ENGINEERING DRAWING

For 1st Year BE/BTech. Students

SIXTH EDITION

K Venugopal
The Best Citizen of India Awardee-2000

V Prabhu Raja
ME, PhD, MIE, MISTE
Associate Professor of Mechanical Engineering
PSG College of Technology
Coimbatore, Tamil Nadu

Publishing Globally

NEW AGE INTERNATIONAL (P) LIMITED, PUBLISHERS
LONDON • NEW DELHI • NAIROBI
Bangalore • Chennai • Cochin • Guwahati • Hyderabad • Kolkata • Lucknow • Mumbai
Visit us at **www.newagepublishers.com**

Published by New Age International (P) Ltd., Publishers
First Edition: 1992
Sixth Edition: 2011
Reprint: 2020

GLOBAL OFFICES

- **New Delhi** **NEW AGE INTERNATIONAL (P) LIMITED, PUBLISHERS**
  7/30 A, Daryaganj, New Delhi-110002, (INDIA)
  Tel.: (011) 23253771, 23253472, **Telefax:** 23267437, 43551305
  E-mail: contactus@newagepublishers.com • Visit us at www.newagepublishers.com
- **London** **NEW AGE INTERNATIONAL (UK) LTD.**
  27 Old Gloucester Street, London, WC1N 3AX, UK
  E-mail: info@newacademicscience.co.uk • Visit us at www.newacademicscience.co.uk
- **Nairobi** **NEW AGE GOLDEN (EAST AFRICA) LTD.**
  Ground Floor, Westlands Arcade, Chiromo Road (Next to Naivas Supermarket)
  Westlands, Nairobi, KENYA, **Tel.:** 00-254-713848772, 00-254-725700286
  E-mail: kenya@newagepublishers.com

BRANCHES

- **Bangalore** 37/10, 8th Cross (Near Hanuman Temple), Azad Nagar, Chamarajpet, Bangalore- 560 018
  Tel.: (080) 26756823, **Telefax:** 26756820, **E-mail: bangalore@newagepublishers.com**
- **Chennai** 26, Damodaran Street, T. Nagar, Chennai-600 017, **Tel.:** (044) 24353401, **Telefax:** 24351463
  E-mail: chennai@newagepublishers.com
- **Cochin** CC-39/1016, Carrier Station Road, Ernakulam South, Cochin-682 016, **Tel.:** (0484) 2377303, **Telefax:** 4051304
  E-mail: cochin@newagepublishers.com
- **Guwahati** Hemsen Complex, Mohd. Shah Road, Paltan Bazar, Near Starline Hotel, Guwahati-781 008,
  Tel.: (0361) 2513881, **Telefax:** 2543669, **E-mail: guwahati@newagepublishers.com**
- **Hyderabad** 105, 1st Floor, Madhiray Kaveri Tower, 3-2-19, Azam Jahi Road, Near Kumar Theater, Nimboliadda
  Kachiguda, Hyderabad-500 027, **Tel.:** (040) 24652456, **Telefax:** 24652457
  E-mail: hyderabad@newagepublishers.com
- **Kolkata** RDB Chambers (Formerly Lotus Cinema) 106A, 1st Floor, S N Banerjee Road, Kolkata-700 014
  Tel.: (033) 22273773, **Telefax:** 22275247, **E-mail: kolkata@newagepublishers.com**
- **Lucknow** 16-A, Jopling Road, Lucknow-226 001, **Tel.:** (0522) 2209578, 4045297, **Telefax:** 2204098
  E-mail: lucknow@newagepublishers.com
- **Mumbai** 142C, Victor House, Ground Floor, N.M. Joshi Marg, Lower Parel, Mumbai-400 013
  Tel.: (022) 24927869, **Telefax:** 24915415, **E-mail: mumbai@newagepublishers.com**
- **New Delhi** 22, Golden House, Daryaganj, New Delhi-110 002, **Tel.:** (011) 23262368, 23262370, **Telefax:** 43551305
  E-mail: sales@newagepublishers.com

ISBN: 978-81-224-3149-0
C-19-08-12200

Printed in India at Ajit Printing Press, Delhi.
Typeset by Author.

NEW AGE INTERNATIONAL (P) LIMITED, PUBLISHERS
7/30 A, Daryaganj, New Delhi-110002
Visit us at **www.newagepublishers.com**
(CIN: U74899DL1966PTC004618)

PREFACE TO THE SIXTH EDITION

This book **ENGINEERING DRAWING** is intended to be used as a classroom textbook by the beginners of the Engineering Curriculum, both of Degree and Diploma level, including AMIE Course.

The latest **BIS Drawing Codes SP 46 : 2003** are followed in this text. Numerous standard solved problems, from various leading Universities all over India, with detailed constructional Step-By-Step procedure and Side-By-Side simplified drawings are presented in this book such that the book is a self-learning tool.

We are highly indebted to the learned Professors, Faculty Members and Students of various engineering colleges for their magnificent response to this student-friendly textbook for the past 25 years.

We are immensely thankful to our Publishers Shri R.K. Gupta, B.Sc. (Engg.), Chairman and Shri Saumya Gupta, B.E., M.S. (U.S.), Managing Director, M/s. New Age International (P) Ltd., New Delhi, for their persistent effort for bringing out the book in the present form with a nice getup.

We solicit comments and suggestions for further improvement.

K. VENUGOPAL
V. PRABHU RAJA

PREFACE TO THE FIRST EDITION

"Engineering Drawing is the Universal Language of Engineers." This book ENGINEERING DRAWING is intended to be used as a classroom textbook by the beginners of the engineering curriculum. All the drawings as well as the text matter are in conformity with the specification of Bureau of Indian Standards (B.I.S.), adopted from I.S.O.

"A neat diagram with clarity is worth pages of explanation!" Sincere efforts have been made to simplify the drawings that they will be more clear and striking and give a clear concept of the subject.

Although brevity is the keynote throughout the text, it has not been gained at the expense of omitting essential details. This book will serve as an excellent guide for an average student, because it highlights the important Points to Remember, Note, Rule, Hint, etc. I am quite confident that this book will dispel the common illusion that the subject Engineering Drawing is a tough one.

I am very thankful to my publisher, M/s. New Age International (P) Ltd., New Delhi, for publishing this book in the best possible form.

Suggestions to improve the usefulness of the book will be gratefully received.

K. VENUGOPAL

ACKNOWLEDGEMENTS

I. We must express a deep sense of gratitude to the following renowned Technical Universities for prescribing this treatise as a Text Book / Reference Book in their syllabi:

- ANNA University (Chennai, Coimbatore, Trichirapalli, Madurai and Tirunelveli) Tamil Nadu
- JNTU, Andhra Pradesh
- Kerala University, Kerala
- West Bengal University of Technology, West Bengal
- Karunya University, Coimbatore
- SRM University, Chennai
- Maharashtra State Board of Technical Education
- Many reputed Deemed Universities and Autonomous Institutions

II. We most sincerely thank the following Technical Universities, the examination papers of which have been included in the text by way of illustrations:

- ANNA University, Tamil Nadu
- JNTU, Andhra Pradesh
- Kerala University, Kerala
- West Bengal Technical University, West Bengal
- Mumbai University
- Pune University
- Gujarat Technological University
- Visveswaraiah Technological University, Karnataka
- Punjab Technological University

AUTHORS

DETAILED CONTENTS

SOLID GEOMETRY

1

DRAWING INSTRUMENTS

1. IMPORTANCE OF ENGINEERING DRAWING AS A GRAPHIC COMMUNICATION

"Scientists study the world as it is!
Engineers create the world that never has been!!"

Graphic Communication: Communication of ideas through sketches is probably the oldest form of communication among humans. Pre-historic man used a stick as equipment to scratch out a message in the sand. Early Egyptians used pictures as a form of communication with the equipment of paint and brush. Whether the tools are a simple pencil and sketch pad or a computer controlled plotting device – the basic need to draw pictures is still present.

Language of Engineers: "*A neat drawing is worth pages of explanation.*" We easily grasp the information about an object, if it is illustrated with drawings. For example, by drawing one or two sketches of a building, the details such as shapes and sizes of various rooms, etc., can be conveyed precisely.

Engineering Drawing is the Language of Engineers: Engineering Drawing is the drawing of an object containing all necessary information like actual shape, size, etc., required for the manufacturing of an engineering component. It is drawn well before it is manufactured. Therefore, the person who draws the drawing should have a clear picture of the shape and size of the component in his / her mind.

Universal Language of Engineers: This is an era of globalization and establishment of Multi National Companies (MNCs). Technological development of a country is the deciding factor for its export potential. Engineering drawings prepared in a country are adopted in other countries. So, engineering drawing is called the Universal Language of Engineers.

Visualization Concept: Engineering Graphics is the graphic language, from which an engineer can visualize the object and then effectively communicate his/her ideas and designs to others in the form of pictorial drawings. Reading, understanding and interpreting an engineering drawing requires a lot of skill of imagination and the ability of visualization of the actual object.

Communication Skill: The engineer must have perfect knowledge and communication skill in drawing. As the bad language is unpleasant to read and communicate, so also a wrong drawing will worry a trained eye. It is doubtful if any engineer can fulfill his / her functions completely and efficiently without the ability to read or to prepare, i.e., to communicate others in the form of drawings. Yes! Many good technical ideas of yours may get rejected because of poor communication skill in the graphic language.

1.1 CLASSIFICATIONS

(i) **Plane Geometry:** Plane Geometry is the art of representation of geometrical objects having Two Dimensions (2D), *i.e.*, length and breadth on a Two Dimensional (2D) drawing sheet. Chapters 3 – 6 deal with the concept of Plane Geometry, where the constructions of the geometrical figures having only two dimensions (2D) are discussed.

(ii) **Solid Geometry:** Solid Geometry is the representation of geometrical objects of Three Dimensions (3D), *i.e.*, Length, Breadth/Width and Height on a Two Dimensional (2D – Length and Breadth/Width) drawing sheet. Chapters 7 – 24 deal with the concept of Solid Geometry. Solid Geometry deals with the representation of Points, Lines, Plane Surfaces and Solids in the form of projections on your 2D drawing sheet.

1.2 APPLICATIONS

Engineering Drawing and Graphics is a ***core subject*** for all the branches of engineering. It is used in different engineering disciplines to design and draw:

(i) Machine components, Transmission systems, CNC machine tools, Mechanics of solids, Robotics, etc., in *Mechanical and Production Engineering.*

(ii) Structures, Plan of buildings, Multi-storied complex, Bridges, Stadium, etc., Town planning, Mapping, Contour plotting in *Civil and Architectural Engineering.*

(iii) Circuit layouts, Electrical-wiring diagrams, Panel design, Control schematics, etc., in *Electrical Engineering.*

(iv) Measuring instruments, Sensors, etc., in *Instrumentation Engineering.*

(v) Schematic diagrams of Printed circuits, Microprocessors, Integrated circuits, etc., in *Electronics including Computer Science Engineering.*

(vi) Communication network, Satellite transmitting pictures, etc., in *Communication Engineering.*

(vii) I.C. Engines, Kinematics, Steering, Hydraulic circuits, etc., in *Automobile Engineering.*

(viii) Spacecraft, Jet propulsion, Flight simulator, Lofting, etc., in *Aeronautical Engineering.*

(ix) Fashion design of garments in Textile Technology

(x) Metallurgical Engineering, Mining Engineering, Marine Engineering, etc.

1.3 CAN COMPUTER AIDED DRAFTING (CADr) REPLACE THE MANUAL DRAFTING TOOLS FOR THE LEARNING PROCESS?

Compared to Manual Drafting, Computer Aided Drafting is faster as well as accurate. No doubt, much time can be saved by using the computer to make even complicated engineering drawings. However, remember, the learning process is accomplished through the traditional manual drafting tools only. The computer cannot replace the drafting board and equipment as a learning tool. Yes! Computer Aided Drafting is not a substitute for Manual Drafting. It is only a tool that can be used to supplement traditional drafting tools. The underlying basic concepts of construction of 2D and 3D drawings remain the same regardless of the tools chosen to create the graphics.

It is ideal to first master the fundamentals of engineering drawings. Having grasped the fundamentals and acquired sufficient practice on the drawing board, then try your hands on your user- friendly computer. The computer would then come handy in doing the task.

Parallel Example: A somewhat parallel example is the usefulness of calculators. Calculators are used to solve math(s) problems, but knowledge of math(s) is still required by the user to solve problems. What is eliminated is the tedious task of performing longhand mathematical manipulations!

2. DRAWING INSTRUMENTS [DRAFTING TOOLS]

"Plan your work and work your plan!"

Drawing Instruments are used to prepare neat and scaled drawings. They are listed below:

1.	Drawing Board	**7.**	Protractor-cum-Procircle
2.	Drawing Sheets	**8.**	Drawing Pencils (HB, H and 2H)
3.	Standard Set-Scales	**9.**	Sharpener and Emery Paper
4.	Mini-Drafter	**10.**	Adhesive Tape (*Cello-Tape*)/Clips
5.	Instrument Box	**11.**	Pencil Eraser
6.	Set-Squares (45° & 30°-60°)	**12.**	French Curves

Note: Check your instruments while buying.

2.1 DRAWING BOARD (IS 1444 : 1989)

Drawing board is of standard size 650 mm × 470 mm as recommended by IS 1444 : 1989.

2.2 DRAWING SHEETS (IS 10711 : 2001)

Standard sizes of *trimmed* drawing sheets recommended by IS 10711 : 2001 are given in the Table Shown: (Area of A0 size is 1 sq. m. of length to width ratio $\sqrt{2}$: 1. Successive smaller sizes are obtained by halving previous size sheets, with the above constant length to width ratio.)

Designation	Dimension, mm Trimmed size
A0	841 × 1189
A1	594 × 841
A2*	420 × 594
A3	297 × 420
A4	210 × 297

** A2 size drawing sheet is preferred for class use.*

Note: Always fix the drawing sheet over a padding sheet on the drawing board.

2.3 STANDARD SET-SCALES

BIS recommends the following set-scales (plastic/cardboard), eight in number with designations Ml, M2 and so on, with the scales graduated on each edge. *See Table below:*

	M1	M 2	M 3	M 4	M 5	M 6	M 7	M 8
Scale on one edge	1:1	1:2.5	1:10	1:50	1:200	1:300	1:400	1:1000
Scale on other edge	1:2	1:5	1:20	1:100	1:500	1:600	1:800	1:2000

Important Note: ***Do not use the set-scale as a straight edge for drawing straight lines.***

2.4 MINI-DRAFTER: (*Fig. 1*)

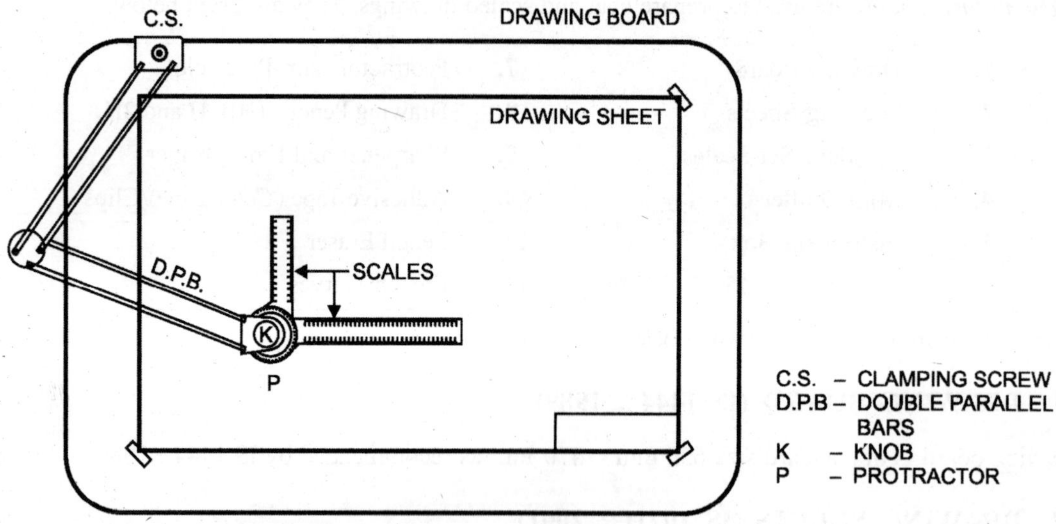

FIG. 1 MINI-DRAFTER CLAMPED TO DRAWING BOARD

A Drafting Machine is used by professional draftsman to prepare drawings. It combines the functions of T-Square, Set-Squares, Scales, Clinograph and Protractor. A miniature version of the drafting machine known as Mini-Drafter is used for drafting by students.

Mini-Drafter: *See Fig. 1.* One end of the mini-drafter is clamped by means of a Clamping Screw (C.S.) to the longer edge of the Drawing Board. At its other end, an adjustable Knob (K) having Protractor (P) markings is fitted. Two Scales of transparent celluloid, set at right angles to each other are attached to the knob.

How to use your Mini-Drafter?

1. *Clamping the Mini-Drafter:* Set the reference mark on the protractor head to read 0° and tighten the knob. Then, clamp the drafter to the left-side top edge of the drawing board using the clamping screw, such that the horizontal scale is parallel to the horizontal edge of the Drawing Sheet.
2. Now, the scales along with the protractor head as a unit can be moved easily to any location on the drawing sheet without altering the position of the two scales. The Double Parallel-Bar (D.P.B.) mechanism, pivoted as shown, keeps the scales in every position to remain parallel to their previous position.
3. To draw horizontal and vertical lines, the reference mark on the protractor head has to be set to 0°. To draw inclined lines, the scales should be fixed with the reference mark, indexing the required angle. Do all the positioning by one hand and draw the lines by the other.

In short, the mini-drafter is used for different drafting operations, viz., to draw horizontal, vertical and inclined lines, parallel and perpendicular lines and also for measuring lines and angles.

2.5 INSTRUMENT BOX

It contains the following instruments:

2.5.1 COMPASSES

Compasses are used to draw circles and arcs.

(i) *Large Compass*

See Fig. 2(a). Large Compass is used to draw circles from 30 mm to 120 mm diameter.

See Fig. 2(b). It shows the lead stick L projecting approximately 1 mm less than the needle point N of the compass. Reason is that the needle tip gets inserted in the drawing sheet while drawing.

Also, keep the inclined cut of the lead *turned outside.*

See Fig. 2(c). Circles upto 50 mm radius can be drawn with the legs of the compass kept straight. For drawing circles of more than 50 mm radius, both the legs should be bent at the knee joints so that they are perpendicular to the surface of the paper as shown.

(ii) *Lengthening Bar*

See Fig. 2(d). For drawing larger circles of more than 125 mm radius, *Lengthening Bar* attachment is used.

For this, the lower part of the pencil leg is detached and lengthening bar is inserted in its place. The detached pencil leg is then fitted at the end of the lengthening bar. Thus, length of the pencil leg is increased.

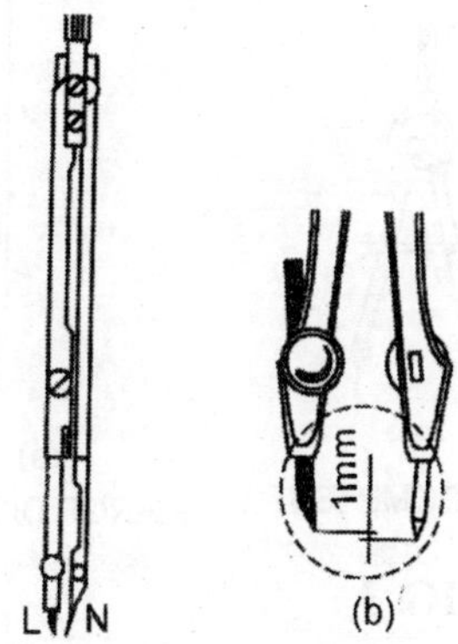

(a) LARGE COMPASS

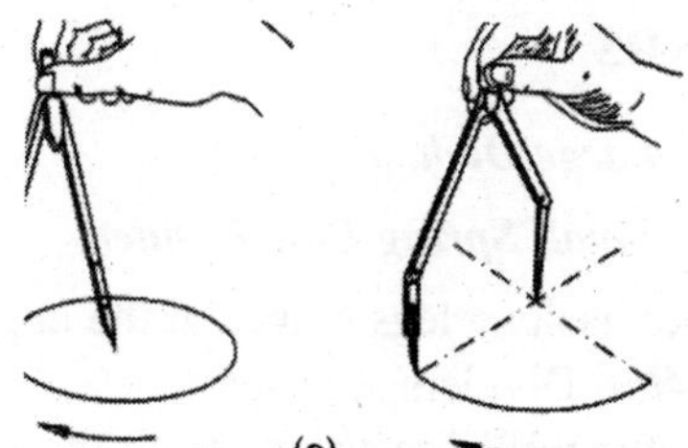
(c)

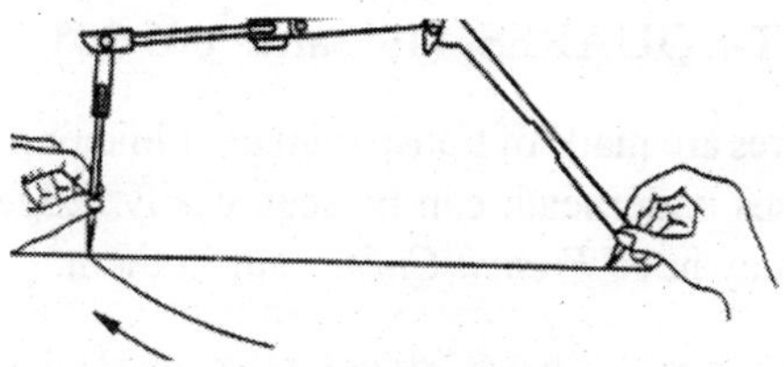
(d) LENGTHENING BAR

FIG. 2

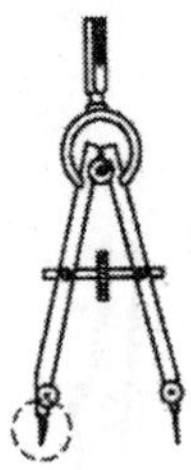

BOW COMPASS

FIG. 3

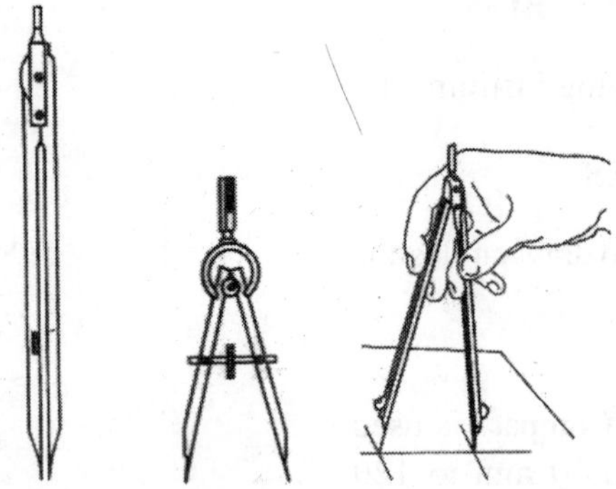

(a) LARGE DIVIDER (b) BOW DIVIDER (c) SETTING-OFF DIMENSION

FIG. 4

(iii) *Small Spring Bow Compass*

See Fig. 3. Bow Compass is used for drawing small circles upto 30 mm diameter. In this, the inclined cut of the lead is turned inside as shown.

2.5.2 DIVIDERS

Fig. 4(a) shows ***Large Divider.***

Fig. 4(b) shows ***Small Spring Bow Divider.***

The divider has two legs hinged at the upper end. It is provided with steel pins at both the lower ends. *See Fig. 4(c).* Dividers are used:

- For setting-off dimensions from the set-scale to the drawing,
- For transferring dimensions from one part of the drawing to another part and
- For dividing straight or curved lines into a number of equal parts.

2.6 SET-SQUARES (45° and 60°-30° Triangles)

Set-Squares are made of transparent celluloid or plastic. Those made of transparent celluloid are preferred as the lines underneath can be seen easily. Edge of the set-squares may be bevelled or levelled. Set-squares may have *French Curves* cut in them.

2.7 PROTRACTOR-CUM-PROCIRCLE

Protractor is divided into 0.5° division. It is used to mark or measure accurately such angles that cannot be precisely drawn or measured with the mini-drafter. Pro-circle is used to draw small circles of different diameters without using the compass.

2.8 DRAWING PENCILS

Drawing pencils are different from ordinary pencils and are in grades HB, H, 2H, etc. Grade HB denotes medium soft. Grades H, 2H, etc., denote the degree of hardness in increasing order. Grades B, 2B, etc., denote the degree of softness in increasing order. The grade of the pencil is decided by the amount of graphite mixed with clay and is used as follows:

Grade		Use
HB - (*Soft grade*)	...	Used for drawing Border Lines, Lettering and Freehand sketching.
H - (*Medium grade*)	...	Used for drawing Visible outlines and Visible edges.
2H - (*Hard grade*)	...	Used for Construction lines, Dimension lines, Leader lines, Extension lines, Centre lines, Hatching lines and Hidden lines.

Note: *Never use ordinary cheap quality pencils on your drawing sheet.*

Keep your pencil sharp: A dull pencil produces fuzzy lines. Only a sharp pencil produces black and sharp lines that sparkle with clarity. See Figs. 5(a) and (b). Lead is sharpened to *conical point* or *chisel edge* by rubbing on an emery paper. *Conical point* is used for lettering and freehand sketching. *Chisel edged pencil* is used for drawing lines.

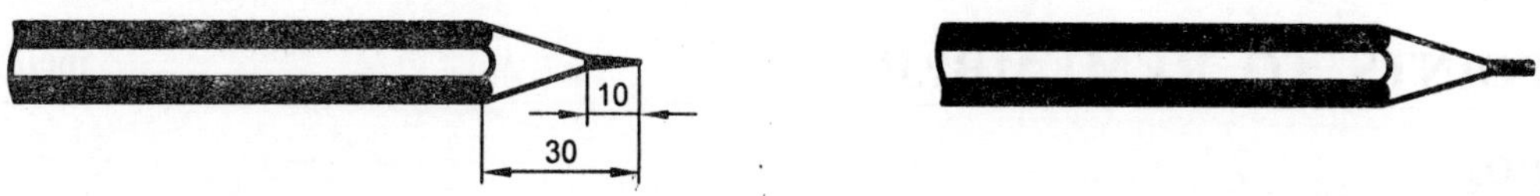

(a) CONICAL POINT (b) CHISEL EDGE

FIG. 5

Micro - tip pencils / Clutch pencils *with 0.5mm thick leads of grades HB, H and 2H are preferred than wooden pencils, as they need no sharpening.*

2.9 SHARPENER AND EMERY PAPER: Drawing pencils are sharpened by the pencil sharpener. Emery paper is used to obtain conical or chisel edge of the lead.

2.10 ADHESIVE TAPE (*CELLO-TAPE*) or CLIPS: A thin adhesive tape (***cello tape***) is used to fix the drawing sheet on the drawing board. Best quality clips can also be used.

2.11 PENCIL ERASER: Pencil Eraser (*It is not 'rubber'! Call it 'eraser'!*) is used to rub-off unwanted lines, etc. A non-dusting good quality eraser called *Eraser Soft* should be used.

2.12 FRENCH CURVE

See Fig. 6. French Curve is made of transparent plastics. It is used to draw irregular curves. It is available in different sizes and shapes. French Curve is also inscribed in the setsquares.

French Curve is used to draw non-circular curves, which cannot be drawn with a compass. The edge of the French Curve is lined up, by trial and error method, so that it passes through at least 3 points of the required curve.

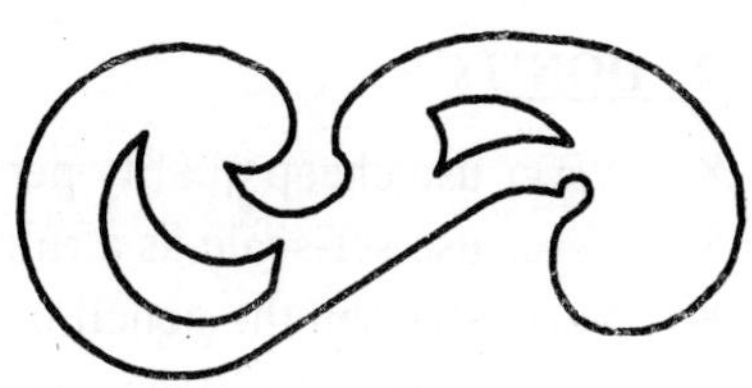

FIG. 6

3. HOW TO BEGIN YOUR DRAWING ?

1. Clean the drawing board and all your drawing instruments using handkerchief.
2. Fix your drawing sheet over a **padding sheet** on the drawing board.
3. Clamping the Mini-drafter : Set the reference mark on the protractor head to read 0° and tighten the knob. Then, clamp the drafter to the left-side top edge of the drawing board using the clamping screw, such that the horizontal scale is parallel to the horizontal edge of the Drawing Sheet (D.S.).
4. Check that scales and protractor head of mini-drafter move to any position on drawing sheet.
5. Draw the **Border Lines** and the **Title Block** on the drawing sheet using HB pencil.
6. Print the **Problem Number** on the **left top side** of the figure. Enclose the number in a small circle of about 10 mm diameter. Then commence your drawing work.
7. Captions like the Name of the Object, Views, etc., should be written below the drawing.
8. Spacing of drawings between two problems / views is to be planned before the commencement of the drawing.

4. POINTS TO REMEMBER

4.1 DOs

✓ *Always* clean hands are the first necessity; ("*Neatness is a personal habit*").

✓ *Always* begin the work after wiping-off the drawing board.

✓ *Always* use *appropriate* drawing pencils / leads. (The grade of the pencil / lead is decided by the amount of graphite mixed with clay.)

✓ *Always* maintain the sharpness of the pencil lead by periodically sharpening it. (Micro-tip pencils are recommended for fast drafting.)

✓ *Always* keep a padding sheet under the drawing sheet in use.

✓ *Always* keep your handkerchief at hand to clean any dirt on the drawing sheet.

✓ *Always* use the divider for transferring measurements from the set-scale to the drawing for getting dimensional accuracy.

✓ *Always* remember that you are NOT writing, but you are printing the text on your drawing sheet.

✓ *Always* remember: *"Small things make perfection, but perfection is no small thing."*

✓ *Always* remember: *"Time is money in industries."* You should not only draw your drawings accurately, but also swiftly to save time.

4.2 DON'Ts

× *Never* use cheap quality pencil / lead and eraser on the drawing sheet.

× *Never* use set-scale as a ruler for drawing lines or for taking measurements directly.

× *Never* sharpen the pencil over the drawing board or drawing sheet.

× *Never* forget to clean every time, the sliding instruments (mini-drafter, set-squares and protractor) placed on the drawing sheet. Sliding of the instruments on the drawing sheet should be reduced to a minimum.

*

2

LINES, LETTERING AND DIMENSIONING

1. BIS ENGINEERING DRAWING PRACTICE SP 46 : 2003

"Standards connect the world !"

Engineering Drawing is the Universal Language of Engineers, spoken, read and written in its own way. Every language has its own rules of grammar. Engineering drawing also has its grammar in the types of lines and lettering, its idioms in the conventional practices, its punctuations in the types of dimensioning, its abbreviations in the symbols and its descriptions in the theory of projections and constructions.

BIS Drawing Codes: The representation of any matter by some sign or mark on the drawing is known as Convention or Code. These conventions are specified by Bureau of Indian Standards (BIS). BIS was formerly known as Indian Standards Institute (ISI). The latest BIS drawing codes are ***SP 46 : 2003 – Engineering Drawing Practice for Colleges***. BIS adopts the ISO standards in *toto* for the various topics in engineering drawing as listed below:

Sl. No.	BIS Codes	Topics
1.	IS 1444 – 1989	Drawing Boards – Specification
2.	IS 10711 – 2001 and SP 46 – 2003	Technical products Documentation – Size and Layout of Drawing Sheets
3.	IS 11665 – 1985 (Reaffirmed 2000)	Technical Drawings – Title Block
4.	IS 11664 – 1986 (Reaffirmed 1999)	Folding of Drawing Sheets
5.	IS 10714 (Part 20) – 2001 and SP 46 – 2003	Lines for Technical Drawings
6.	IS 9609 (Parts 0 & 1) – 2001 and SP 46 – 2003	Technical products Documentation – Lettering
7.	IS 11669 – 1986 and SP 46 – 2003	General Principles of Dimensioning of Technical Drawings
8.	IS 15021 (Parts 1 to 4) – 2001 and SP 46 – 2003	Technical Drawings – Projection Methods

2. LAYOUT and FOLDING OF DRAWING SHEET [IS 10711 : 2001]

Fig. 1 shows the layout of A2 size drawing sheet (420 mm x 594 mm) for class work.

Borders

Border is the drawing space left all around and in between the Trimmed Edges of the drawing sheet and the Frame. Border has a minimum width of 10 mm for A2, A3 and A4 drawing sheets.

Filing Margin

It is provided for taking perforations for the purpose of filing the drawing sheets. It shall have a minimum width of 20 mm *with the border included* therein.

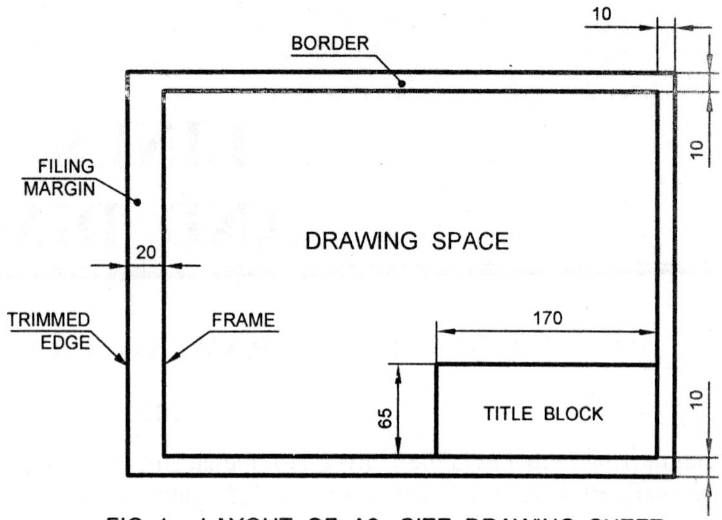

FIG. 1 LAYOUT OF A2 SIZE DRAWING SHEET

2.1 TITLE BLOCK FOR CLASS WORK: Fig. 2

Title Block is provided at the bottom right-hand corner of the drawing sheet. It is a rectangle of 170 mm x 65 mm size. It should be visible when folded. Title Block contains the following particulars as shown in *Fig. 2*.

See Fig. 2. Left Hand Side is the identification zone with the following details:

- **R**oll No.
- **C**lass and **B**ranch
- **S**heet No.*
- **D**ate of Submission
- **S**ymbol for First Angle Projection

Right Hand Side is the information zone with the following details:

- **N**ame of the Student
- **N**ame of the Institution
- **T**itle of the Chapter

**Sheet No.:* If more than one drawing sheet is used for a particular TITLE OF THE CHAPTER, Sheet No. is n / p, where n is the sequential sheet number and p is the total number of sheets, *i.e.*, 1/2, 1/3, etc.

If only one sheet is used, Sheet No. is 1/1.

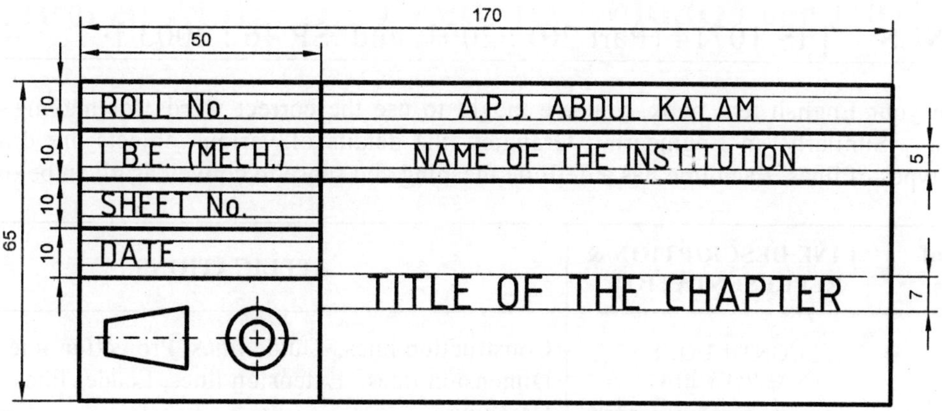

FIG. 2 SUGGESTED TITLE BLOCK FOR CLASS USE

2.2 FOLDING OF DRAWING SHEETS

Fig. 3 shows the method of folding of A2 and A3 size drawing sheets for filing.

<table>
<tr><th>SHEET SIZE</th><th>FOLDING DIAGRAM</th><th>LENGTHWISE FOLDING</th></tr>
<tr><td>A2
420 x 594</td><td>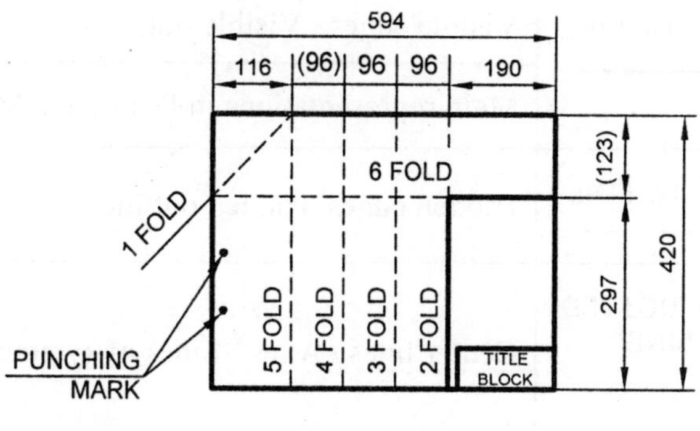
</td><td>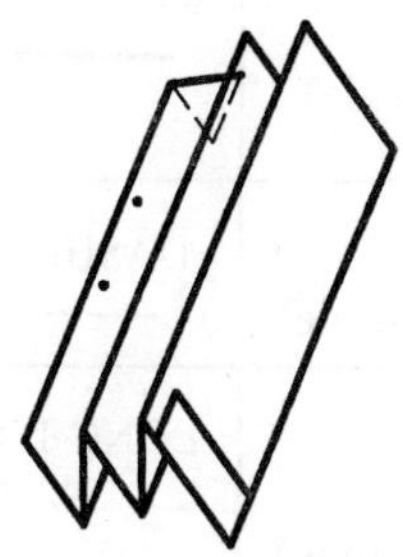</td></tr>
<tr><td>A3
297 x 420</td><td>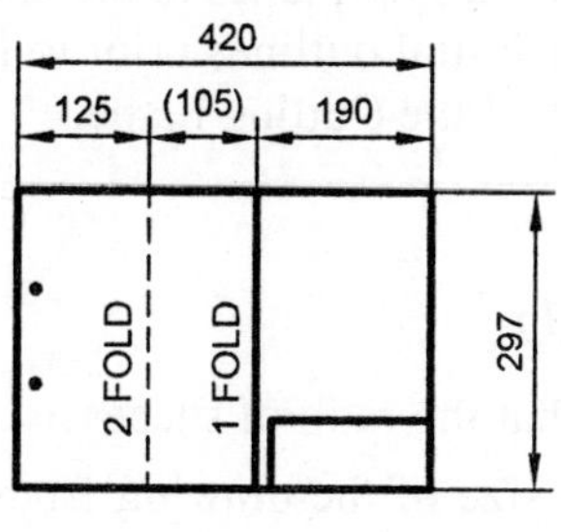
</td><td>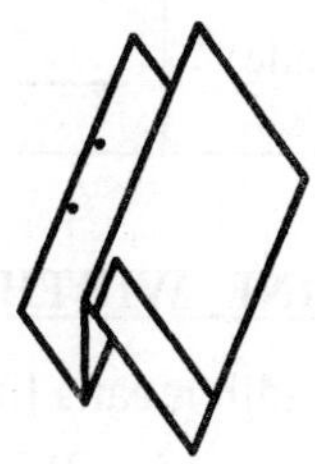</td></tr>
</table>

All dimensions are in millimetres

FIG. 3 FOLDING OF DRAWING SHEETS FOR FILING OR BINDING

3. LINES [IS 10714 (Part 20) : 2001 and SP 46 : 2003]

Just as in your English text book, you are taught to use the correct words for making correct sentences. Similarly, in Engineering Graphics, the details of various objects are drawn by different types of lines. Each line has a definite meaning and sense to convey as given below:

Type of Lines	LINE DESCRIPTION & REPRESENTATION	APPLICATIONS
01.1 (2H Grade)	CONTINUOUS NARROW LINE	Construction lines, Guide lines, Projection lines, Dimension lines, Extension lines, Leader lines, Hatching.
01.1 (2H Grade)	CONTINUOUS NARROW FREEHAND LINE	Preferably manually represented termination of partial or interrupted views, cuts and sections, if the limit is not a line of symmetry or a center line.
01.1 (2H Grade)	CONTINUOUS NARROW LINE WITH ZIGZAGS	Preferably mechanically represented termination of partial or interrupted views, cuts and sections, if the limit is not a line of symmetry or a center line.
01.2 (H Grade)	CONTINUOUS WIDE LINE	Visible edges, Visible outlines.
		Main representations in Diagrams, Maps, Flow charts.
02.1 (2H Grade)	DASHED NARROW LINE - - - - - - - - - - - -	Hidden edges, Hidden outlines.
04.1 (2H Grade)	LONG-DASHED DOTTED NARROW LINE —— . —— . ——	Center lines / Axes, Lines of symmetry.
04.2 (H Grade)	LONG-DASHED DOTTED WIDE LINE —— . —— . ——	Cutting Planes at the ends and changes of direction. Initial outlines prior to forming parts situated in front of the Cutting Plane.

3.1 LINE WIDTHS (IS 10714 : 2001)

Line Width means line thickness. Note that the ratio of successive line widths is 1 : √2.

Choose Line Widths according to the size of the drawing from the following range: **0.13, 0.18, 0.25, 0.35, 0.5, 0.7** and **1 mm.**

BIS recommends *two* Line Widths on a drawing. Ratio between the *thin (narrow)* and *thick (wide)* lines on a drawing shall not be less than **1 : 2.**

4. LETTERING [IS 9609 (Parts 0 & 1) : 2001 and SP 46 : 2003]

Lettering is defined as writing of Titles, Sub-titles, Symbols, Dimensions, etc., on a drawing.

Main Features of Lettering are Legibility, Uniformity and Rapidity of execution. Use of drawing instruments for lettering consumes more time. Also, time-consuming ornamental letters shall not be used. Lettering should be done freehand with speed using HB grade (conical end) pencil. *Yes*, **Lettering is freehand drawing and not writing.**

Practice with continuous efforts would certainly improve the lettering skill and style. Poor lettering will spoil the appearance of an otherwise acceptable drawing.

4.1 SINGLE STROKE LETTERS – Uppercase and Lowercase Letters

The word ***single-stroke*** does not mean that the lettering should be made in one stroke without lifting the pencil. It means that the thickness of the letter should be uniform such as is obtained in one stroke of the pencil. Single stroke letters may be:

1. Capital Letters (Uppercase) and 2. Small Letters (Lowercase)

Note: Lettering in drawing should be in CAPITALS (i.e., Uppercase letters). Lowercase (small) letters are used for abbreviations like mm, cm, etc.

Line Strokes: Line Strokes refer to the directions of drawing straight and curved lines. See Fig. 4(i). Horizontal lines of the letters are drawn from left to right. Vertical and inclined lines are drawn from top to bottom. Curved lines are drawn from left to right or top to bottom. Right or upper half of a circle is drawn clock-wise while left or lower half is drawn anti-clock-wise.

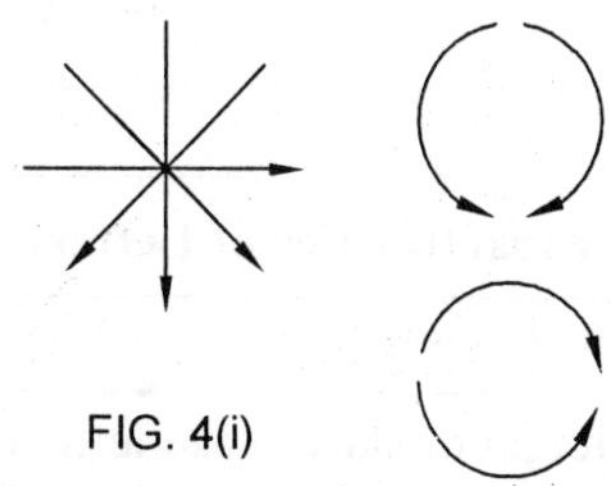

FIG. 4(i)

4.2 TYPES OF SINGLE STROKE LETTERS

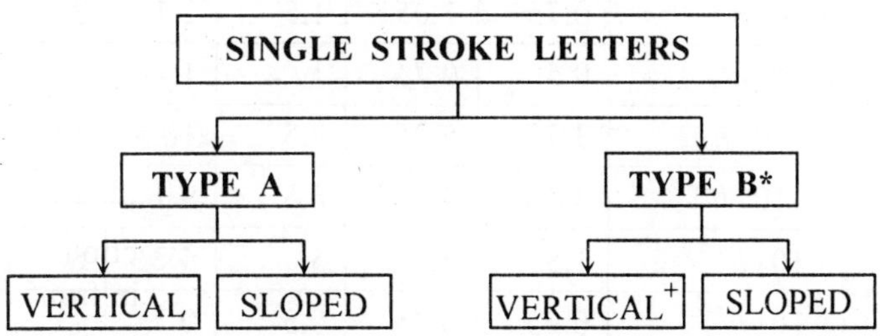

* **Type B preferred:** In Type B, height of capital letter is divided into 10 equal parts. *Type B is preferred for easy and fast execution, because of the division of height into 10 equal parts.* [Refer Table 1] In Type A, height of capital letter is divided into 14 equal parts.

\+ **Vertical Letters of Type B preferred:** Vertical letters are preferred for easy and fast execution, instead of sloped letters. [Sloped letters are inclined at 75° to the horizontal.]

Note:

1. Size of letters used for lettering depends upon the size of the drawing.
2. Draw thin **Guide Lines** (2H pencil) at a distance of **height of letters h** apart.
3. After lettering has been completed, don't erase the guide lines.
4. Maintain proper space between words and also between base lines. (*Refer: Table 1*)

4.3 DIMENSIONING OF TYPE B LETTERS [Refer Fig. 4(ii) and Table 1]

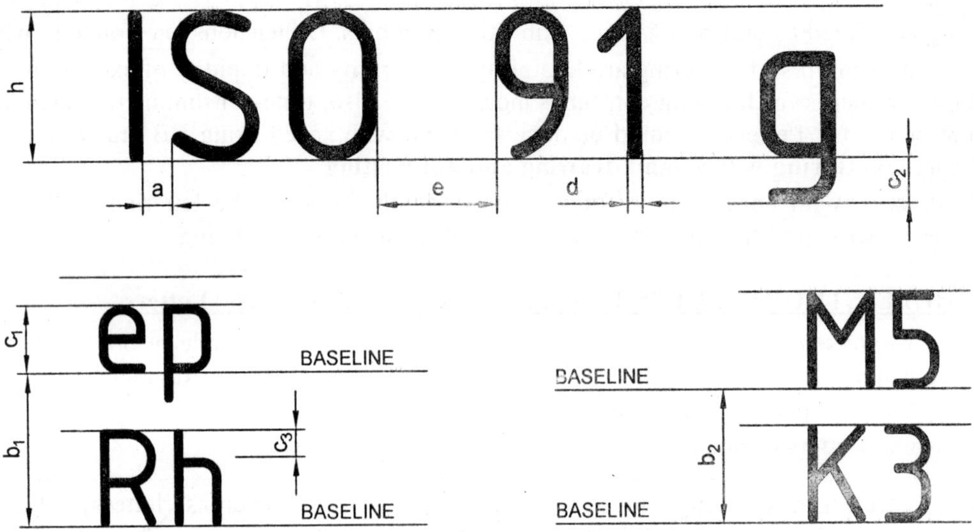

FIG. 4(ii)

Table 1

Characteristics of Lettering		Multiple of *h*	Dimensions (mm)				
Lettering height	*h*	(10/10) *h*	2.5	3.5	5	7	10
Height of lower-case letters	c_1	(7/10) *h*	1.75	2.5	3.5	5	7
Tail of lower-case letters	c_2	(3/10) *h*	0.75	1.05	1.5	2.1	3
Stem of lower-case letters	c_3	(3/10) *h*	0.75	1.05	1.5	2.1	3
Spacing between characters	*a*	(2/10) *h*	0.5	0.7	1	1.4	2
Min. space between baselines[1)]	b_1	(15/10) *h*	3.75	5.25	7.5	10.5	15
Min. space between baselines[2)]	b_2	(13/10) *h*	3.25	4.55	6.5	9.1	13
Spacing between words	*e*	(6/10) *h*	1.5	2.1	3	4.2	6
Line width	*d*	(1/10) *h*	0.25	0.35	0.5	0.7	1

1) *Lettering style: Upper-case & lower-case letters* 2) *Lettering style: Upper-case letters only*

4.4 HEIGHT OF LETTERS (h)

Recommended height h of Letters / Numerals by BIS		
MAIN TITLE OF DRAWING	:	**5 mm** *or* **7 mm** *or* **10 mm**
SUB-TITLES (FRONT VIEW, etc.)	:	**3.5 mm** *or* **5 mm**
Dimensions, Numerals, Notes, etc.	:	**2.5 mm, 3.5 mm** *or* **5 mm**

4.5 VERTICAL CAPITAL & Lower-case LETTERS AND NUMERALS OF TYPE B

Use a graph sheet and mark 10 units of vertical spacing for height **h** of the letters as shown below:

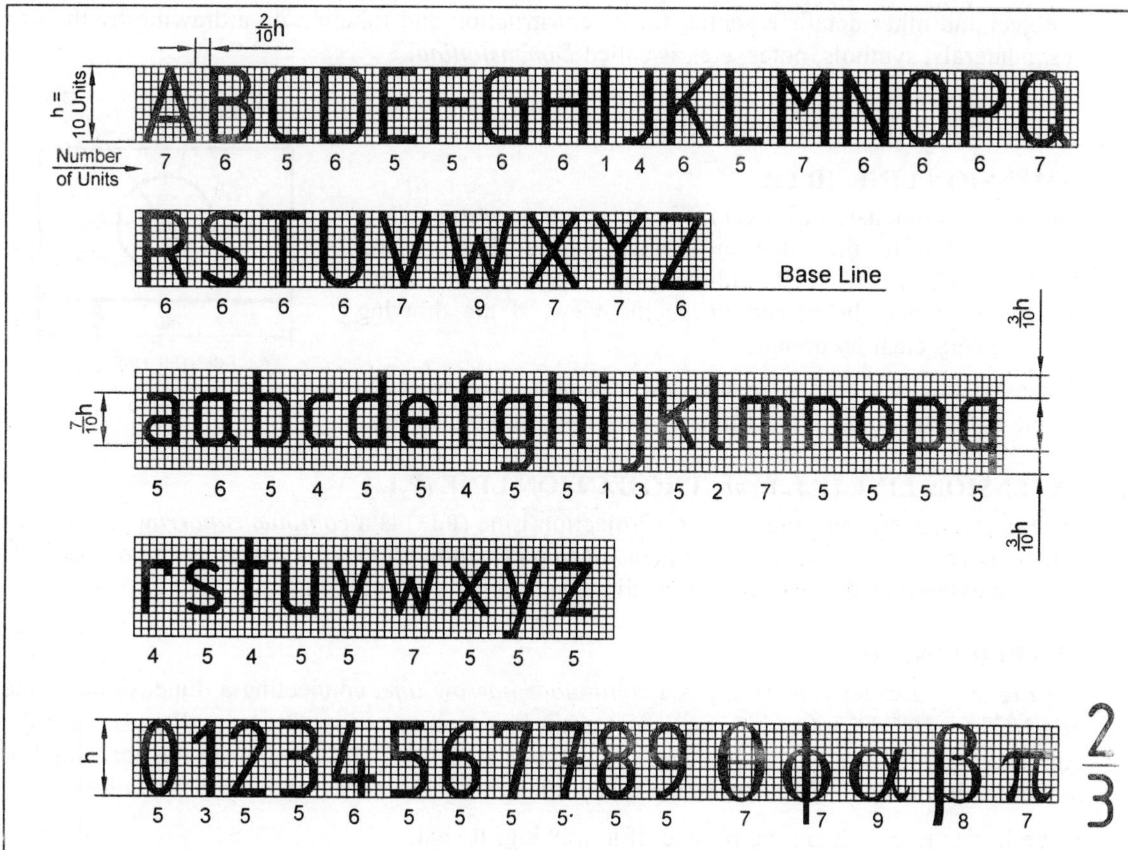

VERTICAL CAPITAL & Lower-Case LETTERS AND NUMERALS OF TYPE B

Example: Write freehand in single stroke vertical CAPITAL letters of 5 mm (h) size:

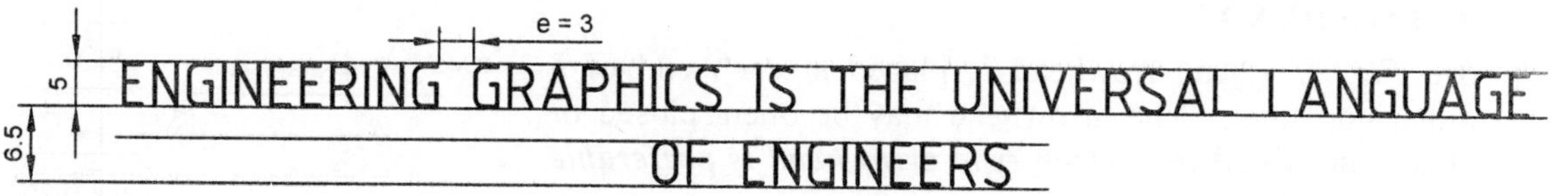

Minimum spacing between baselines (Uppercase letters only) = $\mathbf{b_2}$ = **(13 / 10) h** = **6.5**
Spacing between words = **e** = **(6/10) h** = **3**

Exercise: Write freehand the following, in single stroke (CAPITAL and lower-case) letters:

1. Alphabets (Upper-case & Lower-case) and Numerals 0 to 9 (h = 3.5, 5 and 7 mm)
2. SMALL THINGS MAKE PERFECTION, BUT PERFECTION IS NO SMALL THING. (h = 5 and 7 mm)
3. LETTERING IS FREEHAND DRAWING AND NOT WRITING. (h = 5 mm)

5. DIMENSIONING [IS 11669 : 1986 and SP 46 : 2003]

Indicating the sizes of the *features* (In *Fig. 5,* Circle is a feature; C.L. means Center Line or Axis) of the object and other details essential for its construction and function on a drawing by the use of lines, numerals, symbols, notes, etc., is called *Dimensioning*.

5.1 ELEMENTS OF DIMENSIONING

1. DIMENSION LINE (D.L.)

See Fig. 5. Dimension Line (D.L.) is a *continuous narrow line,* drawn parallel to the edge or surface whose measurement should be shown. It should be placed *outside* the view. Sometimes it may be placed *inside* the view, if the drawing becomes very clear by doing so.

Dimension line is placed at least 6 to 10 mm away from the outline of the drawing. It is terminated by arrowheads at its ends.

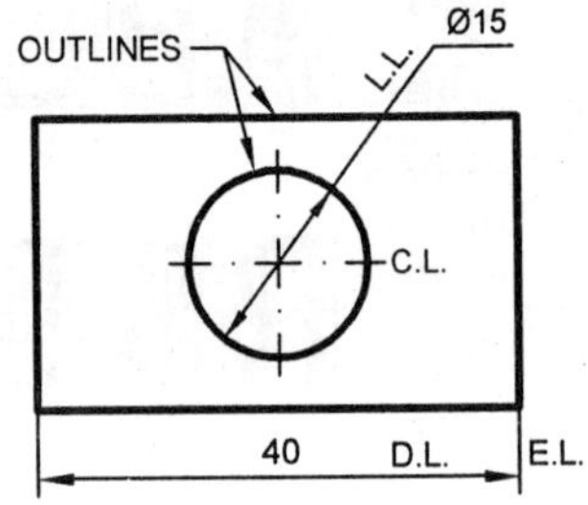

C.L. - CENTER LINE (AXIS)

FIG. 5

2. EXTENSION LINE (E.L.) *or* PROJECTION LINE (P.L.)

See Fig. 5. Extension Line (E.L.) or Projection Line (P.L.) is a *continuous narrow line* drawn perpendicular to the outline to be dimensioned and without leaving a gap from the outline. It is drawn extending slightly beyond the dimension line by about 2 mm.

3. LEADER LINE (L.L.)

See Fig. 5. Leader Line (L.L.) is a *continuous narrow line*, connecting a dimensional value or a NOTE [NOTE gives information relating to a feature, *e.g.,* ∅ 15 in *Fig. 5*] with the corresponding *feature* on the drawing. It is drawn at 30° or 45° or 60° to the horizontal or to the line it touches. When pointing to a circle or an arc, leader line is drawn radially.

If the leader line ends on the outline of a drawing, it shall have an arrowhead [*Fig. 6(i)*].

If it ends within the outline of the drawing, it shall have a dot at the end [*Fig. 6(ii)*].

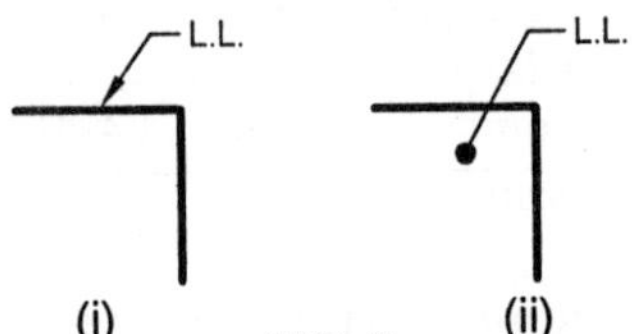

FIG. 6

4. ARROWHEADS

See Fig. 7. An arrowhead is placed at each end of a dimension line. The arrowhead may be open, closed or closed and filled in. ***Open type arrowhead is preferable for fast execution.***

In a drawing, only one style of arrowhead termination should be used.

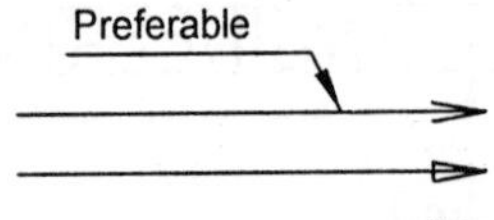

FIG. 7

Shape of the arrowhead

See Fig. 8. Length **l** of an arrowhead is about three times the width **w**. The size of the arrowheads should be proportionate to the size of the drawing.

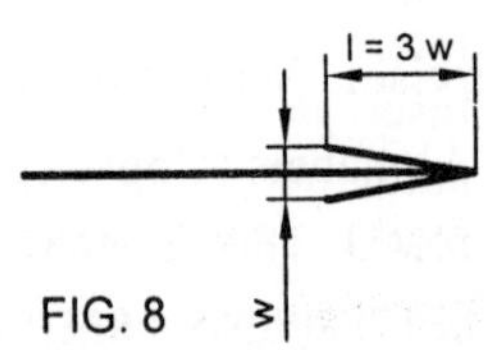

FIG. 8

See Fig. 9. Place the dimensional value parallel to and at the middle of Dimension Line. When the length of the dimension line is not sufficient, place arrowheads outside the Extension Line (*e.g.,* dimensional value 5) ***or*** place the dimensional value (e.g., 3) above the extended portion of the dimension line beyond the arrowhead.

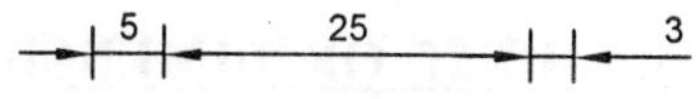

FIG. 9

Oblique stroke

See Fig. 10. Where space is too small for an arrowhead, the *oblique stroke* (drawn as a short line inclined at an angle of 45°) or a dot may be substituted as shown.

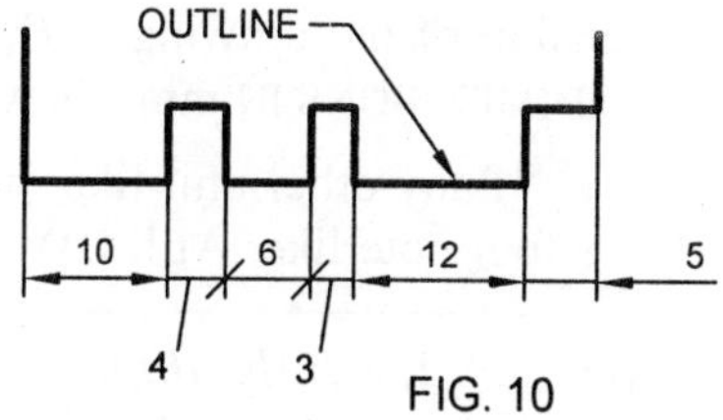

FIG. 10

5.2 METHODS OF DIMENSIONING

Method I: Aligned Method

- *See Fig. 11.* Dimensional value is placed above the dimension line without breaking it. It is placed parallel to and at the middle of the dimension line without touching it.

  Dimensional values are marked so that they can be read *either from the bottom or from the right hand side of the drawing.*
- *Fig. 12* shows the method of placing dimensional values at different positions on the dimension lines.
- ***Angular dimensions***

  Angular dimensions are oriented either as in *Fig. 13* or *Fig. 14* and *Fig. 15.*

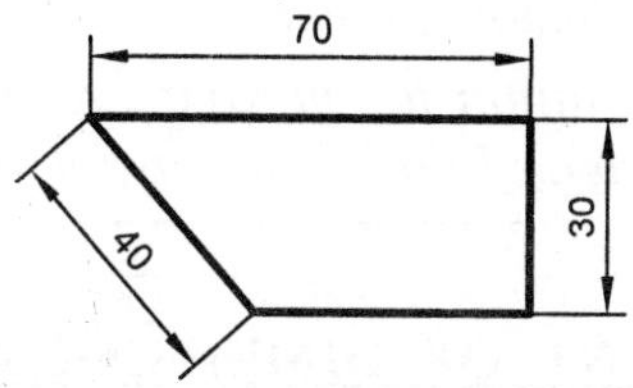

FIG. 11

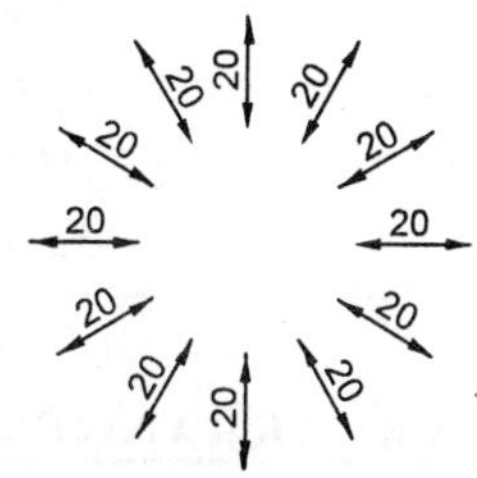

FIG. 12

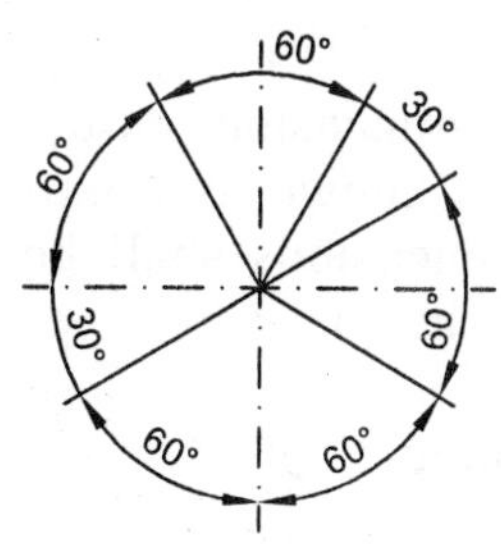

FIG. 13

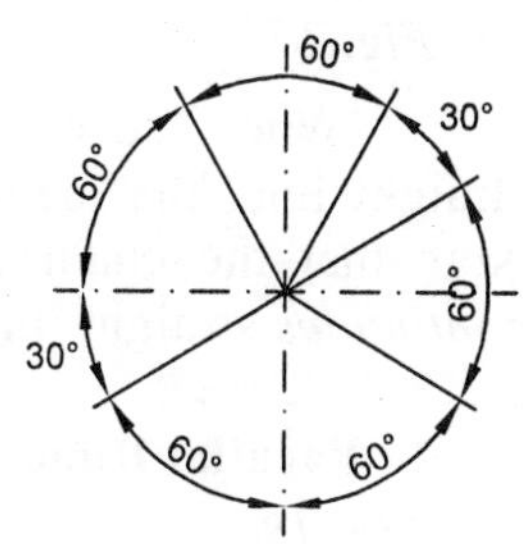

FIG. 14

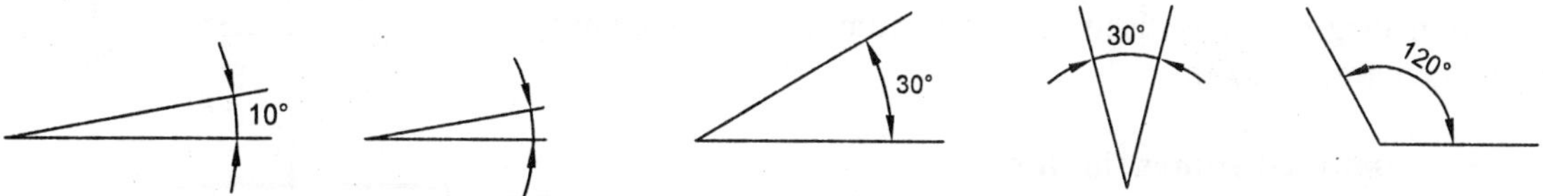

FIG. 15

Method II: Unidirectional Method: Fig. 16

- Dimensional values are indicated so that they may be read from the bottom of the drawing. Non-horizontal dimension lines are interrupted near the middle and the dimensional values are inserted. This method is used on large drawings of automobiles, aircrafts, etc., where it may not be convenient to read the dimensions from right side.

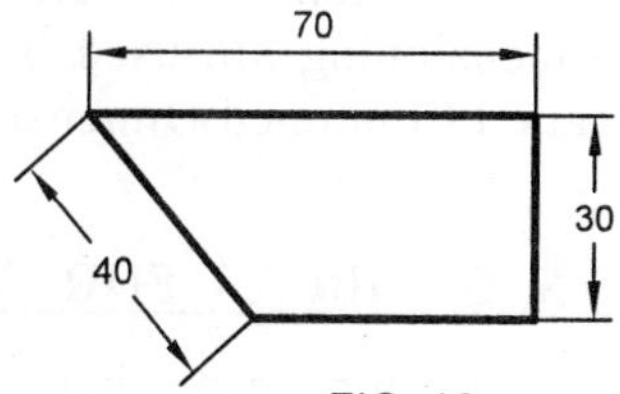

FIG. 16

Note: ***Method I is preferable.*** *Only one method should be used on a drawing.*

5.3 UNIT OF DIMENSIONING

All dimensions in one particular drawing should be expressed in *one unit* only. B.I.S. recommended unit is *millimeter.* Abbreviation **mm** for millimeter need not be shown on each dimension of the drawing. *Example:* A dimension **25** means **25 mm**. A foot note stating 'ALL DIMENSIONS IN mm' is written in a prominent place near Title Block.

If any other unit like m or cm is used, even then only the value is written as dimension. But, a foot-note like 'ALL DIMENSIONS IN cm' is inserted in a prominent place near Title Block.

Note:

i. *If the dimension is less than 1, a zero is placed before the decimal point. Example:* ***0.5 and not .5***

ii. *Decimal point in a dimension should be written in line with the bottom line of the dimension value. Example:* ***0.5 and not 0˙5***

iii. *Abbreviation for millimeters is* ***not mms,*** *but* ***mm only.*** *(s denotes seconds). In other words, s is not added for the plural of any units.*

5.4 ARRANGEMENT OF DIMENSIONS

5.4.1 Chain Dimensioning

See Fig. 17.

When successive dimensions are arranged in a straight line, the method is known as *Chain Dimensioning.* Note that the chain of dimensions shall be arranged in a *continuous* straight line as shown.

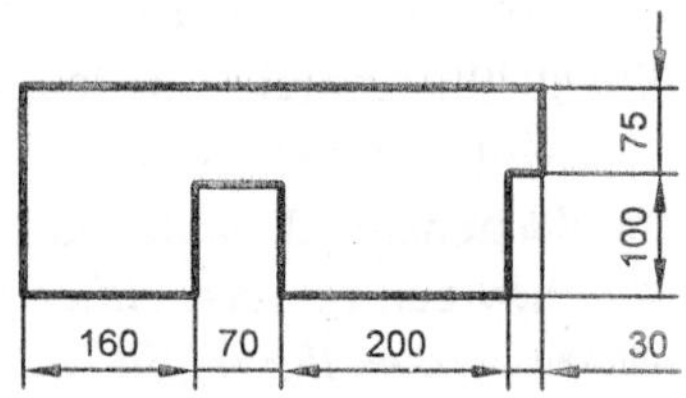

FIG. 17

5.4.2 Parallel Dimensioning

See Fig. 18.

Parallel Dimensioning is the placement of a number of single dimension lines parallel to one another from a common origin. It is used where a number of dimensions have a common origin.

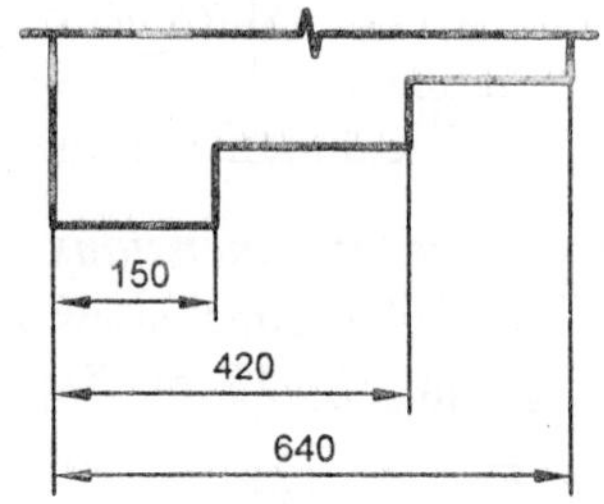

FIG. 18

5.4.3 Combined Dimensioning

See Fig. 19.

When chain dimensioning and parallel dimensioning are used in a single drawing, it is called Combined Dimensioning.

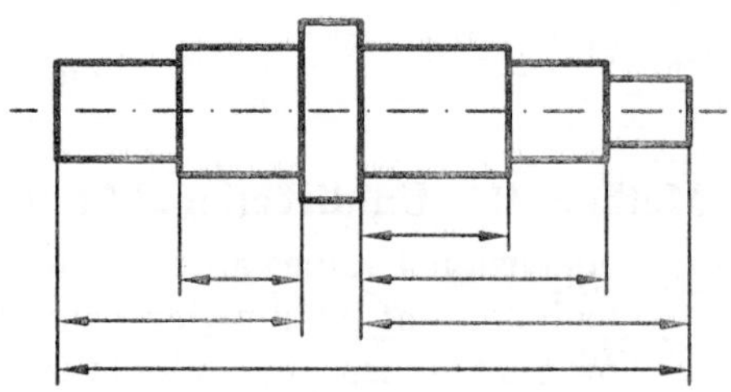

FIG. 19

5.5 SYMBOLS FOR SHAPE INDICATION

Symbol for shape indication shall precede the dimension value. Symbols used are:

Ø – Diameter; **R** – Radius; □ – Square; **SØ** – Spherical Diameter; **SR** – Spherical Radius.

5.5.1 Circle

See Fig. 20. A *circle* is dimensioned using the symbol ∅, by any one of the ways shown. ∅ precedes the dimension value.

Always mark the diameter of a circle and not its radius.

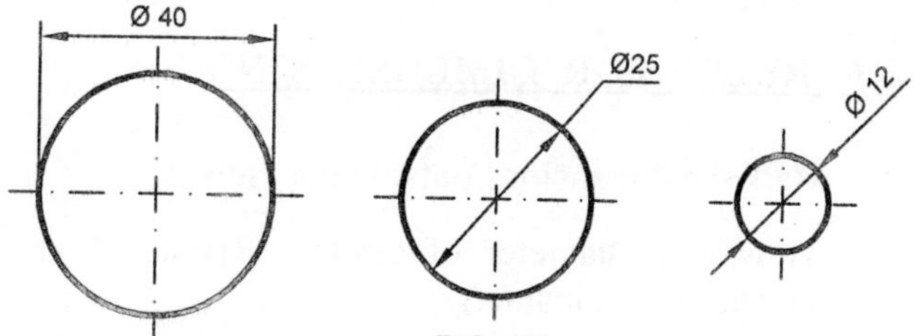

FIG. 20

5.5.2 Cylindrical Diameter

See Fig. 21. Cylindrical Diameter uses symbol ∅ as shown.

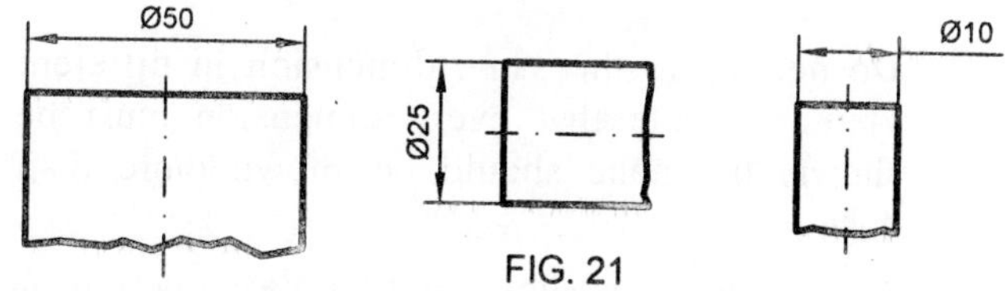

FIG. 21

5.5.3 Arc Radius

See Fig. 22. An *arc* is dimensioned by its Radius R by any one of the ways shown. Only one arrowhead with its point on the arc end of the dimension line is used.

Dimension line of a radius should pass through the center of the arc. Mark the center with a *small cross*.

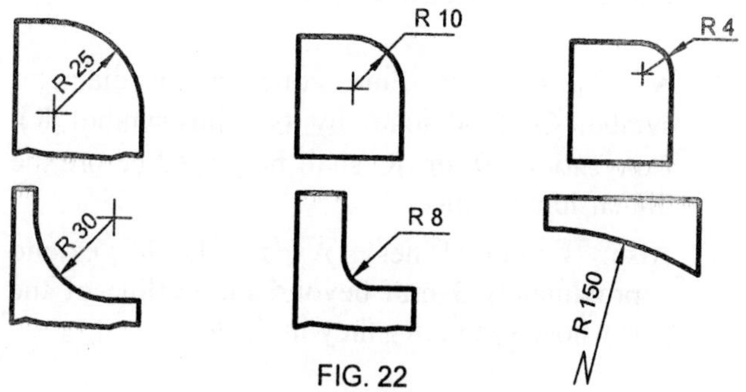

FIG. 22

Note: (i) *Dimensions of small radii are shown preferably outside the view, due to want of space inside the view. Examples: R10 and R4*

(ii) *When the arc center falls outside the limits of space available, dimension line of radius is broken or interrupted according to whether or not it is necessary to locate the center. e.g., R150*

5.5.4 Sphere

See Fig. 23.

Use symbol S∅ to represent Spherical Diameter and SR for Spherical Radius. The symbols S∅ and SR should precede the dimensional value.

Here, S stands for Sphere and R stands for Radius.

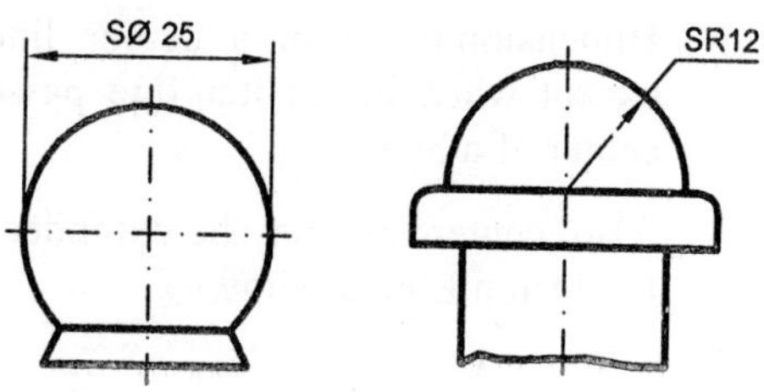

FIG. 23

5.5.5 Square

See Fig. 24.

Symbol □ is used to represent a square and it should precede the dimensional value. Square ends may also be indicated by diagonals drawn as *thin lines*.

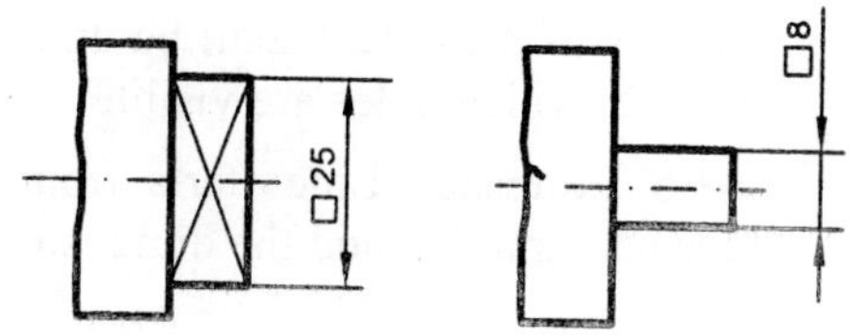

FIG. 24

5.6 RULES FOR DIMENSIONING

INCORRECT | CORRECT

1. **M**ark the dimensions outside the view.

   (However, Diameter of circle / Radius of arc may be shown inside.)

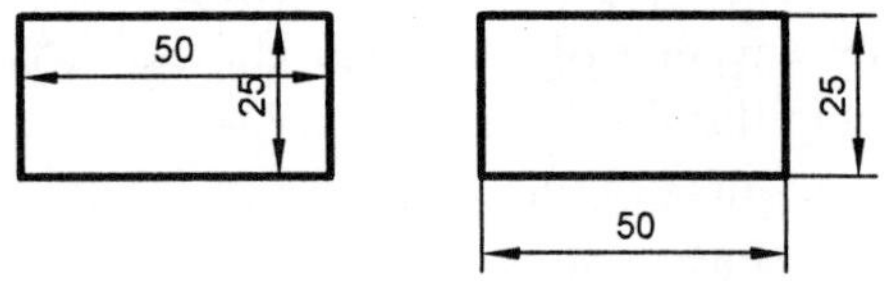

2. **D**o not repeat the same dimension in different views. Note that every dimension must be shown, but none should be shown more than once.

   Also, dimensions should not be placed very near to the parts being dimensioned.

FRONT VIEW | SIDE VIEW

3. **A** circle shall be dimensioned by its diameter symbol ∅ (and an arc by its radius symbol R). Convention ∅ or R shall be placed before the dimensional value.

   Also, Center Lines (Axes) should extend approximately 3 mm beyond the outline of the part whose symmetry they indicate.

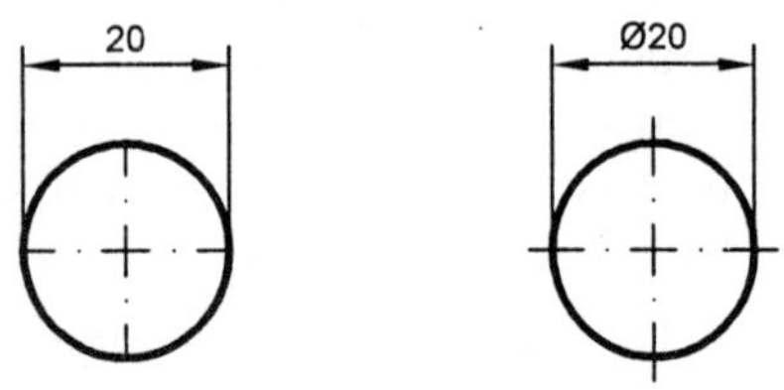

4. **C**enter line (axis) itself shall not be used as a dimension line with arrowheads at its ends.

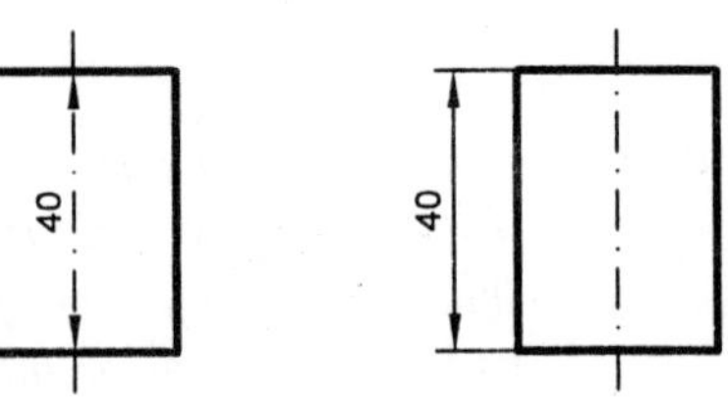

5. **D**imensioning from a center line is incorrect, except when the center line passes through the center of a hole.

   Also, center line may be extended to serve as an Extension Line as shown.

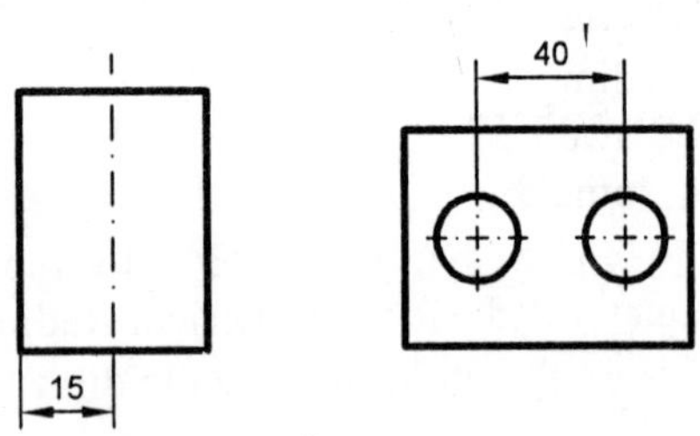

6. **L**ocation of holes (*i.e.*, distance between the centers of the holes) shall be dimensioned in the view in which holes are visible.

   Also, Extension Line starts from the view and extend 2 mm beyond the dimension line.

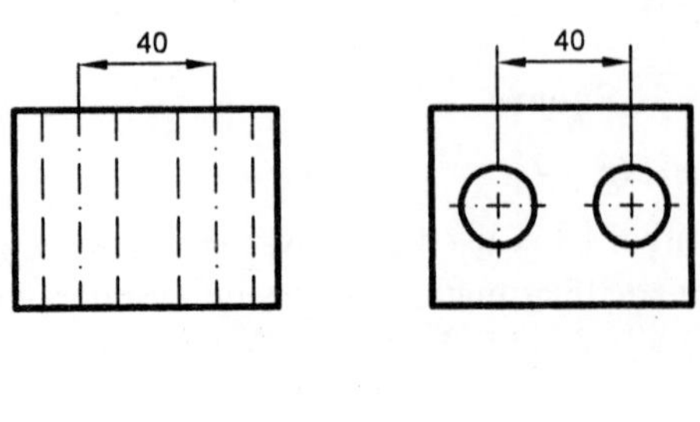

INCORRECT CORRECT

7. **D**imensions shall be given to visible lines and not to hidden (invisible) lines.

   Also, dimension value should be placed little (approximately 2 mm) above the dimension line and not on the dimension line.

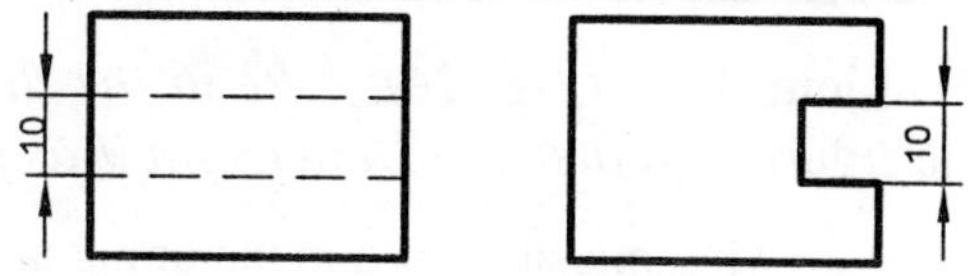

8. **T**aper on diameter shall be dimensioned as shown.

   Example: Taper $= \dfrac{D - d}{L} = \dfrac{1}{10}$

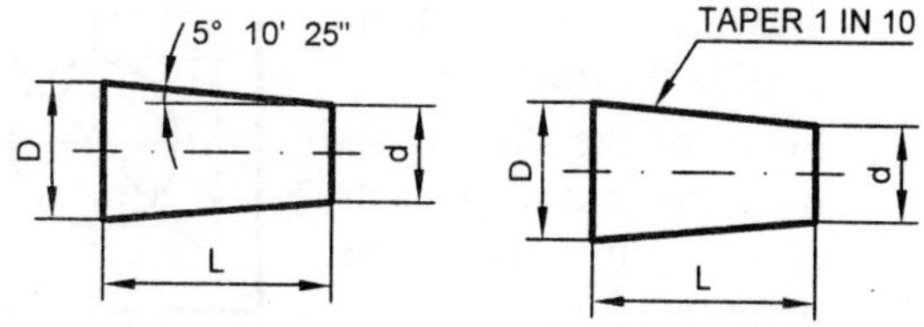

9. **I**f dimensioning inside a hatched portion of a drawing is unavoidable, the hatching lines should not cut the dimensional text.

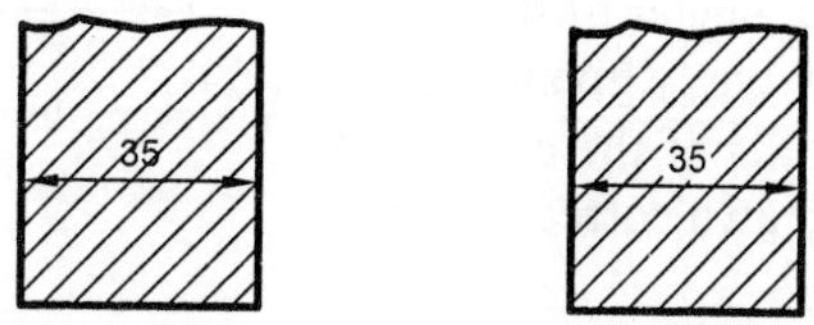

10. **R**epeated Features: Repeated features of the same size are marked to avoid repeating the same dimensional value.

    Notes should always be written horizontally. *Leader line* shall be inclined at an angle of 30°, 45° or 60° to the horizontal as shown.

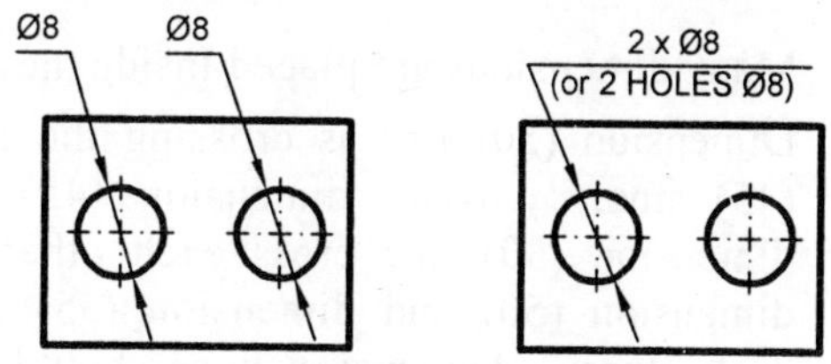

11. **O**verall Dimensioning: **O**verall dimension shall be placed outside the intermediate dimensions. i.e., smaller dimensions shall be placed nearer the view and the larger farther away so that extension lines do not cross dimension lines. (Extension lines may cross each other or the outlines of the drawing, such as the case of leader lines.)

    When an overall dimension is shown, one of the intermediate dimensions should not be marked. See Fig. 25.

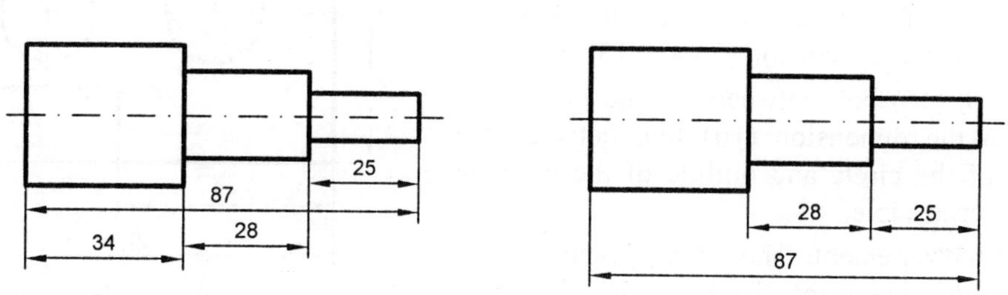

INCORRECT CORRECT

FIG. 25

5.7 PROBLEMS ON DIMENSIONING

Problem 1: *(Fig. 26) Read the dimensioned drawing of GASKET shown. Redraw the figure in full size. Dimension it as per BIS Code.*

The following mistakes of dimensioning are noted in *Fig. 26(i)*.

1. Most of the dimensions are placed inside the drawing. Place them outside the drawing.
2. Some of the dimension lines are crossing each other. It is not permitted.
3. Diameter of the circle is wrongly given as 30Ø.
4. Axis of the circle is not shown correctly.

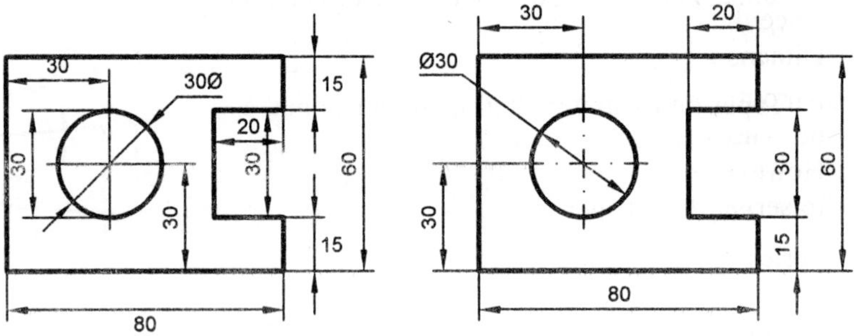

(i) INCORRECT (ii) CORRECT

FIG. 26

Problem 2: *(Fig. 27) Redraw the LOCK PLATE. Mark the dimensions as per BIS.*

1. Many dimensions are placed inside the drawing. Mark them outside the drawing.
2. Dimension (50) line is crossing the dimension (45) line. Also, dimension (45) line and dimension (30) line cross each other. Also, dimension (50) and dimension (35) line cross each other. Dimension lines should not cross other dimension lines.
3. Dimensional value (50) should be placed in the middle of the dimension line.
4. Center lines of both the holes should be drawn. Distance between center lines should be marked.
5. Diameters of the holes are not shown properly. Follow the convention of *Repeated Features* for the two holes using leader line and *Note* (2 x Ø 20 or 2 HOLES Ø 20) written horizontally.
6. Extensions of outlines of the drawing are wrongly used as dimension (35 and 15) lines.
7. Dimension (15) line between outside the circles and also the dimension (10) line between the outline of the circle and outline of the drawing are not permissible.
8. Overall measurements should be given. But, only overall width of the drawing is shown. Mark the overall length too.

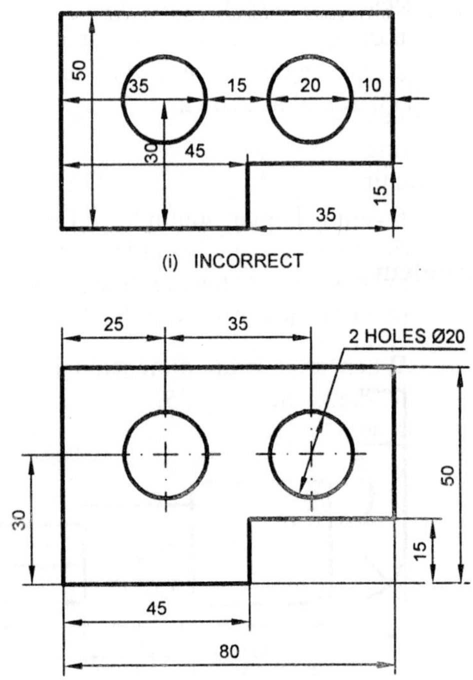

(ii) CORRECT

FIG. 27 LOCK PLATE

Problem 3: *(Fig. 28) Redraw the TEMPLATE to half size. Dimension it as per BIS code.*

Note: Draw the figure to Half Size (Scale 1 : 2). Mark actual dimensions on the drawing.

1. Dimension (12) is wrongly marked, since the outline of the object is used as Dimension Line.

   An outline or a center line should never be used as a Dimension Line. However, a center line may be extended to serve as an Extension Line.
2. Series of dimensions (45, 45 and 12) are not in a continuous line, which is not proper. They should be arranged on a continuous line.
3. The longer dimension (180) is wrongly placed inside the shorter dimension (75).
4. Arc dimension (R23) is wrongly marked as (Ø46).
5. Angle (60°) is not properly dimensioned in both the places. The same should be placed by either of the two methods.
6. Unit of dimension **mm** is marked for the dimensional values (75 and 180), which is not correct. A foot-note stating 'ALL DIMENSIONS IN mm' is written in a prominent place near the Title Block.

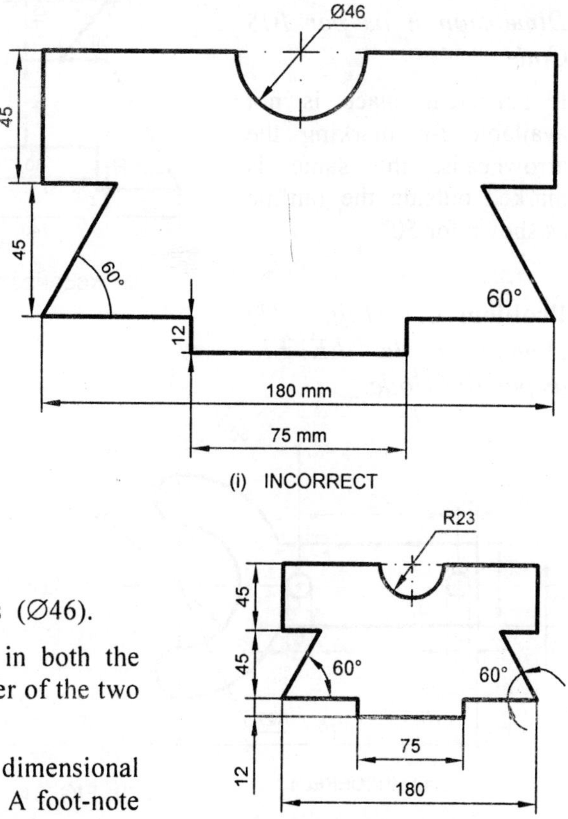

(i) INCORRECT

(ii) CORRECT SCALE 1 : 2

FIG. 28

Problem 4: *(Fig. 29) Redraw the SLIDING PLATE. Dimension it as per BIS Code.*

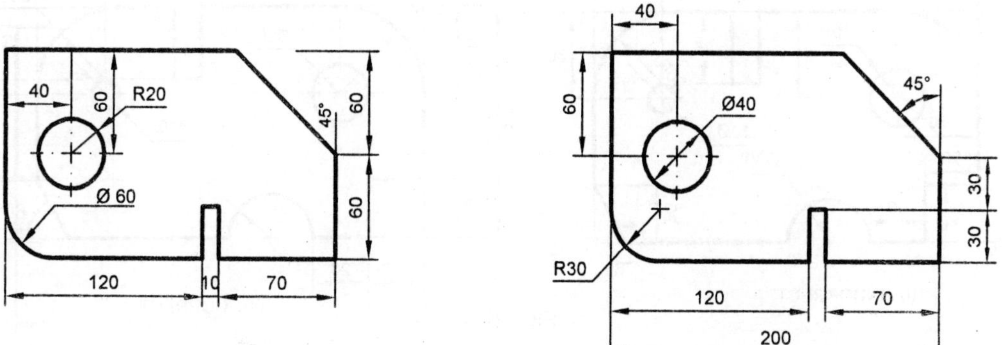

(i) INCORRECT FIG. 29 SLIDING PLATE (ii) CORRECT

Problem 5: *(Fig. 30) Redraw the M.S. PLATE. Dimension it as per BIS Code.*

If sufficient space is not available for marking the arrowheads, the same is marked outside the outline as shown for 50°.

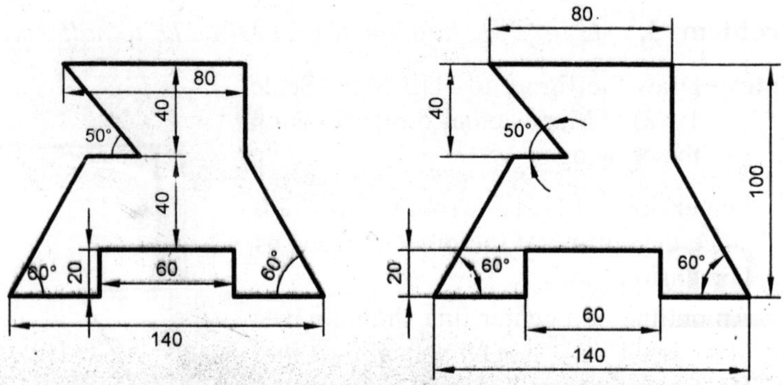

(i) INCORRECT FIG. 30 (ii) CORRECT

Problem 6: *(Fig. 31) Dimension the LEVER. as per BIS Code.*

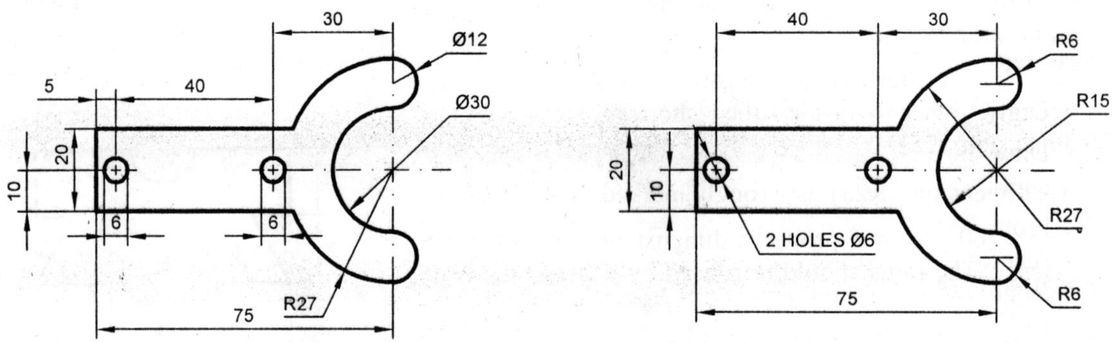

(i) INCORRECT FIG. 31 (ii) CORRECT

Problem 7: *(Fig. 32) Redraw Fig. 32(i). Dimension it as per BIS Code.*

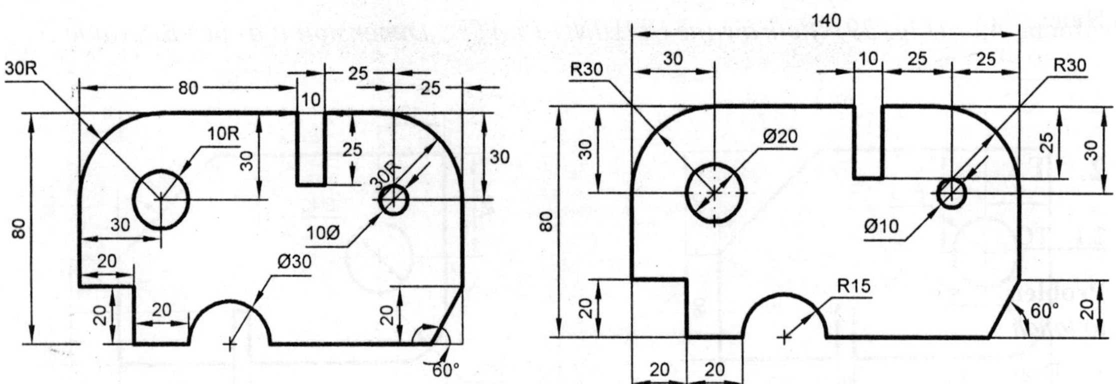

(i) INCORRECT FIG. 32 (ii) CORRECT

3

GEOMETRICAL CONSTRUCTIONS

This Chapter deals with some of the important 2D Geometrical Constructions which will be frequently used in Engineering Drawings.

1. TO DIVIDE A LINE INTO ANY NUMBER OF EQUAL PARTS

Problem 1: *(Fig. 1) Divide a straight line of 80 mm length into six equal parts.*

1. Draw a line AB of given length 80 mm.
2. Draw a *thin* line AC of any length inclined at a *convenient acute angle*, say, 30° or 45° to AB.
3. From A and along AC, step-off (i.e., mark successively) 6 equal divisions of any *convenient length* using the divider. Name the division points as 1, 2, 3, …, 6 on AC. Join B and 6.
4. Draw lines parallel to B6 through 1, 2, 3, 4 and 5 to cut AB at 1′, 2′, 3′, 4′ and 5′. The points 1′, 2′, etc., are division points dividing AB into six equal parts.

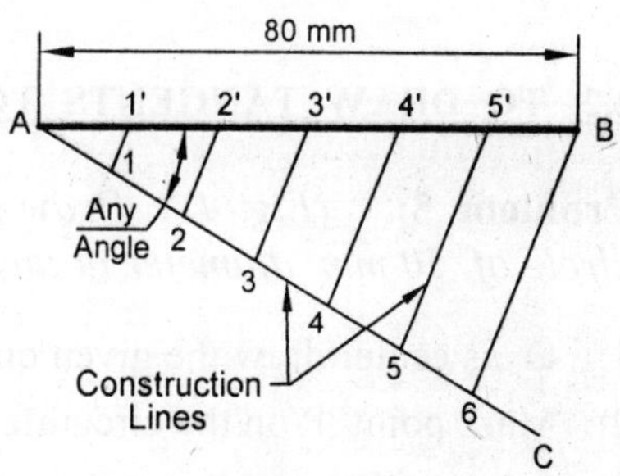

FIG. 1

Note: In the application of any Geometrical Construction, use 2H pencil for the construction lines. Never rub-off these construction lines after completing the problems.

2. TANGENT ARCS TO LINES AND TANGENTS TO CIRCLES

2.1 TO DRAW ARCS TANGENTIAL TO LINES

Problem 2: *(Fig. 2) Draw an arc of radius R = 25 mm to touch 2 given lines AB and AC at 90° to each other.*

1. Draw AB and AC at right angles. A as center and given radius R, draw arcs to cut AB at M and AC at N.
2. M and N as centers and radius R, draw arcs to intersect each other at O.
3. O as center and the same radius R, draw the *tangent arc* NM to touch AC and AB.

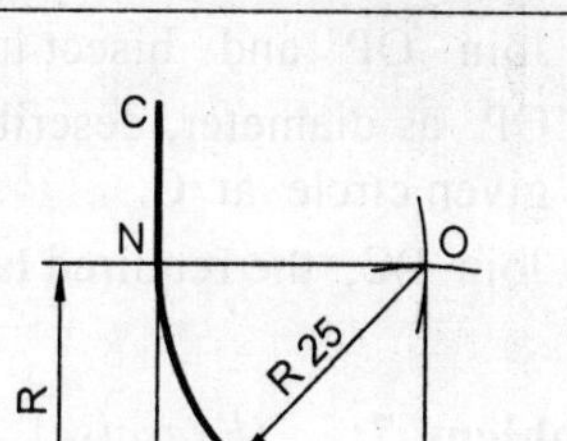

FIG. 2

Problem 3: *(Fig. 3) Draw an arc of a given radius R = 20 mm touching two given straight lines AB and AC which make an acute angle of 59° between them.*

1. Draw the given acute angle CAB.
2. Draw a line PQ parallel to and at a distance = R (given radius) from AB. That is, draw PQ tangential to the two arcs drawn at a distance of R from AB.
3. Draw a line DE parallel to and at a distance of R from AC meeting PQ at O as shown.
4. O as center and R as radius, draw the *tangent arc* PD to touch AB and AC.

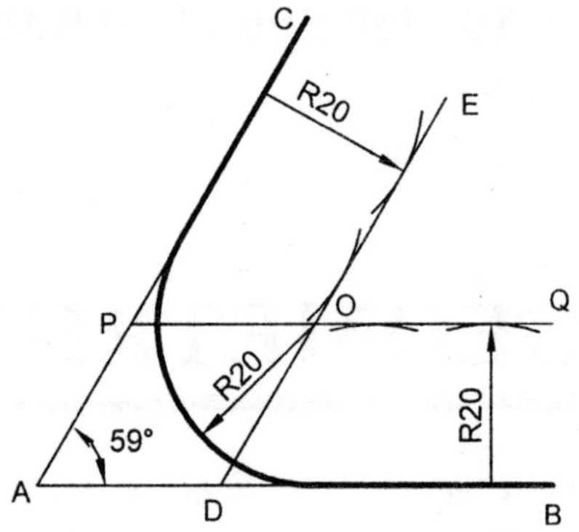

FIG. 3

Problem 4: *(Exercise) Draw an arc of radius 22 mm touching two given straight lines which make an obtuse angle of 135° between them.*

Hint: Follow the *same* procedure given in Problem 3 for *obtuse angle* also.

2.2 TO DRAW TANGENTS TO CIRCLES

Problem 5: *(Fig. 4) Draw a tangent to a given circle of 50 mm diameter at any point P on it.*

1. O as center draw the given circle.
2. Mark point P on the circumference of the circle.
3. Join OP. Through P draw line RS perpendicular to PO. RS is the tangent at P.

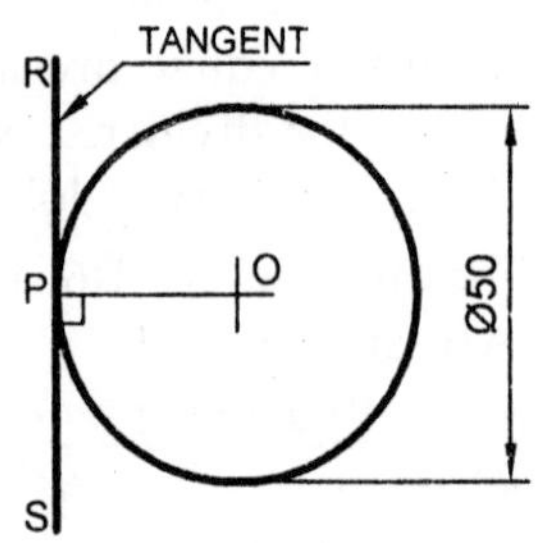

FIG. 4

Problem 6: *(Fig. 5) Draw a tangent to a circle of 48 mm diameter from a point P 58 mm away from its center.*

1. O as center draw the given circle. Mark P outside the circle such that OP = 58 mm.
2. Join OP and bisect it.
3. OP as diameter, describe a semi-circle to cut the given circle at C.
4. Join PC, the required tangent.

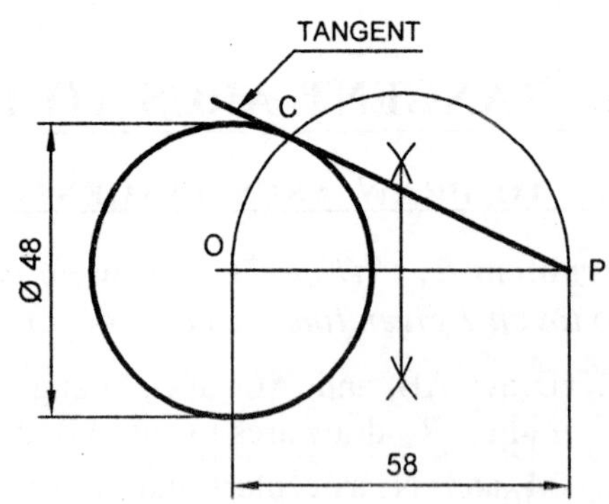

FIG. 5

Problem 7: *(Exercise) Draw tangents to a circle of 52 mm diameter*
i. at any point on it, and
ii. at a point 52 mm away from its center.

2.3 TO FIND THE CENTER OF AN ARC

Problem 8: *(Fig. 6) Find the center of a given arc.*

1. Draw the given arc AB.
2. Draw two chords PQ and RS of any length on AB.
3. Draw perpendicular bisectors of PQ and RS to cut at O. O is the required center of the given arc.

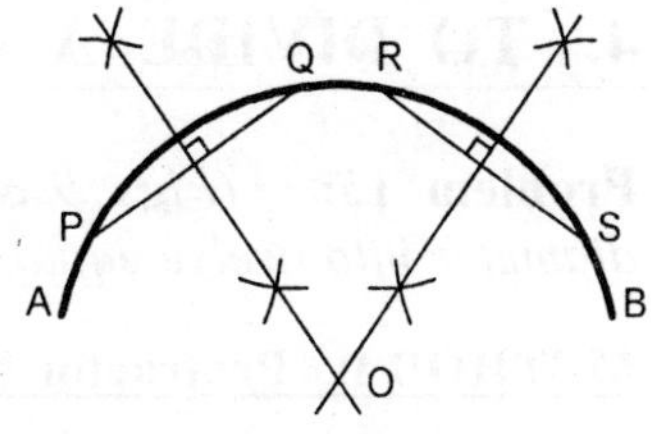

FIG. 6

2.4 TO DRAW AN ARC PASSING THROUGH THREE GIVEN POINTS

Problem 9: *(Fig. 7) Draw an arc passing through three given points not in a straight line.*

1. Mark the given points A, B and C.
2. Join AB and BC. Draw perpendicular bisectors of AB and BC to meet at O.
3. O as center and radius OA or OB or OC, draw the required arc, passing through the three points.

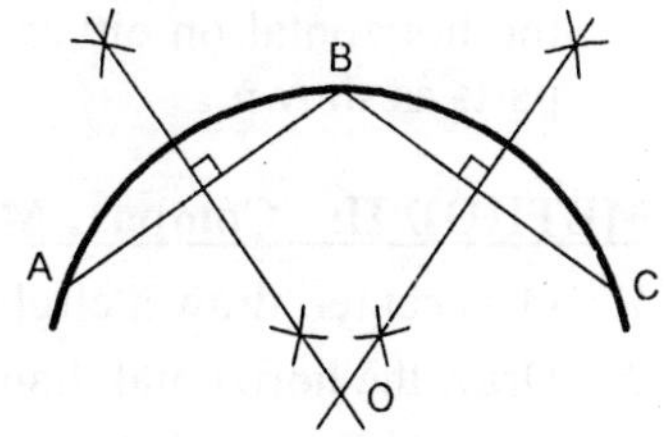

FIG. 7

Note: *Above construction is known as **"Describing a Circle about a ΔABC"**.*

Problem 10: *(Exercise) Describe a circle about a triangle of sides 60, 70 and 75 mm.*

3. LENGTH OF THE CIRCUMFERENCE OF A CIRCLE

Problem 11: *(Fig. 8) Find the length of circumference of a circle of 50 mm diameter.*

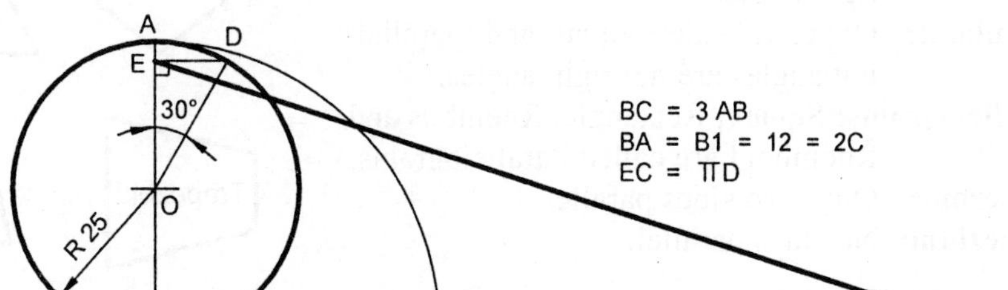

FIG. 8

1. O as center, draw the given circle. Draw a vertical diameter AB through O.
2. At B draw a tangent BC equal to 3 times AB as shown.
3. Draw a radius OD at 30° to OA. From D, draw DE perpendicular to AB.
4. Join E and C. Now EC is approximately = Circumference of the given circle.

Problem 12: *(Exercise) Find the circumference of a circle of 52 mm diameter.*

4. TO DIVIDE A CIRCLE INTO TWELVE EQUAL PARTS

Problem 13: *(Figs. 9 & 10) Divide a circle of 50 mm diameter into twelve equal parts.*

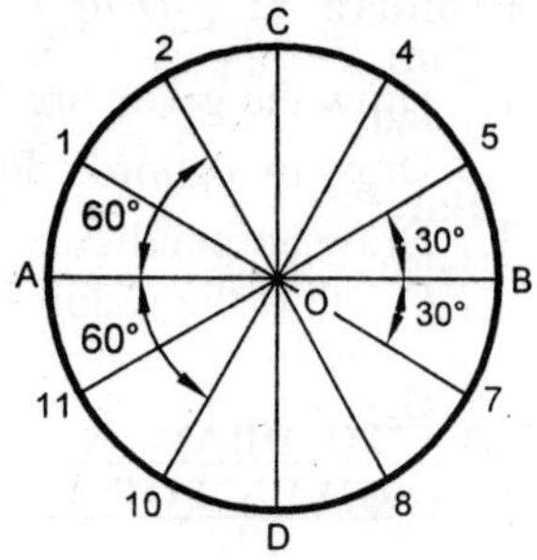

FIG. 9

METHOD I: Protractor Method (Fig. 9)

1. O as center, draw a circle of 50 mm diameter.
2. Draw the horizontal diameter AB and vertical diameter CD.
3. Through O, draw the diameters inclined at 30° and 60° to the horizontal on either side to cut the circle into 12 equal parts as shown.

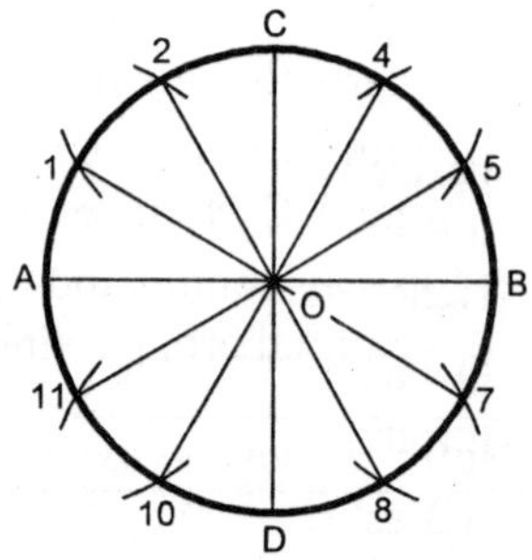

FIG. 10

METHOD II: Compass Method (Fig. 10)

1. O as center, draw a circle of 50 mm diameter.
2. Draw the horizontal diameter AB and vertical diameter CD.
3. With A, B, C and D as centers and radius equal to the radius of the circle, draw arcs on either side to cut the circle into 12 equal parts as shown.
4. Connect the division points 1, 2, 3, ..., 12 to the center O to obtain the radial lines.

5. IMPORTANT GEOMETRICAL TERMS

1. **QUADRILATERALS (Fig. 11)**
   - i) **Rhombus:** All sides equal, but angles are not right angles.
   - ii) **Rhomboid:** Opposite sides equal and parallel, but angles are not right angles.
   - iii) **Parallelograms:** Square, Rectangle, Rhombus and Rhomboid are called Parallelograms.
   - iv) **Trapezoid:** Only two sides parallel.
   - v) **Trapezium:** No sides parallel.

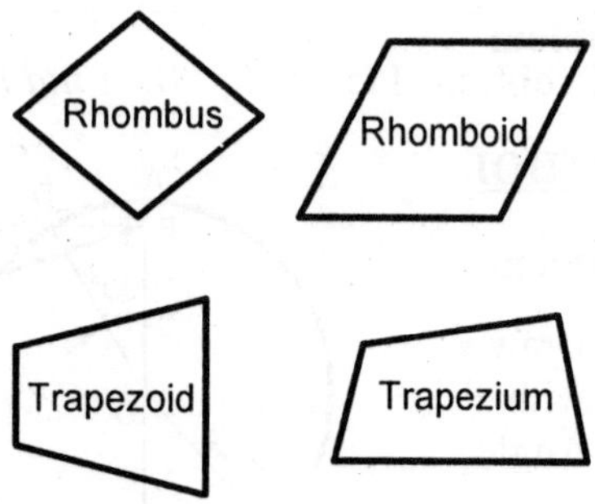

FIG. 11

2. **PARTS OF A CIRCLE (Fig. 12)**
   - i) **Chord** AB is a straight line joining the ends of an arc.
   - ii) A **Segment** is the part of a circle bounded by an arc and its chord.
   - iii) A **Sector** is the part of a circle contained by two radii and the arc between them.

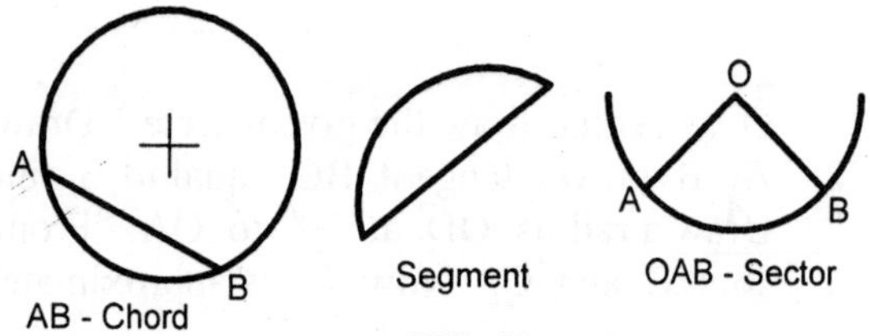

FIG. 12

3. **POLYGONS – Next Page**

6. POLYGONS

Polygon: *Polygon* is a plane figure bounded by more than four lines. Regular polygons include five-sided pentagon, six-sided hexagon, seven-sided heptagon and eight-sided octagon.

Regular Polygon: When all the sides of a polygon are equal, it is known as a *regular polygon*, otherwise *irregular*. **Unless specified as irregular, *polygon* means *regular polygon* only.**

6.1 CONSTRUCTION OF REGULAR POLYGONS, Given Length of Side

METHOD I – Using Base Angle of a Side

Problem 14: *(Fig. 13) Construct a polygon (say, pentagon) of 30 mm side.*

1. Draw AB = length of side = 30 mm.
2. Draw an isosceles triangle OAB with AB as base and base angles of 54°. [For the pentagon, Angle subtended at the center of the isosceles triangle = 360°/5 = 72°. Hence, its Base Angle = (180° – 72°) / 2 = 54°.]
3. O as center and OA as radius, draw a circle passing through A and B.
4. AB as radius, cut the circle successively at C, D and E.
5. Join BC, CD, DE and EA. ABCDE is the required pentagon.

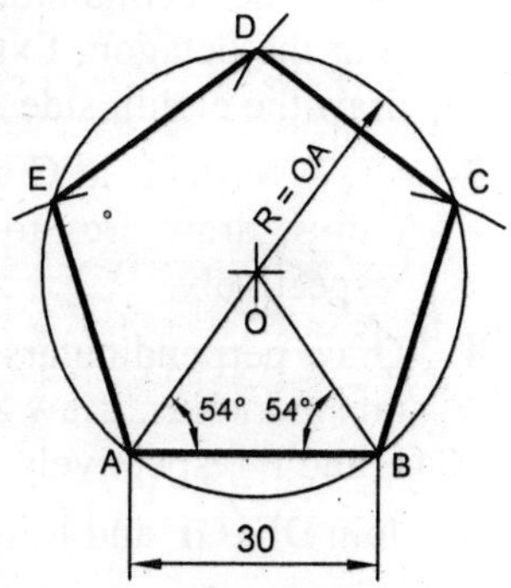

FIG. 13

Note: ***Above Method is Common for All the Polygons.***

Problem 15: *(Exercise) Construct a hexagon of side 28 mm when one side is vertical.*

METHOD II – Exterior Angle Method

Problem 16: *(Fig. 14) Construct a polygon, say, hexagon of length of side 25 mm.*

1. Draw AB = 25 mm. Draw the second side BC = 25 mm with the Exterior Angle of 60°. [For hexagon, Exterior Angle = 360°/6 = 60°.]
2. Similarly, draw AF = 25 mm with the Exterior Angle of 60°.
3. Draw perpendiculars at A and B. C and F as centers and 25 mm as radius, draw arcs to cut the perpendiculars at the points D and E respectively.
4. Join CD, DE and EF. ABCDEF is the required hexagon.

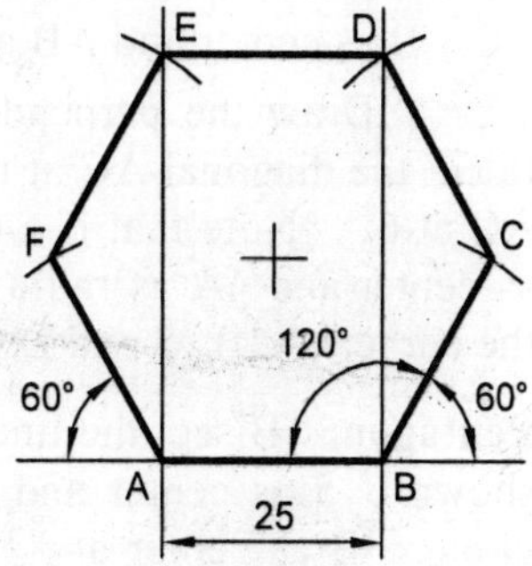

FIG. 14

Note: ***Above Method is Common for All the Polygons.***

Problem 17: *(Fig. 15) Construct a pentagon of length of side 28 mm.*

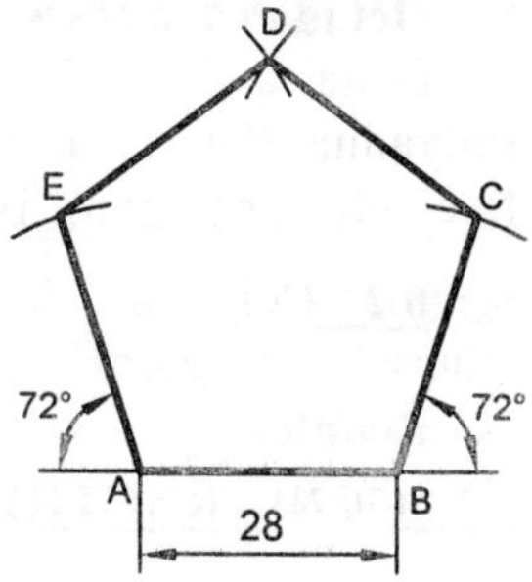

FIG. 15

1. Draw AB = 28 mm.
2. Draw the second side BC = AB with the exterior angle of 72°. [For pentagon, Exterior Angle = 360°/5 = 72°.] Similarly, draw the fifth side AE = AB with the exterior angle of 72°.
3. C and E as centers, draw two arcs of radius AB to intersect each other at D.
4. Join CD and ED. ABCDE is the required pentagon.

Problem 18: *(Fig. 16) Draw an* ***octagon*** *given the length of side 25 mm.*

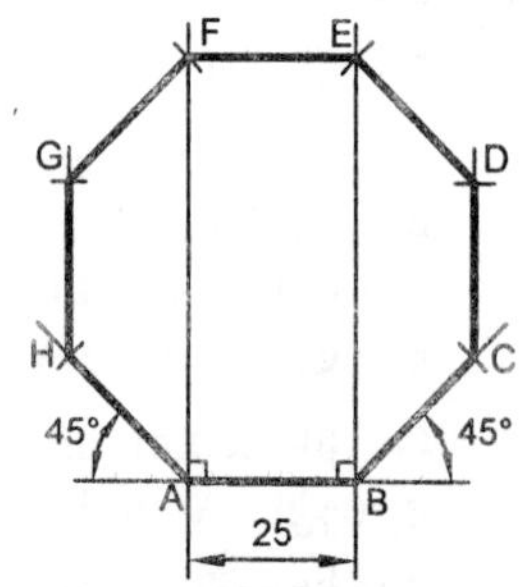

FIG. 16

1. Draw AB = 25 mm.
2. Draw the second side BC = AB with the exterior angle of 45°. [For the octagon, Exterior Angle = 360°/8 = 45°.] Similarly, draw the eighth side AH = AB with the exterior angle of 45°.
3. Draw verticals at C and H. C and H as centers and 25 mm as radius, draw arcs to cut the verticals at the points D and G respectively.
4. Draw perpendiculars at A and B. D and G as centers and 25 mm as radius, draw arcs to cut the perpendiculars at the points E and F respectively.
5. Join DE, GF and EF. ABCDEFGH is the required octagon.

METHOD III

Problem 19: *(Fig. 17) Construct a regular polygon of any number of sides, given the length of its sides equal to 25 mm.*

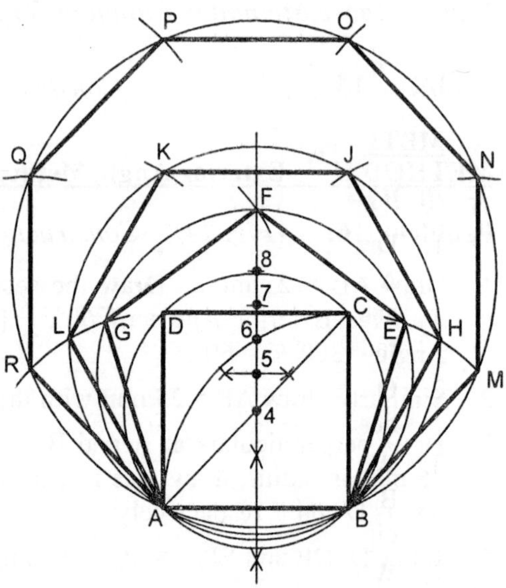

FIG. 17

Square: Draw a line AB of length 25 mm. Construct the square ABCD. Join the diagonal AC. B as center and AB as radius, draw an arc.

Draw the perpendicular bisector of AB to cut the diagonal AC at the point 4 and the arc AC at 6. Note that if a circle is drawn with 4 as center and 4A as radius, it will pass through the corners A, B, C and D of the square.

Pentagon: Bisect the line 46 at the point 5 as shown. 5 as center and 5A as radius, draw a circle. B as center and BA as radius, draw an arc to cut the circle at the point E.

Similarly, fix F and G. ABEFG is the required pentagon.

Hexagon: 6 as center and 6A as radius, draw a circle. AB as radius cut this circle successively to obtain the required hexagon ABHJKL.

Octagon: 6 as center and 45 as radius, cut the perpendicular bisector of AB at the point 7. 7 as center and with the same radius, cut the bisector of AB at the point 8. 8 as center and 8A as radius, draw a circle. AB as radius, cut this circle successively to obtain octagon ABMNOPQR.

6.2 CONSTRUCTION OF REGULAR HEXAGON, Given Length of Side

Problem 20: *(Figs. 18 & 19) Construct a regular hexagon of side 28 mm when one side is (a)* ***horizontal*** *and (b)* ***vertical****.*

METHOD I (Fig. 18)

(a) When one side is Horizontal

1. O as center and the given side length 28 mm as radius, draw a circle. Mark the radius OA *horizontally*.
2. A as center and given side length as radius, draw an arc to cut the circumference of the circle at B.
3. Similarly, mark points C, D, E and F. Now, six division marks are obtained on the circumference of the circle.
4. Join the above division marks to obtain the required hexagon ABCDEF.

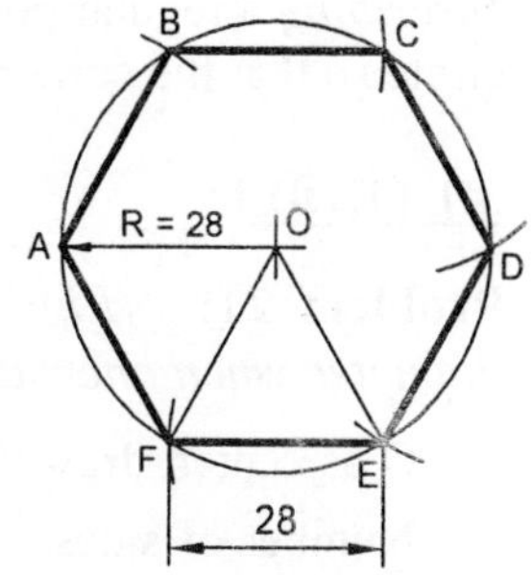

OFE - Equilateral Triangle

FIG. 18

Note: For hexagon, Δ OFE is an Equilateral Triangle. Therefore, Radius of the circle = Side of the hexagon.

(b) When one side is Vertical: Mark the radius OA vertically and complete the hexagon.

METHOD II (Fig. 19)

(i) When one side is Horizontal: Fig. (i)

1. Draw AB = 28 mm. For hexagon, Δ OAB is equilateral. Hence, A and B as centers and AB as radius, draw two arcs to meet at O.
2. O as center and OA as radius describe a circle.
3. With the same radius and A and B as centers, draw arcs to cut the circle at F and C respectively.
4. With the same radius and C and F as centers, draw arcs to cut the circle at D and E respectively.
5. Join BC, CD, DE, EF and FA. ABCDEF is the required hexagon.

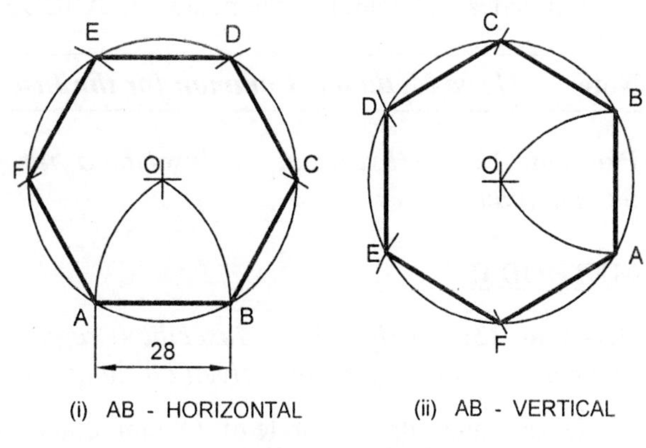

(i) AB - HORIZONTAL (ii) AB - VERTICAL

FIG. 19 HEXAGON

(i) When one side is Vertical: Fig. (ii). Mark radius AB vertically and complete the hexagon.

7. INSCRIPTION OF POLYGONS

Note: 1. *If a polygon has all its vertices on a circle, it is said to be* ***inscribed*** *in the circle. The above circle is said to be* ***circumscribed*** *about the polygon.*

2. *If a polygon has its sides tangential to a circle, it is said to be* ***circumscribed*** *about the circle. The circle is said to be* ***inscribed*** *in the polygon.*

Inscribing a regular polygon *in* a given circle is the construction of the largest polygon inside that circle so that the corners of the polygon are on the circumference of the circle.

METHOD I

Problem 21: *(Fig. 20) Inscribe a regular polygon of sides, say 5, in a circle of 64 mm diameter when one side is* ***horizontal****.*

1. O as center, draw the given circle of 64 mm diameter.
2. Number of sides = n = 5. Angle subtended at the center by one side of the pentagon = $360^\circ / 5 = 72^\circ$.
3. To draw one side of the pentagon to be *horizontal*, draw a *vertical* radial line OP. On either side of this vertical line, mark 36° at the center and draw the corresponding two radial lines from the center.
4. Mark the points of intersection of the radial lines with the circumference of the circle as AB. Join the two intersecting points to obtain the horizontal side of the pentagon.
5. Similarly, mark the remaining 4 subtended angles of 72° at the center. Obtain the other four sides and complete the pentagon ABCDE.

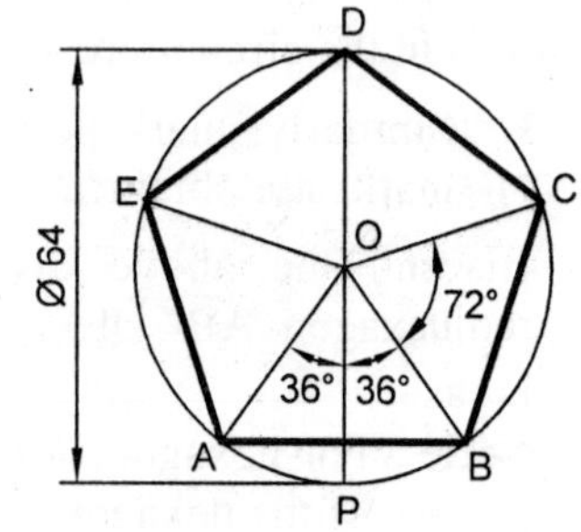

FIG. 20

Note: ***Above Method is Common for the Inscription of All Polygons.***

Problem 22: *(Exercise) Inscribe a hexagon in a circle of 68 mm diameter.*

METHOD II

Problem 23: *(Fig. 21) Inscribe a regular polygon of any number of sides, say 5, in a given circle of 68 mm diameter.*

1. O as center, draw a circle of 68 mm diameter.
2. Draw the horizontal diameter PN. Divide it into 5 equal parts (same number of parts as the number of sides). Name them as 1, 2, 3 and 4.
3. P and N as centers and PN as radius, draw arcs cutting each other at M.

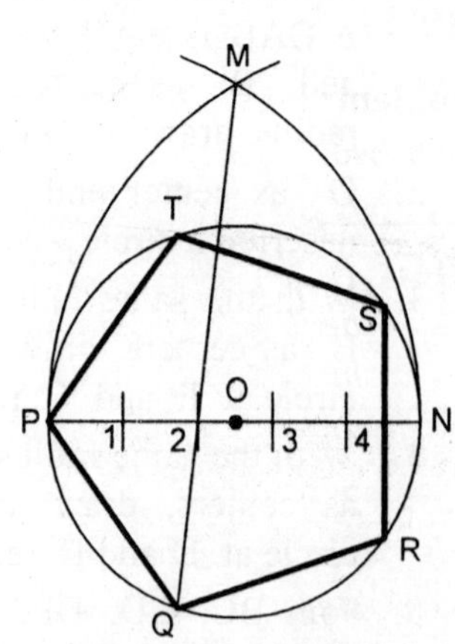

FIG. 21

4. Join M2. Extend it to meet the circle at Q. PQ = a side of the pentagon.
5. Starting from Q, mark the points R, S and T on the circle such that QR = RS = ST = TP = **PQ**.

Note: ***Above Method is Common for the Inscription of All Polygons.***

Problem 24: *(Exercise) Inscribe a square in a circle of 60 mm diameter.*

Hint: O as center, draw the given circle. Draw the vertical and horizontal diameters AB and CD. Draw lines AC, CB, BD and DA to obtain the square.

Problem 25: *(Exercise) Inscribe a regular hexagon, heptagon and octagon in a circle of 72 mm diameter.*

7.1 INSCRIPTION OF PENTAGON

Problem 26: *(Fig. 22) Inscribe a regular pentagon in a circle of 70 mm diameter.*

1. O as center, draw a circle of 70 mm diameter.
2. Draw diameters DP and TS perpendicular to each other.
3. Bisect TO at point M. M as center and MD as radius, draw an arc to cut the diameter TS at N.
4. D as center and DN as radius, draw an arc to cut the circle at points C and E.
5. C and E as centers and DN as radius, draw arcs to cut the circle at the points B and A respectively.
6. Join AB, BC, CD, DE and EA, thus completing the required pentagon.

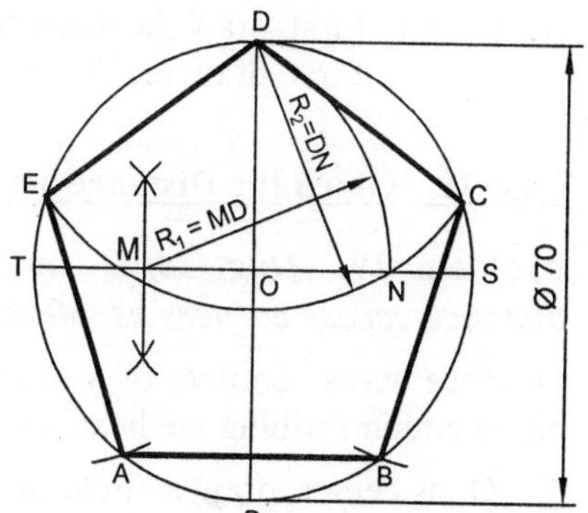

FIG. 22

7.2 INSCRIPTION OF HEXAGON

Case I : Given the Diameter of the Circle

Problem 27: *(Fig. 23) Inscribe a regular hexagon in a given circle of 70 mm diameter, when two sides of the hexagon are (a) horizontal and (b) vertical.*

Note: *1. For one side of the hexagon to be **horizontal**, the starting point for marking equal divisions should be on an end of the **horizontal** diameter.*

*2. For one side of the hexagon to be **vertical**, the starting point for stepping-off equal divisions should be on an end of the **vertical** diameter.*

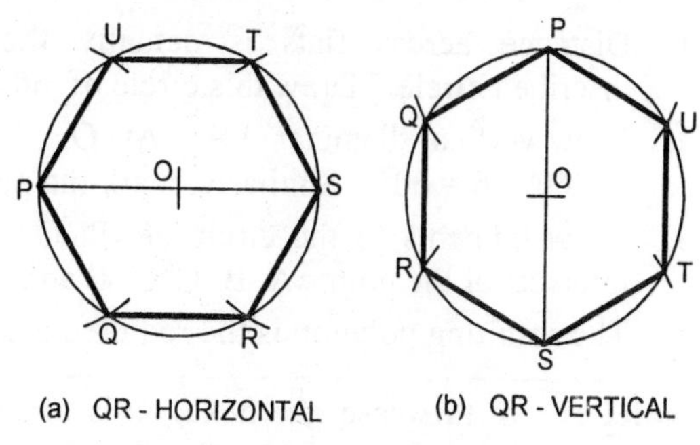

(a) QR - HORIZONTAL (b) QR - VERTICAL

FIG. 23 INSCRIPTION OF HEXAGON

(a) When two sides to be Horizontal

1. O as center, draw a circle of 70 mm diameter. Mark the horizontal diameter PS.
2. With the same radius of 35 mm and starting from point P on the circle, set-off six divisions on the circumference of the circle using the divider. *See Fig. (a).*
3. Join the division points in proper sequence to obtain the hexagon P Q R S T U.

(b) When two sides to be Vertical: *See Fig. (b).*

Important Note: ***Above method may also be used for the construction of hexagon, given the Length of a Side. In this case, draw the Circle of Radius = Given Length of Side.***

Problem 28: *(Exercise) Inscribe a hexagon in a circle of 34 mm radius, when two sides of the hexagon are required to be (i) horizontal and (ii) vertical.*

Hint: (i) First, draw the diameter of the given circle horizontally.
(ii) First, draw the diameter of the given circle vertically.

Case II: Given the Distance across Corners of the Hexagon

Problem 29: *(Fig. 24) Construct a regular hexagon given the distance across corners as 60 mm.*

Distance across corners of a hexagon is actually the diameter of the circle circumscribing the hexagon.

1. O as center, draw a circle of diameter 60 mm equal to the given distance across corners of the hexagon.
2. Draw the horizontal diameter AD. With centers A and D and radius equal to 30 mm (= radius of the circle), draw arcs to cut the circumference of the circle at points B, F and C, E respectively.
3. Join AB, BC, CD, DE, EF and FA to obtain the required hexagon.

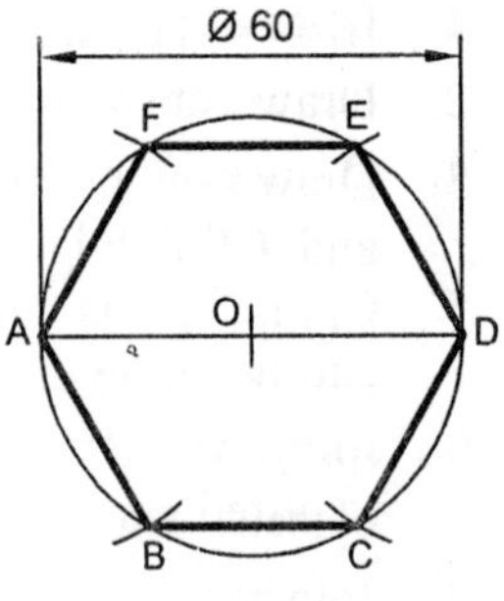

FIG. 24

Case III : Given the Distance across Flats of the Hexagon

Problem 30: *(Fig. 25) Construct a regular hexagon given the distance across flats as 60 mm.*

1. Distance across flats is actually the diameter of the inscribed circle. Draw this circle of 60 mm diameter.
2. Draw vertical diameter PS. At O, draw the angle POQ = 360° / 6 = 60°. Similarly, mark the points R, T and U.
3. Draw tangents to the circle at P, Q, R, S, T and U to intersect at the points A, B, C, D, E and F.
4. The resuiting polygon is the required hexagon.

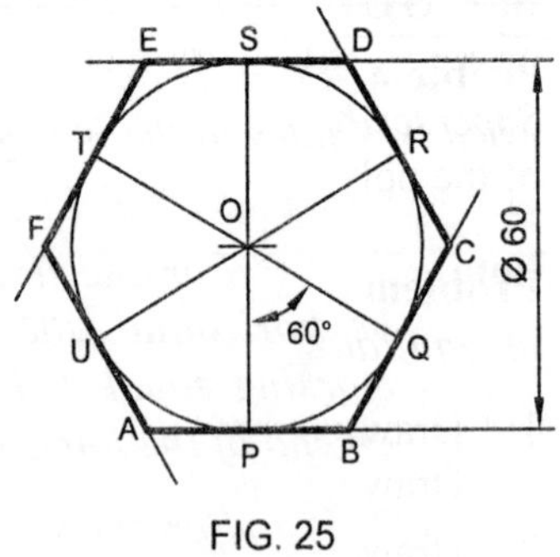

FIG. 25

Note: In this case, the hexagon is ***described*** about a circle, whose Diameter is equal to the given Distance Across Flats of the hexagon.

7.3 INSCRIPTION OF HEPTAGON

Problem 31: *(Fig. 26) Inscribe a regular heptagon inside the given circle of 68 mm diameter.*

1. O as center, draw the given circle of 68 mm diameter.
2. Draw the horizontal diameter AB.
3. A as center and AO as radius, draw an arc cutting the circle at the points E and F.
4. Join EF to meet AO at the point P. Now EP = FP = Length of the side of the heptagon.
5. From any point, say A, step-off divisions equal to EP on the circumference of the circle.
6. Join the division points A1, 12, 23, 34, 45, 56 and 6A. Thus obtain the heptagon.

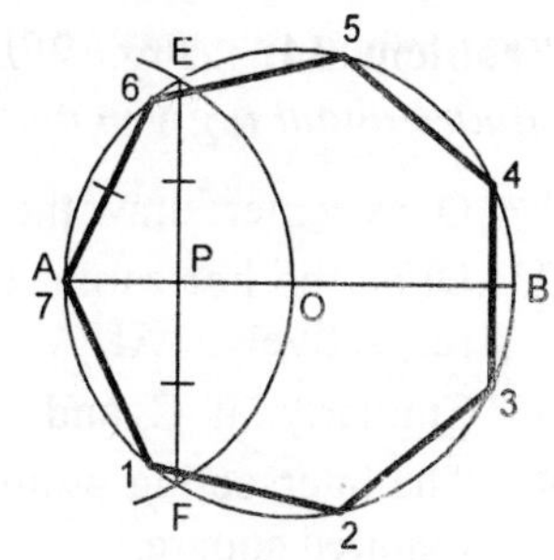

FIG. 26

5.4 INSCRIPTION OF OCTAGON

Problem 32: *(Fig. 27) Inscribe an octagon in a circle of 66 mm diameter.*

1. O as center , draw a circle of 66 mm diameter.
2. Draw the horizontal diameter AB and vertical diameter CD.
3. Draw the diameters EF and GH bisecting the angles AOC and COB respectively as shown.

   [O as center and any length as radius, draw arcs cutting OA and OC at 1 and 2 respectively. 1 and 2 as centers and the same or any length as radius, draw arcs to cut at 3. O3 bisects the angle AOC.]
4. Join AE, EC, CH, HB, BF, FD, DG and GA to obtain octagon.

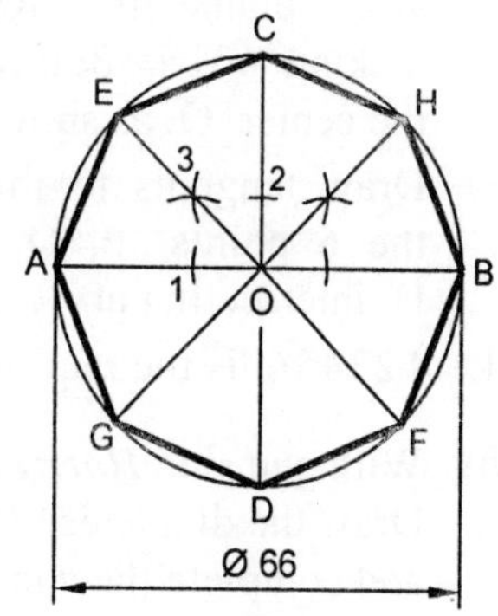

FIG. 27

8. SUPERSCRIPTION / CIRCUMSCRIPTION OF POLYGONS

Superscribing / Describing / Circumscribing a regular polygon about a circle is the construction of the polygon, keeping the sides tangential to the given circle.

Problem 33: *(Fig. 28) Superscribe / Describe / Circumscribe an equilateral triangle about a circle of 50 mm diameter.*

1. Draw the given circle of 50 mm diameter with center O. Draw vertical radius OA.
2. Draw radii OB and OC such that angles AOB and AOC are each equal to 120°.
3. At A, B and C, draw tangents to the circle to obtain the required equilateral triangle PQR.

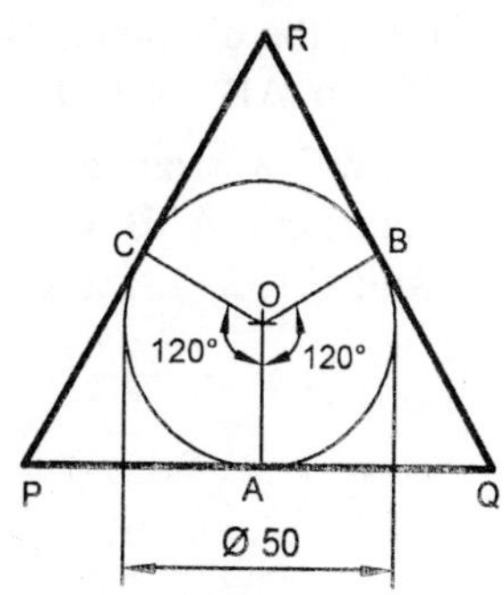

FIG. 28

Problem 34: *(Fig. 29) Superscribe / Describe / Circumscribe a square about a given circle of diameter 58 mm.*

1. O as center, draw the given circle of 58 mm diameter.
2. Draw the horizontal and vertical diameters of the circle AB and CD respectively. At A and B, draw vertical lines tangential to the circle.
3. Similarly, at C and D draw horizontal lines tangential to the circle.
4. The intersecting points obtained are P, Q, R and S. PQRS is the required square.

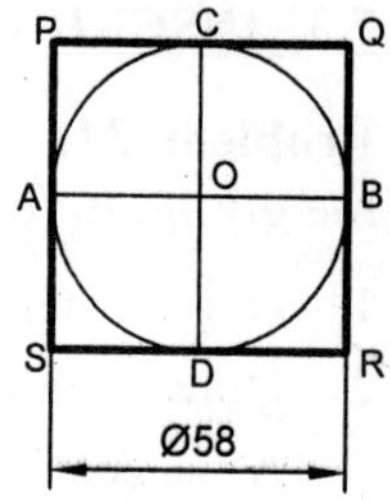

FIG. 29

Problem 35: *(Fig. 30) Superscribe / Describe / Circumscribe a hexagon about a circle of 60 mm diameter,* with one side (a) *vertical and* (b) *horizontal.*

a) With one side ***Vertical*** *[Fig. (a)]*

1. Draw the given circle with center O.
2. Draw the *horizontal* diameter PQ and diameters RS and TU making 60° (= 360°/6) with PQ at the center O as shown.
3. Draw tangents to the circle at all the 6 points P, Q, R, S, T and U intersecting at 1, 2, ..., 6.
4. 123456 is the required hexagon.

b) With one side ***Horizontal*** *[Fig. (b)]* Draw the diameter PQ to be vertical and complete the construction.

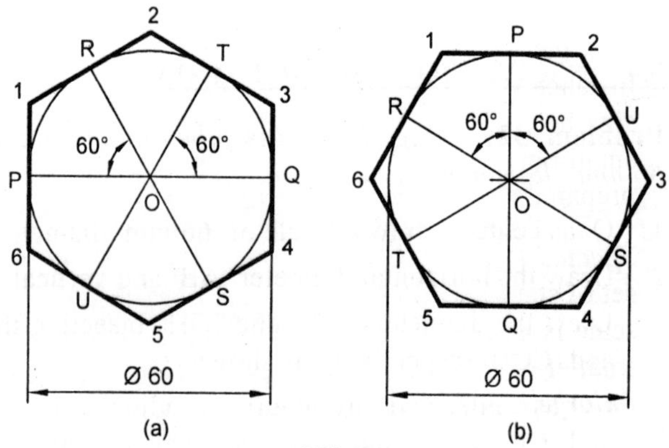

FIG. 30

Problem 36: *(Fig. 31) Superscribe a regular octagon about a given circle of 70 mm diameter.*

1. Draw the given circle with center O and diameter equal to 70 mm.
2. Draw the horizontal diameter AB and the vertical diameter CD.
3. Draw the diameters EF and GH inclined at an angle of 45° to AB or CD.
4. Draw 8 tangents to the circle at the ends of the diameters A, B, C, ..., H to intersect one another at the points 1, 2, 3, ..., 8. Complete the octagon 1 2 3 ... 8.

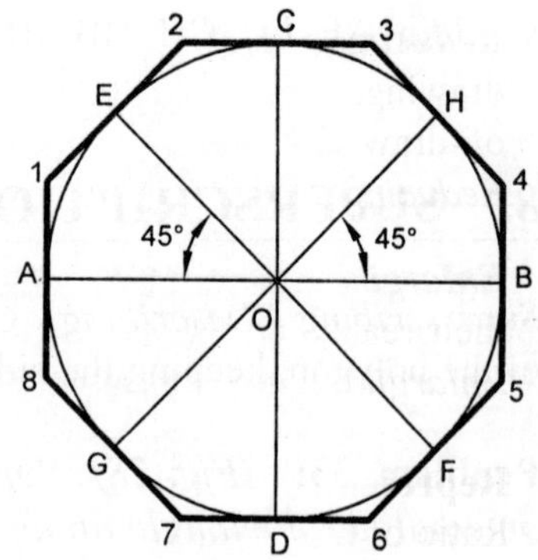

FIG. 31 OCTAGON

*

4

ENGINEERING SCALES

1. TERMINOLOGY

It is not always possible or convenient to draw drawings of an object to its actual size. For instance, drawings of very big objects like buildings, machines, etc., cannot be prepared in full size because they would be too big to accommodate on the drawing sheet. Drawings of very small objects like precision instruments, *namely*, watches, electronic devices, etc., also cannot be prepared in full size because they would be too small to draw and to read.

Special Unusual Scales: When you require to make measurements other than those given by the set of standard scales M1 to M8, *Special Unusual Scales* have to be constructed. A convenient scale is always chosen to prepare the drawings of big as well as small objects in proportionately smaller or larger sizes. So, the scales are used to prepare a drawing at *Full Size, Reduced Size* or *Enlarged Size*.

Full Size Scale: If we show the actual length of an object on a drawing, then the scale used is *Full Size Scale*. *Designation of Full Size Scale:* Scale 1 : 1.

Reducing Scale: If we reduce the actual length of an object so as to accommodate that object on drawing, then the scale used is called *Reducing Scale*. Such scales are used for the preparation of drawings of large machine parts, buildings, architectural drawings, etc. *Designation of Reducing Scale:* Scale 1 : x. (Example : Scale 1 : 20)

Enlarging or Increasing Scale: Drawings of instruments, watches, etc., are made larger than their real size. These are said to be drawn in an *Enlarging* or *Increasing Scale*. *Designation of Enlarging Scale:* Scale x : 1. (Example : Scale 20 : 1)

Representative Fraction (R.F.)
Ratio of the drawing size of an object to its actual size is called *Representative Fraction.*

$$\text{i.e.,}\quad \text{R.F.} = \frac{\text{Drawing size of an object}}{\text{Its actual size}} \quad (\text{ in same units })$$

Reducing Scale: For *Reducing Scale*, the drawings will have R.F. values of less than unity. For example, if 1 centimeter on a drawing represents 1 meter length of an object,

$$\text{then}\quad \text{R.F.} = \frac{1\text{ cm}}{1 \times 100\text{ cm}} = \frac{1}{100} < 1$$

Enlarging Scale: For drawings using *Enlarging Scale*, R.F. values are > 1. For example, when 1 mm length of an object is shown by a line of length of 1 cm on the drawing,

$$\text{then} \quad \text{R.F.} = \frac{1 \times 10 \text{ mm}}{1 \text{ mm}} = \frac{10}{1} = 10 > 1$$

Metric Measurements:

10 decimeters (dm)	=	1 meter (m)
10 meters (m)	=	1 decameter (dam)
10 decameters (dam)	=	1 hectometer (hm)
10 hèctometer (hm)	=	1 kilometer (km)

Types of Scales

1. PLAIN SCALES
2. DIAGONAL SCALES
3. VERNIER SCALES
4. COMPARATIVE or CORRESPONDING SCALES
5. SCALE OF CHORDS

2. PLAIN SCALES

It is a line divided into a number of equal parts; the first part is further sub-divided into small parts. It represents either two units or a unit and its fraction such as km and hm, etc.

Note: *1. Before constructing a scale, it is necessary to know: (a) Its R.F., (b) Maximum length to be measured and (c) Divisions it has to show.*

2. If the length of scale and distance to be marked are not given in the problem, then assume the scale length = 15 cm.

Problem 1: *(Fig. 1) Construct a plain scale to show meters when 1 centimeter represents 4 meters and long enough to measure up to 50 meters. Find the R.F. and mark on your scale a distance of 36 meters.*

1. $\text{R.F.} = \frac{\text{Drawing size}}{\text{Actual size}} \text{ (in same units)} = \frac{1 \text{ cm}}{4 \times 100 \text{ cm}} = \frac{1}{400}$

2. Length of scale = L = R.F. x Maximum length to be measured (50 m)

$$= \frac{1}{400} \times 50 \text{ m} = \frac{1}{400} \times 50 \times 100 \text{ cm} = 12.5 \text{ cm}$$

3. Draw a horizontal line of length L = 12.5 cm. Draw a rectangle of size 12.5 cm x 0.5 cm on the horizontal line drawn above.

Note: *Width of the scale is usually taken as 5 mm.*

4. **Main Unit:** Total length to be measured is 50 m. Therefore divide the rectangle into 5 (= n) equal divisions, each division representing 10 m. [For dividing the length L into n number of equal parts, use the Geometrical Construction Method in **Chapter 3.**]

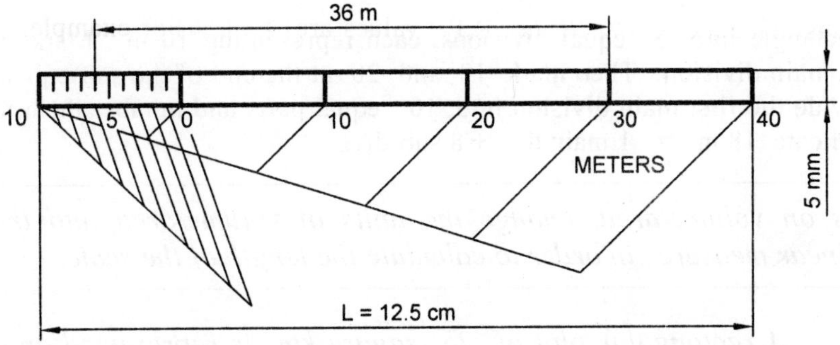

Note: Never rub-off the construction lines.

R.F. = 1/400

FIG. 1 PLAIN SCALE

5. Mark 0 (zero) at the end of the first main division. From 0, number 10, 20, 30 and 40 at the end of subsequent main divisions towards **right** as shown.

Note: *In general purpose scales, zero is marked at the beginning. But, in engineering scales zero will be marked at the end of the first main division. This will enable sub-dividing the first main unit into sub-divisions.*

---

6. **Sub-divisions of Main Unit :** Then sub-divide the first main division into 10 sub-divisions to represent meters. Number the sub-divisions *i.e.*, meters to the **left** of 0.
7. Write the names of the main units and sub-units in capitals (METERS) below the scale. Also indicate the R.F. value as shown.
8. Indicate on the scale a distance of 36 meters
= 3 main divisions to the **right** side of 0 (zero) + 6 sub-divisions to **left** of 0 (zero).

Problem 2: *(Exercise) Construct a plain scale of R.F. 1 : 50,000 to show kilometers and hectometers and long enough to measure up to 7 kilometers. Measure a distance of 54 hectometers on your scale.*

Hint: Length of scale $L = \frac{1}{50,000} \times 7 \text{ km} = \frac{1}{50,000} \times 7 \times 1000 \times 100 \text{ cm} = 14 \text{ cm}.$

Problem 3: *(Fig. 2) A room of 1000 m^3 volume is represented by a block of 125 cm^3 volume. Find the R.F. and construct a plain scale to measure up to 30 m. Measure a distance of 18 m on your scale.*

1. 125 cm^3 represents 1000 m^3, i.e., 5 cm represents 10 m.
2. $\text{R.F.} = \frac{125 \text{ cm}^3}{1000 \text{ m}^3} = \frac{1 \text{ cm}}{2 \text{ m}} = \frac{1 \text{ cm}}{2 \times 100 \text{ cm}} = \frac{1}{200}$
3. Length of scale = L = R.F. x 30 x 100 cm = 15 cm.

R.F. = 1/200

FIG. 2 PLAIN SCALE

4. Draw and divide the rectangle into 3 equal divisions, each representing 10 m. Mark 0 (zero) at the end of first main division. Then mark 10 and 20 at the end of the subsequent main divisions. Sub-divide the first main division into 10 equal parts and number them to the left of 0 (zero). Indicate 18 m = 1 main div. + 8 sub-div.

Note: *In the problems on volume/area, change the units of volume/area into the corresponding linear measures in order to calculate the length of the scale.*

Problem 4: *(Exercise) A rectangular plot of 25 square km is represented on a certain map by a similar rectangle of area of 1 square cm. Find the R.F. Draw a plain scale to show kilometers. Indicate a distance of 55 mm on it.*

Problem 5: *(Exercise) Draw a plain scale of R.F. 1/35,000 to show kilometers and hectometers and long enough to measure 5 km. Mark a distance of 3.3 km.*

1. L = R.F. x 5 x 1000 x 100 cm = 14.3 cm. This length *cannot be measured accurately* using ordinary scale. Hence increase the maximum measurement of the scale to obtain a length of the scale that can be measured accurately. In this problem a maximum measurement of 7 km will give a length of scale of 20 cm.

2. Draw a line of 20 cm length (L) and divide it into 7 equal divisions, each division representing 1 km. The scale should be long enough to measure 5 km (*Given*). Hence construct the scale only for 5 km and leave the remaining 2 km divisions as *thin* lines.

3. DIAGONAL SCALES

Plain Scales are used to read lengths in two units or to first decimal accuracy. *Diagonal Scales* are used either to measure very minute distances such as 0.1 mm, etc., or to measure in three units such as dm, cm and mm.

Principle of Diagonal Scale (Fig. 3)

1. Let AB be the small length to be divided into 10 equal parts. Draw the rectangle ABCD of, say, 5 cm height. Divide AD into 10 equal divisions. Join AC. Draw horizontals through the division points to meet AC at 1′, 2′, ..., 9′.

2. In the similar Δs ADC and A88′, $\dfrac{88'}{DC} = \dfrac{A8}{AD}$

   But $A8 = \dfrac{8}{10} AD$; $\therefore \dfrac{88'}{DC} = \dfrac{8}{10}$

   i.e. $88' = 0.8\ DC = 0.8\ AB$.

3. Thus horizontal lengths 11′, 22′, 33′, etc., are equal to 0.1 AB, 0.2 AB, 0.3 AB, etc., respectively.

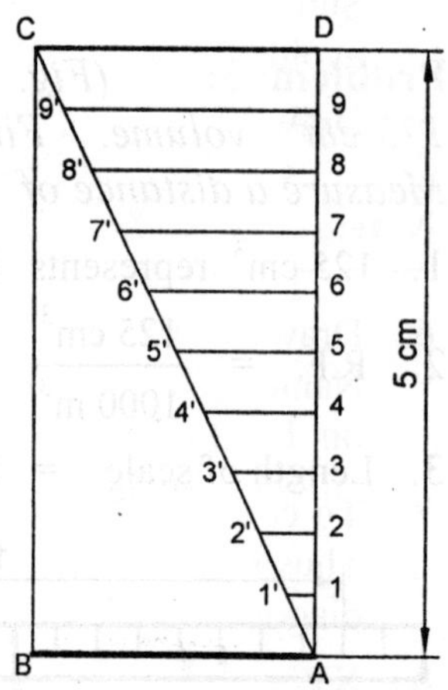

FIG. 3

Problem 6: *(Fig. 4) Construct a diagonal scale of R.F. = 1 : 32,00,000 to show kilometers and long enough to measure upto 400 km. Show distances of 257 km and 333 km on your scale.*

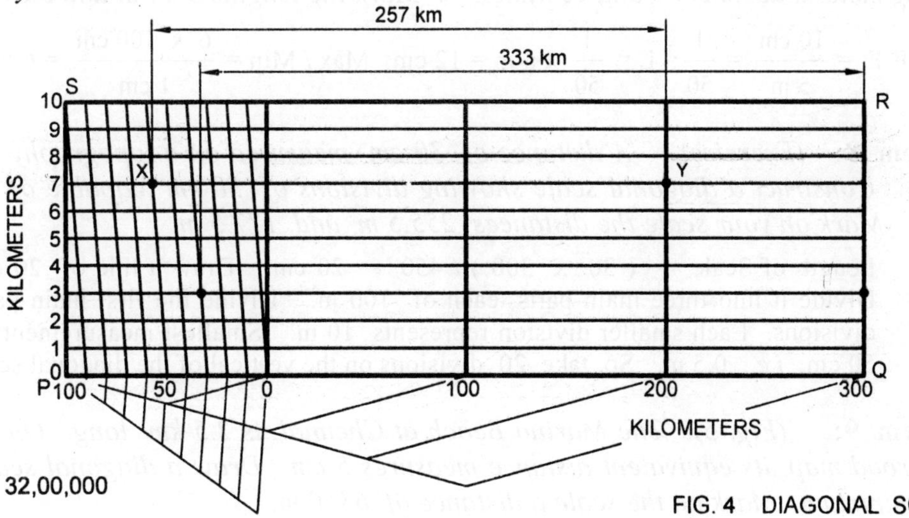

R.F. = 1 / 32,00,000

FIG. 4 DIAGONAL SCALE

1. Length of the scale = L = R.F. x Maximum distance to be measured

$$= \frac{1}{32,00,000} \times 400 \text{ km} = \frac{1}{32,00,000} \times 400 \times 1000 \times 100 \text{ cm.} = 12.5 \text{ cm.}$$

2. Draw a line PQ of 12.5 cm long (L). Maximum length to be measured is 400 km. Minimum distance to be measured = 1 km (which is obtained from data 257 / 333 km).

   ∴ Max / Min = 400 km / 1 km = 400. This can be obtained in 3 steps as 4 x 10 x 10.

3. Divide PQ into 4 main divisions, each representing 100 km. Mark 0 (zero) at the end of the first main division. Also mark 100, 200 and 300 towards the right of zero.
4. Sub-divide the main division into 10 equal sub-divisions, each representing 10 km. To avoid crowding of numbers mark only 50, 100 towards the left of zero.
5. Draw a line PS of 5 cm long perpendicular to PQ. Complete the rectangle PQRS.

Note: *Height PS may generally be taken as 5 cm for convenience.*

---

6. Draw vertical lines from each main division on PQ. Divide PS into 10 equal divisions. Name the divisions as 0, 1,...., 10 from P to S. Draw horizontal lines from each division on PS.
7. **To construct the Diagonal:** Join S to the first sub-division from P on the main scale PQ. Thus first diagonal line is drawn. Similarly draw remaining 9 diagonals parallel to the first diagonal. Thus each 10 km is divided into 10 equal parts.
8. **XY** = 257 km = 200 km on Main division + 50 km on Sub-division + 7 km on Diagonal.

Problem 7: *(Exercise) On the plan of a shopping complex, a line 10 cm long represents a distance of 5 m. Draw a diagonal scale for the plan to read up to 6 m, showing meters, decimeters and centimeters. Mark the lengths 3.24 m and 5.57 m.*

Hint: $\text{R.F.} = \dfrac{10 \text{ cm}}{5 \text{ m}} = \dfrac{1}{50}$; $L = \dfrac{1}{50} \times 6 \text{ m} = 12 \text{ cm}$; $\text{Max / Min} = \dfrac{6 \times 100 \text{ cm}}{1 \text{ cm}} = 6 \times 10 \times 10$

Problem 8: *(Exercise) A distance of 30 cm measured on a topography represents 450 m. Construct a diagonal scale showing divisions of 50 cm capable of measuring 300 m. Mark on your scale the distances 255.5 m and 177.5 m.* (UQ)

Hint: Length of Scale = (30 · x 300) / 450 = 20 cm. Draw a line of 20 cm length. Divide it into three main parts, each of 100 m. Divide the first main part into 10 divisions. Each smaller division represents 10 m. Smallest measurement required is 50 cm, *i.e.*, 0.5 m. So, take 20 divisions on the vertical of the diagonal scale.

Problem 9: *(Fig. 5) The Marina Beach at Chennai, is 2.5 km long. On inspection of the road map, its equivalent distance measures 5 cm. Draw a diagonal scale to read 50 m minimum. Mark on the scale a distance of 6350 m.* (UQ)

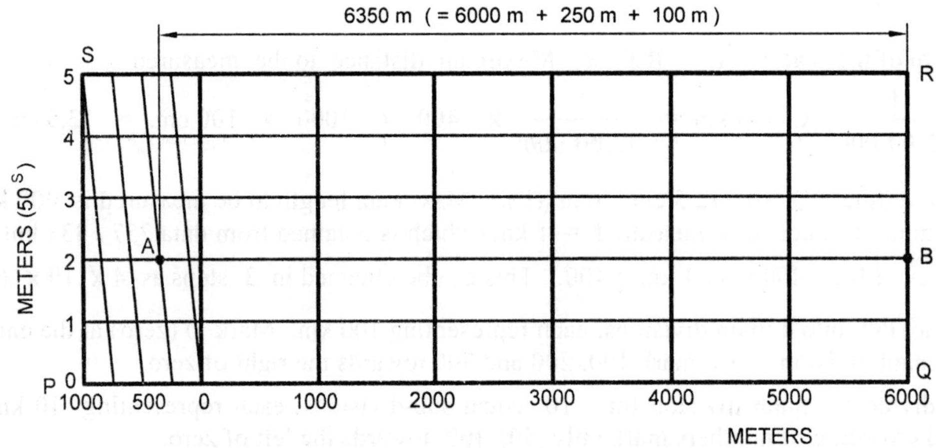

R.F. = 1 / 50,000

FIG. 5 DIAGONAL SCALE

1. $\text{R.F.} = \dfrac{5 \text{ cm}}{2.5 \text{ km}} = \dfrac{1}{50{,}000}$
2. L = R.F. x **7000** x 100 cm = 14 cm (since a distance of **6350** m is to be marked).
3. Draw a rectangle PQRS of 14 cm x 5 cm. Maximum = 7000 m; Minimum = 50 m; Max / Min = 140 = 7 x 4 x 5. Divide PQ into 7 equal parts, each representing 1000 m. Sub-divide the first main division into 4 equal parts, each representing 250 m. Divide PS into 5 equal parts and draw horizontal lines.
4. Complete the diagonals. (Each sub-division = 250 m, divided into 5 equal parts). Hence , 250 m / 5 = 50 m can be read. Mark **AB** = 6000 m + 250 m + 100 m.

Problem 10: *(Exercise) The distance between Delhi and Agra is 200 km and its equivalent distance on map measures 10 cm. Draw a diagonal scale to indicate 223 km and 135 km.* (UQ)

1. R.F. = 10 cm / 200 km = 1 / 20,00,000

Note: *Select the length of scale such that the actual distance corresponding to this selected length is greater than that of the distance of 223 km to be measured.*

---

2. Hence, take the length of the scale as 12.5 cm to represent the actual distance 12.5 cm x (200 km/10 cm) = 250 km (within which the given distance of 223 km can be measured).
3. Max = 250 km ; Min = 1 km ; Max / Min = 250 = 5 x 5 x 10.

Problem 11: *(Fig. 6) Construct a diagonal scale to read up to 1 / 100 th of a meter given R.F. = 1 / 50 and to measure up to 7 m. Indicate a distance of 5.45 m.*

L = 14 cm ; Max / Min = 700 = 7 x 10 x 10.

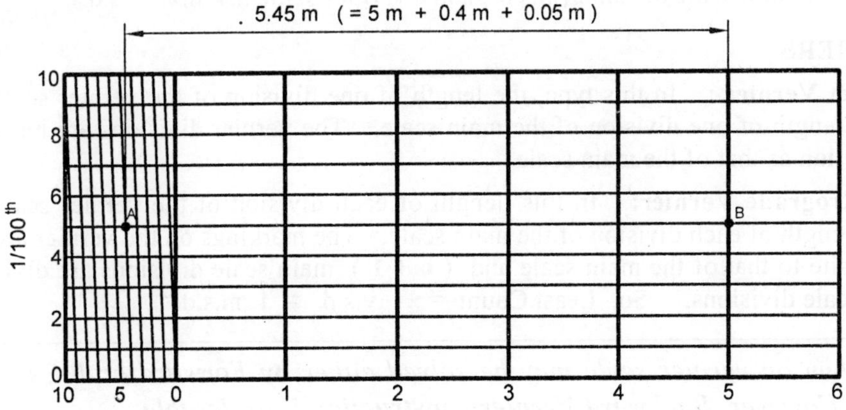

FIG. 6

Problem 12: *(Exercise) An IT Park in a land of area 0.45 hectare is represented on a map by a similar rectangle of 5 sq. cm. Construct a diagonal scale to read up to a maximum of single km from the map. The scale should be long enough to measure up to 500 m.* (UQ)

1 Hectare = 10,000 sq. m.; Therefore, 0.45 Hectare = 45,000,000 sq. cm. 5 sq. cm. on the map represents 0.45 Hectare = 45,000,000 sq. cm. ∴ 1 cm on the map represents 3000 cm. L = (1 / 3000) x 500 x 100 cm = 16.6 cm; Note that 16.6 cm cannot be drawn accurately using the *ordinary scale*. Hence, L is recalculated for the revised value of maximum measurement of 600 m, which will give L = 20 cm.

Problem 13: *(Exercise) A car is moving at a speed of 360 km / hour. Draw a diagonal scale to represent 6 km by 1 cm to show a maximum distance of 60 km. Measure the distance traveled by the car at 6 minutes 10 seconds.* (UQ)

1. L = 100 mm. 360 km is traveled in 1 hour. Hence, 100 m is covered in 1 sec.

2. Draw a line PQ = 100 mm to represent 60 km or 10 minutes. Divide PQ into 10 equal parts, each representing 6 km or 1 minute.
3. Divide the first main division into 6 equal parts, each sub-division representing 1 km or 10 seconds. Draw PR perpendicular to PQ and draw the rectangle PQRS. Divide PR into 10 equal parts. Join the first sub-division with R diagonally.

4. VERNIER SCALES

Vernier Scale is a short scale used when a diagonal scale is inconvenient to use due to lack of space. It consists of two parts, *i.e.*, ***Main Scale*** (which is a Plane Scale fully divided into minor divisions) and a ***Vernier Scale***. Vernier scale slides on the side of the main scale and both of them are used to measure small divisions up to 3 divisions like diagonal scales.

Least Count: It is the smallest distance that is measured accurately by the vernier scale. It is the difference between a *main scale division* (m.s.d.) and a *vernier scale division* (v.s.d.).

TYPES OF VERNIERS

1. **Forward / Direct Vernier :** In this type, the length of one division of the vernier scale is smaller than the length of one division of the main scale. The vernier divisions are marked in the same direction as that of the main scale.
2. **Backward / Retrograde Vernier:** In this, length of each division of the vernier scale is greater than the length of each division of the main scale. The markings on the vernier are in a direction opposite to that of the main scale and (**n + 1**) main scale divisions are divided into **n** vernier scale divisions. So, Least Count = 1 v.s.d. – 1 m.s.d.

Note: *Any problem on vernier scale may be solved either by Forward or Backward Vernier. However,* ***Backward Vernier construction is preferable.*** *Take the width of both plain and vernier scales as 5 mm each.*

Problem 14: *(Fig. 7) Construct a vernier scale to read meters, decimeters and centimeters and long enough to measure up to 4 m. R.F. of the scale is 1/20. Mark on your scale a distance of 2.28 m.* ***(Backward / Retrograde Vernier)*** (UQ)

1. Least Count = Smallest distance to be measured = 1 cm (given) = 0.01 m
2. L = R.F. x Maximum distance to be measured = (1/20) x 4 m = 20 cm
3. **Main Scale:** Draw a line of 20 cm length. Complete the rectangle of 20 cm x 0.5 cm.

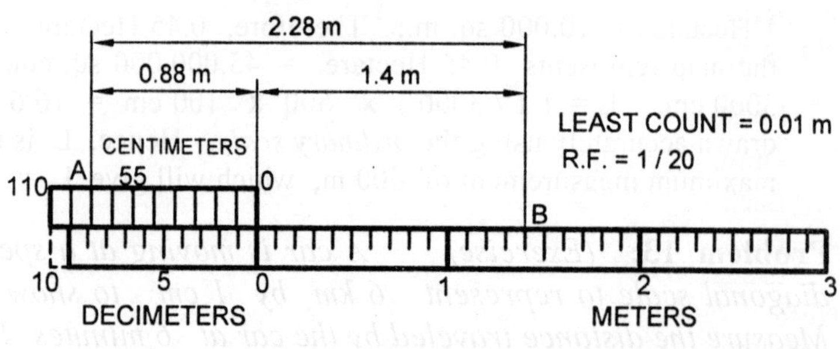

FIG. 7 BACKWARD / RETROGRADE VERNIER SCALE

Divide it into 4 equal parts each representing 1 meter. Sub-divide each part into 10 main scale divisions. Hence 1 m.s.d. = 1m / 10 = 0.1 m = 1 dm.

4. **Backward Vernier:** Take 11 divisions on main scale. Divide it into 10 equal parts on vernier scale. So 1 v.s.d. = 11 m.s.d. / 10 = 11 x 1 dm / 10 = 0.11 m = 1.1 dm = 11 cm.

Mark 0, 55, 110 towards left from 0 on the vernier scale. The units of main divisions is METERS, sub-divisions is DECIMETERS and vernier divisions is CENTIMETERS.

Note: *Given distance is split into two parts, one to be read on main scale and other on vernier scale. Sum of these two parts give the required length.*

---

5. **AB** = 2.28 m = (v.s.d. x 8) + (m.s.d. x 14) = (0.11 m x 8) + (0.1 m x 14) = (**0.88 + 1.4**) m

Problem 15: *(Exercise) Construct a vernier scale of R.F. = 2 to show cm, 1/10th of cm and 1/100th of cm to read up to 9 cm. Mark on the scale the lengths 7.02 cm and 2.25 cm.* ***(Backward / Retrograde Vernier)*** (UQ)

1. Least Count = 1 cm / 100 = 0.01 cm ; L = (2/1) x 9 cm = 18 cm.
2. Draw the main scale of 18 cm length. Divide this into 9 equal parts. Each part is divided into 10 main scale divisions. 1 m.s.d. = 1 cm/10 = 0.1 cm.
3. To construct the vernier, take 11 divisions on the main scale and divide it into 10 equal parts on the vernier. Hence, 1 v.s.d. = 11 x 0.1 cm / 10. = 0.11 cm.
4. Mark **7.02 cm = (0.11 x 2) cm + 6.8 cm**

Problem 16: *(Exercise) Construct a vernier scale to show readings of 1/10th of a meter when 3 cm represents 10 m. Construct the scale to read up to 60 m and mark the distances of 35.3 m and 47.3 m on you scale.* ***(Backward Vernier)***

1. L.C. = 1 m / 10 = 0.1 m.
2. L = (3 cm / 10 m) x 60 m = 18 cm. Draw main scale of 18 cm length. Divide it into 6 equal parts each representing 10 m. Each part is further divided into 10 main scale divisions. So, 1 m.s.d. = 1 m. 1 v.s.d. = 11 x (1 m / 10) = 1.1 m

Problem 17: *(Fig. 8) Construct a vernier scale of R.F. = 1: 2.5 to show decimeters, centimeters and millimeters. The scale should be capable of reading up to 4 decimeters. Mark on your scale the following distances : (a) 3.23 dm and (b) 3.65 dm.* ***(Backward / Retrograde Vernier)*** (UQ)

1. Least Count = 1 dm / 100 = 1 mm; L = (1 / 2.5) x 40 cm = 16 cm.
2. Draw the main scale of 16 cm length. Divide it into 4 equal parts each of 1 dm.
3. Each part is divided into 10 main divisions. Hence, 1 msd = 1/10 x 1 dm = 10 mm.
4. Divide 11 divisions on main scale into 10 equal parts on vernier. 1 vsd = 11 mm.

Note: *To mark the given distances, add* ***one more part*** *on the main scale as shown.*

---

5. Mark **AB = 3.23 dm** and **CD = 3.65 dm.**

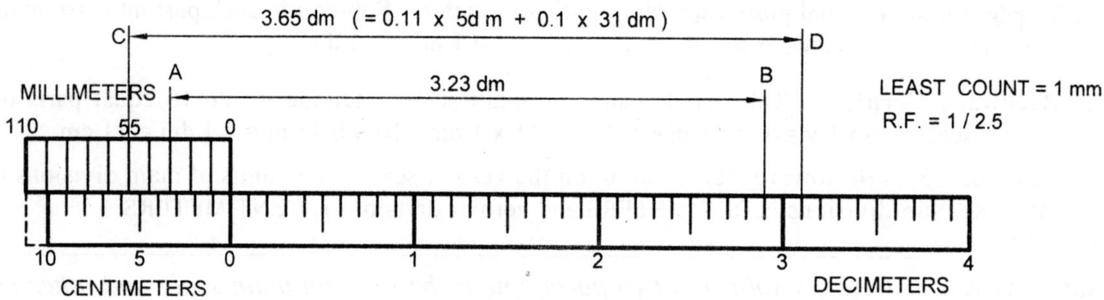

FIG. 8 BACKWARD / RETROGRADE VERNIER SCALE

Problem 18: *(Fig. 9) The actual length of 300 m of an Auditorium is represented by a line of 10 cm on a drawing. Draw a vernier to read upto 400 m. Mark on it, a length of 343 m.* ***(Backward / Retrograde Vernier)*** (UQ)

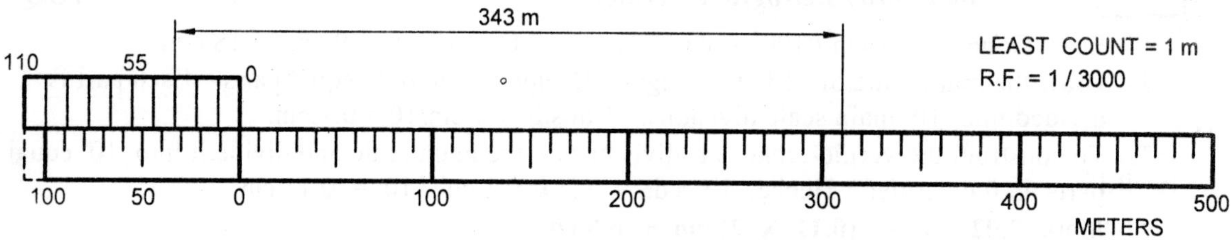

FIG. 9 BACKWARD / RETROGRADE VERNIER SCALE

1. Least Count = 1 m (since the length to be measured is 343 m.)

2. Length of the scale $= \dfrac{10}{300 \times 100} \times 400 \times 10$ cm = 13.333 cm.

Note: *This length of 13.333 cm cannot be measured accurately using ordinary scale. Therefore, to obtain the length of scale as an integer, take the maximum length to be measured as 600 m instead of 400 m.*

Length of Scale for 600 m = L = 600 x 100 x 10 / 30,000 = 20 cm.

3. Draw the main scale as shown. 1 msd = 100 m / 10 = 10 m.
4. Draw the vernier scale as shown. 1 vsd = 11 x 10 m / 10 = 11 m.
5. **343 m** = 3 vsd + 31 msd = (3 x 11 m) + (31 x 10 m).

Problem 19: *(Exercise) Draw a vernier scale of R.F. = 3. Show on it the distance of 3.55 cm.*

Hint: R.F. = 3 / 1 ; Least Count = 0.01 cm; Assume the maximum length to be measured as 4 cm in order to mark the given distance of 3.55 cm.

5. COMPARATIVE OR CORRESPONDING SCALES

When you have to make measurements in one system of measurement from a drawing in another system of measurement, Comparative or Corresponding Scales are used. Both the scales are constructed with the *same* R.F. These scales may be Plain Scales or Diagonal Scales or Vernier Scales and may be constructed either *separately* or *one above the other.*

Problem 20: *(Fig. 10) Draw a comparative scale to be used with a road map drawn to a scale of 1 cm = 5 km. The scale is to measure upto (i) 40 knots and (ii) 60 km. (1 knot = 1.85 km)* (UQ)

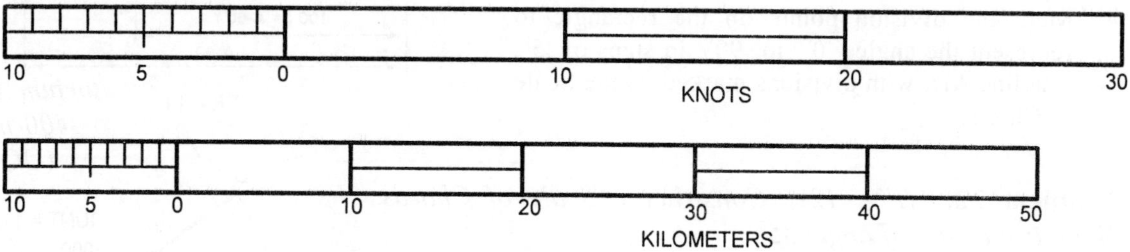

FIG. 10 COMPARATIVE SCALE

1. R.F. = 1 cm / 5 km = $1 / (5 \times 10^5)$
2. Length of scale to represent 40 knots = $1 / (5 \times 10^5) \times 40 \times 1.85 \times 10^6$ = 148 mm.
3. Length of scale representing 60 km = $1 / (5 \times 10^5) \times 60 \times 10^6$ = 120 mm.
4. Construct both the plain comparative scales separately as shown.

Problem 21: *(Exercise) The distance between Mumbai and Pune is 180 km. A car covers this distance in 6 hours. R.F. is 1 / 200,000. Draw a comparative scale to measure time up to a single minute. Show the distance travelled by it in 38 minutes.*

6. SCALE OF CHORDS

When a protractor is not available, a scale of chords is used to set-out or measure angles.

Construction of Scale of Chords showing 5° Divisions [Fig. 11]

1. Draw AB of any length, say, 100 mm. At B, erect a perpendicular. B as center and BA radius, draw the chord AC cutting the perpendicular at C. Thus, the arc AC subtends 90° at B.
2. Divide AC into 18 equal parts, so that each part subtends 5° at B as follows:
   - Divide AC into 3 equal parts, by drawing arcs with centers A and C and radius AB. Divide each of these parts further into 3 equal parts using the divider and by trial and error method. Each of these 9 equal parts on the chord subtends 10° at the center B.
   - A as center and radius A-10°, A-20°, ..., A-90°, draw arcs to obtain 10, 20, ..., 90 on AB and AB produced up to D.

- Subdivide the arcs A-10°, A-20°, ..., A-90°. Each of these 18 equal parts on the chord subtends 5° at B.
- A as center and radius A-5°, A-15°, A-25°, ..., A-85°, draw arcs to obtain the division points 5, 15, 25, ..., 85 on the line AD. Thus, all the division points from the chord are transferred on to the line AD.

3. Draw a rectangle below AD taking any width, say, 10 mm.
4. Mark the division points on the rectangle to represent the angles 0° to 90° in steps of 5°. The line AD, with divisions marked, is the Scale of Chords.

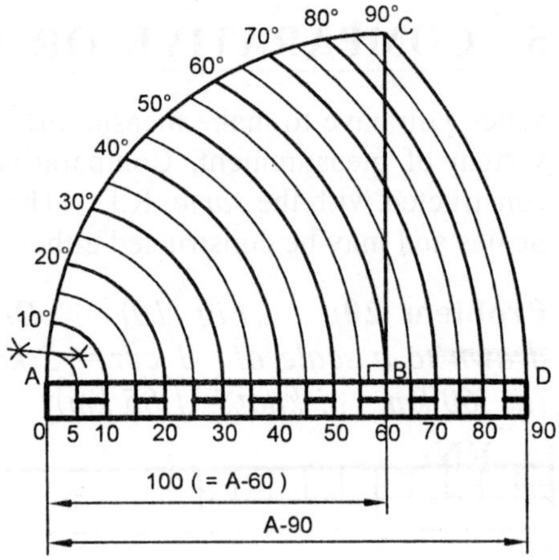

FIG. 11

Problem 22: *(Fig. 12) Construct a Scale of Chords. With its aid, set-off angle 35°.*

1. Construct the Scale of Chords showing 5° divisions.
2. Draw PO of length 100 mm (= AB on the Scale of Chords). O as center and OP as radius, draw an arc PX.
3. P as center, A-35 (measured from the Scale of Chords) as radius, draw an arc to cut the arc PX at Q.
4. Join OQ. The angle POQ is the required angle of 35°.

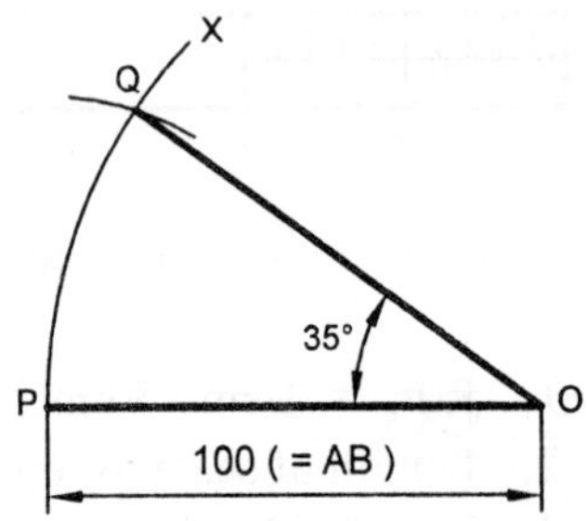

FIG. 12

O is the point where angle is constructed. OP is the line from which angle is measured.

Problem 23: *(Fig. 13) Construct a Scale of Chords. With its aid, set-off angle 145°.*
If the angle to be constructed is greater than 90°, it should be taken into two parts in such a way that the sum of the two parts is equal to the given angle. Angle 145° can be set-off in two parts as given below: (i) 145° = 180° – 35° and (ii) *145° = 90° + 55° (Exercise).*

(i) Construct the Scale of Chords showing 5° divisions for 145° = 180° – 35°

1. Draw a line PO of length 100 mm (= AB on the Scale of Chords).
2. Extend the line PO to OS such that OS = PO.
3. O as center and OP as radius, draw the semi-circle.
4. S as center, A-35 (measured from the Scale of Chords) as radius, draw an arc to cut the semi-circle at Q.
5. Join OQ. The angle POQ is the required angle of 145°.

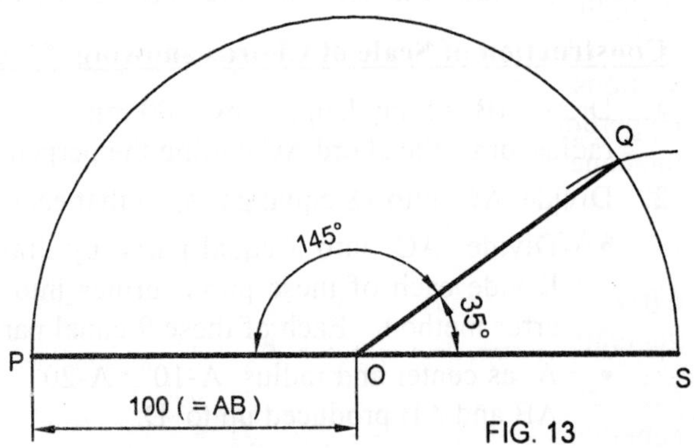

FIG. 13

*

5

CONIC SECTIONS

1. ENGINEERING CURVES

In engineering practice, we come across various types of *Non-Circular Curves.* These are used in finding the path of space vehicles (aircrafts, missiles, etc.), machine components (elliptical gears, etc.), civil structures (bridges, arches, etc.), electrical equipment (reflectors, etc.), electronic transmitters and receivers (radar antenna, etc.), solar concentrators, etc.

These non-circular curves cannot be drawn by compass alone. Sufficient number of points are first located and then a smooth curve passing through them are drawn *freehand* or by using French Curve. *Examples:* Conic Sections, Cycloids, Involutes, Spirals, Helices, etc.

2. CONIC SECTIONS

See Fig. 1. A *cone* is a surface generated by the rotation of a straight line whose one end is in contact with a fixed point while the other end is in contact with a closed curve, not lying in the plane of the curve. A *right circular cone* is a cone having its axis perpendicular to its base.

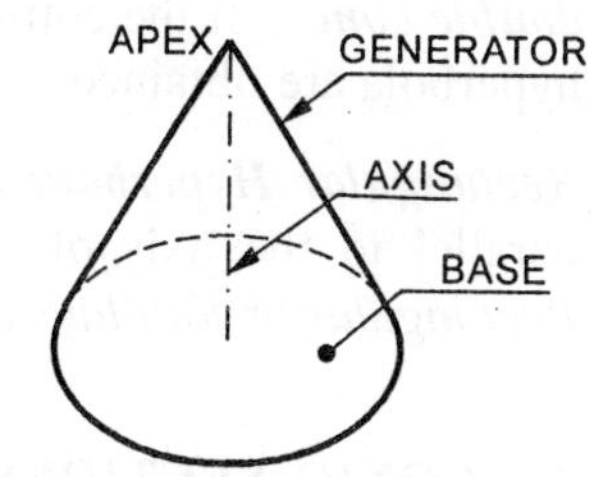

FIG. 1 CONE

Apex or Vertex is the top point of the cone. *Axis* is an imaginary line joining apex and center of base. *Generator* is the straight line which is generating the surface of the cone. *Base of the cone* is the closed curve.

2.1 CONIC SECTIONS : Definition I

A right circular cone can be cut by *Section* (*Cutting*) *Planes* in different positions relative to the axis of the cone to obtain cut sections of different shapes, namely, Circle, Ellipse, Parabola, Hyperbola and Rectangular Hyperbola. These cut sections seen on the cone are called Conic Sections or simply Conics.

In *Fig. 2* the cutting plane is shown in different positions, *viz.*, AA, BB, CC, DD and EE.

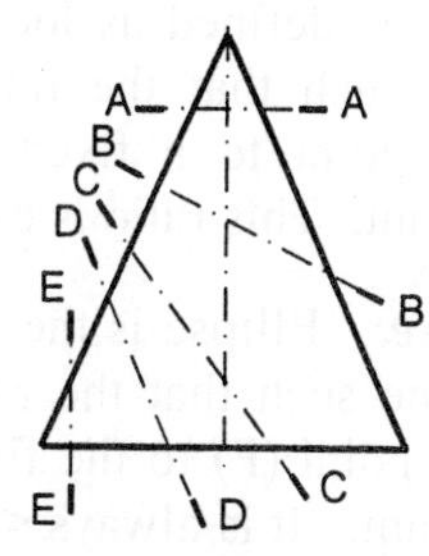

FIG. 2

Circle: *(Fig. 3)* When the cutting plane AA is perpendicular to the axis and cuts all the generators, the section obtained is a *Circle.*

Ellipse: *(Fig. 4)* When the cutting plane BB is inclined to the axis of the cone and cuts all the generators on one side of the apex, the section obtained is an *Ellipse.* Only when the cutting plane cuts all the generators of the cone, the elliptical section formed is a closed one.

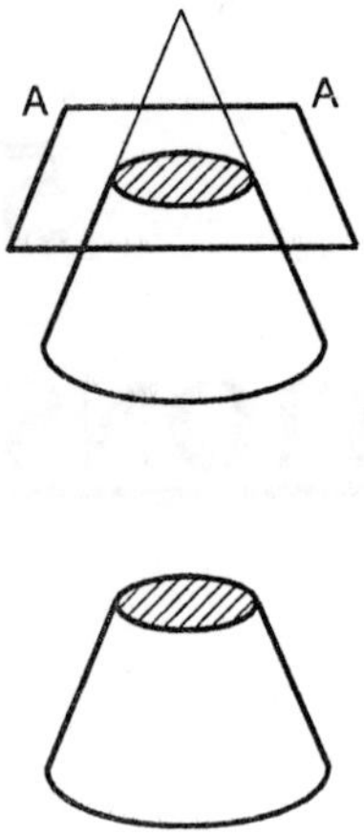

FIG. 3
CIRCLE

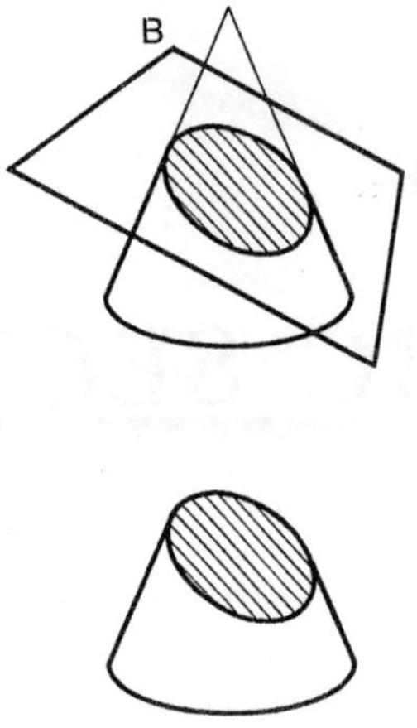

FIG. 4
ELLIPSE

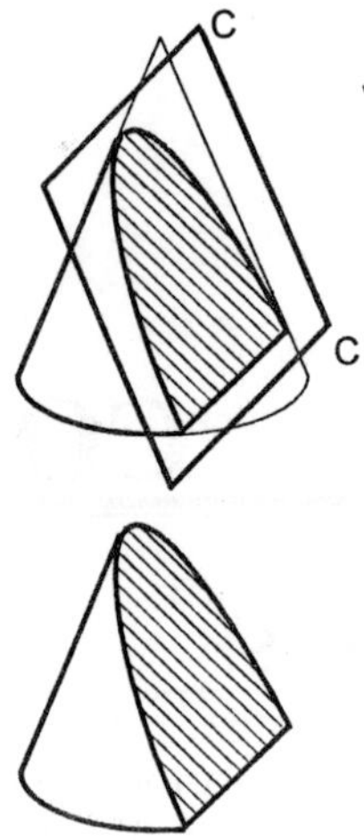

FIG. 5
PARABOLA

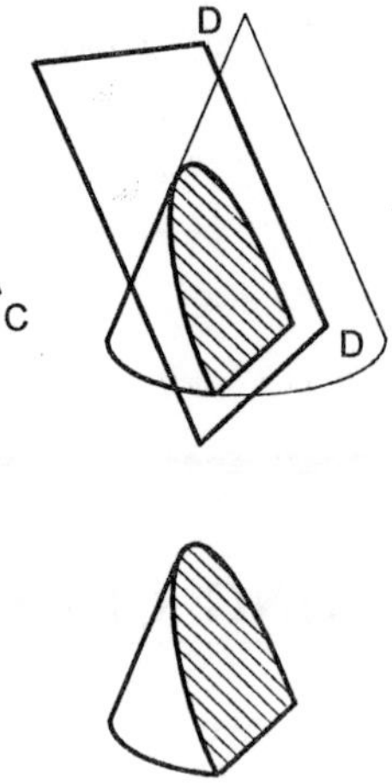

FIG. 6
HYPERBOLA

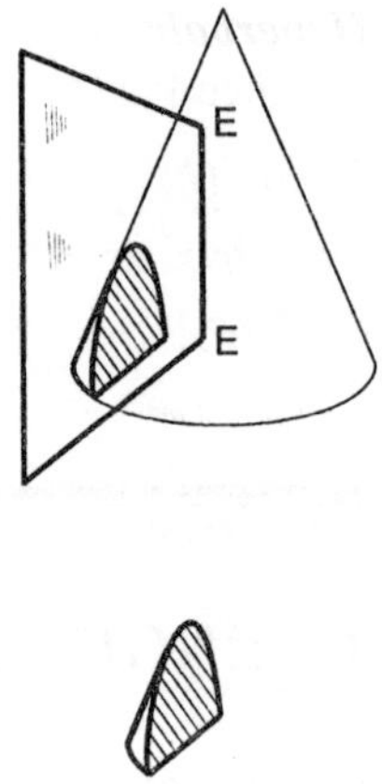

FIG. 7
RECTANGULAR
HYPERBOLA

Parabola: *(Fig. 5)* When the cutting plane CC is inclined to the axis of the cone and parallel to one of the generators, the section obtained is a *Parabola*.

Hyperbola: *(Fig. 6)* When the cutting plane DD makes a smaller angle with the axis than that of the angle made by the generator of the cone, the section obtained is a *Hyperbola*. Consider a ***double cone***. If the cutting plane lies on one side of the axis of the double cone, two branches of hyperbola are obtained.

Rectangular Hyperbola or Equilateral Hyperbola: *(Fig. 7)* When the cutting plane EE is parallel to the axis of cone, but does not pass through the apex, the section obtained is a *Rectangular* or *Equilateral Hyperbola*.

2.2 CONIC SECTIONS DEFINED AS LOCI : Definition II [Fig. 8]

Conic is defined as locus of a point moving in a plane such that the ratio of its distance from a fixed point to a fixed straight line is always a constant. This ratio is called ***eccentricity*** (e).

Ellipse: Ellipse is the locus of a point moving in a plane such that the ratio of its distance from a fixed point (F) to the fixed straight line (DD) is a constant. It is always < 1.

Parabola: It is the locus of a point moving in a plane such that the ratio of its distance from a fixed point (F) to the fixed straight line (DD) is a constant and is always = 1.

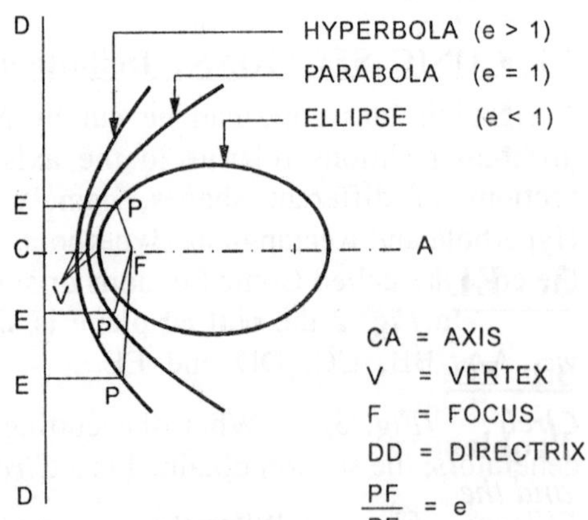

FIG. 8

Hyperbola: It is the locus of a point moving in a plane in such a way that the ratio of its distance from a fixed point (F) to the fixed straight line (DD) is constant and is > 1.

Focus (F): The fixed point is called the Focus.

Directrix (DD): The fixed line is called the Directrix.

Axis (CA): Axis is the line passing through focus and perpendicular to the directrix.

Vertex (V): Vertex is a point at which the conic cuts its axis.

Eccentricity: $e = \dfrac{\text{Distance of the moving point from the focus}}{\text{Distance of the moving point from the directrix}}$

Note: If the eccentricity (e) is < 1, the curve obtained is called *Ellipse*.
If the eccentricity (e) is = 1, the curve obtained is called *Parabola*.
If the eccentricity (e) is > 1, the curve obtained is called *Hyperbola*.

Engineering Applications

Ellipse : Elliptical shape is used in the construction of arches, bridges, dams, elliptical gears of textile machines and printing presses, ends of cylindrical tanks, tops or bottoms of utensils, man-holes, glands, stuffing boxes, flanges of pipes, monuments, etc.

Parabola is used for suspension bridges, reflectors for parallel beams such as head lights of automobiles, solar concentrators, including machine tool structures, etc. Path of a thrown object or missile and path of a jet of water issuing from vertical orifice are of parabolic shape.

Hyperbola is used in the design of cooling towers, hydraulic channels, electronic transmitters and receivers like radar antenna, etc. ***Rectangular Hyperbola*** is used to represent the Boyle's Law expansion curve or the Theoretical Indicator Diagram of an Engine.

2.3 METHODS OF CONSTRUCTION

- Eccentricity Method .... (*Ellipse / Parabola / Hyperbola*)
- Intersecting Arcs or Arc of Circles or Foci Method / Concentric Circles Method / Parallelogram Method using Conjugate Diameters / Four Centers Method ... ... (*Ellipse*)
- Rectangle or Oblong Method .... (*Ellipse / Parabola*)
- Parallelogram Method / Tangent Method .... (*Parabola*)
- Foci & Transverse Axis Method / Ordinate, Abscissa & Transverse Axis Method (*Hyperbola*)
- Asymptotes and a Point on the Curve Method .... (*Rectangular Hyperbola*)

3. ELLIPSE

3.1 ECCENTRICITY METHOD

Problem 1: *(Fig. 9) (a) Construct an ellipse when the distance between the focus and the directrix is 30 mm and the eccentricity is 3/4.*

(b) Draw the tangent and normal at any point P on the curve using directrix.

1. Draw a vertical line DD to represent the directrix. At any point C on it draw a line perpendicular to the directrix to represent the axis CC′.
2. Distance between the focus and the directrix is 30 mm. Hence mark F_1, the focus such that CF_1 = 30 mm.
3. e = 3/4. So construct a right-angled Δ CXY such that $\frac{XY}{CX} = \frac{3 \text{ units}}{4 \text{ units}}$ (X is any point on the axis).
4. From F_1 draw a 45° line to cut CY at S. From S erect vertical to intersect CF_1 at V_1, the vertex. Now $SV_1 = F_1V_1$.
5. Δ CXY is similar to Δ CV_1S.

$$\therefore \frac{SV_1}{CV_1} = \frac{XY}{CX} = \frac{F_1V_1}{CV_1} = 3/4$$

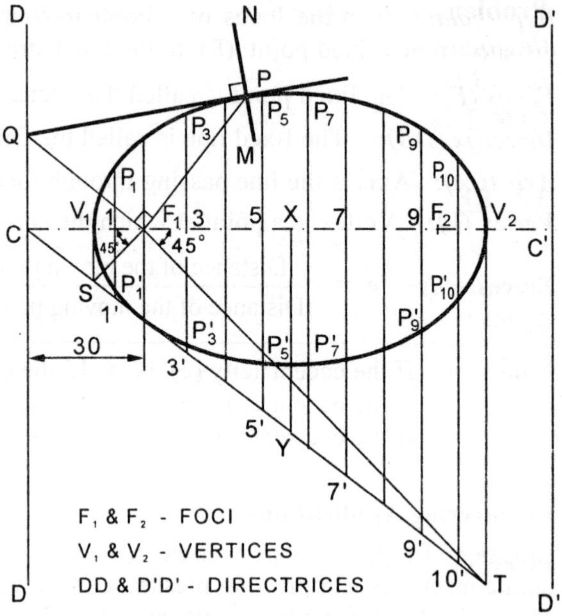

FIG. 9 ELLIPSE - ECCENTRICITY METHOD

6. From F_1 draw another 45° line to intersect the extension of CY at T. From T erect vertical to intersect the axis at V_2, another vertex. $V_1 V_2$ = Major Axis.
7. Along the major axis, mark points 1, 2, ..., 10 at approximately equal intervals. Through these points erect verticals to intersect CY (produced if necessary) at 1′, 2′, ..., 10′.
8. With 11′ as radius and F_1 as centre draw two arcs on either side of the axis to intersect the vertical line drawn through 1 at P_1 and P_1'.
9. With 22′ as radius and F_1 as centre draw two arcs on either side of the axis to intersect the vertical line drawn through 2 at P_2 and P_2'.
10. Repeat the above and obtain P_3 and P_3',....., P_{10} and P_{10}' corresponding to 2, 3,..., 10 respectively. Draw a *smooth ellipse* passing through V_1, P_1, ..., P_{10}, V_2, P_{10}' ,..., P_1', V_1.
11. *To mark another Focus* F_2 : Mark F_2 on the axis such that $V_2 F_2 = V_1 F_1$.
12. *To mark another Directrix* D′D′: Mark C′ along the axis such that $C' V_2 = CV_1$. Through C′ draw a vertical line D′D′.

Note: *First draw thin freehand curve followed by smooth thick curve using French Curve.*

Tangent and Normal at any point P on the ellipse using Directrix

1. Mark the given point P and join PF_1. At F_1 draw a line perpendicular to PF_1 to cut DD at Q. Join QP and extend it. **QP is the tangent at P.**
2. Through P, draw a line NM perpendicular to QP. **NM is the normal at P.**

Problem 2: *(Exercise) A fixed point F is 7.5 cm from a fixed straight line. Draw the locus of a point P moving in such a way that its distance from the fixed straight line is 2/3 times its distance from F. Name the curve. Draw normal and tangent at a point 6 cm from F.*

3.2 RELATIONSHIP BETWEEN MAJOR AXIS, MINOR AXIS AND FOCI

Ellipse may also be defined as the locus of a point which moves such that the sum of its distances from two fixed points is a constant and is equal to the length of the major axis.

See Fig. 10. F_1 and F_2 are foci. The line AB which passes through F_1 and F_2 is the Major Axis. CD which bisects the major axis at 90° is called the Minor Axis. Now, $CF_1 = 1/2$ AB.

Using this, if major axis and minor axis are given, the two fixed points F_1 and F_2 can be located.

Note : $CF_1 = CF_2 = DF_1 = DF_2 = AB/2$

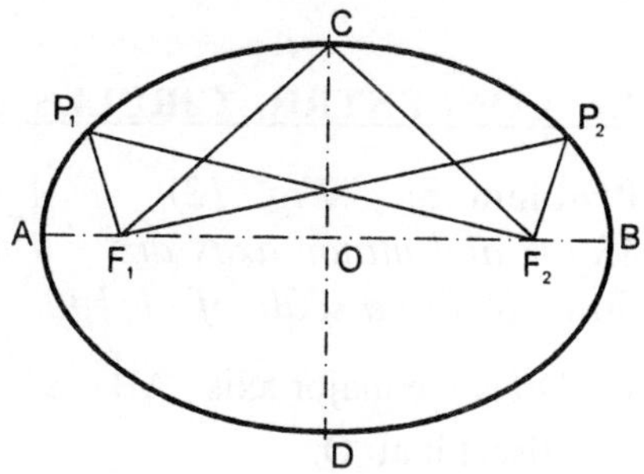

RESULT : $CF_1 = CF_2 = DF_1 = DF_2 = \frac{1}{2}$ AB

FIG. 10

3.3 INTERSECTING ARCS or ARC OF CIRCLES or FOCI METHOD

Problem 3: *(Fig. 11) Construct an ellipse when the major axis is 120 mm and the distance between the foci is 108 mm. Determine the length of the minor axis.* (UQ)

1. Draw the major axis AB = 120 mm and bisect it at O.
2. Mark the foci F_1 and F_2 on AB such that $OF_1 = OF_2 = 108/2 = 54$ mm.
3. Mark any number of points 1, 2, ..., (say up to) 10 at approximately equal intervals on AB in between F_1 and F_2.
4. F_1 as centre and A1 as radius draw two arcs on either side of AB.
5. F_2 as centre and B1 as radius draw 2 arcs on either side of AB to cut the previous two arcs at P_1 and P_1'.
6. Again F_1 as centre and A2 as radius draw two arcs on either side of AB.
7. F_2 as centre and B2 as radius draw two arcs on either side of AB to cut the previous two arcs at P_2 and P_2'.
8. Repeat the above and mark P_3 & P_3', ..., P_{10} & P_{10}' and draw a *smooth ellipse.*

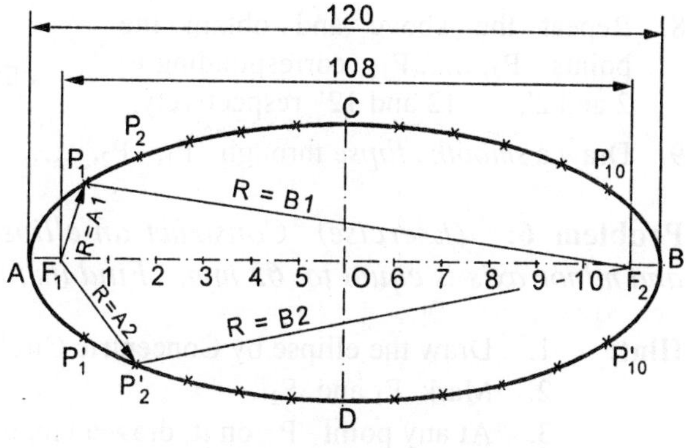

RESULT : MINOR AXIS CD = 50 mm

FIG. 11

ELLIPSE - INTERSECTING ARC METHOD

9. **Minor Axis:** Take half of the major axis (OA) as radius and any one of the foci (F_1 or F_2) as centre draw two arcs to cut the ellipse at C and D. Join CD which is the **Minor Axis = 50 mm**.
10. **Check:** CD should pass through O. **[CF_1 = CF_2 = ½ AB]**

Problem 4: *(Exercise) The major axis of an ellipse is 100 mm and the minor axis is 55 mm. Find the foci and construct the ellipse by Intersecting Arcs Method.*

3.4 CONCENTRIC CIRCLES METHOD

Problem 5: *(Fig. 12) A flowerbed in a botanical garden is elliptical in shape. Major and minor axes are 9 m and 5.5 m respectively. Draw the profile of the flowerbed to a scale of 1:100.* (UQ)

1. Draw the major axis AB = 90 mm.
2. Bisect it at O.
3. Through O draw a vertical line CD = 55 mm, the minor axis.
4. O as centre and major and minor axes as diameters draw two concentric circles.
5. Divide the circles into 12 equal parts.
6. Draw the radial lines to intersect the major and minor axes circles at 1, 2,...,12 and 1′, 2′, ..., 12′ respectively.
7. From 1 draw a vertical line parallel to CD. From 1′ draw a horizontal line parallel to AB. Both meet at P_1.
8. Repeat the above and obtain the points P_2, ..., P_{12} corresponding to 2 and 2′, ..., 12 and 12′ respectively.
9. Draw a *smooth ellipse* through P_1, P_2, ... , P_{12}, P_1.

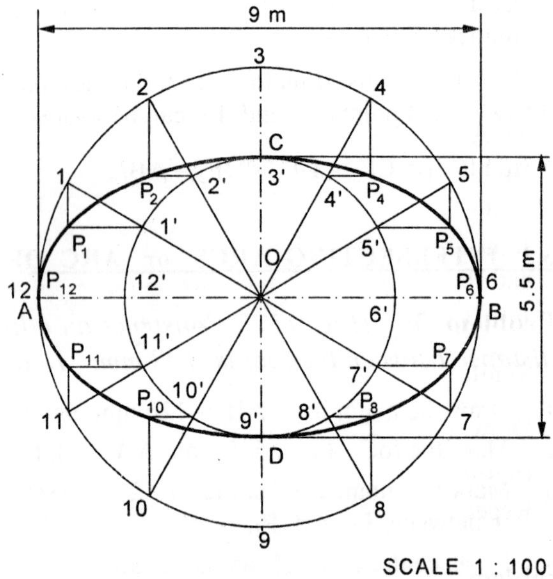

FIG. 12

ELLIPSE - CONCENTRIC CIRCLES METHOD

Problem 6: *(Exercise) Construct an ellipse when its major axis is equal to 100 mm and minor axis is equal to 65 mm. Find its foci, directrix and eccentricity.* (UQ)

Hint:
1. Draw the ellipse by Concentric Circles Method / Rectangle Method in Section 3.5.
2. Mark F_1 and F_2.
3. At any point P on it, draw a tangent using F_1 and F_2.
4. At F_1 draw a perpendicular to PF_1 to cut the tangent at Q.
5. Draw a vertical line through Q to obtain the directrix.
6. Measure e = $V_1 F_1$ / $V_1 C$ = 2/3.

3.5 RECTANGLE METHOD or OBLONG METHOD

Problem 7: *(Fig. 13) A plot of ground is in the shape of a rectangle 110 m x 50 m. Inscribe an elliptical lawn in it. Take a suitable scale.* (UQ)

1. Select scale 1 : 1000. Draw major axis AB = 110 mm and minor axis CD = 50 mm. Both bisect at O.
2. Through A and B draw lines parallel to CD.
3. Through C and D draw lines parallel to AB and construct the rectangle PQRS. Now PS = AB and SR = CD.
4. Divide AQ and AP into any number of equal parts (*say 4*) and name the points as 1, 2, 3 and 1′, 2′, 3′ respectively starting from A on AQ and AP.

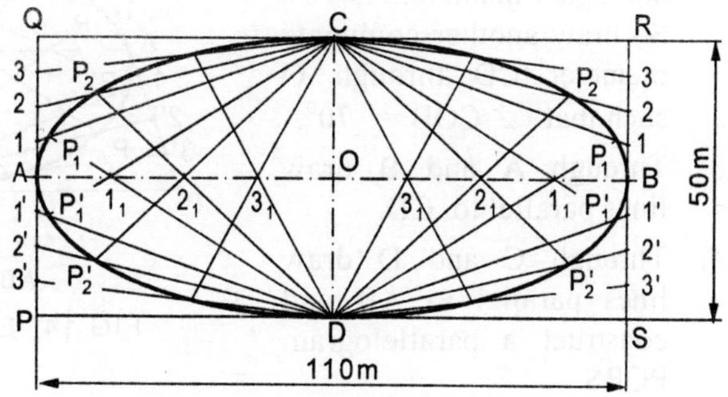

SCALE 1 : 1000

FIG. 13

ELLIPSE - RECTANGLE or OBLONG METHOD

5. Divide AO into same (4) number of equal parts, and name the points as 1_1, 2_1, 3_1 starting from A on AO.
6. Join 1, 2, 3 with C. Join $D1_1$ and extend it to intersect C1 at P_1. Similarly extend $D2_1$ and $D3_1$ to intersect C2 and C3 at P_2 and P_3 respectively. Join 1′, 2′, 3′ with D. Join $C1_1$ and extend it to intersect D1′ at P_1'.
7. Similarly extend $C2_1$ & $C3_1$ to intersect D2′ & D3′ at P_2' & P_3' respectively.
8. Draw a smooth curve through C, P_3, P_2, P_1, A, P_1', P_2', P_3', D and obtain *one half (left-half)* of the ellipse.
9. Repeat the above and *draw the right-half of ellipse, symmetrical to the left-half.*

Problem 8: *(Exercise) The major and minor axes of an elliptical fish pond are 10 m and 6 m respectively. Draw one-half of the ellipse by Oblong Method and the other half by Concentric Circles Method.* (UQ)

3.6 PARALLELOGRAM METHOD

Conjugate Axes: *Conjugate Axes* are two lines intersecting each other at the centre of an ellipse when they are parallel to the tangents drawn at their extremities. An ellipse may have an unlimited number of pairs of *Conjugate Axes*. Given the Conjugate Axes (Conjugate Diameters), the ellipse may be constructed by *Parallelogram Method.*

Problem 9: *(Fig. 14) (a) Construct an ellipse when a pair of conjugate axes are equal to 110 mm and 50 mm respectively. The angle between the conjugate axes is 70°. (b) Determine the lengths of the major and minor axes.* (UQ)

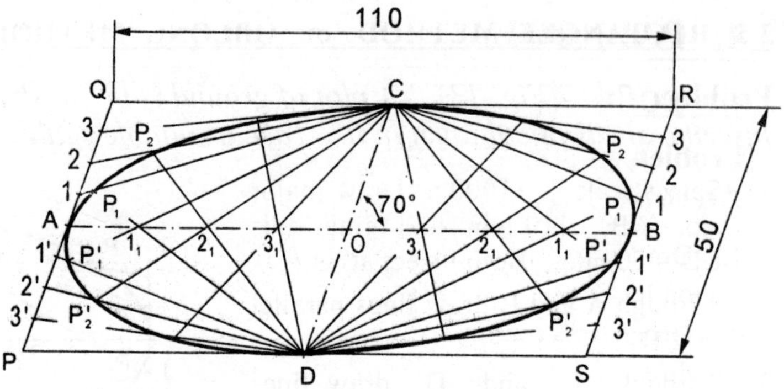

AB & CD : CONJUGATE AXES

FIG.14 ELLIPSE - PARALLELOGRAM METHOD

1. Draw a conjugate diameter AB = 110 mm and bisect it at O.
2. Angle between the conjugate diameters is 70°. So draw another conjugate diameter CD through O such that ∠ COB = 70°.
3. Through A and B draw lines parallel to CD.
4. Through C and D draw lines parallel to AB and construct a parallelogram PQRS.
5. Repeat the procedure given in *Step Nos. 4 to 9 in Problem 7 (Rectangle Method)* and complete the construction of the ellipse inside the parallelogram PQRS.

Hint: To find Major and Minor axes of the above ellipse drawn using Conjugate Axes

1. O as center and OC as radius, draw an arc to cut the ellipse at E. Join CE.
2. Through O, draw FG parallel to CE. **FG** is the Minor Axis.
3. Through O, draw the Major Axis **JK** perpendicular to FG.
4. Measure the major axis = **JK = 112 mm** ($>$ 110 mm) and minor axis = **FG = 47 mm** ($<$ 50 mm).

Problem 10: *(Exercise) A plot of land in a Five Star Hotel is in the shape of a parallelogram of 20 m x 12 m. Angle between the sides is 55°. Inscribe an elliptical swimming pool in it.* (UQ)

3.7 FOUR CENTERS (approximate) METHOD (GIVEN MAJOR AND MINOR AXES)

Problem 11: *(Fig. 15) Construct an ellipse when the major and minor axes are 90 mm and 50 mm respectively.*

1. Draw AB = 90 mm and CD = 50 mm. Join AC. O as centre and OA as radius draw an arc to cut the extension of CD at K.
2. C as centre and CK as radius draw an arc to cut AC at L.
3. Draw perpendicular bisector to AL. Extend it to cut AB at M and the extension of CD at N. Locate Q along AB such that QB = AM.
4. Locate R along the extension of CD such that CR = DN. Now M, N, Q and R are the *four centers* for the ellipse.
5. N and R as centers and NC as radius draw two large arcs. M and Q as centers and AM as radius draw two arcs. Obtain the *ellipse*.

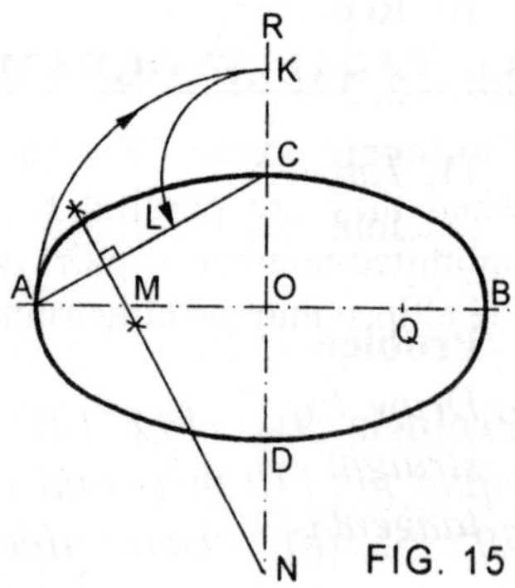

FIG. 15

4. PARABOLA

4.1 ECCENTRICITY METHOD

Problem 12: *(Fig. 16) Construct a parabola when the distance between focus and the directrix is 40 mm. Draw tangent and normal at any point P on your curve.*

1. Draw directrix DD. At any point C on it draw axis CA ⊥ʳ to DD.
2. Distance between the focus and the directrix is 40 mm. So mark F the focus, such that CF = 40 mm.
3. For parabola, **e** is 1; so construct the right-angled Δ CXY such that $\frac{XY}{CX} = 1$ (X is any point on axis)
4. From F draw a 45° line to intersect CY at S.
5. From S erect vertical to cut CF at vertex V. Now SV = FV.
6. From similar Δs CXY and CVS, $\frac{SV}{CV} = \frac{XY}{CX} = \frac{FV}{CV} = 1.$
7. Along the axis CA mark the points 1, 2, ..., 5 at approximately equal intervals.
8. Through 1, 2,.., 5 erect verticals to intersect the line CY (produced if necessary) at 1′, 2′, ..., 5′ respectively.
9. With 11′ as radius and F as center, draw two arcs on either side of the axis to intersect the vertical line drawn through 1 at P_1 and P_1'.
10. Repeat the above and obtain P_2 & P_2', P_3 & P_3', ..., P_5 & P_5' corresponding to points 2, 3, 4, 5 respectively. Draw a smooth *parabola* through P_5, ..., P_1, V, P_1', etc.
11. *Tangent and Normal:* Join PF. At F, draw a perpendicular to PF to meet CD at Q. Join QP, the tangent. Draw NM perpendicular to QP. NM is the normal.

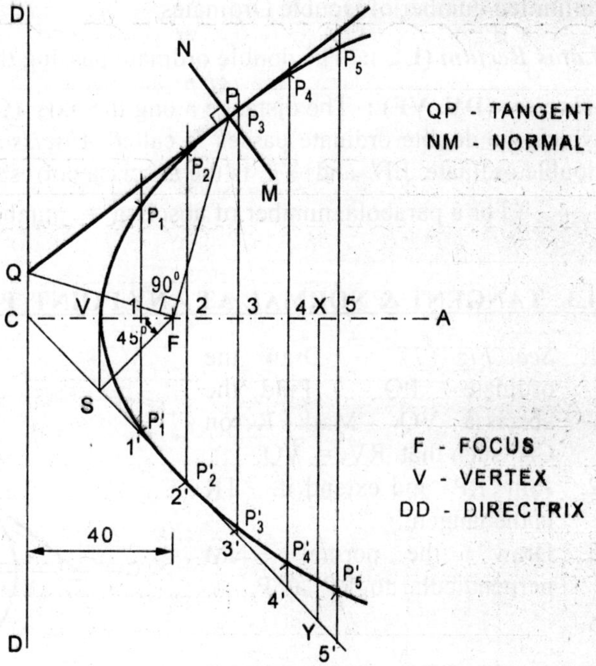

FIG. 16
PARABOLA - ECCENTRICITY METHOD

Problem 13: *(Exercise) A fixed point F is 7.5 cm from a fixed straight line. Draw the locus of a point P moving in such a way that its distance from the fixed straight line is equal to its distance from F. Name the curve. Draw normal and tangent at a point 6 cm from F.*

4.2 PARABOLA – TERMINOLOGY

See Fig. 17. Parabola is an open unlimited curve. CA is the Axis. V is the Vertex.

Ordinate **(ED)** : Any line ED drawn from E on the parabola such that it is perpendicular to the axis CA is called *Ordinate*.

Double Ordinate **(EE′)** : The ordinate ED is extended to cut the parabola at E′. EDE′ is *Double Ordinate,* which is twice the length of the ordinate, *i.e.*, EE′ = 2ED. A parabola has unlimited number of Double Ordinates.

Latus Rectum **(LL′):** The double ordinate passing through the focus is *Latus Rectum*.

Abscissa **(DV, VF)** : The distance along the axis (CA) between vertex (V) and a point through which the double ordinate passes is called *Abscissa.* DV is the abscissa corresponding to the double ordinate EE′ and VF is the abscissa corresponding to latus rectum LL′.

For a parabola, number of abscissa = number of double ordinates.

4.3 TANGENT & NORMAL AT ANY POINT P WHEN F & DD ARE NOT KNOWN

1. *See Fig. 17.* Draw the ordinate PQ. Find the abscissa VQ. Mark R on CA such that RV = VQ.
2. Join RP and extend it. TR is the tangent.
3. Draw the normal NM perpendicular to RP at P.

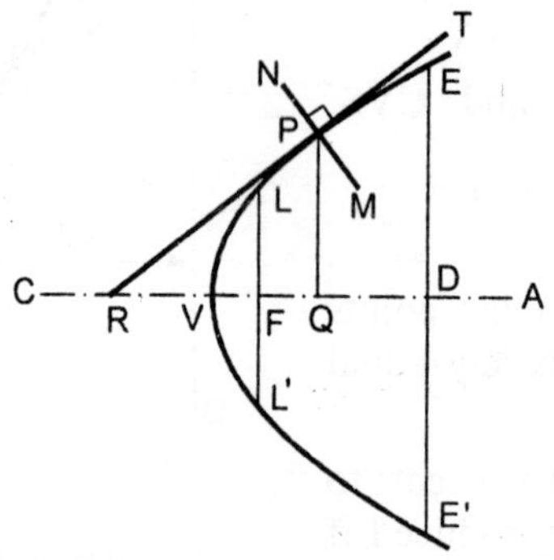

FIG. 17

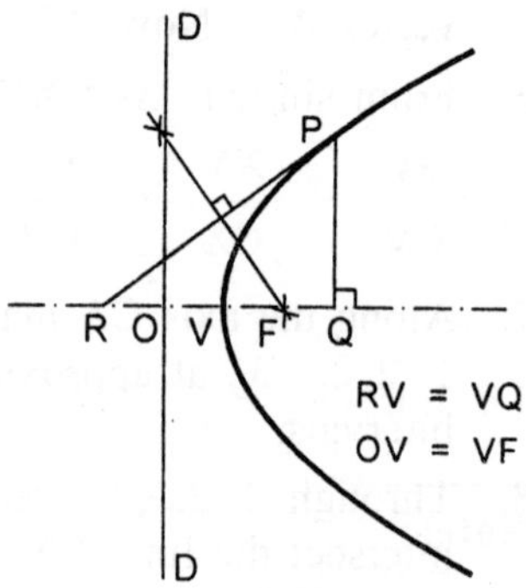

FIG. 18

4.4 TO FIND THE FOCUS AND THE DIRECTRIX OF A PARABOLA GIVEN ITS AXIS

1. *See Fig. 18.* Mark any point P on the parabola.
2. Draw a perpendicular PQ to the given axis.
3. Mark a point R on the axis such that RV = VQ.
4. *Focus :* Join RP. Draw a $\perp^r$ bisector of RP cutting the axis at F. F is the Focus.
5. *Directrix :* Mark O on the axis such that OV = VF. Through O draw the *Directrix* DD perpendicular to the axis.

4.5 RECTANGLE METHOD GIVEN THE BASE (DOUBLE ORDINATE) AND AXIS

Problem 14: *(Fig. 19) A cricket ball thrown from the ground level reaches the Wicket Keeper's gloves. Maximum height reached by the ball is 5 m. The ball travels a horizontal distance of 11 m from the point of projection. Trace the path of the ball.* (UQ)

1. Take a scale of 1: 100.
2. Draw PS = 11 cm to represent the double ordinate. Bisect PS at O.
3. The ball reaches a maximum height of 5 m. Hence from O erect vertical and mark the vertex V such that OV = 5 cm. Now OV is the abscissa.
4. Draw rectangle PQRS such that PS is double ordinate and PQ = RS = VO (abscissa).
5. Divide PQ and RS into 8 equal parts as 1, 2, ..., 8 and 1′, 2′, ..., 8′ respectively, starting from P on PQ and S on SR.

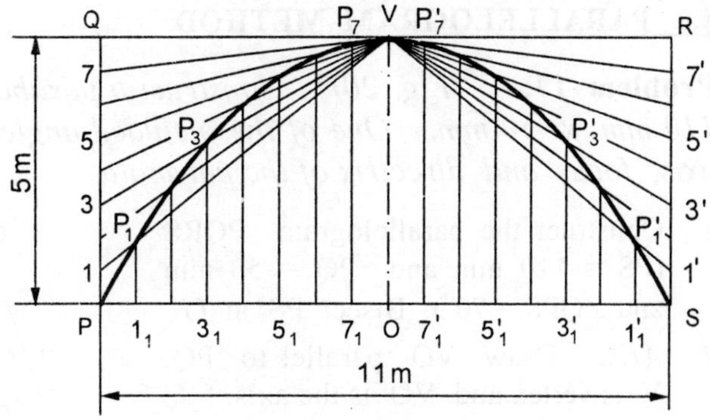

SCALE : 1:100

FIG. 19

PARABOLA - RECTANGLE METHOD

6. Join 1, 2, ..., 8 and 1′, 2′, ..., 8′ with V.
7. Divide PO and OS into 8 equal parts as 1_1, 2_1,..., 8_1 and $1_1'$, $2_1'$, ..., $8_1'$ respectively, starting from P on PO and from S on SO.
8. From 1_1 erect vertical to meet the line V1 at P_1.
9. Similarly from 2_1, ..., 8_1 erect verticals to meet the lines V2,..., V8 at P_2 ,..., P_8 respectively.
10. Also erect verticals from $1_1'$, $2_1'$, ..., $8_1'$ to meet the lines V1′, V2′ ,..., V8′ at P_1', P_2', ..., P_8' respectively.
11. Join P, P_1, P_2, ..., P_7, V, P_7', ..., P_1' and S to represent the path of the ball *(parabola).*

Note: *Path of a projectile when thrown / falling due to gravitational force is Parabola.*

Problem 15: *(Exercise) A rocket shot up in the air, reaches a maximum height of 800 m and falls on the ground at a distance of 1800 m from the point of projection. Trace the path of the rocket in space. What is the angle of the projectile?* (UQ)

Hint: Draw the parabola by *Rectangle Method.* Angle of the projectile is obtained from the direction of motion of the rocket with reference to the horizontal. Measure the angle of projectile by drawing a tangent at the point of projection. It is equal to **58°**.

Problem 16: *(Exercise) The headlight reflector of a four-wheeler has a maximum rim diameter of 115 mm and a maximum depth of 90 mm. Draw the shape of the reflector. Draw a tangent and normal at any point on the curve.* (UQ)

Hint: Shape of the reflector is parabolic; Maximum rim diameter of the four-wheeler is equal to Base of the parabola; Maximum depth of the reflector is equal to Axis of the parabola; Use *Rectangle Method.*

4.6 PARALLELOGRAM METHOD

Problem 17: *(Fig. 20) Construct a parabolic arch within a parallelogram of sides 110 mm x 50 mm. One of the included angles between the sides is 70°. Locate the axis, focus and directrix of the parabola.* (UQ)

1. Construct the parallelogram PQRS (PS = 110 mm and PQ = 50 mm and ∠QPS = 70°). Bisect PS at O.
2. *Axis:* Draw VO parallel to PQ. V is vertex and VO is the axis.
3. Divide PQ and SR into any number of (4) equal parts as 1, 2, 3 and 1′, 2′, 3′ respectively starting from P on PQ and from S on SR.
4. Join V1, V2 and V3. Also, join V1′, V2′ and V3′.

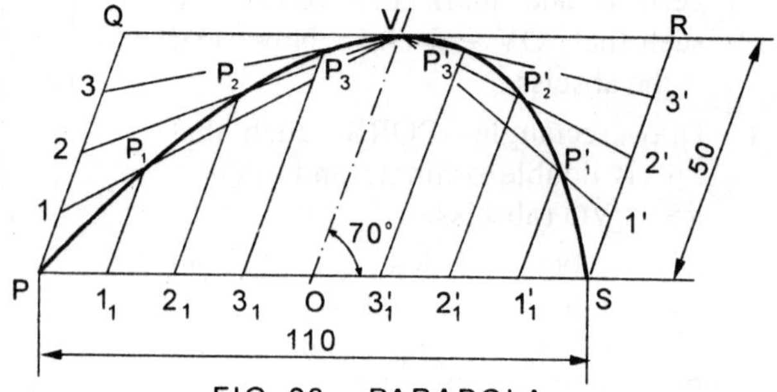

FIG. 20 PARABOLA
PARALLELOGRAM METHOD

5. Divide PO and OS into 4 equal parts as 1_1, 2_1, 3_1 and $1_1'$, $2_1'$, $3_1'$ respectively starting from P on PO and from S on SO.
6. From 1_1 draw a line parallel to PQ to meet V1 at P_1. Similarly obtain P_2 and P_3.
7. Also from $1_1'$, $2_1'$, $3_1'$ draw lines parallel to RS to meet the lines V1′, V2′ and V3′ at P_1', P_2' and P_3' respectively and draw a *smooth parabola*.

Note: *Refer Fig. 18 to find the focus and directrix of the parabola.*

Problem 18: *(Exercise) Construct a parabola within a parallelogram of sides 90 mm and axis 40 mm. The acute angle is 65°.*

4.7 TANGENT METHOD or TRIANGLE METHOD

Problem 19: *(Fig. 21) A fountain jet discharges water from ground level at an inclination of 50° to the ground. The jet travels a horizontal distance of 9 m from the point of discharge and falls on the ground. Trace the path of the jet.* (UQ)

1. Select a scale 1 : 100. Draw PQ = 9 cm.
2. Jet discharges water at an angle of 50° to the ground. Hence, at P and Q draw 50° lines to intersect at R. PQR is an isosceles triangle.
3. Bisect PQ at O. At O, erect vertical to pass through R. Bisect OR at V, the vertex.
4. Divide PR into, *say*, 8 equal parts as 1, 2, ..., 7 starting from P on PR.
5. Divide RQ into the *same* number of (8) equal parts as 1′, 2′, ..., 7′ starting from R on RQ.
6. Join 11′ and also 77′. Both will meet the vertical OR at a point.
7. Join 22′ and also 66′. Both will meet the vertical OR at *another* point.

8. Join 33′ and also 55′. Both will meet the vertical OR at a *third* point.
9. Join 44′ to meet OR at V.
10. Draw a *smooth parabola* through P, V, Q such that the curve is tangential to the lines 11′, 22′, ……, 77′.

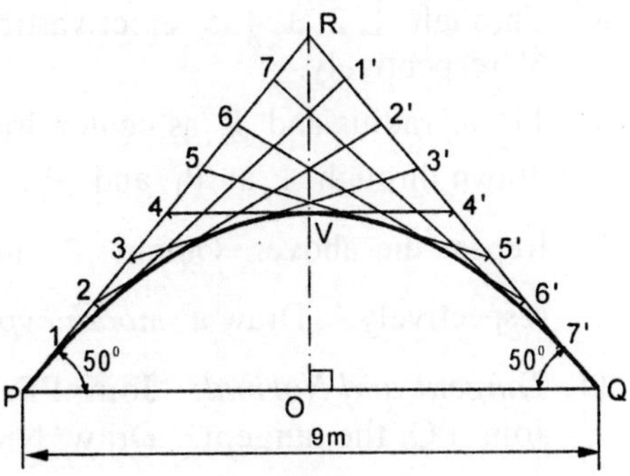

SCALE : 1:100

FIG. 21 TANGENT METHOD

Problem 20: *(Exercise)*
(a) An aircraft is shot at an angle of 45° *to the horizontal from the ground level. It is hit at a maximum height of 255 m of the trajectory. Trace the path of the shot in space using Tangent Method or Parallelogram Method. Name the curve.*

(b) Draw a tangent and normal at any point on the curve. (UQ)

5. HYPERBOLA

5.1 ECCENTRICITY METHOD

Problem 21: *(Fig. 22) Construct a hyperbola when the distance between the focus and directrix is 40 mm. The eccentricity is 4/3. Draw a tangent and normal at any point on the hyperbola.* (UQ)

1. Draw directrix DD. At any point C on it draw CA perpendicular to DD to represent the axis.
2. Mark F the focus, such that CF = 40 mm.
3. e = 4/3. So construct right-angled Δ CXY such that $\frac{XY}{CX} = \frac{4 \text{ units}}{3 \text{ units}}$ (X is any point on axis)
4. From F draw a 45° line to cut CY at S.
5. From S erect vertical to intersect CF at V, the vertex. Now SV = FV.
6. From similar Δs CXY and CVS,

$$\frac{SV}{CV} = \frac{YX}{CX} = \frac{FV}{CV} = \frac{4}{3}$$

7. Along CA mark 1, 2, 3, 4, 5 at approximately equal intervals.

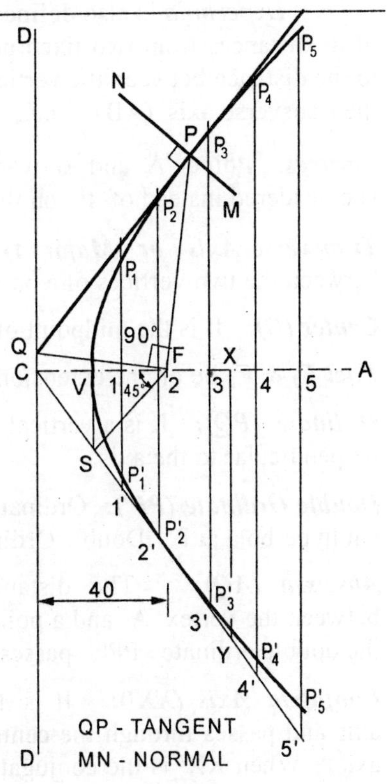

FIG. 22

ECCENTRICITY METHOD

8. Through 1, 2, 3, 4, 5 erect verticals to intersect CY (produced if necessary) at 1′, 2′, ..., 5′ respectively.
9. 11′ as radius and F as center draw two arcs on either side of the axis to cut the vertical line drawn through 1 at P_1 and P_1'.
10. Repeat the above. Obtain P_2 and P_2', ..., P_5 and P_5' corresponding to 2, 3, 4 and 5 respectively. Draw a *smooth hyperbola* through P_5, P_4, ..., P_1, V, P_1', ..., P_5'.
11. *Tangent and Normal:* Join PF. At F, draw a perpendicular to PF to meet CD at Q. Join PQ, the tangent. Draw NM perpendicular to PQ. NM is the normal.

Problem 22: *(Exercise) A fixed point F is 7.5 cm from a fixed straight line. Draw the locus of a point P moving in such a way that its distance from the fixed straight line is 3/2 times its distance from F. Name the curve. Draw normal and tangent at a point 6 cm from F.*

5.2 HYPERBOLA – TERMINOLOGY [Fig. 23]

Hyperbola has two branches. Each branch has the same eccentricity and therefore this curve is also known as Central Conic. *A hyperbola has two foci and two directrices.*

Hyperbola is also defined as the locus of a point P which moves such that the difference of its distances from two fixed points called Foci (F_1 and F_2) is always a constant and is equal to the distance between the vertices of the *two branches of Hyperbola*, *i.e.*, equal to the length of the transverse axis (AB). *i.e.*, $PF_2 \sim PF_1 = AB$.

Vertices: Points A and B where the hyperbola cuts the axis are the Vertices. When point A is considered instead of P on the curve, $F_2A - F_1A = AB$ (since $F_1A = F_2B$).

Transverse Axis or Major Axis (AB): It is a horizontal axis and is equal to the distance between the two vertices of a pair of hyperbolas. Thus by definition, $AB = PF_2 \sim PF_1$.

Center (O): It is the midpoint of transverse axis AB.

Foci F_1 & F_2 lie on the extension of transverse axis.

Ordinate (PQ): It is a vertical line drawn from P perpendicular to the axis.

Double Ordinate (PP′): Ordinate PQ is extended to cut hyperbola at P′. Double Ordinate = PP′ = 2PQ.

Abscissa (AQ): The distance along the axis between the vertex A and a point Q through which the double ordinate PP′ passes is called *Abscissa*.

Conjugate Axis (XX′): It is ⊥ʳ to the transverse axis and passes through the centre of the transverse axis. When XX′ is the conjugate axis, AX = AX′ = OF_1. Conjugate Axes are situated such that each is parallel to tangents at extremities of other.

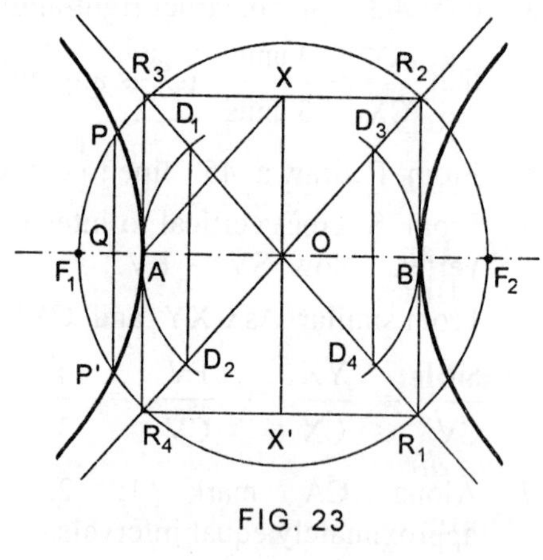

FIG. 23

Asymptotes ***(R_1 R_3 and R_2 R_4):*** Draw a circle with O as centre and radius OF_1. Draw perpendicular to AB at A and B cutting the circle at R_3, R_4, R_1 and R_2. The diagonals R_1 R_3 and R_2 R_4 are the *Asymptotes*. When they are extended on either side, they never touch the curve, but approach nearer and nearer to the curves and will be tangential to the curves at infinity.

Directrices ***(D_1 D_2 and D_3 D_4) :*** O as centre and OA or OB as radius, draw arcs to cut the asymptotes at D_1, D_2, D_3, D_4. D_1 D_2 and D_3 D_4 are the *Directrices*.

5.3 FOCI AND TRANSVERSE AXIS METHOD

Problem 23: *(Fig. 24) The distance between two fixed points is 90 mm. A point P moves such that the difference of its distances from two fixed points always remains constant and is equal to 60 mm. Draw the loci of P. Draw the tangent and normal at any point on the hyperbola.* (UQ)

1. Draw F_1 F_2 = 90 mm. Bisect it at O.
2. A point P moves such that the difference of its distances from F_1 and F_2 = 60 mm. But PF_2 – PF_1 = AB = Transverse axis.
3. Hence mark transverse axis AB = 60 mm along F_1 F_2 such that OA = OB = 1/2 AB = 30 mm.
4. Take any number of points 1, 2, ..*(say 5)* along extension of AF_1.
5. A1 as radius and F_1 as centre, draw the first set of arcs on either side of the axis.
6. Keep the same A1 as radius and F_2 as centre, draw the second set of arcs on either side.

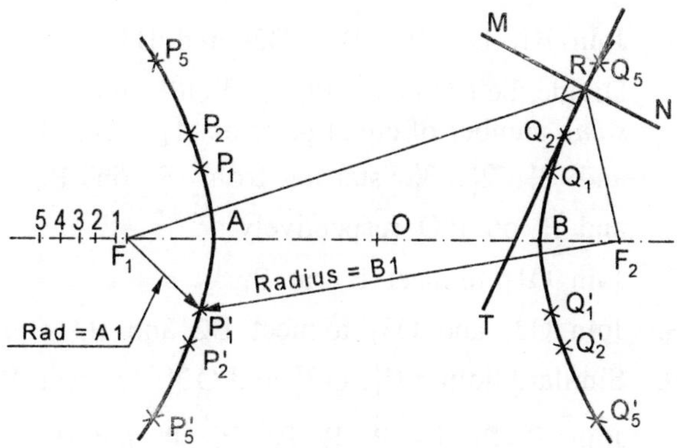

F_1F_2 = 90mm, $PF_2 \sim PF_1$ = AB = 60mm

FIG. 24 A PAIR OF HYPERBOLAS

7. B1 as radius and F_2 as center, cut the first set of arcs at P_1 and P_1'.
8. Keep the same B1 as radius and F_1 as center, cut the second set of arcs at Q_1 and Q_1'.
9. Repeat the above. Obtain P_2, P_2', P_3, P_3', Q_2, Q_2', Q_3, Q_3', etc., corresponding to 2, 3, etc. Join P_5, ..., P_1, A, P_1', ..., P_5' and Q_5, ..., Q_1, B, Q_1', ..., Q_5'. Obtain a *pair of hyperbolas*.
10. **Tangent and Normal:** Mark R on the hyperbola. Join RF_1 and RF_2. Bisect $\angle F_1 RF_2$. Bisector TR is the tangent at R. Draw the normal MN perpendicular to TR through R.

Problem 24: *(Exercise) Draw a pair of hyperbolas when the distance between the foci is 70 mm and the length of the transverse axis is 45 mm. Locate the asymptotes and directrices. Find the eccentricity of the curve.* (UQ)

Hint: Length of transverse axis = Distance between vertices = 45 mm; e = 1.56.

5.4 ORDINATE, ABSCISSA AND TRANSVERSE AXIS METHOD

Problem 25: *(Fig. 25) Draw a hyperbola when its double ordinate is 90 mm, abscissa is 35 mm and half the transverse axis is 45 mm.* (UQ)

1. Draw OBQ such that OB = half the transverse axis = 45 mm and BQ = abscissa = 35 mm.
2. Through Q erect vertical such that PP′ = double ordinate = 90 mm = 2PQ.
3. Construct the rectangle PP′R′R.
4. Divide PR and P′R′ into any number of equal parts *(say 4)* as 1, 2, 3 and 1′, 2′, 3′ starting from P on PR and P′ on P′R′ respectively.
5. Join B1, B2, B3, B1′, B2′ and B3′.
6. Divide the ordinates PQ and QP′ into the *same* number of equal parts as 1_1, 2_1, 3_1 and $1'_1$, $2'_1$, $3'_1$ starting from P on PQ and P′ on P′Q respectively.
7. Join $O1_1$ to meet B1 at P_1.
8. Join $O2_1$ and $O3_1$ to meet B2 and B3 at P_2 and P_3 respectively.
9. Similarly join $O1'_1$, $O2'_1$ and $O3'_1$ to meet B1′, B2′ and B3′ at P'_1, P'_2 and P'_3 respectively.

Join P, P_1, P_2, P_3, B, P'_3, P'_2, P'_1 and P′ by a *smooth hyperbola*.

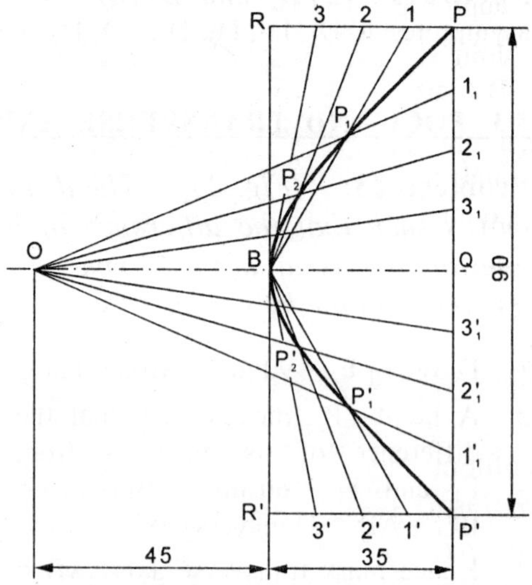

FIG. 25 HYPERBOLA

Problem 26: *(Exercise) Construct a hyperbola given half the major axis equal to 40 mm, single ordinate 60 mm and abscissa 48 mm.* [Major axis means Transverse axis.]

5.5 RECTANGULAR HYPERBOLA or EQUILATERAL HYPERBOLA

When angle between the asymptotes is **90°**, the curve is called *Rectangular* or *Equilateral Hyperbola* of **e** = **√2**.

Problem 27: *(Fig. 26) Construct a rectangular hyperbola when a point P on it is at a distance of 15 mm and 20 mm respectively from the two asymptotes.* (UQ)

1. Draw OR_1 and OR_2 such that $\angle R_1OR_2$ is 90°.
2. Mark A and B along OR_2 and OR_1 respectively such that OA = 20 mm and OB = 15 mm.
3. From A draw AX parallel to OR_1. From B draw BY parallel to OR_2. Both meet at P.
4. Along BP mark 1, 2 and 3 at approximately equal intervals.
5. Join O1, O2 and O3 and extend them to meet AX at 1_1, 2_1 and 3_1 respectively.
6. From 1_1 draw a line parallel to OR_2. From 1 draw a line parallel to OR_1. Both meet at P_1.

7. From 2_1 and 3_1 draw lines parallel to OR_2. From 2 and 3 draw lines parallel to OR_1. They intersect at P_2 and P_3 respectively.
8. Then along PA mark points 4_1 and 5_1 approximately at equal intervals.
9. Join $O4_1$ and $O5_1$ and extend them to meet BY at 4 and 5 respectively.
10. From 4_1 and 5_1 draw lines parallel to OR_2 and from 4 and 5 draw lines parallel to OR_1 to intersect at P_4 and P_5 respectively.
11. Join P_1, P_2, P_3, P, P_4, P_5 by a *smooth Rectangular Hyperbola.*

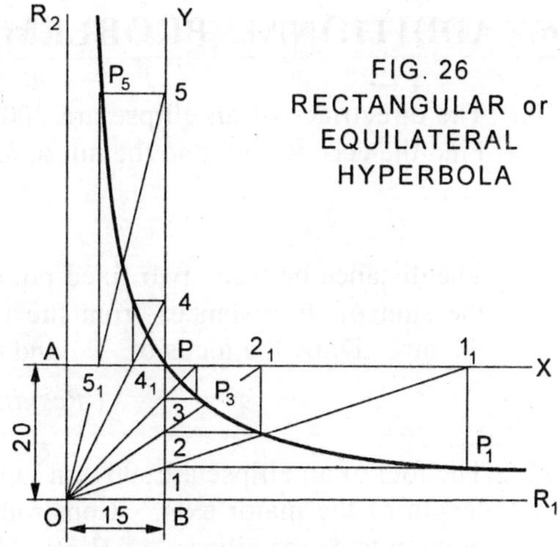

FIG. 26
RECTANGULAR or EQUILATERAL HYPERBOLA

Note: *Extend the curve on either side such that it does not touch the asymptotes.*

Problem 28: *(Exercise) The asymptotes of a hyperbola make 80° with each other. A point P on the curve is at a distance of 30 mm from the horizontal asymptote and 45 mm from the inclined asymptote when measured horizontally. Plot the hyperbola. Draw tangent and normal at any point R on the curve.*

Hint: Procedure for this problem is exactly similar to that of Rectangular Hyperbola.

Problem 29: *(Fig. 27) An ideal gas expands according to Boyle's Law pv = a constant. Pressure p is $3\ N/cm^2$. and volume v is $0.06\ cm^3$. Select a suitable scale and plot the expansion curve.* (UQ)

1. Mark p and v on two mutually perpendicular axes OX and OY.
2. Scale: X axis: 1 mm = $0.1\ N/cm^2$
   Y axis: 1 mm = $0.001\ cm^3$
3. Draw asymptotes OX and OY.
4. Locate P such that OA = $3\ N/cm^2$ and OB = $0.06\ cm^3$.
5. Along AP mark 1, 2 and 3 at approximately equal distances.
6. Join O1. Extend it to cut $B1_1$ at 1_1.
7. From 1_1 draw a line parallel to OY. From 1 draw a line parallel to OX. Both intersect at P_1.
8. Similarly obtain P_2, P_3, P_4 and P_5 as explained in Problem 26.
9. Join P_1, P_2, P_3, P, P_4 and P_5 by a *smooth Rectangular Hyperbola.*

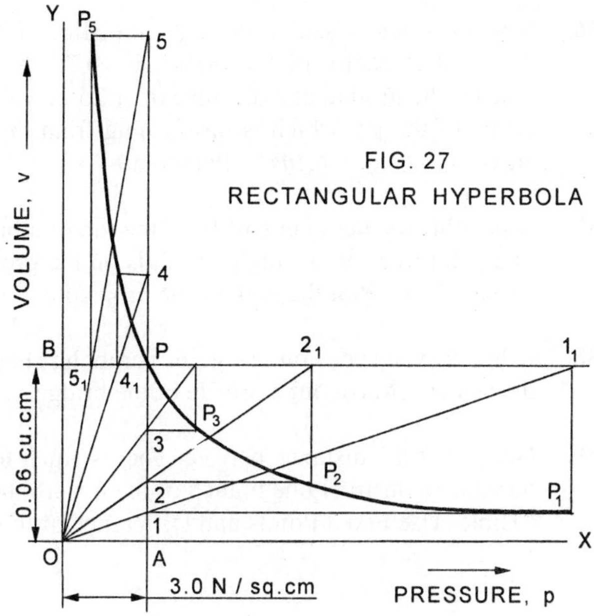

FIG. 27
RECTANGULAR HYPERBOLA

6. ADDITIONAL PROBLEMS

30. The directrices of an ellipse are 200 mm apart. Distance between the vertices is 150 mm. Find the eccentricity and the minor axis. Use Concentric Circles Method. (UQ)

(*Results :* e = 3/4 ; Minor Axis = 94 mm)

31. The distance between two fixed points is equal to 75 mm. A point P moves such that the sum of its distances from the two fixed points is always a constant and is equal to 90 mm. Draw the locus of P and determine the minor axis. (UQ)

(*Results :* Intersecting Arcs Method; Minor axis = 50 mm)

32. The foci of an ellipse are 90 mm apart and the minor axis is 72 mm long. Determine the length of the major axis. Construct the ellipse. Draw a tangent to the ellipse from any point outside the ellipse. [Hint: Use Intersecting Arcs Method] (UQ)

33. The span and rise of a parabola are 1800 mm and 1320 mm respectively. Construct the parabola. Draw a tangent and normal to the curve at a point on it, located 45 mm from the vertex on the right side. [Hint: Use Rectangle Method] (UQ)

34. The distance between a fixed point and fixed line is 54 mm. Trace the path of a point moving in the same plane such that its distance between the fixed point and the fixed line is always equal. If the point moves upto 72 mm from the fixed point, plot the curve. [Hint: Use Eccentricity Method] (UQ)

35. A roof arch is in the shape of a parabola. Span of the arch is 40 m and the rise is 8 m. Plot the profile. [Hint: Span = Base and Rise = Height of the parabola]

36. A jet of water is issuing through an orifice of 25 mm diameter fitted to a vertical side of a tank. The centre of the orifice is 0.75 m above the ground and the centre of the jet touches the ground at a distance of 1.25 m from the vertical orifice. Draw the locus of the centre of the jet, which is just issuing from the orifice till it reaches the ground. Name the curve. (*Ans :* Parabola) (UQ)

37. A boy throws a cricket ball from the top of a building 4 m high. The ball crosses the top of a palm tree 9 m high and falls on the ground. Distance between the building and the tree is 3 m. Plot the path of the projectile. [Hint: Use Rectangle Method]

38. A highway suspension bridge of parabolic shape has a span of 25 m and a dip of 5 m at the center. Make out a profile of the bridge.

39. Given that the distance between foci is equal to 90 mm and the transverse axis is equal to 50 mm. Construct one branch of the hyperbola and mark its asymptotes and directrix. [Hint: Use Fixed Points and Difference of the Distances Method.] (UQ)

*

6

CYCLOIDS, INVOLUTES, SPIRALS AND HELICES

1. SPECIAL CURVES

Apart from the curves obtained under Conic Sections, a number of other types of *Special Curves* are also used in engineering applications. A knowledge of the methods of constructing such curves is required for the manufacture of machine components made up of standard shapes. Special curves are classified as: *Plane Curves* and *Space Curves*

A *Curve* is a line traced out by a moving point by constantly changing its direction. A *Plane Curve* is a curve traced by a point which moves in a constantly changing direction on a plane surface. It is also known as *Single Curved Line*. *Examples* : Cycloidal Curves, Involutes and Spirals. A *Space Curve* is a curve traced by a point which moves in a constantly changing direction in space. It is also known as *Double Curved Line*. *Examples*: Helices. Table below gives the classifications of the family of Special Curves:

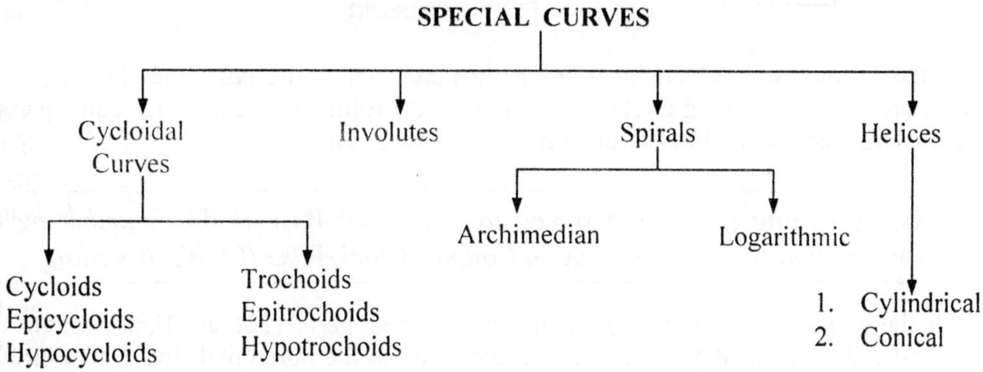

2. CYCLOIDAL CURVES

A *Cycloidal Curve* is traced by a fixed point on the circumference of a circle which rolls without slipping along a fixed straight line or a circle. It is also known as *Roulette*.

The circle which rolls along the line is called *Rolling Circle,* also known as *Generating Circle*. The fixed line *or* circle is called *Directing Line / Base Line or Directing Circle / Base Circle*. When a circle makes one revolution on the base line, it would have moved through a distance equal to the circumference of the rolling circle.

2.1 CYCLOIDS

Cycloid is a curve traced by a point on the circumference of a circle which rolls along a straight line without slipping. *Engineering Applications:* Cycloid is used for small gears in instruments like dial gauges and watches.

Problem 1: *(Fig. 1) A coin of 40 mm diameter rolls over a horizontal table without slipping. A point on the circumference of the coin is in contact with the table surface in the beginning and after one complete revolution. Draw the path traced by the point. Draw a tangent and normal at any point on the curve.* (UQ)

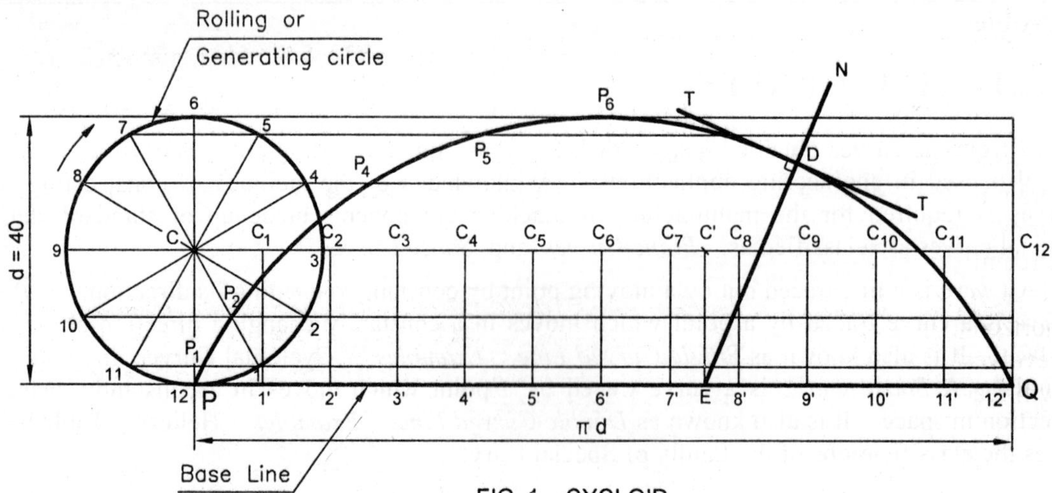

FIG. 1 CYCLOID

1. Draw the rolling circle of diameter (2r) 40 mm. Draw the base line PQ equal to the circumference of the rolling circle at P. Divide the rolling circle into 12 equal parts as 1, 2, etc. Draw horizontal lines through the points 1, 2, etc.

Note: *As the rolling circle is assumed to roll Clock-Wise (CW), numbering of the division points on it should be in Counter Clock-Wise (CCW) direction.*

---

2. Divide the base line PQ into the same number of equal parts (12) at 1′, 2′, ..., etc. Draw lines perpendicular to PQ at 1′, 2′, ..., etc., to cut the horizontal line drawn through C (called *Locus of Center*) at C_1, C_2, ..., etc., respectively.
3. C_1, C_2, ..., etc., as centers and radius equal to the radius of rolling circle (20 mm), draw arcs to cut the horizontal lines through the points 1, 2, ..., etc., at P_1, P_2, ..., etc. Draw a smooth *cycloid* curve through the points P, P_1, P_2, ..., etc.

Note: *When the circle rolls forward CW by 1/12 of a revolution, the division point 1 on the rolling circle will come in contact with 1′ and the center of the rolling circle moves to the new position C_1 which is vertically above 1′.*

To draw the Normal and Tangent at a point D

4. D as center and radius = radius of the rolling circle, cut the line of locus of center at C′. From C′, draw a perpendicular line to PQ to get E on the base line.
5. Connect DE, the *Normal*. At D, draw a line perpendicular to DE to get *Tangent* TT.

Problem 2: *(Exercise) A roller of 40 mm diameter rolls on a straight line without slip. In the initial position the diameter AB of the circle is parallel to the line on which it rolls. Draw the locus of the points A and B for one complete revolution of the roller. Name the curve. Draw a tangent and normal at any point on the curve.*

Problem 3: *(Fig. 2) A circle of 40 mm diameter rolls on a horizontal line. Draw the curve traced out by a point R on the circumference for one half revolution of the circle. For the remaining half revolution the circle rolls on the vertical line. The point R is vertically above the center of the circle in the starting position.* (UQ)

Draw the rolling circle of 2r = 40 mm. Draw the base line = half the circumference of the rolling circle, since it revolves half revolution on a horizontal line. As rolling circle revolves ½ revolution, the point 12 will reach the bottom position.

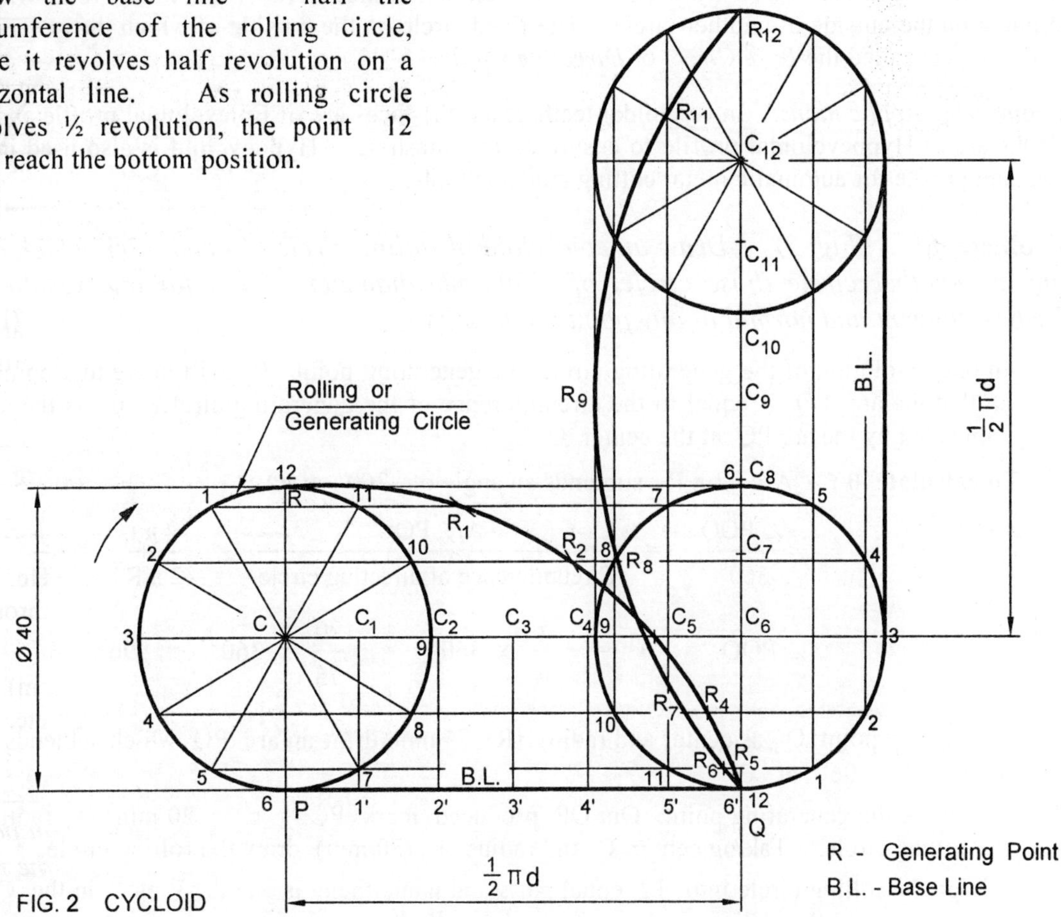

FIG. 2 CYCLOID

Problem 4: *(Exercise) A wheel of a bike of diameter 500 mm rolls without slipping on a level road through a distance of 1025 mm. Trace the path of a point P on the wheel which is initially in contact with the road. Name the curve. Find the angle through which the wheel is turned.* (UQ)

Hint: Distance moved by P when the wheel rotates through $180^\circ = \pi \times 250 = 785$ mm.
Angle turned through the remaining distance of (1025 – 785 =) 240 mm = 55°.
Therefore, the total angle turned through by the wheel = $180^\circ + 55^\circ$ = **235°.**
Construct the ***cycloid*** for the given length of (785 + 240) mm = 1025 mm.

Problem 5: *(Exercise) Construct a cycloid generated by a rolling circle of 45 mm in diameter. Draw a tangent and a normal at a point on the curve 30 mm above the directing line.*

2.2 EPICYCLOIDS

Epicycloid is a curve traced by a point on the circumference of a circle which rolls without slipping on the outside of another circle. The fixed circle on the outside of which the generating circle rolls is called the *Base Circle* or *Directing Circle.*

Engineering Applications: In Cycloidal teeth gears, the faces are of Epicycloidal profile and the flanks are of Hypocycloidal profile to ensure correct meshing. Hypocycloid is also used in the flat cam profile for automatic metal cutting machine tools.

Problem 6: *(Fig. 3) Draw an epicycloid of rolling circle 40 mm (2r), which rolls outside another circle (base circle) of 150 mm diameter (2R) for one revolution. Draw a tangent and normal at any point on the curve.* (UQ)

1. In one revolution of the generating circle, the generating point P will move to a point Q, so that the arc PQ is equal to the circumference of the generating circle. θ is the angle subtended by the arc PQ at the center O.

   To calculate θ : Arc $2\pi R$ subtends an angle of 360° at O.

$$\frac{\angle POQ}{360^\circ} = \frac{\text{Arc PQ}}{\text{Circumference of directing circle}} = \frac{2\pi r}{2\pi R} = \frac{r}{R}$$

$$\therefore \angle POQ = \theta = \frac{r}{R} \times 360^\circ = \frac{20}{75} \times 360^\circ = 96^\circ$$

2. Taking any point O as center and radius (R) 75 mm, draw an arc PQ which subtends an angle $\theta = 96^\circ$ at O.
3. Let P be the generating point. On OP produced, mark PC = r = 20 mm = radius of the rolling circle. Taking center C and radius r (20 mm) draw the rolling circle.
4. Divide the rolling circle into 12 equal parts and name them as 1, 2, 3, etc., in the CCW direction, since the rolling circle is assumed to roll *clockwise*.

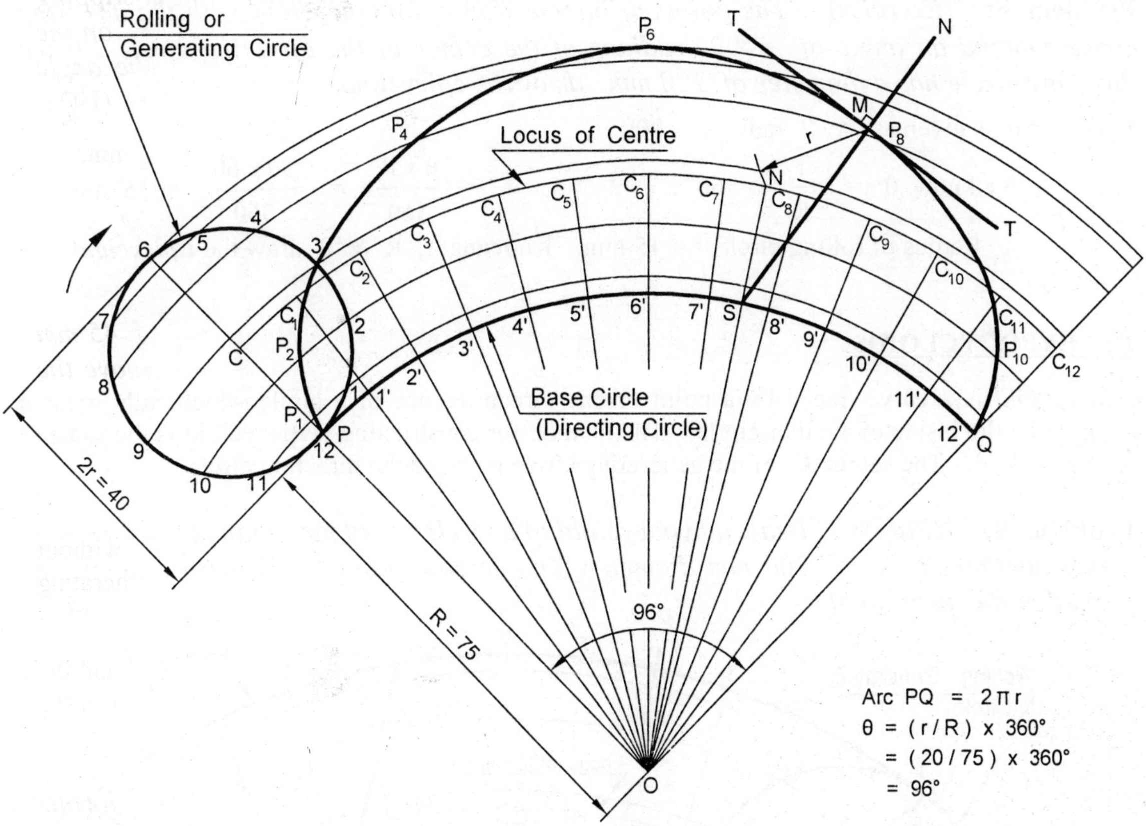

FIG. 3 EPICYCLOID

5. O as center, draw concentric arcs passing through 1, 2, 3,..., etc.
6. O as center and OC as radius draw an arc to represent the locus of center.
7. Divide arc PQ into same number of equal parts (12) and name them as 1′, 2′, ..., etc.
8. Join O1′, O2′, ..., etc., and extend them to cut the locus of center at C_1, C_2, ..., etc.
9. Taking C_1 as center and radius equal to r, draw an arc cutting the arc through 1 at P_1.
10. Similarly, obtain the other points and draw a ***smooth Epicycloid*** through them.

To draw a Tangent and Normal at a given point M

11. M as center, and radius r = CP cut the locus of center at the point N.
12. Join NO which intersects the base circle arc PQ at S.
13. Join MS, the normal and draw the tangent perpendicular to it.

Problem 7: *(Exercise) Draw an epicycloid, the directing circle of which is 160 mm in diameter and the generating circle 40 mm in diameter. Draw a tangent and normal to the curve at any point on it.*

Problem 8: *(Exercise) The points of intersection of an epicycloid with the directing circle subtend an angle of $\pi / 2$ radians at the center of the directing circle. If the directing circle has a diameter of 120 mm, draw the epicycloid.* (UQ)

Hint: θ is given as $\pi / 2$ radians = 90°.

$$\text{We know } \theta = \frac{r}{R} \times 360^\circ ; \qquad \therefore \; r = \frac{\theta \times R}{360} = \frac{90 \times 60}{360} = 15 \text{ mm.}$$

$\therefore$ Radius of rolling circle r = 15 mm. Knowing r, R & θ draw the ***Epicycloid***.

2.3 HYPOCYCLOIDS

Hypocycloid is a curve traced by a point on the circumference of a circle which rolls without slipping on the inside of another circle. The method for constructing Hypocycloid is the same as for Epicycloid. The center C of the generating circle is inside the directing circle.

Problem 9: *(Fig. 4) Draw a hypocycloid of a circle of 40 mm diameter which rolls inside another circle of 200 mm diameter for one revolution. Draw a tangent and normal at any point on it.* (UQ)

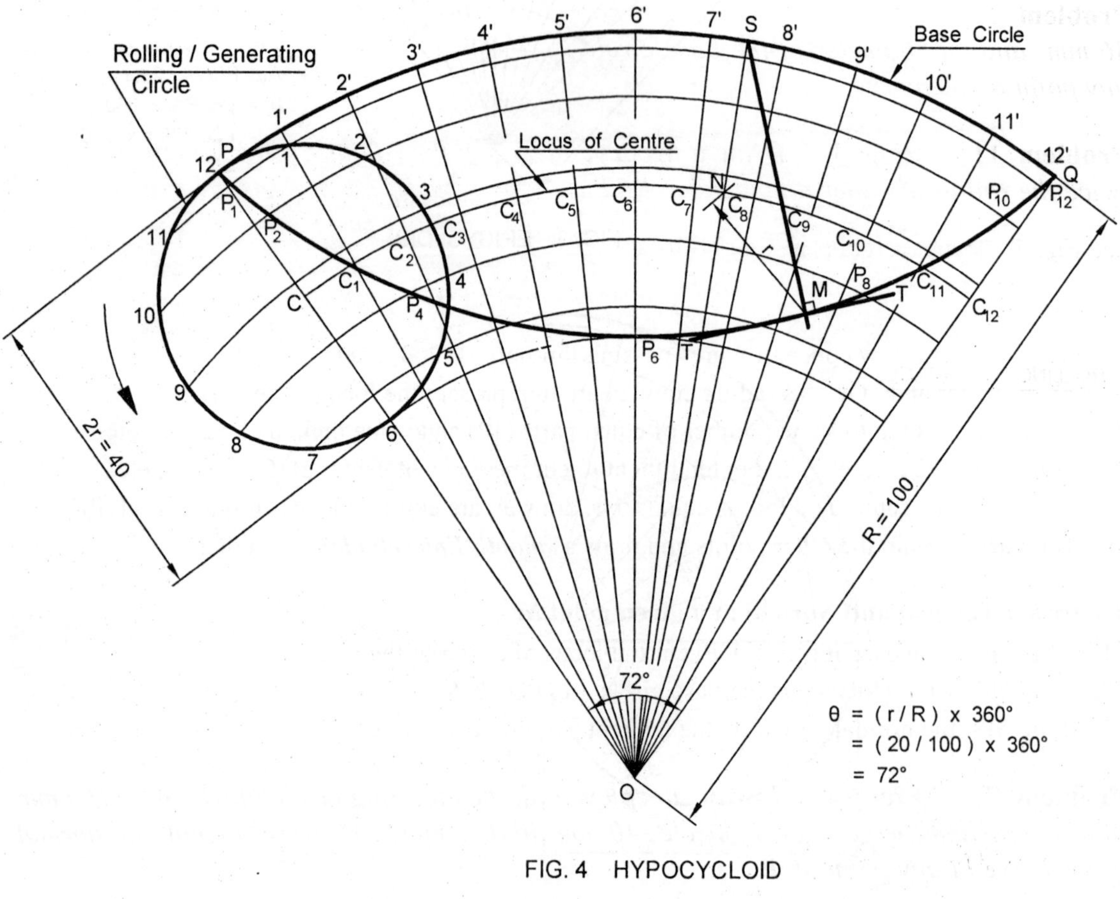

FIG. 4 HYPOCYCLOID

1. **To calculate θ :** $\theta = \frac{r}{R} \times 360^\circ = \frac{20}{100} \times 360^\circ = 72^\circ$.
2. Taking any point O as center and radius (R) 100 mm, draw an arc PQ which subtends an angle $\theta = 72^\circ$ at O.
3. Let P be the generating point. On OP mark PC = r = 20 mm = radius of rolling circle.
4. With C as center and radius r (20 mm) draw the rolling circle.
5. Divide the rolling circle into 12 equal parts as 1, 2, 3, etc., in CW direction, since the rolling circle is assumed to roll CCW.
6. **Repeat Step Nos. 5 to 10 in *Problem 6* and draw a smooth Hypocycloid curve.**

Tangent and Normal at a given point M

7. M as center and radius r = CP cut the locus of center at the point N.
8. Join ON and extend it to intersect the base circle at S.
9. Join MS, the **Normal**.
10. At M, draw a line perpendicular to MS to get the required **Tangent**.

Problem 10: *(Exercise) Draw a hypocycloid. The diameter of the rolling circle is 36 mm and the diameter of the base circle is 108 mm. Draw a tangent and normal at any point on the curve.*

Problem 11: *(Fig. 5) Draw a hypocycloid when the radius of the directing circle is twice the radius of generating circle. Radius of the generating circle is 35 mm.* (UQ)

See Fig. 5. Hypocycloid is a straight line.

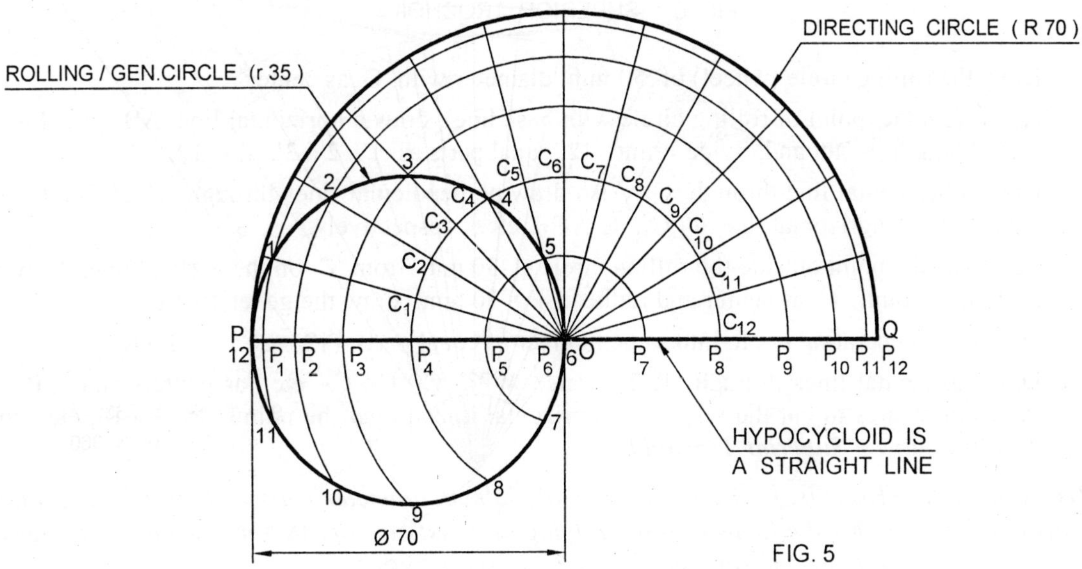

FIG. 5

3. TROCHOIDS

Trochoid is a curve traced by a point fixed to a circle, inside or outside its circumference, as the circle rolls without slipping along a straight line. If the point lies outside the rolling circle, the curve traced by the point is ***Superior Trochoid.*** If the point lies inside the rolling circle, the curve traced by the point is ***Inferior Trochoid.***

Problem 12: *(Fig. 6) A wheel 50 mm diameter rolls on a straight road surface without slip. Trace the path of a point which is at a radius of 30 mm from the center of the wheel, for one complete revolution of the wheel.* (UQ)

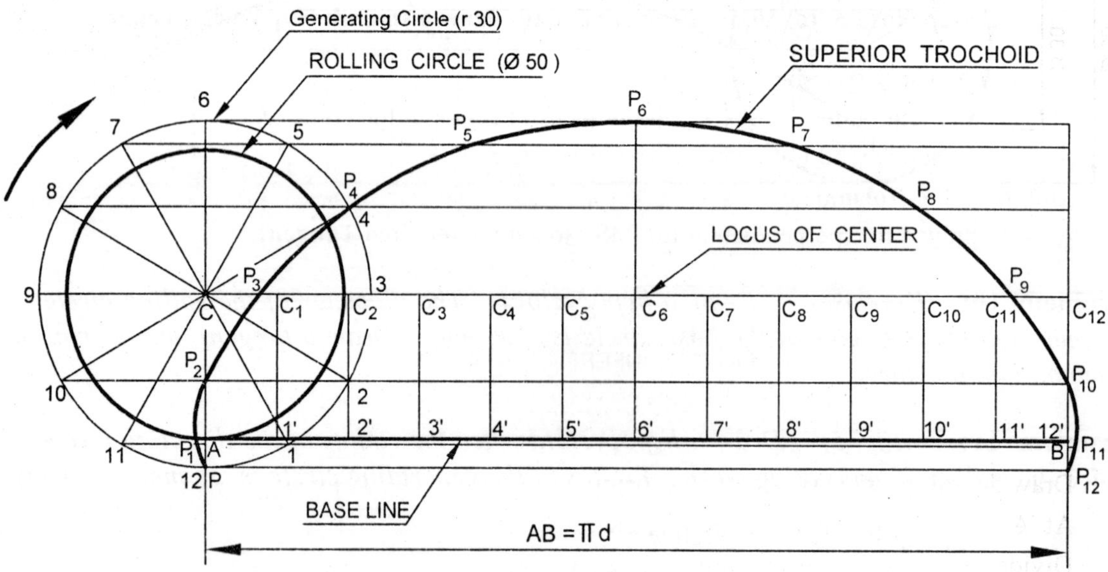

FIG. 6 SUPERIOR TROCHOID

1. Draw the rolling circle (wheel) of 50 mm diameter with C as center.
2. At A (contact point of rolling circle with base line), draw a horizontal line AB called base line of length = πd and divide it into 12 equal parts as 1′, 2′, 3′, ..., 12′.
3. Draw a horizontal line through C. Also draw perpendicular lines through 1′, 2′, 3′, etc., to cut the above horizontal line at C_1, C_2, C_3, etc., respectively.
4. Let P be the point outside the rolling circle at 30 mm from C on the vertical line. Now P is starting point. C as center and radius CP = 30 mm, draw the generating circle.
5. Divide the generating circle into same number of equal parts (12) as 1, 2, 3, etc.
6. Draw horizontal lines through 1, 2, 3, etc. With C_1, C_2, C_3, etc., as centers and CP as radius draw arcs to cut the respective horizontal lines to get the points P_1, P_2, P_3, etc., and draw the curve *(**Superior Trochoid**).*

Problem 13: *(Exercise) A circular disc of 55 mm diameter rolls without slipping along a straight line. Trace the locus of a point lying at a distance of 38.5 mm from the center of the disk, when it rolls by one revolution.*

Problem 14: *(Fig. 7) A wheel 50 mm diameter rolls on a straight road surface without any slip. Trace the path of a point on one of the spokes 10 mm from the rim towards the center of the wheel for one complete revolution.* (UQ)

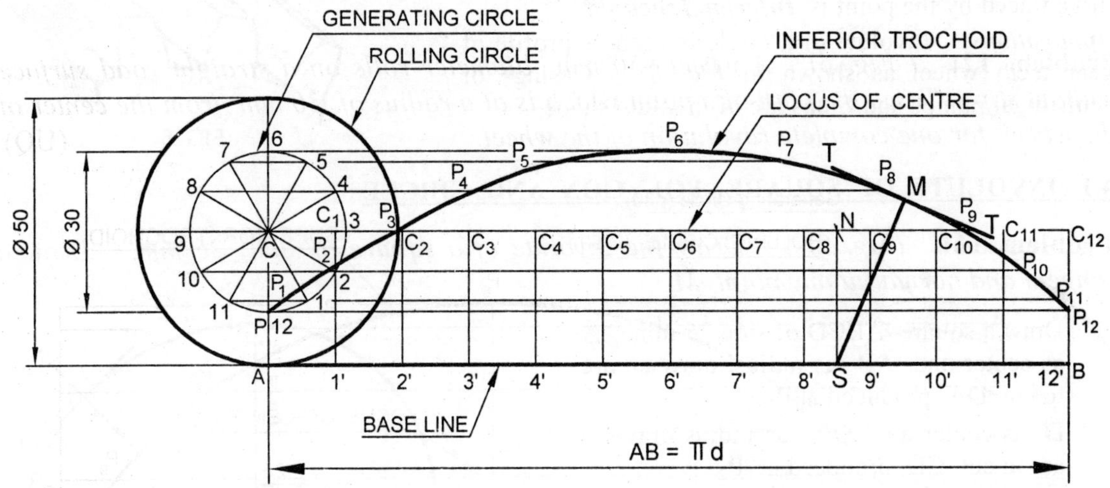

FIG. 7 INFERIOR TROCHOID

1. Draw the rolling circle (wheel) of 50 mm diameter with C as center.
2. At A draw a horizontal line AB = circumference of the rolling circle.
3. Divide AB into 12 equal parts as 1′, 2′, 3′, etc.
4. Draw a horizontal line through C. Also draw perpendiculars through 1′, 2′, 3′, etc., to cut the above horizontal line at C_1, C_2, C_3, etc., respectively.
5. Let P be the starting point inside the rolling circle at 10 mm from A on AC.
6. C as center and radius CP = 15 mm draw the generating circle.
7. Divide the generating circle into same (12) number of equal parts as 1, 2, 3, etc.
8. Draw horizontal lines through 1, 2, 3, etc.
9. With C_1, C_2, C_3, etc., as centers and CP as radius, draw arcs to cut the respective horizontal lines to get the points P_1, P_2, P_3, etc.
10. Join P_1, P_2, P_3, etc., by a smooth curve ***(Inferior Trochoid)***.

Note: *Draw Tangent and Normal on Trochoids similar to the Cycloids.*

Problem 15: *(Exercise) A bicycle wheel of 1 m diameter rolls on a straight surface. A point P is on a spoke of the wheel at a distance of 0.2 m from the rim. Draw the locus of P for one revolution of the wheel. Name the curve.*

4. INVOLUTES

Involute is a single-curved line traced out by an end of a string when unwound itself from a straight line or a circle or a polygon, the string being always kept taut.

Applications: Involute of a circle is used as profile of gear teeth wheel as shown in *Fig. 8*. Casings of centrifugal pumps and Cams are of involute shape.

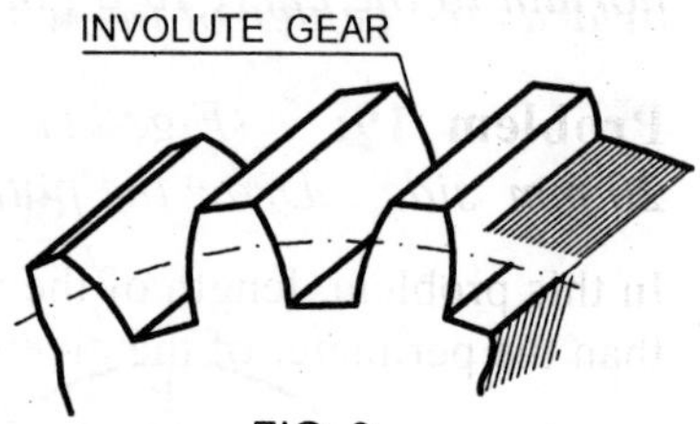

FIG. 8

4.1 INVOLUTE OF SQUARE, POLYGON AND CIRCLE

Problem 16: *(Fig. 9) Draw the involute of a* ***square*** *of side 25 mm. Draw a tangent and normal at any point M.*

1. Draw a square ABCD of side 25 mm. A as center and AB as radius, draw an arc to cut DA produced at P_1.
2. D as center and DP_1 as radius draw an arc to cut CD produced at P_2.
3. C as center and CP_2 as radius draw an arc to cut BC produced at P_3.
4. Similarly, B as center and BP_3 as radius draw an arc to cut AB produced at P_4.
5. **Normal and Tangent:** The given point M lies in the arc $P_3 P_4$. The center of the arc $P_3 P_4$ is point B. Join B and M.

   Extend BM which is the required *Normal*. At M draw a perpendicular to the normal to obtain the *Tangent* TT.

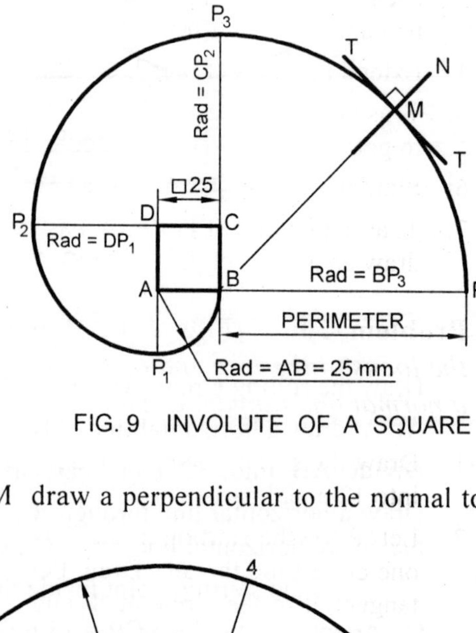

FIG. 9 INVOLUTE OF A SQUARE

Problem 17: *(Fig. 10) Draw the involute of a* ***pentagon*** *of 22 mm side.*

1. Draw the pentagon ABCDE. Produce the sides CB, DC, ED, AE and BA.
2. B as center and BA as radius, draw an arc to intersect CB produced at 1. C as center and C1 as radius, draw an arc to intersect DC produced at 2.
3. Similarly, obtain 3, 4 and 5. The curve A12345 is the required involute.

The above procedure is common for the construction of involute of any other polygon.

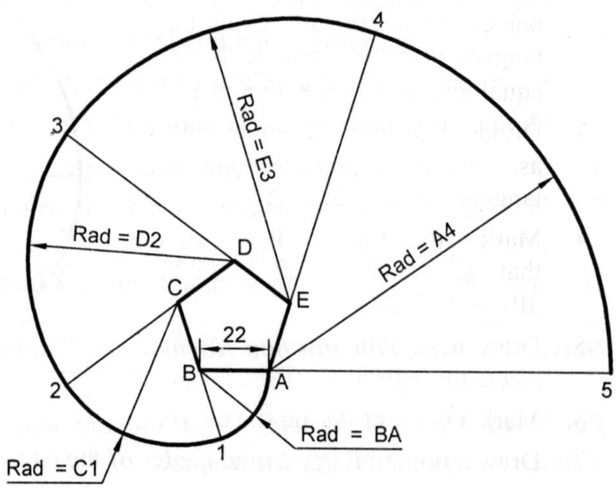

FIG. 10 INVOLUTE OF A PENTAGON

Problem 18: *(Exercise) Draw the curve traced out by an end of a thin wire unwound from a regular hexagon of side 15 mm, the wire being kept taut. Draw a tangent and normal to the curve at a point 80 mm from the center of the hexagon.*

Problem 19: *(Fig. 11) A string of length 108 mm is wound round a pentagon of 20 mm side. Draw the path traced by the end of the spring.* (UQ)

In this problem, length of the string is more than the perimeter of the given pentagon.

1. Draw pentagon PQRST of side 20 mm.
2. At T draw a horizontal line TA of 108 mm length.
3. On TA, mark 5 divisions 1, 2, ..., 5, each of length 20 mm (= side of pentagon) starting from T.
4. Extend the lines TP, PQ, QR and RS.
5. T as center TA as radius, draw an arc to get P_1.
6. Similarly obtain Q_1, R_1, S_1 and T_1.
7. T as center and TT_1 (=5A) as radius, draw an arc to get U_1.

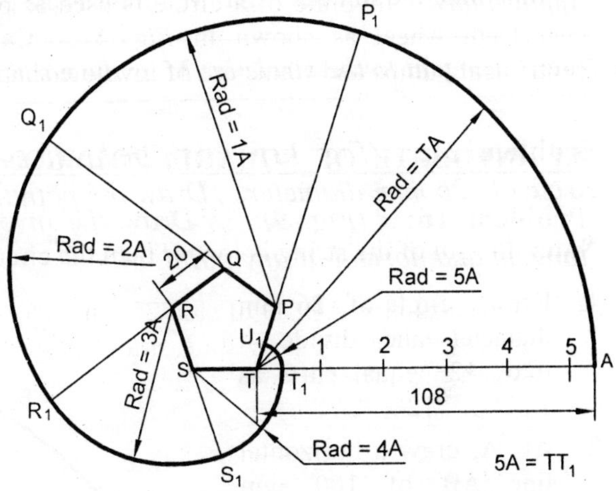

FIG. 11 INVOLUTE OF A PENTAGON

Problem 20: *(Figs. 12) A coir is unwound from a drum of 30 mm diameter. Draw the locus of the free end of the coir for unwinding through an angle of 360°. Draw also a normal and tangent at any point on the curve.* (UQ)

1. Draw the given circle. Divide it into 12 equal parts as 1, 2, ..., 12.
2. Let P be the starting point *i.e.*, one end of the thread. Draw PQ tangential to the circle at P and equal to π d (= π x 30 mm).
3. Divide PQ into 12 equal parts as 1′, 2′, ..., 12′. Draw tangents at 1, 2, 3,..., etc.
4. Mark P_1, P_2, ..., P_{12} such that $1P_1$ = P1′, $2P_2$ = P2′, $3P_3$ = P3′, etc.
5. Draw a smooth *involute of the circle* through P, P_1, P_2, ..., P_{12}.
6. Mark any point M on the curve.

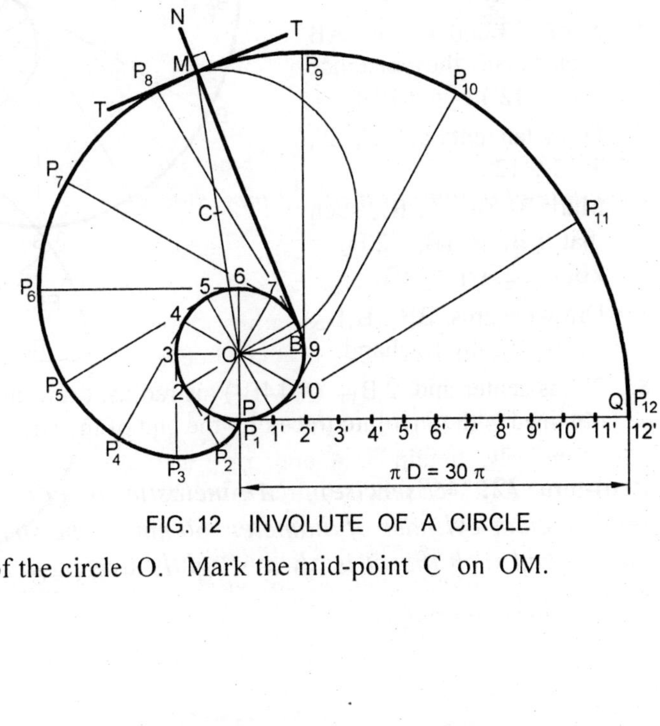

FIG. 12 INVOLUTE OF A CIRCLE

7. Draw a line joining M and center of the circle O. Mark the mid-point C on OM.

8. C as center and MC as radius, draw a semi-circle to cut the given circle at B.

Note: *Two semi-circles can be drawn, one on the converging side of the involute and another on the diverging side. It is important to note that the semi-circle on the diverging side of the involute alone should be considered.*

---

9. Join MB, the *Normal*. At M, draw a line perpendicular to MB, to get *Tangent* TT.

Note: *Tangent to the circle at M is the normal to the involute at the same point M.*

Problem 21: *(Fig. 13) An inelastic string of length 100 mm is wound around a circle of 26 mm diameter. Draw the path traced by the end of the string.* (UQ)

Note: Length of the string is longer than the circumference of the circle on which it is wound.

1. Draw a circle of 26 mm diameter and divide it into 12 equal parts as 1′, 2′, ..., 12′.
2. At A, draw a horizontal line AB of 100 mm length.
3. Mark A 12 = Circumference of the circle = π x 26 mm. Divide the length A 12 into 12 equal parts as 1, 2, ..., 12.
4. Mark 13 and 14 on AB such that the distance A1 = 12 13 = 13 14.
5. Draw tangents at 1′, 2′, 3′, ..., 12′.
6. Mark $B_1, B_2, \ldots, B_{14}$ such that $1'B_1$ = 1B, $2'B_2$ = 2B, ..., $2'B_{14}$ = 14B.
7. Draw the arcs $BB_1, B_1B_2, \ldots, B_{13}B_{14}$ by freehand.

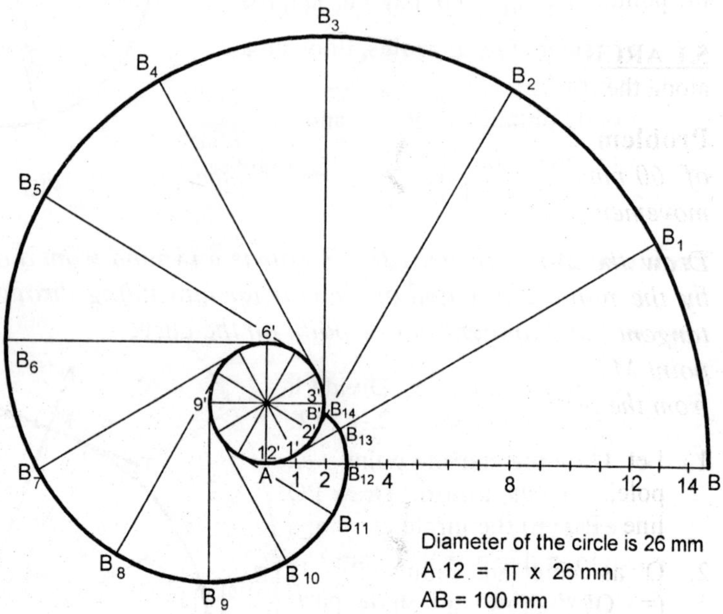

Diameter of the circle is 26 mm
A 12 = π x 26 mm
AB = 100 mm

FIG. 13 INVOLUTE OF A CIRCLE

8. 2′ as center and $2'B_{14}$ (= 14B) as radius, draw an arc to meet the circle at B′. The curve obtained is the involute traced by the end of the string wound round the circle.

Problem 22: *(Exercise) An inelastic string is wound around the circumference of a semi-circular cylinder of diameter 66 mm. The string is unwound completely by holding its free end such that it is always tightly stretched. Draw the path of the free end of the string.* (UQ)

Problem 23: *(Exercise) Trace the path of the end point of a cable unwinding itself from a drum of 500 mm diameter, such that the unwound cable is taut during unwinding. Draw a tangent and normal at any point on the curve.* (UQ)

5. SPIRALS – ARCHIMEDIAN AND LOGARITHMIC SPIRALS

Spiral is defined as the locus of a point which moves around a center, called *Pole*, while moving towards or away from the center. The point will move along a line called Radius Vector while the line itself rotates about one of its end points. The locus of the point for one complete revolution of the radius vector is called *Spiral for one convolution.*

Pole is the fixed point about which the line rotates. *Radius Vector* is the line joining any point on the curve with the pole. *Vectorial Angle* is the angle between the line, at any instant, and its initial position. The curve traced by the moving point for one complete revolution of the line is called a *Convolution*. For a spiral curve, there may be any number of convolutions before the point reaches final position. *There are two types of spirals:*

5.1 ARCHIMEDIAN SPIRAL: It is the locus of a point moving with uniform linear velocity along the radius vector, while the radius vector itself rotates with uniform angular velocity.

Problem 24: *(Fig. 14) A point moves away from the pole O and reaches a distance of 60 mm while moving around it once, its movement from O being uniform with its movement around it.*

Draw the curve traced out by the point P. Draw a tangent and normal at a point M on the curve 30 mm from the pole. (UQ)

1. Let O be the center of the pole. Draw a horizontal line PO = 60 mm.
2. O as center and radius R (= OP) draw a circle. Divide it into 12 equal parts as 1, 2, etc.
3. Divide OP into same number of parts (12) as 1′, 2′, 3′, ..., 12′.
4. Draw radial lines connecting 1, 2, 3, ..., 12 with O.
5. O as center and O1′ as radius, draw an arc to cut the radial line O1 at P_1.

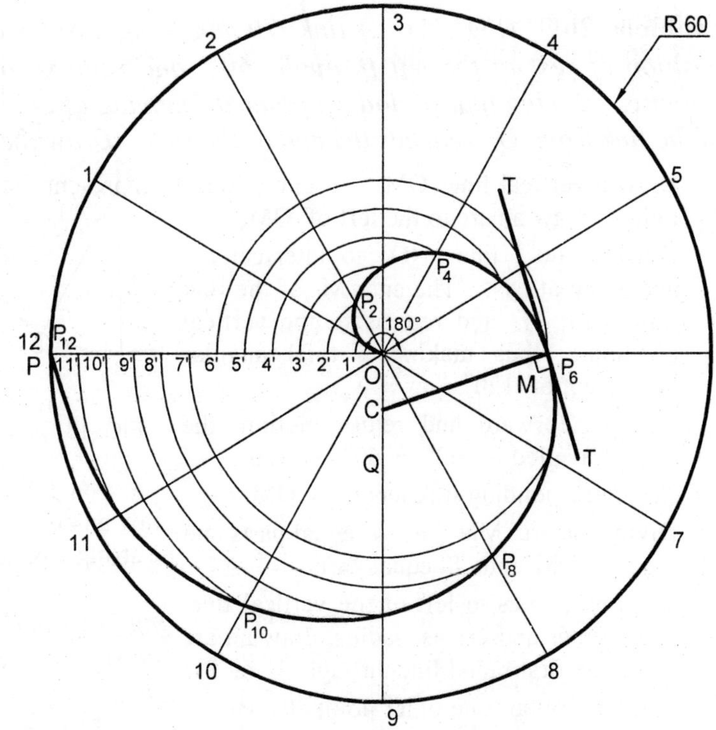

FIG. 14 ARCHIMEDIAN SPIRAL

6. Similarly obtain $P_2, P_3, ..., P_{12}$ corresponding to radii O2′, O3′, ..., O12′ and radial lines O2, O3, ..., O12. Draw a *smooth Archimedian Spiral* through the above points.

Tangent and Normal at the given point M

7. With pole O as center and 30 mm as radius draw an arc to cut the spiral at M.
8. Join MO. Measure the angle MOP = θ, which is equal to 180°. Draw OQ ⊥^{r} to OM.
9. The polar equation of Archimedian Spiral is $r = c + a\theta$.

a	=	constant of the curve		
r	=	radius vector at any point M	=	30 mm (given)
c	=	initial radius vector	=	0 mm
θ	=	corresponding vector angle	=	180° (measured) = π radians

$$\therefore a = \frac{r - c}{\theta} = \frac{30 - 0}{\pi} = 9.55$$

10. Mark C on OQ such that OC = a = 9.55 mm. Join MC, the **Normal**.
11. At M, draw a perpendicular to CM to obtain the **Tangent** TT.

Problem 25: *(Exercise) Draw an Archimedian Spiral of 1 ½ convolutions when the greatest and the shortest radii are 75 mm and 15 mm respectively.*

Engineering Applications: *Spirals* are used to provide spring action of watches, clocks, etc. *Archimedian Spiral* is used in cam design and lathe chucks. *Logarithmic Spiral* is used in the construction of loud speakers, audio amplifiers, etc.

Problem 26: *(Fig. 15) A link 70 mm long, swings on a pivot O from its vertical position of rest to the left through 60° and returns to its initial position at uniform velocity. During that period, a point P moving at uniform speed along the center line of the link from O, reaches the end of the link. Draw the locus of P.* (UQ)

1. Draw a vertical line OM 70 mm long to represent the link. O as center and OM as radius, draw an arc to the **left** of OM.
2. Draw a 60° line NO to intersect the above arc at N. The end M of the link travels to N and returns to the vertical position, thus making a total angular movement of 120°.
3. For one forward and return motion, the angle travelled is 60° + 60° = 120°.
4. The corresponding link length is OM.
5. Divide the arc MN into 4 equal parts and the link OM into 8 equal parts.
6. The link moves to left of the vertical line. O as center and O1 as radius, draw an arc to cut the first radial line through 1′ at P_1.
7. Similarly obtain the other points $P_2, P_3, ..., P_8$ and draw a *smooth Archimedian Spiral.*

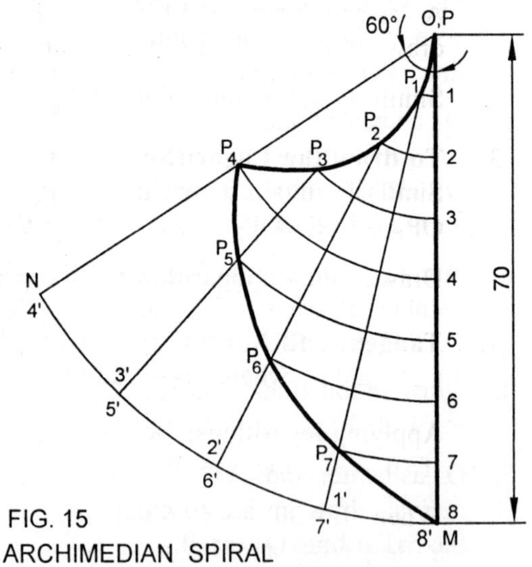

FIG. 15
ARCHIMEDIAN SPIRAL

Problem 27: *(Exercise) A sportsman was asked to slide down a rope tied to a beam at a height of 9 m above the ground. He takes 8 seconds to slide down with uniform velocity. The rope oscillates, in a vertical plane, through an angle of 25° on either side of the plumb line and takes 4 seconds for one complete oscillation. Draw the path traced out by the sportsman (assumed as a point).* (UQ)

Hint: Sportsman takes 8 seconds to slide down and 4 seconds are required for one oscillation. Hence, the rope will make two oscillations. For one oscillation, the rope moves angularly by 100°.

5.2 LOGARITHMIC SPIRAL or EQUIANGULAR SPIRAL

In a *Logarithmic Spiral,* the ratio of the lengths of consecutive radius vectors enclosing equal angles is always constant. In other words, the values of vectorial angles are in arithmetical progression and the corresponding values of radius vectors are in geometrical progression.

The logarithmic spiral is also known as *Equiangular Spiral* because of its property that the angle which the tangent at any point on the curve makes with the radius vector at that point is constant.

Problem 28: *(Fig. 16) Draw a* ***logarithmic spiral*** *for one convolution such that the angle between consecutive radii is 30°, the successive radii are of ratio 6 : 5 and final radius vector is 90 mm. Draw tangent and normal at any point on the curve.*

1. *See Fig. 16(b).* Mark point O (pole). Draw number of radial lines at $\theta = 30^\circ$ to each other.

2. **To determine the lengths of Radius Vectors :** [*Fig. 16(a)*] Draw a line AB and mark A 12 (= greatest radius 90 mm) on it and draw AC at 30°.

   A as center and A 12 as radius draw an arc to cut AC at 12′. Mark A 11 (= 5/6 of greatest radius) on AB. Join 12′ and 11.

   A as center and A 11 as radius, draw an arc to cut AC at the point 11′. Through 11′ draw a line parallel to 12′ 11 to cut AB at 10.

   Similarly obtain the points 9, 8, 7, . . . , 1 and 0.

3. **To draw the Logarithmic Spiral:** Transfer the distance A0 on OP to get point P_0. Similarly obtain the points P_1, P_2, P_3, ..., P_{12} on the respective radial lines such that $OP_1 = A1$, $OP_2 = A2$, $OP_3 = A3$, etc.

   Draw a smooth logarithmic spiral passing through these points.

4. **Tangent and Normal at a given point M:** Polar equation of the logarithmic spiral is $r = a\theta$, where r = radius vector, θ = vectorial angle, a = constant.

   Applying logarithms, $\log r = \theta \times \log a$
   If $\theta = 0$, $\log r = 0$ $\quad \therefore r = 1$

   When $\theta = 30^\circ = \pi / 6$ radians, $r = 6/5 \times 1 = 6/5$
   $\therefore \log 6/5 = \pi / 6 \log a.$

From the property of spiral, $\angle \beta$ is found by the equation

$\tan \beta = \log_{10} e / \log_{10} a$

$= [\log_{10} 2.718] / [(6/\pi) \log_{10} (6/5)]$

$= (\pi/6) \times (0.4343/0.0792)$

$= 2.871$; Hence $\beta = 70^\circ 48'$

Connect OM. AT M draw a line at angle $\beta = 70^\circ 48'$ to OM, the required Tangent.

Draw MN, the Normal.

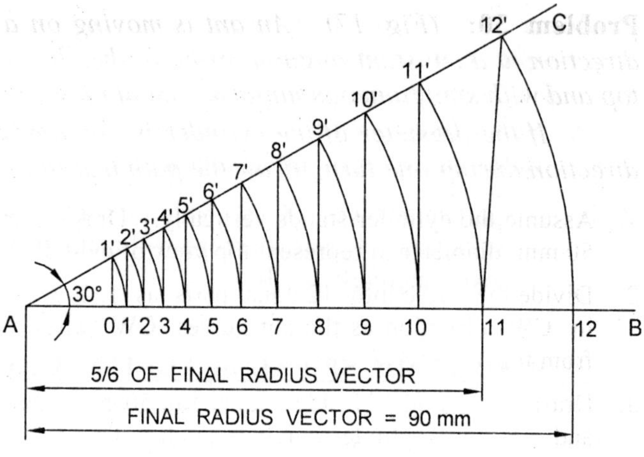

(a)

Problem 29: *(Exercise) Construct a Logarithmic Spiral for one convolution, given the length of the shortest radius as 30 mm and the ratio of the lengths of the radius vector enclosing an angle of 30° as 8 / 7. Draw a normal and a tangent at any point on the logarithmic spiral.*

Hint: Initial radius vector will be unity. Therefore take the initial radius vector of 30 mm as one unit.

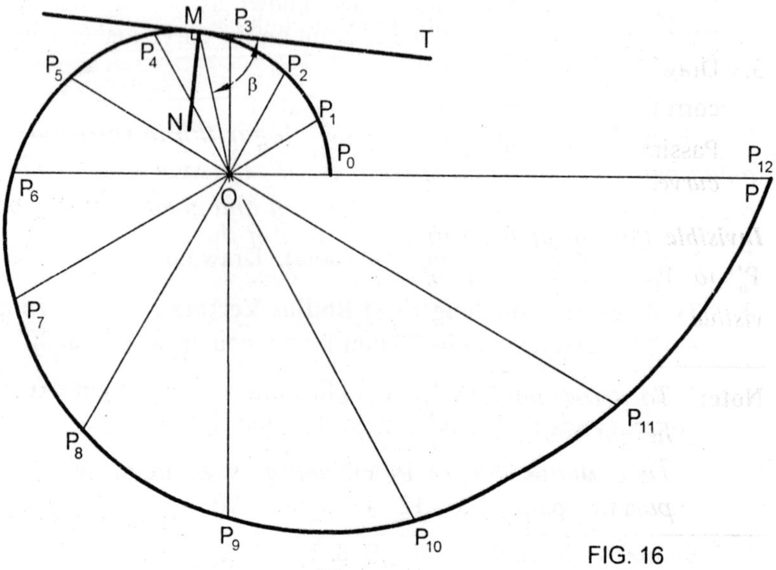

(b)

FIG. 16
LOGARITHMIC SPIRAL

6. HELICES – CYLINDRICAL AND CONICAL

Helix is defined as a curve generated by a point moving around and along the surface of a right circular ***cylinder*** or ***cone*** with uniform angular velocity about the axis and with a uniform linear velocity in the direction of the axis. Lead or Pitch is the axial distance moved by the generating point in one revolution.

Types of Helices: 1. Cylindrical Helix, simply known as Helix and 2. Conical Helix

Problem 30: *(Fig. 17) An ant is moving on a vertical cylinder in a circumferential direction at a constant angular speed in the clock-wise direction as seen axially from the top and with simultaneous uniform rate of advance in the axial direction.*

If the diameter of the cylinder is 50 mm and the ant moves 75 mm in the axial direction during one turn, draw the path traced by the ant.

1. Assume the cylinder stands vertically. Draw a circle of 50 mm diameter to represent top view of cylinder.
2. Divide the circle into 12 equal parts as 1, 2, 3, . . . , 12, in CW direction as the ant moves CW when viewed from top.
3. Draw a rectangle of sides equal to 50 mm (diameter) and 75 mm (cylinder height = pitch). This is the front view of the cylinder.
4. Divide the pitch also into same (12) equal parts as 1′, 2′, ..., 12′ and draw horizontal lines through each point.
5. Draw vertical projectors from 1, 2,..., 12 to intersect the corresponding horizontal projectors at P_1', P_2', ..., P_{12}'. Passing through these points, draw a *smooth helical curve*.

Invisible Portion of the Curve: *Portion of the curve from* P_6' *to* P_{12}' *is on rear side of cylinder / cone and will not be visible.*

Note: *To draw helix for two revolutions, the cylinder height shall be taken as equal to twice the pitch. Then divide the cylinder height into 24 equal parts.*

Problem 31: *(Exercise) Plot a right-hand single start helix for one convolution on a cylinder of base diameter 60 mm, height 72 mm and pitch 72 mm.*

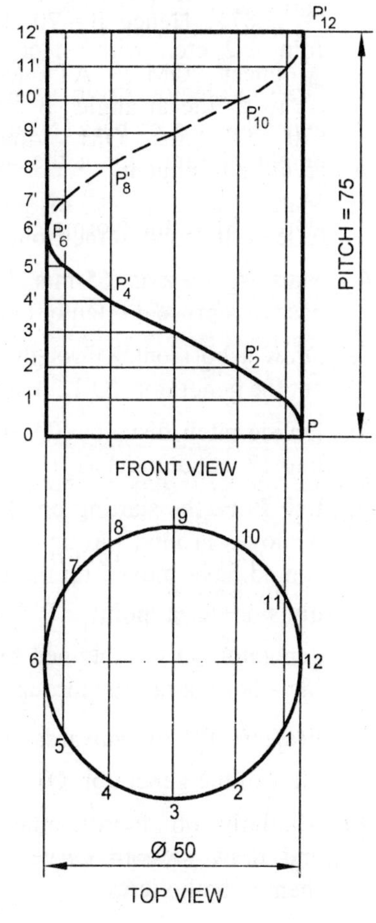

FIG. 17
L.H. CYLINDRICAL HELIX

6.1 CONICAL HELIX

Conical helix construction is similar to cylindrical helix. The difference is that each point is constructed on a different radius. It is generated as the point moves on the surface of a cone such that it moves around the axis and simultaneously moves towards vertex.

Applications: Helix is widely used in helical springs, worm gears, screw threads of V and square types, spiral stairways, etc. Helical springs are used in instruments, etc., for various purposes. Helical grooves of square shape is used on cylindrical cams, screw conveyors, etc.

Problem 32: *(Fig. 18) Draw a helix of one convolution around a cone, base 50 mm diameter, height 80 mm and pitch 65 mm.*

1. Draw the top view of the cone as a circle of diameter 50 mm.
2. Divide the circle into 12 equal parts as 1, 2, 3, . . . , 12.
3. Join 1, 2, etc., with center O.
4. Draw projectors from these points and draw a Δ of base 50 mm and height 80 mm. O′ is the apex.
5. Mark 1′, 2′, ..., 12′ on the above base of the cone in the front view.
6. Mark A on axis 65 mm from the base to represent pitch of helix.
7. Draw a horizontal through A to cut the generator (O′12′) at A′.
8. Divide pitch distance 3′A (65 mm) into 12 equal parts as 1″, 2″, etc.
9. Let P be the starting point. When it moves around through 30°, it should have moved up through one division to a point P_1' on the generator O′1′ obtained by drawing a horizontal line through 1″.
10. Project P_1' downwards to cut top view of the generator O1 at P_1.
11. Similarly obtain all other points and draw smooth curves through them in both views.

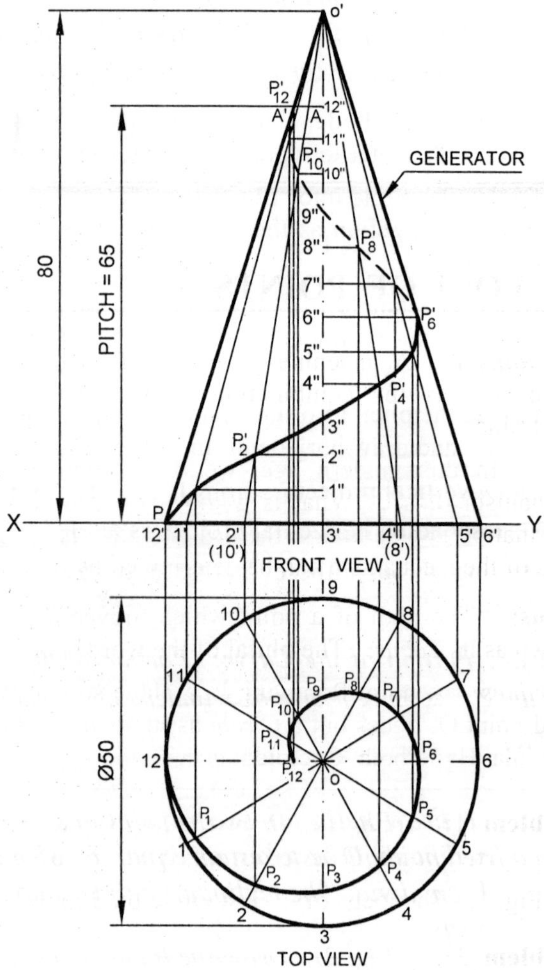

FIG. 18 CONICAL HELIX

Problem 33: *(Exercise) Draw a helix of one convolution around a cone of diameter 50 mm. Height and pitch of the helix are 60 mm.*

7

LOCI OF POINTS

1. LOCI OF POINTS

Mechanism: A machine is made of different types of mechanisms to transmit motion. Mechanism is the combination of various links so paired that the motion is completely constrained.

In this modern age, we often come across machines and mechanisms. How the mechanism runs? What is going to be the path of various points of the mechanism? This information is essential in the design and force analysis of the mechanisms. The *lengths* of the links of the mechanism can be determined by plotting the *path* of the required points.

Locus: The path of a point which moves according to mathematically defined conditions is known as its *Locus*. The plural of the word 'Locus' is *Loci*.

Examples: A point P moving in a plane so that it is always at a constant distance from another fixed point O, traces out a *Circle* as its locus. Many important geometrical curves, viz., Ellipse, Parabola, Hyperbola, Cycloidal curves are typical examples of loci of points.

Problem 1: *(Fig. 1) Draw the locus of a point P moving in a plane, keeping its distance from a fixed point O as constant, equal to 30 mm.*

See Fig. 1. The locus of P is a circle of 30 mm radius with the fixed point O as center.

Problem 2: *(Fig. 2) Draw the locus of a point P moving in a plane, keeping its distance from a fixed straight line AB as constant, equal to 35 mm.*

See Fig. 2. The locus of P is a line parallel to the fixed line AB at a distance of 35 mm.

Problem 3: *(Fig. 3) Draw the locus of a point P moving in a plane, keeping its distance from a fixed circle as constant equal to 10 mm and outside it. Radius of fixed circle is 20 mm.*

See Fig. 3. The locus of P is another circle of radius 30 (= 20 + 10) mm with the same center O.

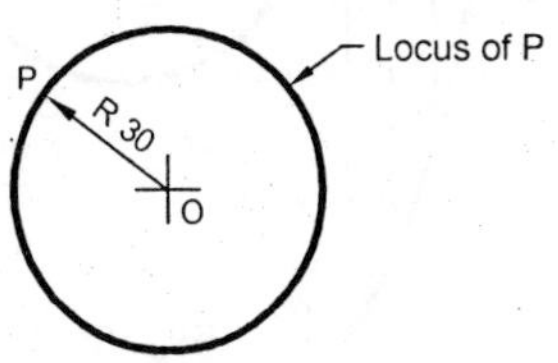

FIG. 1

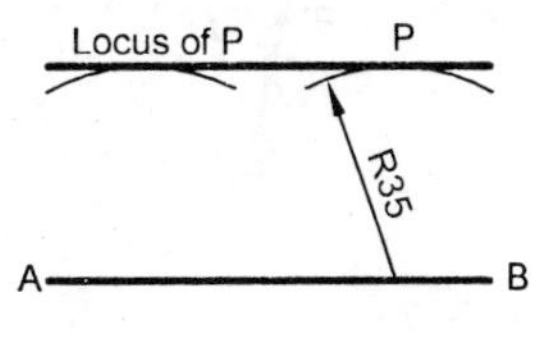

FIG. 2

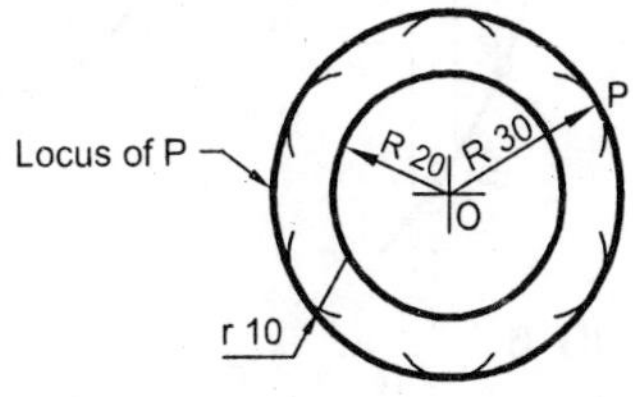

FIG. 3

Problem 4: *(Fig. 4) Draw the locus of P equidistant from two fixed points A and B in the same plane.*

See Fig. 4. The locus of P equidistant from fixed points A and B in the same plane, is the perpendicular bisector of the line joining the two points.

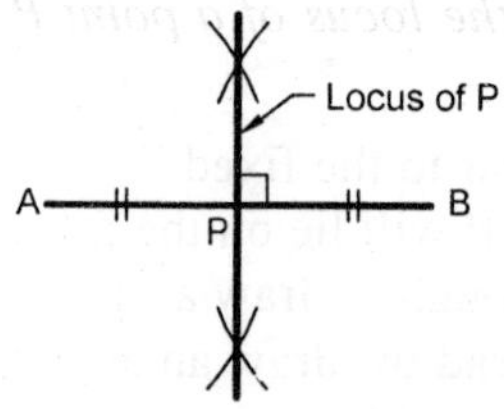

FIG. 4

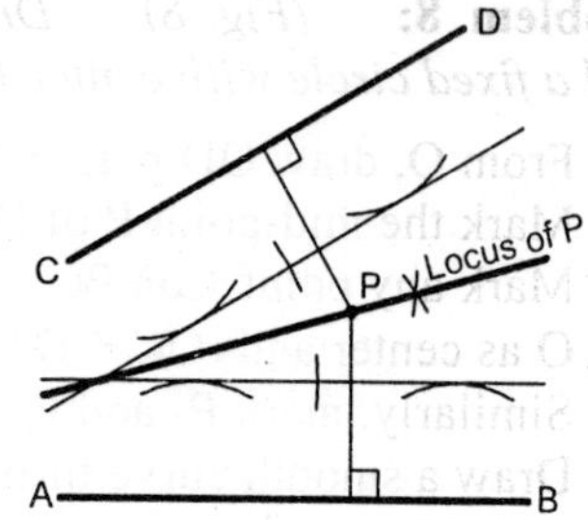

FIG. 5

Problem 5: *(Fig. 5) Draw the locus of P equidistant from two fixed non-parallel straight lines AB and CD.*

See Fig. 5. The locus of P equidistant from two fixed non-parallel lines AB and CD is a line bisecting the angle between them.

Problem 6: *(Fig. 6) Draw the locus of a point P equidistant from a fixed straight line AB and a fixed point C.*

1. From fixed point C, draw CD perpendicular to AB. The mid-point P of CD is equidistant from fixed line AB and C. Hence, P lies on the locus.
2. Mark 1, 2, 3, etc., on PC. Through 1, 2, 3, etc., draw lines parallel to AB.
3. C as center and D1 as radius, draw an arc to cut the line through 1 at P_1 and P_1'.
4. Similarly, mark P_2 and P_2', P_3 and P_3', etc. Through P_3, P_2, P_1, P, P_1', P_2', P_3', etc., draw a smooth curve to obtain the locus of P.

Problem 7: *(Fig. 7) Draw the locus of P equidistant from a fixed circle with center O and a fixed point C.*

1. Join OC to cut the fixed circle at A. Mid-point P of AC will lie on the locus.
2. Mark 1, 2, 3, etc., on PC. Through 1, 2, 3, etc., draw arcs with O as center.
3. C as center and A1 as radius, draw an arc to cut the arc through 1 at P_1 and P_1'.
4. Similarly, mark P_2 and P_2', P_3 and P_3', etc. Draw a smooth curve through P_3, P_2, P_1, P, P_1', P_2', P_3', etc., to obtain the locus of P.

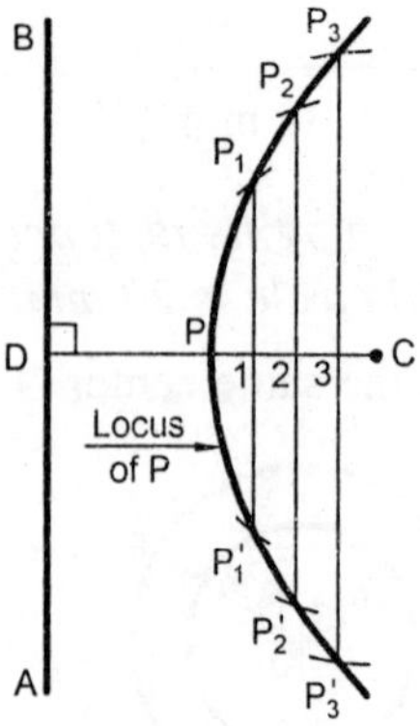

FIG. 6

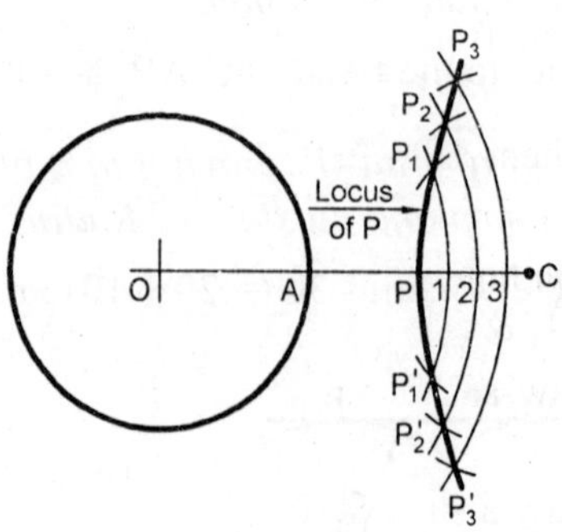

FIG. 7

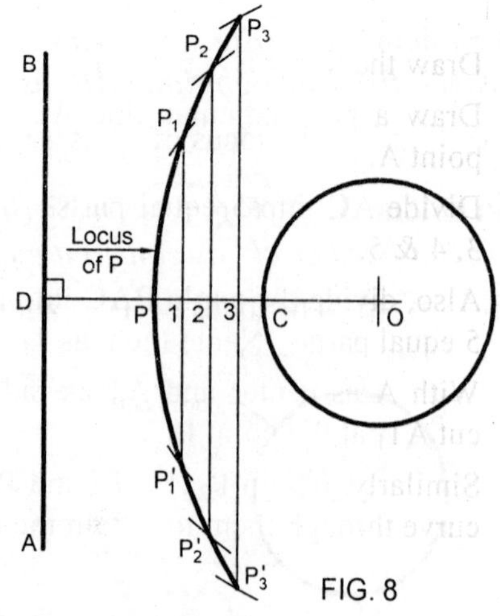

FIG. 8

Problem 8: *(Fig. 8) Draw the locus of a point P equidistant from a fixed straight line AB and a fixed circle with center O.*

1. From O, draw OD perpendicular to the fixed line AB to cut the circle at C.
2. Mark the mid-point P of DC. P will lie on the locus.
3. Mark any point 1 on PC. Through 1, draw a line parallel to AB.
4. O as center and (OC + D1) as radius, draw an arc to cut the line through 1 at P_1 and P_1'.
5. Similarly, mark P_2 and P_2', P_3 and P_3', etc.
6. Draw a smooth curve through P_3, P_2, P_1, P, P_1', P_2', P_3', etc., to obtain the locus of P.

Problem 9: *(Fig. 9) Draw the locus of a point P equidistant from two given unequal circles with centers A and B.*

1. Join AB. AB cuts the two unequal circles at C and D. Mark the mid-point P of CD.
2. Mark any point 1 on PD. Through 1, draw an arc with center A.
3. B as center and (BD + C1) as radius, draw an arc to cut the arc through 1 at P_1 and P_1'.
4. Similarly, mark P_2 and P_2', P_3 and P_3', etc.
5. Draw a curve through P_3, P_2, P_1, P, P_1', P_2', P_3', etc., to obtain the locus of P.

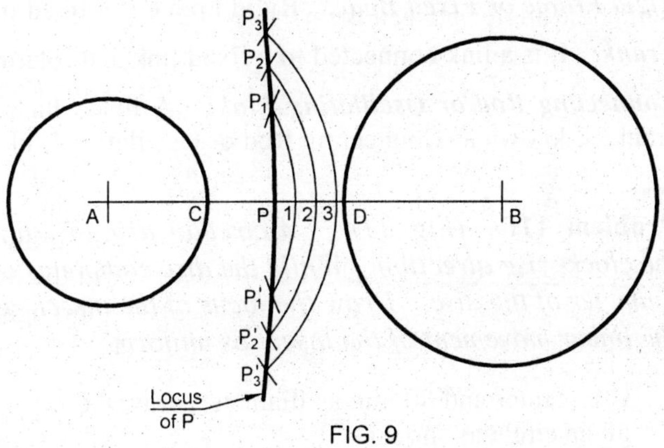

FIG. 9

Problem 10: *(Fig. 10) Draw the locus of a point P moving at uniform rate along a straight line AB, 55 mm long, pivoted at the end A. The point P moves from end A to end B during the revolution of the straight line AB through 90°.*

1. Draw the line AB = 55 mm.
2. Draw a perpendicular line AC = 55 mm from the point A.
3. Divide AC into 5 equal parts. Name them as 1, 2, 3, 4 & 5.
4. Also, divide the angle BAC into the same number of 5 equal parts. Name them as 1_1, 2_1, 3_1, 4_1 & 5_1.
5. With A as center and A1 as radius, draw an arc to cut $A1_1$ at the point P_1.
6. Similarly, obtain P_2, P_3, P_4 and P_5. Draw a smooth curve through them to obtain the required locus.

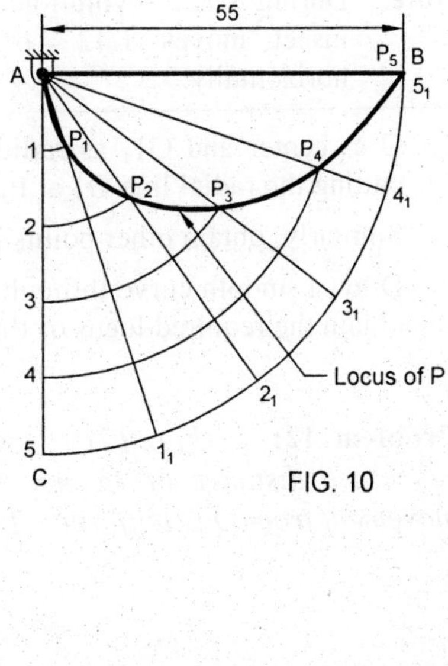

FIG. 10

2. SIMPLE MECHANISMS

2.1 LOCUS OF A POINT ON A MECHANISM

Locus of a Point in a Mechanism is the path traced by the point when the mechanism moves through a complete cycle of operation. The exact nature of the motion (locus) of a mechanism is required while designing the mechanism so that the forces present may be determined.

In general, the method for drawing the locus of a particular point in a mechanism is to construct the mechanism in several positions. The point is plotted for each position. Its locus is obtained by drawing a smooth curve through these plotted points.

Link: The element in a mechanism is called Link. A link may be a rod, a rigid frame or a slider.

Rigid Frame** or **Fixed Body: Rigid Frame is a fixed link with pivoted end or pivoted joint.

Crank: It is a link connected to a fixed link and rotating or oscillating about its pivoted end.

Connecting Rod or Oscillating Link: A link which connects a crank and another link or two cranks is known as Connecting Rod or Oscillating Link.

Problem 11: *(Fig. 11) A circular disc of 80 mm diameter rotates about its center in the clock-wise direction. While the disc completes one revolution, an insect walks across the diameter of the disc. Draw the locus of the insect, assuming both the rotation of the disc and the linear movement of the insect as uniform.* (UQ)

1. O as center and 80 mm as diameter, draw the given circle as shown.
2. Divide the circle into 12 equal parts. Name them as 1, 2, 3, ..., 12.
3. Divide the horizontal diameter (12-6) also into the same 12 number of equal parts. Name them as 1_1, 2_1, 3_1, etc.

Note: During 1/12th revolution of the disc, the insect moves 1/12th of the diameter horizontally.

4. O as center and $O1_1$ as radius, draw an arc cutting the radial line O1 at P_1.
5. Similarly, obtain other points P_2, P_3, ..., P_{12}.
6. Draw a smooth curve through these points to obtain the required locus of the insect.

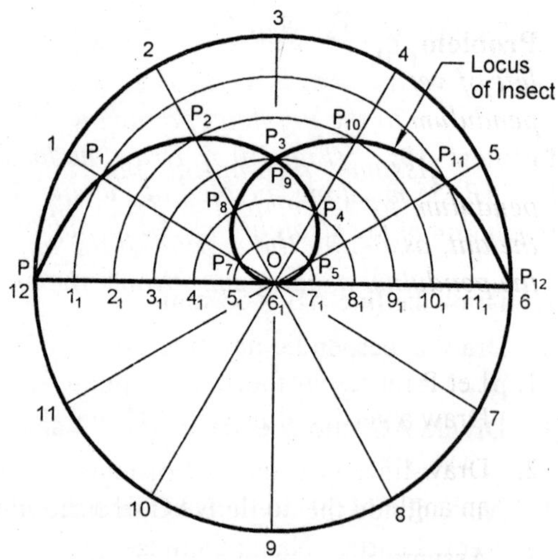

FIG. 11

Problem 12: *(Fig. 12) A point P starts moving 15 mm away from the pole O and reaches a distance of 75 mm from O while moving around it for 1½ convolutions. Its movement from O is uniform. Trace the locus of P. Name the curve.* (UQ)

1. O as center and horizontal radius OB = 75 mm, draw a circle. Divide it into 12 equal parts.
2. Mark A on OB such that OA = 15 mm.
3. Divide AB into 18 equal parts for 1½ convolutions.
4. O as center and O1 as radius draw an arc to intersect the radial line $O1_1$ at the point P_1.
5. Similarly, obtain the points P_2, P_3, …, P_{12} for one revolution.
6. O as center and O-13 as radius draw an arc to intersect the radial line $O1_1$ at the point P_{13}.
7. Similarly, obtain the points P_{14}, P_{15}, …, P_{18} for the next half revolution.
8. Draw a smooth curve through the points A, P_1, P_2, …, P_{18}.

Locus: Archimedean Spiral.

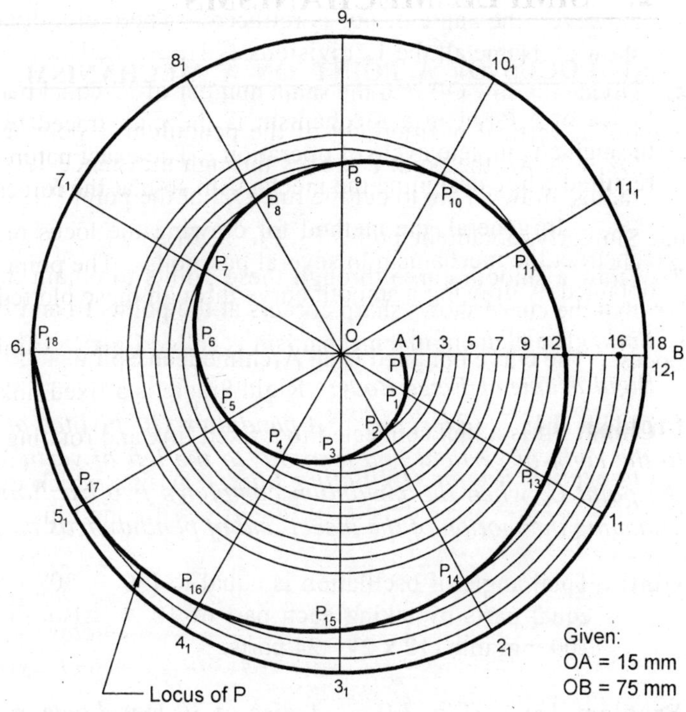

FIG. 12

Problem 13: *(Fig. 13) A pendulum of a clock, 80 mm long, swings through 30° to the left of vertical and 30° to the right of vertical. An ant moves from the moving end of the pendulum to the pivoted end during one oscillation.*

Assume the initial position of the pendulum to be vertical. Trace the path of the ant, assuming the motion of the ant and of the pendulum as uniform. Name the curve.

(UQ)

1. Let P represent the starting point of the ant. Draw a vertical line OP = 80 mm.
2. Draw lines OA and OB of length 80 mm at an angle of 30° to the vertical line OP.
3. Assume that the pendulum swings first to left from vertical position, then to right and again to left to vertical position for one oscillation. Hence, during one oscillation, the pendulum moves through an angle of 120° (= 30° + 60° + 30°). Divide this angle into 12 equal parts.

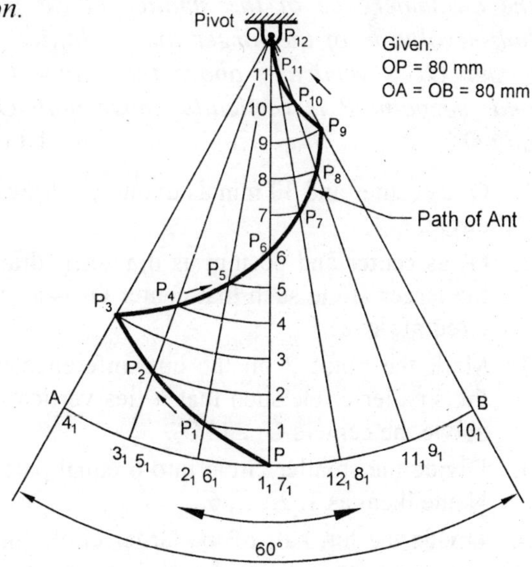

FIG. 13

However, the angle of 60° is retraced. Therefore, divide the angle 60° into 6 equal parts as shown. Name all the 12 divisions as $1_1, 2_1, \ldots, 12_1$.

4. Divide the line OP into the same number of 12 equal parts. Name them as 1, 2, ..., 12.
5. Initial vertical position OP of the pendulum is point 1_1. The pendulum moves to position $O2_1$. Now, the point P moves through a distance $1_1 1$. Hence, with O as center and O1 as radius, draw an arc to cut the line $O2_1$ at the point P_1.
6. Similarly, obtain the points $P_2, P_3, \ldots, P_{12}$.
7. Draw a smooth curve through these points to obtain the locus of the path of the ant. Note that the curve shows sharp corners at the points P_3 and P_9, since it changes the direction.

Locus: The curve obtained is an Archimedean Spiral.

Problem 14: *(Exercise) A pendulum OC pivoted at O is 100 mm long. It oscillates 30° to the right of vertical and also 30° to the left of vertical. An insect, initially at O reaches the point C, when the pendulum completes two oscillations. Trace the path of the insect, assuming the motion of the insect and of pendulum as uniform.* (UQ)

Hint: Total angle of oscillation is equal to [30° + 30° + 30° + 30°] = 120°. Divide it into 12 equal parts by taking each part of 10°. Also, divide the total motion of the insect of 100 mm into (12 x 2) = 24 parts.

Problem 15: *(Fig. 14) A disc of 30 mm diameter revolves in anti-clockwise direction about its center O_1, which is at the circumference of another disc of 60 mm diameter. The larger disc revolves in clockwise direction about its center O_2 such that it completes one revolution while the smaller disc completes two revolutions.*

Trace the locus of a point P lying at the circumference of the smaller circle for half-revolution of the larger disc. Initially, center O_1 is vertically above the center O_2 and the point P is vertically above both O_1 and O_2. (UQ)

1. O_1 as center and 30 mm as diameter, draw a circle.
2. O_2 as center and 60 mm as diameter, draw the larger circle such that center O_1 is at its circumference.
3. Mark the point P on the circumference of the smaller circle such that it lies vertically above the centers O_1 and O_2.
4. Divide the smaller circle into 6 equal parts. Name them as 1, 2, ..., 6.
5. Divide the left half of the larger circle into same number of 6 equal parts. Name them as $1_1, 2_1, \ldots, 6_1$.

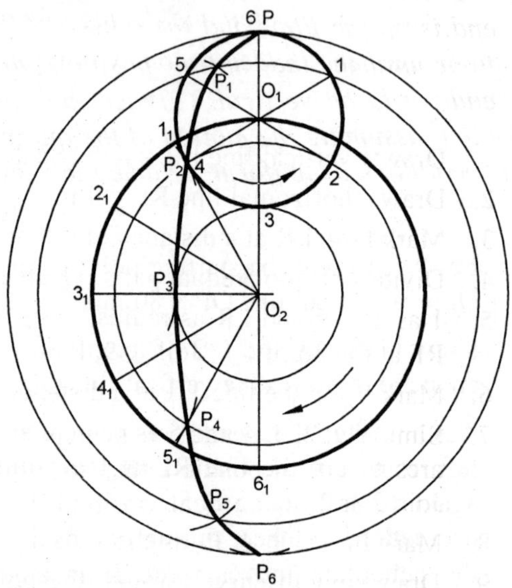

FIG. 14

6. *Given:* During half-revolution of the larger disc, the smaller disc makes one revolution. With O_2 as center, draw circles through the points 6, 1, 2 and 3.
7. With 1_1 as center and O_1P as radius, draw an arc to cut the circle through 1 at the point P_1. Note that two arcs are obtained. Since the smaller disc revolves in anti-clockwise direction, select the arc nearer to the line O_1O_2.
8. Similarly, with 2_1, 3_1, ..., 6_1 as centers and O_1P as radius, draw the arcs to cut the circles through the points 2, 3, ..., 6 at the pints P_2, P_3, ..., P_6.
9. Draw a smooth curve through the points P, P_1, P_2, ..., P_6 to obtain the required locus.

Problem 16: *(Fig. 15) A link MN of length 50 mm rotates clockwise about its end M for one complete revolution. During this period, a point P moves along the link from M to N and returns back to the point M.*

If the rotary motion of the link about point M and the linear motion of point P along MN are both uniform, trace the path of the point P. (UQ)

1. Draw the link MN = 50 mm. M as center and MN as radius, draw a circle. Divide it into 12 equal parts.
2. Divide the line MN into 6 equal parts. Name them as 1, 2, ..., 12 as shown.
3. M as center and M1 as radius, draw a circle to obtain point P_1. Similarly obtain the points P_2, P_3, ..., P_{12}.
4. Draw a smooth curve through these points.

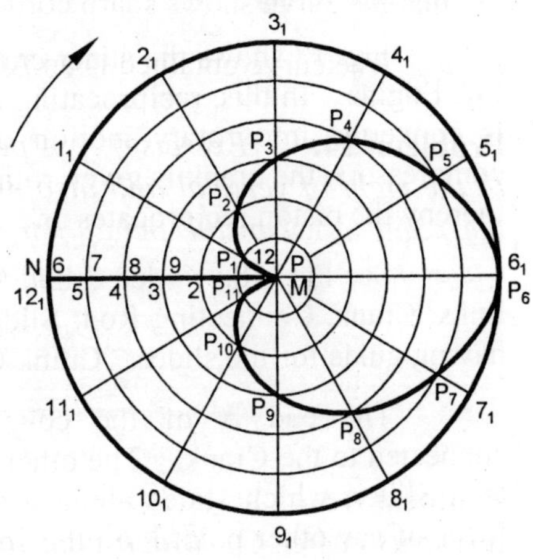

FIG. 15

Problem 17: *(Fig. 16) A steel ladder of 4 m height rests vertically on a wall. Its one end is on the floor and the other end is on the wall. The ends slide along the wall and the floor until the ladder becomes horizontal. Trace the locus of a point P, 0.8 m from the top end.* (UQ)

1. Draw a vertical line LR = 4 m (Scale 1 : 100).
2. Draw a horizontal line RF = 4 m.
3. Mark P on LR at a distance of 0.8 m from L.
4. Divide LR into 5 equal parts. Name them as 1, 2, ..., 5.
5. 1 as center and LR as radius, draw an arc to cut the line RF at the point 1_1. Join 1 & 1_1.
6. Mark P_1 on the line $1_1 1$ at a distance of 0.8 m from 1.
7. Similarly, 2, 3, 4 and 5 as centers and LR as radius, draw arcs to cut the line RF at the points 2_1, 3_1, 4_1 and 5_1. Join 2 & 2_1, ..., 5 & 5_1.
8. Mark P_2, P_3, P_4 & P_5 as shown.
9. Draw smooth curve through the points P, P_1, P_2, ..., P_5 to obtain the required locus.

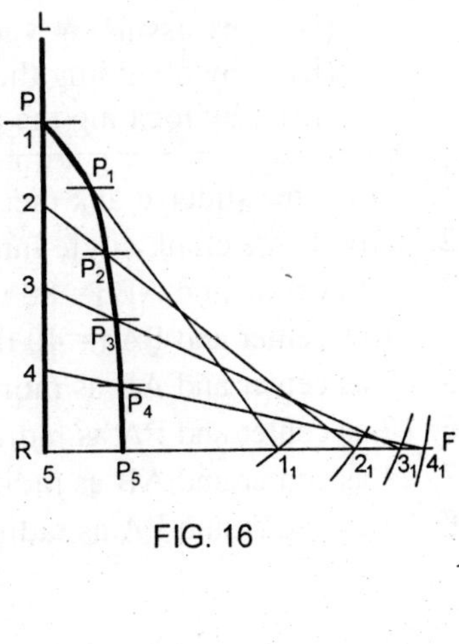

FIG. 16

3. SLIDER-CRANK MECHANISM

Since one of the elements in the crank mechanism is a sliding element (piston), the mechanism is termed as Slider-Crank Mechanism.

Slider-Crank Mechanism is used in Internal Combustion (I.C.) Engines, Pumps, Compressors, Printing Press, etc. This mechanism converts rotary motion into reciprocating motion *or* reciprocating motion into rotary.

Fig. 17 shows the slider crank mechanism of an I.C. Engine. In this, reciprocating motion of the piston is converted into rotary motion of the crank. [In compressors, the crank is given rotation by a motor and thereby the piston reciprocates and compresses the air.]

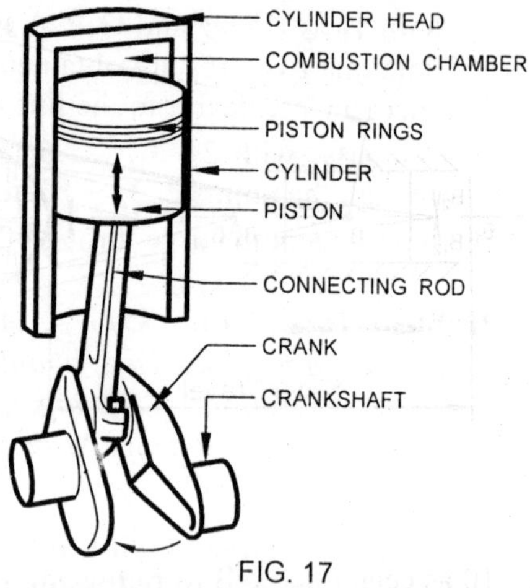

FIG. 17
I.C. ENGINE - CUT SECTIONAL VIEW

See Fig. 18. Slider-Crank Mechanism has four links: Crank, Connecting Rod, Slider and Rigid Frame having guide for the slider. Crank OA rotates about O.

The end A of the connecting rod AB is connected to the Crank. The other end B is connected to a Slider, which slides along a straight line. The locus of any other point P on the connecting rod has to be determined.

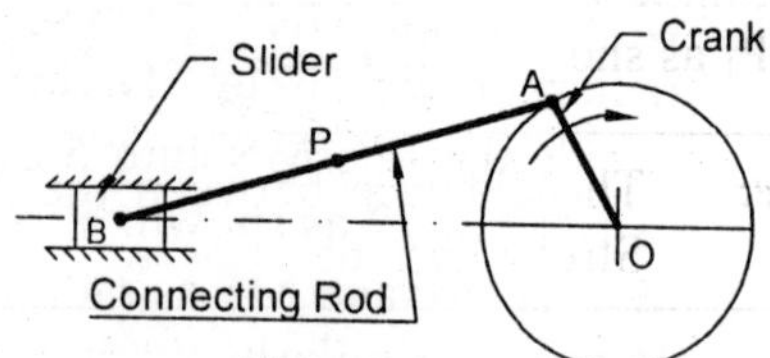

FIG. 18

Problem 18: *(Fig. 19) The slider-crank mechanism of a reciprocating engine has a crank OA of length 20 mm and a connecting rod AB of length 80 mm. The end B moves along a straight line passing through O. Draw the locus of a point P located on the connecting rod at a distance of 40 mm from the point A for one revolution of the crank.* (UQ)

Note: Locus of crank-end A is a ***Circle***. Locus of the end B is a *Straight Line*. The locus of any other point P on the connecting rod is determined:

(i) by assuming various positions of A,
(ii) by obtaining the corresponding positions of the end B and
(iii) by locating the positions of P on the lines joining the first two positions.

1. Draw the slider-crank mechanism of the given engine as shown.
2. Divide the crank circle into 12 equal parts to represent 12 positions of A. Name them as shown.
3. 1 as center and AB (= 80 mm) as radius, draw an arc cutting the path of B at B_1.
4. 1 as center and PA (= 40 mm) as radius, cut the line $1B_1$ at the point P_1.
5. 7 as center and AB as radius, draw an arc cutting the path of B at B_7.
6. 7 as center and PA as radius, draw an arc to cut the line $7B_7$ at P_7.
7. 4 as center and AB as radius, draw an arc cutting the path of B at B_4. Join $4B_4$.
8. 4 as center and PA as radius, draw an arc to cut the line $4B_4$ at P_4.

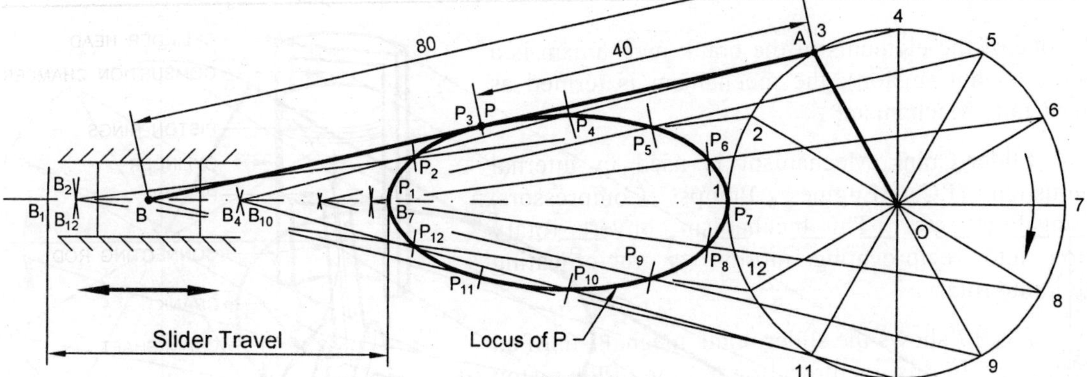

FIG. 19

9. 10 as center and AB as radius, draw an arc cutting the path of B at B_{10}. Join $10B_{10}$.
10. 10 as center and PA as radius, draw an arc to cut the line $10B_{10}$ at P_{10}.
11. Similarly, obtain the other points. Draw a smooth curve through the points P_1, P_2, ……, P_{12}, P_1 as shown. This curve is the locus of the point P. It is known as Coupler Curve.

Note: The distance between the two extreme positions, i.e., B_1B_7 is called Slider Travel or Stroke Length.

Problem 19: *(Exercise) In a slider crank mechanism, the crank OA is 20 cm long and the connecting rod AB is 90 cm long. The end B moves along a straight line passing through O. Trace the locus of a point P, 40 cm from A along the rod for one revolution of the crank.*

3.1 OFFSET SLIDER-CRANK MECHANISM

If the slider moves along a straight line that is offset to the axis of rotation of the crank, the slider crank mechanism is called Offset-Slider Crank Mechanism.

Problem 20: *(Fig. 20)*
In an offset-slider crank mechanism OAB, crank OA = 24 mm, connecting rod BA = 102 mm and offset = 32 mm. Trace the locus of a point P at a distance of 36 mm from A for one complete revolution of the crank. (UQ)

See Fig. 20(i) for the Given Data.

1. See Fig. 20(ii). O as center and crank OA (= 24 mm) as radius, draw a circle, which is the locus of A.

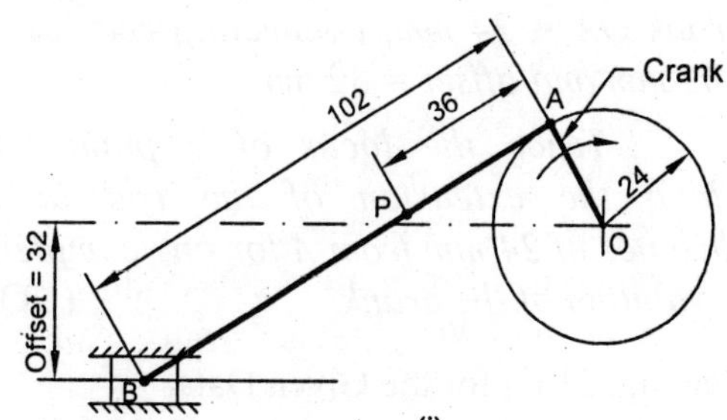

(i)

FIG. 20

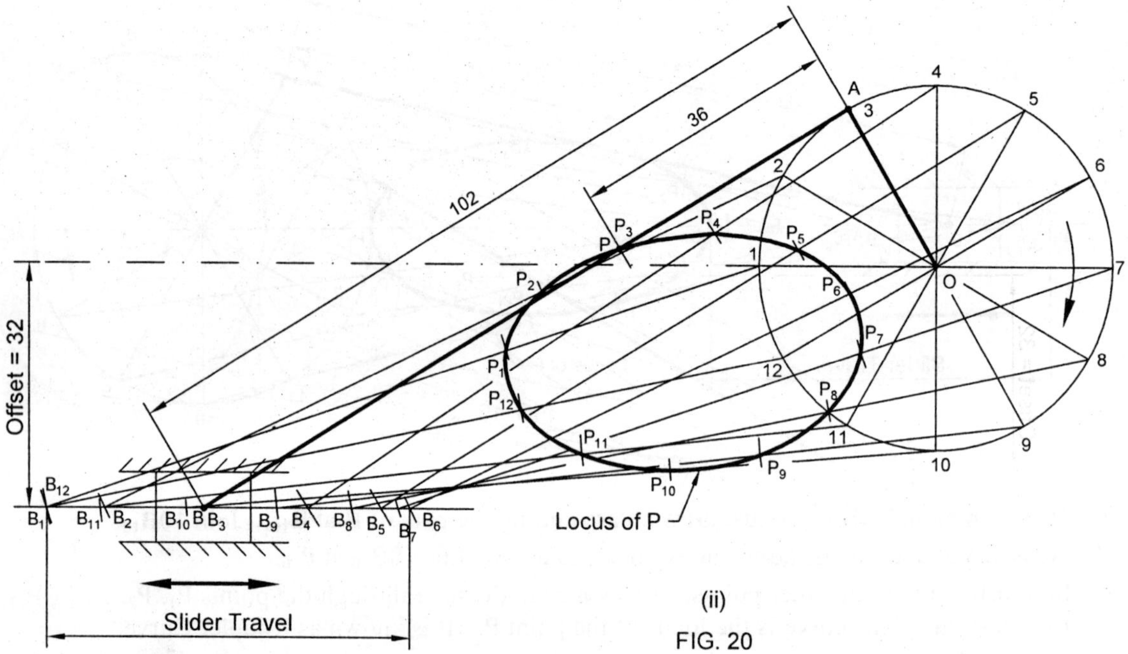

(ii)

FIG. 20

2. Divide the circle into 12 equal parts. Name them as 1, 2, 3, ..., 12.
3. Through O, draw a horizontal line. Then draw a line parallel to and 32 mm (*Offset*) away from it. This gives the locus of the slider B.
4. 1 as center and AB (= 102 mm) as radius, draw an arc cutting the path of B at B_1. Join $1B_1$.
5. 1 as center and PA (= 36 mm) as radius, draw an arc to cut the line $1B_1$ at the point P_1.
6. Similarly, obtain the other points. Draw a smooth curve passing through the points P, P_1, P_2, ..., P_{12}, P as shown. This curve shows the locus of the point P.

Problem 21: *(Fig. 21)*

In an offset-slider crank mechanism OAB, crank OA = 24 mm, connecting rod BA = 102 mm and offset = 32 mm.

Trace the locus of a point P along the extension of the rod at a distance of 24 mm from A for one complete revolution of the crank. (UQ)

See Fig. 21 (i) for the Given Data.

See Fig. 21 (ii) for Solution.

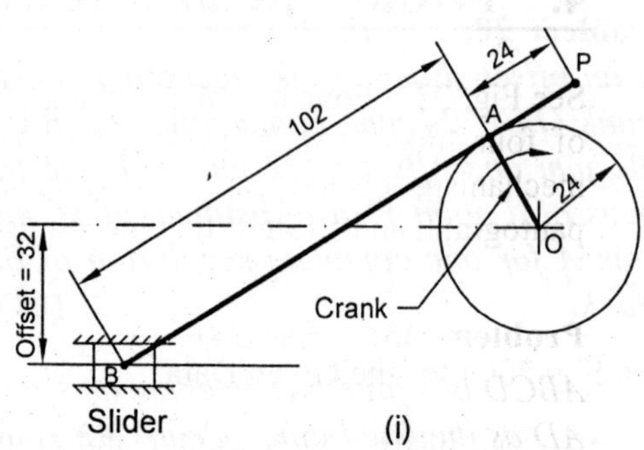

(i)

FIG. 21

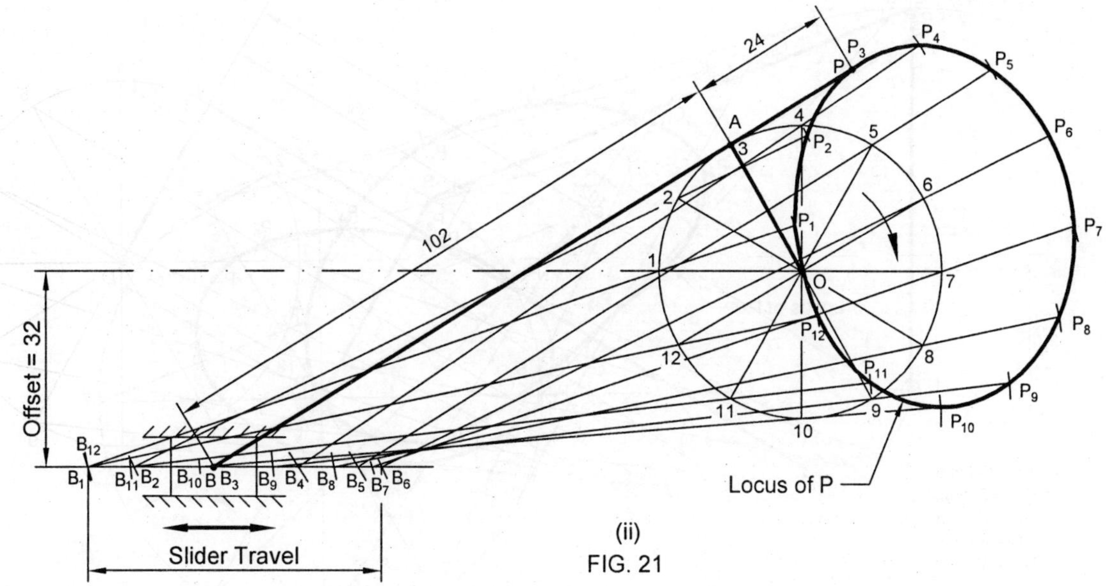

(ii)
FIG. 21

Problem 22: *(Exercise)* *In an offset-crank mechanism, crank OA = 38 cm and connecting rod AB = 150 cm. The slider-end B moves in guides along the line CD, 24 cm below the axis O of the crankshaft.*

Trace the locus of a point P along the extension of the rod, 30 cm beyond B. Also, determine the slider travel.

See Fig. 22 for the Given Data.

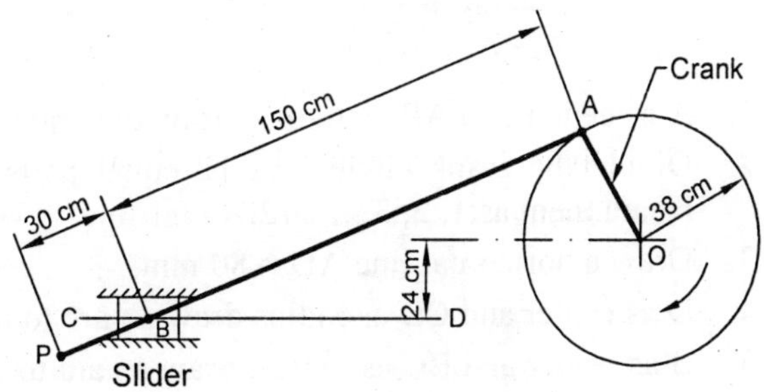

FIG. 22

4. FOUR-BAR MECHANISM

See Fig. 23. Four-bar Mechanism consists of four links. It is used in the steering mechanism of the car, sewing machine, pantograph and straight line mechanisms.

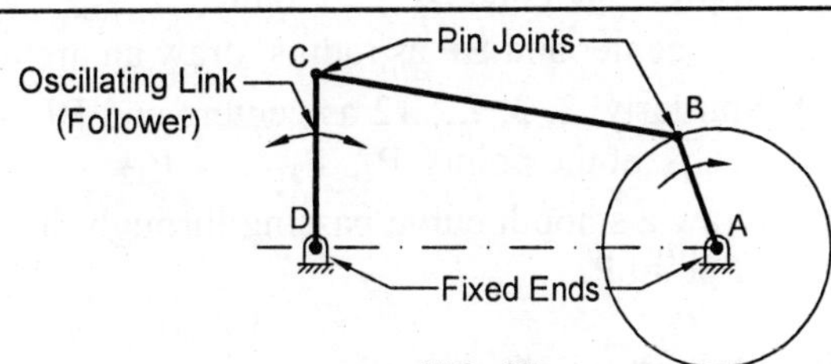

FIG. 23

Problem 23: *(Fig. 24)*

ABCD is a four-bar mechanism with link AD as the fixed link. Crank AB rotates about A while the follower CD oscillates about D.

Draw the locus of a point P on the oscillating link BC for one revolution of the crank. AB = 30 mm, CD = 50 mm, BC = 70 mm, AD = 80 mm and BP = 30 mm. (UQ)

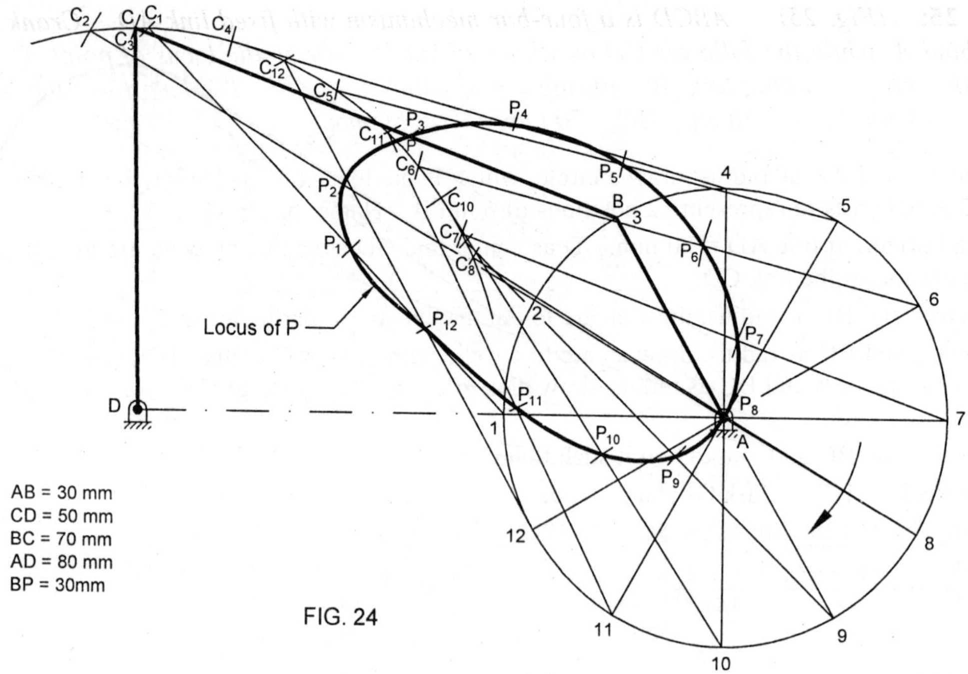

FIG. 24

1. A as center and AB as radius, draw the crank circle, which is the locus of B.
2. Divide the crank circle into 12 equal parts to represent 12 different positions of point B. Name them as 1, 2, 3, …, 12.
3. Draw a horizontal line AD = 80 mm.
4. D as center and CD as radius draw an arc to represent the oscillation of the link CD.
5. B as center and BC as radius, draw an arc to cut the above oscillating arc at point C. Join CB and CD.
6. 1 as center and BC as radius, draw an arc to cut the path of C at C_1. Join $1C_1$.
7. Similarly, 2, 3, …, 12 as centers and BC as radius, draw arcs to cut the path of C at C_2, C_3, …, C_{12}. Join $2C_2$, $3C_3$, …, $12C_{12}$.
8. Mark P such that BP = 30 mm.
9. 1 as center and BP as radius, draw an arc to cut the line $1C_1$ at the point P_1.
10. Similarly, 2, 3, …, 12 as centers and BP as radius, draw arcs to cut the lines $2C_2$, $3C_3$, …, $12C_{12}$ at the points P_2, P_3, …, P_{12}.
11. Draw a smooth curve passing through the points P_1, P_2, P_3, …, P_{12}, P_1 to obtain the locus of point P.

Problem 24: *(Exercise)*

Two cranks AB and CD are connected by a link BD. AB rotates about A, while CD oscillates about C. Trace the locus of the mid-point of the link BD for one complete revolution of the crank AB. AB = 45 cm, CD = 75 cm, BD = 105 cm. Distance between A and C is equal to 90 cm.

(UQ)

Problem 25: *(Fig. 25) ABCD is a four-bar mechanism with fixed link AD. Crank AB rotates about A, while the follower CD oscillates about D. Draw the locus of point Q on the extension of the floating link BC during one revolution of AB. The lengths of the links are: AB = 30 mm, CD = 50 mm, BC = 70 mm, AD = 80 mm and CQ = 30 mm.* (UQ)

1. A as center and AB as radius, draw a circle, which is the locus of B. Divide the crank circle into 12 equal parts to represent 12 positions of point B. Name them as 1, 2, 3, ..., 12.
2. Draw a horizontal line AD = 80 mm. D as center and CD as radius draw an arc to represent the oscillation of the link CD.
3. B as center and BC as radius, draw an arc to cut the above oscillating arc at C. Join CB & CD.
4. 1 as center and BC as radius, draw an arc to cut the path of C at C_1. Join $1C_1$. Similarly, 2, 3, ..., 12 as centers and BC as radius, draw arcs to cut the path of C at C_2, C_3, ..., C_{12}. Join $2C_2$, $3C_3$, ..., $12C_{12}$.
5. Extend the link BC and mark Q on it such that CQ = 30 mm as shown.
6. Extend the line $1C_1$. Mark Q_1 at a distance of 30 mm from C_1.
7. Simiiarly, extend the lines $2C_2$, $3C_3$, ..., $12C_{12}$. Mark Q_2, Q_3, ..., Q_{12} at a distance of 30 mm from C_2, C_3, ..., C_{12}.
8. Draw a smooth curve through the points Q_1, Q_2, Q_3, ..., Q_{12}, Q_1 to obtain the locus of Q.

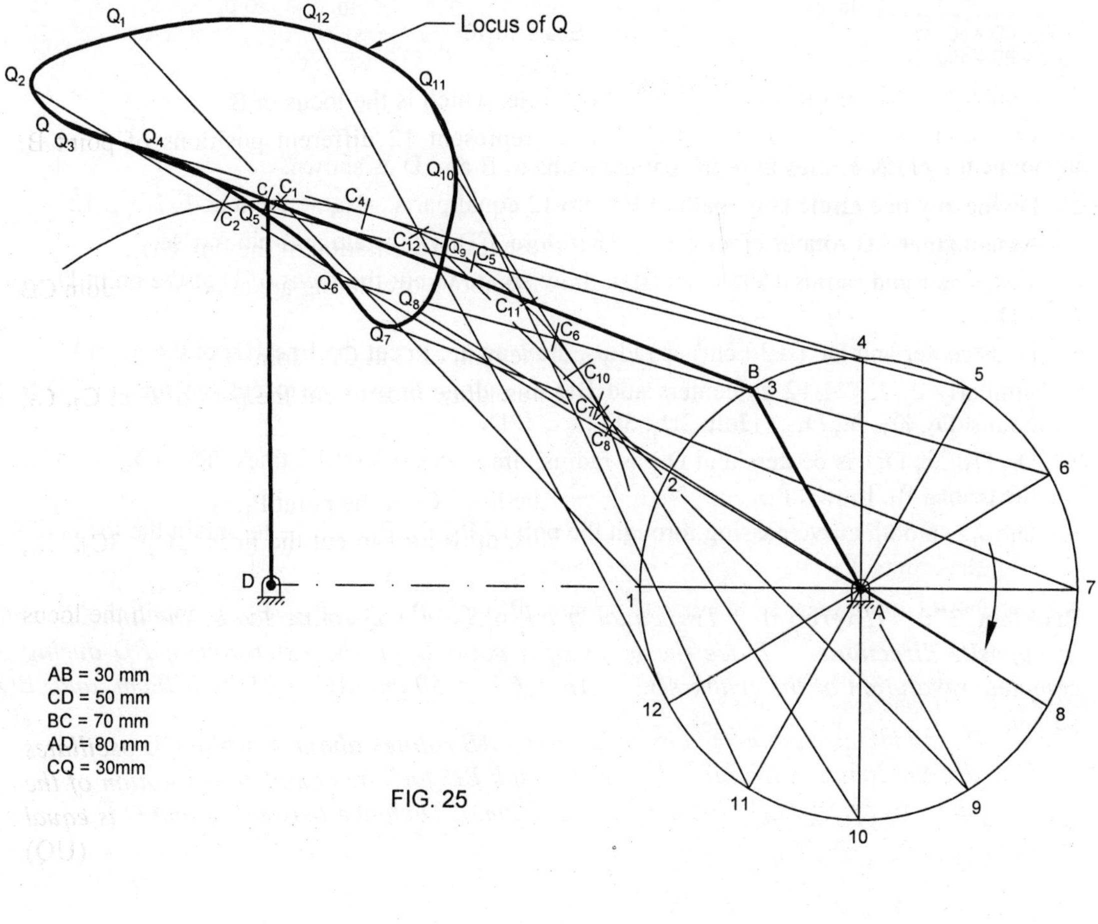

FIG. 25

Problem 26: *(Fig. 26) Two equal cranks AB and CD connected by the link BD, rotate in* ***opposite directions****. Trace the locus of a point P on BD during one complete revolution of the crank AB. AB = CD = 50 cm, AC = BD = 150 cm and PD = 30 cm.* (UQ)

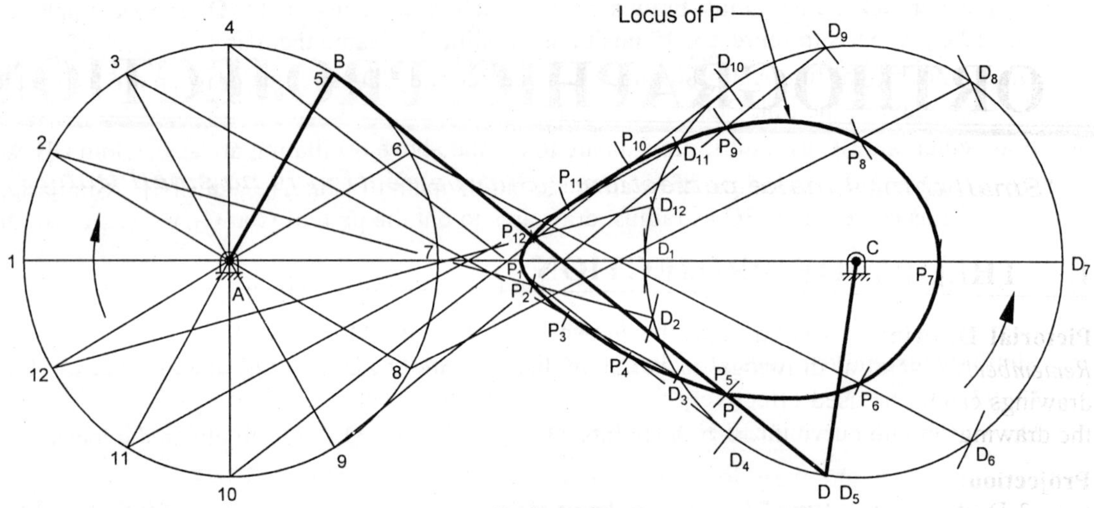

AB = CD = 50 cm
AC = BD = 150 cm
BP = 50 cm

Scale 1 : 10

FIG. 26

1. Draw the crank circles to represent the paths of B and D as shown.
2. Divide any one circle (say, path of B) into 12 equal parts. Name them as 1, 2, ..., 12.
3. Assume that AB rotates clockwise. Therefore, CD will rotate anti-clockwise.
4. 1 as center and radius 150 cm (= BD), draw an arc to cut the locus of D at the point D_1. Join 1 D_1.
5. D_1 as center and DP (= 30 cm) as radius, draw an arc to cut the line 1D_1 at the point P_1.
6. Similarly 2, 3, ..., 12 as centers and BD as radius, draw arcs to cut the locus of D at the points $D_2, D_3, \ldots, D_{12}$. Join 2D_2, 3D_3, ..., 12D_{12}.
7. $D_2, D_3, \ldots, D_{12}$ as centers and DP as radius, draw arcs to cut the lines 2D_2, 3D_3, ..., 12D_{12} at the points $P_2, P_3, \ldots, P_{12}$.
8. Draw a smooth curve passing through the points $P_1, P_2, P_3, \ldots, P_{12}$ to obtain the locus of P.

Problem 27: *(Exercise) Two equal cranks AB and CD connected by the link BD, rotate in* ***opposite directions****. Trace the locus of a point Q on the extension of BD during one complete revolution of the crank AB. AB = CD = 50 cm, AC = BD = 150 cm and BQ = 25 cm.* (UQ)

*

8

ORTHOGRAPHIC PROJECTION

"Small things make perfection; But, perfection is no small thing."

1. THEORY OF PROJECTION

Pictorial Drawings: Solids may be drawn on a drawing paper pictorially in its 3-D form. *Remember!* Pictorial drawings do *not* show the true shape of the solid. Therefore, pictorial drawings cannot be used effectively for the manufacturing technology, for the simple reason that the drawings of engineering components have to be necessarily in their true shapes and sizes.

Projection: Each and every information about the shapes and sizes of the various components (i.e., 3-D objects) has to be prepared in detail in an industry. The task of recording the shapes and the sizes of 3-D objects on your 2-D drawing sheets is done by the *Method of Projection.*

The concept of *Projection* was invented by the French scientist Monge in the year 1800. The word *projection* is of Latin origin. It means to *throw forward.* Thus, *Projection* is defined as an image of an object thrown forward on to a plane by means of straight lines called Visual Rays. The figure formed by joining various points thus obtained on the plane, is the image of the object. It is termed as *Projection of the object.*

Observer, Object and Visual Rays: In the study of Optics in Physics, you are already familiar with the terms, namely, *Observer*, *Object* and *Visual Rays.* An object is seen when the light rays called *Visual Rays* coming from the object strike the observer's eye. The size of the image formed in the retina depends on the distance of the observer from the object.

See Fig. 1. Suppose, now an *Imaginary Transparent Plane* is introduced such that the object is in-between the observer and the plane. The image obtained on the plane is as shown. This is called ***Perspective View***.

Here, straight lines (*rays*) are drawn from various points on the contour of the object to meet the transparent plane. Thus the object is said to be *projected* on that plane from the eye of the observer.

Plane of Projection: The transparent plane on which the projections are drawn is known as *Plane of Projection.*

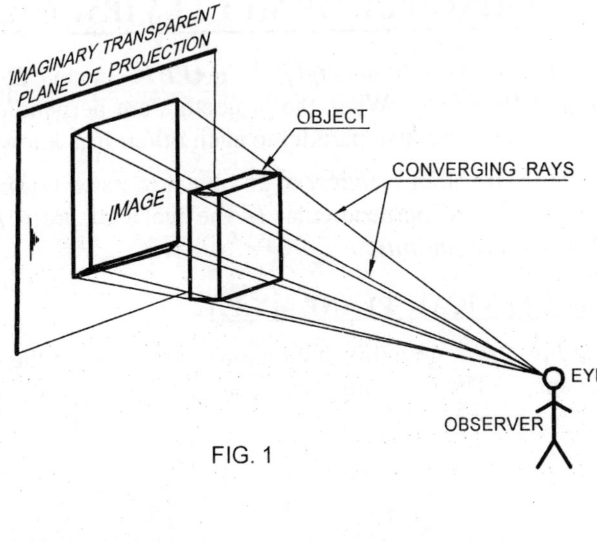

FIG. 1

View: The figure formed on the plane by the projection of the faces of an object is called a *View*. The size of the view depends on the distance from the observer to the plane and the distance from the plane to the object.

Projection: The figure or view formed by joining, in correct sequence, the points at which the rays meet the plane is called the *Projection of the object*. (It is obvious that the outlines of the shadow on the plane show the projection of the object).

Projectors: The lines or rays drawn from the object to the plane are called *Projectors*.

1.1 METHODS OF PROJECTION

1. **Pictorial Projections:** The projection in which the description of the object is completely understood in one view is known as *Pictorial Projection*. It has advantage of conveying an immediate impression of the shape of the object, but not its true sizes. Its types are:
   - (i) **Axonometric Projection:** In this projection, only one view showing all the three dimensions of an object is drawn on a Plane of Projection. The orientation of object is kept in such a way that its three mutually perpendicular edges will remain inclined to the plane of projection. Isometric Projection is a type of Axonometric Projection.

     Isometric Projection: Isometric Projection gives the true shape of the object, but not the true size. It shows three sides of the object in one projection. *(Discussed later)*
   - (ii) **Oblique Projection:** *Oblique* means *Slanting*. In this, projectors are not perpendicular to the plane of projection. The front face of the object is placed parallel to the plane of projection and hence it is shown in its true size and shape. *(Discussed later)*
   - (iii) **Perspective Projection:** In Perspective Projection, the projecting lines or visual rays converge at a point as shown in *Fig. 1*. Hence it is termed as *Convergent Projection*. Perspective Projections are *not* used by engineers for manufacturing, because the perspective view does not reveal the exact size and shape. Perspectives may be used in marketing where a natural view of a product is desirable. *(Discussed later)*
2. **Orthographic Projection:** In Orthographic Projection, an object is represented by two or three views on the mutually perpendicular projection planes.

2. PRINCIPLE OF MULTI-VIEW ORTHOGRAPHIC PROJECTION

ORTHO means ***Right-angle;*** ***GRAPHIC*** means ***Drawing***; ***ORTHO-GRAPHIC*** means ***Right-angled Drawing***. When the projectors are perpendicular to the plane on which the projection is obtained and are also parallel to each other, it is known as *Orthographic Projection*.

To obtain the view of an object to its *same size*, the visual rays should be made parallel to each other and perpendicular to the plane of projection. This will be possible only when the observer is at *infinity*.

2.1 VERTICAL PLANE (V.P.)

See Fig. 2(a). Imagine that the observer looks at the object (*a rectangular block*) from an *infinite* distance. The rays will now be *parallel* to each other and perpendicular to both the front surface of the object and the plane.

(When the observer is at a *finite* distance from the object, the rays converge to the eye as in the case of Perspective Projection.)

The observer looks at the rectangular block (object) from the front as shown. The front face **F** of the block of Length L and Height H, is seen in its true shape and size.

Extend the rays, called *projectors* further to meet a *transparent Vertical Plane (denoted as V.P.)* located behind the object.

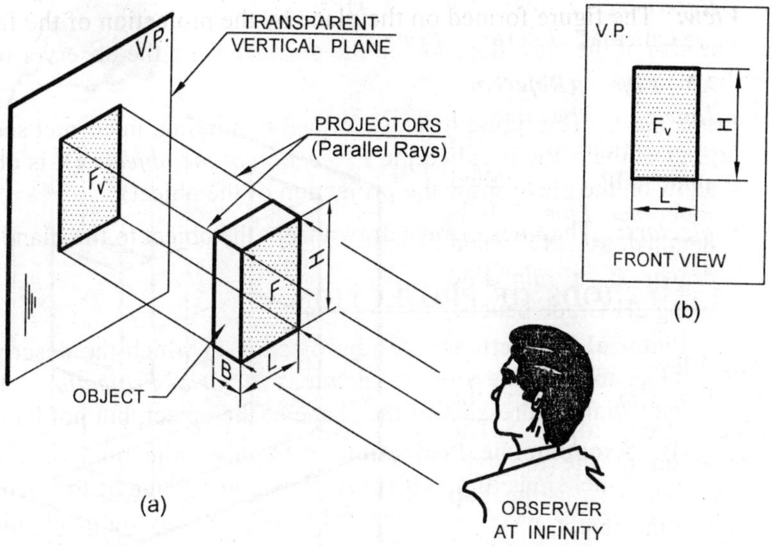

FIG. 2 ORTHOGRAPHIC PROJECTION - FRONT VIEW

Join the points at which the projectors meet the plane, in correct sequence, to obtain the orthographic projection.

Orthographic View: Since the visual rays, called projectors are perpendicular, *i.e.*, orthogonal to the plane of projection, the projected view is called *Orthographic View*.

Front View

See Fig. 2(a). The orthographic view is obtained by projecting the object when the observer is viewing it from the front. Hence it is called the *View from the front* or **Front View** ($\mathbf{F_v}$) of the object. Both F and F_V are of exactly the same shape and size.

See Fig. 2(b). Front View is shown separately with its Length (L) and Height (H).

Front View shows only two dimensions of the object, *viz.*, Length (L) and Height (H). It does not show the Breadth (B), since it lies along the direction of sight while looking from the front. Thus, Front View alone cannot describe the shape and size of the object completely, as it shows only two out of three dimensions, *viz.*, L and H.

Note: 1. ***Front View:*** *Projection of the object on VP is known as Front View.*
2. *Front view always shows Length (L) and Height (H) of the object.*

2.2 HORIZONTAL PLANE (H.P.)

One more view is necessary to show the Breadth (B) of the object. Generally, the View from above or View from top, called Top View is projected to show the breadth of the object.

See Fig. 3(a). Let us assume another transparent plane, called Horizontal Plane (H.P.) hinged perpendicular to VP such that the object is in front of VP and above HP.

* *Reference Planes:* VP and HP are called as *Reference Planes.*
* *XY Line (Reference Line):* The line of intersection of VP and HP is called *Reference Line.* It is denoted as XY and drawn as a *thin* line using 2H Grade pencil.

Top View

See Fig. 3(a). Look at the object from top. The projection of the top face T is T_H. T_H is the **Top View** of the object.

Both T and T_H are of exactly same shape and size. Thus T_H gives Length (L) and Breadth (B) of the block, but not the Height (H).

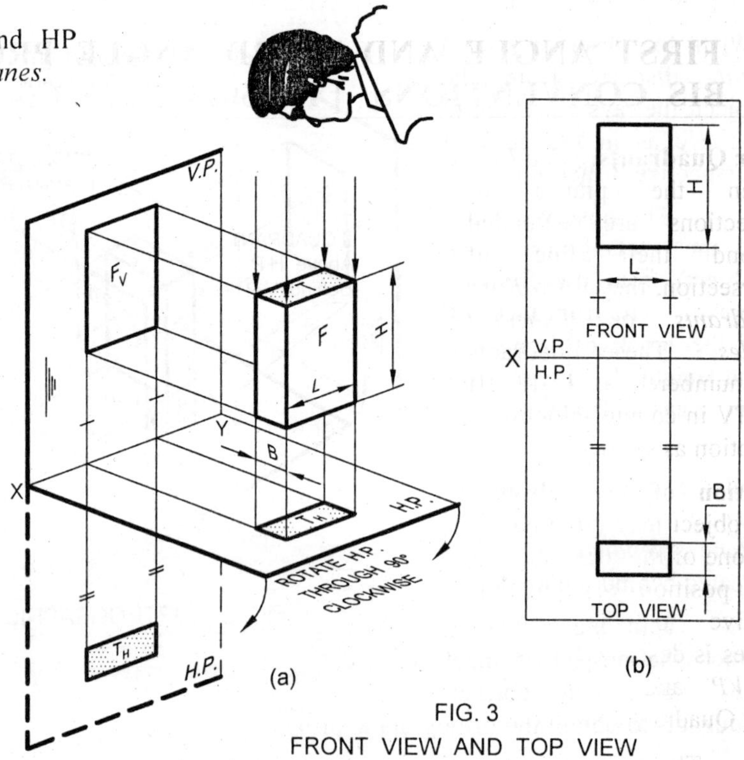

FIG. 3
FRONT VIEW AND TOP VIEW

Rule: *Each projection shows that face of the object which is nearer to the observer and far away from the plane.*

2.3 HOW TO OBTAIN THE PROJECTIONS ON THE DRAWING SHEET ?

See Fig. 3(a). It is convention to rotate HP through 90° in clockwise direction about XY line so that it lies in the extension of VP as shown in dashed line. Thus the Top View T_H along with HP is rotated through 90° and occupies the position vertically below XY line.

See Fig. 3(b). Now the two projections F_V and T_H are drawn on your two dimensional drawing sheet as shown.

Thus all the details regarding the shape and size, *viz.,* Length (L), Height (H) and Breadth (B) of the object may be represented by means of the orthographic projections, *viz.,* Front View and Top View.

Note:
1. ***Top View:*** *Projection on the HP is known as Top View.*
   Top View shows Length (L) and Breadth (B) of the object.
2. *Front and Top Views are drawn as wide (thick) lines (H Grade Pencil).*

Orthographic Projection is the standard drawing form of the industrial world. The form is unreal in that we do not see an object as it is drawn orthographically. Pictorial drawing, however, has photographic realism.

3. FIRST ANGLE AND THIRD ANGLE PROJECTIONS – BIS CONVENTIONS [IS 15021 (Parts 1 to 4) : 2001]

Four Quadrants: *See Fig. 4.* When the planes of projections are extended beyond their line of intersection, they form *Four Quadrants* or *Dihedral Angles*. These quadrants are numbered as I, II, III and IV in counter-clockwise direction as shown.

Position of the Observer: The object may be situated in any one of the four quadrants. The position of the object relative to the reference planes is described as *in front of VP* and *above HP* in First Quadrant and so on.

The observer will always be on the right side of the four quadrants and in front of VP to obtain the Front View and above HP to obtain the Top View, irrespective of the position of the object in any of the four quadrants.

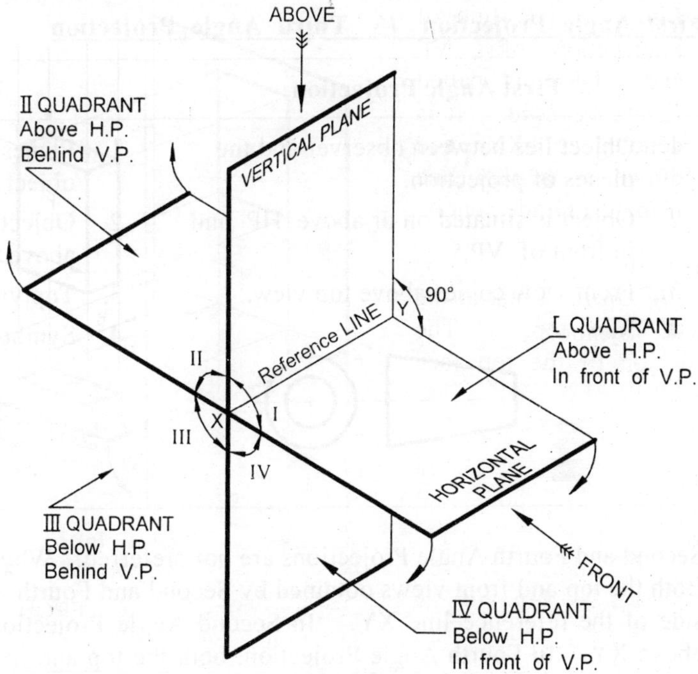

FIG. 4 FOUR QUADRANTS OR DIHEDRAL ANGLES

ALWAYS ROTATE H.P. CLOCKWISE TO OPEN-OUT I QUADRANT

First Angle Projection: *Fig. 3* shows *First Angle Projection*, in which the object lies in front of VP and above HP, i.e., in *First Quadrant*. ***Figs.** 2 and 3* explain the *concept* of First Angle Projection as follows:

1. Object lies in between the observer and the planes of projection.
2. Object is then projected on VP to obtain front view and on HP to obtain top view.
3. First and Third Quadrants are always opened out while rotating the HP.
4. When the views are drawn in their relative positions, Top View comes below Front View.
5. Front View is drawn above the XY line and Top View is drawn below the XY line. (Here, above XY line represents the VP and below XY line represents the HP).
6. In the Front View, HP coincides with XY and in Top View, VP coincides with XY.
7. Each projection shows the view of that face of the object which is far away from the plane on which it is projected and which is nearest to the observer.
8. Front View always shows the Length (L) and Height (H) of an object.
9. Top View always shows Length (L) and Breadth (B) / Width (W) / Thickness (T) of it.

Third Angle Projection: *See Fig. 4.* In Third Angle Projection, the object is situated in the *Third Quadrant.* The planes of projection, namely, HP and VP lie in between the object and the observer. The front view comes below the XY line and the top view above it.

First Angle Projection *Vs* Third Angle Projection

First Angle Projection	Third Angle Projection
1. Object lies between observer and the planes of projection.	1. Planes of projection lie between the object and observer.
2. Object is situated on or above HP and in front of VP.	2. Object is situated below HP (i.e., on or above the ground) and behind VP.
3. Front view comes above top view.	3. Top view comes above front view.
4. Symbol: 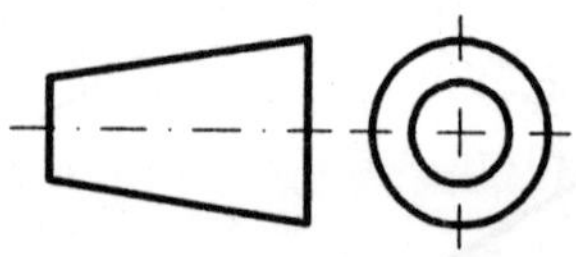	4. Symbol: 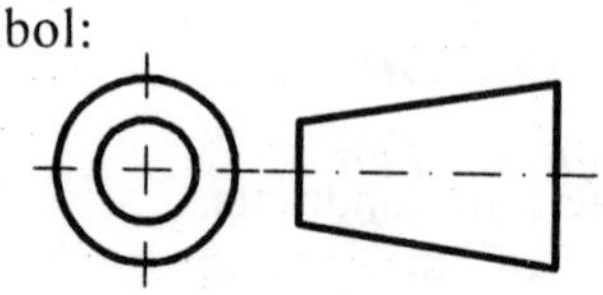

Second and Fourth Angle Projections are not preferred. Why?
Both the top and front views obtained by Second and Fourth Angle Projection Methods lie on one side of the reference line XY. In Second Angle Projection, both top and front views appear above XY. In Fourth Angle Projection, both the top and front views appear below XY. Thus, showing the projections on a flat plane surface becomes inconvenient.

B.I.S. has recommended the use of only First Angle Projection in line with the specifications of I.S.O., being adopted by all the countries of the world. The adopted version of B.I.S. and I.S.O. with regard to the method of projection reads as follows:

"B.I.S. specifies that the use of First Angle Projection is to be adopted in Engineering Drawing. However, the Third Angle Projection may also continue till 1990."

Note: (i) Projectors as well as the XY line are drawn as *narrow (thin) lines*.
(ii) HP and VP may also be written on either side of XY line in orthographic views.

4. PROFILE PLANES AND SIDE VIEWS

Two orthographic views, *viz.*, front and top views, are necessary to represent a symmetrical object. However, complex objects require three views, namely, front view, top view and also a side view. Side view is obtained by looking at the object from either left or right side.

Profile Plane (PP)
See *Fig. 5(a)*. Sometimes some features of an object lie on its left or right side. Top and Front Views will not show these features in their true shape. In such cases, you are required to project the object either on a Left or Right Side Plane, known as *Profile Plane (PP)* and the view projected on it is known as *Profile View* or *Side View*.

Auxiliary View
Projection on an Auxiliary Plane is called Auxiliary View. Side view is a special case of auxiliary view. PP is also known as Auxiliary Vertical Plane (AVP).

Right Side PP and Left Side PP
In First Angle Projection, the object lies in between the observer and HP & VP.

Therefore, Left Side View will be projected on the Right side Profile Plane (RPP) and drawn on the right side of the front view. Right Side View is projected on the Left side Profile Plane (LPP) and drawn on the left side of the front view.

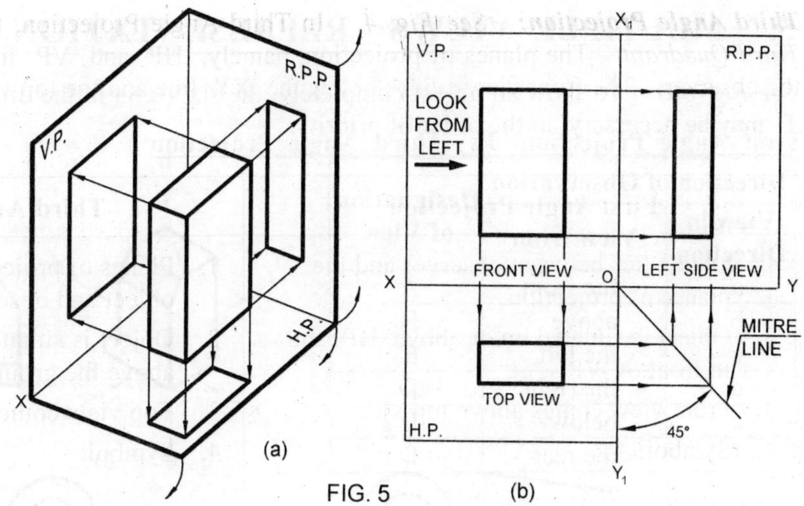

FIG. 5 (b)

FRONT VIEW, TOP VIEW AND LEFT SIDE VIEW

Left Side View
See Fig. 5(a). It shows HP, VP and a Right Profile Plane (RPP) drawn perpendicular to both VP and HP. RPP is used to project the Left Side View as shown. *See Fig. 5(b).* Draw the Front and Top Views. Look from left. Project Left Side View on RPP.

Rotate the RPP in the direction of the arrow shown in *Fig. 5(a)*, so as to make it to coincide with the VP [*Fig. 5(b)*]. Look at the object from the left to obtain the Left Side View and draw it on the right side of the Front View.

How to draw Left Side View using 45° Mitre Line ?
1. Draw $X_1 Y_1$ perpendicular to XY to meet at O. Through O draw a 45° (Mitre) line.
2. Draw horizontal projectors from all corners of the Top View to intersect the 45° line.
3. Through these points of intersection, draw vertical projectors as shown in *Fig. (b)*.
4. Also, draw horizontal projectors from all the corners of the Front View to intersect the previously drawn vertical projectors.
5. Join the corresponding points and complete the Left Side View as shown in *Fig. (b)*.

4.1 SYSTEM OF NOTATION

System of Notation is the conventions used for the orthographic views as explained below:
1. In this text, the actual corners of solids are denoted by capital letters A, B, C, etc.
2. **Front View:** Their front views are denoted by corresponding lower case letters with dashes a′, b′, c′, etc.
3. **Top View:** Their top views are denoted by corresponding lower case letters a, b, c, etc.
4. **Side View:** In the side view, the corresponding points are denoted as a″, b″, c″, etc.

5. MULTIPLE VIEWS – REPRESENTATION OF 3-D OBJECTS

See Fig. 6(i). To show an object completely, the six views in the directions a, b, c, d, e and f may be necessary, in the order of priority.

Direction of Observation		Designation of View
View in Direction	**View from**	
a	the front	A
b	above	B
c	the left	C
d	the right	D
e	below	E
f	the rear	F

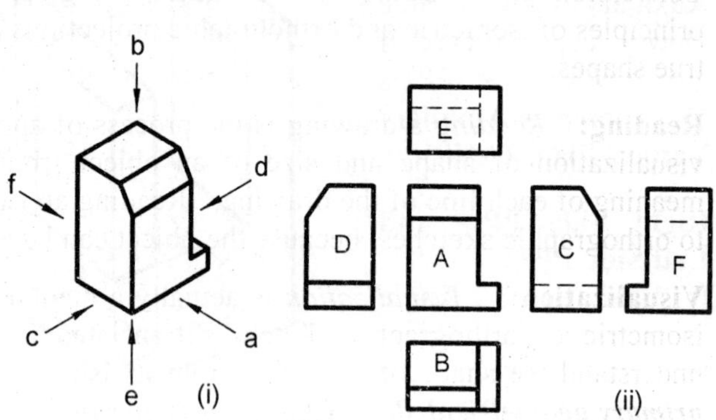

FIG. 6

See Fig. 6(ii).

View A : Front View known as Principal View.
View B : Top View obtained by looking from above and placed below Front View.
View E : View obtained by looking from below and placed above Front View.
View C : View obtained by looking from left and placed on the right of Front View.
View D : View obtained by looking from right and placed on the left of Front View.
View F : View obtained by looking from the rear and placed on the right or left.

In practice, not all the above six views are needed. View E (Bottom View) and View F (Rear View) are not used. Limit the number of views to the minimum necessary and sufficient to fully represent the object without ambiguity.

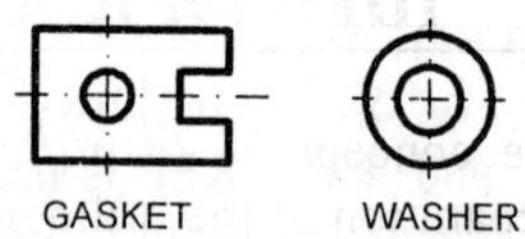

FIG. 7 ONE VIEW DRAWINGS

One View Drawings: *See Fig. 7.* A single view is sufficient for simple objects, if a *note* is added. Flat objects like washers, shims, gasket, etc., are represented with thickness mentioned.

Two View Drawings: *See Fig. 8.* Two views can explain objects *symmetrical* about their axes, provided they are correctly chosen. Careful analysis of the object will enable you to determine whether top or side view is necessary after front view is selected. The front and left side views of the objects shown in Fig. 8 are sufficient to describe them completely. Note that the front and top views will *not* be sufficient to describe the objects in the figure.

Three View Drawings: Irregular or unsymmetrical objects require three views for complete shape description.

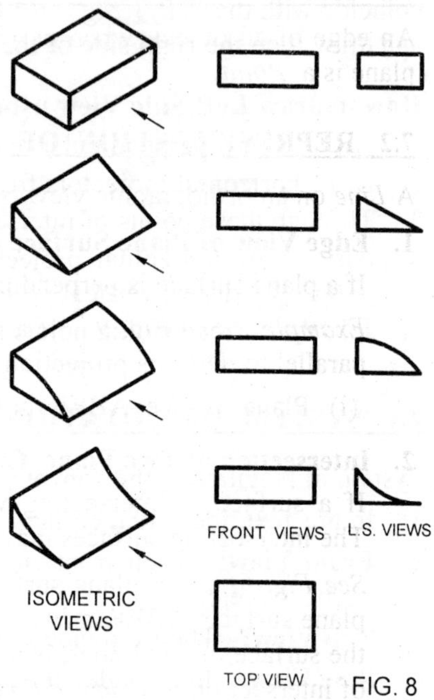

FIG. 8

6. VISUALIZATION OF OBJECTS FROM PICTORIAL VIEWS

Conversion of an isometric view into orthographic views requires sound knowledge of the principles of isometric and orthographic projections and a little bit of imagination, to visualize the true shapes.

Reading: *Reading* a drawing is the process of applying the principles of projection. It means visualization of shape and size of an object from its orthographic views by interpreting the meaning of each line of the drawing. Reading an isometric view is rather easy and fast compared to orthographic sketches, because the object can be easily visualized.

Visualization: *Visualization* is actually a mental process by which the details furnished in an isometric or orthographic sketch is translated to understand the shape of the object. To understand the shape of an object from its isometric view, you may break the object down into ***primary geometrical shapes like prisms, pyramids, cylinders, cones, etc.***

These shapes may be additions in the form of projections or subtractions in the form of holes, grooves, cavities, etc. The entire object is obtained by adding or subtracting the above geometrical shapes. Assemble them mentally to obtain the final shape.

7. IDENTIFICATION OF SURFACES

The concept of *Identification of Surfaces* of an object for its visualization followed by the preparation of the orthographic views is discussed below:

7.1 REPRESENTATION OF A POINT

An edge of an object, when perpendicular to HP or VP, its orthographic projection on that plane is a *Point.*

7.2 REPRESENTATION OF A LINE

A *Line* on an orthographic view is used to represent the following:

1. **Edge View of Plane Surface of an Object**

   If a plane surface is perpendicular to VP or HP, its projection on that plane is a ***Line***.

   Example: *See Fig. 9.* If a plane surface ABEF of an object is perpendicular to VP and parallel to HP, its projection on VP is a *line*. *This line represents the following:*

   (i) Plane surface ABEF in Front View, and (ii) Edge AB.

2. **Intersection of Two Plane, Curved or Spherical Surfaces of an Object**

   If a surface is intersecting with another surface, a ***line*** is formed. The intersecting surfaces may be *plane, curved* or *spherical* in shape.

   See Fig. 9. If a plane surface ABCD is intersecting with another plane surface ABEF, a *line* AB is formed. If a curved surface, *say,* the surface of a cylinder, is intersecting with its plane surface, the line of intersection is a *curved line* (here, *a circle*).

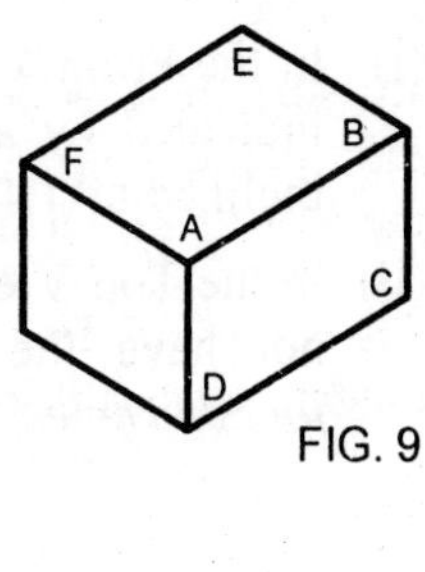

FIG. 9

If a spherical surface, *say*, the surface of a hemi-sphere, is intersecting with its plane surface, the line of intersection is a *curved line* (here, *a circle*).

3. Boundary of the Surfaces of an Object

Boundary of a plane surface A B C D is its outer *edges* AB, BC, CD and DA. Boundary of a circular plane is a *circle*.

7.3 REPRESENTATION OF AN AREA

An area on an orthographic view represents the projected surface area of a part of an object. The meaning of an area in an orthographic view can be understood only after analyzing the corresponding part in the other view or views.

1. Parallel Surfaces (Area parallel or perpendicular to HP/VP)

See Fig. 10. In the Pictorial View of the L-BLOCK, surfaces H_1 and H_2 are horizontal and are perpendicular to VP. Hence,

(i) in the Top View, horizontal surfaces H_1 and H_2 have ***true shape*** and ***size***.

(ii) in the Front View, the horizontal ***surfaces*** H_1 and H_2 are indicated by horizontal ***lines*** H_1 and H_2.

(iii) in the Right Side View, the horizontal ***surfaces*** H_1 and H_2 are indicated by horizontal ***lines*** H_1 and H_2, as they are perpendicular to the Left Profile Plane.

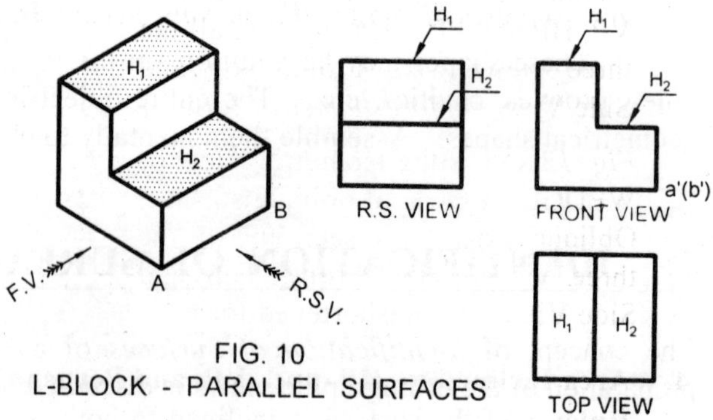

FIG. 10
L-BLOCK - PARALLEL SURFACES

Rule:
1. If a surface of an object is parallel to HP / VP, the projection of the surface on that plane to which it is parallel will have its true size and shape.
2. Projection of a surface normal to HP / VP is represented by a straight line.
3. Parallelogram and rhombus in the isometric view are rectangle and square surfaces respectively in the orthographic views.
4. Lines / Edges of an object parallel to the direction of viewing will be seen as points in the orthographic views. [Edge AB is seen as a point a′ (b′).]

---

2. Inclined Surface (Area Inclined to VP or HP)

See Fig. 11. In the Pictorial View, the surface I is inclined to HP. Hence,

(i) in the Front View, this inclined ***surface*** is indicated by an inclined ***line*** I, as it is inclined to HP.

(ii) in the Top View, this inclined surface will not have the true shape and size, but *foreshortened* and is indicated as I.

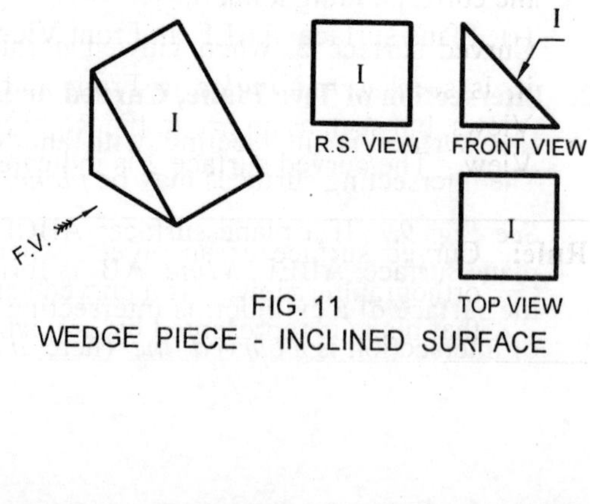

FIG. 11
WEDGE PIECE - INCLINED SURFACE

(iii) in the Right Side View, this inclined surface will *not* have the true shape and size, but *foreshortened* and is indicated as I.

Rule: 1. If a surface of an object is perpendicular to either HP or VP and inclined to the other, the projection of the surface to which it is perpendicular is a *straight line*.
2. Projection of surface on HP or VP to which it is inclined appears *foreshortened.*

---

3. Oblique Surface (Area Inclined to VP, HP and PP)

If a surface of an object is inclined to the three reference planes, the surface is called as *Oblique Surface*.

Rule: An oblique surface is inclined to all the three planes, *viz.*, VP, HP and PP. It is seen in all the three views, *viz.*, Front, Top and Side Views in *foreshortened* form.

Fig. 12 shows the Isometric View of WEDGE BLOCK of oblique surface. Oblique surface is seen in all the three views, viz., Front. Top and Side Views in foreshortened form.

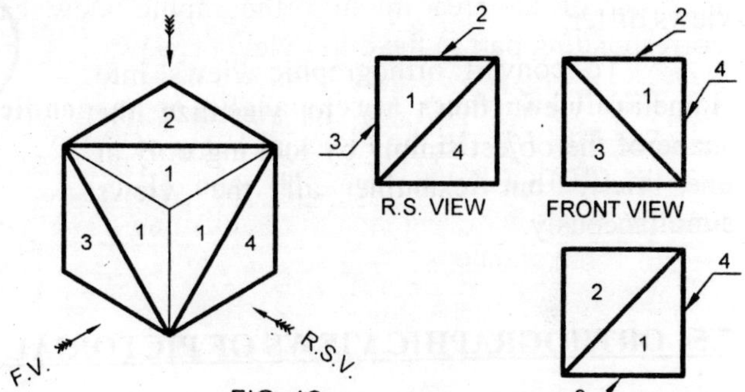

FIG. 12
WEDGE BLOCK (OBLIQUE SURFACE)

4. Area Inclined to HP and VP and Perpendicular to PP

Rule: If the surface is inclined to both HP and VP and perpendicular to PP, its Front and Top Views will get *foreshortened* while the Side View is a *straight line*.

5. Curved Surface (Curved Area Perpendicular to VP or HP)

Fig. 13 shows the Isometric View of a STEEL BLOCK having curved surface. Identify the visible surfaces and name them using numerals or alphabets as below:

Surface 1 is perpendicular to HP. Hence, its true shape is seen in the Right Side View. But it is seen as *lines* in Front and Top Views. These lines are indicated by the corresponding numeral 1.

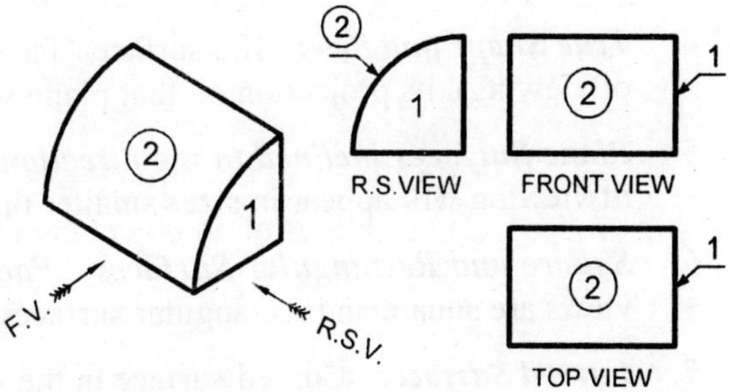

FIG. 13 STEEL BLOCK - CURVED SURFACE

Curved surface 2, when viewed normal to it, is seen as *rectangles* in Front and Top Views, but seen as an *arc* in the Right Side View. The curved surface 2 is indicated by the corresponding numeral 2 enclosed in circles.

Rule: Curved surface of an object, when viewed normal to it, appears as a *rectangle* in the orthographic view. If a curved surface is perpendicular to HP or VP, its projection on that plane is represented by *arc*, while its projection on other plane is *foreshortened* area.

7.4 READING A DRAWING

See Fig. 14. Objects I and II have the same front view, but the top views differ. Similarly, objects III and IV have the same front view, but the top views differ.

Also, it is interesting to note that objects I and III have the same top view, but the front views differ. Similarly, objects II and IV have the same top view, but front views differ.

To convert orthographic views into isometric view, don't try to visualize the shape of the object simply by looking only at one view, but examine all the views simultaneously.

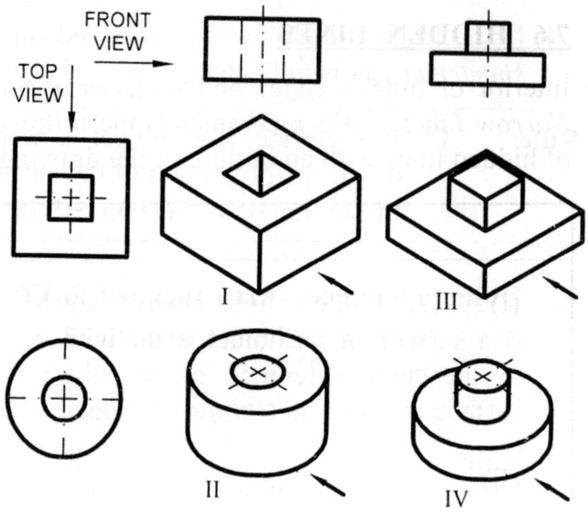

FIG. 14

7.5 ORTHOGRAPHIC VIEWS OF PICTORIAL VIEWS – POINTS TO REMEMBER

1. In Isometric View, only the overall shape of the object can be visualized. But the Orthographic View gives the true dimensions or size of the object.
2. Each view gives *two* dimensions, *i.e.*, front view gives length and height, top view gives length and width and side view gives height and width.
3. ***Lines and Points:*** Surfaces / Faces of the object, parallel to the direction of vision will be seen as ***lines*** in orthographic views. Also, edges of the object, parallel to the direction of vision will be seen as ***points*** in orthographic views.
4. ***True Shape and Size:*** If a surface / face is parallel to a plane (*i.e.*, normal to the direction of viewing), its projection on that plane will show its ***true shape and size.***
5. ***Plane Surfaces inclined to the direction of view:*** Plane surfaces inclined to the direction of viewing will appear in sizes smaller than true sizes in orthographic view.
6. ***Square and Rectangular Surfaces / Faces:*** Rhombus and parallelogram in the isometric views are square and rectangular surfaces / faces respectively in the orthographic views.
7. ***Curved Surface:*** Curved surface in the isometric view of an object when viewed normal to it, appears as rectangle in the orthographic view. If a curved surface is perpendicular to HP or VP, its projection on that plane is represented by arc, while its projection on other plane is foreshortened area.
8. ***Circle:*** Ellipse in isometric view should be visualized as a circle in orthographic view.
9. ***Semi-circle or Arc of a circle:*** Semi-circle or Arc of a circle in orthographic view is shown as semi-ellipse or a portion of an ellipse in the isometric view.
10. Dimensions which are parallel to the direction of vision will ***not*** be seen.

7.6 HIDDEN LINES

Interior or hidden edges and surfaces of an object are shown by *Hidden Lines* called *Dashed Narrow Lines*. BIS recommends the following conventions to indicate the points of intersection of hidden lines with an outline of the drawing or another hidden line:

		CORRECT	INCORRECT
(i)	A hidden line always starts with a dash and ends with a dash.		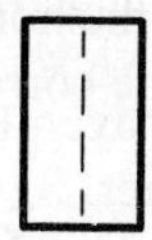
(ii)	A hidden line, when drawn in continuation of a visible line, always starts with a space.	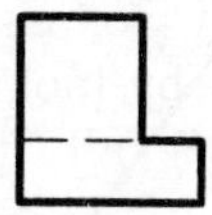	
(iii)	Hidden line jumps the visible line.	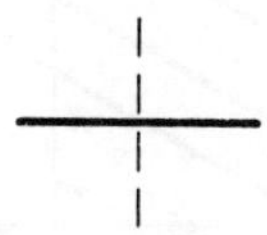	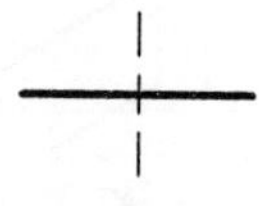
(iv)	Hidden lines always meet in dashes at intersections or corners.	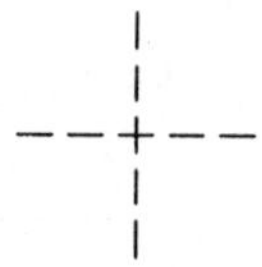	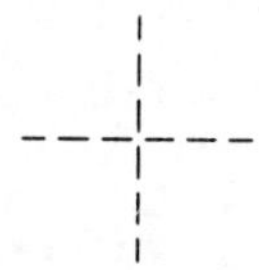
(v)	Hidden arc always starts in dashes and ends in dashes.	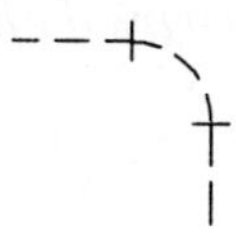	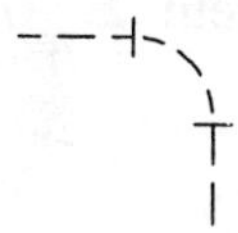

7.7 PRECEDENCE OF LINES

Lines such as visible line, hidden line and center line (axis) at times coincide on the drawings. It is necessary to know which line should be shown under such circumstances. The order of precedence of lines in a drawing is as follows: Full Line – Dashed Line – Center Line (Axis) / Section Plane Line – Dimension / Extension Lines.

1. When a visible line coincides with a hidden line (*i.e.*, dashed line), draw the visible line.
2. When a visible line coincides with a center line (axis), draw the visible line and extend the center line (axis) beyond the outlines of the view.
3. When a section plane line coincides with a center line (axis), show the center line and draw the section plane line outside the outlines of the object at the ends of the center line by *thick* dashes.
4. When a dashed (invisible) line coincides with the center line, draw the dashed line.

8. SIDE VIEWS USING MITRE LINE

Problem 1: *(Fig. 15) Draw Right Side View of the STOP BLOCK using mitre line.*

1. Draw the given front and top views. Draw the projection lines to the left of the top view.
2. Construct a mitre line at 45° to the horizontal passing through the point O. From the points of intersection of mitre line with the above projection lines, draw vertical projection lines.
3. Draw the horizontal projection lines from the front view to intersect the above lines.
4. Join the corresponding points of intersection in proper sequence and obtain the side view.

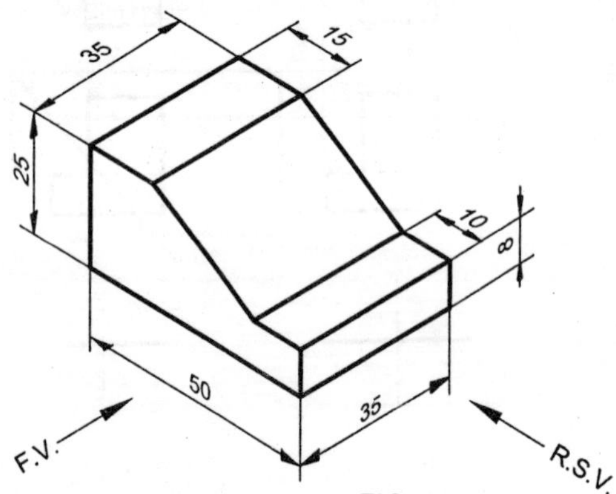

FIG. 15

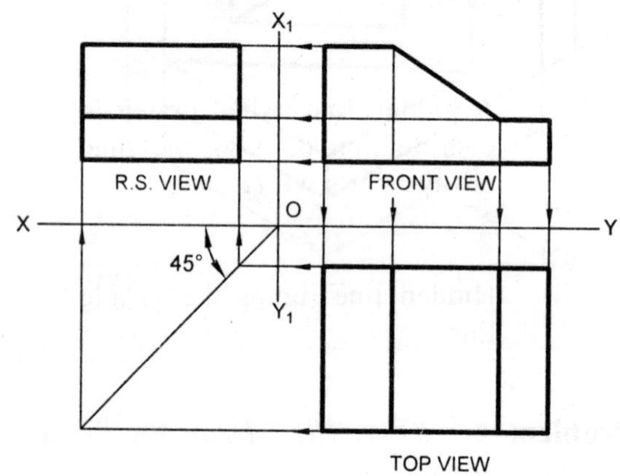

Problem 2: *(Fig. 16) Draw Left Side View of the ANGLE PLATE using mitre line.*

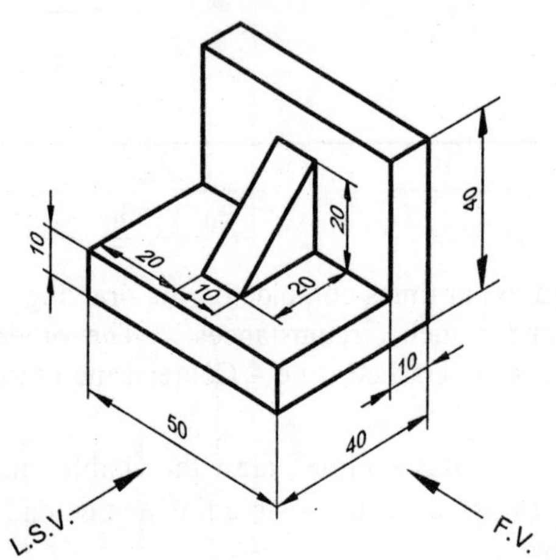

FIG. 16

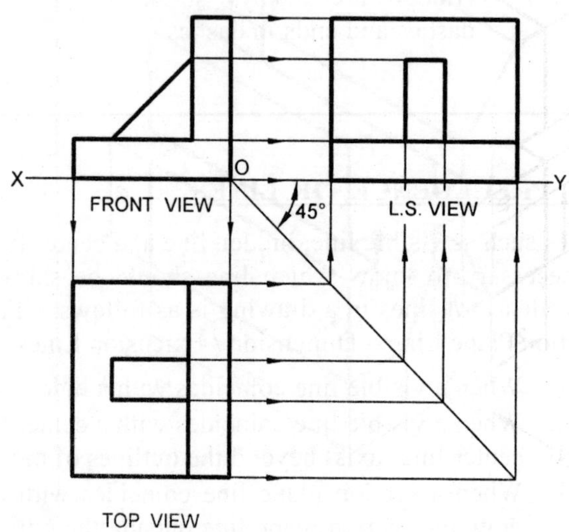

Problem 3: *(Fig. 17) Draw the Right Side View of the object shown below, using mitre line.*

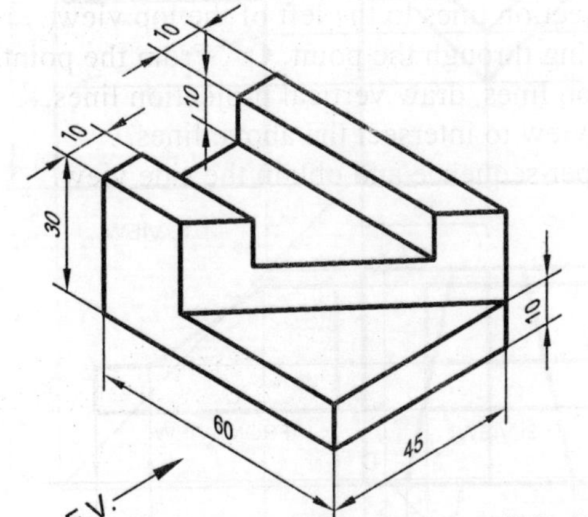

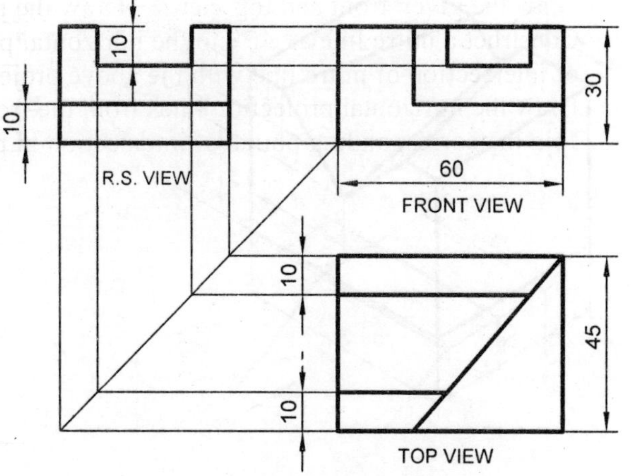

FIG. 17

Problem 4: *(Fig. 18) Draw the Right Side View of the object using Mitre Line.* (UQ)

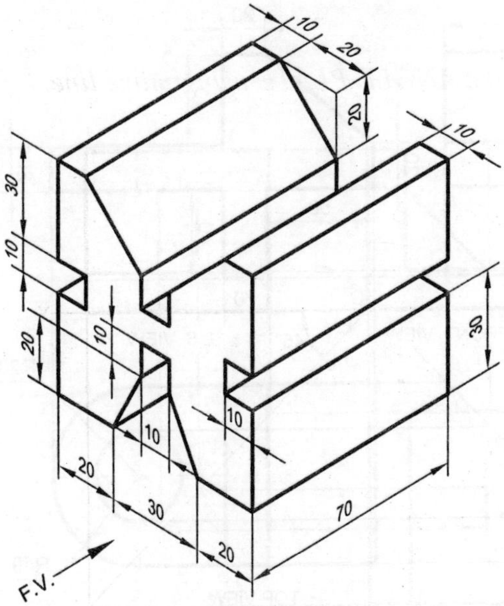

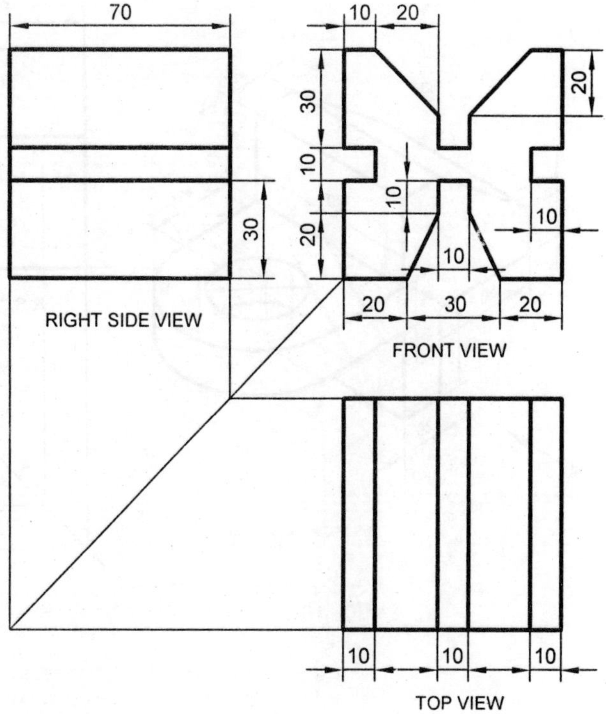

FIG. 18

Problem 5: *(Fig. 19) Draw the Left Side View of the object using Mitre Line.* (UQ)

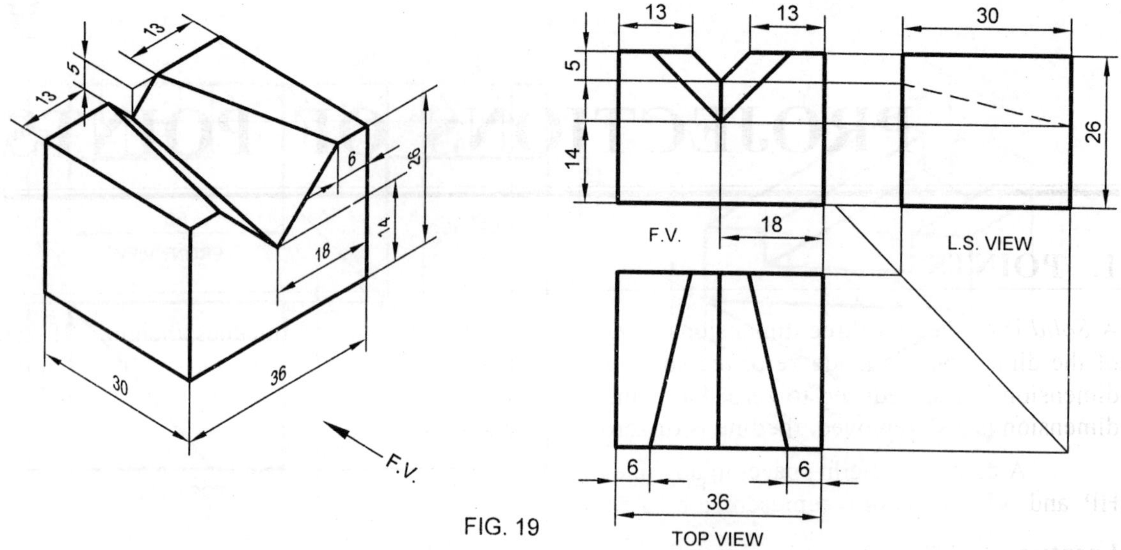

FIG. 19

Problem 6: *(Fig. 20) Draw the Right Side View of the object using Mitre Line.* (UQ)

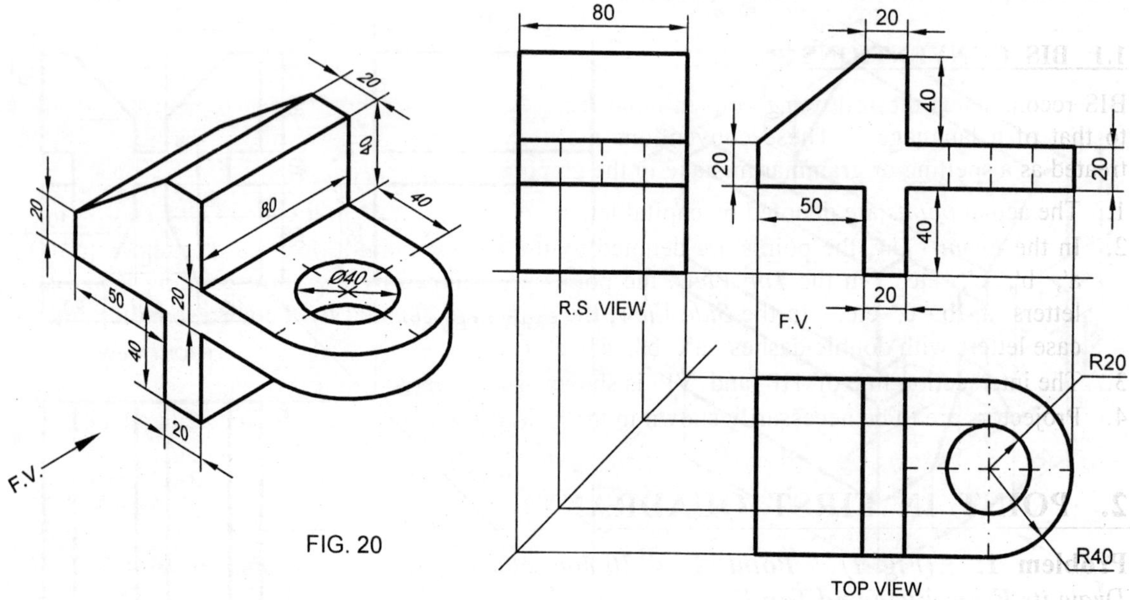

FIG. 20

9

PROJECTIONS OF POINTS

1. POINTS

A *Solid* is formed by three dimensions, viz., length, width/breadth and thickness/height. If one of the dimensions is made zero, the solid is converted into a *2-D Plane Surface*. If one more dimension is also reduced to zero, the plane is converted into a *Line*. Finally, if the remaining dimension is also removed, the line is of zero length and becomes a *Point*.

A point may lie in space, in any one of the four quadrants formed by two reference planes HP and VP. A *point* is represented by a *dot* (.) in the drawings.

Locations of a Point

When a point lies in the **First Quadrant**, it will be **above HP** and **in front of VP**.
When the point lies in the **Second Quadrant,** it will be **above HP** and **behind VP**.
When the point lies in the **Third Quadrant,** it will be **below HP** and **behind VP**.
When the point lies in the **Fourth Quadrant,** it will be **in front of VP** and **below HP**.

1.1 BIS CONVENTIONS

BIS recommends the following conventional representations in orthographic projections, similar to that of a language. These conventions are obeyed internationally and a variation will be treated as a spelling or grammar mistake in the graphics language:

1. The actual *points* are denoted by capital letters A, B, C, etc., in the *pictorial view*.
2. In the *Front View*, the points are denoted by the corresponding lower-case letters with dash a′, b′, c′, etc. In the *Top View*, the points are denoted by the corresponding lower-case letters a, b, c, etc. In the *Side View*, the points are denoted by the corresponding lower-case letters with double dashes a″, b″, c″, etc.
3. The intersection line of HP and VP is shown as the reference line XY.
4. Projectors are to be necessarily shown in the orthographic views in *thin lines* (2H pencil).

2. POINT IN FIRST QUADRANT

Problem 1: *(Fig. 1) Point A is 30 mm above HP and 45 mm in front of VP. Draw its Front View and Top View.*

1. *See Fig. (i).* Look at the Pictorial View. Point **A** lies in the first quadrant.
2. **To obtain the Front View a′, look at the Pictorial View from the front**
   Point **A** is 30 mm above HP. **Aa′** is the projector perpendicular to the VP.
   Hence **a′** is the Front View of the point **A** and it is 30 mm above the XY line.

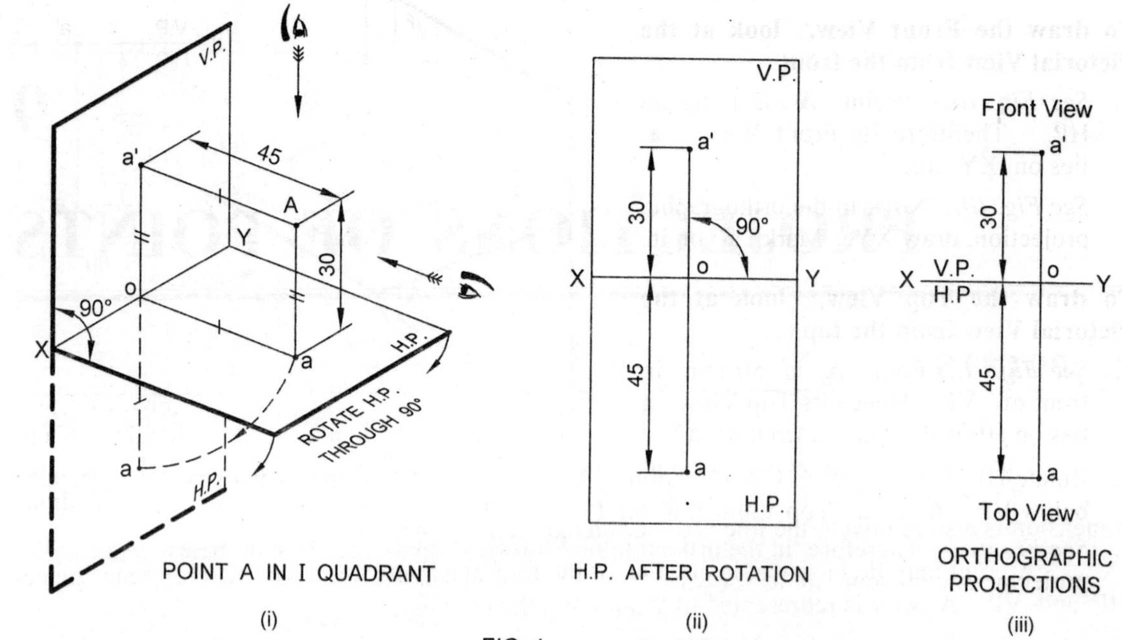

FIG. 1

3. **To obtain the Top View a , look at the Pictorial View from the top**
   Point **A** is 45 mm in front of VP. **Aa** is the projector perpendicular to the HP.
   Hence **a** is the Top View of the point **A** and it is 45 mm in front of XY.
4. **To convert projections a′ and a from Pictorial View into Orthographic Projections**
   - Rotate HP about XY line through 90° in the CW direction as shown in *Fig. (i)*.
   - After rotation, the first quadrant is opened out and the HP occupies the position (dashed lines) vertically below the VP.
   - Also, the point **a** on HP will trace a quadrant of a circle with **o** as center and **oa** as radius. Now, **a** occupies the position just below **o**.
   - *See Fig .(ii)*. Line joining **a′** and **a**, called *projector*, is perpendicular to XY.

How to draw the orthographic projections on the drawing sheet: *Fig. (iii).*

5. **Front View :** *See Fig. (iii).* Draw the XY line and draw a projector at any point on it. Mark the Front View **a′** 30 mm above XY on the projector.
6. **Top View :** On the *same* projector, mark the Top View **a** 45 mm below XY.

Rule: *XY represents HP in Front View and VP in Top View. So, while drawing Front and Top Views on your drawing sheet, the squares or rectangles for the above individual imaginary planes are unnecessary and are to be discarded. Draw only the orthographic views shown in Fig. (iii) and not Figs. (i) and (ii).*

Problem 2: *(Fig. 2) Draw the projections of a point A lying on HP and 50 mm in front of VP.*

To draw the Front View, look at the Pictorial View from the front

1. *See Fig. (i).* Point **A** is lying on HP. Therefore its Front View **a′** lies on XY line.
2. *See Fig. (ii).* Now, in the orthographic projection, draw XY. Mark **a′** on it.

To draw the Top View, look at the Pictorial View from the top

3. *See Fig. (i).* Point **A** is 50 mm in front of VP. Hence its Top View **a** lies on HP itself and in front of XY.

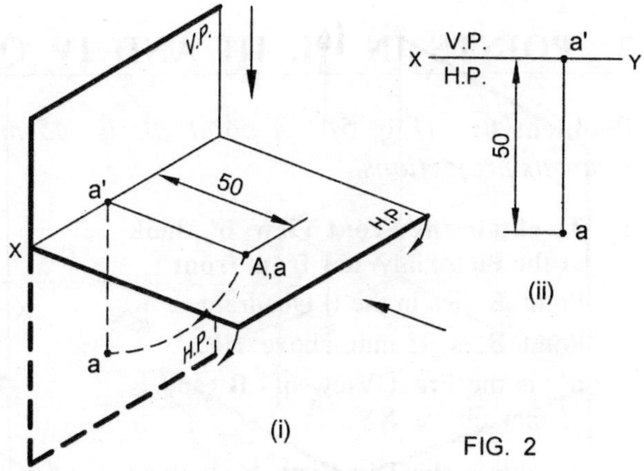

FIG. 2

4. Rotate HP through 90° in CW direction. After rotation, HP occupies the position vertically below VP. Also, **a** (Top View of **a**) on HP comes vertically below **a′**.
5. *See Fig. (ii).* Therefore, in the orthographic projection, mark **a** 50 mm below XY on the projector drawn from **a′** as shown.

Problem 3: *(Fig. 3) Draw the projections of a point A lying on VP and 55 mm above HP.*

To draw the Front View, look at the Pictorial View from the front

1. *See Fig. (i).* Point **A** is 55 mm above HP. Hence **a′** is 55 mm above XY.
2. Therefore in the orthographic projection *(Fig. ii),* mark **a′** 55 mm above XY.

To draw the Top View, look at the Pictorial View from the top

3. Point **A** is on VP. Therefore, its Top View lies on XY itself.
4. Top View **a** does not lie on HP. Thus, in this case, the HP need *not* be rotated. Therefore mark **a** on XY on the projector drawn from **a′**.

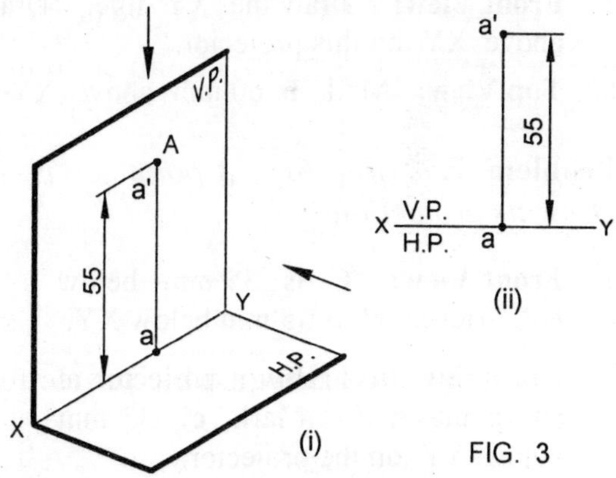

FIG. 3

Problem 4: *(Exercise) Draw the projections of a point Q lying on VP and 58 mm above HP.*

Problem 5: *(Fig. 4) Draw the projections of a point F which lies in both the HP and the VP.*

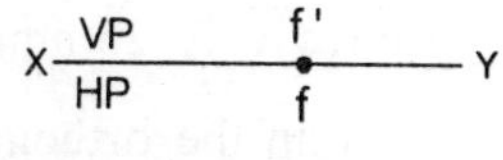

FIG. 4

Note: 1. *Distance of Front View from XY = Distance of the point from HP.*
2. *Distance of Top View from XY = Distance of the point from VP.*

---

3. POINTS IN II, III AND IV QUADRANTS

Problem 6: *(Fig. 5) A point B is 45 mm above HP and 60 mm behind VP. Draw its projections.* [II Quadrant]

1. **To obtain the Front View b′, look at the Pictorial View from front**
   Point **B** lies in the II Quadrant.
   Point **B** is 45 mm above HP.
   b′ is the Front View of **B** and is 45 mm above XY.
2. **To obtain the Top View b, look at the Pictorial View from the top**
   Point **B** is 60 mm behind VP.
   b is the Top View of **B** and is 60 mm behind XY.

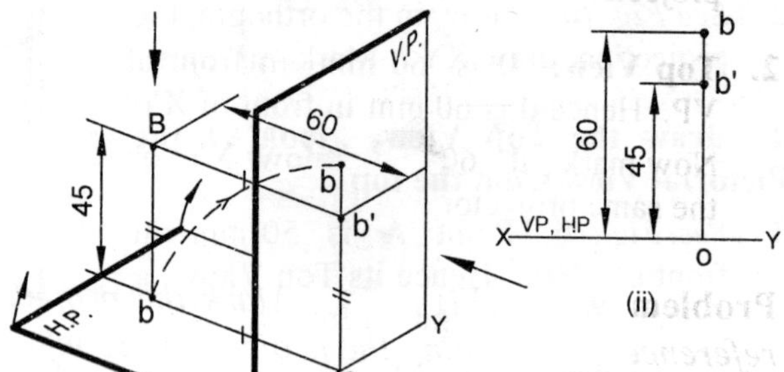

FIG. 5
II QUADRANT

3. To convert the projections **b′** and **b** obtained in the Pictorial View into orthographic projections, rotate the HP about the XY line through 90° in the CW direction.
4. After rotation, HP coincides with VP and both the Front and Top Views are seen above XY. Now, **b** on HP will trace a quadrant of a circle with **o** as center and **ob** as radius. Now **b** occupies the position above **o**.
5. **Front View:** Draw the XY line. Draw a projector at any point on it. Mark **b′** 45 mm above XY on this projector.
6. **Top View:** Mark **b** 60 mm above XY on the same projector.

Problem 7: *(Fig. 6) A point C is 35 mm below HP and 25 mm behind VP. Draw its projections.* [III Quadrant]

1. **Front View:** **C** is 35 mm below HP. Hence **c′** is 35 mm below XY.

   Draw XY and draw a projector at any point on it. Mark **c′** 35 mm below XY on the projector.

2. **Top View:** **C** is 25 mm behind VP. Hence **c** is 25 mm behind XY.

   Now in the orthographic projections mark **c** 25 mm above XY on the above projector.

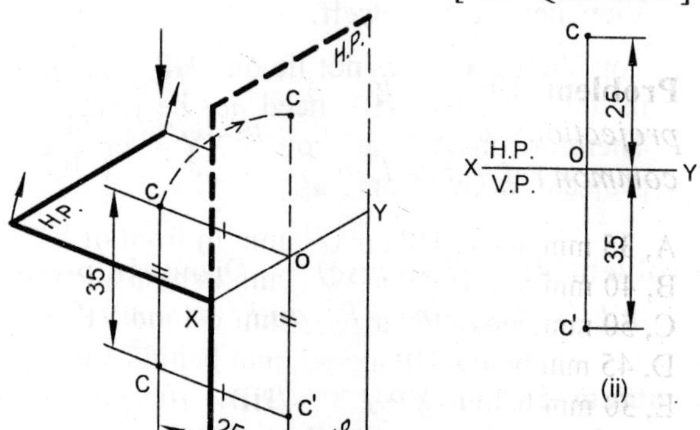

FIG. 6
III QUADRANT

Problem 8: *(Fig. 7) A point D is 45 mm below HP and 60 mm in front of VP. Draw its projections.* [IV Quadrant]

1. **Front View:** **D** is 45 mm below HP. Hence **d′** is 45 mm below XY.

   Draw XY line and draw a projector perpendicular to it.

   Mark **d′** 45 mm below XY on the projector.

2. **Top View:** **D** is 60 mm in front of VP. Hence **d** is 60 mm in front of XY.

   Now mark **d** 60 mm below XY on the same projector.

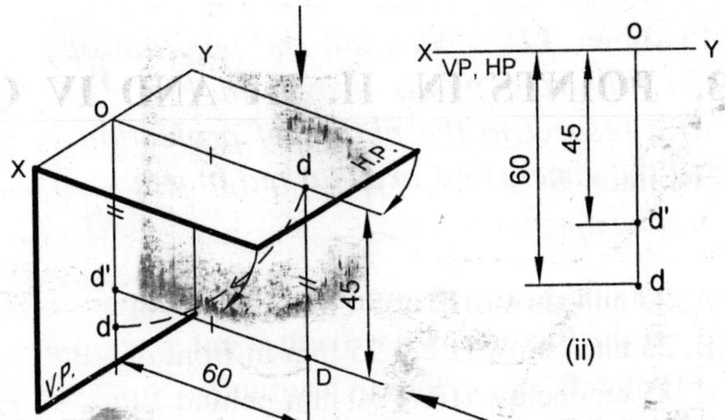

(i)

FIG. 7
IV QUADRANT

Problem 9: *(Fig. 8) Mark the projections of the following points on a common reference line, keeping the projectors 35 mm apart.* (UQ)

1. A, 25 mm above HP and 35 mm in front of VP
2. B, 25 mm above HP and 40 mm behind VP
3. C, 30 mm below HP and 45 mm behind VP
4. D, 30 mm below HP and 40 mm in front of VP
5. E, 25 mm above HP and on VP
6. F, 35 mm below HP and on VP
7. G, 25 mm in front of VP and on HP
8. H, 20 mm behind VP and on HP

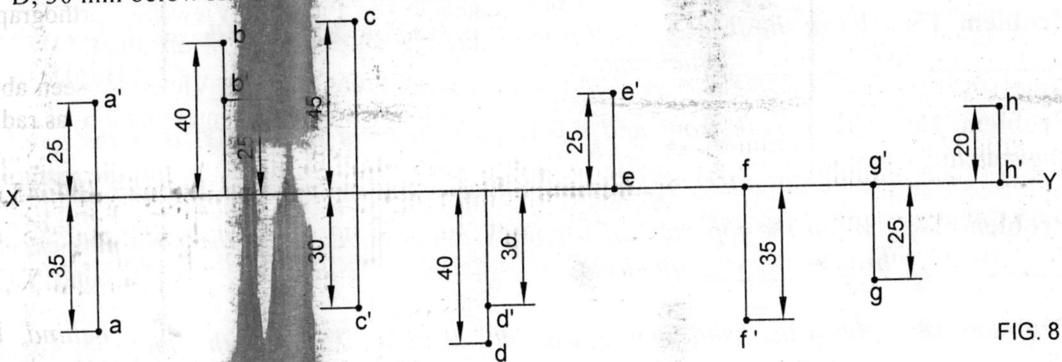

FIG. 8

Problem 10: *(Fig. 9) Draw the projections of the following points on a common reference line.* (UQ)

A, 35 mm above HP and 25 mm in front of VP
B, 40 mm below HP and 15 mm behind VP
C, 50 mm above HP and 25 mm behind VP
D, 45 mm below HP and 25 mm behind VP
E, 30 mm behind VP and on HP

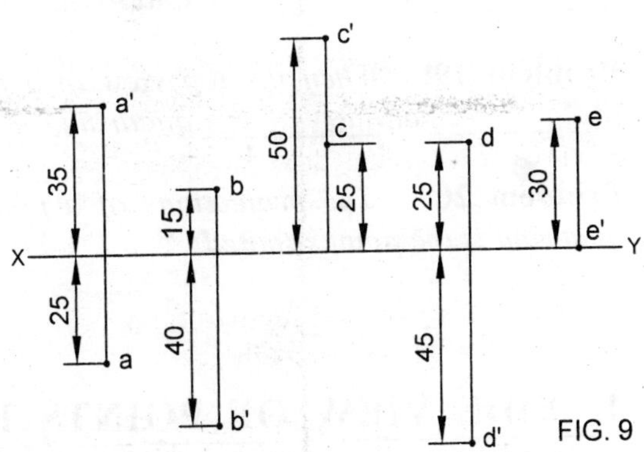

FIG. 9

Problem 11: *(Exercise) A point 25 mm below XY is the top view of two points A and B. A is in HP and B is 35 mm below HP. Mark the projections of A and B.*

Problem 12: *Mention the position of the following points shown in Fig. 10, with respect to the planes of projection. All the dimensions are marked in mm.* (UQ)

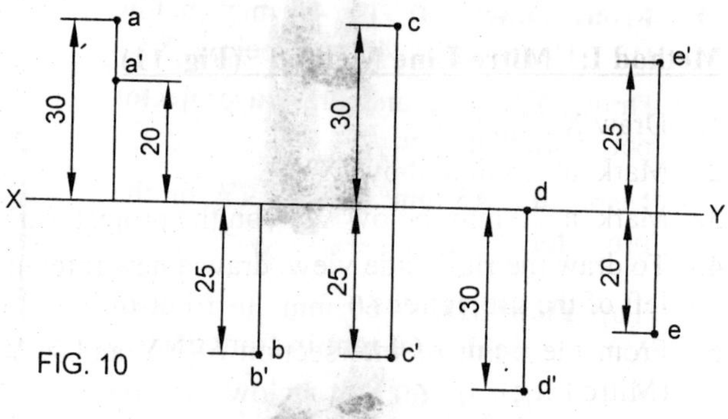

FIG. 10

A, 20 mm above HP & 30 mm behind VP;
B, 25 mm below HP & 25 mm in front of VP;
C, 25 mm below HP & 30 mm behind VP;
D, 30 mm below HP & in VP;
E, 25 mm above HP & 20 mm in front of VP.

Problem 13: *When the front view of a point lies on XY and its top view lies below XY, then the point lies either in* __________ *or* __________ *quadrant.* [First, Fourth]

Problem 14: *When the top view of a point lies on XY and its front view lies above XY, then the point lies either in* __________ *or* __________ *quadrant.* [First, Second]

Problem 15: *When the front view of a point lies below XY, then the point lies either in* __________ *or* __________ *quadrant.* [Third, Fourth]

Problem 16: *When both front and top views of a point lie on XY, then the point lies common to* ________________. [all the quadrants]

Problem 17: *When the top view of a point lies above XY, then the point lies either in* __________ *or* __________ *quadrant.* [Second, Third]

Problem 18: *When the front view of a point lies above XY, then the point lies either in* __________ *or* __________ *quadrant.* [First, Second]

Problem 19: *When the top view of a point lies below XY, then the point lies either in* __________ *or* __________ *quadrant.* [First, Fourth]

Problem 20: *The projections of a point coincide and both fall below XY. In which quadrant is the point situated?* [Fourth]

4. SIDE VIEW OF POINTS IN FIRST QUADRANT

Problem 21: *(Figs. 11 & 12) A point A is 35 mm above HP and 20 mm in front of VP. Draw the front, top and right side view.*

Method I: Mitre Line Method (Fig. 11)

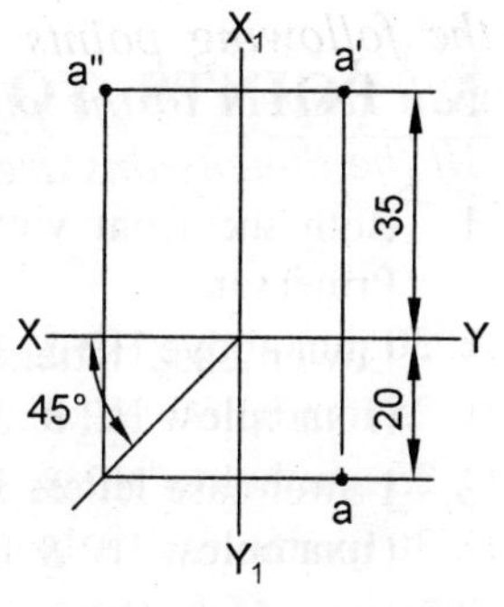

FIG. 11

1. Draw XY line.
2. Mark a′ 35 mm above XY.
3. Mark a 20 mm below XY on the projector of a′.
4. To draw the right side view, draw a new reference line X_1Y_1 to the left of the projections.
5. From the point of intersection of XY and X_1Y_1, draw a 45° line (Mitre Line).
6. Draw a horizontal line from a till it cuts the mitre line.
7. Now, draw a vertical projector from the point on the mitre line. Also, draw a horizontal line from a′. These two lines will intersect at a point a″ which is the required right side view.

Method II: Arc Method (Fig. 12)

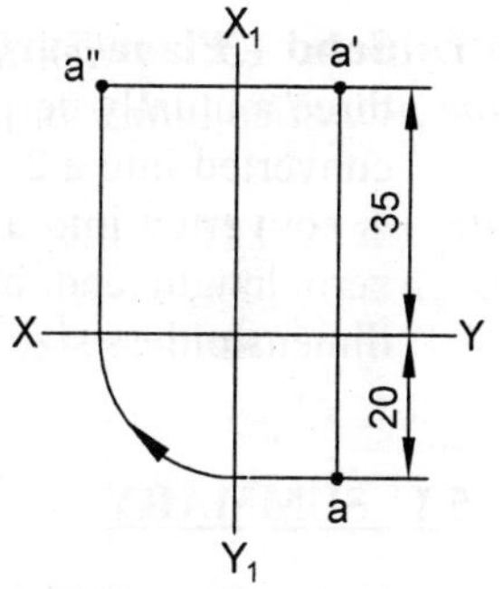

FIG. 12

1. Draw the front and top view of the point A as shown.
2. To draw the right side view, draw a new reference line X_1Y_1 to the left of the projections.
3. Through a, draw a horizontal line till it cuts the X_1Y_1 line.
4. Then, from that point draw an arc with the point of intersection of XY and X_1Y_1 as center till it cuts the XY line.
5. Draw a vertical projector from the above point to meet the horizontal line drawn from a′.
6. These two lines will intersect at a point a″ which is the right side view.

Problem 22: *(Exercise) A point P is 35 mm above HP and 20 mm in front of VP. Draw the front, top and left side view by Mitre Line Method and Arc Method.*

NOTE :

1. *Students are advised to make their own paper / cardboard / perspex model of HP and VP. The model will facilitate developing a good concept of the relative positions of the points lying in any of the four quadrants.*
2. *Since the Projections of Points, Lines and Planes are the basic Chapters for the forthcoming topics on Solids (viz., Projections of Solids, Sections, Development of Surfaces, etc.), the students should follow these basic Chapters carefully to draw the orthographic projections.*
3. *In this Chapter, a series of pictorial views are illustrated to visualize easily the different positions occupied by a point with respect to HP and VP.*

5. POINTS TO REMEMBER

1. Both the front view and top view of a point must lie in a single straight line called the Projector.
2. Lower case letters with a prime should be used to represent the points of front view. (Examples: a′, b′, c′, etc., and 1′, 2′, 3′, etc.)
3. Lower case letters should be used to represent the points of top view. (Examples: a, b, c, etc., and 1, 2, 3, etc.)
4. Lower case letters with double prime should be used to represent the points of side views. (Examples: a″, b″, c″, etc., and 1″, 2″, 3″, etc.)
5. **Solid – Plane Surface – Line – Point:** A *solid* is formed by three dimensions measured in three mutually perpendicular directions. If one of the dimensions is made zero, the object is converted into a 2-D *plane surface*. If one more dimension is also reduced to zero, the plane is converted into a *line*. Finally, if the remaining dimension is also removed, the line is of zero length and becomes a *point*. Therefore, a point in 3-D geometry is the smallest dimensionless size of a solid.

5.1 SUMMARY

Sl.No.	Location of Point	Front View	Top View	Quadrant
1.	Above HP In Front of VP	Above XY	Below XY	First
2.	Above HP Behind VP	Above XY	Above XY	Second
3.	Below HP Behind VP	Below XY	Above XY	Third
4.	Below HP In Front of VP	Below XY	Below XY	Fourth
5.	Above HP On VP	Above XY	On XY	First or Second
6.	Below HP On VP	Below XY	On XY	Third or Fourth
7.	On HP In Front of VP	On XY	Below XY	First or Fourth
8.	On HP Behind VP	On XY	Above XY	Second or Third

*

10

PROJECTIONS OF STRAIGHT LINES

STRAIGHT LINES

A *straight line* is the shortest distance between two given points. Thus, the two ends of a straight line are points. Projection of a Straight Line can therefore be drawn by joining the respective projections of its *end points.*

Note: *Straight Lines are also called simply as Lines.*

The shape of a 3-D object is formed by different *surfaces*. These surfaces are formed by moving or rotating *lines* in different ways. Thus, *line* forms the basic geometrical entity using which any engineering component is made. Therefore, clear understanding of Projections of Lines is a necessary requirement for an engineer to understand the projections of 3-D objects.

True Length: Actual length of a line is termed as its *True Length.*

BIS CONVENTIONS FOR LINE THICKNESS

- XY line, Projector lines, Locus line, Construction line, Dimension line: *Narrow Thin Lines.* [*2H Grade Pencil*]
- Front View, Top View, True Length line: *Thick Lines.* [*H Grade Pencil*]

POSITIONS OF A LINE WITH RESPECT TO HP AND VP

A line may occupy an infinite number of positions in space with respect to HP and VP. The various positions may be classified into the following types:

1. Parallel to Both the Planes HP and VP
   - 1.1 Contained in One Plane and Parallel to the Other
   - 1.2 Contained in Both the Planes HP and VP
2. Perpendicular to One Plane and Parallel to the Other
   - 2.1 Perpendicular to HP and Parallel to VP
   - 2.2 Perpendicular to VP and Parallel to HP
3. Parallel to One Plane and Inclined to the Other
   - 3.1 Parallel to VP and Inclined to HP
   - 3.2 Parallel to HP and Inclined to VP
   - 3.3 Contained in One Plane and Inclined to the Other
4. Inclined to Both the Planes HP and VP

1. LINE PARALLEL TO BOTH HP AND VP

Fig. 1(i) shows the Pictorial View of a line CD parallel to both HP and VP.

1. Look from front to get Front View. As CD is parallel to VP, c′ d′ = True Length of CD.
2. Look from top to get the Top View. As CD is parallel to HP, cd = True Length of CD.
3. The position of the line CD and its projections on HP and VP are shown in *Fig. 1(i)*.
4. Rotate HP (along with the projection on it) through 90° in CW direction as shown.
5. *See Fig. (ii)*. Projection of CD on VP is Front View and projection on HP is Top View.

Problem 1: *(Fig. 1) A line CD 30 mm long is parallel to both the planes. The line is 40 mm above HP and 25 mm in front of VP. Draw its projections.*

1. *See Fig. (ii)*. Draw reference line XY. Draw a projector at any point on it (*thin line*).

Front View

2. *See Fig. (i)*. Line CD is 40 mm above HP, *i.e.*, point C is 40 mm above HP. So mark c′ 40 mm above XY in *Fig. (ii)*.
3. Line CD is parallel to HP. Front View will be in True Length and parallel to XY. *See Fig. (ii)*. Draw c′ d′ parallel to XY such that c′ d′ = True Length CD = 30 mm.

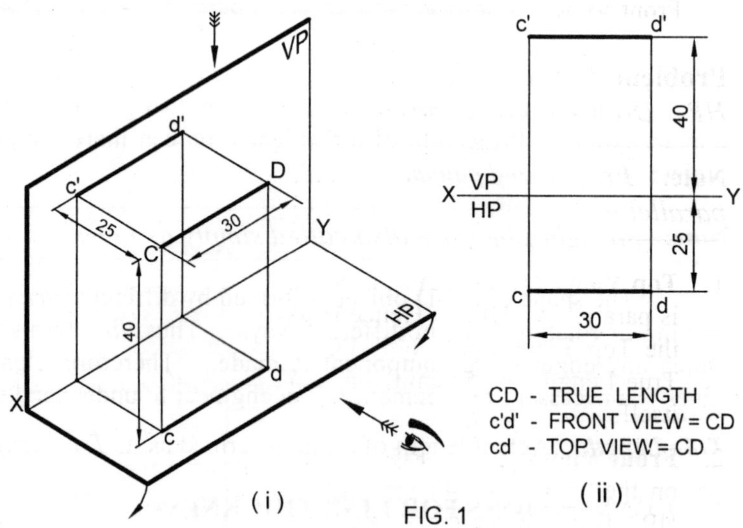

FIG. 1

Top View

4. *See Fig. (i)*. Line CD is also parallel to VP and 25 mm in front of VP. Hence on the projector from c′, mark c at a distance of 25 mm below XY as in *Fig. (ii)*.
5. *See Figs. (i) & (ii)*. Top View is also in True Length and parallel to the XY line. Hence, draw cd parallel to XY such that cd = True Length = CD = 30 mm.

Note: Draw the orthographic projections only as in *Fig. 1(ii)* and not the pictorial projection.

1.1 LINE CONTAINED IN ONE PLANE AND PARALLEL TO THE OTHER

Problem 2: *(Fig. 2) A line RS 60 mm long lies in HP and 45 mm in front of VP. Draw its projections.*

Note: ***From the given data, decide the view to be drawn first.*** *In this Problem, since the line is parallel to VP, its True Length is seen in Front View. Hence, draw* ***Front View first.***

1. **Front View:** Draw XY. Line RS lies on HP. Hence mark Front View r′ s′ (= RS = True Length = 60 mm) on XY itself.
2. **Top View:** Line RS is 45 mm in front of VP. Rotate HP through 90° in CW direction. Hence mark the Top View rs (= RS = True Length = 60 mm) at a distance of 45 mm below XY on the projectors drawn from the Front View.

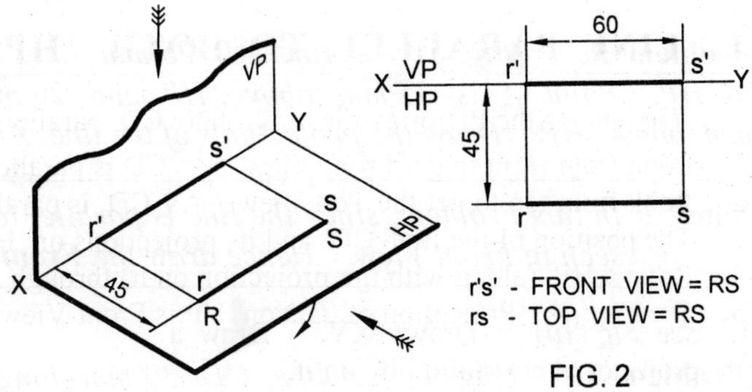

r's' - FRONT VIEW = RS
rs - TOP VIEW = RS

FIG. 2

Problem 3: *(Fig. 3) A line PQ 55 mm long is lying in VP and 45 mm above HP. Draw its projections.*

Note: ***From the given data, decide the view to be drawn first.*** *In this Problem, line is parallel to HP. Hence, its True Length is seen in Top View. Draw* ***Top View first.***

---

1. **Top View:** Draw XY. Line PQ is parallel to HP. Hence mark the Top View pq (= PQ = True Length = 55 mm) on XY itself.
2. **Front View:** Line PQ is lying on the VP and 45 mm above HP. Hence mark the Front View p′q′ (= PQ = True Length = 55 mm) at a distance of 45 mm above XY on the projectors drawn from the Top View.

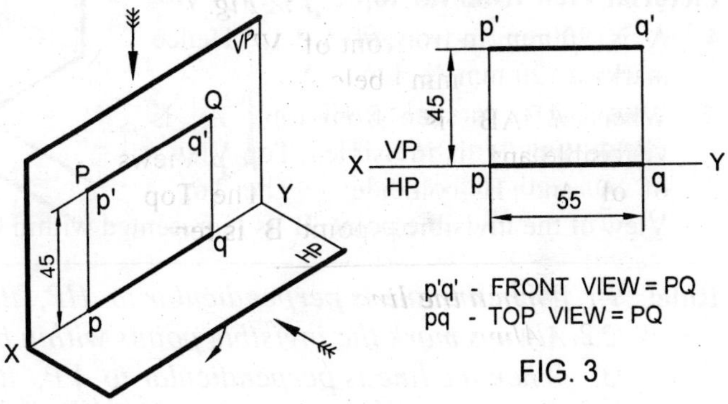

p'q' - FRONT VIEW = PQ
pq - TOP VIEW = PQ

FIG. 3

1.2 LINE CONTAINED IN BOTH HP AND VP

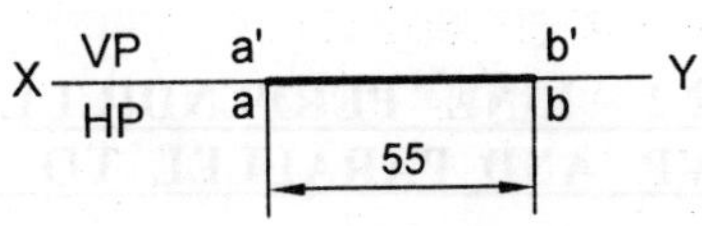

Problem 4: *(Fig. 4) A line AB 55 mm long is lying on both HP and VP. Draw its projections.*

See Fig. 4 for Solution.

a'b' - FRONT VIEW = AB
ab - TOP VIEW = AB

FIG. 4

2. LINE PERPENDICULAR TO ONE PLANE AND PARALLEL TO THE OTHER

2.1 LINE PERPENDICULAR TO HP AND PARALLEL TO VP

Fig. 5(i) shows the Pictorial View of a line AB perpendicular to HP and parallel to VP.

Problem 5: *(Fig. 5) A line AB 25 mm long is parallel to VP and perpendicular to HP. Point A is 35 mm above HP and 20 mm in front of VP. Point B is 10 mm above HP. Draw the projections of the line AB.*

Note: *In this Problem, since the line is parallel to VP, the True Length of the line is seen in Front View. Hence draw the* ***Front View first****.*

---

1. *See Fig. (ii).* Draw XY. Draw a projector at any point o on it.

To draw Front View a′ b′, look at the Pictorial View from the front: *Fig. (i)*

2. B is 10 mm above HP. Hence mark b′ 10 mm above XY.
3. A is 35 mm above HP. So mark a′, 25 mm (= 35–10) above b′ on the *same* projector.

To draw the Top View a b, look at the Pictorial View from the top : *Fig. (i)*

4. A is 20 mm in front of VP. Hence mark a 20 mm below XY.
5. When AB is seen from top, A is visible and B is invisible. Top Views of A and B coincide. The Top View of the invisible point B is represented within bracket as (b).

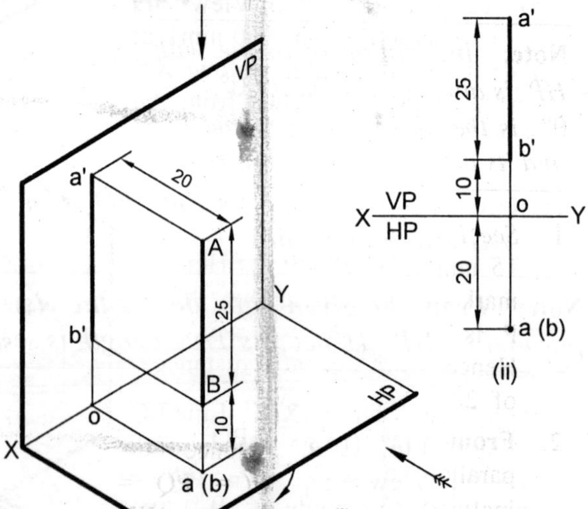

(i) FIG. 5

Rule: *1. When the line is perpendicular to HP, its top view will be a point.*
2. Always mark the invisible points within brackets.
3. When the line is perpendicular to VP, its front view will be a point.

<u>2.2 LINE PERPENDICULAR TO VP AND PARALLEL TO HP</u>

Problem 6: *(Fig. 6) A line AB 25 mm long is perpendicular to VP and parallel to HP. Its end A is 10 mm in front of VP and the line is 20 mm above HP.*

Draw the projections of the line.

See Fig. 6 for Solution.

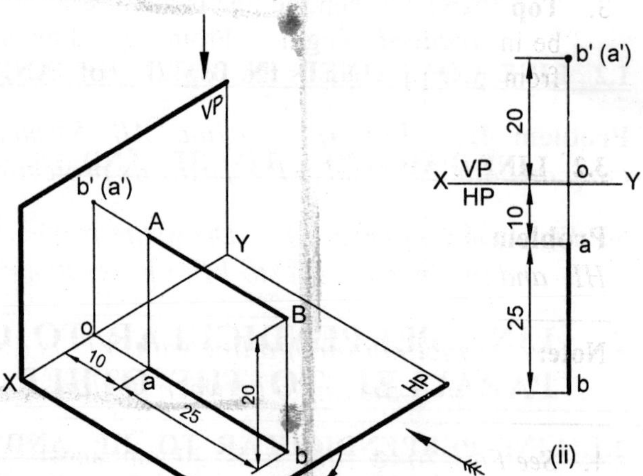

(i) FIG. 6

3. LINE PARALLEL TO ONE PLANE & INCLINED TO OTHER

3.1 LINE PARALLEL TO VP AND INCLINED TO HP

Problem 7: *(Fig. 7) A line PQ 40 mm long is parallel to VP and inclined at an angle of 30° to HP. The lower end P is 15 mm above HP and 20 mm in front of VP. Draw the projections of the line.*

Note: *Inclination of a line with HP is always denoted as θ°. θ° is the angle between line PQ and its projection pq on HP.*

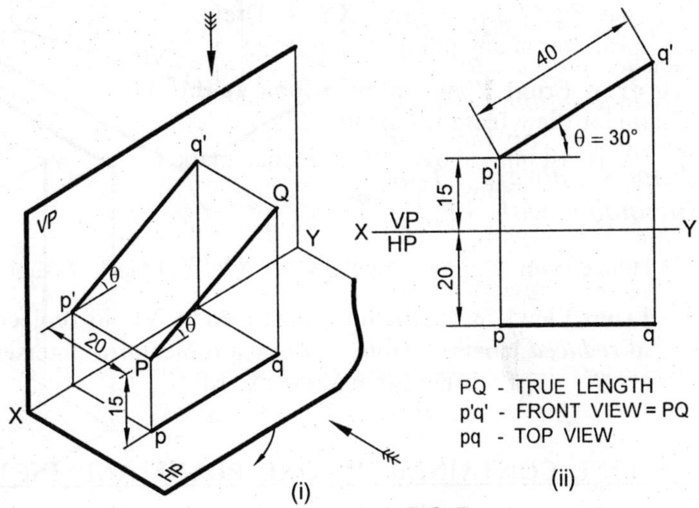

FIG. 7

1. *See Figs. (i) and (ii).* P is 15 mm above HP. Hence mark p′ 15 mm above XY. P is 20 mm in front of VP. Hence, mark p at a distance of 20 mm below XY.
2. **Front View:** As PQ is parallel to the VP and inclined at an angle of θ° to HP, p′q′ will be equal to its **True Length** and inclined at an angle of θ° to XY.

Rule: *When a line is parallel to VP and inclined at θ° to HP, this inclination θ° is seen in Front View.* ***θ° denotes always True Inclination with HP.***

So from p′ draw a line at 30° to XY and mark q′ such that p′ q′ = 40 mm = True Length.

3. **Top View:** When the line is inclined to HP, its projection on HP, *i.e.*, Top View will be in ***reduced*** length. From q′ draw a projector to intersect the horizontal line drawn from p at q. pq is the Top View of PQ.

3.2 LINE PARALLEL TO HP AND INCLINED TO VP

Problem 8: *(Fig. 8) Draw the projections of a line EF 40 mm long parallel to HP and inclined at 35° to VP. E is 20 mm above HP and 15 mm in front of VP.*

Note: *Inclination of a line with VP is always denoted as ϕ. ϕ is the angle between the line EF and its projection e′ f′ on VP.*

1. *See Figs. (i) & (ii).* E is 20 mm above HP. Therefore mark e′ 20 mm above XY. E is 15 mm in front of VP. Therefore mark e 15 mm below XY.

2. **Top View:** As EF is parallel to HP and inclined at ϕ to VP, ef will be equal to **True Length** EF and inclined at ϕ to XY.

Rule: ***When a line is parallel to HP and inclined at ϕ to VP, this inclination ϕ will be seen in the Top View. ϕ denotes always True Inclination with VP.***

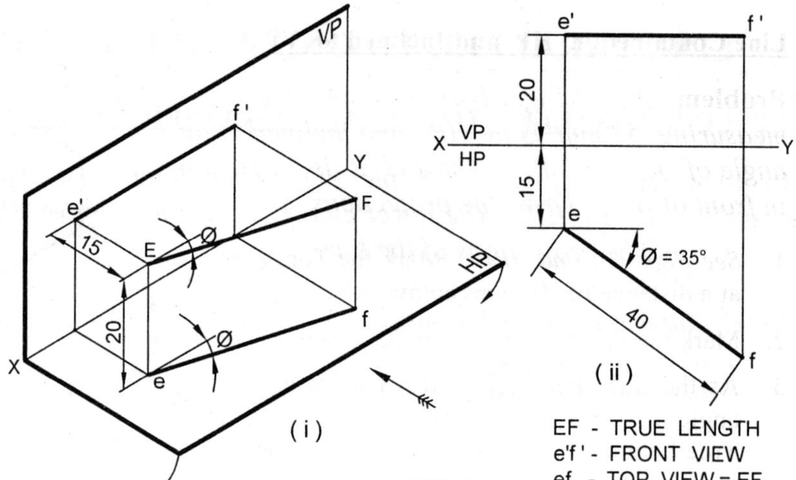

FIG. 8

Hence from e draw a line 35° to XY and mark f such that ef = 40 mm = True Length.

3. **Front View:** When the line is inclined to VP its projection on VP, *i.e.*, Front View will be in ***reduced*** length. From f draw a projector to intersect the horizontal line drawn from e′ at f′. e′ f′ is the Front View of EF.

3.3 LINE CONTAINED BY ONE PLANE AND INCLINED TO THE OTHER

Line Contained in VP and Inclined to HP

Problem 9: *(Fig. 9) A line AB 50 mm long is in VP and inclined at 35° to HP. End A is 10 mm above HP. Draw the projections.*

1. *See Figs. (i) & (ii).* A is 10 mm above HP. Hence mark a′ 10 mm above XY.
2. Line AB is in VP. So mark a on XY itself.
3. As the line AB is in VP and inclined at 35° to HP, a′ b′ = True Length = AB and a′ b′ will be inclined at θ = 35° to XY.

   So, from a′ draw a line at 35° to XY and mark b′ such that a′ b′ = 50 mm = **True Length**.
4. Since the line is inclined to HP, the Top View will be in ***reduced length***. Since the line is in VP, Top View will lie on XY itself.

   Therefore from b′ draw a projector to intersect the XY line at b. ab is the Top View of AB lying on XY.

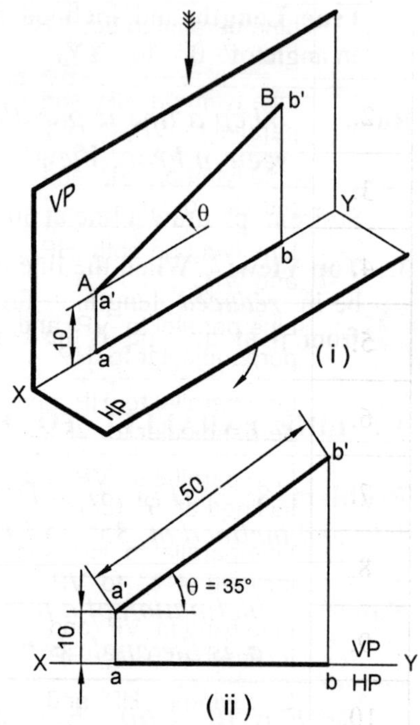

FIG. 9

Line Contained in HP and Inclined to VP

Problem 10: *(Fig. 10) A line RS measuring 52 mm is in HP and inclined at an angle of 45° to VP. The end R is 10 mm in front of VP. Draw the projections.*

1. *See Figs. (i) and (ii).* Mark the Top View r at a distance of 10 mm below XY.
2. Mark the Front View r′ on XY itself.
3. As the line RS is in HP and inclined at an angle of 45° to VP, rs will be equal to the True Length RS (= 52 mm) and inclined at an angle of $\phi = 45°$ to XY.
4. As the line is in HP, its Front View r′ s′ will lie on XY.

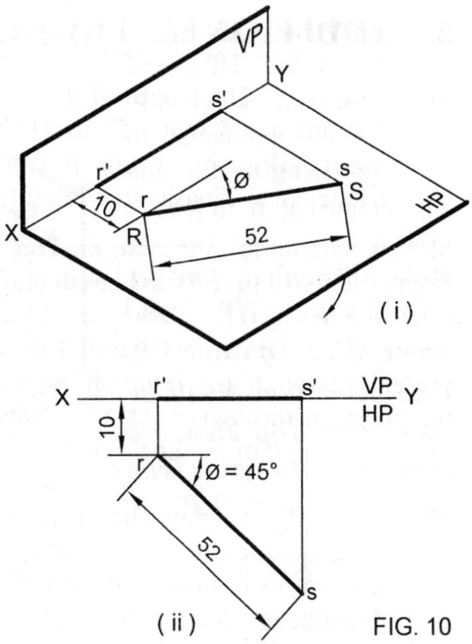

FIG. 10

4. POINTS TO REMEMBER

Sl. No.	Position of Line	Front View	Top View
1.	Line parallel to both HP and VP.	Parallel to XY and having True Length of the line.	Parallel to XY and having True Length of the line
2.	Line lying on HP and parallel to VP.	Lies in XY line and equal to True Length.	Lies below XY line and equal to True Length.
3.	Line lying on VP and parallel to HP.	Lies above XY and equal to True Length.	Lies in XY line and equal to True Length.
4.	Line lying on both HP and VP.	Lies in XY line and equal to True Length.	Lies in XY line and equal to True Length.
5.	Line parallel to VP and perpendicular to HP	Perpendicular to XY and equal to True Length.	A point.
6.	Line parallel to HP and perpendicular to VP	A point.	Perpendicular to XY and equal to True Length.
7.	Line parallel to VP and inclined at θ° to HP.	Inclined at θ° to XY and is equal to the True Length.	Parallel to XY line and shorter than the True Length.
8.	Line parallel to HP and inclined at ϕ° to VP	Parallel to XY line and shorter than the True Length.	Incline at ϕ° to XY line and is equal to the True Length.
9.	Line lying in VP and inclined at θ° to HP.	Inclined at θ° to XY and is equal to the True Length.	Lies in XY line and shorter than the True Length.
10.	Line lying in HP and inclined at ϕ° to VP.	Lies in XY line and shorter than the True Length.	Inclined at ϕ° to XY and is equal to the True Length.

5. ADDITIONAL PROBLEMS – I

11. *(Fig. 11)* The length of the top view of a line MN parallel to VP and inclined at 45° to HP is 50 mm. Point M is 12 mm above HP and 25 mm in front of VP. Draw the projections of the line. Find its true length.

MN is parallel to VP. So its Top View will be parallel to XY and Front View will give True Length and True Inclination with HP.

See Fig. 11. Mark m 25 mm below XY and m′ 12 mm above XY. Draw mn parallel to XY and 50 mm long. Draw a projector through n. From m′ draw a line at 45° with XY to meet the projector through n at n′. Now m′n′ is the Front View giving the *True Length* of the line. m′n′ = MN = 71 mm

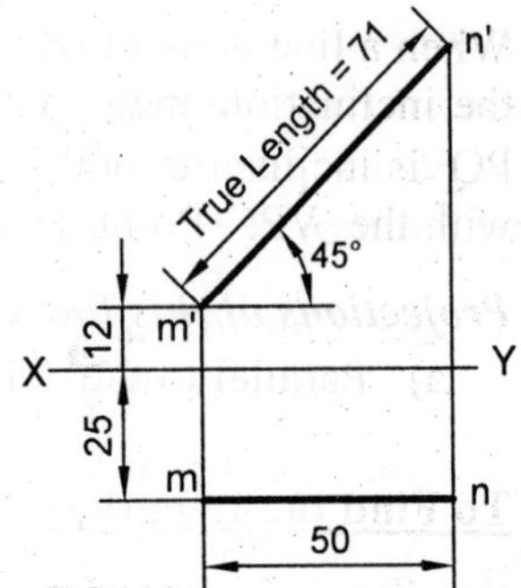

FIG. 11

12. *(Fig. 12)* Draw the projections of a line CD 50 mm long, parallel to HP and inclined to VP. The end C is 10 mm in front of VP and D is 30 mm in front of VP. The line is 15 mm above HP.

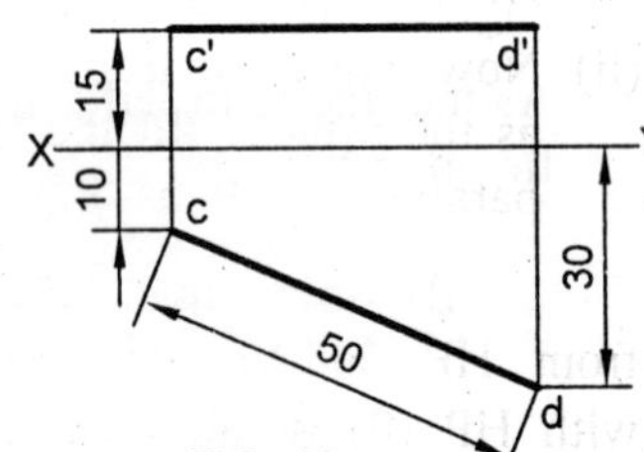

FIG. 12

13. A line CD is parallel to VP and inclined at 40° to HP. C is in HP and 25 mm in front of VP. Top view is 50 mm long. Find its true length. *(Result : CD = 65 mm)*

14. *(Fig. 13)* A line EF 60 mm long is in VP and inclined to HP. The top view measures 45 mm. The end E is 15 mm above HP. Draw the projections of the line. Find its inclination with HP. (UQ)

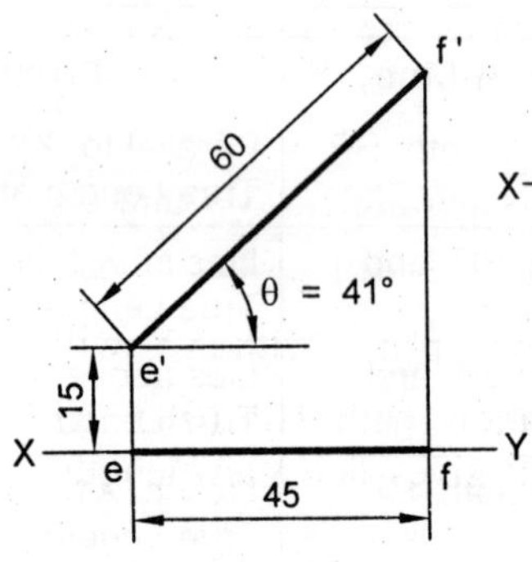

FIG. 13

15. *(Fig. 14)* A line GH 45 mm long is in HP and inclined to VP. The end G is 15 mm in front of VP. Length of front view is 35 mm. Draw the projections of the line. Determine its inclination with VP.

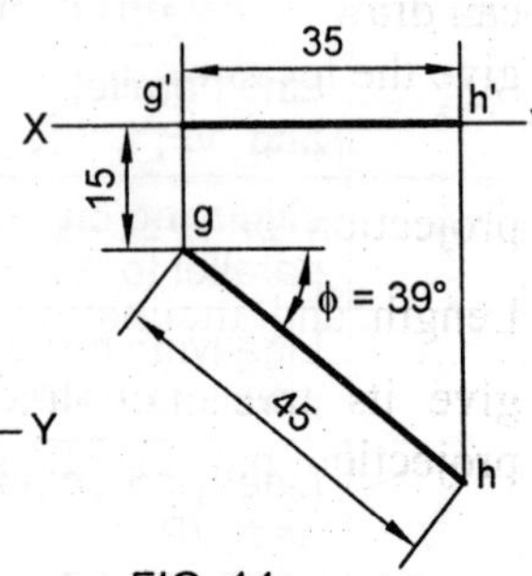

FIG. 14

16. A line AB 60 mm long is parallel to HP. The point A is 20 mm above HP and 35 mm in front of VP. The length of the front view is 50 mm. Determine its true inclination with VP. *(Result : ϕ = 35°)*

17. *(Fig. 15)* A line MN 50 mm long is parallel to VP and inclined at 30° to HP. The end M is 20 mm above HP and 10 mm in front of VP. Draw the projections of the line.

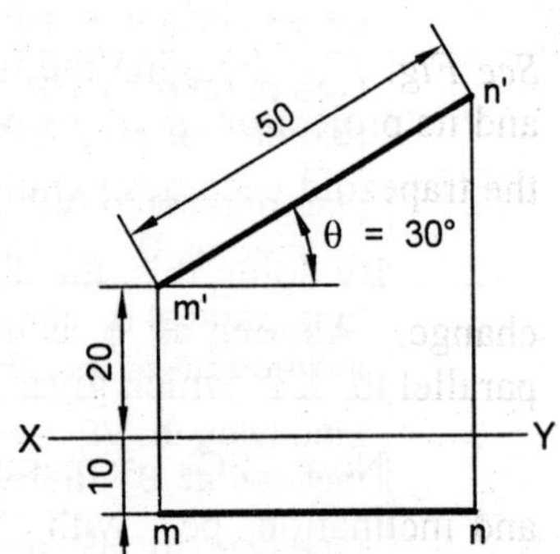

FIG. 15

6. LINE INCLINED TO BOTH HP AND VP

When a line is inclined to both HP and VP, the projections are *shorter* than the True Length and the inclination angles with XY are *greater* than the true inclinations of the line. *See Fig. 16.* PQ is inclined to both the planes in first quadrant. It makes an angle of θ with the HP and ϕ with the VP. Point P is above the HP and in front of the VP.

Projections of this line can be obtained by rotating it into two simple positions, *viz.*,
a) Parallel to the VP and b) Parallel to the HP

To Find the Length of the Top View pq_1 and the Locus of Q in the Front View

(i) *See Fig. 16.* Imagine a trapezoid made up of line PQ, its projectors Pp, Qq to HP and its projection pq on HP.

(ii) Now rotate this trapezoid PQqp with Pp as the hinge, till the trapezoid becomes parallel to VP as PQ_1q_1p.

By doing this, height of the point Q from HP and the inclination of the line PQ with HP (θ) do not change.

As long as θ is unaltered, the length of the Top View remains the same. So we can draw a line cd parallel to XY which will give the locus of q′.

Now PQ_1 is parallel to VP and its projection on VP, $p'q_1'$ will give its True Length and inclination $p'q_1'$ with XY will give its true inclination with HP (θ). Its projection pq_1 on HP gives Top View length.

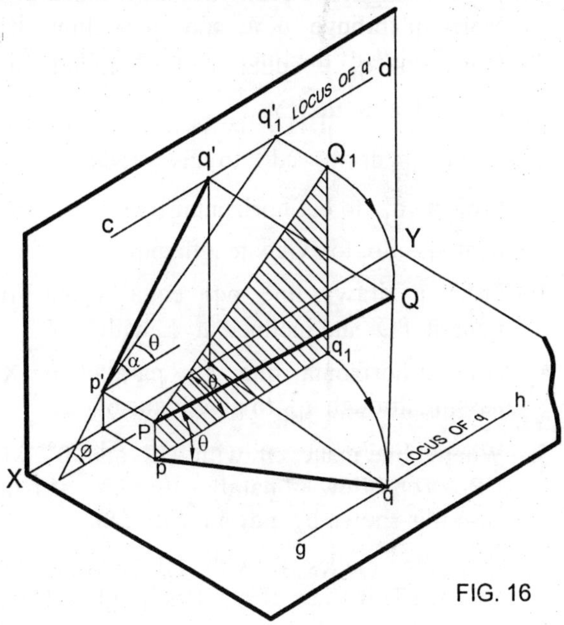

FIG. 16

To find the Front View Length $p'q_2'$ and the Locus of Q in the Top View

See Fig. 17. Imagine the trapezoid made up of the line PQ, its projectors Pp′, Qq′ to VP and its projection p′ q′ on VP. Now rotate this trapezoid PQ q′ p′ with Pp′ as the hinge, till the trapezoid becomes parallel to HP as $PQ_2\ q_2'\ p'$.

By doing this, the distance of Q from VP and inclination of PQ with VP (ϕ) do not change. As long as ϕ is unaltered, Front View length remains the same. Now draw a line gh parallel to XY which gives the locus of q.

Now PQ_2 is parallel to HP and its projection on HP, pq_2 will give its True Length and inclination pq_2 with XY gives its true inclination (ϕ) with VP. Its projection on VP, $p'\ q_2'$ gives Front View length.

Note: i) *See Fig. 16. If PQ makes θ to HP, for various values of ϕ the Front View of Q always lies on cd and Top View length pq_1 is unaltered.*

ii) *See Fig. 17. If PQ makes ϕ to VP, for various values of θ the Top View of Q always lies on gh and Front View length $p'q_2'$ remains unaltered.*

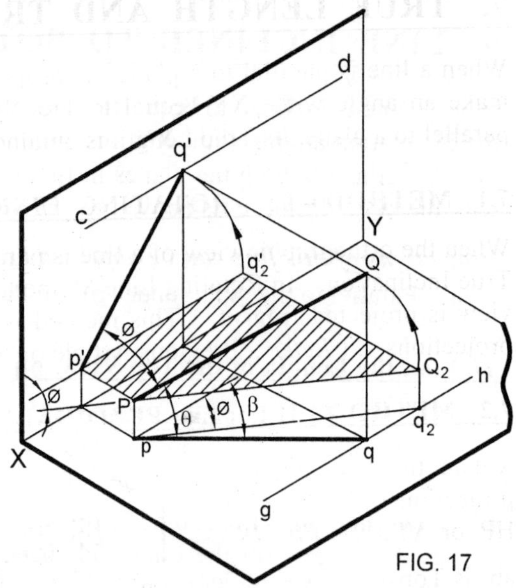

FIG. 17

Above two cases are integrated as in *Fig. 18.*

1. Mark p' above XY and p below XY. Draw $p'q_1'$ equal to True Length and inclined at θ to XY.
2. From q_1' draw cd to give locus of q'. Project q_1' to cut horizontal line drawn from p at q_1. So, top view length pq_1 is obtained.
3. From p draw a line pq_2 equal to the True Length PQ and inclined at ϕ with XY.
4. Draw a horizontal line gh parallel to XY passing through q_2 to give locus of q.
5. When a line makes θ with HP and 0° with VP, Top View is parallel to XY (in this case it is shown by pq_1 in *Fig. 18*).

   If the line makes θ with HP and ϕ with VP, the Top View of Q lies in gh. Hence draw an arc with p as centre and pq_1 as radius to cut gh at q. Join pq.
6. Draw a projector from q to meet cd at q'. Join p'q' to give its Front View.

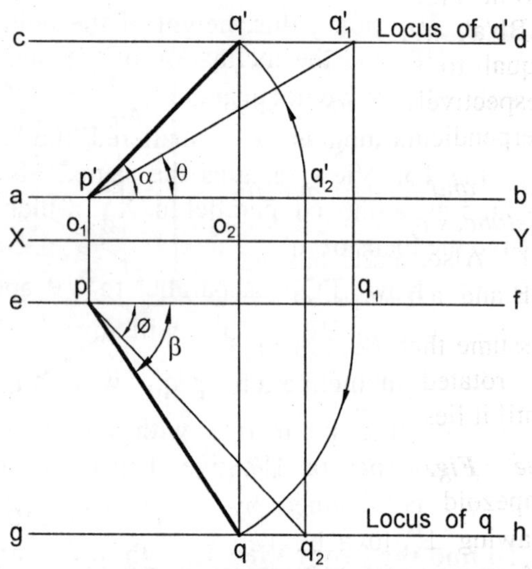

FIG. 18

Projections of the line can also be obtained as follows:

7. When the line makes an angle of ϕ with VP and 0° with HP the Front View ($p'q_2'$) is parallel to XY. If the line makes ϕ with VP and θ with the HP, q' will lie on cd. Hence draw an arc with p' as centre and $p'q_2'$ as radius to meet cd at qq'.
8. From q' draw a projector to intersect gh at q. Join pq which is Top View of the line.

Note: 1. *Angle made by Front View p'q' with XY is called* ***Apparent angle α.*** *Angle made by the Top View pq with XY is called* ***Apparent angle β.***
2. *Apparent Inclinations are always greater than the True Inclinations.*

7. TRUE LENGTH AND TRUE INCLINATIONS

When a line is parallel to a plane, its projection on that plane will give its True Length and will make an angle with XY equal to True Inclination with other plane. After making the line parallel to a plane, its True Length is obtained by any one of the following *two* methods:

7.1 METHOD – I : ROTATING LINE METHOD

When the orthographic view of a line is parallel to XY, its other view will have True Length and True Inclination. In Rotating Line Method, each view is made parallel to XY line and the other view is projected from it. This method is just the reverse of the method used for drawing the projections. (*Refer* : Previous Article on ***Lines Inclined to Both the Planes.***)

7.2 METHOD – II : TRAPEZOIDAL PLANE METHOD

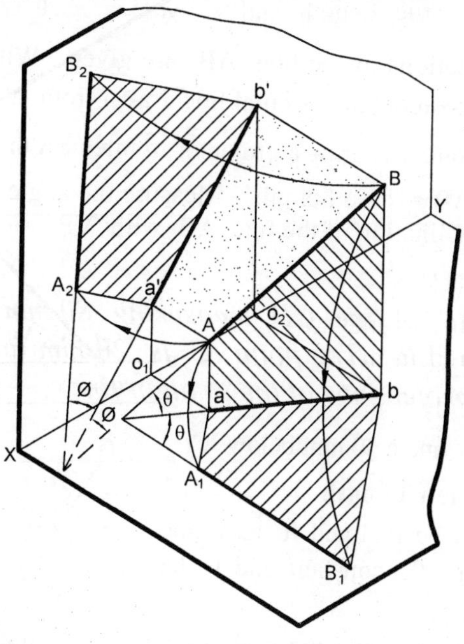

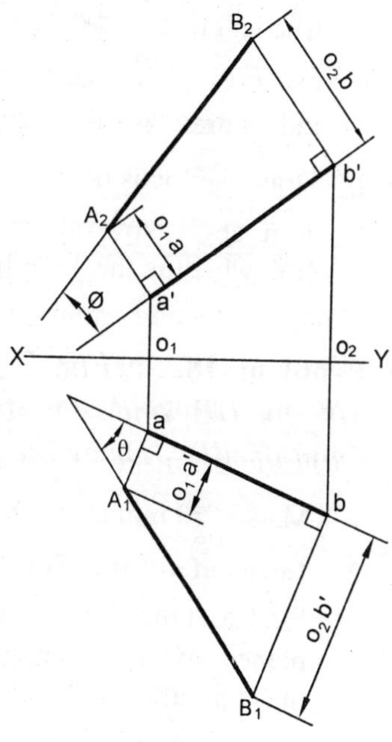

FIG. 19 TRAPEZOIDAL METHOD FIG. 20

Rotate the line about its projections till it lies in HP or VP. *See **Fig. 19**.*

ab is Top View. a′ b′ is Front View. In trapezoid ABb′a′, a′A and b′B are equal to ao_1 and bo_2 respectively and also perpendicular to a′b′.

ao_1 and bo_2 are the distances of a and b from XY. Also, angle between AB and a′b′ is ϕ.

Assume that the trapezoid is rotated about a′b′ until it lies in VP.

*See **Fig. 20**.* This trapezoid is obtained by drawing $\perp^{rs}$ to a′ b′, *i.e.*, $a'A_2 = ao_1$ and $b'B_2 = bo_2$ and joining A_2B_2. A_2B_2 is True Length of AB. Its inclination with a′ b′ is ϕ. Similarly abBA is assumed rotating about ab as axis, until it lies in HP. In *Fig. 20* this trapezoid is obtained by drawing perpendiculars to ab, *i.e.*, $aA_1 = a'o_1$ and $bB_1 = b'o_2$ and joining A_1B_1. [= True Length of AB] Inclination of A_1B_1 with ab = θ.

8. GIVEN FRONT AND TOP VIEWS, TO FIND TRUE LENGTH AND TRUE INCLINATIONS

Make the top view parallel to XY. Draw corresponding front view to obtain true length and θ.

1. In *Fig. 21*, top and front views are given. a as centre and ab as radius draw an arc to cut the horizontal drawn from a at b_1.
2. Draw the locus of b′. From b_1 draw a projector to intersect the locus of b′ at b'_1. Join $a'b'_1$. Now $a'b'_1$ = True Length AB.
3. Measure angle θ between $a'b'_1$ and XY. Make the front view parallel to XY. Draw the corresponding top view to obtain the true length and ϕ.

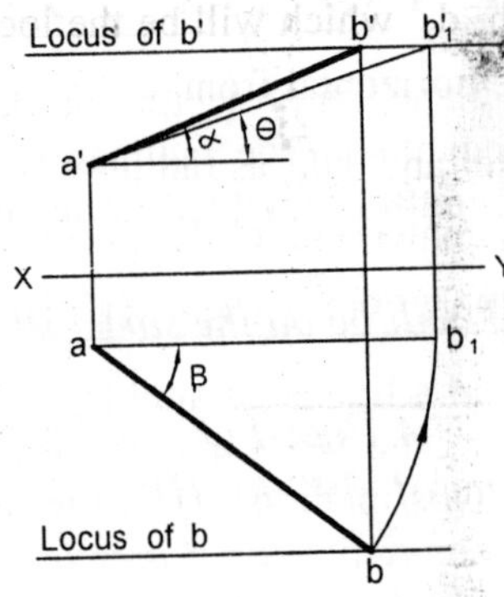

FIG. 21

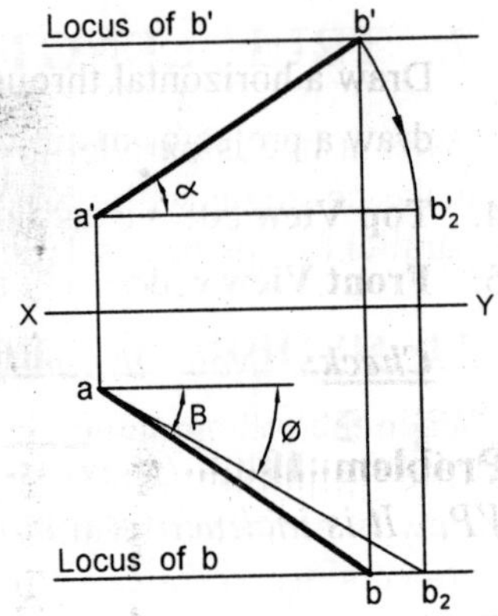

FIG. 22

1. *See Fig. 22.* The projections of the line AB are given. With a′ as centre and a′ b′ as radius draw an arc to intersect the horizontal line drawn from a′ at b'_2.
2. Draw the locus of b. From b'_2 draw a projector to intersect the locus of b at b_2.
3. Join ab_2. Now ab_2 = AB = True Length. Measure the angle ϕ between the line ab_2 and XY which is the True Inclination of the line AB with VP.

Problem 18: *(Fig. 23) A line CD measuring 80 mm is inclined at an angle of 30° to HP and 45° to VP. The point C is 20 mm above HP and 30 mm in front of VP. Draw the projections of the straight line.* (UQ)

1. Mark c′ 20 mm above XY and c 30 mm below XY.
2. **Locus of d′ and Top View Length**

   First assume the line to be inclined to only one plane, say, HP, keeping θ constant and to be made parallel to VP.

   From c′ draw 30° line to XY and mark d'_1 such that $c'd'_1$ = **True Length** = 80 mm. Draw a horizontal through d'_1 which is the locus of d′ (i.e., D in Front View). Draw a horizontal from c.

   From d'_1 draw a projector to cut the horizontal drawn from c at d_1. **cd_1 is Top View Length.**

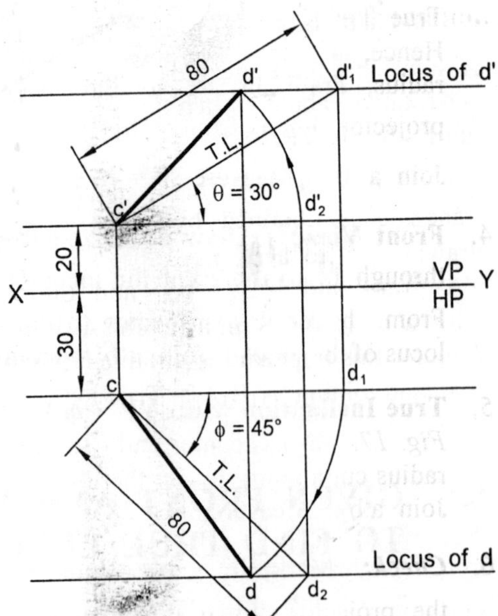

FIG . 23

3. **Locus of d and Front View Length**

   Next assume the line to be parallel to HP and inclined at ϕ to VP. From c draw a line 45° to XY and mark d_2 such that cd_2 = **True Length** = 80 mm.

Draw a horizontal through d_2 which will be the locus of d (i.e., D in Top View). From d_2 draw a projector to cut the horizontal from c′ at d_2'. $c'd_2'$ is **Front View Length**.

4. **Top View cd:** c as center and cd_1 as radius draw an arc to cut the locus of d at d.
5. **Front View c′d′:** c′ as center and $c'\ d_2'$ as radius draw an arc to cut the locus of d′.

Check: *Now d′ and d will be on the same projector.*

Problem 19: *(Exercise) A line PQ 75 mm long has its end P in both HP and VP. It is inclined at an angle of 30° to HP and 45° to VP. Draw the projections.*

Problem 20: *(Fig. 24) A line AB is 75 mm long. A is 50 mm in front of VP and 15 mm above HP. B is 15 mm in front of VP and is above HP. Top View of AB is 50 mm long. Find the front view length and the true inclinations.* (UQ)

1. Mark a′ 15 mm above XY and a 50 mm below XY.
2. **Top View:** B is 15 mm in front of VP. So, draw a horizontal 15 mm below XY to represent the locus of b. Top View length is 50 mm. Hence a as center and 50 mm as radius, draw an arc to cut the locus of b at b. **ab** is the Top View.
3. **True Inclination with HP (θ):** Now, the Top View is made parallel to XY. i.e., the line AB is made parallel to VP. A is fixed, θ is fixed. So, the Front View corresponding to this Top View ab_1 gives True Length and True Inclination θ with HP.

   a as center and ab as radius draw an arc to meet the horizontal drawn through a at b_1.

   True Length is given as 75 mm. Hence, a′ as center and 75 mm as radius, draw an arc to intersect the projector drawn from b_1 at b_1'.

   Join $a'\ b_1'$. Measure **θ = 48°**.
4. **Front View:** Draw a horizontal line through b_1' to represent the locus of b′. From b draw a projector to cut the locus of b′ at b′. Join **a′b′ = 66 mm.**
5. **True Inclination with VP (φ):** *See Fig. 17.* a as center and 75 mm as radius cut a point b_2 on the locus of b. Join a b_2. Measure **φ = 28°.**
6. ***Check:*** Mark b_2'. b_2' should lie on the projector drawn through b_2 as shown.

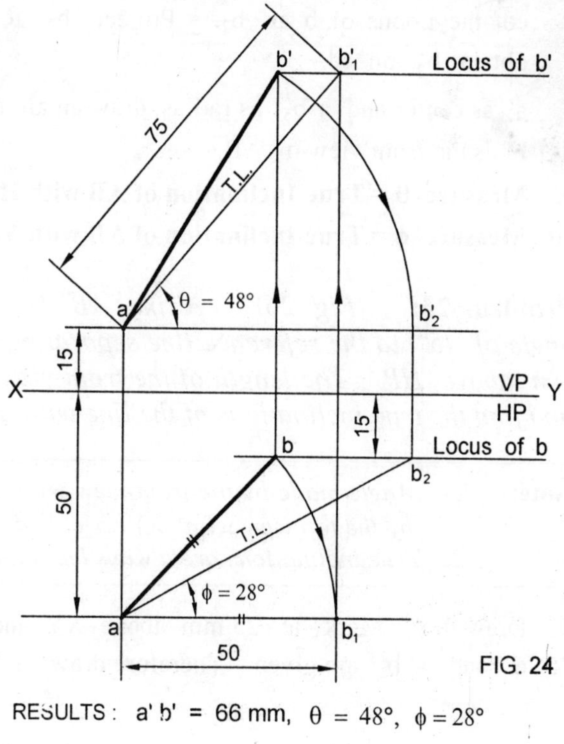

FIG. 24

RESULTS : a′b′ = 66 mm, θ = 48°, φ = 28°

Problem 21: *(Exercise) A line measuring 80 mm long has one of its ends 60 mm above HP and 20 mm in front of VP. The other end is 15 mm above HP and in front of VP. The front view of the line is 60 mm long. Draw the top view.*

Problem 22: *(Fig. 25) A line AB, 65 mm long, has its end A 20 mm above HP and 25 mm in front of VP. End B is 40 mm above HP and 65 mm in front of VP. Draw the projections of AB. Find its inclinations with HP and VP.* (UQ)

1. Mark a′ 20 mm above XY. Draw the Locus cd of a′ (front view of A).
2. Mark a 25 mm below XY. Draw the Locus ef of b (top view of B).
3. Draw the locus of b′ (front view of B) at a height of 40 mm from XY.
4. Similarly, draw the Locus of b (top view of end B) 65 mm below XY.
5. a′ as center and 65 mm as radius, draw an arc to cut the Locus of b′ at b'_1. Project b'_1 to fix b_1 on ef.
6. a as center and a b_1 as radius, draw an arc to cut the Locus of b at b. Join a b. a b is the top view of the line AB.
7. a as center and radius 65 mm, draw an arc to cut the Locus of b at b_2. Project b_2 to obtain b'_2 on cd.
8. a′ as center and a′b'_2 as radius, draw an arc to cut the Locus of b′ at b′. Join a′ b′. a′ b′ is the front view of AB.
9. **Measure θ = True Inclination of AB with HP = 18°.**
10. **Measure φ = True Inclination of AB with VP = 38°.**

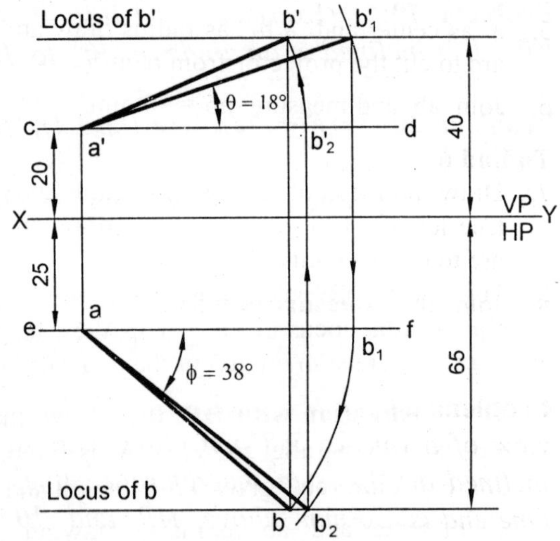

FIG. 25 RESULTS: θ = 18° and φ = 38°

Problem 23: *(Fig. 26) A line AB 100 mm long has its front view inclined at an angle of 45° to the reference line separating the views. The end A is in VP and 25 mm above HP. The length of the front view is 60 mm. Draw the top view of the line and find the true inclinations of the line with HP and VP.* (UQ)

Note: 1. *Angle made by the front view with XY is called Apparent Angle α. Angle made by the top view with XY is called Apparent Angle β.*
2. *True inclinations are always less than the apparent inclinations.*

---

1. Draw XY. Mark a′ 25 mm above XY and a on XY.
2. α and a′ b′ are given. Therefore draw a′ b′ = 60 mm at α = 45° to XY.

To obtain the Top View and θ

3. Draw the locus of b'. a' as center and true length 100 mm as radius draw an arc to cut the locus of b' at b'_1.

   Join $a'b'_1$. **Measure θ = 25°.**
4. From b'_1 draw a projector to cut XY at b_1. ab_1 is the top view of $a'b'_1$.
5. a as center and ab_1 as radius draw an arc to cut the projector from b' at b.
6. Join ab and measure ab = 90 mm.

To find ϕ

7. Draw the locus of b. a as centre and true length 100 mm as radius draw an arc to cut the locus of b at b_2.
8. Join ab_2. **Measure ϕ = 53°.**

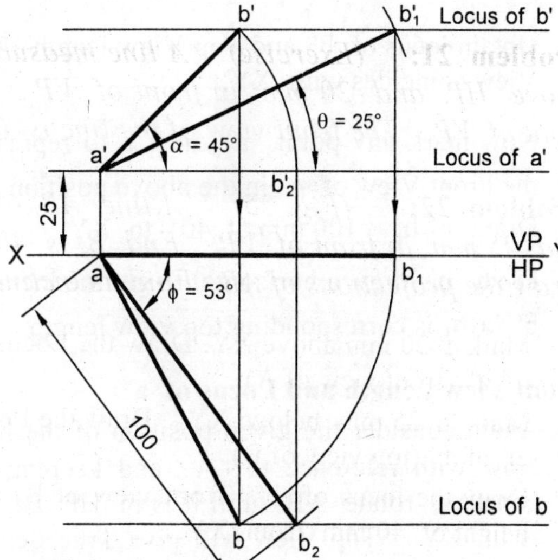

FIG. 26

RESULTS: θ = 25°, ϕ = 53° and ab = 90 mm

Problem 24: *(Fig. 27) The top view of a line is 65 mm long and is inclined at 30° to the reference line. One end is 20 mm above HP and 10 mm in front of VP. The other end is 60 mm above HP and is in front of VP.*

Draw the projections and find the true length of the line and its true inclinations to HP and VP. (UQ)

Results: TL = 76 mm, θ = 32° and ϕ = 25°.

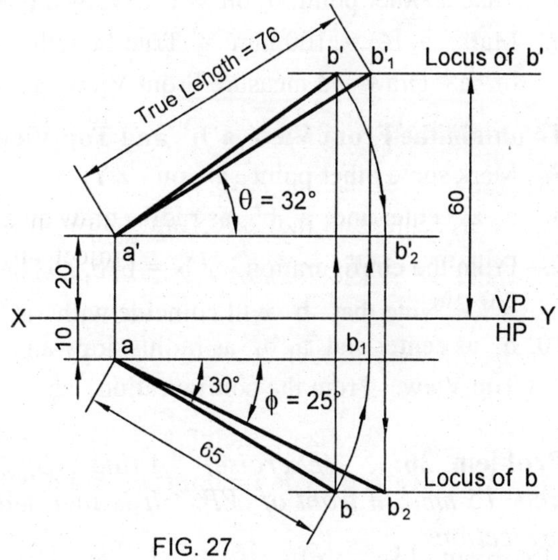

FIG. 27

Problem 25: *(Fig. 28) Draw the projections of a straight line AB of 100 mm length when one of its ends is touching the VP and the other end touching HP. The angles of inclination with HP and VP are 40° and 50° respectively.* (UQ)

Note: *When θ + ϕ = 90°, the projections of the line lie in the same projector.*

Top View Length and Locus of b′

- Let A be touching HP and B touching VP. So a' and b is on XY.
- With reference to B, rotate the line parallel to VP, keeping θ constant. Now end A will be on XY. Front View gives the True Length and θ.

- Obtain locus of b′ and Top View length. Top View coincides with XY.

1. Now mark any point a_1' on XY to represent the Front View of A in the above position.
2. Draw $a_1' b_1'$ = 100 mm at 40° to XY.
3. From b_1' draw a horizontal to get the locus of b′. $\mathbf{a_1b_1}$ is corresponding top view length.

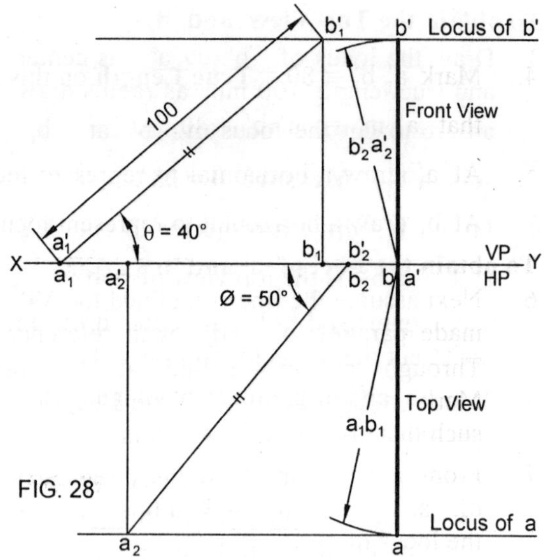

FIG. 28

Front View Length and Locus of a

4. Again consider the given position of the line. Now with reference to A and keeping ϕ constant, rotate AB until it is in HP *i.e.,* B is in XY. Top View will give True Length and ϕ. Front View coincides with XY. Now obtain the locus of a and Front View length.
5. Take another point b_2 on XY. Draw a line from b_2 at an angle of 50° (ϕ) to XY.
6. Mark $a_2 b_2$ = 100 mm = True Length. Draw a horizontal through a_2 to represent locus of a. Draw and measure Front View $a_2' b_2'$ on XY corresponding to $a_2 b_2$.

To obtain the Front View a′b′ and Top View a b

7. Mark some other point a′ on XY.
8. a′ as center and $a_2' b_2'$ as radius draw an arc to cut the locus of b′ at b′.
9. From the configuration, a′ b′ = $b_1 b_1'$. Therefore draw **a′ b′** (Front View) perpendicular to XY. Note that b will coincide with a′.
10. b as center and $a_1 b_1$ as radius draw an arc to cut the locus of a at a. Now **a b** is the Top View. From the configuration, ab = $a_2' a_2$. So ab is perpendicular to XY.

Problem 26: *(Exercise) A line AB, 65 mm long has its end A, 15 mm above HP and 15 mm in front of VP. It is inclined at 55° to HP and 35° to VP. Draw its projections.* (UQ)

Problem 27: *(Fig. 29) The* ***mid-point M*** *of a straight line AB is 60 mm above HP and 50 mm in front of VP. The line measures 80 mm long and inclined at an angle of 30° to HP and 45° to VP. Draw its projections.* (UQ)

1. Mark m′ above XY and m below XY.

To obtain the loci of a′ and b′

2. First assume the line inclined to only one plane HP and made parallel to VP with reference to M. Thus obtain the loci of b′ and a′. The whole line is now 50 mm in front of VP. Top View corresponding to this is parallel to XY and passes through m.
3. Through m′ draw a line 30° (θ) to XY.

4. Mark $a_1' b_1'$ = 80 = **True Length** on this line such that $a_1' m' = m' b_1'$ = 40.

5. At a_1' draw a horizontal to represent locus of a'.

   At b_1' draw a horizontal to represent locus of b'.

To obtain the Loci of a and b

6. Next assume the line is inclined to VP only and made parallel to HP with reference to M. Through m draw a line at 45° (φ) to XY. Mark $a_2 b_2$ = 80 mm = **True Length** on this line such that $a_2m = mb_2$ = 40 mm.

7. From a_2 draw a horizontal to represent the locus of a. From b_2 draw a horizontal to represent the locus of b.

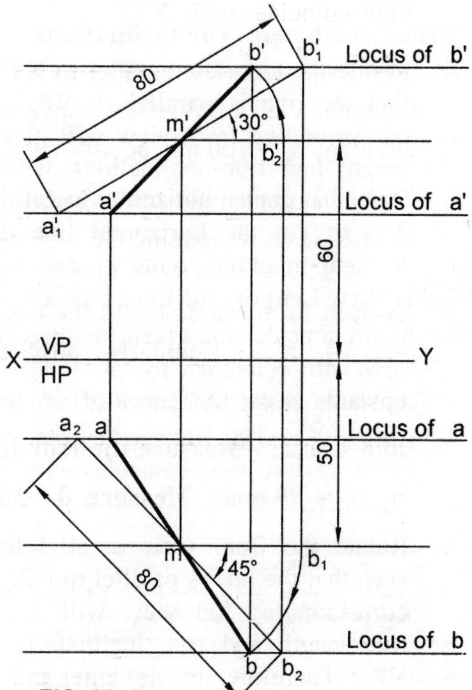

FIG. 29

Consider the projection of right half of the line. Projection of remaining half is symmetrical.

8. Now consider the Front View $m'b_1'$ = 40 mm inclined at 30° to XY. From b_1' draw a projector to cut the horizontal drawn through m at b_1. mb_1 is the Top View length for half of the straight line.

9. m as center and mb_1 as radius draw an arc to cut the locus of b at b. From b draw a projector to cut the locus of b' at b'.

10. Join bm. Extend this till it touches the locus of a at a. (amb is Top View of AB).

11. Join b'm'. Extend this till it touches locus of a' at a'. (a'm'b' is Front View of AB).

Check: *Now a' and a will be on the same projector.*

Problem 28: *(Fig. 30) The projections of a line measure 80 mm in the top view and 70 mm in the front view. The* ***mid-point*** *of the line is 45 mm in front of VP and 35 mm above HP. One end is 10 mm in front of VP and nearer to it. Draw the projections. Find true length and true inclinations with reference planes.* (UQ)

Top and Front Views

1. Draw front view m' of mid-point M, 35 mm above XY. Draw top view m, 45 mm below XY. Draw the Locus of a at a distance of 10 mm below XY.

2. m as center and half the length of top view as radius (40), draw an arc to cut Locus of a at a. Join am. Extend it to b such that am = mb. Through b draw the Locus of b parallel to XY. Draw vertical projectors through a and b.

3. m' as center and half the length of the front view as radius (35 mm), draw arcs to cut the projector through a at a' and to cut the projector through b at b'. Now, ab and a'b' are the top and front views of the line respectively.

True Length and True Inclinations

4. Rotate the top view parallel to XY such that the line is parallel to VP. The corresponding front view will give true length and true inclinations with HP. So, m as center and ma as radius, draw arcs to cut the horizontal line drawn through m at the points a_1 and b_1.

5. Project a_1 upwards to cut the Locus of a′ at a_1'. Similarly, project b_1 upwards to cut the Locus of b′ at b_1'. Join $a_1' b_1'$. **Measure the true length** $a_1' b_1'$ = **99 mm. Measure $\theta = 36^o$.**

6. Rotate the front view parallel to XY such that the line is parallel to HP. The corresponding top view will give the true length and true inclinations with VP. Therefore, m′ as center and m′ a′ as radius, draw arcs to cut the horizontal line drawn through m′ at the points a_2' and b_2'.

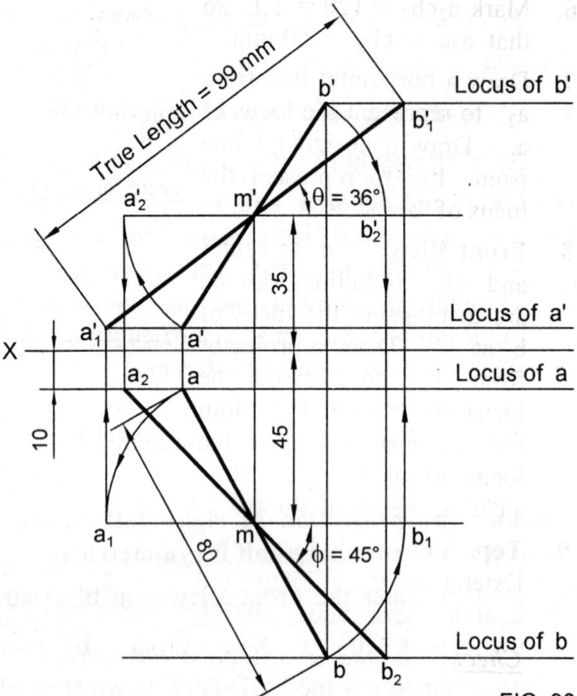

FIG. 30

RESULTS: TL = 99 mm, $\theta = 36°$ and $\phi = 45°$

7. Project a_2' downwards to cut the Locus of a at a_2. Similarly, project b_2' downwards to cut the Locus of b at b_2. Join $a_2 b_2$. Measure the true length $a_2 b_2$ = **99 mm** (***Checked***). Also, **measure $\phi = 45^o$.**

Problem 29: *(Fig. 31) A line AB 120 mm long is inclined at 45° to HP and 30° to VP. Its* ***mid-point*** *C is in VP and 20 mm above HP. The end A is in third quadrant and B is in first quadrant. Draw the projections of the line.* (UQ)

To obtain the Loci of a′ and b′ and Top View length

1. Projections of mid-point C in the VP : Mark c′ 20 mm above XY and c on XY.
2. Keeping θ constant and with reference to C rotate AB to lie on VP. Through c′ draw a line at 45° to XY. Mark $a_1' c' b_1'$ = 120 mm = T.L. such that $a_1' c' = c' b_1'$ = 60 mm.
3. Through b_1' & a_1' draw horizontal lines to represent the loci of b′ & a′ respectively.
4. From b_1' draw a projector to cut XY at b_1. Now cb_1 is the Top View length for half of the straight line.

To obtain the Loci of a and b

5. Keeping ϕ constant and with reference to C rotate the line parallel to HP. Now the line is 20 mm above HP. Through c draw a line at 30° to XY.

6. Mark a_2cb_2 = 120 = T.L. so that a_2c = cb_2 = 60 mm.
7. Draw a horizontal line from a_2 to represent the locus of a. Draw a horizontal line from b_2 to represent the locus of b.
8. **Front View:** c as center and cb_1 as radius draw an arc to intersect the locus of b at b. Draw a projector from b to intersect the locus of b′ at b′. Join b′c′. Extend this to cut the locus of a′ at a′. Now a′c′b′ is the Front View.
9. **Top View:** Join bc. Extend this to cut locus of a at a. Now acb is the Top View.

<u>***Check***</u>: *a and a′ will be on the same projector.*

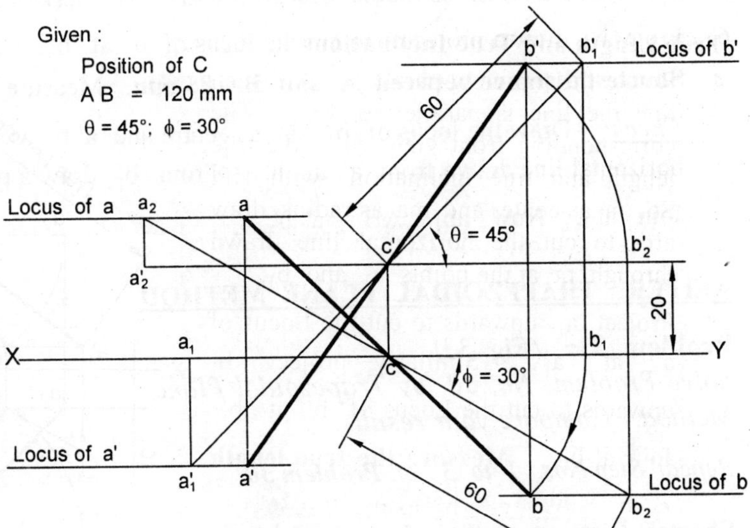

FIG. 31

<u>ROTATING LINE METHOD – True Length and True Inclinations</u>

Problem 30: *(Fig. 32) The distance between the projectors of two points A and B is 70 mm. A is 10 mm above HP and 15 mm in front of VP. B is 50 mm above HP and 40 mm in front of VP. Find the shortest distance between A and B by Rotating Line Method. Find true inclinations of AB with VP and HP.* (UQ)

Front and Top Views

1. Mark a′ above XY and a below XY. Draw another projector 70 mm from the projector of A.
2. Mark b′ 50 mm above XY and b 40 mm below XY on the second projector.
3. Join a′ b′ and ab.

Shortest Distance between A and B (= True Length of AB)

4. Through b′ draw the locus of b′.

Make top view parallel to XY. Draw the corresponding front view to get shortest distance.

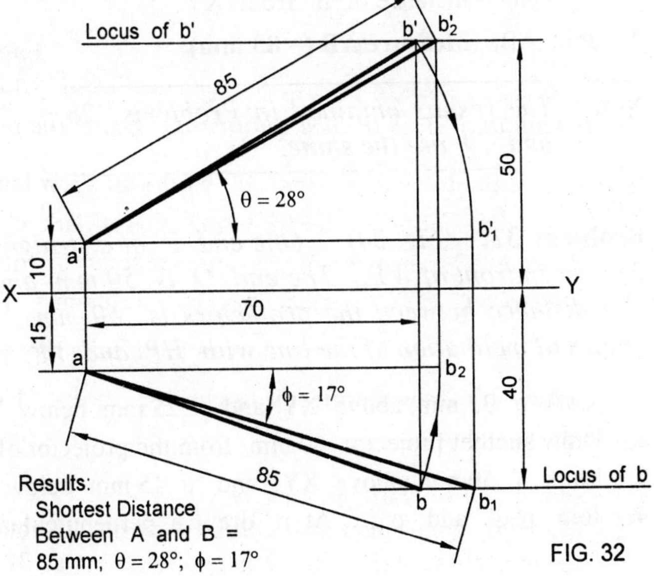

FIG. 32

5. a as center and ab as radius draw an arc to cut the horizontal drawn from a at b_2.
6. From b_2 draw a projector to cut the locus of b′ at b'_2. Join $a'b'_2$. Now $a'\,b'_2$ = AB = **Shortest distance between A and B = 85 mm**. Measure **θ = 28°** and **Φ = 17°**.

Check: Draw the locus of b. a′ as centre and a′ b′ as radius draw an arc to intersect the horizontal line drawn from a′ at b'_1. From b'_1 draw a projector to cut locus of b at b_1. Join ab_1. Now ab_1 = AB = $a'\,b'_2$.

ALITER: TRAPEZOIDAL PLANE METHOD

Problem 31: *(Fig. 33)*
Solve Problem No. 30 by Trapezoidal Plane Method. Compare your results.

*Repeat **Step Nos. 1 to 3** of **Problem 30.***

Shortest Distance (= True Length of AB)

4. At a′ draw a perpendicular to a′ b′.
5. Mark a′A = 15 mm = distance of a from XY. Similarly draw a perpendicular to a′ b′ at b′.
6. Mark b′B = 40 mm = distance of b from XY. Join AB. Now **AB = 85 mm.**

Check

7. At a draw a perpendicular to ab. Mark aA = 10 mm = distance of a′ from XY.
8. At b draw a perpendicular to ab. Mark b B = 50 mm = distance of b′ from XY.
9. Join AB. **Measure AB (= 85 mm).**

Note: *The results obtained in Problems 28 and 29 are the* ***same.***

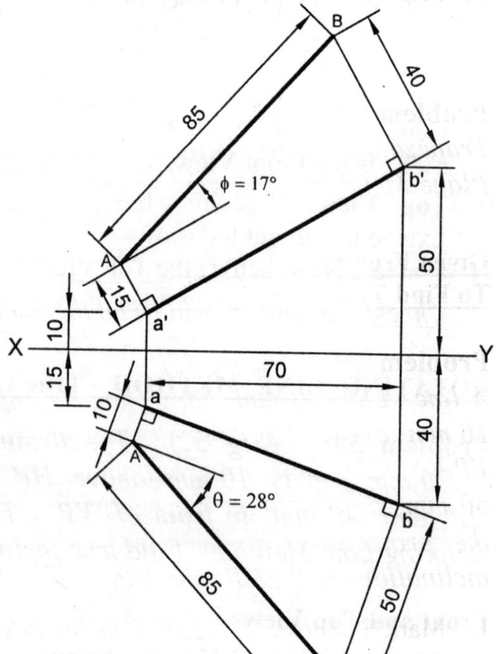

FIG. 33
RESULTS :
True Length = AB = 85 mm; θ = 28° ; ϕ = 17°

Problem 32: *(Fig. 34) One end P of a straight line PQ is 35 mm above HP and 25 mm in front of VP. The end Q is 50 mm above HP and 45 mm in front of VP. The distance between the projectors is 60 mm. Determine the true length and true angles of inclination of the line with HP and VP.* (UQ)

1. Mark p′ 35 mm above XY and p 25 mm below XY.
2. Draw another projector 60 mm from the projector of P.
3. Mark q′ 50 mm above XY and q 45 mm below XY on the second projector.
4. Join p′ q′ and p q. At p′ draw a perpendicular to p′ q′.

5. Mark p′P = 25 mm = distance of p from XY.
6. Similarly draw a perpendicular to p′q′ at q′. Mark q′Q = 45 mm = distance of q from XY.
7. Join PQ. Now **PQ = 65 mm.**

Check

8. At p draw a perpendicular to pq. Mark pP = 35 mm = distance of p′ from XY.
9. At q draw a perpendicular to pq. Mark qQ = 50 mm = distance of q′ from XY.
10. Join PQ. **Measure PQ = 65 mm.** (***Checked***)

Result: True Length = 65 mm; ϕ = 13°; θ = 18°.

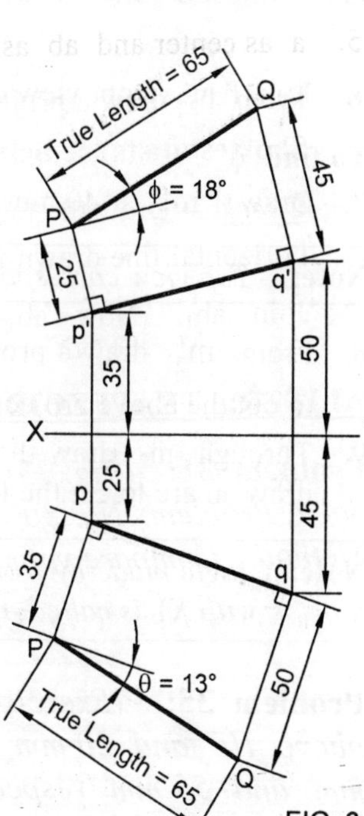

FIG. 34

Problem 33: *(Exercise) The above problem is solved by Trapezoidal Plane Method. Solve this problem by Rotating Plane Method. Compare the results.*

Given Front and Top Views,
To Find True Length and True Inclinations

Problem 34: *(Fig. 35)*
A line LM 70 mm long, has its end L 10 mm above HP and 15 mm in front of VP. Its top and front views measure 60 mm and 40 mm respectively. Draw the projections of the line. Find its inclinations with HP and VP. (UQ)

1. Mark **l′** above XY and **l** below XY.

To find θ

2. From **l** draw a line parallel to XY. Mark lm_1 = 60 mm = Top view length.

Note: *Front view corresponding to this top view will give the true length and θ.*

---

3. From m_1 draw a projector. l′ as center and true length 70 mm as radius draw an arc to cut the above projector at m_1'.
4. Join $l'm_1'$. Now measure θ = **31°**.
5. Draw the locus of m′. l′ as centre and front view 40 mm as radius draw an arc to cut the locus of m′ at m′.

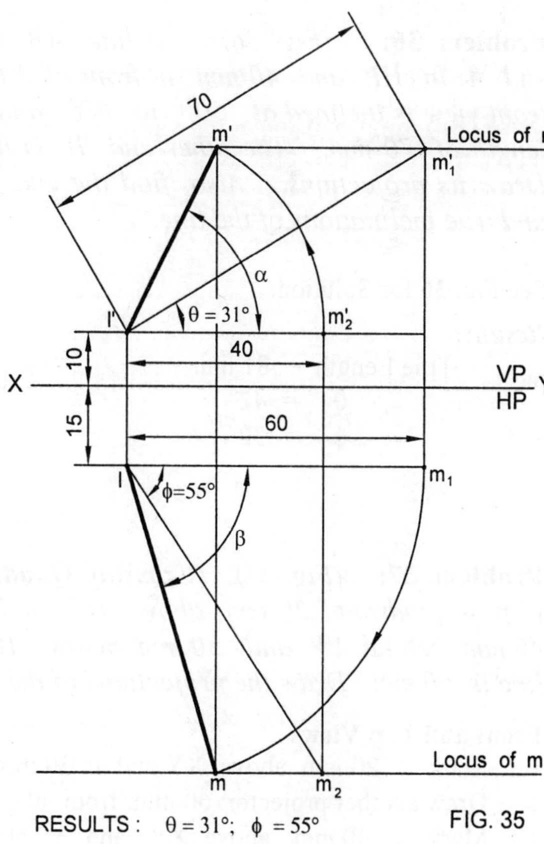

RESULTS : θ = 31°; ϕ = 55°

FIG. 35

6. Join l′ m′, front view of LM.

To find ϕ

7. Draw $l'm_2'$ = 40 mm = Front view length and parallel to XY.

Note: *Top view corresponding to this front view will give the true length and ϕ.*

---

8. From m_2' draw a projector. l as center and true length 70 mm as radius, draw an arc to cut the above projector at m_2. Join lm_2. Measure **ϕ = 55°**.
9. Through m_2 draw the locus of m. l as center and 60 mm (top view length) as radius draw an arc to cut the locus of m at m. Join lm, the top view.

Note: *Angle made by front view with XY is called Apparent Angle α. Angle made by top view with XY is called Apparent Angle β.*

Problem 35: *(Exercise) A line AB measuring 85 mm has its end A 25 mm above HP and 20 mm in front of VP. The front and top views of the line measure 70 mm and 55 mm respectively. Draw the projections of the line and determine its true inclinations.* (UQ) (Results : θ = 49° and ϕ = 35°)

Problem 36: *(Fig. 36) A line AB has its end A in HP and 40 mm in front of VP. Its front view is inclined at 50° to XY and has a length of 70 mm. The other end B is in VP. Draw its projections. Also, find the true length and true inclinations of the line.* (UQ)

See Fig. 36 for Solution.

Results:

True Length = 81 mm
θ = 42°
ϕ = 30°

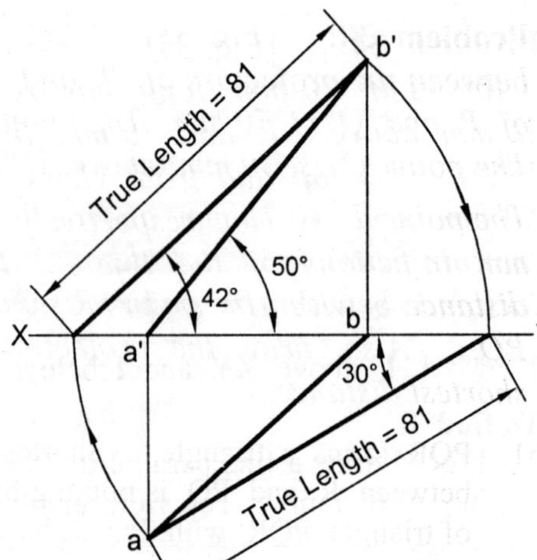

FIG. 36

Problem 37: *(Fig. 37)* ***(Crossing Quadrant Problem)*** *One end of the line UV is in first quadrant, 20 mm above HP and 30 mm in front of VP. The other end is 40 mm behind VP and 50 mm below HP. The distance between the projectors of the line is 60 mm. Draw the projections of the line and find its true length.* (UQ)

Front and Top Views

1. Mark u′ 20 mm above XY and u 30 mm below XY.
2. Draw another projector 60 mm from u′u.
3. Mark v 40 mm above XY and v′ 50 mm below XY.

4. Join u′v′ **(F.V.)** and uv **(T.V.)**.

To find True Length

5. Draw the loci of v, u′, u and v′. Rotate the top view uv to be parallel to XY and obtain uv_1.
6. The corresponding front view $u'v_1'$ gives the true length. It is obtained by drawing the projector from v_1 to cut locus of v′ at v_1'.
7. Join $u'v_1'$. Now $u'v_1'$ = **T.L. = 117 mm.**
8. **To check the True Length:** At u′ draw a perpendicular to u′v′ and mark U such that u′U = 30 mm. At v′ draw a perpendicular to u′v′ in opposite direction. Mark V such that v′V = 40 mm. Join UV. Now UV = **True Length = 117 mm.**

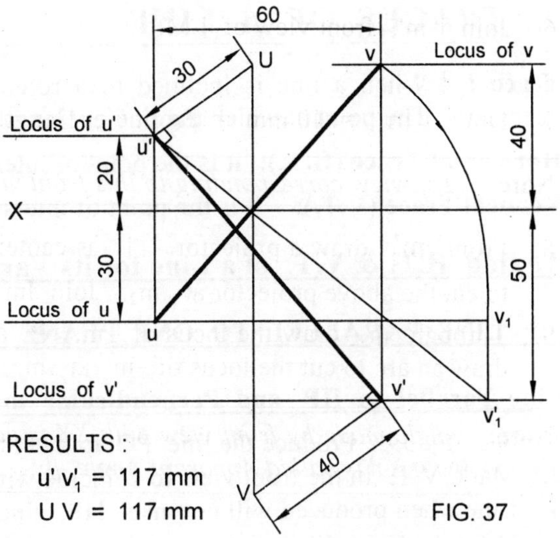

FIG. 37

Note: *True length UV is obtained by both the Trapezoid Method and Rotating Line Method. The results are compared as above.*

Problem 38: *(Fig. 38) A line PQ and a point R is situated in space. The distance between the projectors of P and R is 25 mm and the distance between the projectors of P and Q is 70 mm. The point P is 45 mm above HP and 10 mm in front of VP. The point Q is 20 mm above HP and 40 mm in front of VP.*

The point R is 15 mm above HP and 50 mm in front of VP. Find out the shortest distance between the point R and the line PQ. Also, draw the projections of the shortest distance. (UQ)

1. PQR forms a triangle. Shortest distance between R and PQ is nothing but altitude of triangle PQR with PQ as base.
2. Now the problem reduces to the determination of True Lengths of PQ, QR and RP. For this, draw the projections of each point and determine True Length of each side of the triangle by *Trapezoidal Method*.
3. Construct the triangle in its actual size separately and find the shortest distance between R and PQ.
4. *Result:* The shortest distance between R and the line PQ = RS = **32 mm.**

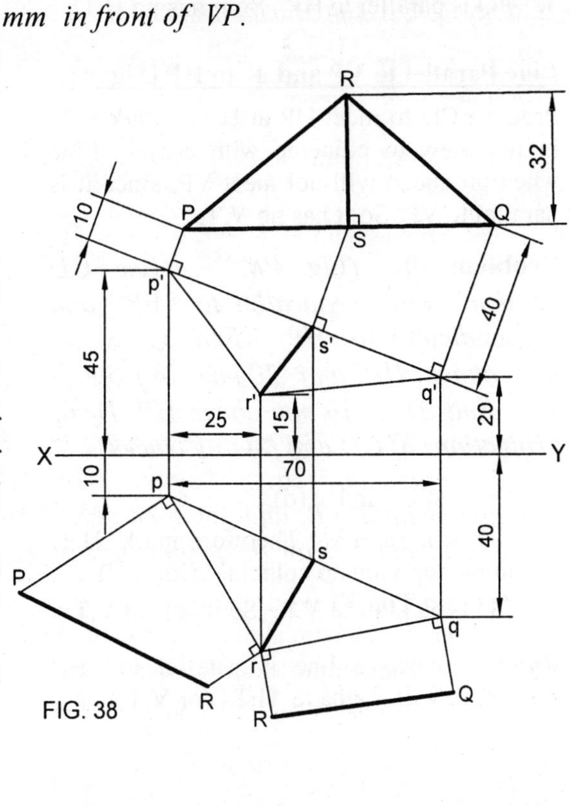

FIG. 38

9. TRACES OF A LINE

Trace : When a line is inclined to a reference plane, it will meet that plane, produced if necessary. The point at which the line or line produced meets the plane is called its *Trace*.

Horizontal Trace (H.T.): It is the point of intersection of the line or line produced with HP.

Vertical Trace (V.T.): It is the point of intersection of the line or line produced with VP.

To find H.T. & V.T. of a Line for its Various Positions with respect to HP & VP

9.1 LINE PARALLEL TO ONE PLANE AND PERPENDICULAR TO OTHER

Line Parallel to HP and Perpendicular to VP

1. *See Fig. 39.* Produce the line PQ to meet the VP at V.T.
2. Mark V.T. in the front view to coincide with p′ (q′).
3. Line when produced will not meet HP, since it is parallel to HP. Hence it has no H.T.

Problem 39: *(Fig. 39) A line PQ 25 mm long is parallel to HP and perpendicular to VP. The end Q is 10 mm in front of VP and the line is 20 mm above HP. Draw the projections of the line and find its traces.*

1. Draw p′ (q′) and pq.
2. PQ is ⊥ʳ to VP. Hence mark V.T. in the front view to coincide with p′(q′).
3. PQ is parallel to HP. So it has no H.T.

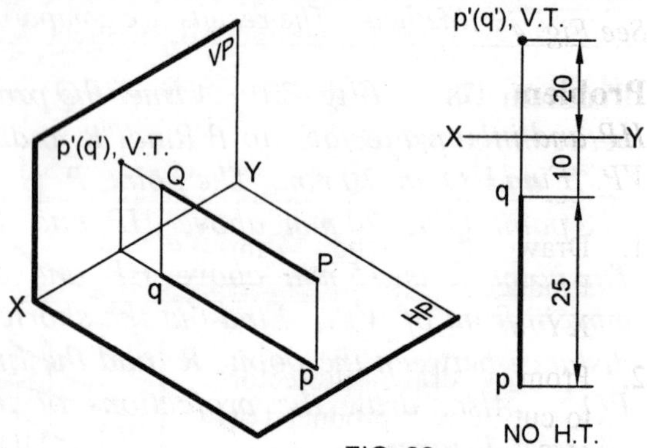

FIG. 39

Line Parallel to VP and ⊥ʳ to HP [Fig. 27]

Produce CD to meet HP at H.T. Mark H.T. in top view to coincide with c(d). Line when produced will not meet VP, since it is parallel to VP. So it has no V.T.

Problem 40: *(Fig. 40) A line CD 25 mm long is parallel to VP and perpendicular to HP. End C is 35 mm above HP and 20 mm in front of VP. End D is 10 mm above HP. Draw projections of CD and find its traces.*

1. Draw c′d′ and c (d).
2. CD is ⊥ʳ to HP. Therefore mark H.T. in the top view to coincide with c(d).
3. CD is parallel to V.P. So it has no V.T.

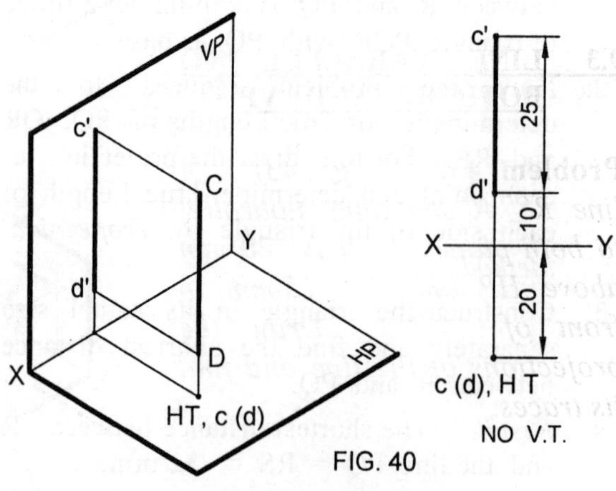

FIG. 40

Rule: When a line is parallel to HP (or VP) it will not have H.T. (or V.T.).

9.2 LINE PARALLEL TO ONE PLANE AND INCLINED TO THE OTHER

Line Parallel to VP and Inclined to HP

See Fig. 41. Produce line BA to meet HP at H.T. Front view of H.T. is h′ on XY.

Problem 41: *(Fig. 41) A line AB 40 mm long is parallel to VP and inclined at 35° to HP. A is 15 mm above HP and 20 mm in front of VP. Draw the projections of AB. Find its traces.*

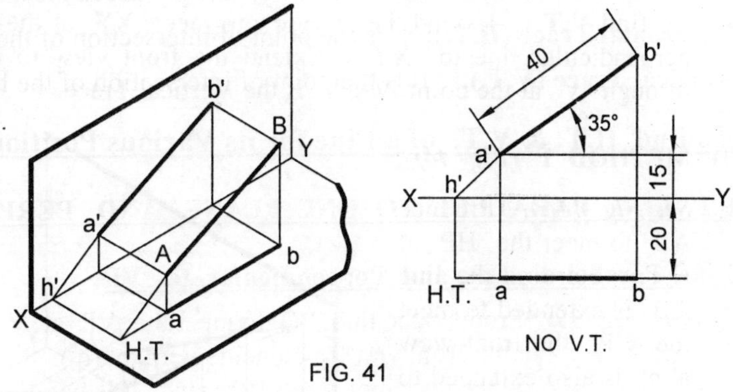

FIG. 41

1. Draw a′b′ and ab. Produce b′a′ to meet XY at h′.
2. From h′ draw a projector to cut ba produced at H.T. AB is parallel to VP. Hence it has no V.T.

Line Parallel to HP and Inclined to VP

See Fig. 42. Produce the line ML to meet VP at V.T. Top View of V.T. is v on XY.

Problem 42: *(Fig. 42) Draw the projections of line LM 40 mm long, parallel to HP and inclined at 35° to VP. The end L is 20 mm above HP and 15 mm in front of VP. Find its traces.*

1. Draw l′m′ and l m. Produce ml to meet XY at v.
2. From v draw a projector to cut l′ m′ produced at V.T. LM is parallel to HP. Hence it has no H.T.

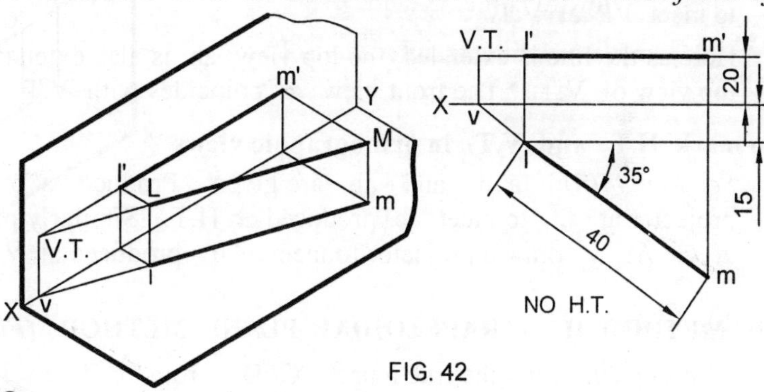

FIG. 42

9.3 LINE PARALLEL TO BOTH HP AND VP

Problem 43: *(Fig. 43) A line RS 40 mm long is parallel to both planes. It is 20 mm above HP and 15 mm in front of VP. Draw the projections of the line and find its traces.*

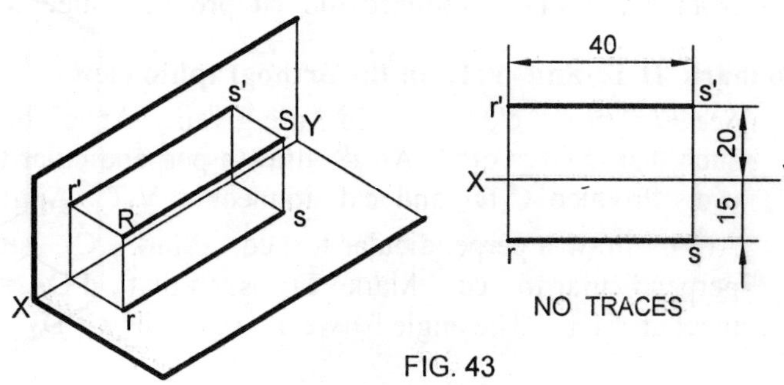

FIG. 43

9.4 LINE INCLINED TO BOTH HP AND VP

- **To find H.T.:** Draw the given top and front views of the line. Extend the front view to meet XY at the point h. At the point h, draw a perpendicular to XY. Extend the top view to meet the perpendicular line drawn through h at H. H is the Horizontal Trace.
- **To find V.T.:** Extend the top view to meet XY at the point v. At the point v, draw a perpendicular line to XY. Extend the front view to meet the perpendicular line drawn through v at the point V. V is the Vertical Trace.

(i) METHOD I *(Fig. 44)*

1. *See Fig. 44(i).* Produce AB to meet the HP at H.T. Here as the line AB is extended to meet the HP, the front view a′ b′ is also extended to meet XY at h′.

   Now h′ is front view of H.T.

   Top view h coincides with H.T. itself and hence not shown.

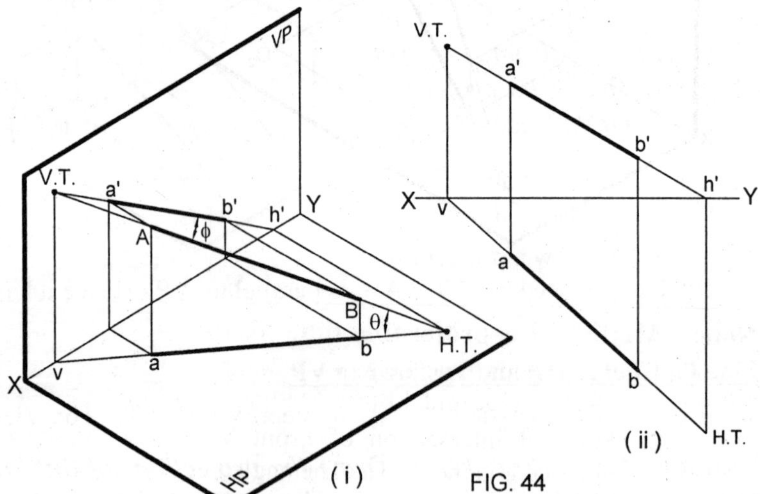

FIG. 44

2. Similarly produce AB to meet VP at V.T.

   Here as the line is extended, the top view ab is also extended to meet XY at v. v is the top view of V.T. The front view v′ coincides with V.T. and hence not shown.

To mark H.T. and V.T. in orthographic views

3. *See Fig. 44(ii).* a′ b′ and ab are given. Produce a′ b′ to meet XY at h′. Draw a projector at h′ to meet ab produced at H.T. Similarly produce ab to meet the XY line at v. At v draw a projector to meet a′ b′ produced at V.T.

(ii) METHOD II TRAPEZOIDAL PLANE METHOD *(Fig. 45)*

1. *See Fig. (i).* Find the true length $C_1 D_1$ from c′ d′ by Trapezoidal Plane Method. $C_1 D_1$ produced and c′ d′ produced meet at the V.T. Similarly find the true length $C_2 D_2$ from cd. $C_2 D_2$ produced and cd produced meet at the H.T.

To mark H.T. and V.T. in the orthographic views

2. *See Fig. (ii).* c′d′ and cd are given. At c′ draw a perpendicular to c′d′. Mark C_1 such that $c'C_1 = o_1c$. At d′ draw a perpendicular to c′d′. Mark D_1 such that $d'D_1 = o_2d$. Produce C_1D_1 and c′d′ to meet at V.T. Angle between C_1D_1 and c′d′ is ϕ.
3. At c draw a perpendicular to cd. Mark C_2 such that $c\,C_2 = o_1c'$. At d draw a perpendicular to cd. Mark D_2 such that $d\,D_2 = o_2\,d'$. Produce $C_2 D_2$ and cd to meet at H.T. The angle between cd and $C_2 D_2$ is θ.

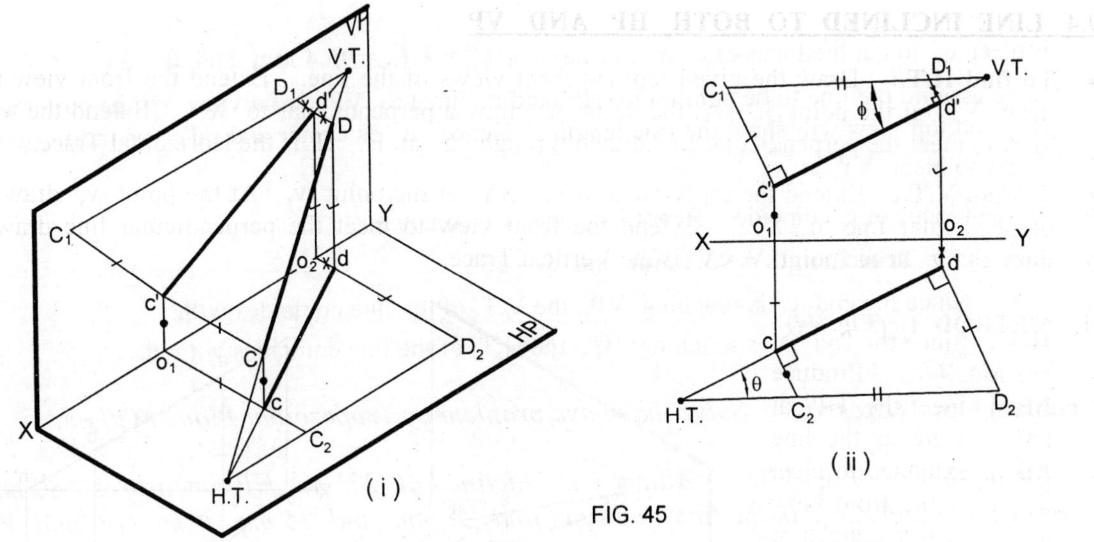

FIG. 45

Note: Method II is preferable than Method I for the following reasons:

1. θ, ϕ and True Length are simultaneously determined along with H.T. and V.T. by the Trapezoidal Plane Method. It is not so in the case of Method I.
2. Point of intersection of Front View and the corresponding True Length line, if produced, is the V.T. The angle between these gives True Inclination ϕ.
3. Point of intersection of Top View and the corresponding True Length line, if produced, is the H.T. The angle between these gives True Inclination θ.
4. When the projections of the line are perpendicular to XY i.e., when the sum of its inclinations with HP and VP is 90°, it is possible to find the traces by Method II only (Refer **Problem 42**).

Problem 44: *(Fig. 46) The top view of a line CD has a length of 80 mm and makes 30° with the horizontal. The end C is in VP and 52 mm above HP. The end D is in HP.*
Draw the projections of the line. Find true length and true inclinations with HP and VP. Mark its traces on the two planes. (UQ)

1. Mark c in XY and c′ 52 mm above XY.
2. Draw cd = 80 mm at 30° to XY.
3. Since D is in HP, project d to d′ in XY.
4. Join c′d′, the front view.
5. Assume CD to be parallel to VP and inclined to HP. d as center and dc as radius, draw an arc to cut the locus of d at c_1.

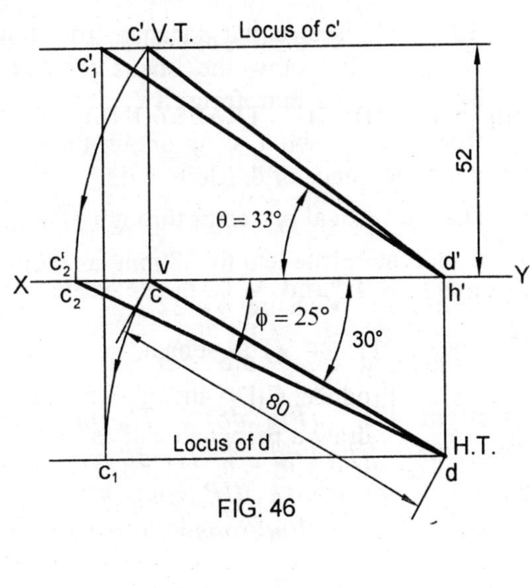

FIG. 46

6. Project c_1 to cut the locus of c' at c_1'. Join $d'c_1'$ = **T.L. = 95.4 mm** and **$\theta = 33^\circ$.**
7. If we assume the line to be parallel to HP and inclined to VP, front view will be parallel to XY and top view will show the true length. Hence, d' as center and $d'c'$ as radius, draw an arc to meet XY at c_2'.
8. XY and locus of c coincide. Hence, c_2' and c_2 coincide.
9. Join c_2d = **True Length = 95.4 mm** and **$\phi = 25^\circ$.**
10. **V.T.:** Since the end C is touching VP, the V.T. of the line coincides with c'.
11. **H.T.:** Since the end D is touching HP, the H.T. of the line coincides with d.

Problem 45: *(Exercise) Solve the above problem by Trapezoidal Plane Method.*

Problem 46: *(Fig. 47) A line CD, inclined at 25° to HP, measures 80 mm in top view. End C is in the first quadrant and 25 mm and 15 mm from HP and VP respectively. End D is at equal distances from both reference planes. Draw the projections, find true length and true inclination with VP. Locate the traces.* (UQ)

1. Mark c' and c. From c draw a line $c\,d_1$ of length 80 mm (top view length) parallel to XY.
2. Through c', draw a line at an angle of 25° to XY to meet the vertical projector from d_1 at the point d_1'.
3. Draw the Locus of d' through d_1'.
4. Joint $c'\,d_1'$ = True Length = 88 mm.
5. Height of C above HP is equal to height of d_1' from XY = 62 mm.
6. Since D is at equal distance from both HP and VP, draw the Locus of d at a distance of 62 mm from XY.
7. Rotate $c\,d_1$ about c to obtain the point d on the Locus of d. Join c d.
8. Draw a vertical projector through d to meet the Locus of d' at d'. Join $c'\,d'$.
9. c as center, true length 88 mm as radius, draw an arc to cut the Locus of d at d_2.
10. Join $c\,d_2$. Measure **$\phi = 32^\circ$.**
11. Locate HT and VT as shown.

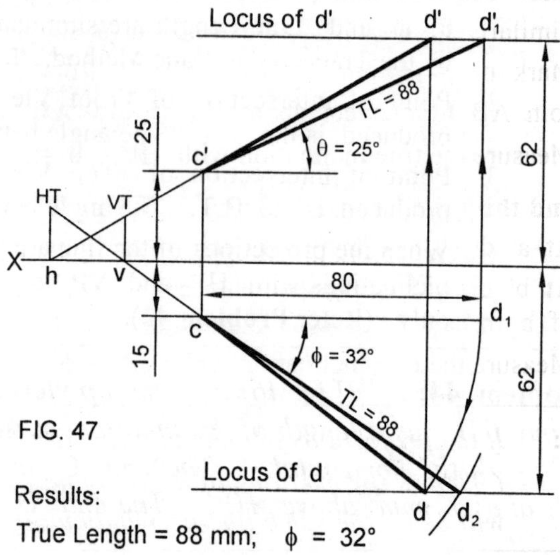

FIG. 47

Results:
True Length = 88 mm; ϕ = 32°

Problem 47: *(Fig. 48) The projections of the ends of a line AB are on the same projector. The end A is 30 mm below HP and 15 mm behind VP. The end B is 35 mm above HP and 40 mm in front of VP. Determine its true length, the traces and the inclinations with the reference planes.* (UQ)

1. Mark a′ and a from the given data.
2. Mark b′ and b on the same projector from the given data. Now ab is the top view and a′ b′ is the front view.

To find True Length, θ and H.T. by Trapezoidal Plane Method

3. From a draw a line perpendicular to ab.

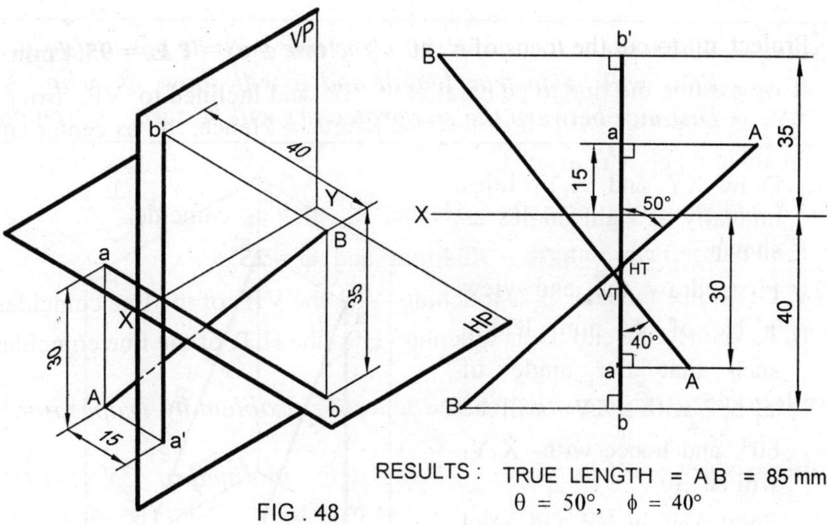

FIG . 48

RESULTS : TRUE LENGTH = A B = 85 mm
θ = 50°, φ = 40°

4. Mark aA = 30 mm = distance of a′ from XY.
5. Similarly, at b draw a perpendicular to ab in opposite direction.
6. Mark bB = 35 mm = distance of b′ from XY.
7. Join AB to intersect ab at H.T. Now **AB = 85 mm = True Length.**
8. Measure the true inclination with HP = **θ = 50°**. (θ is the angle between ab and AB)

To find the True Length, φ and V.T. by Trapezoidal Plane Method

9. At a′ draw a′A perpendicular to a′ b′ and equal to 15 mm = distance of a from XY.
10. At b′ draw b′B perpendicular to a′b′ in opposite direction and equal to 40 mm = distance of b from XY. Join AB to intersect a′ b′ at V.T. Now **AB = 85 mm = True Length.**
11. Measure the true inclination with VP = **φ = 40°**. (φ is angle between a′ b′ and AB).

Rule: *If the projections of ends of a line are on the same projector,* $\theta + \phi = 90^o$.
(That is, projections of a line lying on a plane perpendicular to both HP and VP will be on the same line perpendicular to XY. This happens when $\theta + \phi = 90^o$. *)*

Problem 48: *(Exercise) The projections of the ends of a line EF are on the same projector. The end E is 10 mm above HP and 15 mm in front of VP. The end F is 35 mm above HP and 45 mm in front of VP. Draw the projections. Find the true length, true inclinations and traces.* (UQ)

Results: *True Length. = 39 mm,* $\theta = 40^o$ *and* $\phi = 50^o$.

Problem 49: *(Fig. 49) A line AB 65 mm long has its end A in VP and end B in HP. The line is inclined to HP by* 60° *and to VP by* 30°*. Draw the projections of the line. Locate its traces.* (UQ)

Note: *Whenever (θ + ϕ) = 90°, the line must be parallel to the Profile Plane. Its End View will be in true length and it will show θ with XY and ϕ with X_1Y_1. Distance between the end projectors will be zero. Given data satisfy this.*

1. Draw XY and X_1Y_1 lines mutually at right angles as shown.
2. First, draw the end view $a_1' b_1'$ of 65 mm length such that the angle of $a_1' b_1'$ with XY will be 60° and hence with X_1Y_1 will be 30°.
3. Point A is in VP and point B is in HP (given). Hence, mark a and b′ on XY.
4. Find a′ and b by projection method as shown by the arrows.
5. **V.T.** coincides with point a′. **H.T.** coincides with b.

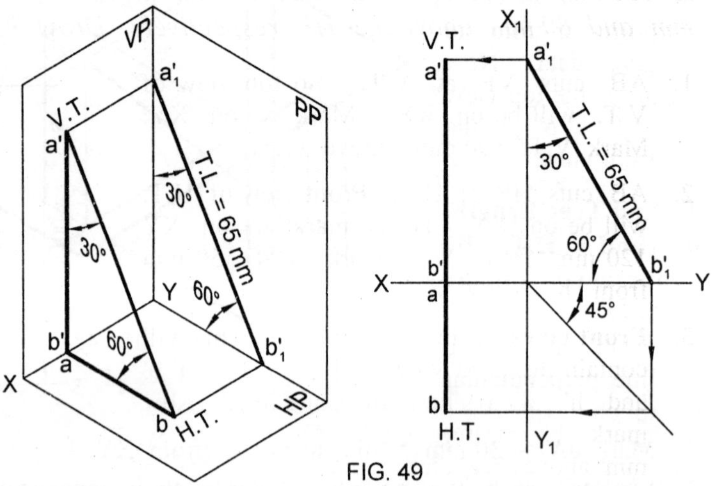

FIG. 49

Problem 50: *(Fig. 50) A straight line AB has its end A 20 mm above HP and 25 mm in front of VP. The other end B is 60 mm above HP and 65 mm in front of VP. The ends of the line are on the same projector. Draw its projections. Find the true length, true inclinations of the line with HP and VP. Also, mark the traces.* (UQ)

See Fig. 50 for Solution.

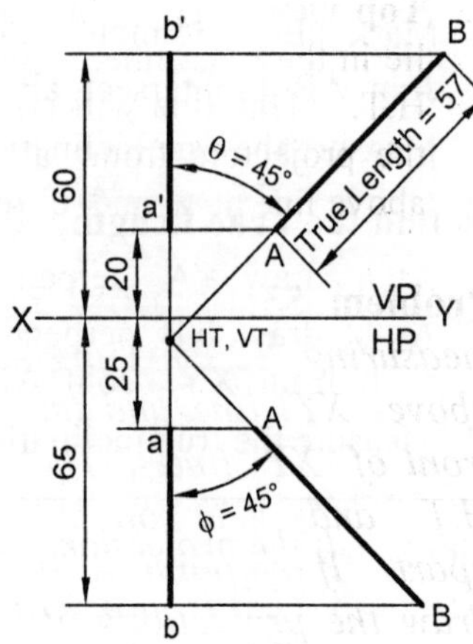

FIG. 50

Result:

True Length = 57 mm; θ = 45°; ϕ = 45°.

Problem 51: *(Fig. 51) A line AB of 75 mm long is inclined at 30° with VP and lies in a plane perpendicular to both VP and HP. Its end A is in VP and the end B is in HP. Draw the projection of the line AB. Locate its traces.* (UQ)

See Fig. 51. Locate the traces and complete the problem.

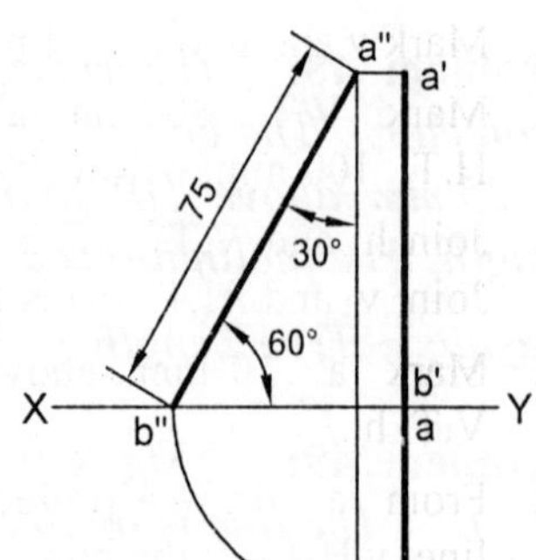

FIG. 51

Given the Traces, to Find the Projections

Problem 52: *(Fig. 52) The projections on the XY line of the horizontal and vertical traces of a straight line AB in the first quadrant are 120 mm apart. The V.T. is 100 mm above XY and H.T. 50 mm in front of XY. The points A and B are 30 mm and 80 mm above the HP respectively. Draw the projections.* (UQ)

1. AB cuts VP at V.T. So top view of V.T. will be on XY. Mark v on XY. Mark V.T. 100 mm above v.
2. AB cuts HP at H.T. Front view of H.T. will be on XY. Hence mark h′ on XY 120 mm from v. Mark H.T. 50 mm from h′ below XY.
3. **Front view:** Join V.T. and h′. This will contain the front view a′ b′. V.T., a′ b′ and h′ always lie in the same line. Now mark b′ 80 mm above XY and a′ 30 mm above XY on this line.
4. **Top view:** H.T., ab and v will always lie in the *same* line. Therefore join v and H.T. This line will contain ab. Draw the projectors from a′ and b′ to cut the above line at a and b respectively.

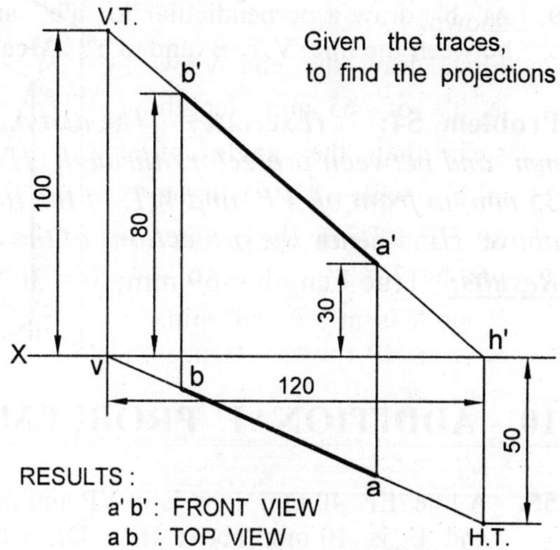

FIG. 52

Problem 53: *(Fig. 53) A line AB measuring 12 cm long has its V.T. 8 cm above XY line and the H.T. 10 cm in front of XY line. The projections of the H.T. and V.T. on XY line are 11 cm apart. If the point A is 1 cm above HP, draw the projections of AB and find its true inclinations.* (UQ)

1. Mark v and h′ on XY line 110 mm apart.
2. Mark V.T. 80 mm above v and mark H.T. 100 mm below h′.
3. Join h′ and V.T. This line contains a′b′. Join v and H.T. This line contains ab.
4. Mark a′ 10 mm above XY on the line V.T. h′.
5. From a′ draw a projector to intersect the line v H.T. at the point a.

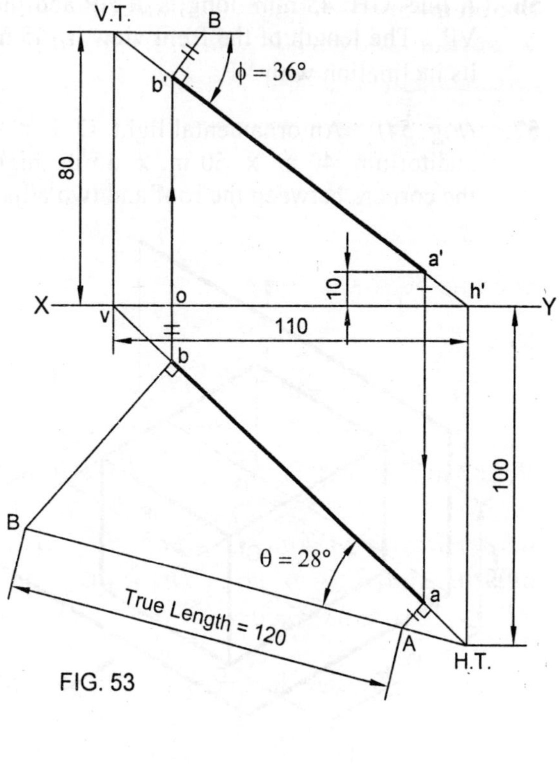

FIG. 53

6. From a draw a perpendicular to v H.T., such that aA = 10 mm = distance of a′ from XY.
7. θ is the angle between ab and AB. ab and AB always meet at H.T. Therefore, join A and H.T. and extend this line. Mark AB = 120 mm = True Length on this line. **Measure θ = 28°.**
8. From B draw a perpendicular to v H.T. line to meet at b. From b draw a projector to meet the V.T. h′ line at b′ and XY at o.
9. At b′ draw a perpendicular to a′b′ and mark B such that b′B = ob. ϕ is the angle between the line V.T. B and a′b′. **Measure ϕ = 36°.**

Problem 54: *(Exercise) The distance between the end projectors of a line PQ is 55 mm and between projectors through H.T. and V.T. is 80 mm. The H.T. of the line is 35 mm in front of VP and V.T. of the line is 50 mm above HP. The end P is 10 mm above HP. Draw the projections of the line and find its inclinations with HP and VP.*
Results: True Length = 69 mm; θ = 30°; ϕ = 20.5°.

10. ADDITIONAL PROBLEMS – II

55. A line EF 40 mm long is in VP and inclined to HP. The top view measures 30 mm. The end E is 10 mm above HP. Draw the projections of the line. Determine its inclination with HP. [*Results :* θ = 42°]

56. A line GH 45 mm long is in HP and inclined to VP. The end G is 15 mm in front of VP. The length of the front view is 35 mm. Draw the projections of the line. Determine its inclination with VP. [*Results :* ϕ = 38°.]

57. *(Fig. 54)* An ornamental light O is placed 10 m above the floor and in the center of an auditorium 40 m x 50 m x 35 m high. Determine graphically its distance from one of the corners between the roof and two adjacent walls. (UQ)

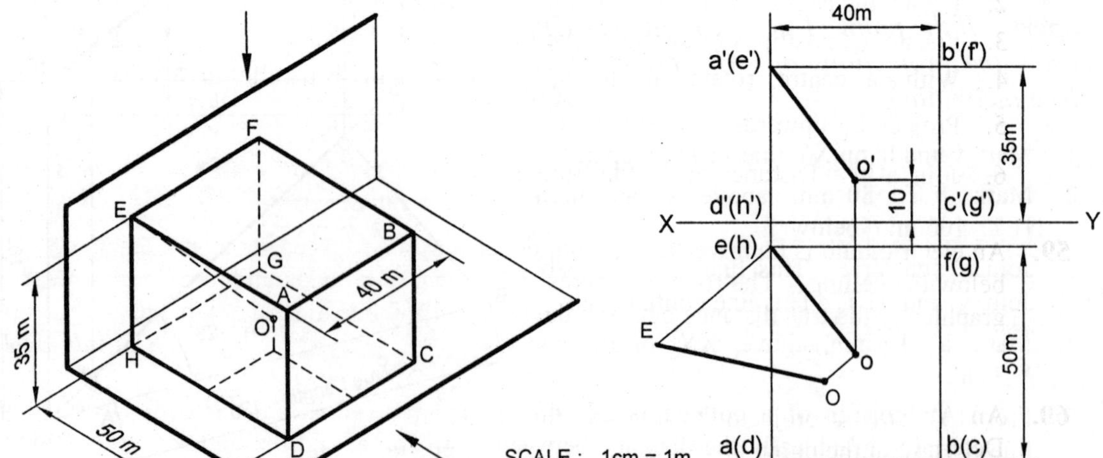

SCALE : 1cm = 1m

FIG. 54

RESULT : OE = 40.5m

1. *See Fig. 54.* The projections of the line joining the light O and a corner between the roof and the adjacent walls are drawn first.
2. Then, by Trapezoidal Plane Method, the distance between the light and that corner is determined as shown.

58. *(Fig. 55)* A fan is hanging in the center of a room 4 m x 4.5 m x 4 m high. The center of the fan is 0.6 m below the ceiling. Determine graphically the shortest distance of the fan from one of the corners of the floor. (UQ)

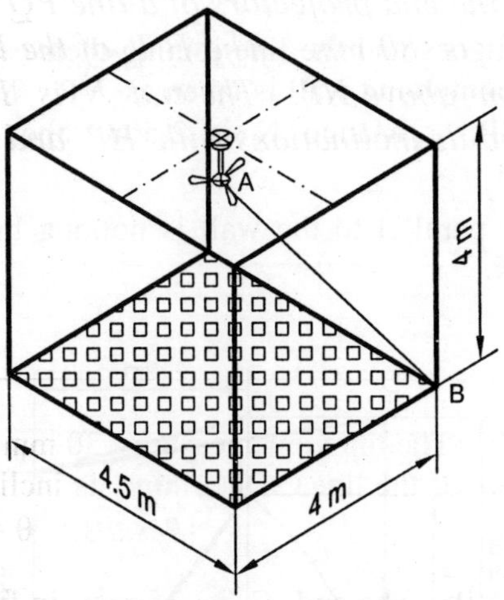

RESULT : SHORTEST DISTANCE OF THE FAN FROM THE CORNER OF THE FLOOR = a' b'$_1$ = 4.54 m

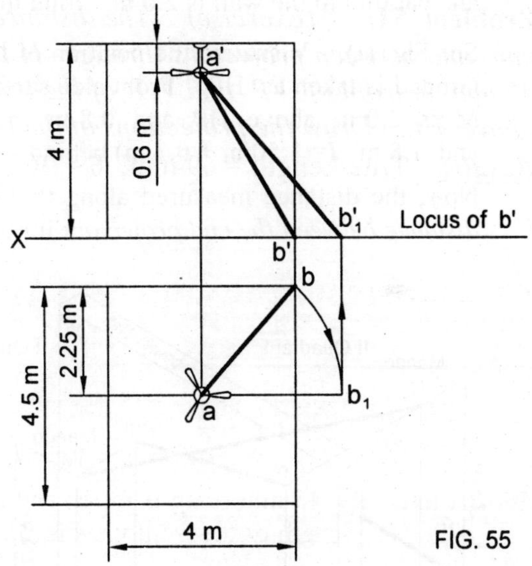

FIG. 55

1. Take a suitable Scale. Draw top and front views of the room as shown.
2. Locate the positions of the fan in the top view and front view as a and a′.
3. Locate the positions of corner of the floor as b and b′. Join ab and a′ b′.
4. With a as centre, rotate ab to b_1 such that a b_1 is parallel to XY.
5. Project b_1 upwards to meet the locus of corner b′ of the floor at b_1' in front view.
6. Join a′ b_1' = Distance of the fan from the corner of the floor = **4.54 m.**

59. An electric bulb is hung vertically from the center of a terraced room at a height of 0.5 m below the ceiling. The room measures 5 m long, 4 m wide and 3 m high. Determine graphically the true distance between the bulb and any one of the corners below. (UQ)

(Result: 4.1 m)

60. An Auditorium of a college is of 48 m length, 42 m width and 36 m height. Determine graphically the distance between a top corner and the bottom corner diagonally opposite to it. (UQ)

(Result: 73 m)

61. A fan is hanging in the center of a room of 4 m x 3 m x 3 m height. The center of the fan is 0.75 m below the ceiling. The switch of this fan is on 3 m x 3 m size wall at the center height and 0.5 m from the adjacent wall. Find the distance between the fan center and the switch. Select a suitable scale. [*Result:* 2.36 *m*] (UQ)

62. *(Fig. 56)* Two mangoes on a tree are 2.0 m and 2.5 m respectively above the ground and 0.8 m and 1.50 m from a 0.3 m thick compound wall of a bungalow, but on the opposite sides of the wall. The distance between the mangoes, measured along the ground and parallel to the wall is 2.0 m. Find the real distance between the two mangoes. (UQ)

See Fig. (i). Visualize the position of the two mangoes and the line joining their centers. Ground is taken as HP. Front side surface of the compound wall is taken as VP. Mango M is 2.0 m above HP and 0.8 m in front of VP. Mango N is 2.5 m above HP and 1.8 m (= 1.50 m + 0.3 m) behind VP.

Now, the distance measured along the ground and parallel to the wall is nothing but the *distance between the end projectors* for the line MN.

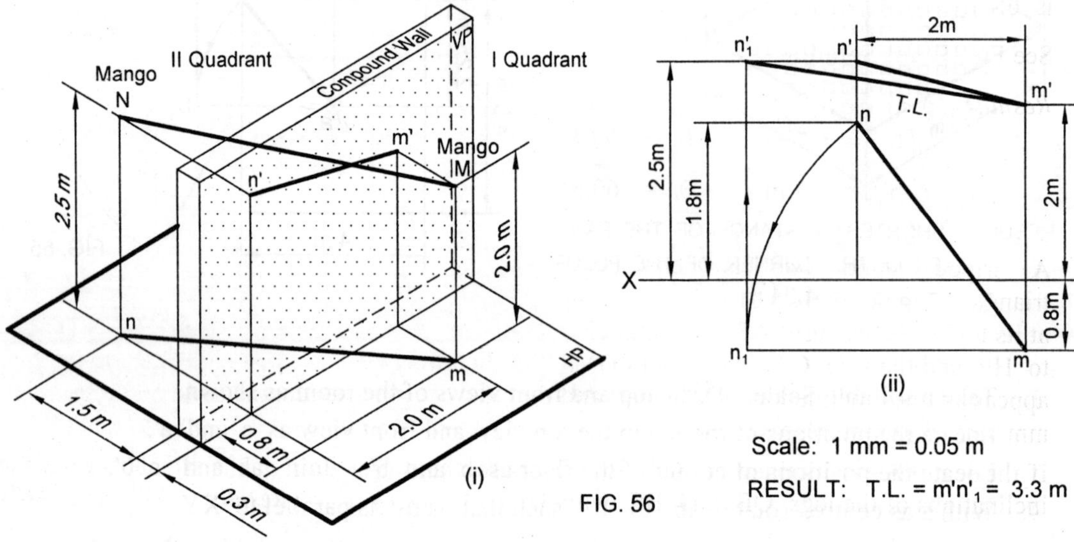

Scale: 1 mm = 0.05 m

FIG. 56 RESULT: T.L. = m'n'$_1$ = 3.3 m

1. *See Fig. (ii).* Draw two vertical projectors 2.0 m apart.
2. Mark m and m′ on one projector and n′ and n on another projector.
3. Make the top view mn parallel to XY to obtain mn$_1$ and get its corresponding front view m′n$_1'$. m′n$_1'$ is the real distance between the two mangoes = **3.3 m.**

63. Two apples A and B on an apple tree are 3 m and 2 m respectively above the ground. Apple A is 0.8 m and 1.2 m from the adjacent compound walls P and Q respectively while the corresponding distances of the apple B are 4 m and 8 m.

Draw the projections of the apples if the two walls are at right angles to each other. Find the true distance between the apples. [*Result :* AB = 7.6 *m]* (UQ)

64. *(Fig. 57)* A chimney of a hostel kitchen is 6.4 m high and 0.8 m in diameter. It is supported by 3 guy wires which appear in the top view to be inclined at equal angles to each other. The ends of the wires are fixed to the top of the chimney. The other ends are anchored to the ground at distances of 2.4 m, 3.2 m and 4.0 m from the center of the chimney.

Determine graphically the true lengths and inclinations of the wires with ground. (UQ)

Draw top view first. c is the center of chimney. cp, cq and cr show top view of the three wires. Project front view. c′ is 6.4 m above XY.

See Fig. 57 for Solution.

Results: $p'a' = 7.0$ m $\quad \theta_p = 66.5^\circ$

$q_1'b' = 6.7$ m $\quad \theta_q = 72.5$

$r_1'b' = 7.3$ m $\quad \theta_r = 60.5^\circ$

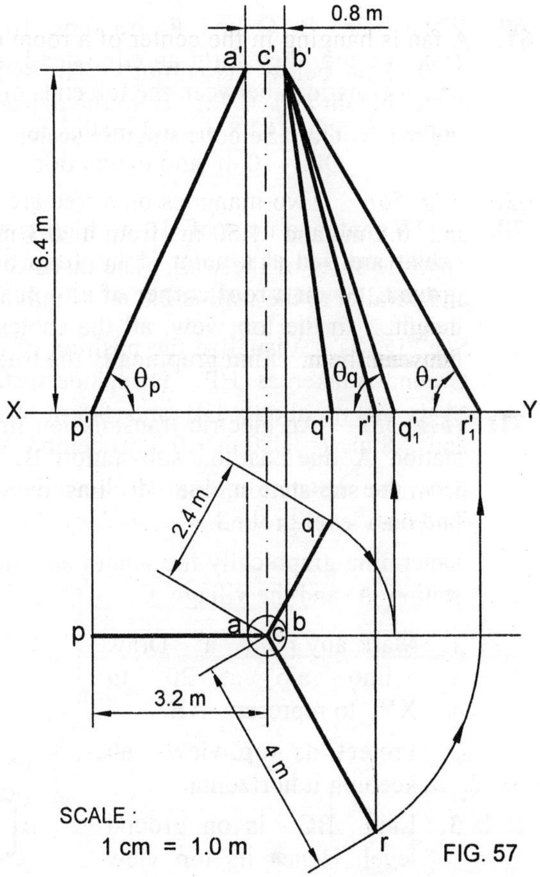

FIG. 57

65. A tripod stand has an equilateral triangular frame ABC of side 100 mm at its top. The triangle ABC is parallel to HP and the side CA is parallel to VP. In the top views the three legs pa, qb and rc appear as lines inclined at 120° apart and they are having true lengths of 500 mm, 550 mm and 600 mm respectively.

If the leg PA is inclined at 60° to HP, draw the projections of the tripod. Find the inclinations of the legs QB and RC with HP. (UQ) (*Results:* 51° and 46°)

66. Two legs of the divider PQ and PR are seen at 135° between them in their end elevation and plane. Leg PQ is perpendicular to VP. Draw the projections and find the real angle between two legs of the divider. (UQ) (*Result: 125°*)

67. Two unequal legs AB and AC, hinged at A, make an angle of 130° between them in their front and top views. Leg AB is parallel to both HP and VP. Determine the real angle between them. (UQ) (*Result: 121°*)

68. A bent lever PQR appears as an angle having 120° in between them in both front and top views. One of its arms PQ is parallel to both HP and VP. Assuming that both the arms PQ and QR have equal lengths in both front and top views, find the true angle between them. (UQ) (*Result:* True Angle between the arms PQ and QR = 111.5°)

69. Three poles P, Q and R are 6 m, 16 m, and 10 m heights respectively. When viewed from the top, they form an equilateral triangle of side 8 m. Determine the distances and true inclinations between the top ends of the poles. (UQ)

Results: PQ = 12.8 m and inclination = 51°
QR = 10 m and inclination = 37° RP = 9 m and inclination = 27°

70. A TV antenna mast of 10 m high is supported by three steel cables R, S and T. The cables are tied at a point 1 m from the top of the mast. Cable R is anchored on the ground, S on a roof corner of a 4 m high building and T on a compound wall of 2 m height. In the top view, all the cables are of the same length of 6 m and make 120° between them. Find graphically the true lengths and inclinations of the cables.

71. *(Fig. 58)* An electric transmission line laid along an uphill from a hydroelectric power station A due East to a sub-station B, is 4 km long and has a slope of 20°. Another line from the sub-station, due 30° East of South (E 30° S) to a village C is 3 km long, but is laid on a level ground.

Determine graphically the length and slope of a proposed telephone line joining the power station A and the village C. (UQ)

1. Mark any point a′. Draw a line a′b′ at 20° to XY, to represent AB.
2. Project its top view ab keeping it horizontal.
3. Line BC is on ground level. Hence its top view bc represents 3 km.
4. Draw bc 30° East of South, that is, (90°+30°) = 120° inclined to ab.
5. Project c to c′ making b′c′ horizontal. a′c′ and ac are the front and top views of the line AC.
6. Find true length $a'c_1'$ and the angle θ which are respectively the length and slope of the telephone line AC. **Measure θ = 13°.**

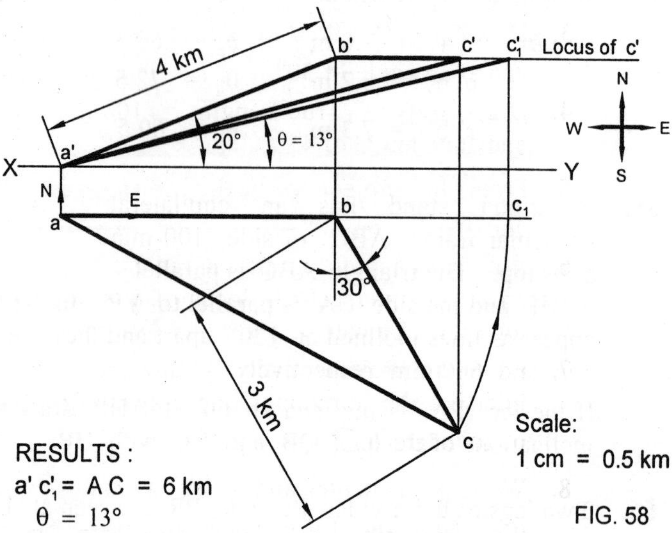

FIG. 58

72. Two roads start at a common point A. Road AB is 3 km long and bears 50° SE on a downward slope of 30°. Road AC is 1.8 km long and bears 65° NE on a downward slope of 20°. What is the true length and slope of the connecting road B and C ?

(Results : BC = 2.82 km and Slope = 18°)

73. (Fig. 59) Draw the projections of a line PQ 100 mm long inclined at 30° to HP and 45° to VP. Point P is 20 mm above HP and in VP. Also, determine the apparent lengths and inclinations. (UQ)

Apparent Length: When a line is inclined to a reference plane, its view or length on that plane appears to be shortened. This shortened length is called Apparent Length. Thus, if a line is parallel to HP and inclined to VP, the top view gives the true length and the front view gives the apparent length.

Apparent Inclination: If the view of a line on a reference plane gives true length, it gives true inclination with respect to the other reference plane. In other words, if the top view of a line gives true length, then its inclination with XY will be the true inclination with VP. On the other hand, if the view of a line on HP gives apparent length, its inclination with VP gives apparent inclination.

Note: Apparent Lengths are always less than the corresponding true length. True inclinations are always less than the corresponding apparent inclinations.

1. Mark p′ 20 mm above XY.
2. Project p′ on XY and mark p.
3. Draw $p'q_1'$ = True Length = 100 mm and at an angle of 30°.
4. Draw $p\,q_2$ = True Length = 100 mm and at an angle of 45°.
5. From q_1' project vertically downwards and mark q_1 on XY.
6. With p as center and pq_1 as radius, draw an arc to intersect the locus of q in the top view to locate q.
7. From q_2 project vertically upwards to intersect the horizontal line drawn from p′ to locate q_2'.
8. With p′ as center and $p'q_2'$ as radius, draw an arc until it intersects the locus of q in the front view to obtain q′.
9. Draw p′ q′ and p q to represent the front and top views respectively.
10. Measure the Apparent Length in Front View = **p′ q′ = 70.7 mm** and Apparent Length in Top View = **p q = 86.6 mm**.
11. Measure the Apparent Inclination in Front View = **α = 45°** and Apparent Inclination in Top View = **β = 55°**.

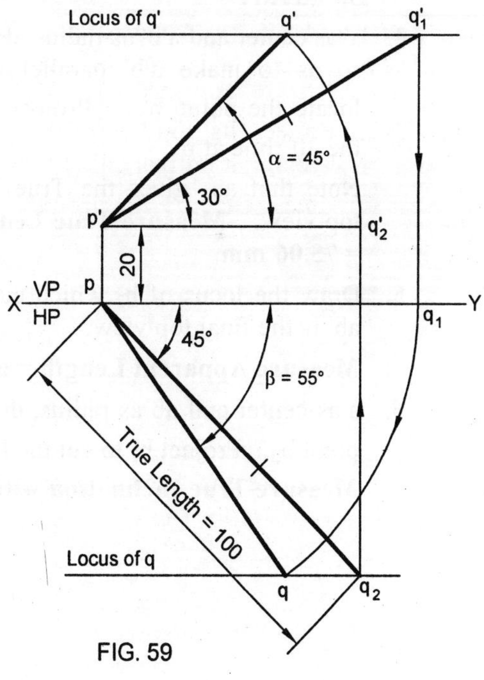

FIG. 59

74. (Fig. 60) A line AB has its end A 25 mm above HP and 20 mm in front of VP. The end B is 50 mm above HP and the line is inclined at 30° to VP. The distance between the end projectors of the line when measured parallel to the line of intersection of HP and VP is 60 mm.

Draw the projections. Find the apparent lengths, apparent inclinations, true length and true inclination with HP. (UQ)

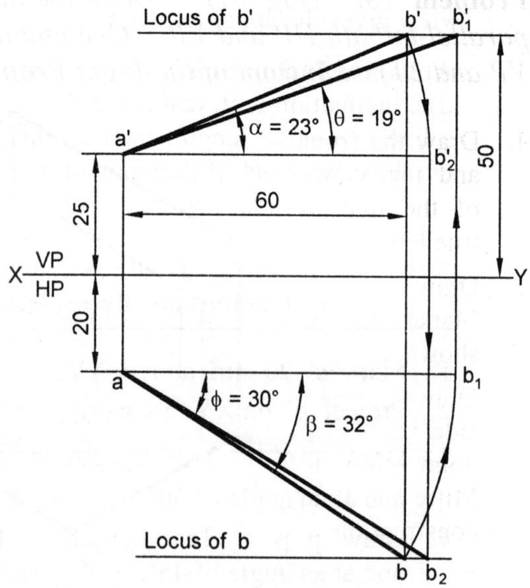

FIG. 60

1. Draw two projectors 60 mm from each other. Mark a 20 mm below XY and a′ 25 mm above XY.
2. Draw the Locus of b′ 50 mm above XY. Right side projector cuts the locus of b′ at b′. Join a′ b′.

   Measure Apparent Length = a′b′ = 65 mm and Apparent Inclination = α = 23°.
3. At the point a, draw a line at an angle of 30° to XY.
4. a′ as center and a′b′ as radius, draw an arc so as to make a′b′ parallel to XY to locate the point b'_2. Project b'_2 to cut the 30° line at b_2.

   Note that ab_2 gives the True Length in top view. **Measure True Length = AB = 75.06 mm.**
5. Draw the locus of b_2 which cuts the right side projector to give the final top view b. ab is the final top view.

   Measure Apparent Length = ab = 70.8 mm and Apparent Angle = β = 32°.
6. a as center and ab as radius, draw an arc so as to make ab parallel to XY to locate the point b_2. Project b_2 to cut the locus of b′ at b'_1. Joint a b'_1.

   Measure True Inclination with HP = θ = 19°.

11. SIDE VIEWS OF STRAIGHT LINES

11.1 LINE PARALLEL TO BOTH HP AND VP

Note: If a line is parallel to both HP and VP, it will be perpendicular to the Profile Plane (PP).

Problem 75: *(Fig. 61) Draw the three principal views of a line 50 mm long, placed parallel to both HP and VP. One end of the line is 40 mm above HP, 35 mm in front of VP and 20 mm in front of the Right Profile Plane (RPP).*

1. Draw the front and top views of the line in true length.
2. Draw X_1Y_1. Mark o as shown.
3. Draw the left side view as a point using the Mitre Line construction.

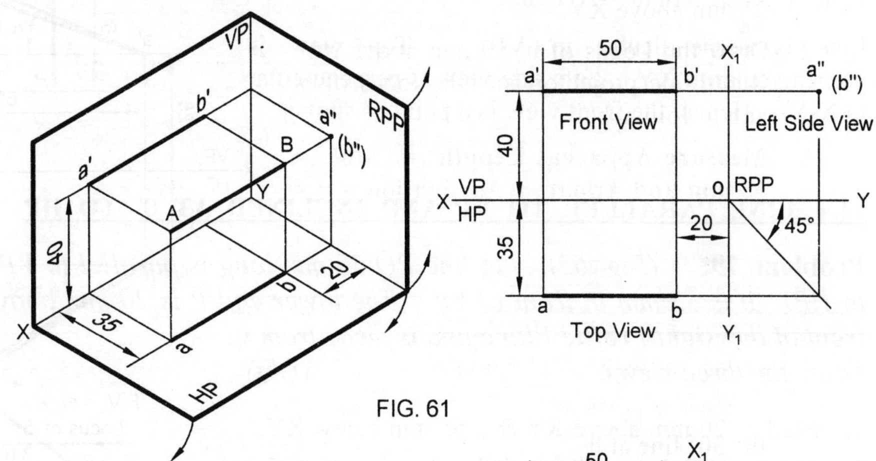

FIG. 61

Problem 76: *(Fig. 62) A line AB, 40 mm long, is perpendicular to the PP. The end A is 20 mm below HP, 30 mm behind VP and 10 mm to the left of PP. Draw the three views.*

Note: Line AB is in Third Quadrant. See Fig. 62.

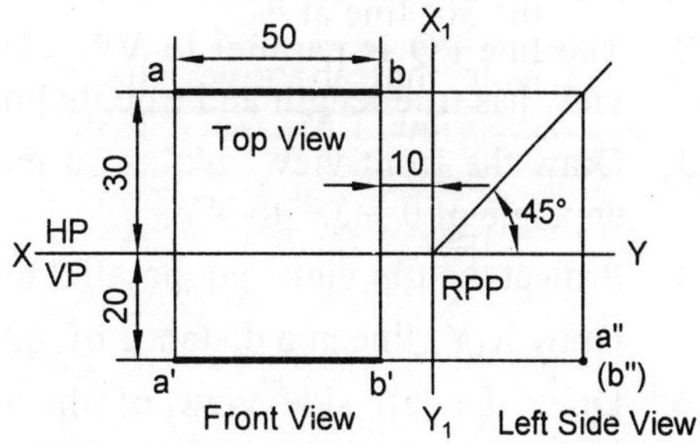

FIG. 62

11.2 LINE PARALLEL TO VP AND PERPENDICULAR TO HP

Problem 77: *(Fig. 63) A line 50 mm long is parallel to VP and perpendicular to HP. It is 40 mm in front of VP and 35 mm in front of R.P.P. The lower end of the line is 20 mm above HP. Draw the three views.*

Line is parallel to VP and RPP. Hence, the front and left side views are in true length = 50 mm.

Line is perpendicular to HP. Front view is perpendicular to XY. Left side view is parallel to X_1Y_1. Hence, the top view is a point as shown.

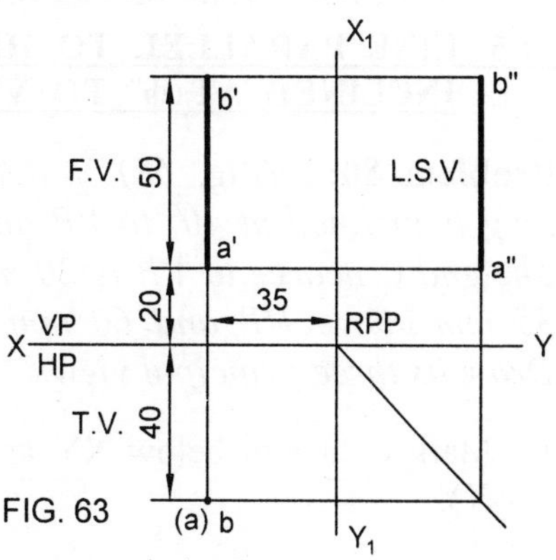

FIG. 63

11.3 LINE PARALLEL TO HP AND PERPENDICULAR TO VP

Problem 78: *(Fig. 64) A line 50 mm long is parallel to HP and perpendicular to VP. Its rear end is 20 mm in front of VP. The line is 35 mm above HP and 30 mm in front of Right Profile Plane. Draw the three views.*

Line is parallel to both HP and RPP. Hence, the top view is in true length = 50 mm.

Line is perpendicular to VP. Top view is perpendicular to XY. Left side view is perpendicular to X_1Y_1. Hence, the front view is a point as shown.

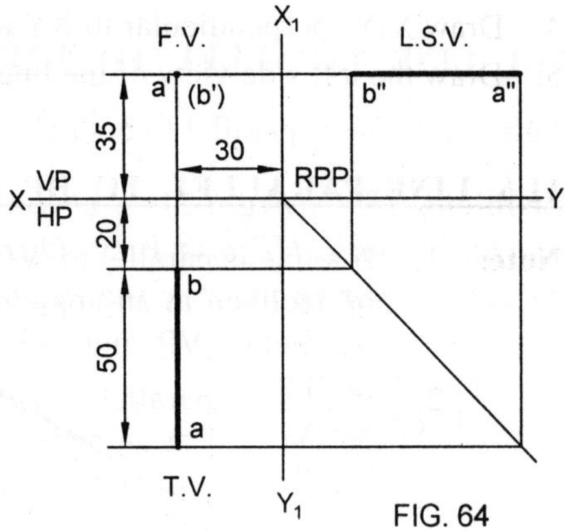

FIG. 64

11.4 LINE PARALLEL TO VP AND INCLINED AT θ° TO HP

Problem 79: *(Fig. 65) A line PQ 50 mm long is parallel to VP and inclined at 35° to HP. It is 55 mm in front of VP. The lower end P is 20 mm above HP and 60 mm in front of the Right Profile Plane and is away from it. Draw the three views.* (UQ)

1. Mark p′ 20 mm above XY & p 55 mm below XY.
2. The line PQ is parallel to VP. Hence, its front view has true length and true inclination.
3. Draw the front view p′q′ = 50 mm, inclined at an angle of θ = 35° to XY.
4. Project the top view pq parallel to XY.
5. Draw X_1Y_1 line at a distance of 60 mm from p′.
6. Draw the left side view of the line, using the Mitre Line construction as shown.

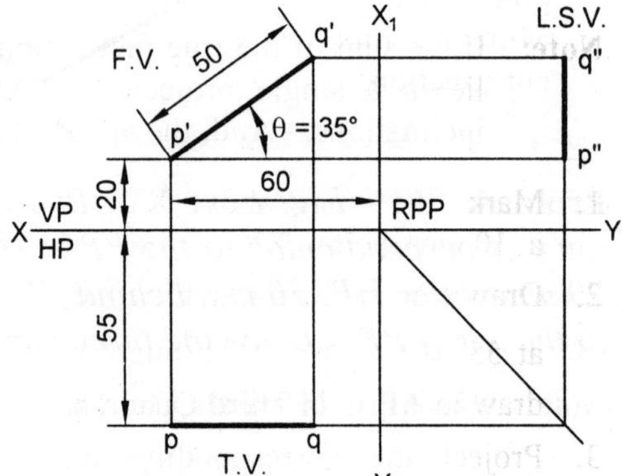

FIG. 65

11.5 LINE PARALLEL TO HP AND INCLINED AT φ° TO VP

Problem 80: *(Fig. 66) A line CD 50 mm long is inclined at 40° to VP and parallel to HP. The end C nearer to VP is 20 mm in front of VP, 35 mm above HP and 60 mm in front of RPP. Draw its three principal views.* (UQ)

1. Mark c 20 mm below XY and c′ 35 mm above XY.

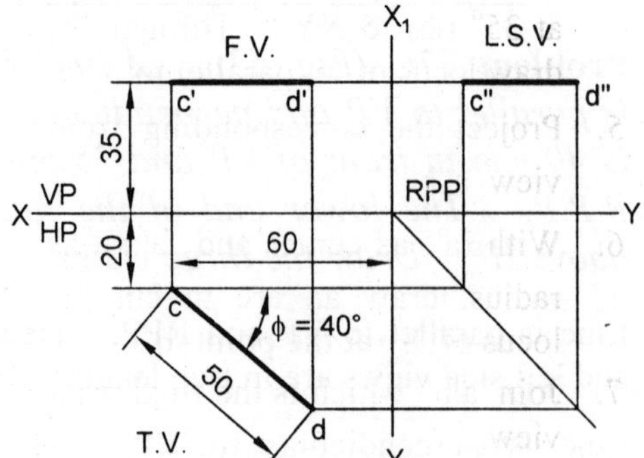

FIG. 66

2. The line is parallel to HP. Hence, its top view gives true length and true inclination.
3. Draw the top view cd = True Length = 50 mm and inclined at 40° to XY.
4. Draw X_1Y_1 perpendicular to XY at a distance of 60 mm from c.
5. Draw the left side view of the line, using the Mitre Line construction as shown.

11.6 LINE PARALLEL TO HP or VP AND INCLINED AT ψ TO PP

Note: 1. If a line is parallel to VP and inclined at an angle of ψ to the PP, obviously it will be inclined at an angle of $(90 - \psi) = \theta$ to HP. This case is the same as a line parallel to VP and inclined at θ° to HP.

2. If a line is parallel to HP and inclined at an angle of ψ to the PP, obviously it will be inclined at an angle of $(90 - \psi) = \phi$ to VP. This case is the same as a line parallel to HP and inclined at ϕ° to VP.

11.7 LINE INCLINED TO BOTH HP AND VP, when $(\theta+\phi) = 90^\circ$

Problem 81: *(Fig. 67) A line AB 55 mm long has its one end 15 mm above HP and 10 mm in front of VP. It is inclined at an angle of 55° to HP and 35° to VP. Draw its three principal views.* (UQ)

Note: If the sum of the true inclinations of a line $(\theta+\phi) = 90^\circ$, then its front and top views will lie in a single projector. Also, the side view will give the true length and true inclinations. And, the apparent angles $\alpha = \beta = 90^\circ$.

1. Mark a′ 15 mm above XY and a 10 mm below XY.
2. Draw $a'b'_1$ = 55 mm and inclined at 55° (θ) to XY. Through b'_1 draw locus of b′ parallel to XY.
3. Project the corresponding top view ab_1.
4. Draw ab_2 = 55 mm and inclined at 35° (φ) to XY. Through b_2 draw locus of b parallel to XY.
5. Project the corresponding front view $a'b'_2$.
6. With a′ as center and $a'b'_2$ as radius, draw an arc to cut the locus of b′ at the point b′.
7. Join a′b′ which is the final front view.

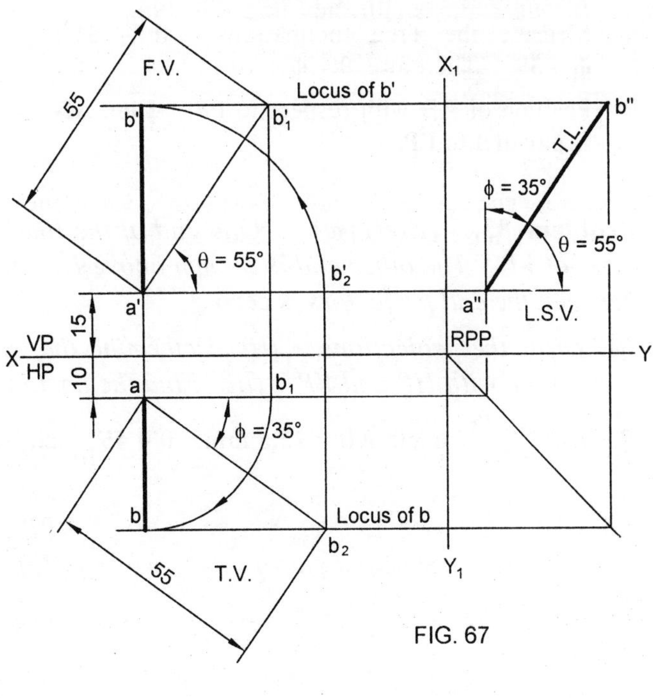

FIG. 67

8. Similarly, with a as center and ab_1 as radius, draw an arc to cut the locus of b at the point b. Join ab which is the final top view.
9. Since $(\theta+\phi) = 90^\circ$, all the points of the projections a, a′, b and b′ lie in the same projector. Therefore, the apparent angles $\alpha = \beta = 90^\circ$.
10. **Left Side View:** Draw X_1Y_1 line perpendicular to XY to the right of the projections at any convenient distance. Draw the left side view a″b″ using the Mitre Line construction.

Problem 82: *(Fig. 68) One end of the line AB is 40 mm above HP and 25 mm in front of VP. The other end is 15 mm above HP and 45 mm in front of VP. The distance between the end projectors is zero.*

(i) Draw the projections. (ii) Determine the true length of the line and its angles of inclination with HP and VP. (iii) Find the position of the line with respect to PP. (UQ)

1. The distance between the end projectors is zero (*Given*). Hence, the line must be parallel to the PP. The side view of the line will give the true length and true inclinations θ and ϕ.
2. Draw a single vertical projector. On it, mark a′, a, b′ and b as per the given data.
3. Draw X_1Y_1 perpendicular to XY.
4. Draw the left side view using the Mitre Line construction as shown.
5. Measure the True Length **AB = 32 mm.**
6. Measure the True Inclinations: **θ = 51°**; **ϕ = 39°**. Therefore, θ + ϕ = 90°.
7. Position of AB with respect to PP: **Line AB is parallel to PP.**

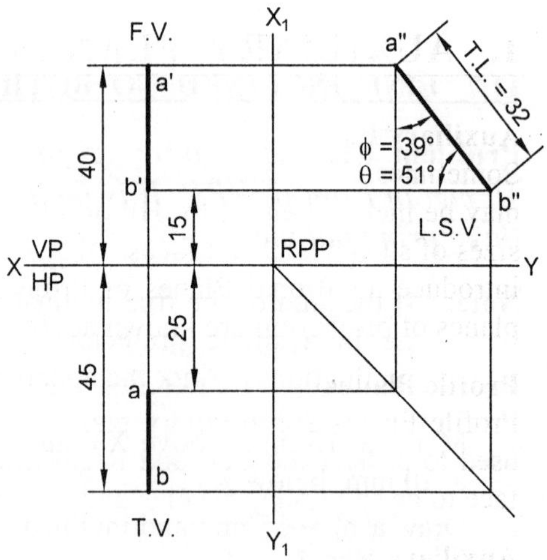

FIG. 68

Problem 83: *(Exercise) One end of the line AB is 50 mm above HP and 30 mm in front of VP. The other end is 20 mm above HP and 60 mm in front of VP. The distance between the end projectors is zero.*

(i) Draw the projections. (ii) Determine the true length of the line and its angles of inclination with HP and VP. (iii) Find the position of the line with respect to PP. (UQ)

Results: True Length **AB = 42.5 mm**, **θ = 45°** and **ϕ = 45°**.

*

11

AUXILIARY PLANE PROJECTION METHOD

1. AUXILIARY PLANES AND AUXILIARY VIEWS

Auxiliary Planes

Sometimes, different edges and faces of an object may not be parallel to either HP or VP. They may be inclined either to HP or VP. In such cases, the front and top views do not show its true sizes of such edges and faces. To show such edges and faces in their true sizes, it is necessary to introduce additional Planes of Projection parallel to these edges and faces. These additional planes of projection are known as ***Auxiliary Planes.***

Profile Planes

Profile Planes are auxiliary planes which are always perpendicular to both HP and VP and are used to project the Left and Right Side Views. But Auxiliary Planes are parallel to the edge or face to be shown in true size and also perpendicular to either HP / VP.

Auxiliary Views

Additional views called **Auxiliary Views,** are obtained by projecting the object on Auxiliary Planes, which are inclined to one of the Principal Planes and perpendicular to the other.

1.1 TYPES OF AUXILIARY PLANES

(i) **Auxiliary Vertical Plane (A.V.P.)**
If the Auxiliary Plane selected is perpendicular to the HP and inclined to the VP, the view of the object projected on Auxiliary Plane is called as ***Auxiliary Front View;*** And the Auxiliary Plane is called ***Auxiliary Vertical Plane*** (A.V.P.).

(ii) **Auxiliary Inclined Plane (A.I.P.)**
If the Auxiliary Plane selected is perpendicular to the VP and inclined to the HP, the view of the object projected on Auxiliary Plane is called as the ***Auxiliary Top View;*** And the Auxiliary Plane is called ***Auxiliary Inclined Plane*** (A.I.P.).

New Reference Lines (X_1Y_1 and X_2Y_2)

Line of intersection of HP with A.V.P. or VP with A.I.P. is called *New Reference Line* X_1Y_1. Line of intersection of A.I.P. and A.V.P. is called *Another Reference Line* X_2Y_2.

2. PROJECTIONS OF POINTS ON AUXILIARY PLANES

2.1 PROJECTION OF A POINT P ON A.V.P.

1. *See Fig. 1.* p′ is the projection of P on VP and p is the projection of P on HP. HP and A.V.P. intersect each other at right angles at the new reference line X_1Y_1.
2. *Refer Fig. (i).* Look at P such that eye ray is perpendicular to A.V.P. The projection of P on A.V.P. is p_1' and is called the ***Auxiliary Front View*** of P. Orthographic views of Auxiliary Projections are obtained by rotating the Auxiliary Plane about HP or VP to which it is perpendicular.
3. *Refer Fig. (ii).* Since A.V.P. is perpendicular to HP, rotate the A.V.P. about X_1Y_1 to coincide with HP. Angle between XY and X_1Y_1 is ϕ^o.

Problem 1: *[Fig. 1(iii)] A point P is 40 mm above HP and 30 mm in front of VP. Draw an Auxiliary Front View of P on an A.V.P. inclined at 30° to VP.*

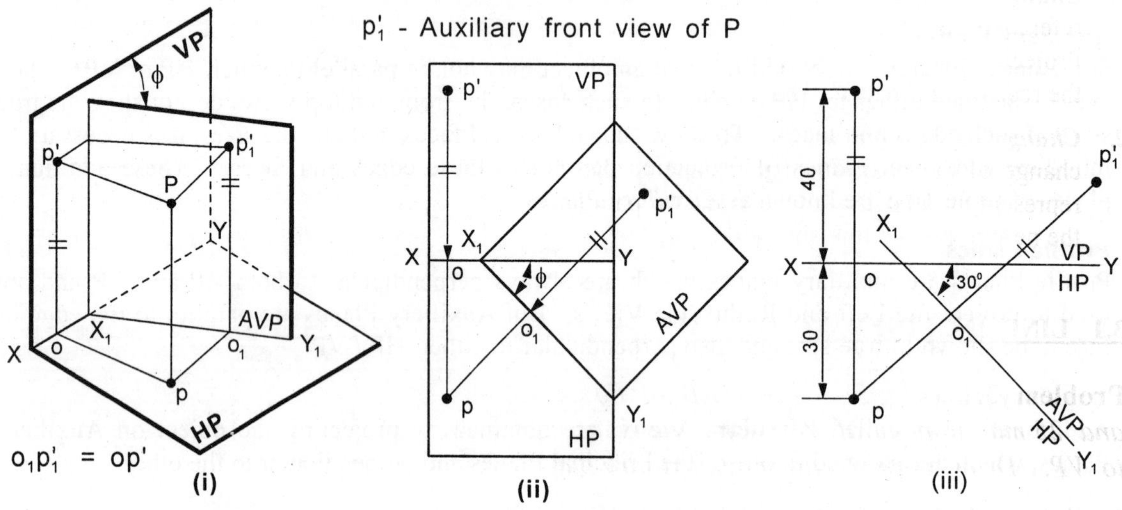

FIG. 1

1. Draw XY line. Draw a projector through o on it.
2. Mark p′ 40 mm above XY and p 30 mm below XY.
3. Draw X_1Y_1 at *any convenient point* on XY at 30° (ϕ) to it.
4. From p draw a projector perpendicular to X_1Y_1 to cut it at o_1.
5. Mark p_1' on this projector such that $p_1'o_1$ = p′ o = 40 mm. p_1' is the Auxiliary Front View.

2.2 PROJECTION OF A POINT P ON A.I.P.

Problem 2: *(Fig. 2) A point P is 25 mm above HP and 35 mm in front of VP. Draw an Auxiliary Top View of P on an A.I.P. inclined at 25° to HP.*

1. Draw XY and a projector perpendicular to it at o.
2. Mark p′ 25 mm above XY and p 35 mm below XY on it.
3. Draw a new reference line X_1Y_1 at any convenient point on XY at 25° (θ) to it.
4. From p′ draw a projector perpendicular to X_1Y_1 to cut it at o_1.
5. Mark p_1 on this projector such that p_1o_1 = po. p_1 is the Auxiliary Top View.

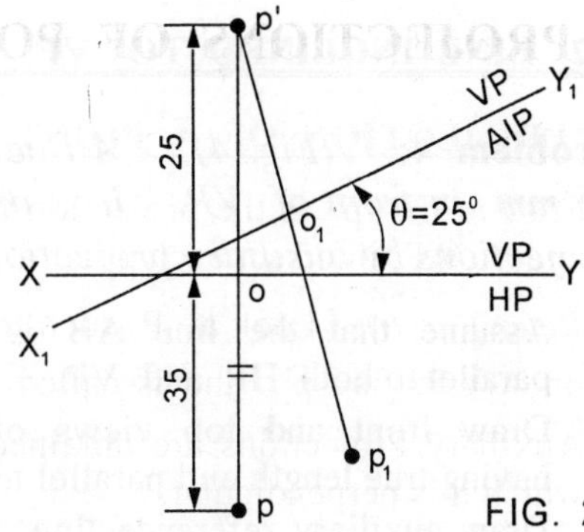

FIG. 2

3. PROJECTIONS OF LINES ON AUXILIARY PLANES

1. ***Change of Position Method:*** In the previous Chapter, line inclined to one or both the reference planes is initially assumed to be in simple position parallel to both HP and VP. Projections are then drawn by changing the position of the line from the simple position to the required position. This is known as Change of Position Method.
2. ***Change of Reference Line Method:*** Instead of changing the position of the projection, change of the position of the plane is also done. In other words, a new reference line to represent an auxiliary plane is drawn parallel to the projection. The auxiliary projection on the new reference line shows the true length and true inclination of the line.

3.1 LINE INCLINED TO HP AND PARALLEL TO VP

Problem 3: *(Fig. 3) A line AB of 60 mm length, has its end A 20 mm above HP and 25 mm in front of VP. The line is inclined at an angle of 45° to HP and parallel to VP. Draw its projections by auxiliary plane method.*

1. Assume the line is kept parallel to both HP and VP. Mark a′ above XY and a below XY.
2. Draw front and top views of AB in true length and parallel to XY.
3. Draw X_1Y_1 inclined at 45° to XY to represent an AIP. From a′ and b′, draw projectors $\perp^r$ to X_1Y_1.
4. Mark a_1 such that the distance of a from XY = 25 mm = distance of a_1 from X_1Y_1. Similarly, mark b_1 such that distance of b from XY = 25 mm = distance of b_1 from X_1Y_1.
5. Join a_1 and b_1 to get Auxiliary Top View corresponding to a′ b′.

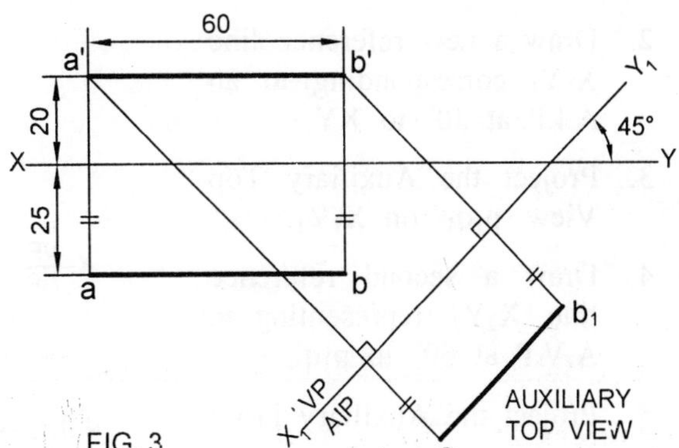

FIG. 3

3.2 LINE INCLINED TO VP AND PARALLEL TO HP

Problem 4: *(Fig. 4) A line 60 mm long has its one end 20 mm above HP and 25 mm in front of VP. It is inclined at 60° to VP and parallel to HP. Draw its projections by auxiliary projection method.*

1. Assume that the line AB is kept parallel to both HP and VP.
2. Draw front and top views of AB having true length and parallel to XY.
3. Draw auxiliary reference line X_1Y_1 at 60° to XY to represent an AVP.
4. From the top view of the ends a and b, draw projectors ⊥r to X_1Y_1.
5. Mark the distance a′ to XY (= 20 mm) from X_1Y_1 to get a_1'.
6. Similarly, mark the distance b′ to XY (= 20 mm) from X_1Y_1 to get b_1'.
7. Join a_1' and b_1', the Auxiliary Front View corresponding to top view ab.

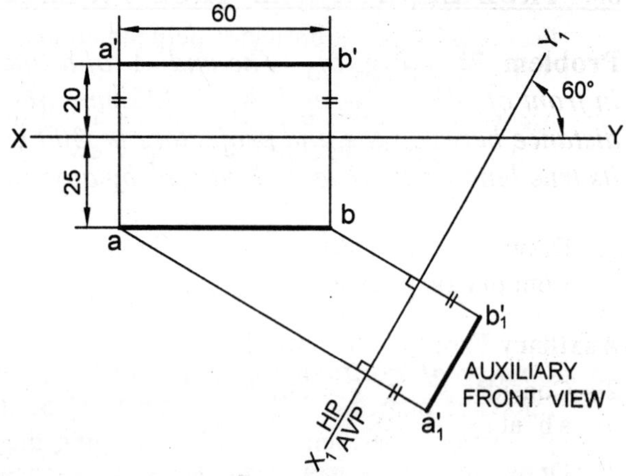

FIG. 4

3.3 LINE INCLINED TO BOTH HP AND VP

Problem 5: *(Fig. 5) A line PQ 48 mm long is inclined at an angle of 30° to HP. Its top view makes an angle 60° with XY. End P is 20 mm above HP and 15 mm in front of VP. Draw the projections of the line by auxiliary plane method.*

(UQ)

1. Assume PQ parallel to both HP and VP. Draw p′ q′ and p q.
2. Draw a new reference line X_1Y_1 corresponding to an A.I.P. at 30° to XY.
3. Project the Auxiliary Top View p_1q_1 on X_1Y_1.
4. Draw a second reference line X_2Y_2 representing an A.V.P. at 60° to p_1q_1.
5. Project the Auxiliary Front View $p_1' q_1'$ on X_2Y_2.

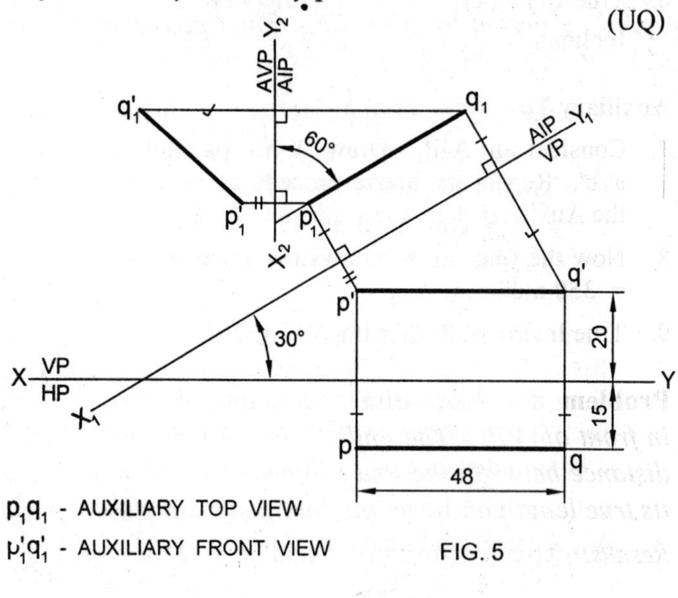

FIG. 5

Problem 6: *(Exercise) A line AB 62 mm long has its end A in HP and 20 mm in front of VP. If the line is inclined at 30° to VP and the front view makes 45° to the reference line, draw its projections by auxiliary plane method.*

3.4 TRUE LENGTH AND TRUE INCLINATIONS

Problem 7: *(Fig. 6) The end A of a line AB is 100 mm above HP and 150 mm in front of VP. The end B is 225 mm above HP and 275 mm in front of VP. The distance between the end projectors is 300 mm. Draw the projections of the line. Find its true length and true inclinations by auxiliary plane method.* (UQ)

1. Draw the front and top views of the line AB from the given data.

Auxiliary Front View and θ

2. Consider an AVP. Draw X_1Y_1 parallel to a b at a convenient position as shown.
3. Draw projectors from a and b and perpendicular to X_1Y_1.
4. Mark a_1' and b_1' on the above projectors such that their distances from X_1Y_1 are equal to the distances of a′ and b′ from XY respectively.
5. Join a_1' and b_1' to obtain the Auxiliary Front View. True Length of the line = **350 mm.**
6. True Inclination θ of the line with HP = Inclination of $a_1' b_1'$ with X_1Y_1 = **21°.**

Auxiliary Top View and ϕ

7. Consider an AIP. Draw X_2Y_2 parallel to a′ b′. Repeat the above procedure and draw the Auxiliary Top View a_1b_1 on X_2Y_2.
8. Now the line a_1 b_1 shows the True Length = **350 mm.**
9. True Inclination Φ with VP = **21°.**

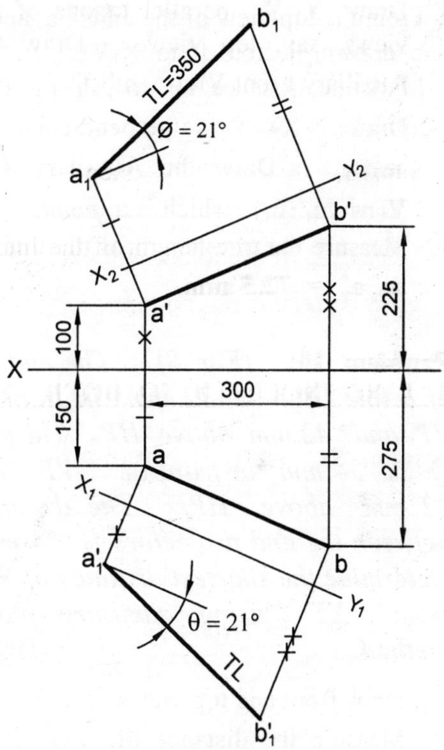

RESULTS : TRUE LENGTH = 350 mm

θ = 21°, Ø = 21°

FIG. 6

Problem 8: *(Exercise) The end A of a line AB is 30 mm above HP and 12 mm in front of VP. The end B is 16 mm above HP and 40 mm in front of VP. The distance between the end projectors is 60 mm. Draw the projections of the line. Find its true length and true inclinations by auxiliary plane method.* (UQ)

Results: True Length = 67.5 mm, θ = 12° and Φ = 24.5°.

3.5 POINT-VIEW OF A LINE

We know that when a straight line is perpendicular to HP or VP, its projection on that plane is a ***point***. Its projection on other reference plane shows its true length.

Problem 9: *(Fig. 7) The front and top views of a line MN are inclined at 20° and 30° to XY respectively. The distance between the end projectors is 60 mm. M is 16 mm above HP and 12 mm in front of VP. Draw the projections of MN on an auxiliary plane so that in the final view it appears as a point. Find the true length of the line.* (UQ)

1. Draw the front and top views of MN.
2. Draw $X_1 Y_1$ parallel to one of the views, say, top view. Draw the Auxiliary Front View $m_1' n_1'$.
3. Draw $X_2 Y_2$ perpendicular to $m_1' n_1'$. Draw the Auxiliary Top View $m_1 (n_1)$ which is a *point.*
4. Measure the true length of the line = $m_1' n_1'$ = **72.5 mm.**

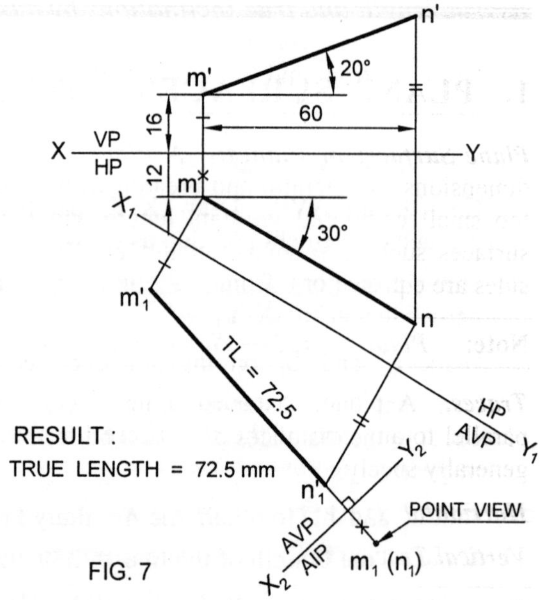

FIG. 7

Problem 10: *(Fig. 8) The end P of a line PQ is 12 mm in front of VP and 42 mm above HP. The end Q is 54 mm in front of VP and 12 mm above HP. The distance between the end projectors is 54 mm. Determine the shortest distance of PQ from XY using auxiliary plane method.* (UQ)

1. Draw front and top views of PQ. Measure the distance of PQ from XY by drawing an auxiliary view on an Auxiliary Plane, on which XY can be represented as a point.
2. Draw X_1Y_1 perpendicular to XY. Project the auxiliary top view p_1q_1. Point X_0Y_0 is point view of XY.
3. Shortest distance of PQ from XY = Length of the perpendicular drawn from X_0 (or Y_0) to p_1q_1 = X_0M = **41 mm.**

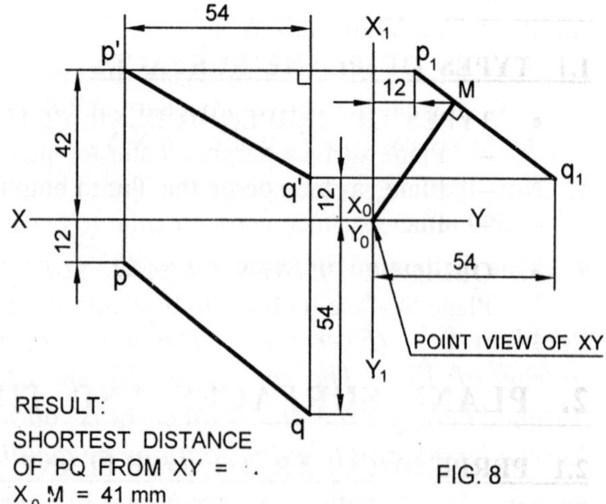

FIG. 8

12

PROJECTIONS OF PLANE SURFACES

1. PLANE SURFACES – Polygonal Surfaces and Circular Lamina

Plane Surfaces or Planes: A *Plane or Plane Surface,* (also known as *Lamina*), has only two dimensions, *viz.,* length and breadth with negligible thickness. Thickness is neglected since it is too small compared with its length and breadth. The plane surfaces may be of polygonal surfaces such as square, rectangular, polyhedral (pentagonal, etc.) or circular lamina, etc. If sides are equal, it is a *Regular Plane Surface,* otherwise an *Irregular Plane Surface*.

Note: ***Plane Surface – Plane – Plane Figure – Lamina – Plate*: All mean the same.**

Traces: A plane, extended if necessary, will meet the reference planes in lines unless it is parallel to any one of them. These lines are called *Traces of the Plane Surfaces*. A plane is generally specified by its Traces.

Horizontal Trace (H.T.) of a plane surface is the line in which plane surface meets the HP.

Vertical Trace (V.T.) of a plane surface is the line in which the plane surface meets the VP.

Note: If a plane is parallel to one of the reference planes, it will not have trace on that reference plane. It means that a plane parallel to HP will not have H.T. and a plane parallel to VP will not have V.T.

1.1 TYPES OF PLANE SURFACES

- **TYPES OF PERPENDICULAR *PLANES / PLANE SURFACES***
  - Plane surface perpendicular to one reference plane and parallel to the other
  - Plane surface perpendicular to both the reference planes
  - Plane surface perpendicular to one reference plane and inclined to the other
- **OBLIQUE *PLANES / PLANE SURFACES***
  Plane Surface inclined to both reference planes is termed as Oblique *Plane Surface*.

2. PLANE SURFACES AND THEIR TRACES

2.1 PERPENDICULAR TO ONE PLANE AND PARALLEL TO THE OTHER

Problem 1: *(Fig. 1) A square lamina ABCD of side 40 mm is perpendicular to HP and parallel to VP. Draw its projections and locate its traces.*

1. From given statement, decide the view to be drawn first. Lamina is parallel to VP.

   When viewed from the front, it is seen in true shape and size. Hence, **draw Front View first.**

2. **Front View:** Lamina ABCD is above HP. So, draw its Front View **a′ b′ c′ d′** in true size above XY.

3. **Top View:** When the lamina is viewed from the top, edge BC is seen in true length and parallel to VP. Hence draw its Top View **bc** (in true length) parallel to and below XY.

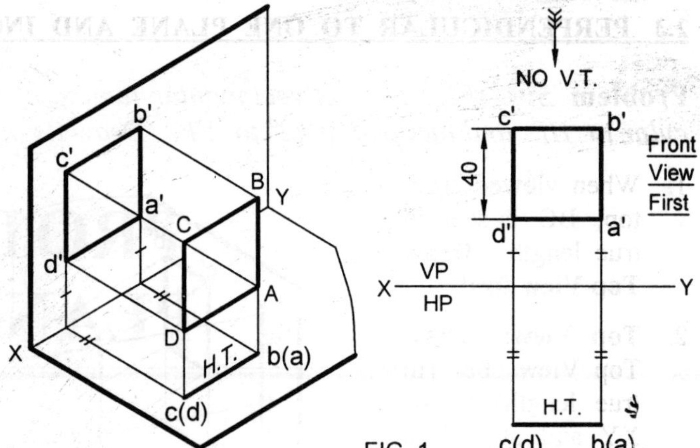

FIG. 1

4. **To mark invisible corners in Top View:** Look at Front View in the direction of arrow. **a′** and **d′** are hidden by **b′** and **c′** respectively. Therefore in the Top View, **ad** coincides with **bc** and is marked as **(a)(d)**. Now **bc(d)(a)** is the Top View.

5. Extend the plane surface to meet the HP. The extension coincides with the Top View. Hence the Top View **bc (d) (a)** itself is the **H.T.** Extend the plane surface to meet VP. The plane surface is parallel to VP and it will not meet. Hence, there is **no V.T.**

Problem 2: *(Fig. 2) A square lamina of side 40 mm is perpendicular to VP and parallel to HP. Draw its projections. Find its traces.*

When the lamina ABCD is viewed from the top, it is seen in true shape and size. Therefore, **draw the Top View first.**

See Fig. 2 for Solution.

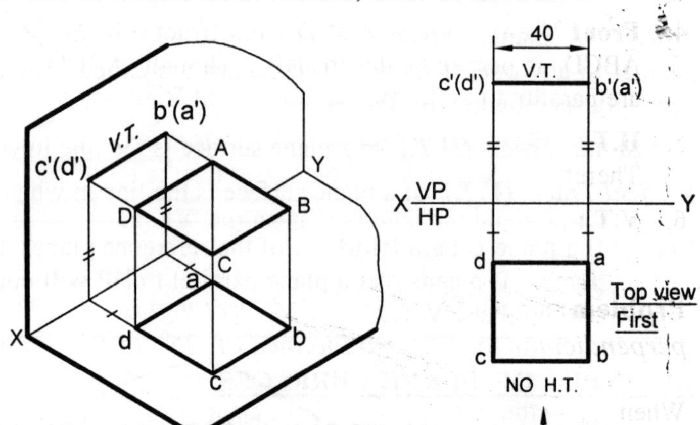

FIG.2

Problem 3: *(Exercise) A rectangular cardboard ABCD of edges AB = 30 mm and BC = 40 mm is placed such that the edge AB is (i) perpendicular to HP and parallel to VP, (ii) parallel to HP and $\perp^r$ to VP. Draw its projections and traces.*

2.2 PERPENDICULAR TO BOTH HP AND VP

Problem 4: *(Exercise) A thin square plate ABCD of side 40 mm is perpendicular to both HP and VP. Draw its projections and find the traces.*

In the given position, the edges are seen in true length both in front and top views. Hence, draw **either the front view or top view first.**

2.3 PERPENDICULAR TO ONE PLANE AND INCLINED TO THE OTHER

Problem 5: *(Fig. 3) A rectangular lamina of sides 30 mm x 40 mm is perpendicular to HP and inclined at 42° to VP. Draw its projections and obtain the traces.*

1. When viewed from top, BC is seen in true length. **Draw Top View first.**
2. **Top View:** Draw Top View bc (in true length) below XY and at 42° to XY.
3. Edge AD is hidden by the edge BC and its Top View (a)(d) coincides with bc. Now bc(a)(d) is required Top View.

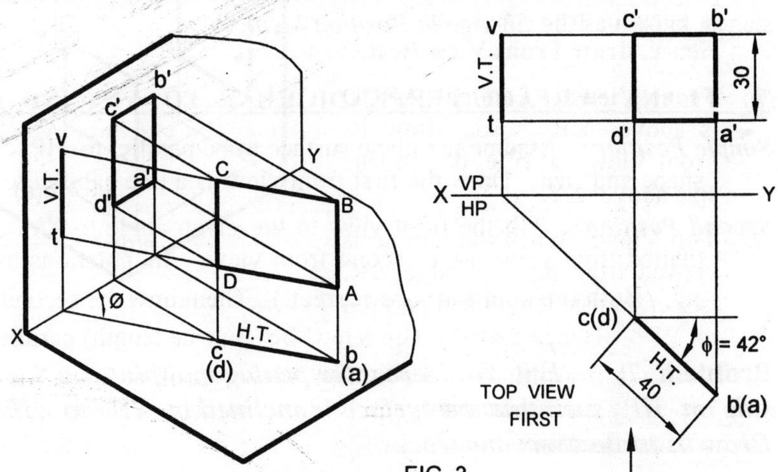

FIG. 3

4. **Front View:** View ABCD from front. Front View a′ b′ c′ d′ is seen in reduced size. ABCD is perpendicular to HP. Though ABCD is inclined to VP, edges CD and AB are parallel to VP. Hence draw Front View a′ b′ c′ d′ such that a′ b′ = c′ d′ = AB.
5. **H.T.:** Extend the lamina to meet the HP. The extension coincides with the Top View. Therefore, the Top View bc(a)(d) itself is the H.T.
6. **V.T.:** Extend the lamina to meet the VP at vt which is the required V.T.

Problem 6: *(Fig. 4) A rectangular lamina of sides 30 mm x 40 mm is perpendicular to VP and inclined at 25° to HP. Draw its projections and traces.*

When the lamina ABCD is viewed from front, the edge BC is seen in true length.

Therefore, ***draw the Front View first.***

See. Fig. 4 for Solution.

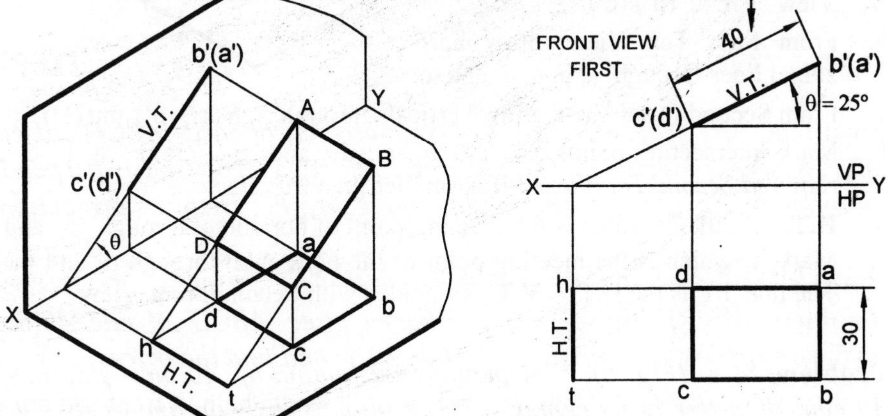

FIG. 4

3. CHANGE OF POSITION METHOD

Orthographic projections of a plane surface inclined to one reference plane are drawn by changing the position of the plane surface from simple position to the required position in *two* stages, known as the *Change of Position Method.*

3.1 PLANE SURFACE PERPENDICULAR TO VP AND INCLINED TO HP

Simple Position: Assume the plane surface to be parallel to HP. Draw the first top view in true shape and size. Draw the first front view as a line parallel to XY.

Second Position: Tilt the front view to the given angle to XY. Draw the second front view. (Tilted front view is the second front view. Its notations may be marked as either a′ or a_1'. Both the notations are correct.) Then draw the second top view.

Problem 7: *(Fig. 5) A regular pentagonal plate of side 28 mm is placed with one side on HP such that the surface is inclined at 45° to HP and perpendicular to VP. Draw its projections and traces.* (UQ)

Simple Position

1. Assume that the pentagonal surface is parallel to HP and $\perp^r$ to VP.
2. Draw First Top View abcde in true shape and size, (side 28 mm), such that side ae is perpendicular to XY.
3. Draw the First Front View a′ b′ c′ (d′)(e′) as a line parallel to XY.

Second Position

4. Tilt the First Front View about a′ (e′) to the given angle of 45° with XY.
5. Redraw it to obtain the Second Front View a′ b′ c′ (d′) (e′).
6. From First Top View, draw horizontal lines through a, b, c, d and e.
7. From Second Front View, draw verticals through a′, b′, c′, (d′) and (e′).
8. Mark intersecting points as a_1, b_1, c_1, d_1 and e_1. Join $a_1\ b_1\ c_1\ d_1\ e_1$ which is the *required Second Top View*. It is smaller than true size.
9. **H.T.:** Mark h which is the meeting point of horizontal through b_1 and vertical through a′. Mark t which is the meeting point of the horizontal through d_1 and the vertical through a′. The line ht is the H.T. **V.T.** coincides with Second Front View.

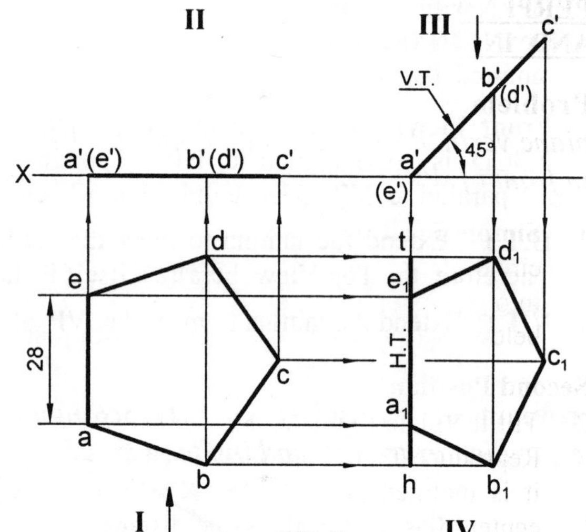

FIG. 5

Problem 8: *(Fig. 6) A pentagonal lamina of 40 mm side has a circular hole of 35 mm diameter in its center. The plane stands on one of its sides on HP with its plane perpendicular to VP and 45° inclined to HP. Draw the projections.* (UQ)

1. Initially, keep the plane of the lamina in HP. First, draw the top view of the pentagon with a circular hole such that a side of the pentagon is perpendicular to XY.
2. Project the corresponding front view.
3. Tilt the front view by 45° to XY such that the point a′(e′) is lying on XY.
4. Project the corresponding final top view as shown.

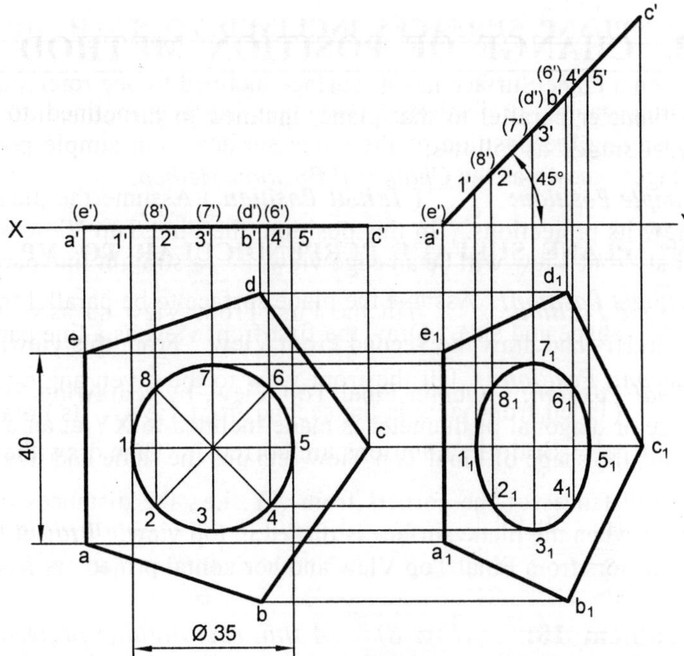

FIG. 6

3.2 PLANE SURFACE PERPENDICULAR TO HP AND INCLINED TO VP

Problem 9: *(Fig. 7)* *A thin circular metal plate of 48 mm diameter, having its plane vertical and inclined at 40° to VP. Its center is 33 mm above HP and 25 mm in front of VP. Draw its projections and locate its traces.* (UQ)

1. **Simple Position:** Assume the circle to be parallel to VP. Draw First Front View as a circle of 48 mm diameter, having its center 33 mm above XY. Divide its circumference into 12 equal parts as 1, 2, ..., 12. Draw First Top View as a line parallel to and 25 mm below XY. Center o coincides with point 4.

Second Position

2. Tilt the plate to make 40° with VP. Reproduce First Top View such that it is inclined at 40° to XY and center o is kept at the same distance 25 mm below XY. Thus, Second Top View is 1 2 3 4 5 6 7.
3. To draw Second Front View, project all the points upwards from Second Top View and horizontally from First Front View. Draw Second Front View as *an ellipse* through the points of intersection $1_1'$, $2_1'$, ..., $12_1'$.
4. H.T. and Second Top View coincide. V.T. is the line vt perpendicular to XY. H.T. and V.T. meet XY at m.

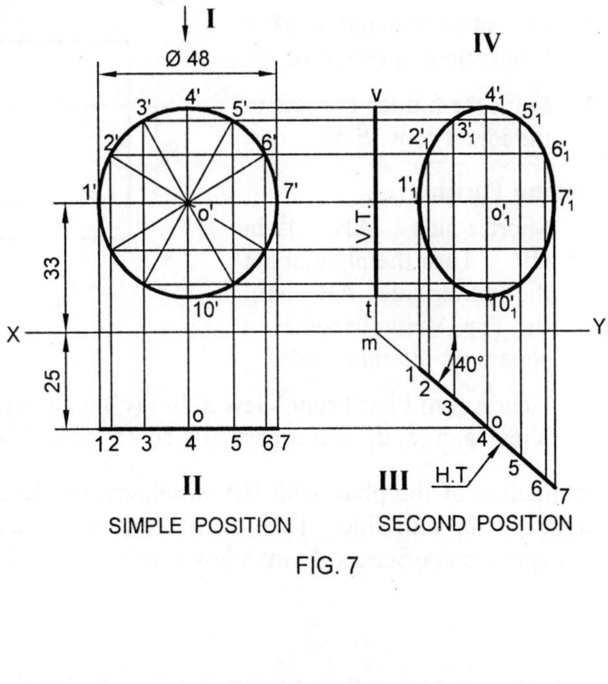

FIG. 7

3.3 PLANE SURFACES INCLINED TO BOTH HP AND VP

When a Plane Surface has its surface inclined to one reference plane and an edge *or* a diagonal *or* a diameter parallel to that plane, inclined to the other reference plane, draw its projections in ***three stages*** as follows:

Simple Position: (***Initial Position***) Assume the plane surface to be parallel to HP and draw its projections. In this position, the First Top View will be in true shape and size and the First Front View will be an edge view, *i.e.*, a straight line parallel to XY.

Second Position: Tilt the First Front View so as to make the given angle of inclination with HP and draw the Second Front View. From this view, project the Second Top View.

Final Position: Obtain Final Top View, by redrawing Second Top View such that the given edge or diagonal or diameter is made inclined to XY at an angle made by plane surface with VP. Now, the shape of Final Top View remains the same and its relative position changes.

The distances of the corners from XY, i.e., the distances of the corners above HP, will be the same when the plane surface is turned in top view. Project Final Front View by drawing vertical projectors from Final Top View and horizontal projectors from Second Front View.

Problem 10: *(Fig. 8) A thin rectangular plate of sides 50 mm x 25 mm has its shorter side in the HP and inclined at an angle of 30° to the VP. Project its front view when its top view is a perfect square of 25 mm side.* (UQ)

Simple Position

1. Assume the rectangular plate ABCD is lying on HP with its shorter side AB = 25 mm ⊥ʳ to VP. Draw First Top View abcd.
2. Project corresponding First Front View a′(b′)(c′)d′.
3. Mark the hidden corners in the First Front View.

Second Position

4. Shorter side AB is in HP. Turn the plate about the shorter side AB until its Top View becomes a square of 25 mm side.

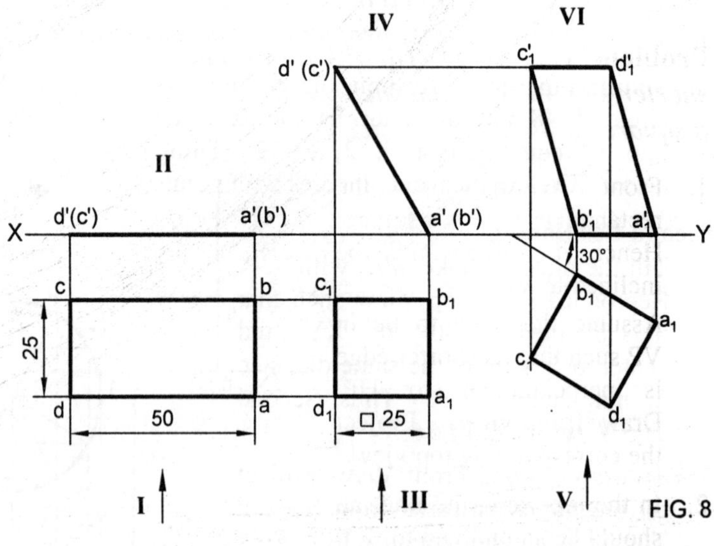

FIG. 8

Hence, turn First Front View a′ (b′) (c′) d′ about edge a′ (b′) such that its corresponding Top View a_1 b_1 c_1 d_1 is a square of side 25 mm. Obtain Second Front View a′ (b′) (c′) d′.

Inclination of the plate with HP is unknown. Hence first draw Second Top View a_1 b_1 c_1 d_1, a square of 25 mm side. Then draw the projectors upwards from Second Top View. On these projectors mark Second Front View a′ (b′) (c′) d′ = length of First Front View = 50 mm.

5. **Final Position:** Shorter edge AB is on HP, making an angle of 30° with VP. Hence turn the Second Top View $a_1\ b_1\ c_1\ d_1$ so that the edge $a_1\ b_1$ is at an angle of 30° to XY. Thus obtain the Final Top View $a_1\ b_1\ c_1\ d_1$. Project all the points upwards from the Final Top View and also horizontally from the Second Front View. Complete the required Final Front View $a_1'\ b_1'\ c_1'\ d_1'$ as shown.

Problem 11: *(Fig. 9) A rectangular lamina of size 60 mm x 30 mm is seen as a square in the top view, when it rests on one of its edges on HP and perpendicular to VP.*

Draw the projections of the lamina. Find the true incli-nation of its surface with HP. Draw the front view of the lamina when the edge about which it is tilted, is inclined at 45° to VP. (UQ)

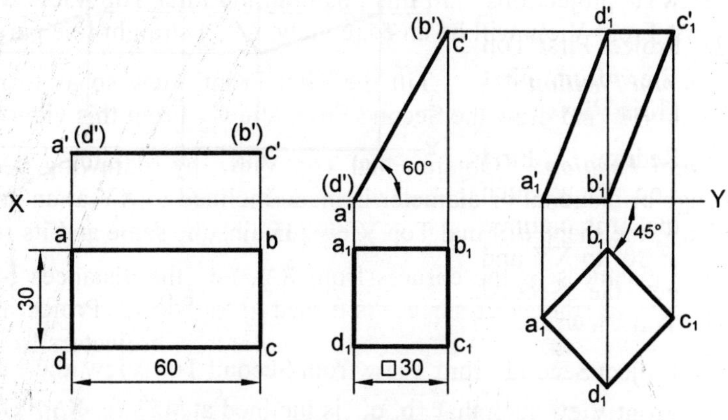

Result: True Inclination of the surface of lamina with HP = 60°

FIG. 9

Problem 12: *(Fig. 10) A thin rectangular plate of sides 60 mm x 30 mm has its shorter side in VP and inclined at 30° to HP. Project its top view, if its front view is a square of 30 mm long sides.* (UQ)

1. Front view of the rectangular plate is a square. Hence, its surface must be inclined to VP.

   Assume the plate to be in VP such that its shorter edge is perpendicular to HP. Draw front view. Project the corresponding top view.

2. In the top view, the line ab should be so inclined to XY that the front view becomes a square.

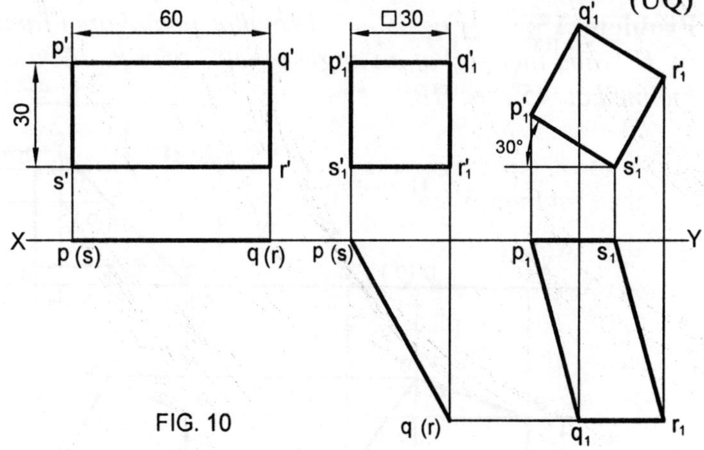

FIG. 10

Problem 13: *(Fig. 11) Draw the projections of a pentagonal sheet of 26 mm side, having its surface inclined at 30° to VP. Its one side is parallel to VP and inclined at 45° to HP.* (UQ)

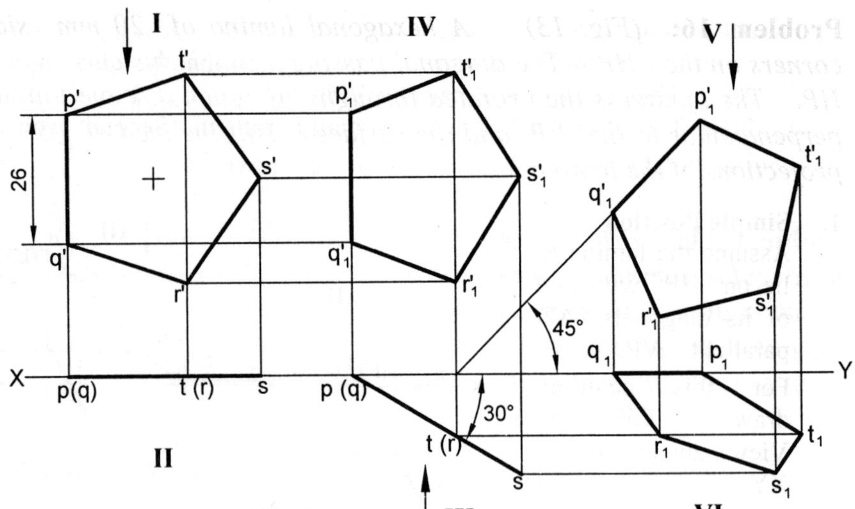

FIG. 11

1. Draw the First Front View as pentagon such that one side (edge) p′q′ is perpendicular to XY.
2. Project First Top View from First Front View.
3. Redraw First Top View such that it is inclined at 30° to XY and get the Second Top View.
4. Project Second Front View from Second Top View and First Front View. Reproduce Second Front View such that $p_1' q_1'$ is inclined at 45° to XY. Obtain Final Front View.
5. Project Final Top View from Final Front View and Second Top View. Note that $p_1 q_1$ is parallel to XY.

Problem 14: *(Exercise) A hexagonal lamina of 26 mm side has a side resting on VP and inclined at 30° to HP. Its surface is inclined at 45° to VP. Draw its projections.*

Problem 15: *(Fig. 12) A regular pentagonal lamina of 30 mm sides has one edge in HP and inclined at an angle of 30° to VP. Draw its projections when its surface is inclined at 45° to HP.* (UQ)

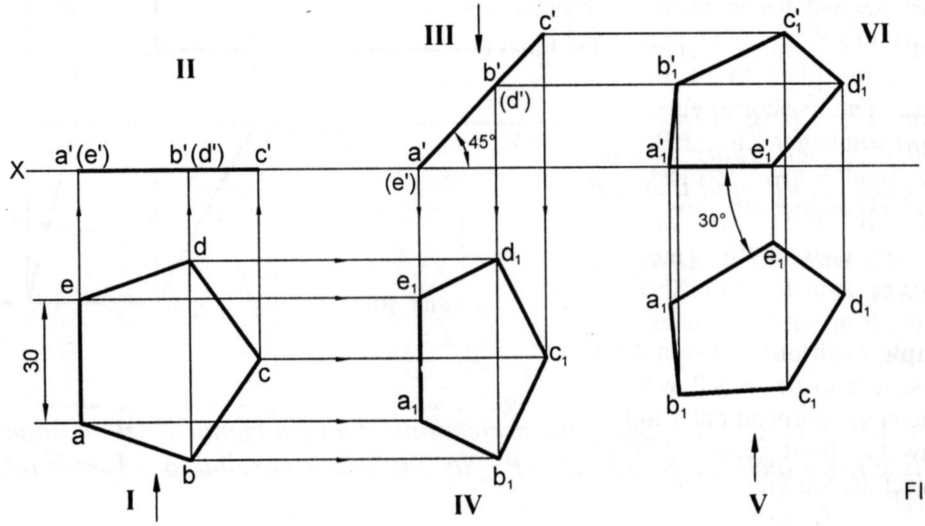

FIG. 12

Problem 16: *(Fig. 13) A hexagonal lamina of 20 mm side rests on one of its corners on the HP. The diagonal passing through this corner is inclined at 45° to the HP. The lamina is then rotated through 90° such that the top view of this diagonal is perpendicular to the VP and the surface is still inclined at 45° to the HP. Draw the projections of the lamina.* (UQ)

1. **Simple Position**
   Assume the lamina to lie on HP with one of its diagonals AD parallel to VP.
   For this position, draw the First Top View abcdef below XY. Project its corresponding First Front View a′ d′ on XY.
   Look at the First Top View in the direction of arrow shown. Mark hidden corners in First Front View.

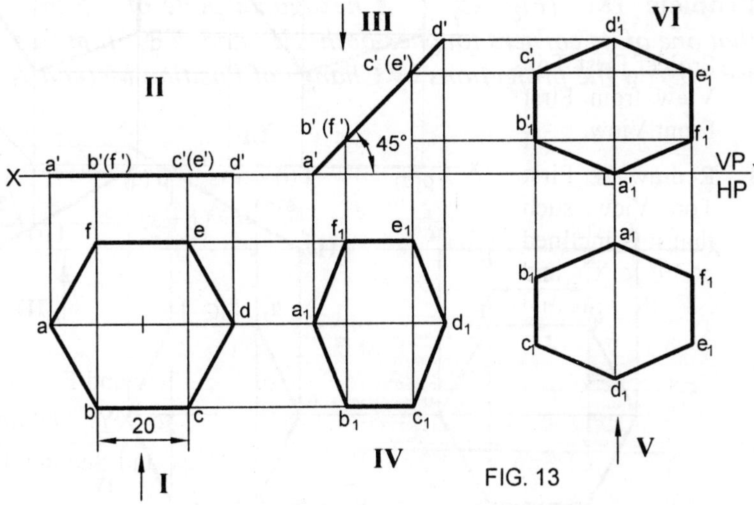

FIG. 13

2. **Second Position:** Diagonal AD of the lamina is inclined at 45° to HP. Hence turn First Front View a′ d′ about the corner a′ through 45° to XY to obtain the *Second Front View* a′ d′. Project the corresponding Second Top View a_1 b_1 c_1 d_1 e_1 f_1.

3. **Final Position:** Then rotate the lamina through 90° such that the Top View a_1 d_1 of diagonal AD is $\perp^r$ to VP. Hence redraw the Second Top View, turning it through 90° such that a_1d_1 is perpendicular to XY. a_1 b_1 c_1 d_1 e_1 f_1 is the required *Final Top View*.

   As per the given condition, diagonal AD of the lamina still makes 45° to the HP. So, from Second Front View draw horizontal lines and from the Final Top View draw the vertical projectors. Complete the Final Front View a_1' b_1' c_1' d_1' e_1' f_1'.

Problem 17: *(Fig. 14) A thin 30°-60° set-square has its longest edge in VP and inclined at 30° to HP. Its surface makes 45° with VP. Draw its projections.* (UQ)

1. **Simple Position:** Assume the set-square to be in VP with its hypotenuse perpendicular to HP. Draw the front view a′ b′ c′. Project the top view.

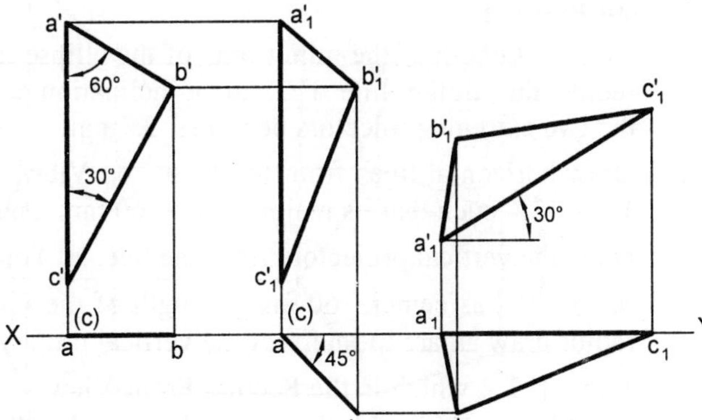

FIG. 14

2. **Second Position:** Tilt the top view ab to make 45° with XY to obtain the second top view. Project the corresponding second front view.
3. **Final Position:** Reproduce the second front view so that the edge a_1' c_1' makes 30° with XY. Project the corresponding final top view as shown.

Problem 18: *(Fig. 15) A hexagonal plate of 25 mm side is resting on HP such that one of its corners touches both HP and VP. It makes 30° with HP and 60° with VP. Draw the projections by Change of Position Method.* (UQ)

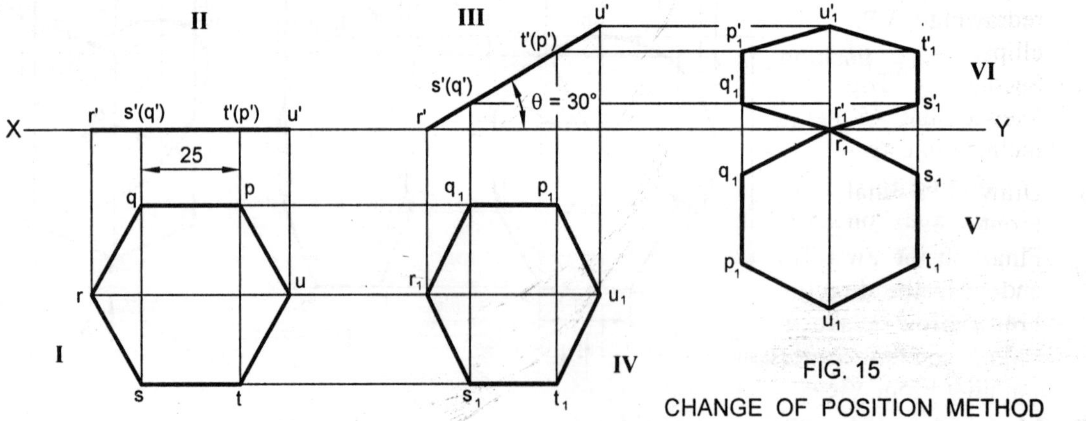

FIG. 15
CHANGE OF POSITION METHOD

Problem 19: *(Fig. 16) A circular lamina of 60 mm diameter rests on HP on a point 1 on the circumference. The lamina is inclined to HP such that the top view of it is an ellipse of minor axis 35 mm. The top view of the diameter through the point 1 makes an angle of 45° with VP. (i) Draw the projections. (ii) Determine the angle made by the lamina with HP.* (UQ)

Simple Position

1. Draw the First Top View of the circular lamina. Divide the circle into 12 equal parts. Draw the corresponding First Front View as a straight line 1′ 7′ lying on XY.

Second Position

Given: Length of the minor axis of the ellipse in the Second Top View is 35 mm. Hence, redraw the straight line 1′ 7′ to an inclination of θ with XY, so that the distance between the two extreme projectors becomes 35 mm.

2. Draw horizontal lines from the First Top View. Draw the Second Top View as an ellipse 1_1 – 12_1 such that its major axis = 60 mm and minor axis = 35 mm.
3. Draw the vertical projectors from the Second Top View. Mark 1′ on XY.
4. With 1′ as center, 60 mm (length of the First Front View = diameter of the circle) as radius draw an arc to intersect the vertical projector from 7_1 at the point 7′.
5. Join 1′ 7′ which is the Second Front View. Measure the angle θ made by the Second Front View 1′ 7′ with XY. **Angle made by the lamina with HP = θ = 54°.**

Final Position

6. The top view of the diameter through the point 1 makes 45° with VP (Given). Therefore, rotate the minor axis of the Second Top View $1_1\,7_1$ such that the Final Top View $1_1 7_1$ makes an angle of 45° with XY.

7. Complete the ellipse of Final Top View by redrawing ellipse of Second Top View on the 45° inclined line.

8. Draw Final Front View from Final Top View and Second Front View.

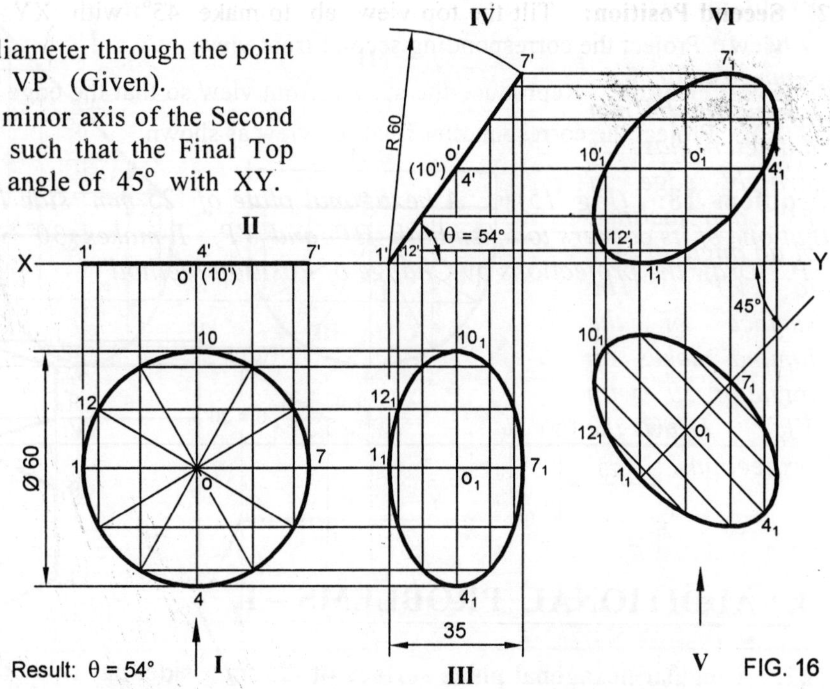

FIG. 16

Problem 20: *(Fig. 17) A regular hexagonal lamina of 26 mm side has a central hole of 30 mm diameter. Draw the front and top views when the surface of the lamina is inclined at 45° to HP. A side of lamina is inclined at 35° to VP.* (UQ)

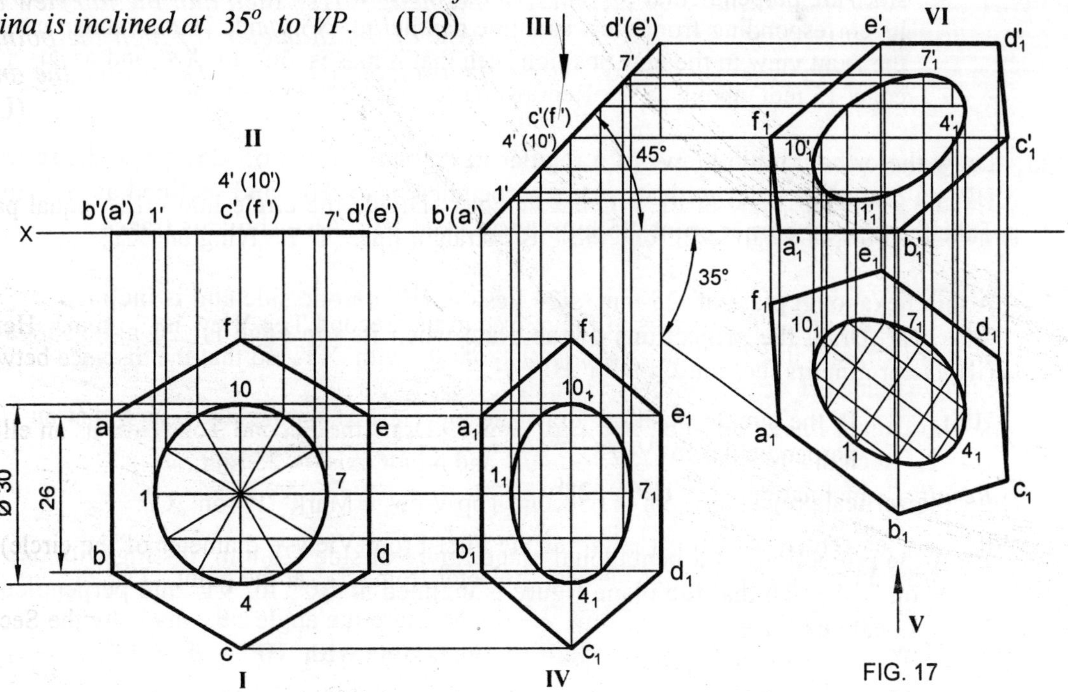

FIG. 17

Problem 21: *(Fig. 18) A semi-circular lamina of 64 mm diameter has its straight edge in VP and inclined at an angle of 45° to HP. The surface of the lamina makes an angle of 30° with VP. Draw the projections.*

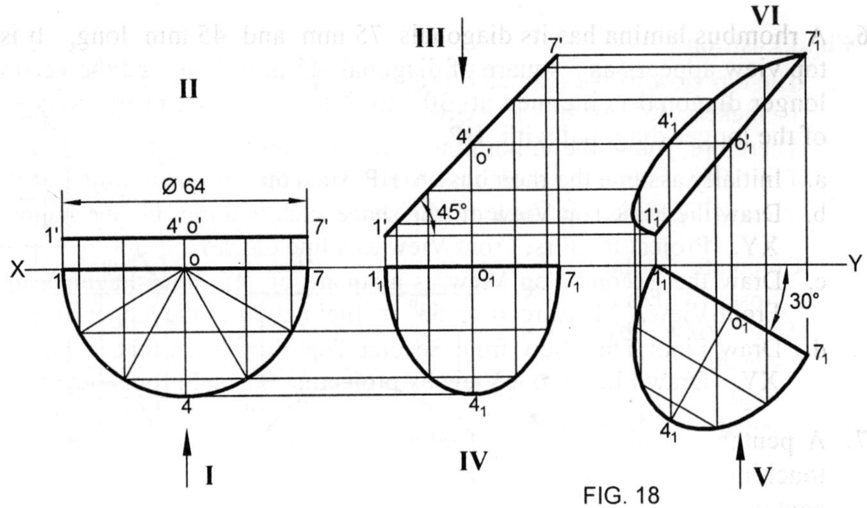

FIG. 18

4. ADDITIONAL PROBLEMS – I

22. A regular hexagonal plane surface of 25 mm side, has two of its edges parallel to both HP and VP and the nearest edge is 15 mm from each plane. The surface is inclined at 60° to HP. Draw the projections. (UQ)

Hint: Draw the top view held parallel to HP and 15 mm above it such that two of its sides are perpendicular to VP. The top view will be in true shape and size. Draw its corresponding front view as a line parallel to XY and 15 mm above it. Draw the front view in the new position such that it makes 60° to XY and has its nearest edge 15 mm above XY. Redraw the second top view as usual.

23. Draw the orthographic views of a regular hexagonal lamina of 25 mm side, resting on HP on one of its sides with its plane perpendicular to HP and inclined at 45° to VP. Take the nearest corner point 25 mm away from VP. Also show its traces.

24. A thin hexagonal plate of 25 mm side lies on HP on one side and is inclined at 30° to VP. (i) Draw the projections of the plate when its top edge is 20 mm above HP. (ii) Determine its inclination with HP. (UQ)

Hint: In the simple position, draw the top view of the hexagon such that an edge is perpendicular to VP.

Result : Inclination with HP = 28°.

25. Draw the projections of a pentagonal plane figure of side 28 mm resting with one of its edges on HP such that the plane figure is inclined at 30° to VP and perpendicular to HP. Locate its traces.

26. A rhombus lamina has its diagonals 75 mm and 45 mm long. It is placed such that its top view appears as a square of diagonal 45 mm long and the vertical plane through the longer diagonal is inclined at 30° to VP. Draw its projections. Find the inclination of the longer diagonal with HP. (UQ)

a. Initially assume the rhombus on HP such that the smaller diagonal is $\perp^r$ to VP.

b. Draw the First Top View of true shape such that the smaller diagonal is perpendicular to XY. Project the First Front View as a line on XY.

c. Draw the Second Top View as a square of 45 mm diagonal and project the Second Front View. Measure **θ = 53°** = Inclination of the longer diagonal with HP.

d. Draw Final Top View from Second Top View such that longer diagonal is at 30° to XY. Draw Final Front View by projecting Second Front View and Final Top View.

27. A pentagonal plane figure of edges 25 mm is lying on HP with one of its corners touching it such that the plane figure makes 60° with HP. Two of the edges containing the corner on which the plane figure rests make equal inclinations with HP. When the edge opposite to this corner makes 45° with VP, draw the top and front views of the plane figure in this position.

Hint: The plane figure is resting with a corner on HP such that the two edges containing the corner on which the plane figure rests, make equal inclinations with HP. Hence, draw the top and front views of the plane figure with its surface completely lying on HP. The front view will be a line on XY line.

28. A circular lamina of 60 mm diameter appears as an ellipse in the top view, having its major axis 60 mm long and minor axis 40 mm long. Draw its front view when the major axis of the ellipse is parallel to both the reference planes.

Hint: Assume the circle parallel to HP. Reproduce front view so that its top view is ellipse of 40 mm minor axis. Reproduce top view with major axis parallel to XY.

29. A thin circular metal plate of 54 mm diameter has a square hole of 27 mm side, cut centrally through it. Draw its projections when the plate is resting on HP with its surface inclined at 30° to HP and an edge of the square hole perpendicular to VP.

30. A circular lamina of 50 mm diameter appears as an ellipse in the front view, its major and minor axes being 50 mm and 30 mm respectively. Draw its top view when the major axis is horizontal. (UQ)

Hint: Circular lamina is seen as an ellipse in front view. Hence its surface must be inclined to VP. Assume the lamina parallel to VP and draw its front & top views.

31. A regular hexagonal lamina of 26 mm side has a central hole of 30 mm diameter. Draw the front and top views when the surface of the lamina is inclined at 45° to HP. A side of the lamina is inclined at 35° to VP.

32. A hexagonal plate of 25 mm side has a corner on HP. Its surface is inclined at 50° to HP. The diagonal through the corner which is on HP, makes 35° with VP. Draw its projections.

5. AUXILIARY VIEWS OF PLANE SURFACES

(PROJECTIONS OF PLANE SURFACES USING AUXILIARY PLANES)

Change of Reference Line Method: In the initial stage, the plane surface is kept parallel to HP / VP and the required view is obtained by projecting on new reference line X_1Y_1.

5.1 PLANE SURFACE PERPENDICULAR TO VP AND INCLINED TO HP

Problem 33: *(Fig. 19) A square plate of 35 mm side has its plane perpendicular to the VP and inclined at an angle of 50° to the HP. Draw the projections of the plate by the auxiliary plane method.*

1. Initially, place the square plate parallel to the HP and draw its top and front views.
2. Draw a new reference line X_1Y_1 inclined at an angle of 50° to XY.
3. Draw the projectors perpendicular to X_1Y_1 from the front view.
4. Mark the distances of the corner points of the top view from XY in the same order from the X_1Y_1 line along the newly drawn projectors.
5. Complete the Auxiliary Top View.

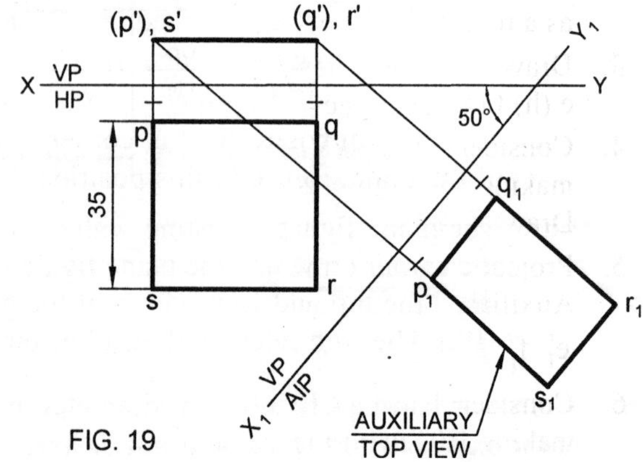

FIG. 19

Problem 34: *(Exercise) A pentagonal lamina of 25 mm side rests on HP on one of its sides with its surface inclined at 55° to HP and ⊥r to VP. Draw its projections by auxiliary plane method.*

5.2 PLANE SURFACE PERPENDICULAR TO HP AND INCLINED TO VP

Problem 35: *(Fig. 20) A pentagonal lamina of 24 mm side is resting on HP with one of its edges ⊥r to it and its surface parallel to VP. Draw an auxiliary front view on AVP ⊥r to HP and inclined at 60° to VP.*

Draw first front view with an edge perpendicular to XY. Draw first top view as a line.

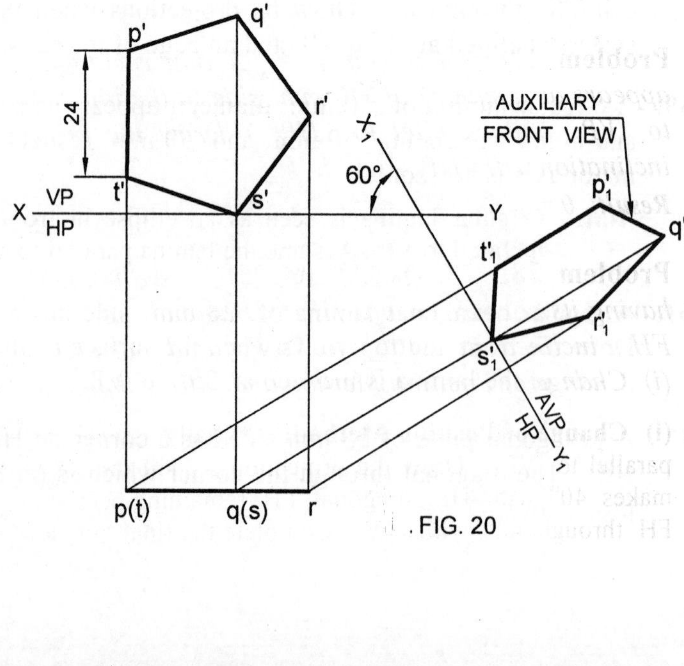

FIG. 20

5.3 PLANE SURFACES INCLINED TO BOTH HP AND VP

Problem 36: *(Fig. 21) A thin rectangular lamina EFGH of 60 mm length and 36 mm width is inclined at an angle of 45° to VP. Its longer edge is making an angle of 30° with HP. Draw the projections by Auxiliary Plane Method.* (UQ)

1. Assume the lamina to be parallel to VP with its longer edge parallel to HP.
2. Draw XY. Draw the front view e′ f′ g′ h′ as a rectangle.
3. Draw the top view e (h) f (g) as a line.
4. Consider an AVP making 45° with XY. Draw X_1Y_1.
5. Project and draw the Auxiliary Front View e_1' f_1' g_1' h_1'.
6. Consider an AIP making 30° with the longer edge g_1' h_1'. Draw X_2Y_2.
7. Project and draw the Auxiliary Top View e_1 f_1 g_1 h_1.

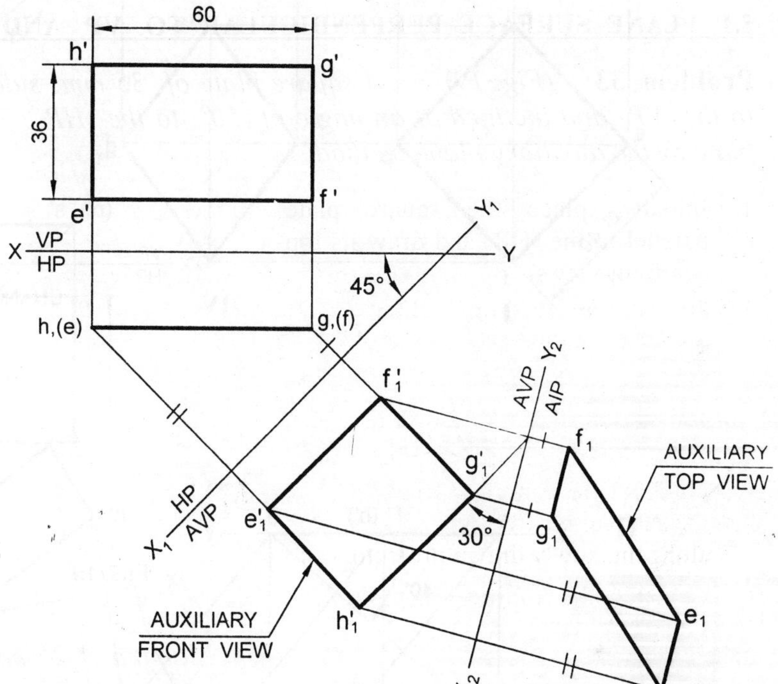

FIG. 21

Problem 37: *(Exercise) A thin rectangular plate of size 70 mm x 40 mm appears as a square of 40 mm side in the top view with one of its sides inclined at 30° to VP and parallel to HP. Draw the projections of the plate. Determine its inclination with HP.* (UQ)

***Result**: $\theta = 55^o$*

Problem 38: *(Figs. 22 & 23) A thin square plate EFGH of 40 mm side is having its corner G on HP. Diagonal GE is inclined at 40° to HP and diagonal FH inclined at 40° to VP and parallel to HP. Draw its projections by (i) Change of Position Method and (ii) Change of Reference Plane Method.* (UQ)

(i) Change of Position Method: Initially assume the square lying on HP with diagonal EG parallel to VP. Draw the projections. Tilt the square about corner G so that diagonal GE makes 40° with HP, diagonal FH remaining perpendicular to VP and parallel to HP. Rotate FH through 40° with XY. Complete the final top view as shown in *Fig. 22.*

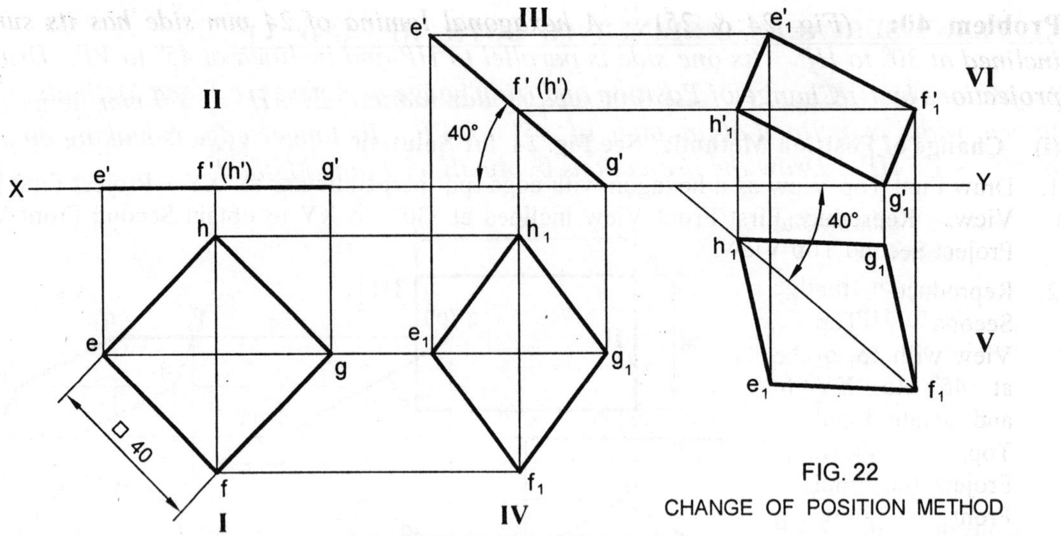

FIG. 22
CHANGE OF POSITION METHOD

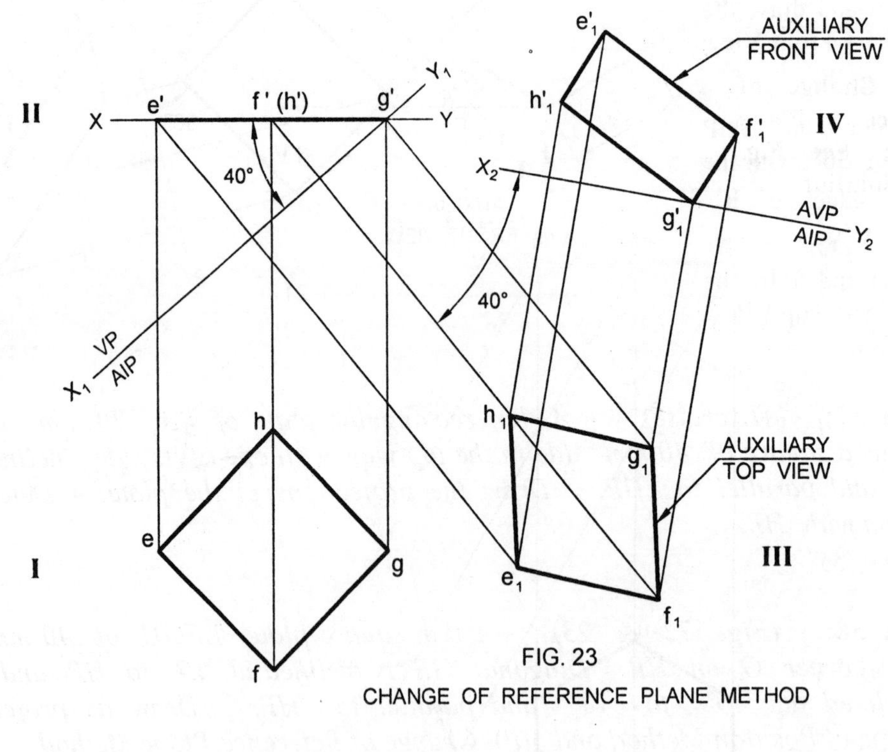

FIG. 23
CHANGE OF REFERENCE PLANE METHOD

Problem 39: *(Exercise) A rhombus has its diagonals 80 mm and 40 mm long. The smaller diagonal is parallel to both HP and VP. The longer diagonal is inclined at 35° to HP. Draw the projections.* (UQ)

Problem 40: *(Fig. 24 & 25) A hexagonal lamina of 24 mm side has its surface inclined at 30° to HP. Its one side is parallel to HP and inclined at 45° to VP. Draw its projections by (i) Change of Position and (ii) Change of Reference Plane Methods.*

(i) Change of Position Method: See Fig. 24 for Solution.

1. Draw First Top View as a hexagon with edge pq perpendicular to XY. Project First Front View. Reproduce First Front View inclined at 30° to XY to obtain Second Front View. Project Second Top View.
2. Reproduce the Second Top View with $p_1 q_1$ at 45° to XY and obtain Final Top View. Project the Final Front View.

   Note that $p_1' q_1'$ is parallel to XY.

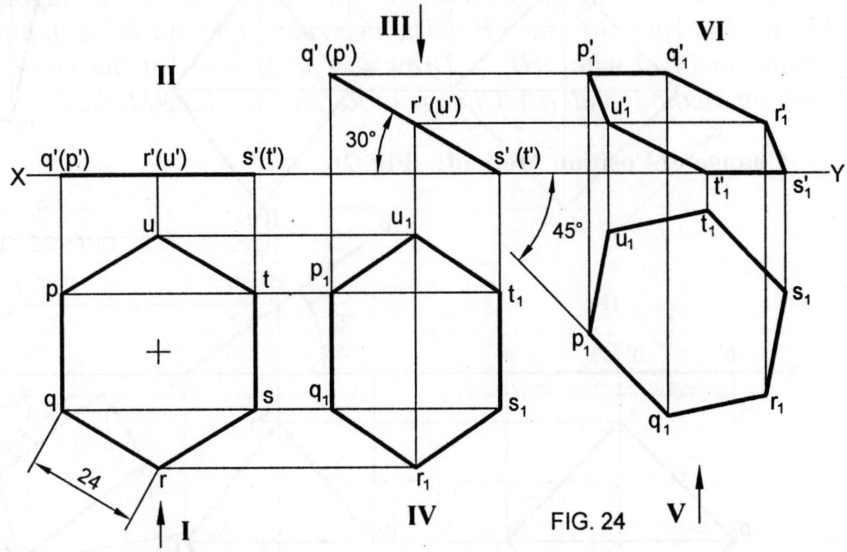

FIG. 24

(ii) Change of Reference Plane Method: See ***Fig. 25*** **for Solution.**

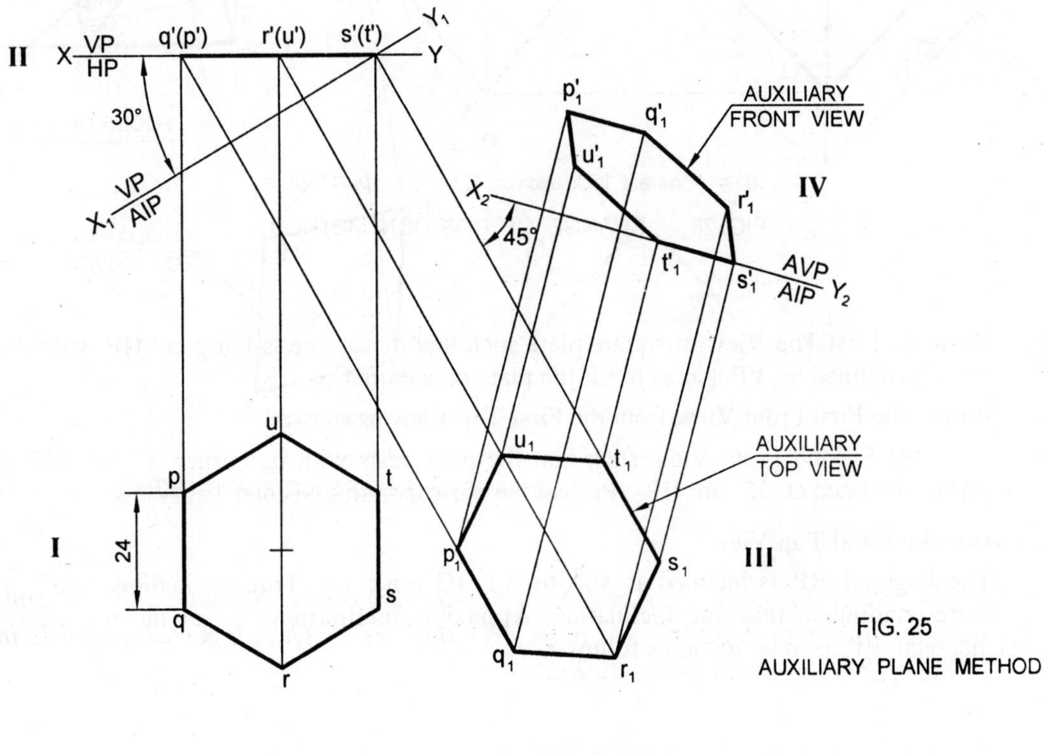

FIG. 25
AUXILIARY PLANE METHOD

Problem 41: *(Exercise) A thin hexagonal plate of 25 mm side lies on HP on one side and is inclined at 30° to VP. (i) Draw the projections of the plate when its top edge is 20 mm above HP. (ii) Determine its inclination with HP.* (UQ)

Hint: Simple Position: Draw top view such that an edge is perpendicular to VP.

Result : Inclination with HP = 28°.

Problem 42: *(Figs. 26 & 27) A square plate PQRS of negligible thickness having 35 mm side, is lying on a corner R on HP. One of the diagonals RP is inclined at 35° to HP and 40° to VP. The two sides QR and RS containing the corner R are equally inclined with HP. Draw the projections of the plate using (i) Change of Position Method and (ii) Change of Reference Plane Method.* (UQ)

(i) Change of Position Method: Fig. 26

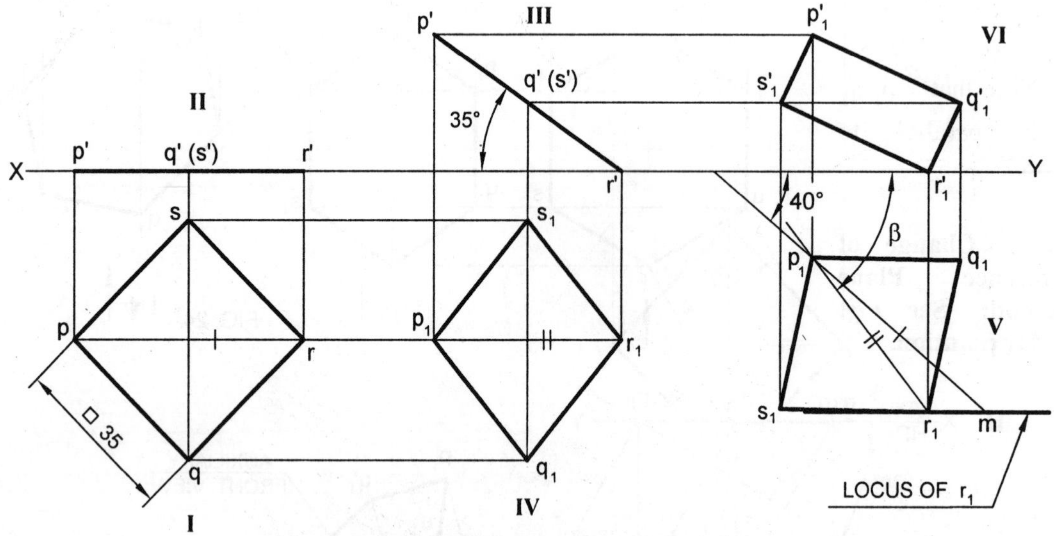

β = Apparent Inclination = 52° $p_1m = pr$

FIG. 26 CHANGE OF POSITION METHOD

1. Draw the First Top View of square plate such that its surface is lying on HP with the sides equally inclined to VP (so as to tilt the plate on a corner r).
2. Project the First Front View from the First Top View as shown.
3. Draw the Second Front View such that the plate rests with its corner r′ on HP and its surface inclined at 35° to HP. Project the corresponding Second Top View.

To draw the Final Top View

The diagonal RP is inclined at 40° to VP (Given); i.e., True Inclination = 40°. Hence, corresponding to this true inclination, **Apparent Inclination β** of the top view of the diagonal RP is to be found as follows:

4. Draw a line inclined at 40° to XY. On this, mark any point p_1. Mark m on this line such that p_1 m = True Length of the diagonal PR = pr from the First Top View.
5. Through m draw the locus of r_1 parallel to XY. With p_1 as center and radius $p_1 r_1$ draw an arc to cut the locus of r_1 at r_1. Join $p_1 r_1$. Redraw the Second Top View to obtain the Final Top View as shown. Project **Final Front View** from Final Top View.

(ii) Change of Reference Plane Method: Fig. 27

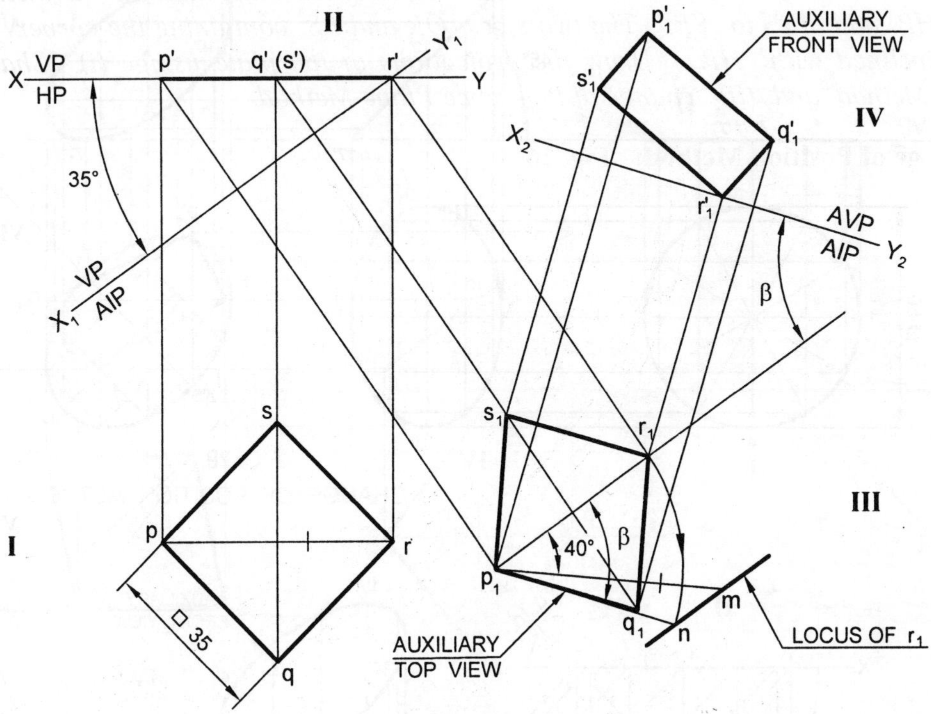

β = Apparent Inclination = 52° p_1m = pr $p_1n = p_1r_1$

FIG. 27 CHANGE OF REFERENCE PLANE METHOD

1. Draw the First Top and Front views in the simple position (Repeat Steps 1 & 2 above).
2. Draw X_1Y_1 passing through r' at 35° to XY. Project Auxiliary Top View $p_1 q_1 r_1 s_1$.

To draw the Final Front View

True Length of the diagonal PR should be inclined at 40° to VP (Given). But, X_2Y_2 cannot be drawn at the True Inclination of 40° to the shortened length of the diagonal in the Auxiliary Top View $p_1 r_1$. Hence, **Apparent Inclination β** of the Auxiliary Top View $p_1 r_1$ of the diagonal PR is to be found as follows:

3. Through p_1 draw a line at 40° to $p_1 r_1$. On this, mark the point m such that p_1m = True Length of the diagonal PR = pr. Through m draw the locus line parallel to X_1Y_1.
4. With p_1 as center and p_1r_1 as radius, draw an arc to cut the locus at n. Join p_1n.
5. Mark β as shown. Draw X_2Y_2 at an angle of β to p_1r_1. Project Auxiliary Front View.

Problem 43: *(Figs. 28 & 29) A thin circular plate of 50 mm diameter lies on HP such that its surface is inclined at 45° to HP. The diameter through the point on which the plate lies on HP appears to be inclined at 45° to VP in the top view.*

Draw the projections by (i) Change of Position Method and (ii) Change of Reference Plane Method. (UQ)

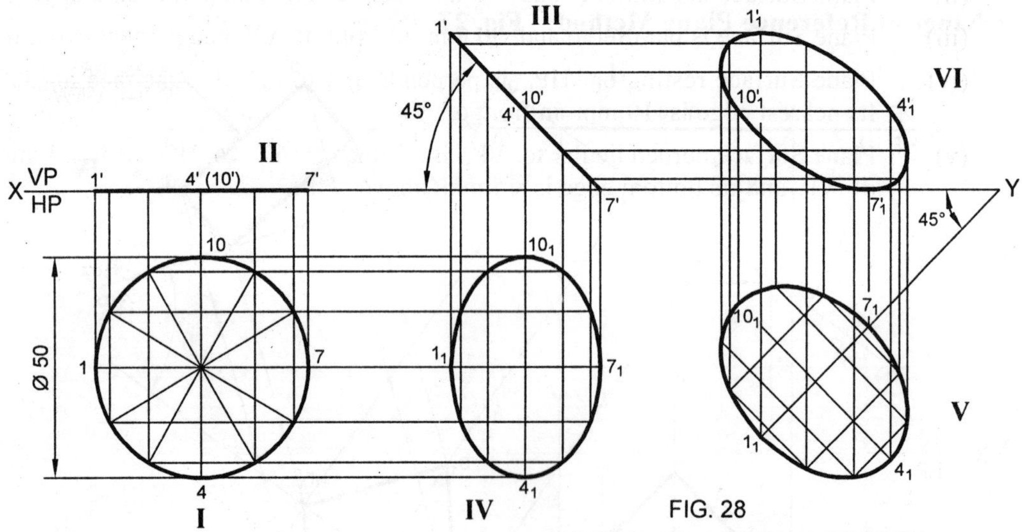

FIG. 28

CHANGE OF POSITION METHOD

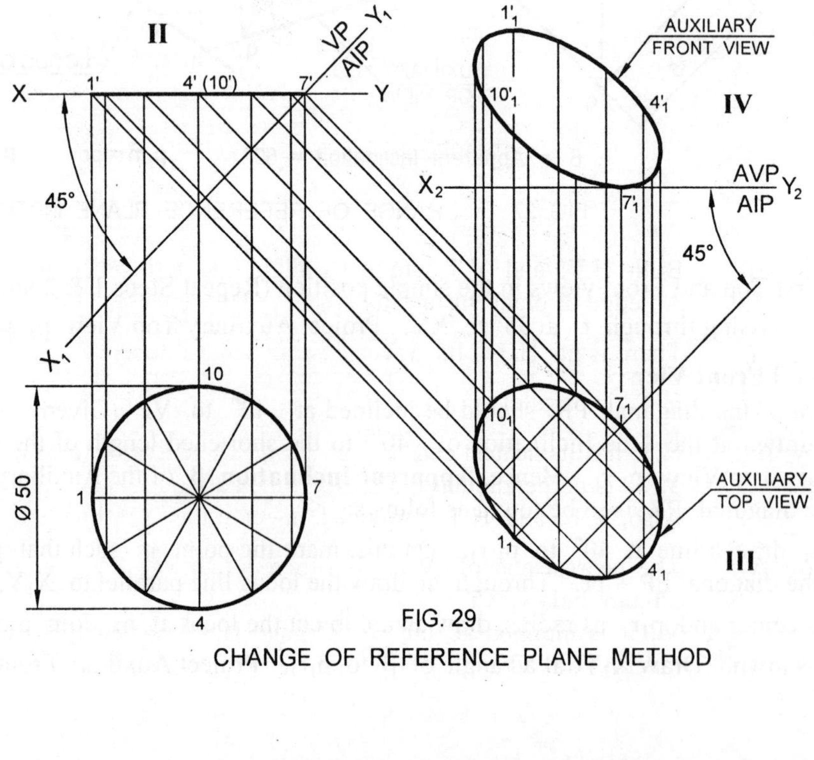

FIG. 29

CHANGE OF REFERENCE PLANE METHOD

6. ADDITIONAL PROBLEMS – II

44. (*Fig. 30*) Represent the following Plane Surfaces by their Traces.

(i) Plane Surface is perpendicular to both HP and VP with an edge on HP.

(ii) Plane Surface is parallel to and 35 mm above HP and perpendicular to VP.

(iii) Plane Surface is parallel to and 30 mm in front of VP and perpendicular to HP.

(iv) Plane Surface resting on HP, is perpendicular to it. It is inclined at 30° to VP. Its nearest edge is 10 mm in front of VP.

(v) Plane Surface perpendicular to VP, is inclined at 60° to HP. It is 15 mm in front of VP and its bottom edge is 15 mm above HP.

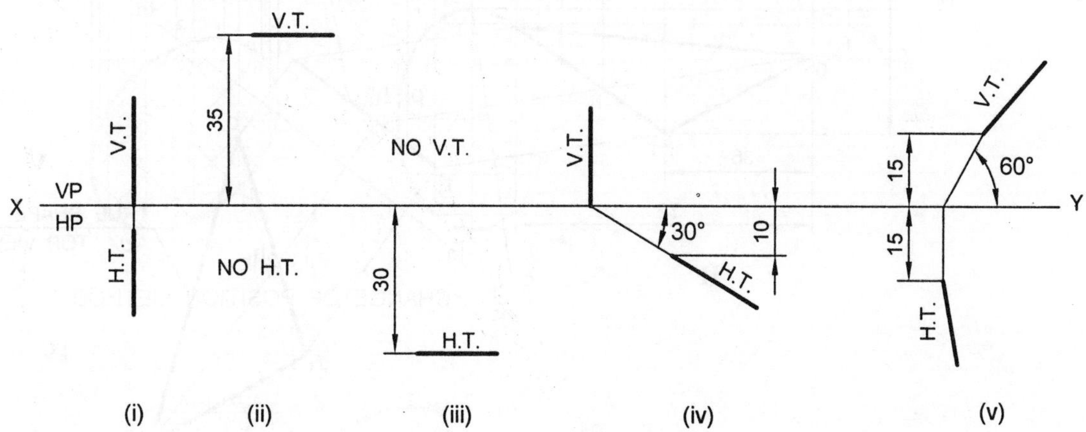

FIG. 30

TRACES OF PLANE SURFACES

(i) *See Fig. (i).* Both H.T. and V.T. are in a straight line perpendicular to XY. An edge of plane surface is on HP. Hence its V.T. is touching XY.

(ii) *See Fig. (ii).* There is no H.T. Its V.T. is a line parallel to XY, 35 mm above it.

(iii) *See Fig. (iii).* H.T. is parallel to XY and 30 mm below it. It has no V.T.

(iv) *See Fig. (iv).* H.T. has its neatest end 10 mm from XY and is inclined at 30° to XY. V.T. is perpendicular to the XY line and touches it. Also, V.T. and H.T. produced, meet at the XY line.

(v) *See Fig. (v).* The nearest end of V.T. is 15 mm above XY and that of H.T. is 15 mm below XY. V.T. is inclined at 60° to XY and H.T. is perpendicular to XY, both meeting on XY.

45. (*Fig. 31*) A thin triangular sheet has its corners P, Q and R above HP by 27 mm, 45 mm and 15 mm and in front of VP by 18 mm, 11 mm and 36 mm respectively. The projectors through P and Q are 72 mm apart and those through Q and R are 36 mm apart. If the projector through R is between those of P and Q, find the true size of the triangle by auxiliary projection method. (UQ)

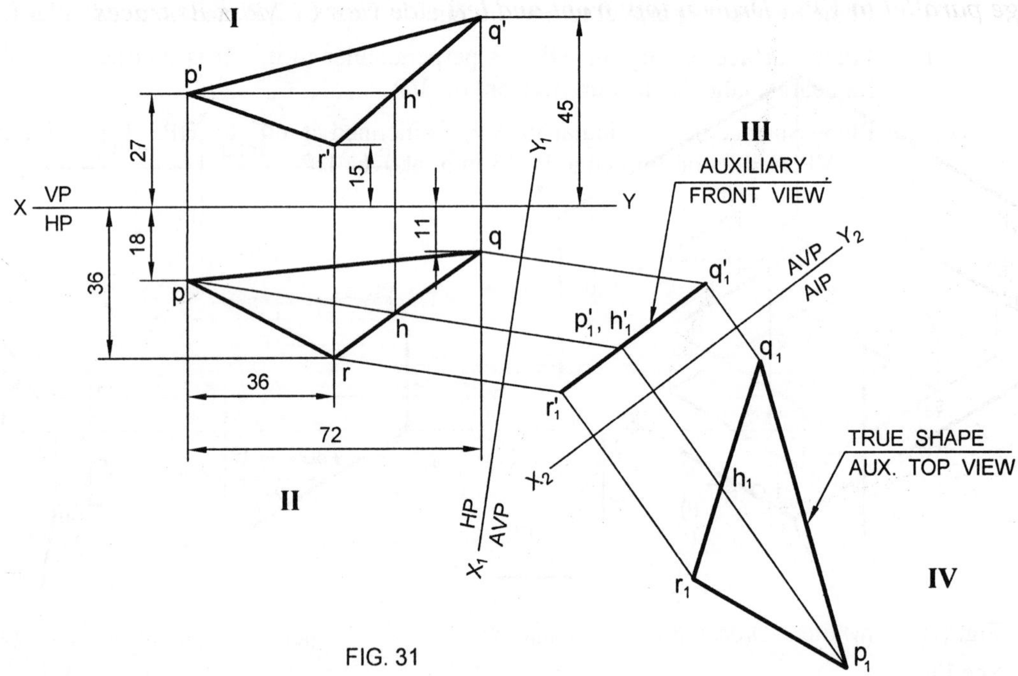

FIG. 31

1. Draw front and top views of the triangle. **To find true shape, setup 2 auxiliary planes.**
2. Draw a horizontal line through any of the corners of the views to cut the opposite side of the triangle. Let p′ h′ be horizontal line drawn in the front view.
3. Draw the same projections of the above line in the top view as p h.
4. Draw X_1Y_1 perpendicular to p h. Draw the auxiliary (second) front view of the triangle as a straight line. That is, the plane of the triangle is perpendicular to AVP.
5. Draw X_2Y_2 parallel to auxiliary (second) front view. Draw auxiliary (second) top view.
6. Since AIP is parallel to the plane of the triangle, the auxiliary (second) top view gives the **True Shape** of the triangle.

7. SIDE VIEWS OF PLANE SURFACES

7.1 PLANE SURFACE PARALLEL TO HP

If a plane surface is parallel to HP, its top view gives true shape. Hence, draw the top view first.

Problem 46: *(Fig. 32) A square lamina of 35 mm side is parallel to HP with an edge parallel to VP. Draw it top, front and* ***left*** *side views. Mark its traces.*

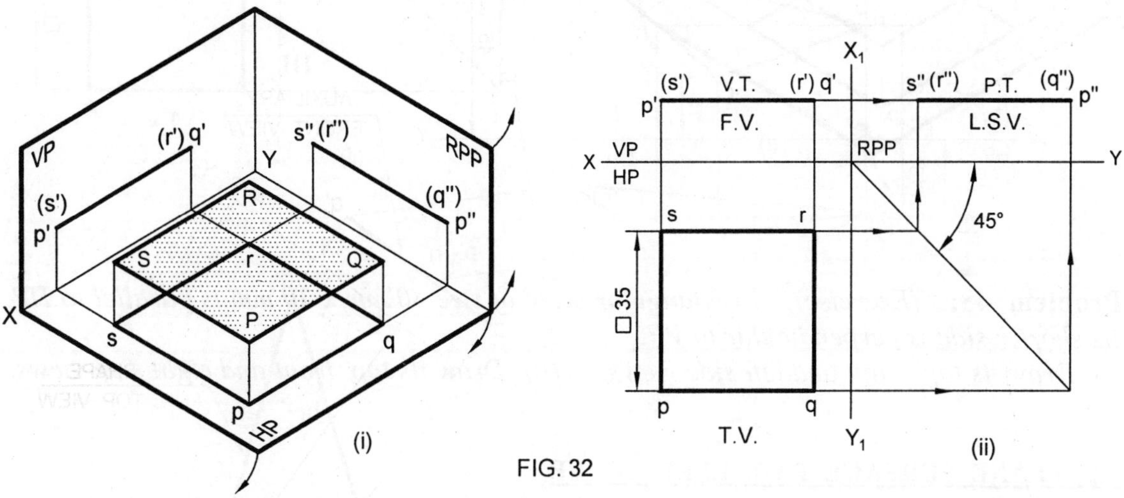

FIG. 32

1. Fig. (i) shows the 3-D drawing of the square lamina parallel to HP. Top View is in true shape.
2. See Fig. (ii). Draw the top view of the lamina with an edge parallel to XY.
3. Project the front view.
4. **Left Side View:** Draw X_1Y_1 perpendicular to XY to the right side of top and front views to represent Right Profile Plane (RPP). Draw the Mitre Line at 45° at the point of intersection of XY and X_1Y_1. Draw the horizontal projector through the corners r and q in the top view to meet the Mitre Line. From these meeting points, draw vertical projectors upwards to meet the horizontal projectors drawn from the front view. Complete the left side view as shown.
5. **Traces:** Front view of the lamina represents the V.T. Left side view represents the Profile Trace (P.T.) of the lamina. As the lamina is parallel to HP, there is no H.T.

Problem 47: *(Fig. 33) A square lamina of 35 mm side is parallel to HP with an edge parallel to VP. Draw it top, front and* ***right*** *side views.*

Right Side View: Draw X_1Y_1 perpendicular to XY to the left of top and front views to represent the Left Profile Plane (LPP). Draw the Mitre Line at 45° at the point of intersection of XY and X_1Y_1. Draw the horizontal projector through the corners r and s in the top view to meet the Mitre Line. From these meeting points, draw vertical projectors upwards to meet the horizontal projectors drawn from the front view. Complete the right side view as shown.

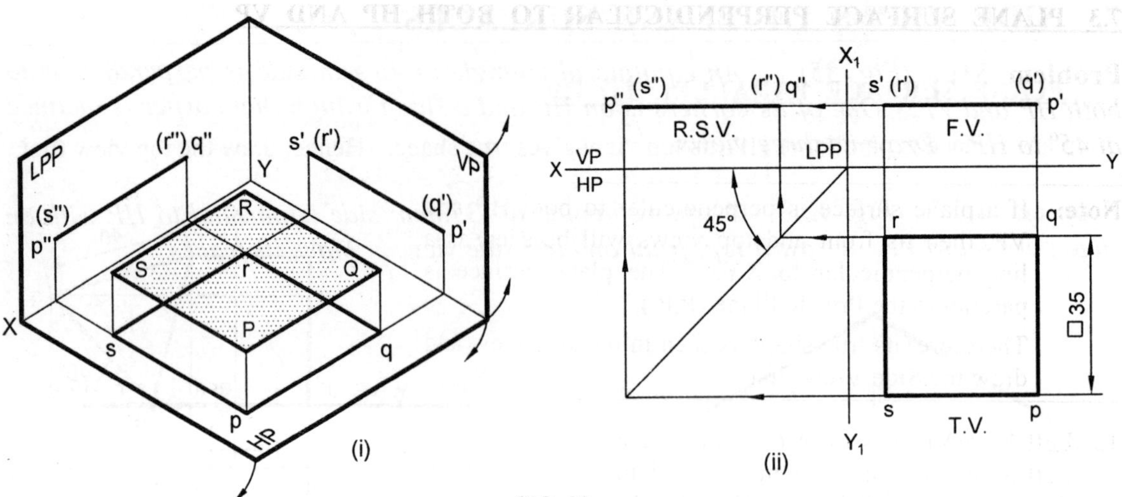

FIG. 33

Problem 48: *(Exercise) A rectangular plate of size 40 mm x 30 mm is parallel to HP. Its shorter side is perpendicular to VP.*
(i) Draw its top, front and left side views. *(ii) Draw its top, front and right side views.*

7.2 PLANE SURFACE PARALLEL TO VP

If a plane surface is parallel to VP, its front view is in true shape. Draw the Front View first.

Problem 49: *(Fig. 34) A thin pentagonal plate of 30 mm side is parallel to VP. One of its sides is perpendicular to HP. Draw its three views.*

1. Draw the front view of the pentagonal plate of side 30 mm with a side perpendicular to HP.
2. Project the top view.
3. Draw the left side view using the Mitre Line construction.
4. **Traces:** Top view of the plane surface represents the H.T. As the plane surface is parallel to VP, there is no V.T.

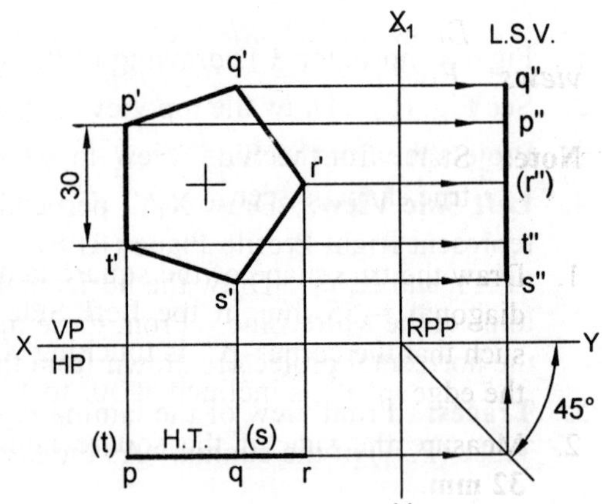

FIG. 34

Problem 50: *(Exercise) A circular lamina of 40 mm diameter is parallel to VP. Its center is 25 mm above HP and 28 mm in front of VP. (i) Draw the top, front and left side views. (ii) Draw the top, front and right side views. Locate its traces.*

7.3 PLANE SURFACE PERPENDICULAR TO BOTH HP AND VP

Problem 51: *(Fig. 35) An equilateral triangle of 40 mm side is perpendicular to both HP and VP. One of its corners is on HP and a side through that corner is inclined at 45° to HP. Draw its three views.*

Note: If a plane surface is perpendicular to both HP and VP, then its front and top views will be seen as a line perpendicular to XY. The plane surface is parallel to the Profile Plane (P.P.).

Therefore, its true shape is seen in the side view and draw the Side View first.

1. Left Side View is in true shape. Therefore, draw the Left Side View of the equilateral triangle p″ q″ r″ such that the corner p″ is on XY and side p″q″ of 40 mm length drawn through p″ is inclined at 45° to XY.
2. Project the front view from the left side view.
3. Project the top view using the Mitre Line construction.

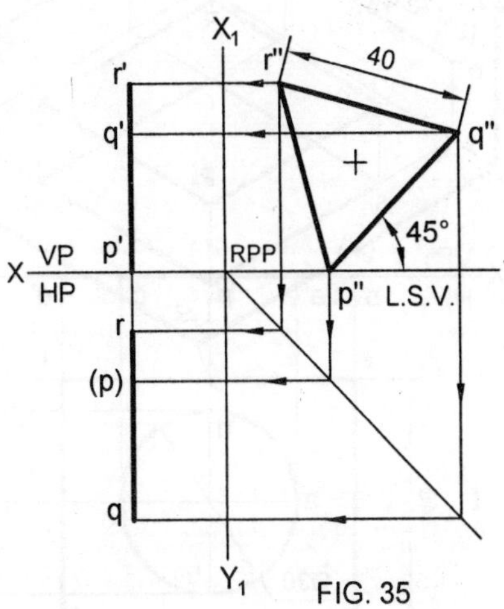

FIG. 35

Problem 52: *(Fig. 36) A square lamina having diagonals 55 mm is resting on HP such that its surface is perpendicular to both HP and VP. One of its sides is inclined at 30° to HP.*

Draw its left side view, top and front views. Find the size of the lamina. (UQ)

Note: Start with the Side View in which the true shape is seen.

1. Draw the true shape of the square lamina of diagonal = 55 mm in the Left Side View such that the corner p″ is touching XY and the edge p″q″ is inclined at 30° to XY.
2. Measure the side of the square lamina = **32 mm.**
3. Project the front view from the left side view.
4. Project the top view, using the Mitre Line construction.

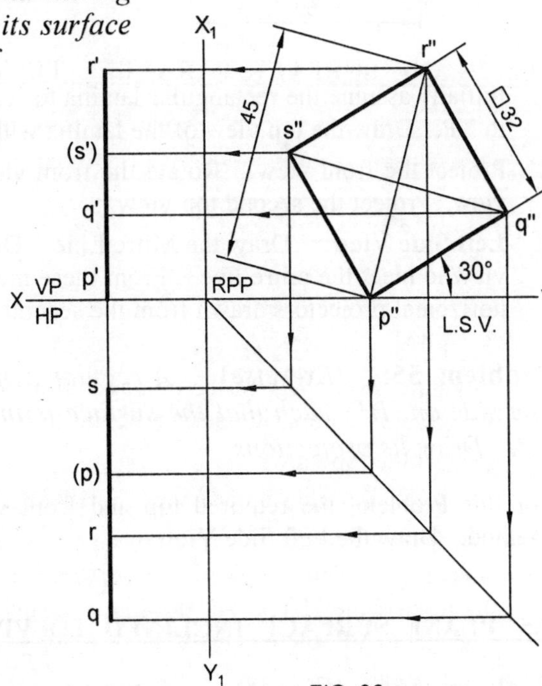

FIG. 36

Problem 53: *(Exercise) In the above Problem, check the result by constructing the right side view of the square lamina.*

7.4 PLANE SURFACE INCLINED TO HP

Problem 54: *(Fig. 37) A rectangular lamina of size 50 mm x 40 mm has a coaxial circular hole of 30 mm diameter. It is resting on HP with a shorter edge perpendicular to VP. The surface of the lamina is inclined at 35° to HP. Draw the top, front and left side views.* (UQ)

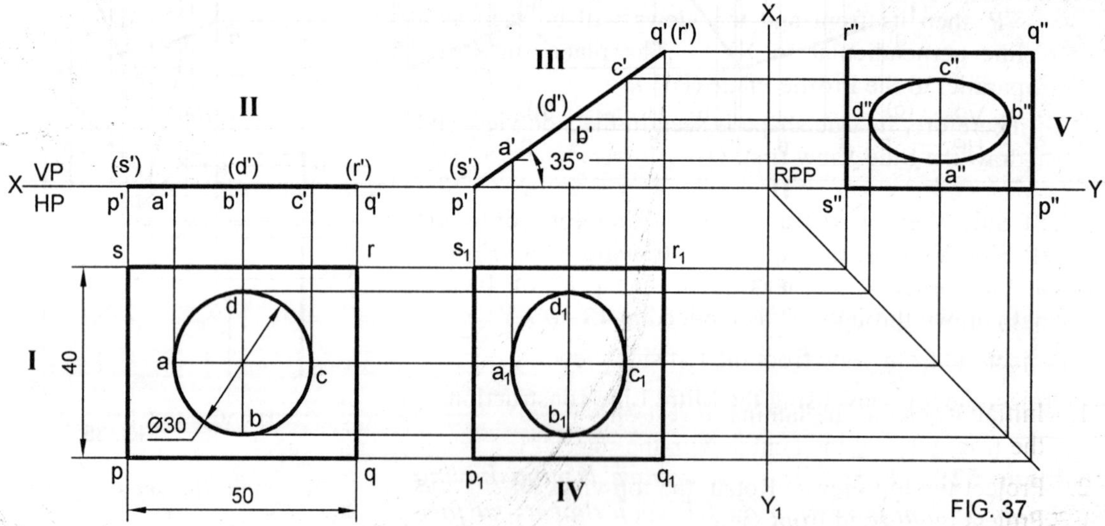

FIG. 37

1. Initially assume the rectangular lamina to rest on HP such that a shorter edge is perpendicular to VP. Draw the top view of the lamina with the circular hole in true shape.
2. Project the front view. Rotate the front view through 35° to XY to obtain the second front view. Project the second top view.
3. **Left Side View:** Draw the Mitre Line. Draw the horizontal projectors from the second top view to meet the mitre line. From these meeting points, draw vertical projectors to meet the horizontal projectors drawn from the second front View. Complete left side view as shown.

Problem 55: *(Exercise) A regular pentagonal plate of side 28 mm is placed with one side on HP such that the surface is inclined at 45° to HP and perpendicular to VP. Draw its projections.*

For this Problem, the required top and front views are already drawn in Change of Position Method. Draw the Left Side View.

7.5 PLANE SURFACE INCLINED TO VP

Problem 56: *(Fig. 38) A pentagonal lamina of 35 mm side has a semi-circular hole of 15 mm radius centrally punched. The lamina is perpendicular to HP with one of its sides touching VP. The surface of the lamina is inclined at 40° to VP. Draw the three views.* (UQ)

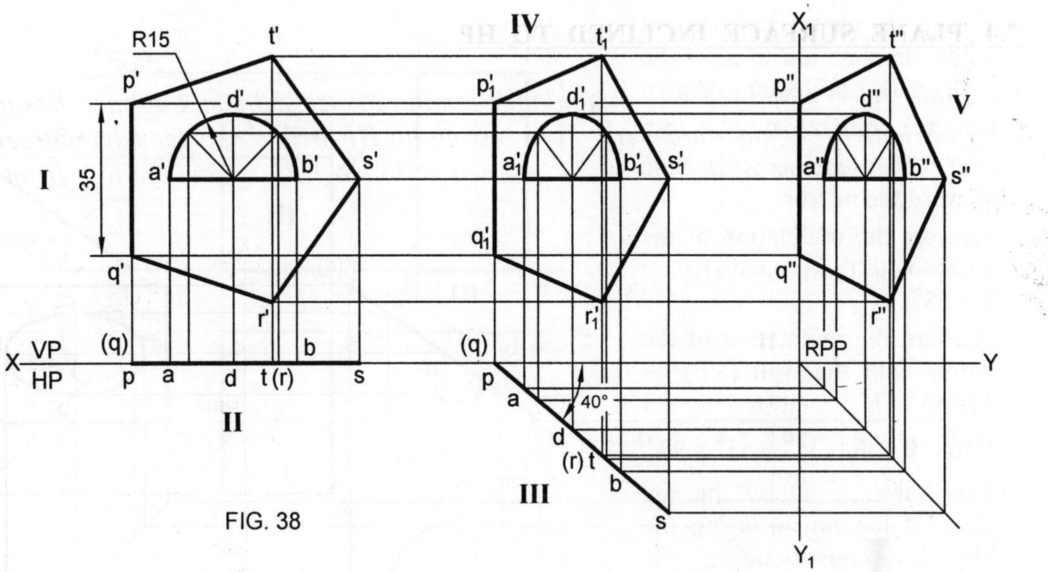

FIG. 38

1. Initially assume the lamina to rest on VP such that one edge is perpendicular to HP. Draw the front view of the lamina with the semi-circular hole in true shape.
2. Project the top view. Rotate the top view through 40° to XY to obtain the second top view. Project the second front view.
3. **Left Side View:** Draw the Mitre Line. Draw the horizontal projectors from the second top view to meet the mitre line. From these meeting points, draw vertical projectors to meet the horizontal projectors drawn from the second front View. Complete left side view as shown.

Problem 57: *(Exercise) A square lamina of 50 mm side has a semi-circular hole of 34 mm diameter. The center of the semi-circle coincides with the center of the lamina. The lamina is perpendicular to HP with one of its sides in VP. The surface of the lamina is inclined at 35° to VP. Draw the three views.* (UQ)

7.6 PLANE SURFACE INCLINED TO BOTH HP AND VP

Problem 58: *(Fig. 39) A mirror of size 560 mm x 320 mm is fixed on a wall on one of its shorter edges. The mirror is so fixed that it appears as a square in the front view. Draw the projections of the mirror. Find its inclinations with the wall and ground.* (UQ)

1. Scale 1 : 10. Initially, assume the mirror resting on the wall. Hence, draw its first front view in its true shape at any distance from XY line. Project the top view as a line.
2. Mirror appears as a square in the front view (*given*). Hence, draw the second front view as a square as shown.
3. Inclination of the mirror to the wall (VP) and ground (HP) is visible only in the side view. Draw X_1Y_1 to the right side of second front view. Draw horizontal projectors from the first front view to meet the X_1Y_1 line at the points a''b''.

4. With b″ as center and b″a″ as radius draw an arc to meet the horizontal projector drawn from the point a_1' at a″. Join a″b″ which represents the Side View of the mirror.
5. Measure the inclination of the mirror with the ground (HP) = **θ = 35°**.
6. Measure the inclination of the mirror with the wall (VP) = **ϕ = 55°**.

   Note: θ + ϕ = 35° + 55° = 90°.
7. **Top View:** Draw the top view of the mirror using the Mitre Line construction.

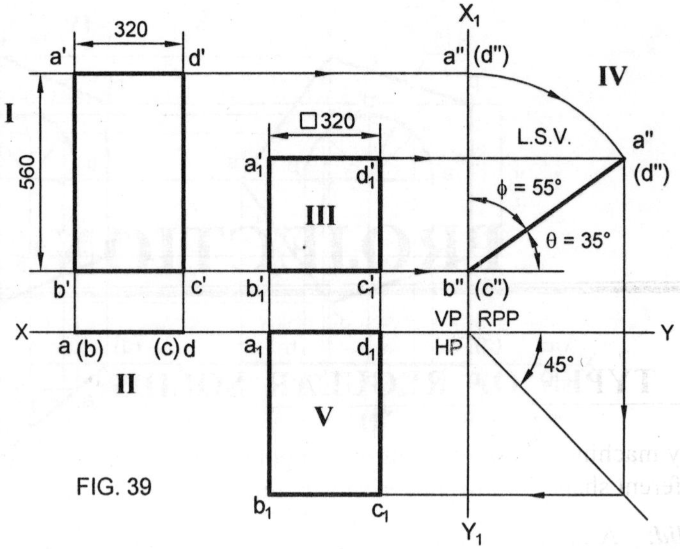

FIG. 39

Problem 59: *(Fig. 40) A hexagonal plate of negligible thickness is of 25 mm side, lying in such a way that one of its corners touches both the reference planes. The plate makes an angle of 60° with HP and 30° with VP. Draw the three views.* (UQ)

Note: The hexagonal lamina is inclined to both HP and VP such that the sum of the inclinations with HP and VP = 60° +30° = 90°. Hence, the Side View appears as a line.

1. Initially, assume the plate to be held parallel to VP with one corner touching XY. In this position, the front view will be a hexagon in true shape. Draw the front and top views.
2. Draw horizontal projectors from the front view to meet X_1Y_1. Draw the side view on X_1Y_1.
3. Rotate the side view from the corner q′ through 30° to X_1Y_1. Then draw the second front view and then the second top view using the Mitre Line construction.

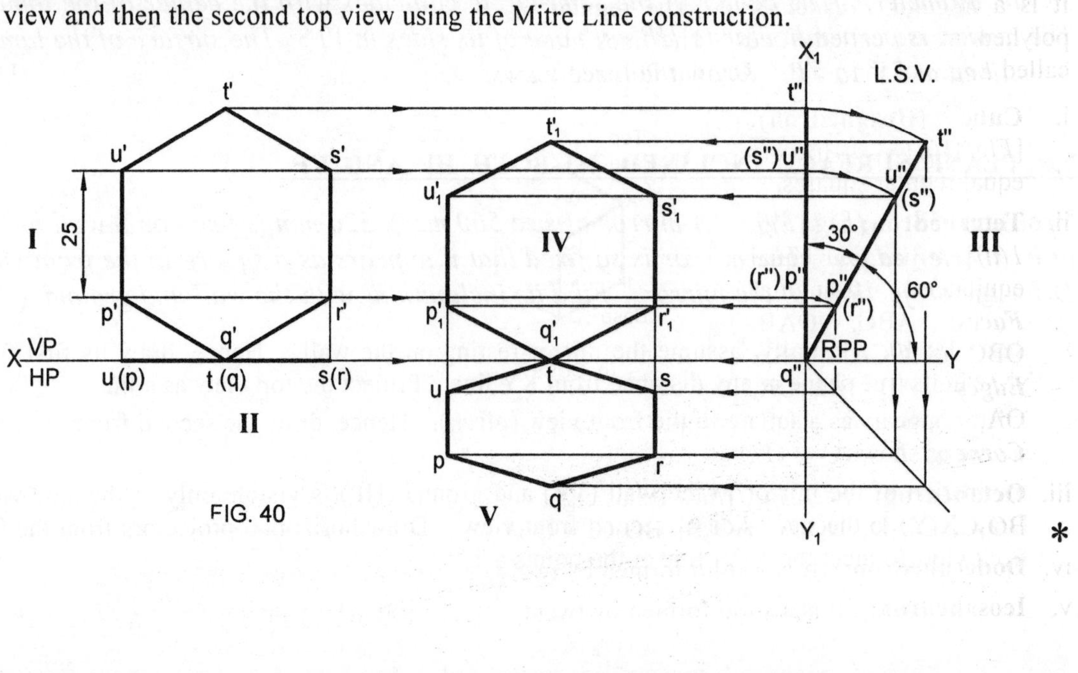

FIG. 40

*

13

PROJECTIONS OF SOLIDS

1. TYPES OF REGULAR SOLIDS

Any machine part or element or component is formed of a single solid or combination of solids of different shapes, known as prisms, cylinders, pyramids, cones and spheres.

Solid: A *Solid* is defined as an object having three dimensions, *viz.*, length, breadth and height or thickness, measured along the three mutually perpendicular axes. It is bounded by plane faces or curved surfaces or combination of both. If the axis of a solid is *perpendicular* to its base, it is a *Right Solid*. If all edges of the base / end faces of a *right solid* are of *equal* lengths, it is known as *Right Regular Solid*; otherwise *Irregular Solid*.

Classification of Solids: 1. Polyhedra 2. Solids of Revolution 3. Frustums & Truncated Solids

1.1 POLYHEDRA

Polyhedron: A solid which is bounded by plane surfaces (*i.e.*, faces) is called *Polyhedron*. Polyhedra are sub-divided into *three* types: 1. Regular Polyhedra, 2. Prisms and 3. Pyramids.

1. **REGULAR POLYHEDRA:** If all the faces of a polyhedron have the same size and shape, it is a *Regular Polyhedron*. The regular plane surfaces which form the surfaces of a polyhedron are called *Faces*. The lines along which two plane faces or surfaces meet are called *Edges* of the solid. Regular Polyhedra are divided into the following *five* types:
   - i. **Cube (Hexahedron):** [*Fig. 1(i)*] It has *six* equal regular squares.
   - ii. **Tetrahedron:** [*Fig. 1(ii)*] It has *four* equal equilateral triangular ***Faces***: ABC, OAB, OBC & OCA. Six ***Edges:*** AB, BC, CA, OA, OB & OC. ***Corners***: A, B, C & O.

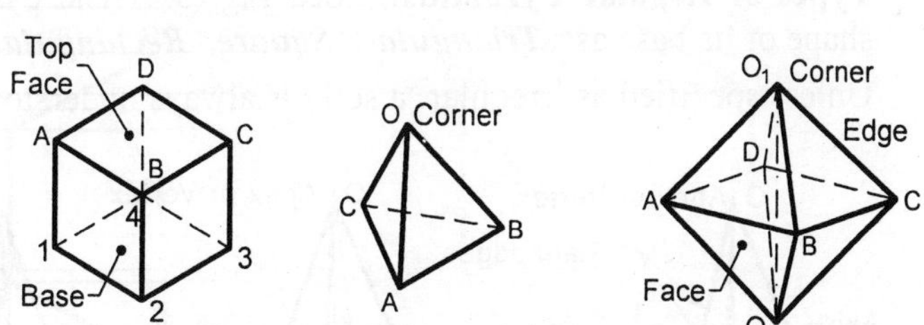

(i) CUBE (ii) TETRAHEDRON (iii) OCTAHEDRON

FIG. 1

   - iii. **Octahedron:** *See Fig. 1(iii)*. It has eight equal equilateral triangular ***Faces***, *viz.*, AO_1B, BO_1C, CO_1D, DO_1A, AO_2B, BO_2C, CO_2D & DO_2A. ***Corners***: A, B, C, D, O_1 & O_2.
   - iv. **Dodecahedron:** It is a solid formed by twelve equal regular pentagonal faces.
   - v. **Icosahedron:** It is a solid formed by twenty equal equilateral triangles.

2. PRISMS: A *Prism* is a polyhedron having two equal and similar faces (called *top face* and *bottom face* or *base*), parallel to each other and joined by other faces which may be *Rectangles* or *Parallelograms*. An imaginary line joining the centers of faces is *Axis*.

Note: *A prism is said to be a **right regular** prism when its axis is perpendicular to its base and all its faces (i.e., lateral surfaces) are equal rectangles.*

Types of Regular Prisms: *See Fig. 2.* Prism is named according to the shape of its base as: Triangular, Square, Rectangular, Pentagonal, Hexagonal, etc.

Nomenclatures of a prism: *Axis, Top face, Bottom face (Base), Rectangular faces and Longer edges.* These are clearly indicated in *Fig. 2.*

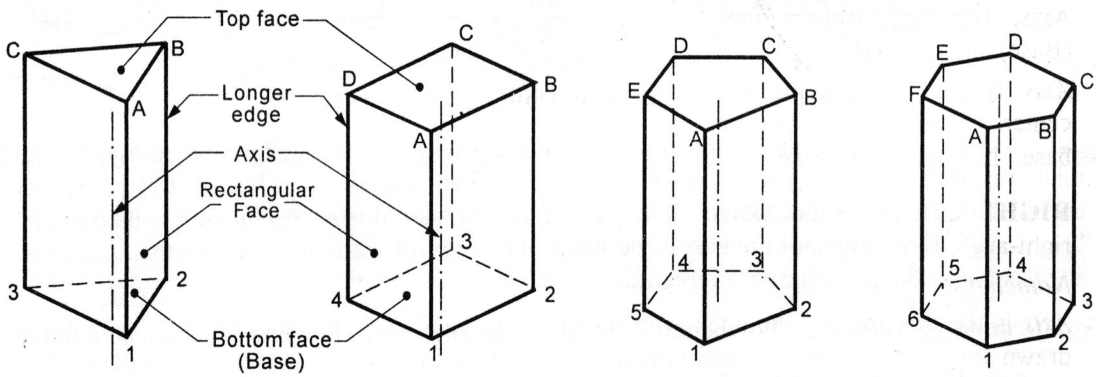

(a) TRIANGULAR (b) SQUARE (c) PENTAGONAL (d) HEXAGONAL

FIG. 2

3. PYRAMIDS: A *Pyramid* is a polyhedron having a plane figure as its ***base*** and a number of ***equal isosceles triangular faces*** meeting at a point called ***Vertex*** or ***Apex***. Pyramid is said to be *Right Regular* when its axis is perpendicular to the base, which is a regular plane figure. The line joining the apex and the corner of the base is called *Slant Edge*.

Types of Regular Pyramids: *See Fig. 3.* The pyramid is also named according to the shape of its base as: ***Triangular, Square, Rectangular, Pentagonal, Hexagonal, etc.*** Unless specified as irregular, a solid is always understood as being **'right'** and **'regular'**.

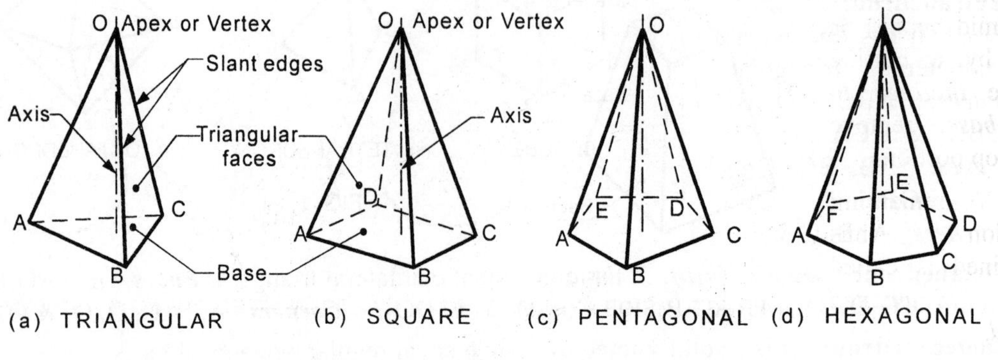

(a) TRIANGULAR (b) SQUARE (c) PENTAGONAL (d) HEXAGONAL FIG. 3

1.2 SOLIDS OF REVOLUTION

Solid of Revolution: *See Fig. 4.* If a plane surface is revolved about one of its edges, the solid generated is called a *Solid of Revolution*. *Examples*: Cylinders, Cones, Spheres, etc.

1. **RIGHT CIRCULAR CYLINDER:** *See Fig. 4(a).* It is a solid generated by the revolution of a rectangle about one of its sides which is fixed. It has top face & bottom base.

   Nomenclatures of a Cylinder: Axis, Top face, Bottom face (Base) and Generators.

   Axis is the line joining the centers of top face and bottom base. It is $\perp^r$ to the base.

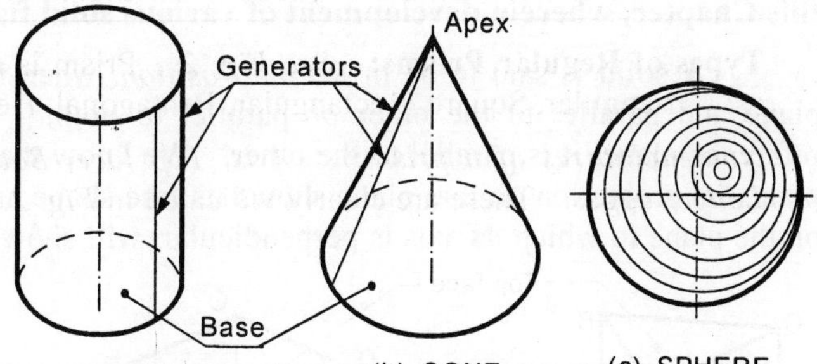

(a) CYLINDER (b) CONE (c) SPHERE

FIG. 4

2. **RIGHT CIRCULAR CONE:** *See Fig. 4(b).* It is a solid generated by the revolution of a right-angled triangle about one of its perpendicular sides which is fixed.

   Nomenclatures of a Cone: Axis, Apex or Vertex, Base, Generators and Slant Height.

   Axis is the line joining the apex with the center of the base. *Generators* are straight lines drawn from the apex to the circumference of the base circle. The length of the generator is the *Slant Height* of the cone.

3. **SPHERE:** *See Fig. 4(c).* A *Sphere* is a solid generated by the revolution of a semi-circle about its diameter as its axis. Mid-point of the diameter is the center of the sphere.

1.3 FRUSTUMS AND TRUNCATED SOLIDS

1. **FRUSTUMS:** When a pyramid or a cone is cut by a cutting plane ***parallel to its base***, remove the top portion. The remaining portion thus obtained is called the *Frustum of Pyramid or Cone. See Fig. 5.*

2. **TRUNCATED SOLIDS** (Fig. 6) When a solid (prism / cylinder / pyramid / cone) is cut by a cutting plane ***inclined to its base***, remove the top portion.

   The remaining portion thus obtained is called the *Truncated Solid.*

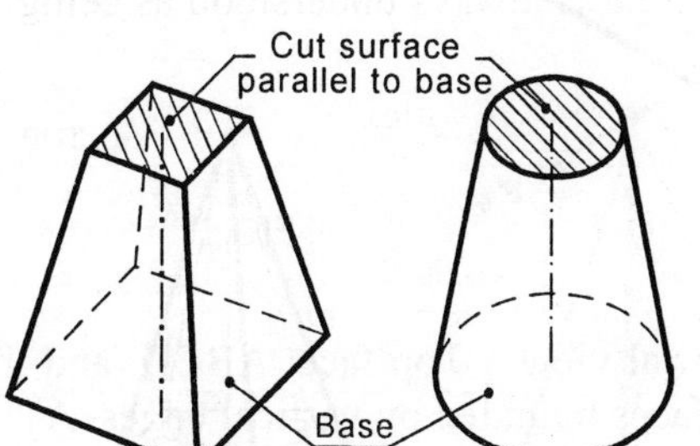

FIG. 5
FRUSTUMS OF PYRAMID AND CONE

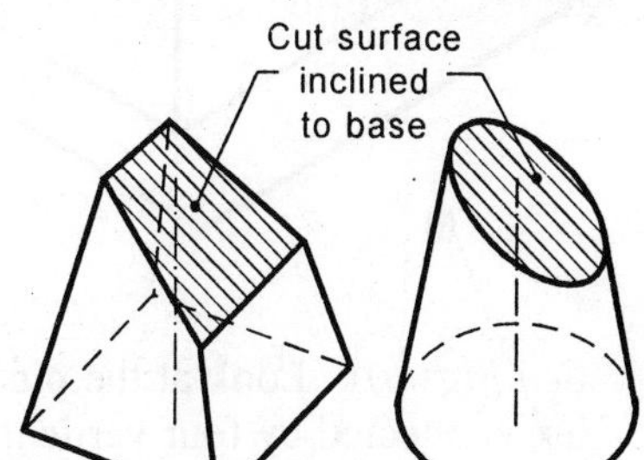

FIG. 6
TRUNCATED PYRAMID AND CONE

2. SOLIDS IN SIMPLE POSITIONS

Students are advised to make their own Padding Sheet (Colour Chart) / Cardboard / Perspex models for various solids. The models will facilitate developing a good concept of Projections of Solids in various positions with respect to HP and VP. Refer to the end of this Chapter, wherein development of various solid figures are given for making models.

A solid is said to be in *Simple Position,* when its axis is perpendicular to one reference plane and parallel to the other *or* parallel to both. When the axis is perpendicular to one reference plane, it is parallel to the other. We know that when a lamina is parallel to a reference plane, its projection on that plane shows its true shape and size. Hence, the projection of a solid on the plane to which its axis is perpendicular, will show the true shape and size of its base.

3. PROJECTIONS OF PRISM, PYRAMID, CYLINDER AND CONE

3.1 AXIS PERPENDICULAR TO HP (Simple Position)

Problem 1: *(Fig. 7) Draw the top and front views of a cube of 40 mm side resting with one of its square faces on HP such that one of its vertical faces is parallel to and 20 mm in front of VP.*

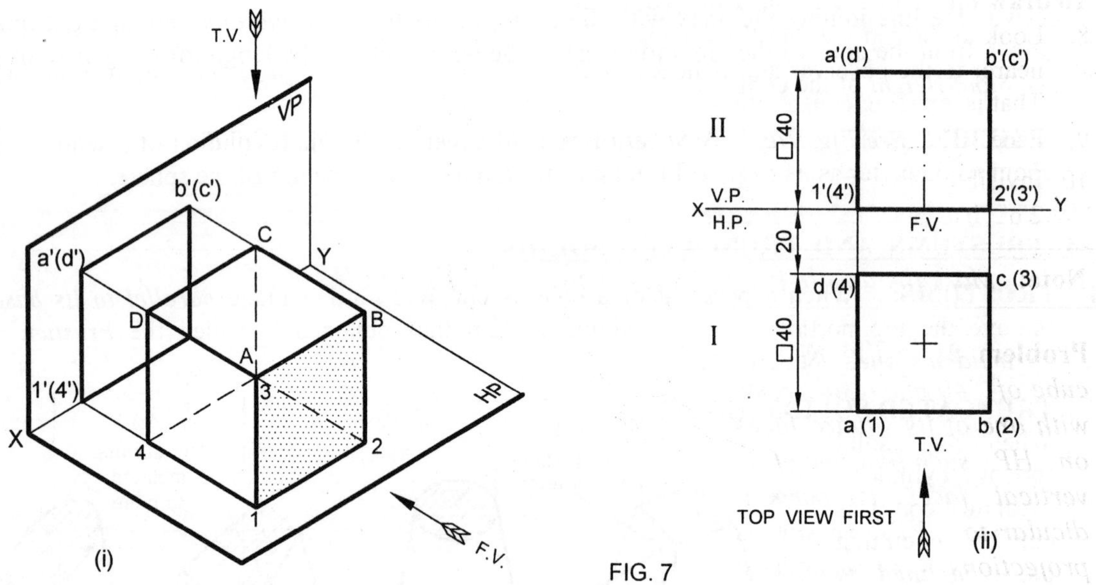

FIG. 7

1. *See Fig. (i).* Look at the pictorial view. Top face ABCD and bottom face (base) 1 2 3 4 are connected by four vertical faces bounded by vertical edges A1, B2, C3 and D4.

From the given statement, decide the view to be drawn first

Rule: The view which gives the true shape of the solid shall be drawn first; Other view(s) shall be projected from it.

2. Cube rests with a face on HP. Thus its axis is perpendicular to HP. Hence, the end face is seen in true size in the top view; And **draw the top view first.**

To draw the top view, look down at the cube in the pictorial view from the top

3. *See Fig. (ii).* Cube rests with a face on HP, *i.e.*, top face ABCD is parallel to HP and an edge CD of a vertical face is parallel to and 20 mm in front of VP. Hence, draw the top view abcd as a square of 40 mm side with its edge cd parallel to and 20 mm below XY.
4. Bottom face 1 2 3 4 is hidden by the top face ABCD. So in the top view mark the bottom face as (1) (2) (3) (4) such that (1) and **a** coincide, (2) and **b** coincide, etc. Thus, complete the top view as shown.
5. **To draw the front view:** Draw the vertical projectors in *thin* lines from the top view. Look at the cube from the front. Bottom face 1 2 3 4 is on HP. Hence draw the front view of the bottom face as 1′ 2′ 3′ 4′ on XY. Height 40 mm is seen in front view. Therefore, draw a horizontal line 40 mm above XY to cut the projectors drawn from a, b, c and d at a′, b′, c′ and d′ respectively.

Rule: *In the Orthographic views of a solid, corners are first located and they are connected by lines. While doing so, the visible and invisible lines are to be determined. The invisible corners are denoted by letters within brackets.*

To draw the visible and invisible edges in the front view

8. Look at the top view in the direction of the arrow (observer) shown. Face ab (2) (1) is nearer to the observer and is hence visible. So draw a′ b′ 2′ 1′ by *thick line* in front view. That is, existing construction *thin* line is drawn *thick.*
9. Face d c (3) (4) is *invisible* and hence mark it as (d′) (c′) (3′) (4′) in the front view.
10. Invisible edges (4′)(3′), (3′)(c′), (c′) (d′) and (d′) (4′) are hidden by the visible edges 1′2′, 2′b′, b′a′ and a′1′ respectively in front view and hence *not* shown in dashed lines.

Note: *Mark the axis in front and top views for all the solids.*

Problem 2: *(Fig. 8) A cube of 40 mm side rests with one of its square faces on HP such that one of its vertical faces is perpendicular to VP. Draw its projections. The nearest edge parallel to VP is 5 mm in front of it.*

When the face 1AD4 is perpendicular to VP, *the face 3CD4 becomes parallel to VP. Now, this Problem is similar to Problem 1. See Fig. 8.*

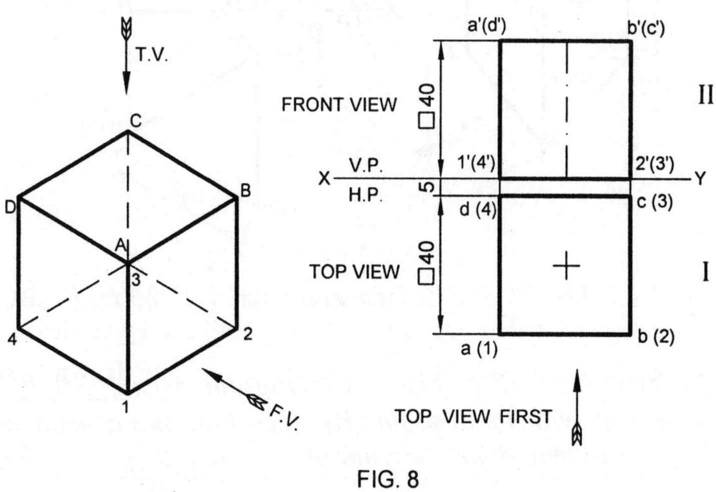

FIG. 8

Problem 3: *(Fig. 9) Draw the projections of a* ***square prism*** *of side of base 30 mm and height 50 mm resting with its base on HP such that one of its rectangular faces is perpendicular to VP. The nearest edge parallel to VP is 5 mm in front of it.*

Base is on HP. So axis is perpendicular to HP. Therefore, the top view gives true size of the end face and **draw the top view first.**

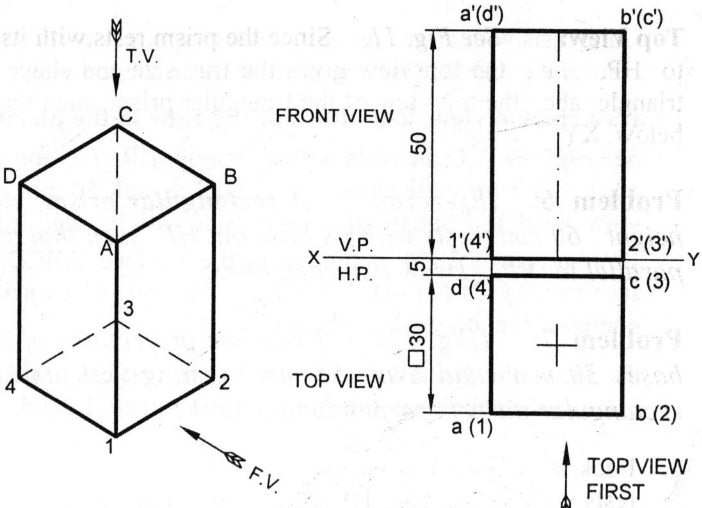

FIG. 9

Problem 4: *(Fig. 10) A* ***triangular prism****, side of base 35 mm and height 50 mm rests with its base on HP such that one of its rectangular faces is perpendicular to VP. Draw its projections. The nearest edge parallel to VP is 10 mm in front of it.*

See Fig. 10. Look at the pictorial view. Prism rests with its base on HP. So its axis is perpendicular to HP, *i.e.*, its top view gives true size of the end face. Hence, **draw the top view first.**

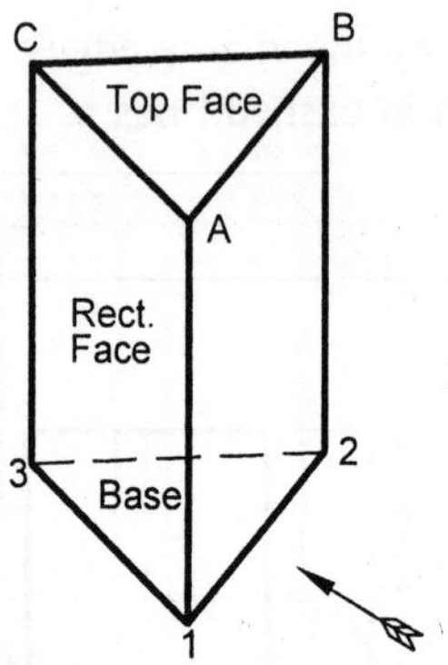

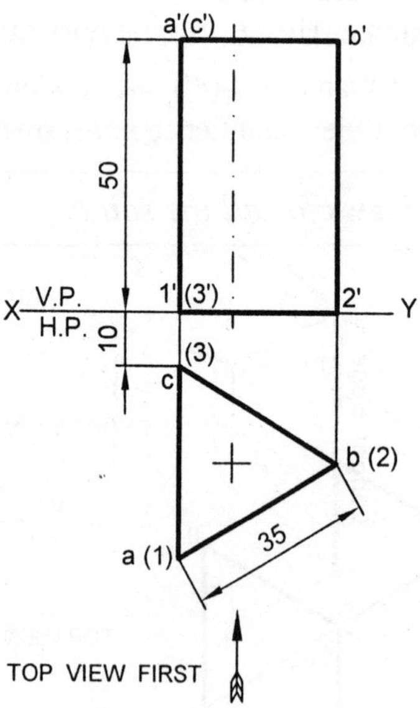

FIG. 10

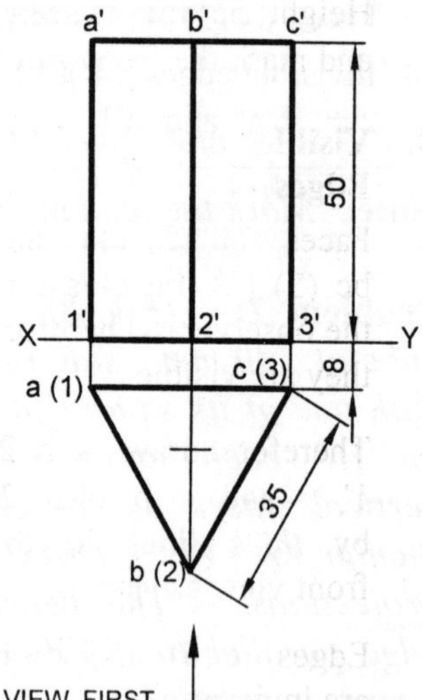

FIG. 11

Problem 5: *(Fig. 11) A* ***triangular prism****, side of base 35 mm and axis 50 mm long rests with its base on HP such that one of its rectangular faces nearer to VP is parallel to and 8 mm in front of it. Draw its projections.*

Top View: *See Fig. 11.* Since the prism rests with its base on HP, its axis is perpendicular to HP, *i.e.*, the top view gives the true size and shape of the end face. So, first draw the triangle abc, the top view of the triangular prism, such that its edge ac is parallel to and 8 mm below XY.

Problem 6: *(Exercise)* *A* ***rectangular prism****, side of base 40 mm x 25 mm and height 60 mm rests with its base on HP such that one of its larger rectangular faces is parallel to VP. Draw its projections.*

Problem 7: *(Fig. 12)* *Draw the projections of a* ***regular pentagonal prism*** *side of base 30 mm and axis 55 mm resting with its base on HP such that one of its rectangular faces is perpendicular to VP.*

1. Look at the pictorial view. ABCDE is the top face and 12345 is base. Since the prism rests with its base on HP, top view gives true size of the end face. Hence, **draw the top view first.**

2. **Top View**
   Look at the prism from the top. Top face ABCDE is parallel to HP. Further the rectangular face AE 51 is perpendicular to VP. Hence draw the top view abcde (a pentagon of 30 mm side) with ae perpendicular to XY. Base is hidden and its top view is marked as (1) (2) (3) (4) (5).

3. **Front View**
   Look at the prism from front and draw the front view of the base 1′ 2′ 3′ 4′ 5′ on XY. Height 55 mm is seen in the front view. Hence, draw a horizontal line 55 mm above XY and mark the front view of the top face as a′ b′ c′ d′ e′.

4. **Visible and Invisible Edges**
   Faces ab (2) (1) and bc (3) (2) are closer to the observer. Therefore they are visible.

   Therefore, draw a′ b′ 2′ 1′ and b′ c′ 3′ 2′ by *thick* lines in the front view as shown.

   Edges e (5) and d (4) are invisible.

   Hence, mark them as (e′) (5′) and (d′) (4′) in the front view.

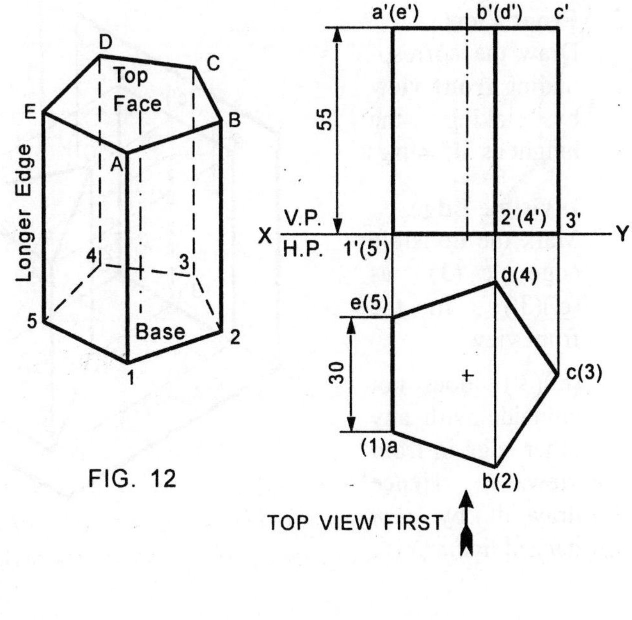

FIG. 12

Problem 8: *(Fig. 13) Draw the projections of a* ***hexagonal prism*** *of side of base 25 mm and height 50 mm resting with its base on HP such that one of its rectangular faces is perpendicular to VP.*

Problem 9: *(Exercise) Draw the projections of a* ***hexagonal prism****, side of base 25 mm and height 60 mm resting with its base on HP such that one of its rectangular faces is parallel to VP.*

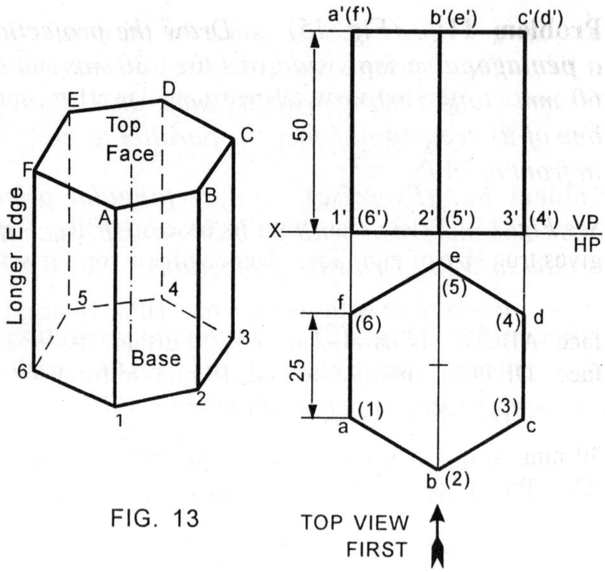

FIG. 13

Problem 10: *(Fig. 14) A* ***triangular prism*** *of side of base 30 mm and height 45 mm rests with its base on HP such that one of its rectangular faces is parallel to and 40 mm farther away from VP. Draw its projections.*

1. Since the prism rests with its base on HP, top view gives true size of the end face. Therefore, **draw the top view first.**
2. **Top View:** Edge AB is parallel to and 40 mm farther away from VP. Hence draw the triangle abc such that ab is parallel to and 40 mm below XY to represent the top view of the top face. Then mark the hidden corners of the base in the top view as (1), (2) and (3).
3. **Front View**
   Draw the corresponding **front view** by taking the height as 45 mm.
4. **Invisible Edge**
   Mark the invisible edge c (3) as (c′)(3′) in the front view.

   (c′)(3′) does not coincide with any other edge in front view. Hence draw it by *thin* dashed line.

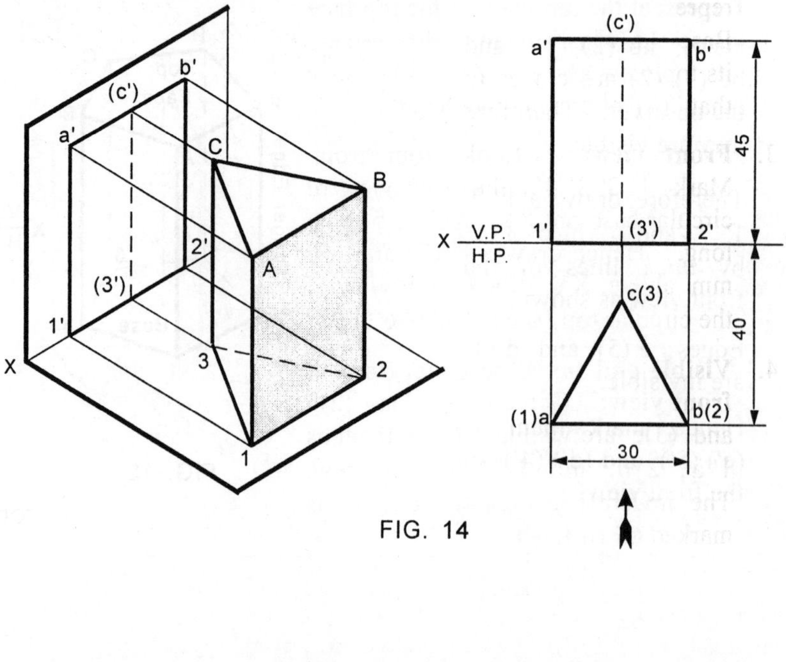

FIG. 14

Problem 11: *(Fig. 15) Draw the projections of a* ***pentagonal prism****, side of base 30 mm and axis 60 mm long resting with its base on HP such that one of its rectangular faces is parallel to and 10 mm in front of VP.*

See Fig. 15. Prism rests with its base on HP. Top view gives true size of end face. Hence, **draw top view first.**

Prism rests with its base on HP. Hence, the top face ABCDE is parallel to HP. Further, its rectangular face DE54 is parallel to and 10 mm in front of VP.

Hence, draw the top view abcde (a pentagon of 30 mm side) with de parallel to and 10 mm below XY. Project the front view as shown.

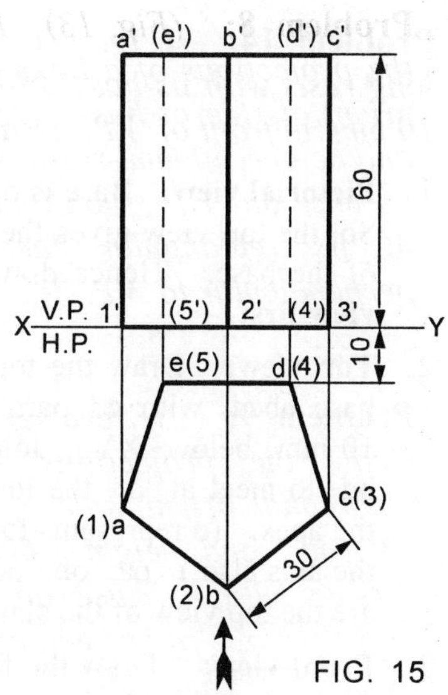

FIG. 15

Problem 12: *(Exercise) Draw the projections of a* ***pentagonal prism****, side of base 25 mm and axis 65 mm long resting with its base on HP such that one of its rectangular faces is parallel to and farther away from VP.*

Problem 13: *(Fig. 16) Draw the projections of a* ***cylinder*** *of base 30 mm diameter and axis 50 mm long resting with its base on HP and axis 25 mm in front of VP.*

1. **Pictorial view:** Both the end faces are connected by a number of generators. Consider 8 of them. Since the cylinder rests with its base on HP, top view gives the true size of the top face. Therefore, **draw the top view first.**
2. **Top view:** Draw a circle abcd of 30 mm diameter such that center is 25 mm below XY to represent the top view of the top face. Base 1234 is hidden. Hence mark its top view as (1) (2) (3) (4) such that (1) and a coincide, etc.
3. **Front view:** Look from front. Mark 1′ 2′ 3′ 4′, the front view of circular base on XY. Axis is 50 mm long. Hence draw a horizontal 50 mm above XY. Mark front view of the circular top face as a′ b′ c′ d′.
4. **Visible and invisible generators in front view:** Generators (1)a, (2)b and (3)c are visible. Mark them as 1′ a′, 2′ b′ and 3′ c′ in front view. The invisible generator d (4) is marked as (d′) (4′).

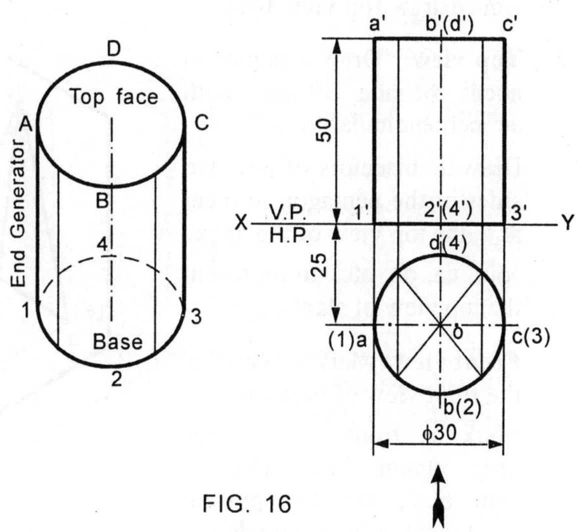

FIG. 16

Problem 14: *(Fig. 17) A **square pyramid**, side of base 30 mm and axis 50 mm long rests with its base on HP such that one of its edges of the base is parallel to and 10 mm in front of VP. Draw its projections.*

1. **Pictorial view:** Base is on HP. Hence the axis is perpendicular to HP. So, the top view gives the true size of the base. Hence draw the **top view first**.
2. **Top view:** Draw the top view of base abcd, with cd parallel to and 10 mm below XY. Join ac and bd to meet at o, the top view of the apex. (o represents top view of the axis also.) oa, ob, oc and od are the top view of the slant edges.
3. **Front view:** Draw the front view of the base as a′ b′ (c′) (d′) on XY. Length of the axis is 50 mm. Hence mark o′ the front view of apex 50 mm above XY on the projector from o. Join o′ with b′ (c′) and (d′) a′.

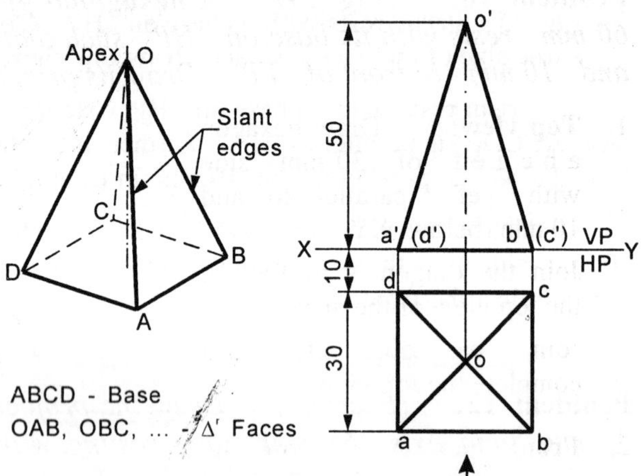

FIG. 17

Problem 15: *(Fig. 18) Draw the projections of a **pentagonal pyramid**, side of base 30 mm and height 60 mm resting with its base on HP such that one of the edges of the base is perpendicular to VP.*

1. **Pictorial view:** Base – ABCDE, apex – O, slant edges are OA, OB, OC, etc. Base is on HP. Hence, axis is perpendicular to HP. True size of the base is seen in the top view. Hence draw **top view first**.
2. **Top view:** Draw a pentagon abcde of side 30 mm with ae perpendicular to XY.

   Draw $\perp^r$ bisectors of any two sides of the pentagon to meet at o, the top view of the apex.

   Join oa, ob, etc., to represent the top view of slant edges.
3. **Front view:** Mark a′ b′ c′ d′ e′ the front view of base on XY.

   Mark o′, front view of the apex 60 mm above XY.

   Join o′ a′, etc., to represent front view of the slant edges.

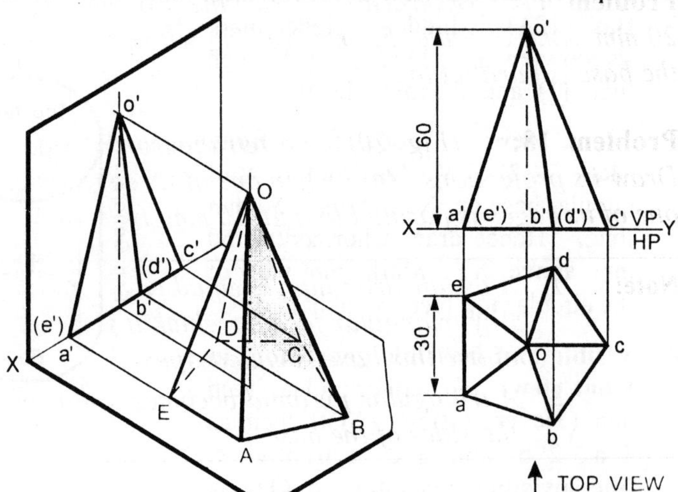

FIG. 18

4. **Visible and invisible edges in front view:** ab and bc are visible edges of the base in the top view. Hence draw a′ b′ and b′ c′ as *thick* lines.
   oa, ob and oc are visible and hence draw o′ a′, o′ b′ and o′ c′ as *thick* lines.
   Since ed, dc are not visible, mark them as (e′) (d′) and (d′) c′.

Problem 16: *(Fig. 19) A **hexagonal pyramid**, side of base 30 mm and height 60 mm rests with its base on HP such that one of the edges of the base is parallel to and 10 mm in front of VP. Draw its projections.*

1. **Top view:** Draw hexagon a b c d e f of 30 mm side with ef parallel to and 10 mm below XY.
   Join the diagonals to fix o, the top view of the apex.
   Join oa, ob, etc., and complete the top view.
2. **Front view:** Follow the procedure given in ***Problem 15*** and draw the front view.
3. **Visible and Invisible:** Draw a′b′, b′c′, c′d′, o′a′, o′b′, o′c′ and o′d′ as *thick* lines. Note that e and f in top view are marked as (e′) and (f′).

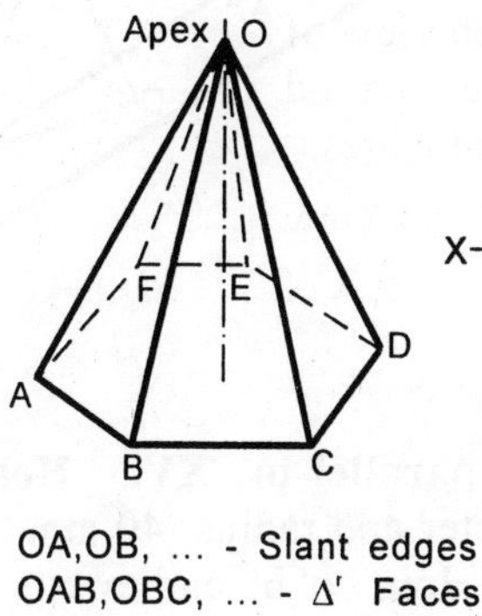

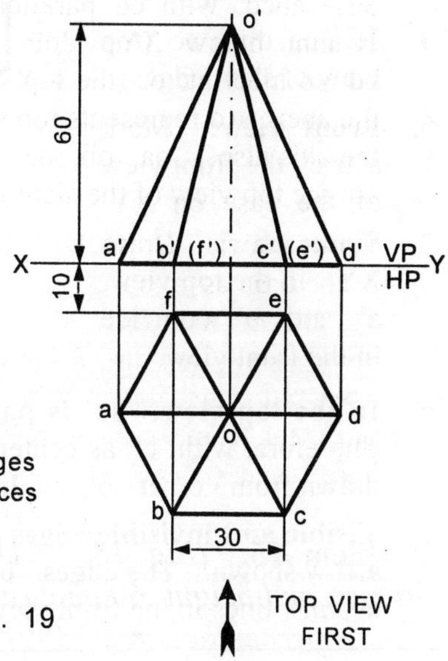

FIG. 19

Problem 17: *(Exercise) Draw the projections of an **octagonal pyramid**, side of base 20 mm and altitude 55 mm resting with its base on HP such that one of the edges of the base is parallel to VP.*

Problem 18: *(Fig. 20) A **tetrahedron** of side 40 mm rests with its base on HP. Draw its projections **(a)** when one of its edges is perpendicular to VP and **(b)** when one of its edges is parallel to and 10 mm in front of VP.*

Note: 1. *A tetrahedron has 6 equal edges and 4 equal equilateral triangular faces. Any triangular face on which it rests on HP is called its **Base**. The corner above this base is called **Apex**.*

2. *A triangular pyramid becomes a tetrahedron when slant edges are equal to the sides of the base.*

---

1. *See Figs. 20(a) and (b).* Look at the pictorial view. Top view gives true size of the base. Hence, draw it first for **Cases (a) and (b).**

CASE (a): When one edge AB is perpendicular to VP [*Fig. (a)*]

2. **Top view:** Draw equilateral Δ abc such that ab is ⊥^{r} to and below XY.
3. Draw ⊥^{r} bisectors for any two sides of Δ abc to cut at o, top view of the apex O.
4. Join oa, ob & oc (top view of slant edges).
5. **Front view:** Mark a′ b′ c′, the front view of the base on XY. Since ab is ⊥^{r} to XY in the top view, a′ and b′ coincide in the front view.

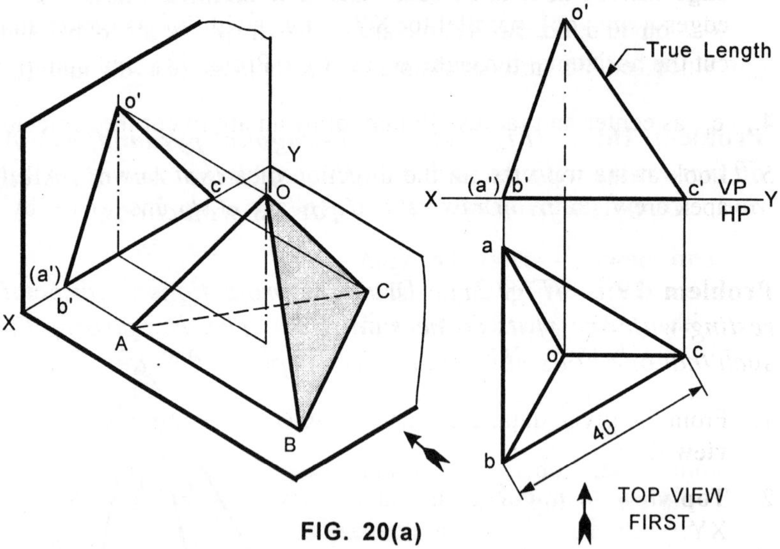

FIG. 20(a)

6. **In the top view, oc is parallel to XY. Hence its front view will be in true length.** Therefore, with c′ as center and radius 40 mm (one side), draw an arc to cut the projector drawn from o at o′. Join o′ b′ and o′ c′.
7. **Visible and invisible edges in the front view:** Look at the top view in the direction of the arrow shown. The edges bc, ob and oc are visible and hence draw b′ c′, o′ b′ and o′ c′ as *thick* lines in the front view. a is hidden by b and is marked as (a′).

Note: *When a base edge of a tetrahedron is perpendicular to VP, one slant edge will be parallel to VP (Property of Tetrahedron).*

CASE (b): When one edge AC is parallel to VP [*Fig. (b)*]

1. **Top view:** Draw the triangle abc with ac parallel to and 10 mm below XY. Fix o as explained in Case (a).
2. **Front view:** Draw a′ b′ c′ (front view of the base) on XY.

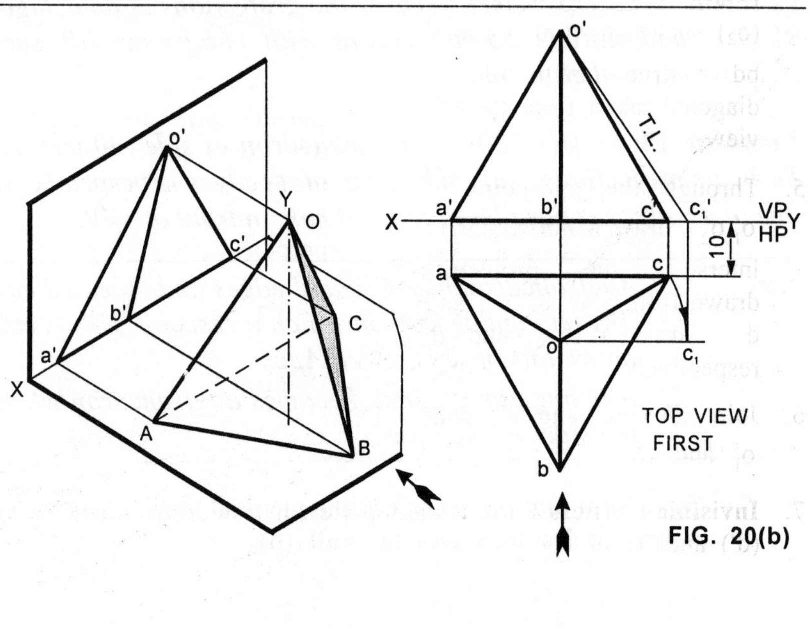

FIG. 20(b)

3. **In the top view, none of the slant edges is parallel to XY. So in the front view no slant edge will be in true length.** Hence to mark o′ turn the top view of any one of the slant edges (*say*, oc) parallel to XY. *i.e.*, with o as center and oc as radius draw an arc to cut the horizontal through o at c_1. Project c_1 and obtain c_1' on XY.

4. c_1' as center and radius 40 mm draw an arc to cut the projector drawn from o at o′.

5. Look at the top view in the direction of arrow shown. All the corners of the base and the apex are visible. Draw a′ b′ c′, o′ a′, o′ b′ and o′ c′ by *thick* lines.

Problem 19: *(Fig. 21) Draw the projections of an* ***octahedron*** *of side 40 mm when resting with one of its corners on HP. The axis passing through this corner is vertical such that one of its horizontal edges is parallel to and 12 mm in front of VP.*

1. From the given data, the square ABCD is seen in true size in top view. Hence, draw the **top view first.**

2. **Top view:** Draw a square abcd of side 40 mm with cd parallel to and 12 mm below XY. Join ac and bd to intersect at o_1, (o_2), the top view of the axis $O_1 O_2$.

Front view

3. One of the axes $O_1 O_2$ is vertical such that O_2 is on HP. Hence mark o_2' on XY.

4. Diagonals AC and BD are parallel to HP. They will be in true length in top view. Hence mark o_1' on the projector drawn from o_1, (o_2) such that $o_1' o_2'$ = ac = bd = true length of the diagonal taken from the top view.

5. Through the mid-point of $o_1' o_2'$ draw a horizontal to intersect the projectors drawn from a, b, c and d at a′ b′ c′ and d′ respectively.

6. Join a′ b′ c′ and d′ with o_1' and o_2'.

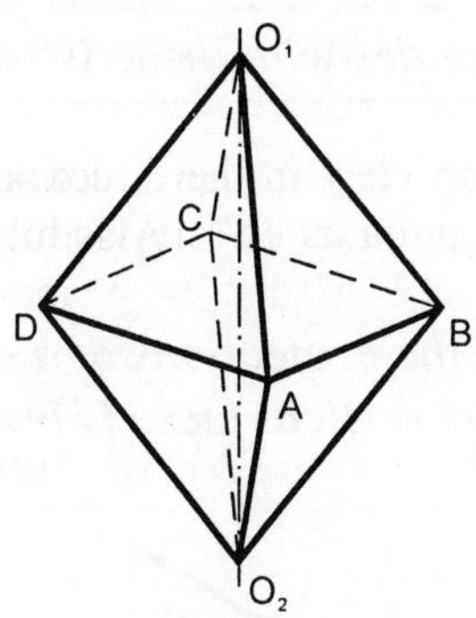

Simple position of an Octahedron means it rests on H.P. with its axis perpendicular to H.P.

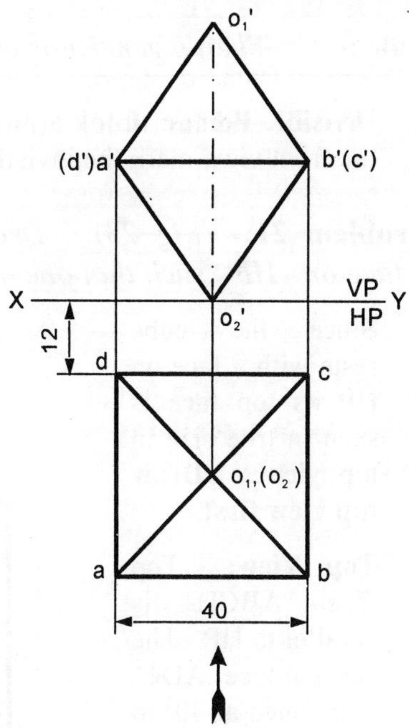

FIG. 21

7. **Invisible Corners** Look at the top view in the direction shown. Mark the hidden corners as (d′) and (c′).

Problem 20: *(Fig. 22) Draw the projections of a right circular* ***cone*** *of base 40 mm diameter and height 60 mm when resting with its base on HP.*

1. Top view gives the true size of the base. Hence, draw the **top view first.**

Top view

2. Draw a circle abcd of 40 mm diameter below XY. Center of the circle o is the top view of the apex and the axis.
3. Join oa, ob, oc and od to represent the top view of the generators.

Front view

4. Draw a′ b′ c′ (d′), the front view of the base on XY.
5. Mark o′ 60 mm above XY. Draw o′ a′ and o′ c′.

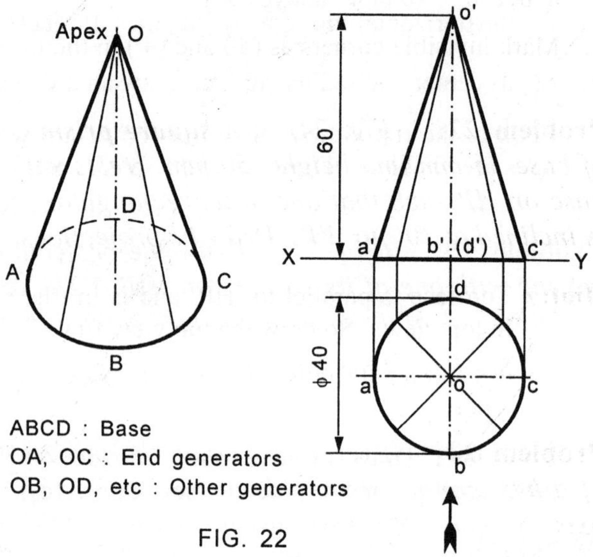

FIG. 22

Rule : *When a generator coincides with the axis, always mark the axis.*

---

6. **Invisible Points:** Look at the top view in the direction shown. od is invisible. Hence, in the front view mark the invisible point as (d′). Also o′ (d′) coincides with o′ b′.

Problem 21: *(Fig. 23) Draw the projections of a* ***cube*** *of 40 mm side, resting with a face on HP such that one of its vertical faces is inclined at 30° to VP.*

1. Since the cube rests with a face on HP its top face is seen in true size in top view. **Draw top view first.**
2. **Top View:** Top face ABCD is parallel to HP. The vertical face AD41 is inclined at 30° to VP *i.e.*, its edge AD is inclined at 30° to VP. Hence draw a square abcd with ad inclined at 30° to XY.

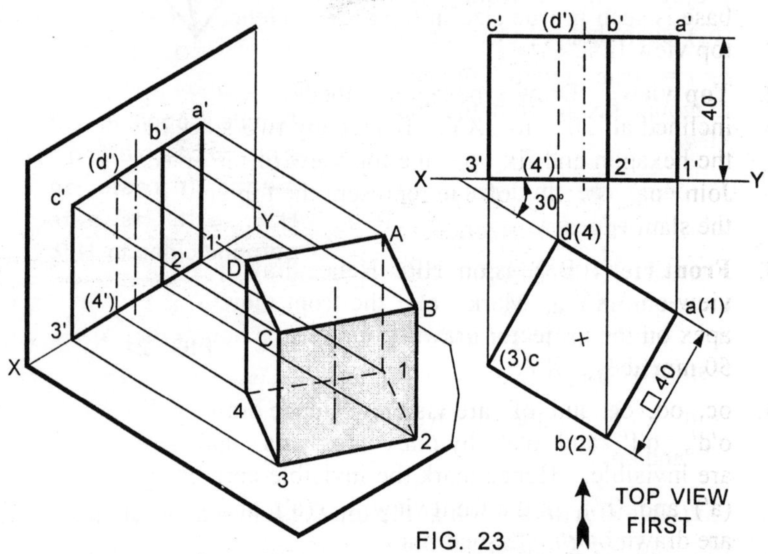

FIG. 23

Base 1234 is hidden. Hence mark (1), (2), (3) and (4) such that (1) and **a** coincide, etc.

3. **Front View:** Base is on the HP. Hence draw its front view 1′ 2′ 3′ 4′ on XY. Mark a′ b′ c′ d′ 40 mm above XY.
4. Mark invisible corners as (d′) and (4′) in the front view. Join them by a *thin* dashed line.

Problem 22: *(Fig. 24) A* ***square prism****, side of base 35 mm and height 50 mm rests with its base on HP such that one of its rectangular faces is inclined at 30° to VP. Draw its projections.*

Hint: Top face is parallel to HP and is in true size in top view. So draw the top view first.
See Fig. 24 for Solution.

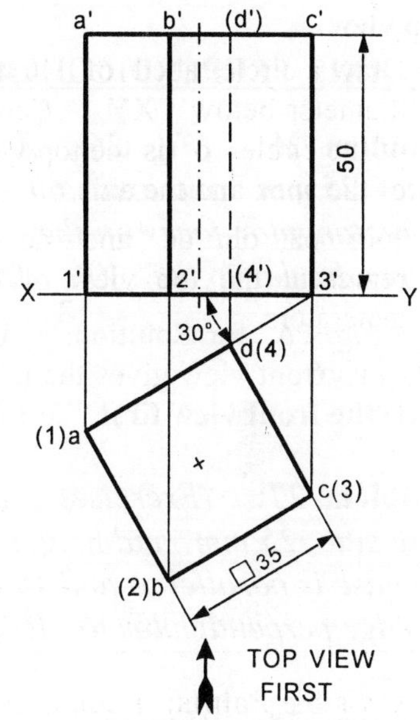

FIG. 24

Problem 23: *(Exercise) Draw the projections of a* ***hexagonal prism****, side of base 25 mm and axis 55 mm long resting with its base on HP. One of its base edges is inclined at 20° to VP.*

Problem 24: *(Fig. 25) A* ***hexagonal pyramid****, side of base 25 mm and height 50 mm rests with its base on HP such that one of the edges of the base is inclined at 20° to VP. Draw the top and front views of the pyramid.*

1. Since the pyramid rests with its base on HP, the base is seen in true size in top view. Hence, draw top view first.
2. **Top view:** Draw a hexagon abcdef with ab inclined at 20° to XY. Bisect any two sides of the hexagon and fix o, the top view of the apex. Join oa, ob, ..., of to represent the top view of the slant edges.
3. **Front view:** Base is on HP. Hence draw its front view on XY. Mark o′, the front view of the apex on the projector drawn from o at a height of 50 mm above XY.
4. oc, od, oe and of are visible. Hence draw o′c′, o′d′, o′f′ and o′e′ by *thick* lines. oa and ob are invisible. Hence mark the invisible corners as (a′) and (b′) in the front view. o′(a′) and o′(b′) are drawn by *thin* dashed lines.

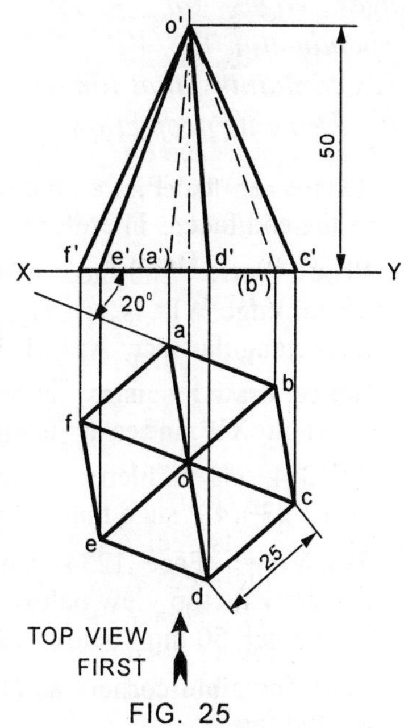

FIG. 25

Problem 25: *(Exercise) Draw the projections of an* ***octahedron*** *of side 35 mm resting on the HP on one end of an axis which is vertical. One of its horizontal edges makes 45° with the VP.*

Hint: Top view is a square with one of its sides making 45° with XY. The diagonals of the square meet at a point, which is the top view of the vertex.

3.2 AXIS PERPENDICULAR TO VP (Simple Position)

Problem 26: *(Fig. 26) A* ***hexagonal prism****, side of base 25 mm and axis 60 mm long, lies with one of its rectangular faces on the HP such that the axis is perpendicular to the VP. Draw its projections.*

See Fig. 26 for Solution. Axis is perpendicular to the VP, *i.e.*, front view gives the true size of the base. Hence, **draw the front view first.**

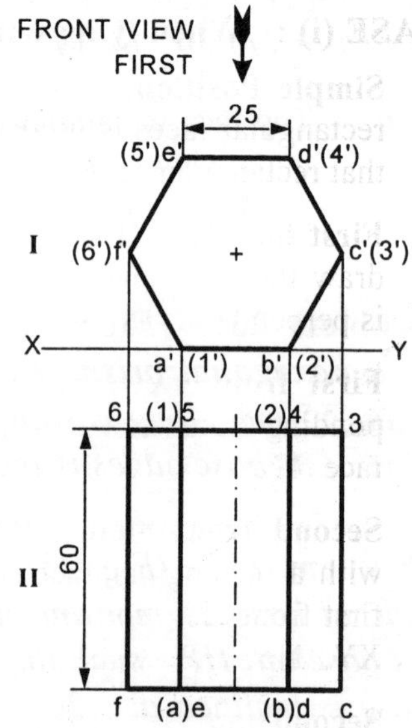

FIG. 26

Problem 27: *(Exercise) A* ***hexagonal pyramid*** *of base side 25 mm and height 65 mm is kept such that the base is parallel to and 15 mm in front of VP with an edge perpendicular to HP. Draw the projections.*

Problem 28: *(Fig. 27) A* ***square prism****, side of base 35 mm and axis 50 mm long lies with one of its longer edges on HP such that its axis is perpendicular to VP. Also one of its rectangular faces containing that longer edge is inclined at 30° to HP. Draw its projections.*

1. Axis is ⊥ʳ to VP, *i.e.*, the front view gives the true size of the end face. Therefore, **draw the front view first.**
2. **Front View:** End face ABCD is parallel to VP. Its longer edge A1 is on HP and perpendicular to VP. Its rectangular face AD41 is inclined at 30° to HP.

   Hence, draw a square a′ b′ c′ d′ such that the corner a′ is on XY and a′d′ is inclined at 30° to XY.

   1 2 3 4 is hidden. Hence, mark its front view (1′)(2′)(3′)(4′) such that (1′) and a′ coincide, etc.
3. **Top View:** Face 1234 is nearer to and in front of VP. So, draw its top view below and nearer to XY. Draw a horizontal 50 mm below 1234. Mark a, b, c and d.
4. Mark invisible corners as (1) and (a). Join (1) (a) by dashed line.

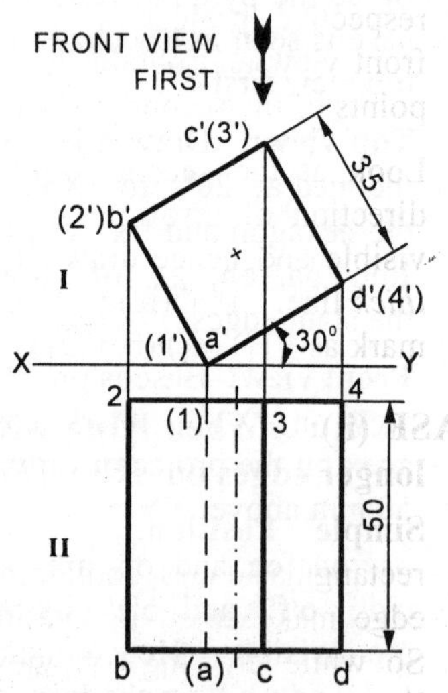

FIG. 27

3.3 AXIS PARALLEL TO BOTH HP AND VP (Simple Position)

Rule: ***As the axis is parallel to both HP and VP, begin with the Side View.***

Problem 29: *(Fig. 28)* ***A square prism**, side of base 30 mm and axis 45 mm long lies on HP such that its axis is parallel to both HP and VP. Draw the top and front views of the prism, when (i) it lies with one of its rectangular faces on HP and (ii) it lies with one of its longer edges on HP.* (UQ)

CASE (i) : When lying with one of its rectangular faces on HP [*Fig. 28(a)*]

1. **Simple Position:** Assume the prism to rest with its base on HP such that one of its rectangular faces is perpendicular to VP to enable to tilt the prism on the right side to lie on that rectangular face.
2. **First top view:** For the simple position draw the top view such that ab (2) (1) is perpendicular to XY and on right side.
3. **First front view:** Project the corresponding front view in which rectangular face 1′ a′ (b′) (2′) is in the right side.
4. **Second front view:** Prism has to lie with a rectangular face on HP. So tilt first front view so that 1′ a′(b′)(2′) is on XY. Now the axis is parallel to XY.
5. **Second top view:** Draw the projectors from second front view. Also draw horizontals from first top view to cut the respective projectors from the second front view and obtain the corresponding points of the second top view.
6. Look at the second front view in the direction of arrow. 4′ d′ (c′) (3′) is visible and hence draw 4_1 d_1 c_1 3_1 as *thick* line. 1′ a′ (b′) (2′) is invisible and mark as (1_1) (a_1) (b_1) (2_1).

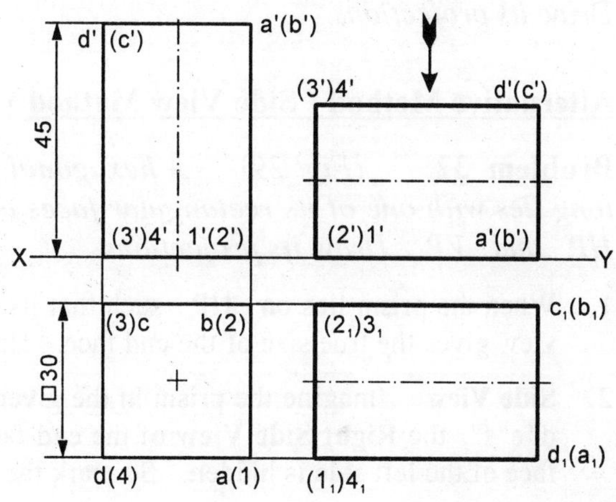

FIG. 28 (a)

CASE (ii): When lying with one of its longer edges on HP [*Fig. 28(b)*]

1. **Simple Position:** Assume two rectangular faces containing the longer edge make equal inclination with HP. So while drawing top view, imaginary line connecting the center of the base with its corner should be parallel to VP.

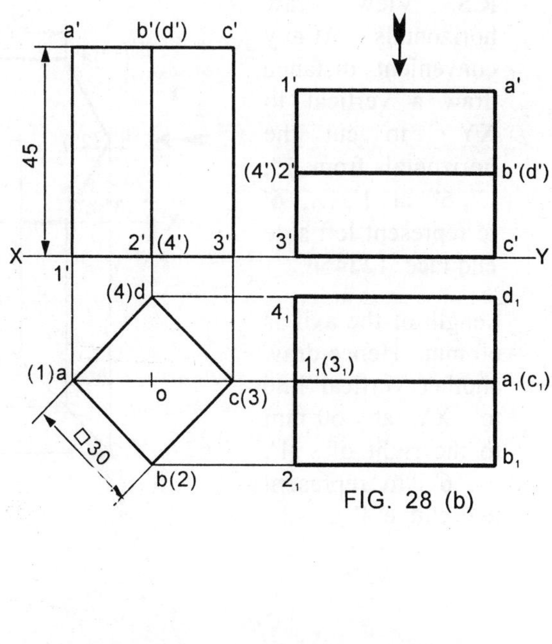

FIG. 28 (b)

2. **First top view** is drawn such that c(3) is on the right side and o3 is parallel to VP.
3. **First front view** is drawn as shown.
4. **Second front view:** Tilt the first front view such that 3′ c′ lies on XY and complete the second front view. **Second top view** is drawn as explained in ***Case (i)***.
5. Look at the second front view in the direction of arrow shown. Mark the invisible corners as (3_1) and (c_1).

Problem 30: *(Exercise) A **triangular prism**, side of base 35 mm and height 60 mm lies with one of its longer edges on HP such that its axis is parallel to both HP and VP. Draw its projections.* (UQ)

Problem 31: *(Exercise) A **cylinder**, base 35 mm diameter and axis 60 mm long lies with one of its generators on HP such that its axis is parallel to both HP and VP. Draw its projections.*

Alternative Method (Side View Method)

Problem 32: *(Fig. 29) A **hexagonal prism**, side of base 20 mm and axis 60 mm long lies with one of its rectangular faces on HP such that its axis is parallel to both HP and VP. Draw its projections.* (UQ)

1. When the prism lies on HP such that its axis is parallel to both HP and VP, the side view gives the true size of the end face. Hence, **draw the side view first**.
2. **Side View:** Imagine the prism in the given position and look from the right. Draw a″ b″ c″ d″ e″ f″, the **Right Side View** of the end face in **left side** such that c″ d″ is on XY. End face of the left side is hidden. So mark the hidden corners as (1″), (2″), …, (6″).
3. **Front View:** From R.S. View draw horizontals. At any convenient distance draw a vertical to XY to cut the horizontals from 1″, …, 6″ at 1′, …, 6′ to represent left side end face 123456.

   Length of the axis is 60 mm. Hence draw another vertical line to XY at 60 mm to the right of 1′, …, 6′ to represent a′ b′ c′ d′ e′ f′.

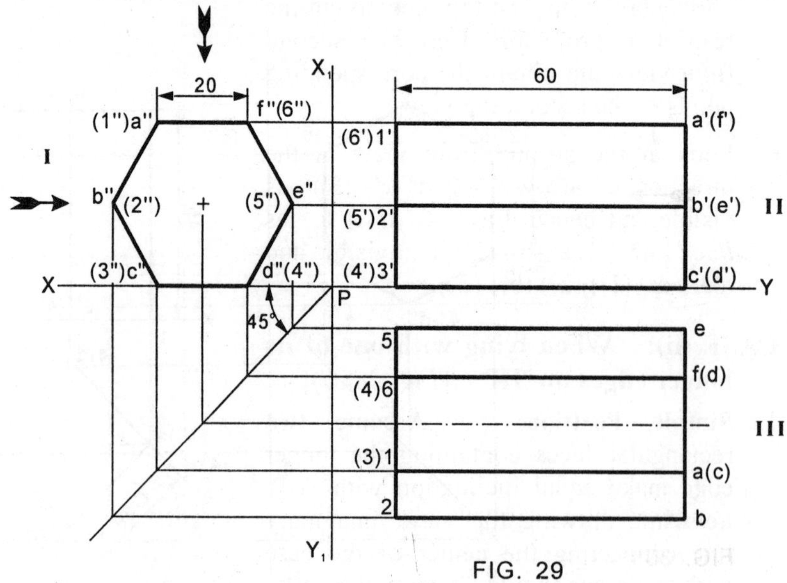

FIG. 29

4. **Look at the Side View from left:** The faces a″ b″ (2″) (1″) and b″ c″ (3″) (2″) are visible. Hence draw a′ b′ 2′ 1′ and b′ c′ 3′ 2′ by *thick* line. Faces f″ e″ (5″) (6″) and e″ d″ (4″) (5″) are invisible and are marked as (f′) (e′) (5′) (6′) and (e′) (d′) (4′) (5′).

5. **Top view:** Draw the projectors from the front view. Draw X_1Y_1 at P ⊥r to XY.

Note: *Distances of all corners of the prism in the top view from XY should be equal to the respective distances of all the corners in the side view from X_1Y_1.*

From P draw a 45° line to XY. Draw the projectors from the side view to cut the 45° line at various points.

From these points, draw horizontals to cut respective projectors drawn from front view.

6. **Look at the Side View from top.** Faces a″f″(6″)(1″), a″b″(2″)(1″) and f″e″(5″)(6″) are visible and hence draw them by *thick* lines in the top view.

   Note that the edges c″ (3″) and d″ (4″) are marked as (c) (3) and (d) (4) in the top view.

Problem 33: *(Fig. 30) A **hexagonal prism**, side of base 20 mm and axis 60 mm long lies with one of its longer edges on HP and its axis is parallel to both HP and VP. Draw its projections.*

Note: *This problem is similar to the previous one, except that the prism lies with one of its longer edges D4 instead of a rectangular face on HP.*

See Fig. 30 for solution.

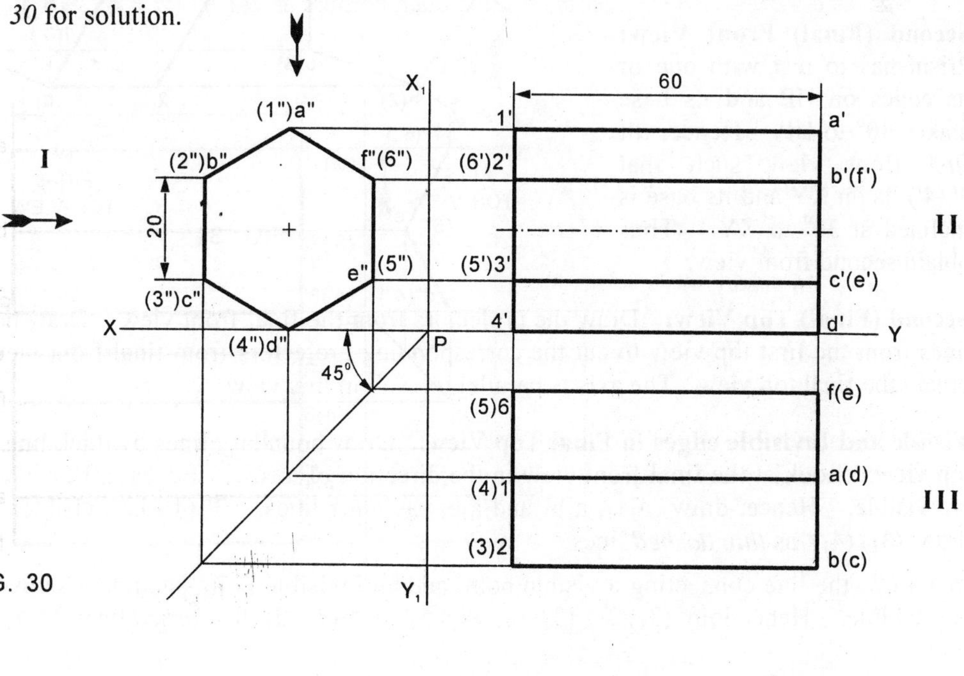

FIG. 30

CHANGE OF POSITION METHOD

4. AXIS INCLINED TO ONE PLANE AND PARALLEL TO OTHER

4.1 AXIS OF SOLID INCLINED TO HP AND PARALLEL TO VP

Problem 34: *(Fig. 31) A **pentagonal prism,** side of base 25 mm and axis 50 mm long, rests with one of its edges on HP such that the base containing that edge makes an angle of 30° to HP and its axis is parallel to VP. Draw its projections.*

1. **Simple Position:** When the prism has to rest with **one of its edges (*i.e.*, the base edge)** on HP, assume the prism in simple position such that its base is on HP and an edge of the base is $\perp^r$ to VP, preferably on the right side for tilting it to the given final position.

   For this simple position, **draw the Top View first.** Now in the top view, (3)(4) is perpendicular to XY and on the right side. Draw the front view, in which the edge 3′ (4′) appears as a point.

2. **Second (Final) Front View:** Prism has to rest with one of its edges on HP and its base makes 30° to HP. Hence, tilt first front view such that 3′ (4′) is on XY and its base is inclined at 30° to XY. Thus obtain second front view.

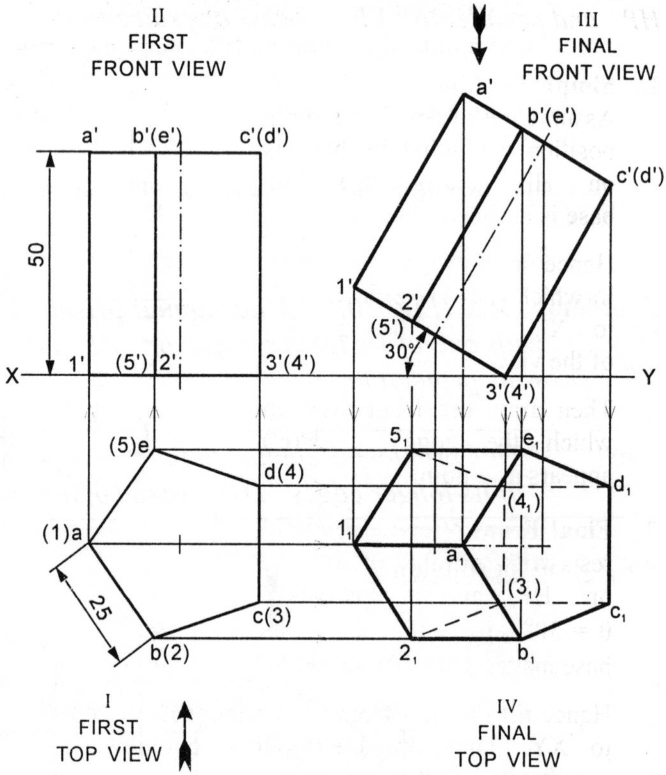

FIG. 31

3. **Second (Final) Top View:** Draw the projectors from the final front view. Draw horizontal lines from the first top view to cut the corresponding projectors from final front view. Thus obtain the final top view. The axis is parallel to XY in this view.

4. **Visible and Invisible edges in Final Top View:** Draw boundary lines by thick lines in final top view. Look at the final front view in the direction of arrow. Edges a′1′, a′b′ and a′(e′) are visible. Hence, draw $a_1 1_1$, $a_1 b_1$ and $a_1 e_1$ as *thick* lines. 3′ (4′) is invisible. Hence, draw $(3_1)(4_1)$ as *thin dashed* lines.

   In a view, the line connecting a visible point and an invisible point should be shown by *thin* dashed line. Hence join $(3_1) 2_1$, $(3_1) c_1$, $(4_1) 5_1$ and $(4_1) d_1$ by *dashed* lines.

Note: *In an orthographic view of a solid, the line connecting an invisible corner and a visible corner should be shown by a dashed line.*

Problem 35: *(Exercise)* *A* ***hexagonal prism,*** *side of base 24 mm and axis 55 mm long, rests with one of its edges on HP such that the base containing that edge makes an angle of 45° to HP and its axis is parallel to VP. Draw its projections.*

Problem 36: *(Fig. 32)* *A* ***hexagonal pyramid,*** *side of base 25 mm and axis 50 mm long, rests with one of the edges of its base on HP and its axis is inclined at 30° to HP and parallel to VP. Draw its projections.* (UQ)

1. **Simple Position:** *See Fig. 32.* Assume the pyramid in simple position such that its base is on HP and an edge of its base is perpendicular to VP.

   Hence draw the top view first, in which bc is perpendicular to XY and on the right side of the view.

   Then draw the front view in which the edge b′(c′) appears as a point.

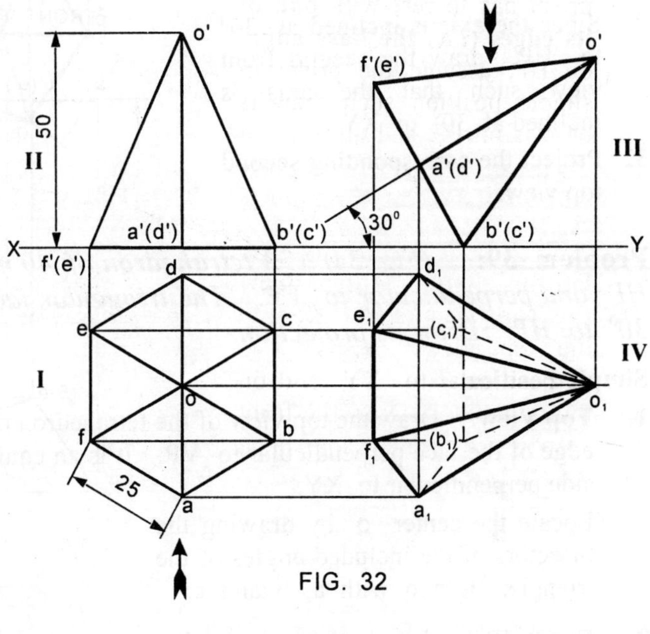

FIG. 32

2. **Final Front View:** Pyramid rests with an edge of its base on HP and its axis makes $\theta = 30°$ to HP, *i.e.*, its base makes $(90 - \theta)$ to HP.

   Hence tilt the first front view such that b′ (c′) is on XY and its base is inclined at 60° to XY. Thus complete the final front view.

3. **Final Top View:** Project the corresponding final top view in which the axis is parallel to XY.

4. **Visible and Invisible Edges in the Final Views:** Draw the boundary lines as *thick* lines in the final top view. Look at the final front view in the direction of the arrow.

   Triangular face o′ f′ (e′) is visible and hence draw $o_1 f_1$ and $o_1 e_1$ as *thick* lines. b′ and (c′) are invisible and hence draw $(b_1)(c_1)$ as *thin* dashed line. Draw also $(b_1) a_1$, $(b_1) o_1$, $(c_1) d_1$ and $(c_1) o_1$ as *thin* dashed lines.

Problem 37: *(Exercise)* *A* ***square pyramid****, side of base 30 mm and height 65 mm, rests with one of the edges of its base on HP such that its base makes 30° to HP. Draw its projections.*

Problem 38: *(Fig. 33) Draw the projections of a* ***pentagonal pyramid*** *of base 25 mm side and axis 60 mm long when it is lying* ***on*** *HP on one of its base edges, such that the axis is parallel to VP and inclined at 30° to HP.* (UQ)

1. Draw the top view of the pyramid such that one of its sides rs, is perpendicular to XY. Draw the corresponding front view.
2. Since the axis is inclined at 30° to HP, draw the second front view such that the axis is inclined at 30° to XY.
3. Project the corresponding second top view.

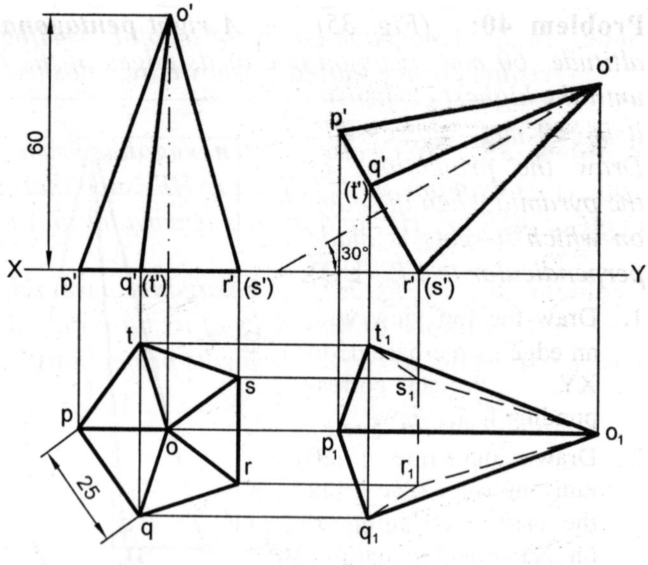

FIG. 33

Problem 39: *(Fig. 34) A* ***tetrahedron*** *of 40 mm side rests with one of its edges on HP and perpendicular to VP. The triangular face containing that edge is inclined at 30° to HP. Draw its projection.* (UQ)

Simple Position

1. **Top View:** Draw the top view of the tetrahedron resting on HP on one of its faces with an edge of the face perpendicular to VP. It is an equilateral triangle of side 40 mm with one side perpendicular to XY.

   Locate the center o by drawing the bisectors of the included angles of the triangle. Join o with a, b and c.

2. **Front View:** Line co representing one of the edges is parallel to XY in top view. Hence it is parallel to VP and its front view shows true length.

   c′ as center and true length (40 mm) as radius draw an arc to meet the projector through o at o′. Join a′ & c′ with o′. Complete the front view.

Second Position

3. Tilt the front view about a′ such that face ABC represented by a′(b′)c′ in the front view is inclined at 30° to XY.
4. Draw the second front and top views.

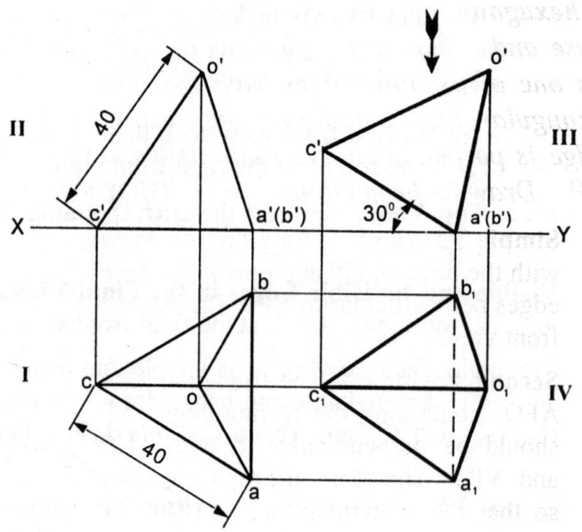

FIG. 34

Problem 40: *(Fig. 35) A right **pentagonal pyramid** of base side 20 mm and altitude 60 mm rests on one of its edges of the base in HP, the base being lifted up until the highest corner in it is 20 mm above HP. Draw the projections of the pyramid when the edge on which it rests is made perpendicular to VP.* (UQ)

1. Draw the top view with an edge of the base $\perp^r$ to XY. Project corresponding front view.
2. Draw a locus line 20 mm above XY. Fix the base edge point r′ on XY.

   Draw an arc with r′ as center and r′ p′ as radius to cut the locus line at p′. Complete second front view.

   Project the corresponding second top view. Mark (s_1) & (r_1).

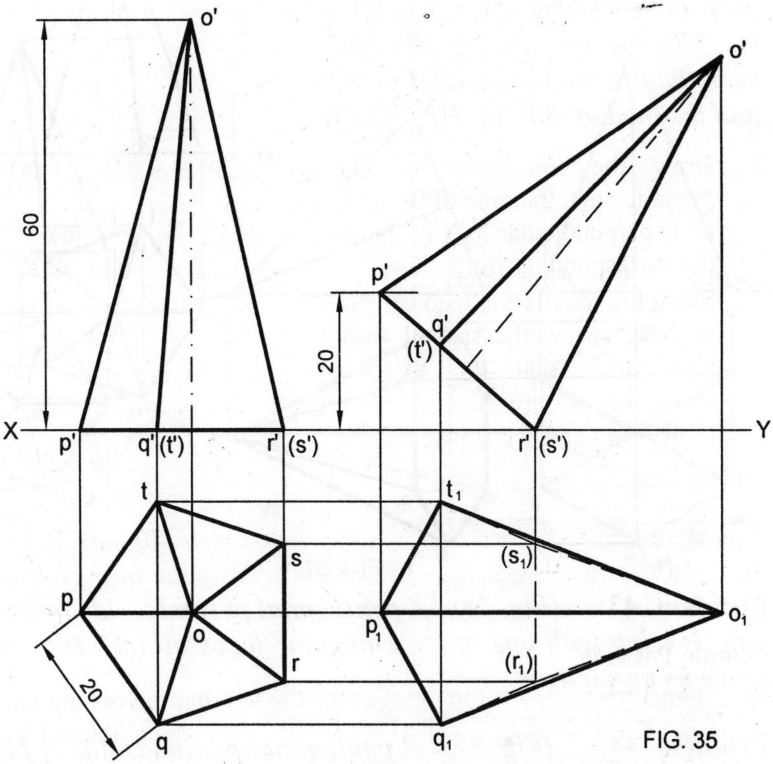

FIG. 35

Problem 41: *(Fig. 36)*
*A **hexagonal pyramid** of 26 mm side of base and 70 mm height rests on HP on one of its base edges such that the triangular face containing the resting edge is perpendicular to both HP and VP. Draw its projections.* (UQ)

1. **Simple Position:** Draw the top view with the base on HP and one of the base edges perpendicular to VP. Project the front view.
2. **Second Position:** The triangular face AFO containing the resting edge AF should be perpendicular to both HP and VP. Therefore, tilt the front view so that the triangular face a′(f′)o′ is vertical, i.e., perpendicular to HP. Complete the projections as shown.

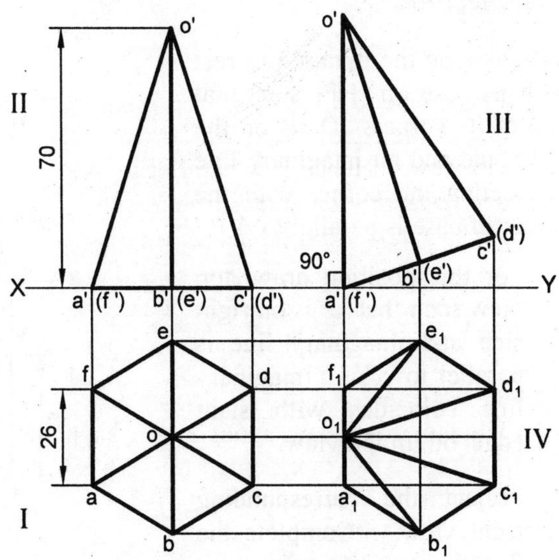

FIG. 36

Problem 42: *(Fig. 37)* ***A hexagonal pyramid** side of base 25 mm, axis 50 mm long lies with one of its triangular faces on the HP and its axis is parallel to the VP. Draw its projections.* (UQ)

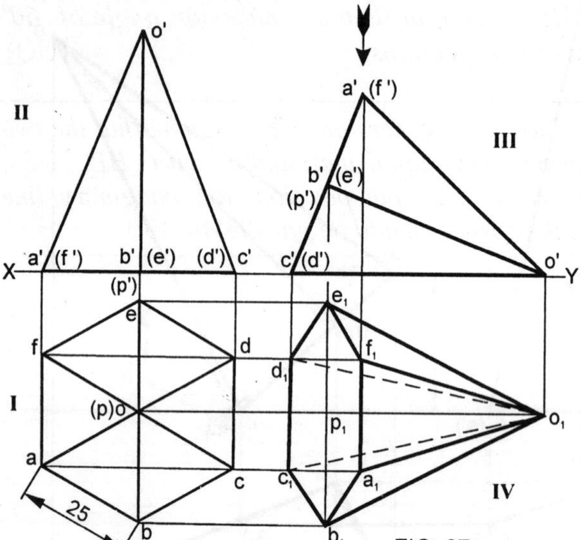

FIG. 37

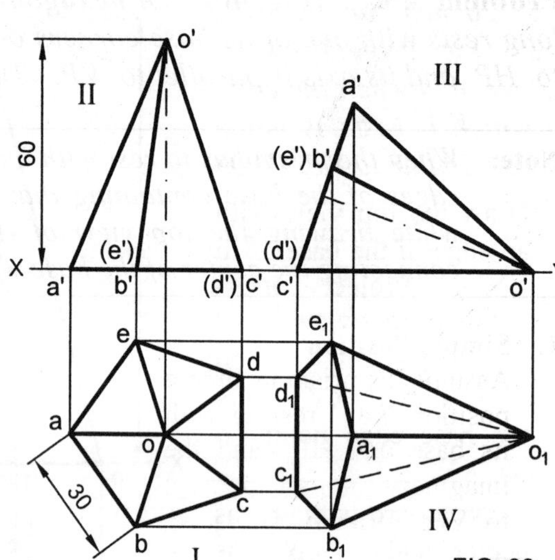

FIG. 38

Problem 43: *(Fig. 38)* ***A pentagonal pyramid**, side of base 30 mm and axis 60 mm long is lying with one of its triangular faces on the HP and axis parallel to the VP. Draw its projections.* (UQ)

Problem 44: *(Fig. 39)* ***A pentagonal pyramid**, side of base 25 mm and axis 55 mm long, lies with one of its slant edges on HP such that its axis is parallel to VP. Draw its projections.* (UQ)

First assume the pyramid to rest with its base on HP such that one of its corners D is on the right side and an imaginary line connecting the corner with the center of base is parallel to VP.

1. For this position draw, top view such that d is on right side and imaginary line is parallel to XY. Imaginary line coincides with slant edge od in top view.

2. Project the corresponding front view. Complete the final front and top views.

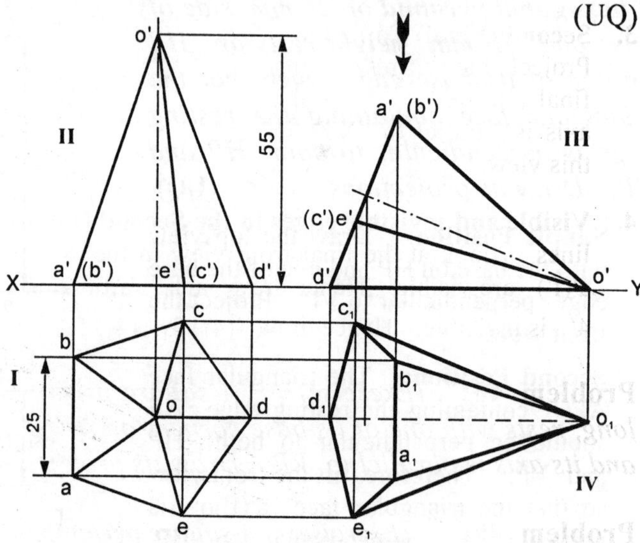

FIG. 39

3. Look in the direction of arrow. The base and apex are visible. Slant edge o′d′ is invisible. So draw the slant edge o_1d_1 by dashed line in final top view.

Problem 45: *(Fig. 40) A* ***hexagonal prism****, side of base 25 mm and axis 50 mm long rests with one of its base corners on HP such that its base makes an angle of 60° to HP and its axis is parallel to VP. Draw its projections.* (UQ)

Note: *When the prism has to rest with one of its base corners on HP, assume that the two edges of the base containing that corner make equal inclinations with HP. So, while drawing the top view of the prism in simple position, the imaginary line connecting the center of the base with the corner should be parallel to VP.*

---

1. **Simple Position**
   Assume the prism in simple position, *i.e.*, resting with its base on HP and the imaginary line o(4) parallel to VP. Draw its top view.

   Project the corresponding front view, in which 4′ is on right side.
2. **Second (Final) Front View**
   Tilt the front view such that 4′ is on XY and its base makes 60° to XY and obtain the final front view.
3. **Second (Final) Top View**
   Project the corresponding final top view. Now the axis is parallel to XY in this view.

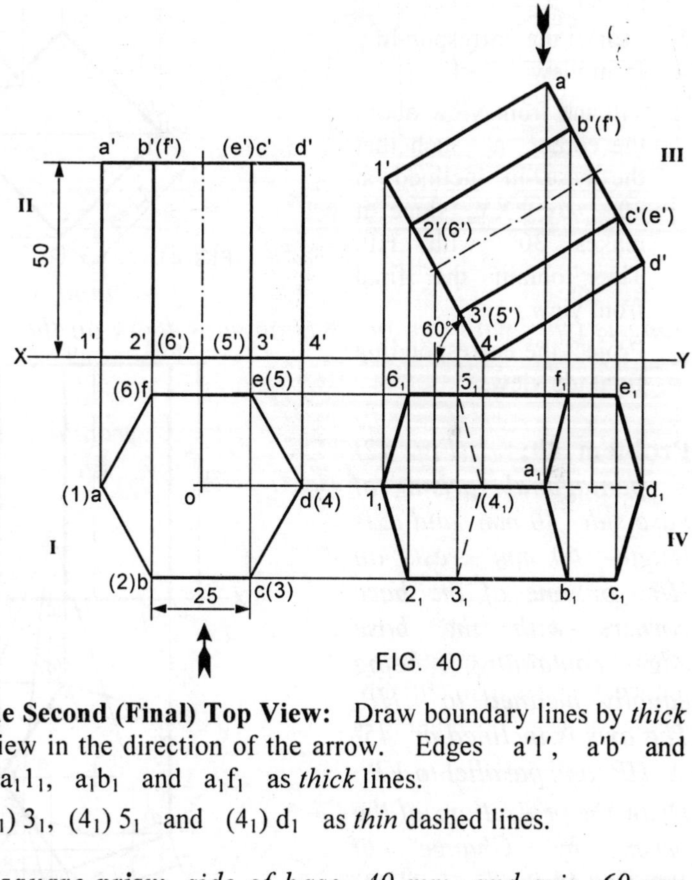

FIG. 40

4. **Visible and invisible edges in the Second (Final) Top View:** Draw boundary lines by *thick* lines. Look at the final front view in the direction of the arrow. Edges a′1′, a′b′ and a′(f′) are visible. Hence draw $a_1 1_1$, a_1b_1 and a_1f_1 as *thick* lines.

   4′ is invisible. Hence draw $(4_1)\,3_1$, $(4_1)\,5_1$ and $(4_1)\,d_1$ as *thin* dashed lines.

Problem 46: *(Exercise) A* ***square prism****, side of base 40 mm and axis 60 mm long, rests with one of its base corners on HP. Its base makes an angle of 45° to HP and its axis is parallel to VP. Draw its projections.*

Problem 47: *(Exercise) A* ***square pyramid*** *of side of base 50 mm and height 75 mm lies on HP on one of its triangular faces with its axis parallel to VP. Draw its projections.*

Problem 48: *(Fig. 41) Draw the projections of a* ***cube*** *of side 40 mm when it rests on the ground on one of its corners and a face containing that corner is inclined at 30° to the ground and perpendicular to VP.* (UQ)

1. The cube rests on one of its corners with the base inclined at 30° to HP and the axis parallel to VP. In this position, the faces are equally inclined to VP. Draw the top view first as a square with sides inclined at 45° to XY.
2. Project the corresponding front view.
3. Tilt the front view about the corner a′ such that the base in inclined at 30° to XY, since it makes 30° with HP. Thus obtain the final front view.
4. Project the corresponding final top view.

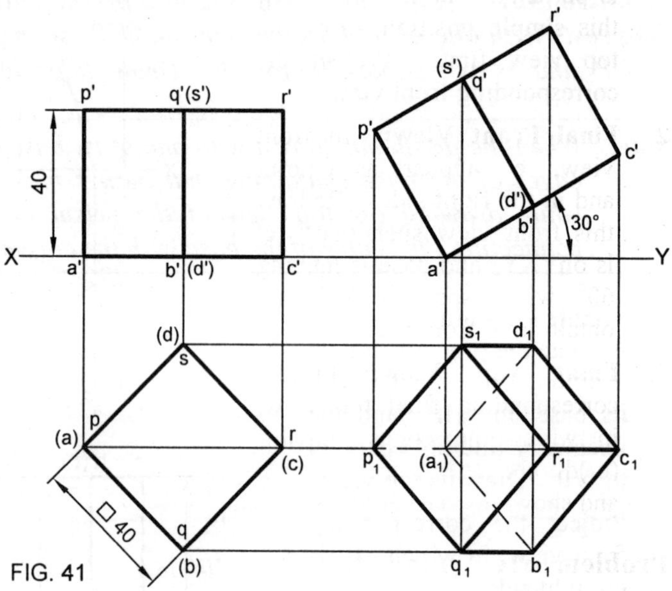

FIG. 41

Problem 49: *(Fig. 42) A* ***pentagonal prism*** *of base side 30 mm and axis length 60 mm rests on HP on one of the base corners with the base edges containing it being equally inclined to HP. The axis is inclined at 45° to HP and parallel to VP. Draw the projections of the prism by Change of Position Method.* (UQ)

FIG. 42

Problem 50: *(Fig. 43) A* ***pentagonal pyramid*** *side of base 20 mm and axis 45 mm long rests with one of its corners on HP such that the base is inclined at an angle of 60° to HP and one side of base is perpendicular to VP. Draw its projections.* (UQ)

1. **Simple Position:** The pyramid rests with its base on HP such that one of its edges of the base is perpendicular to VP. For this simple position, draw the top view first. Project the corresponding front view.
2. **Final Front View:** In front view, c′ appears as a point and in the right side. Redraw this front view such that c′ is on XY and base is making 60° with respect to XY and obtain final front view.
3. **Final Top View:** Project corresponding final top view, in which the edges $(c_1)b_1$, $(c_1)d_1$ and $(c_1)o_1$ are invisible and shown by dashed lines.

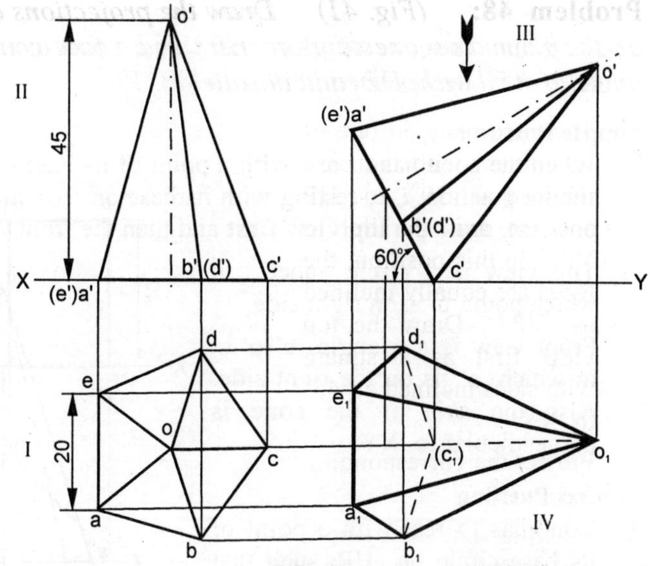

FIG. 43

Problem 51: *(Fig. 44) Draw the projections of a **cylinder**, base 30 mm diameter and axis 40 mm long, resting with a point of its base circle on HP such that the axis is making an angle of 30° with HP and parallel to VP.* (UQ)

1. **Simple position**
   Assume the cylinder, resting with its base on HP and axis perpendicular to it. Draw **top view first** and then the corresponding front view.
2. **Second position**
   Cylinder has to rest with a point of its base circle on HP such that its axis has to make $\theta = 30°$ with HP. Hence tilt the front view such that 3′ is on XY and base is at 60° (*i.e.*, $90° - \theta$) to XY. Thus obtain the second front view with the axis inclined at 30° to XY.

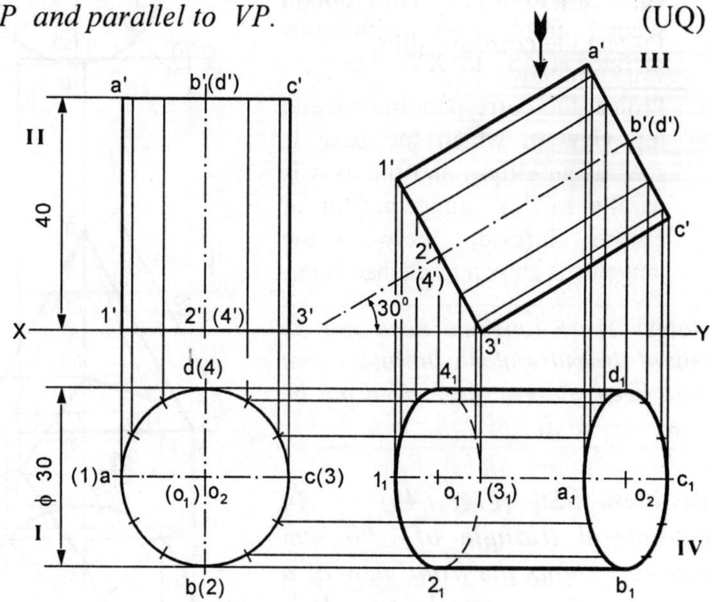

FIG. 44

Project the corresponding second top view, in which the axis is parallel to XY.
Join the ellipses by common tangents in the second top view. Draw dashed line for half of the ellipse for the hidden base.

Problem 52: *(Fig. 45) Draw the projections of a* **cone**, *base 30 mm diameter and axis 50 mm long, resting on HP on a point of its base circle with the axis making an angle of 45° with HP and parallel to VP.* (UQ)

Simple Position

1. When the cone has to rest with a point of its base circle on HP, first assume the cone in its simple position, *i.e.*, resting with its base on HP and the axis perpendicular to HP. For this position, draw the **top view first** and then the front view.
2. Top view is a circle abcd in which point c is on right side.
3. Front view is a triangle a′ o′ c′ in which c′ is on the right side. Also the axis of the cone is perpendicular to XY.

Second Position

4. Cone has to rest with a point of its base circle on HP, such that its axis has to make $\theta = 45^\circ$ with HP. Hence, tilt the first front view such that c′ is on XY and its base is inclined at 45°, (*i.e.*, $90^\circ - \theta$) to XY. Thus obtain second front view with axis inclined at 45° to XY.
5. Project the corresponding second top view in which the base is seen as an ellipse and the axis is parallel to XY. Inner portion of ellipse $b_1(c_1)d_1$ between the tangents is shown as dashed line.

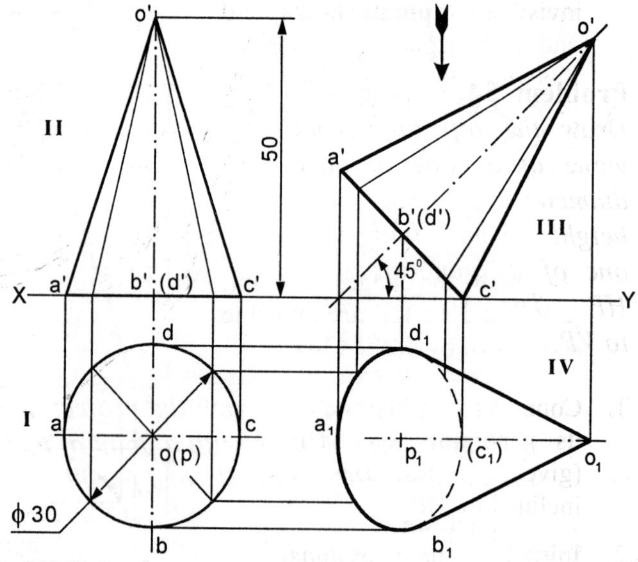

FIG. 45

Note: *These tangents need not be any of the particular generators that have been chosen, i.e., need not be* $o_1 b_1$ *or* $o_1 d_1$.

Problem 53: *(Fig. 46) An equilateral* **triangle** *of 60 mm side represents the front view of a cone standing on its base. It is tilted until its axis makes 30° with HP and top view of the axis is parallel to VP in this position. Draw the projections of cone.* (UQ)

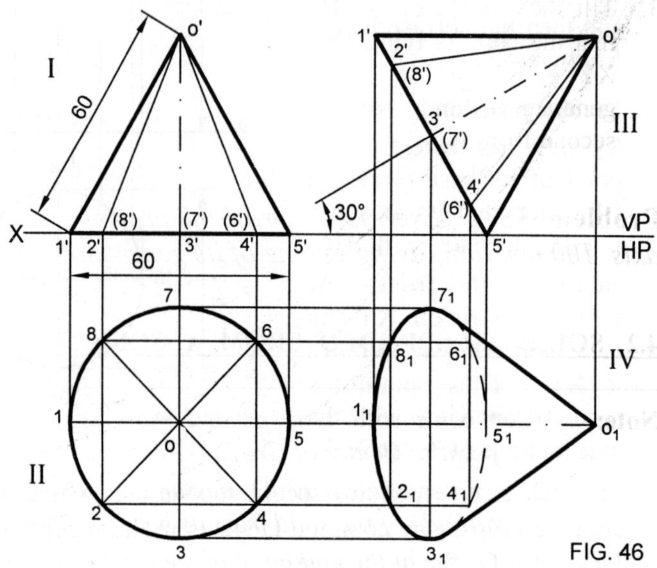

FIG. 46

1. **Simple Position**
   Draw the front view of the cone as an equilateral triangle of side 60 mm. Project the corresponding top view. Divide the circle in the top view into 8 equal parts and project the same to the front view as shown.

2. **Final Position**
   Tilt the front view such that the axis of the cone makes 30° with XY and obtain the final front view. Project the corresponding final top view with axis parallel to XY. Mark the invisible portion as shown.

Problem 54: *(Fig. 47) Draw the top and front views of a cone of base diameter 46 mm and height 65 mm lying with one of its generators on HP. The axis is parallel to VP.* (UQ)

1. Cone lies with its one of its generators on HP (given). So, axis is inclined to HP.
2. Initially assume the axis to be perpendicular to HP and parallel to VP.
3. Tilt the front view such that the cone is lying on XY on one of its generators and draw second front view.

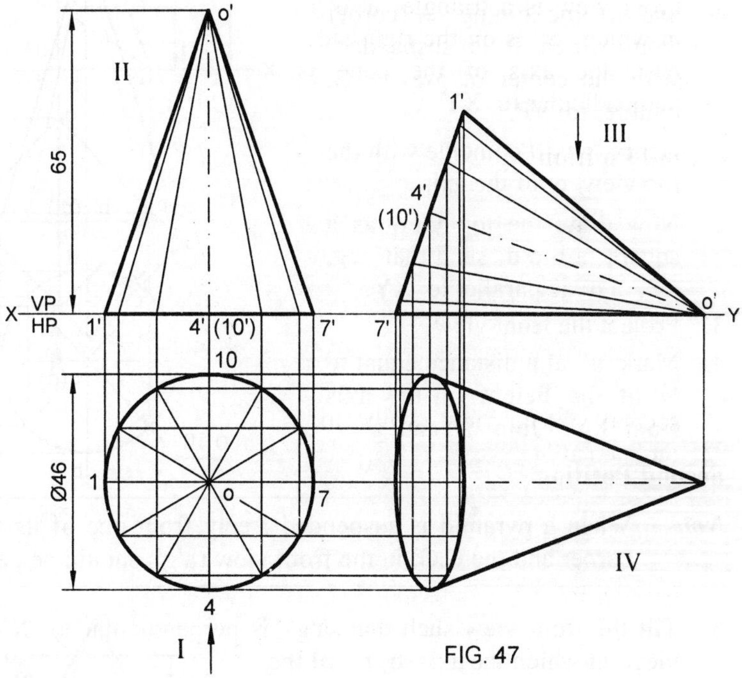

FIG. 47

Problem 55: *(Exercise) Draw the projections of a* **cone**, *base 75 mm diameter and axis 100 mm long lying on one of its generators with the axis parallel to VP.* (UQ)

4.2 SOLIDS SUSPENDED FROM A CORNER

Note:
1. *When a solid is suspended freely about any point, the point of suspension and the point of Center of Gravity G of the solid will be in a vertical line in front view.*
2. *Center of Gravity G lies on the axis of symmetry and at a height of ¼ th of the altitude of the solid from its base for pyramid and cone.*
3. *G lies at the mid point of the axis for prism and cylinder.*

Problem 56: *(Fig. 48) **A square pyramid**, base 32 mm side and axis 60 mm long, is freely suspended from one of the corners of its base with the axis parallel to VP. Draw its projections.* (UQ)

Simple Position

1. Assume the pyramid to be suspended from corner A of its base and axis parallel to VP.

   Initially, keep the pyramid such that its base is on HP and the line joining A (corner from which it is suspended) with the center of gravity G, parallel to VP.

   Hence, g will coincide with the top view o of the axis.
2. Now draw the top view as a square a b c d such that a g, *i.e.*, a o is parallel to XY.
3. Project the front view.
4. Mark g′ at a distance equal to ¼ of the height of the axis from XY. Join a′ and g′.

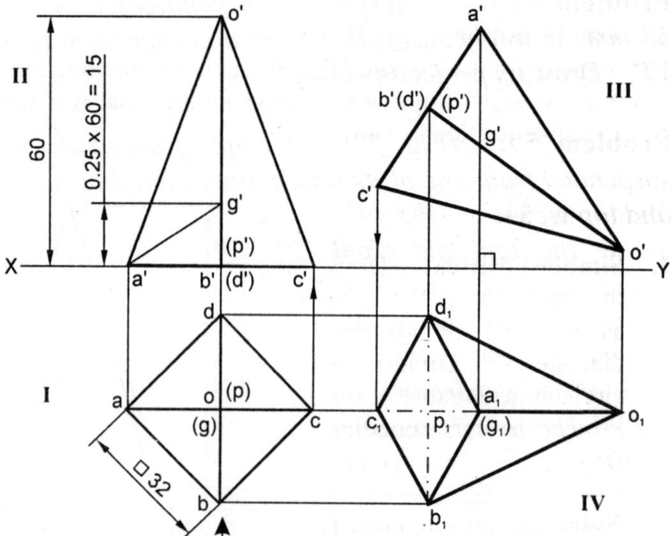

FIG. 48

Second Position

Note: When a pyramid is suspended freely from one of its base corners, the line joining that corner and the C.G. in the front view (a′g′) should be perpendicular to XY.

---

5. Tilt the front view such that a′g′ is perpendicular to XY. Project the corresponding top view, in which the axis o_1 p_1 of the pyramid still remaining parallel to VP.

Problem 57: *(Fig. 49) A **pentagonal pyramid** of side of base 30 mm and altitude 68 mm is suspended freely from a string attached to one corner of the base of the pyramid. The string is held such that the apex of the pyramid is touching HP. The axis of the pyramid is parallel to VP.*

Draw the projections. (UQ)

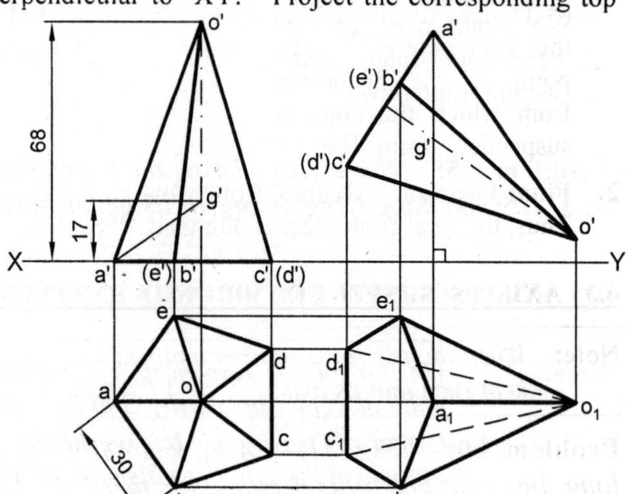

FIG. 49

First, draw the top view such that a is the corner from which the pyramid is suspended. Also, the line joining a and the C.G. of the pyramid (here, o itself) should be parallel to XY in the top view. In the final front view a′ g′ is drawn perpendicular to XY.

Problem 58: *(Exercise) A pentagonal prism of side of base 26 mm and height 58 mm is freely suspended from a corner such that the axis of the prism is parallel to VP. Draw its projections.* (UQ)

Problem 59: *(Fig. 50) A **cone** of diameter 40 mm and height 60 mm is freely suspended from one of its base points such that the axis is parallel to VP. Draw its front and top views.* (UQ)

1. **Simple Position:** Draw the top view of the cone as a circle of 40 mm diameter. Divide the circle into 8 equal parts. Project the corresponding front view.

Note: *When the cone is suspended freely from one of its base points, the line joining that base point and the center of gravity should be* $\perp^r$ *to XY.*

Mark the center of gravity g′ at a distance of ¼ th the height (= 15 mm) of the cone in the front view. The point 1′ is the corner from which the cone is suspended. Join 1′ g′.

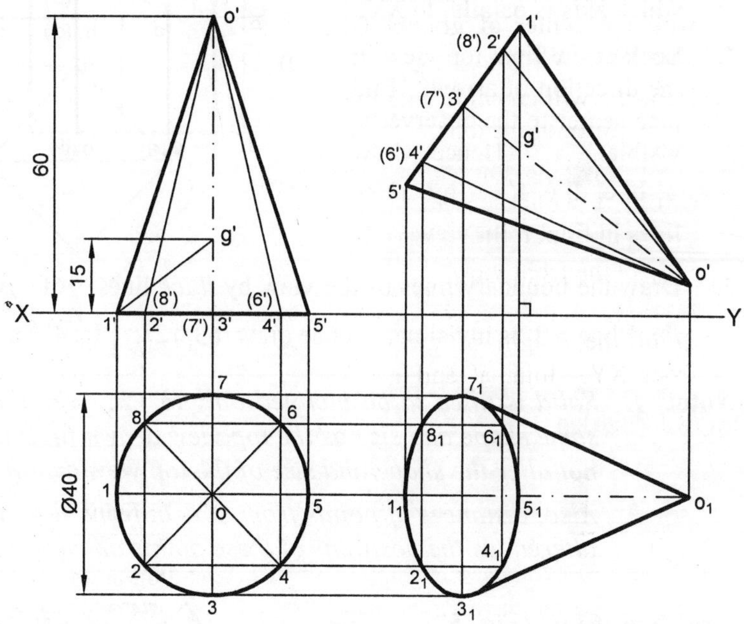

FIG. 50

2. **Final Position:** Redraw front view such that 1′ g′ is $\perp^r$ to XY and o′ touching XY. Thus, the final front view is obtained. Project the corresponding final top view.

4.3 AXIS INCLINED TO VP AND PARALLEL TO HP

Note: *When the axis of a solid is inclined to VP and parallel to HP, assume the solid to be kept such that its axis is* $\perp^r$ *to VP. Draw the projections in this **Simple Position**.*

Problem 60: *(Fig. 51) A **hexagonal prism**, side of base 25 mm and axis 60 mm long, lies with one of its rectangular faces on HP, such that the axis is inclined at 45° to VP. Draw its projections.* (UQ)

1. **Simple Position:** Assume the prism lying with one of its rectangular faces on HP and its axis $\perp^r$ to VP. For this position draw the front view. Project the top view.
2. **Second (Final) Top View:** Redraw top view so that axis is inclined at 45° to XY.
3. **Second (Final) Front View:** Project final front view, in which axis is parallel to XY.
4. Look at the final top view in the direction of arrow. End face nearer to the observer is visible. Hence draw $a_1' \; b_1' \; c_1' \; d_1' \; e_1' \; f_1'$ by *thick* lines in final front view.

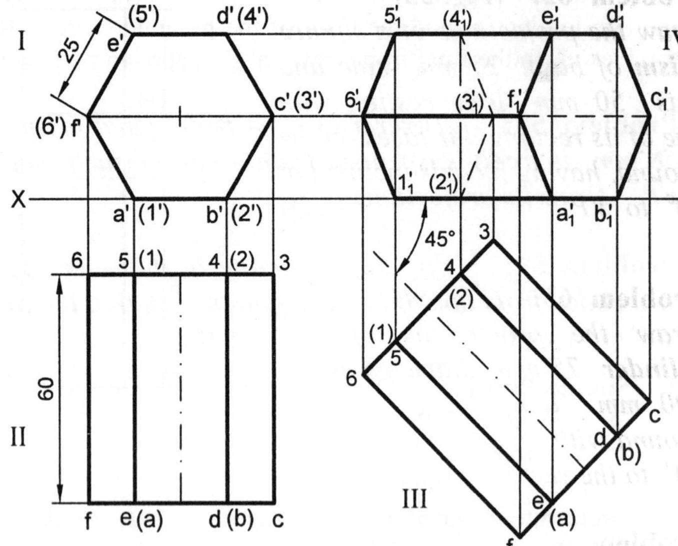

FIG. 51

5. Draw the boundary lines of the view by *thick* lines. 6f is visible. Hence draw $6_1' \; f_1'$ by *thick* line. 3 is invisible. Hence draw $(3_1')\,2_1'$, $(3_1')\,4_1'$ and $(3_1')\,c_1'$ by *dashed* lines.

Note: *1. Solid is tilted to be inclined only to VP. So, the tilted final top view will have same shape and size as the top view drawn first; i.e., mere change in position will not alter the shape and size of the top view drawn in the Simple Position.*

2. Also, distances of points from XY in front view of the simple position will not be altered, as the positions of these points above the HP remain the same.

Problem 61: *(Fig. 52)*

Draw the projections of a ***pentagonal prism*** *of 30 mm side of base and 65 mm long.*

It is lying on one of its longer edges on HP with one rectangular face perpendicular to HP such that the axis makes 60° with VP.

(UQ)

See Fig. 52 for Solution.

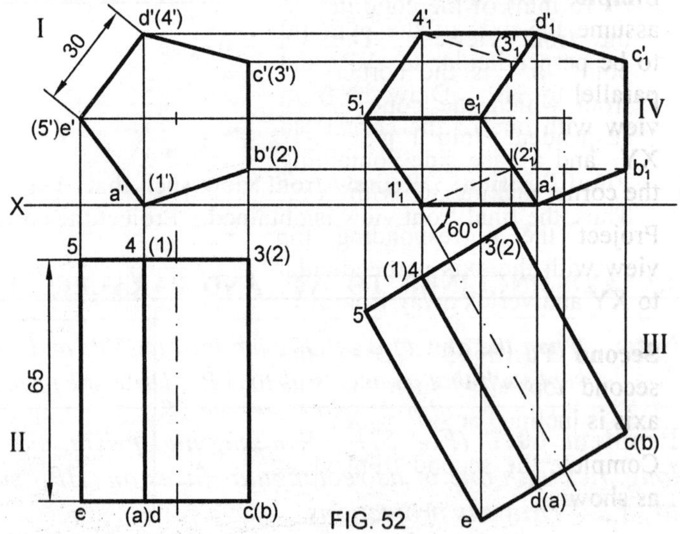

FIG. 52

Problem 62: *(Fig. 53)*
Draw the projections of a ***square prism*** *of base 25 mm side and axis 50 mm long resting with one of its rectangular faces on the ground, having its axis inclined at 30° to VP.* (UQ)

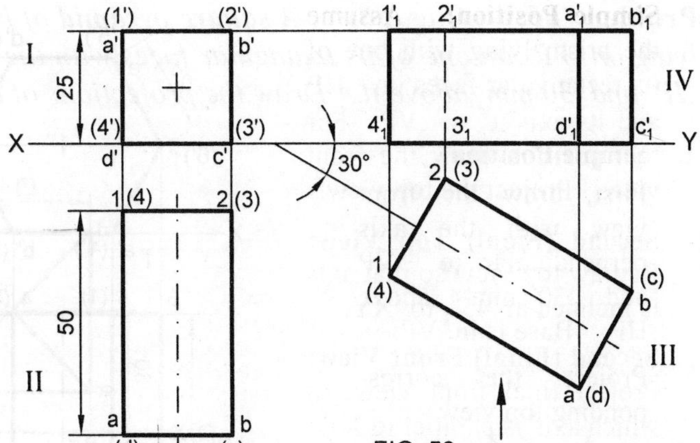

FIG. 53

Problem 63: *(Fig. 54)*
Draw the projections of a ***cylinder*** *75 mm diameter and 100 mm long, lying on the ground with its axis inclined at 30° to the vertical plane.*

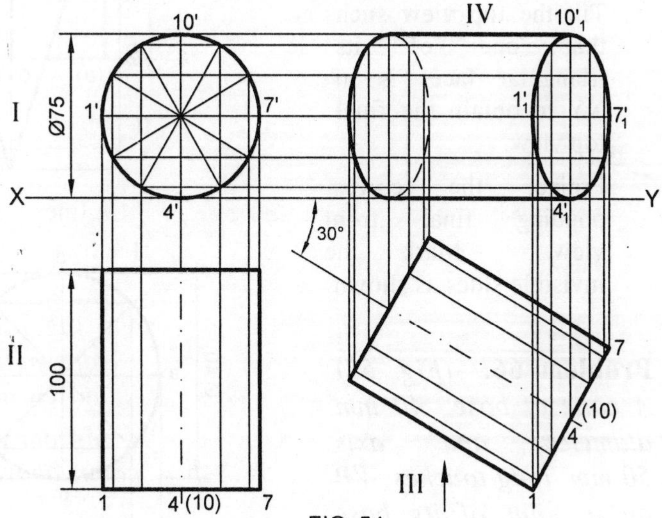

FIG. 54

Problem 64: *(Fig. 55)*
Draw the projections of a ***square pyramid*** *of 32 mm side of base and axis 55 mm. It is resting on HP on one of its base corners with a base side containing the corner making 30° with HP. The axis is inclined at 30° to VP and is parallel to HP. The vertex is away from the VP.* (UQ)

1. **Simple Position:** Initially assume the axis of the pyramid to be perpendicular to VP and parallel to HP. Draw the front view with one of the corners in XY and a base side containing the corner making 30° with XY.

   Project the corresponding top view with the axis perpendicular to XY and vertex away from XY.

2. **Second Position:** Draw the second top view such that the axis is inclined at 30° to XY.

   Complete the second front view as shown.

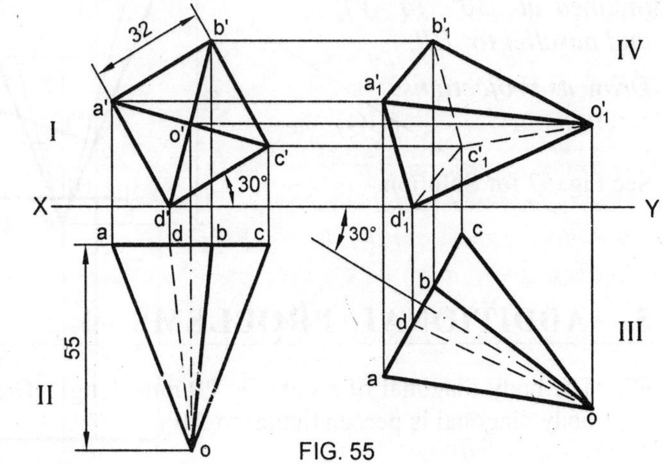

FIG. 55

Problem 65: *(Fig. 56) A* ***square pyramid*** *of base end 40 mm and axis 60 mm is lying on VP on one of its triangular faces with the plane containing the axis parallel to HP and 30 mm above it. Draw the projections of the pyramid.* (UQ)

1. **Simple Position**

   First, draw the front view with the axis perpendicular to VP and 30 mm above HP. Base is in VP.

   Project the corresponding top view.

2. **Final Position**

   Tilt the top view such that one of its triangular faces is in XY to obtain the final top view.

   Project the corresponding final front view. Mark the invisible sides as shown.

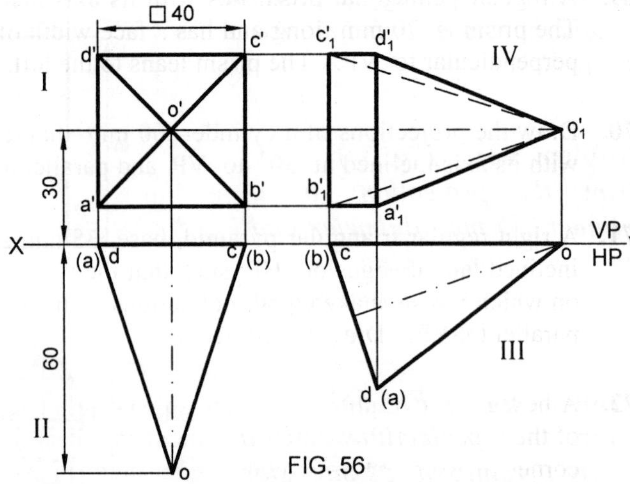

FIG. 56

Problem 66: *(Fig. 57) A* ***cone*** *of base 40 mm diameter and axis 50 mm long touches VP on a point of its base circle. Its axis is inclined at 30° to VP and parallel to HP.*

Draw its projections. (UQ)

See Fig. 57 for Solution.

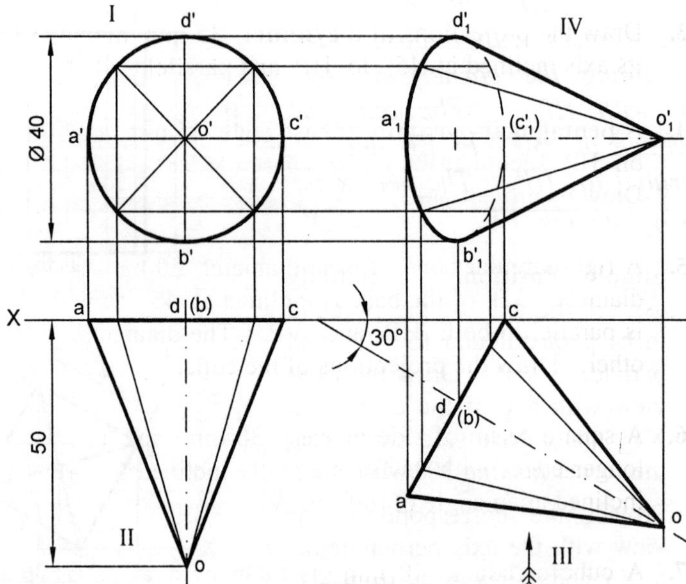

FIG. 57

5. ADDITIONAL PROBLEMS – I

67. The body diagonal of a cube is 70 mm long. Draw the projections of the cube when the body diagonal is perpendicular to HP.

68. A pentagonal prism of 30 mm side of base and height 60 mm is resting on a corner of its base on HP with a longer edge containing that corner inclined at 45° to HP and parallel to VP. Draw its projections. (UQ)

69. A regular pentagonal prism lies with its axis inclined at 60° to HP and parallel to VP. The prism is 70 mm long and has a face width of 25 mm. The highest rectangular face is perpendicular to VP. The prism leans to the left. Draw the top and front views. (UQ)

70. Draw the projections of a cylinder 60 mm diameter and 80 mm long, lying on the ground with its axis inclined at 30° to VP and parallel to HP.

71. A right regular triangular pyramid, base 35 mm side and axis 60 mm rests with one of its inclined lateral edges on HP such that the two triangular faces containing the inclined edge on which it rests make equal inclinations with HP. The projection of the axis on the HP is parallel to VP. Draw the projections. (UQ)

72. A hexagonal pyramid of 25 mm side of base and 50 mm high rests with one of its corners of the base on HP with its axis inclined at 45° to HP. The two base edges containing the corner of which it rests, make equal inclinations with HP. Draw its projections on VP and HP.

73. Draw the projections of a **cylinder** 35 mm diameter and 50 mm long, lying on HP with its axis inclined at 45° to HP and parallel to VP.

74. A **pentagonal pyramid** of base edge 25 mm and altitude 60 mm rests on one side of base on HP such that the highest base corner is 20 mm above HP. Its axis is parallel to VP. Draw its projections. (UQ)

75. A right circular cone of base diameter 50 mm and height 60 mm is placed such that one diameter AB of the base is inclined at 45° to HP and the other diameter CD of the base is parallel to both HP and VP. The diameters AB and CD are perpendicular to each other. Draw the projections of the cone. (UQ)

76. A square prism of side of base 30 mm and axis 55 mm long lies on HP on one of its longer edges with its faces equally inclined to HP. Draw its projections when its axis is inclined at an angle of 60° to VP.

77. A cube of edges 40 mm is hung by a string attached to one of its corners. Draw its projections when two of its edges containing the corner from which it is hung are equally inclined to VP. (UQ)

78. A cylinder of diameter of base 50 mm and height 60 mm is suspended freely from a point on its circular rim. The projection of the axis on HP is parallel to the XY line. Draw the projections of the cylinder. (UQ)

6. POINTS TO REMEMBER

Summary of the ***three simple positions*** *of a solid and the procedure to draw the projections are given. 3-D illustrations in the first quadrant and the corresponding projections for each position of the solid are drawn in the space provided below:*

POSITION OF SOLID	ILLUSTRATION	PROCEDURE TO DRAW THE PROJECTIONS	PROJECTIONS (FRONT & TOP VIEWS)
Axis perpendicular to HP and parallel to VP	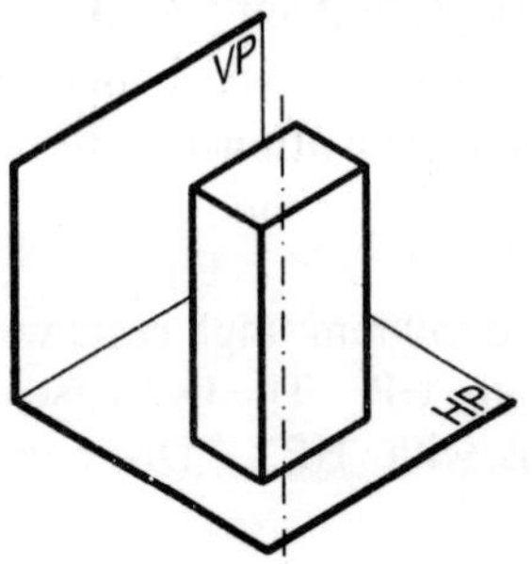	• Axis perpendicular to HP. Hence top view shows true shape and size. • Draw top view first. • Project front view from top view.	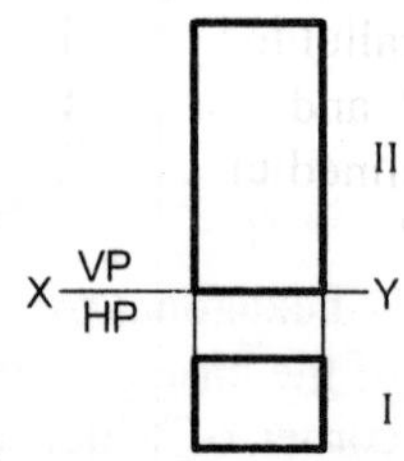
Axis perpendicular to VP and parallel to HP	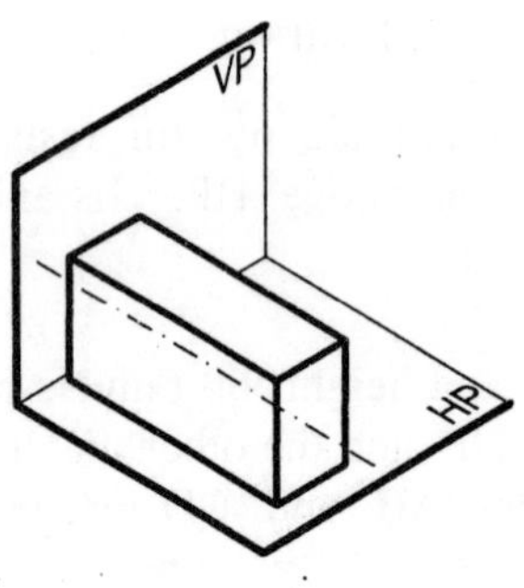	• Axis perpendicular to VP. Hence front view shows true shape and size. • Draw front view first. • Project top view from front view.	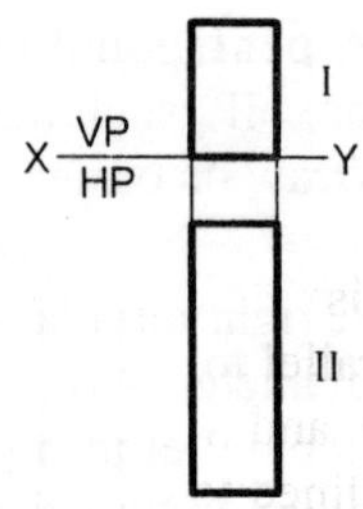
Axis parallel to both HP and VP	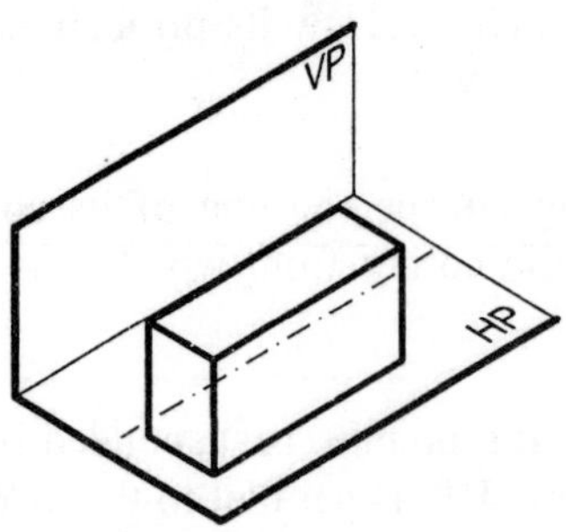	• Draw side view first. • Project front view from side view. • Project top view from front view.	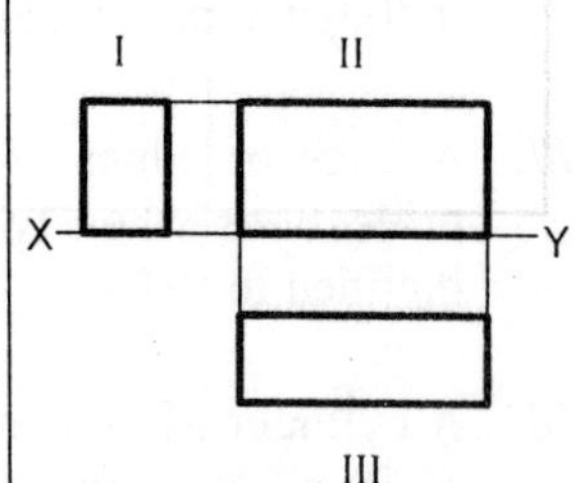

<table>
<tr><th>POSITION OF SOLID</th><th>ILLUSTRATION</th><th>PROCEDURE TO DRAW THE PROJECTIONS</th><th>PROJECTIONS (FRONT & TOP VIEWS)</th></tr>
<tr>
<td>Axis parallel to VP and inclined to HP</td>
<td>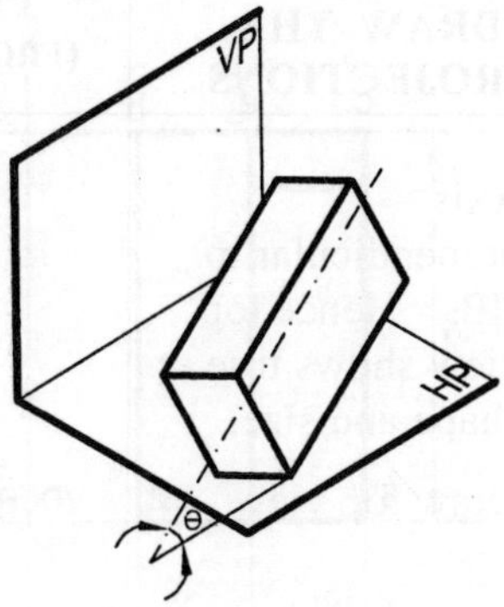
</td>
<td>Simple Position
<ul><li>Draw front and top views in simple position with axis perpendicular to HP.</li></ul>
Second Position
<ul><li>Draw second front view in the tilted position.</li>
<li>Project the second top view from the second front view.</li></ul></td>
<td>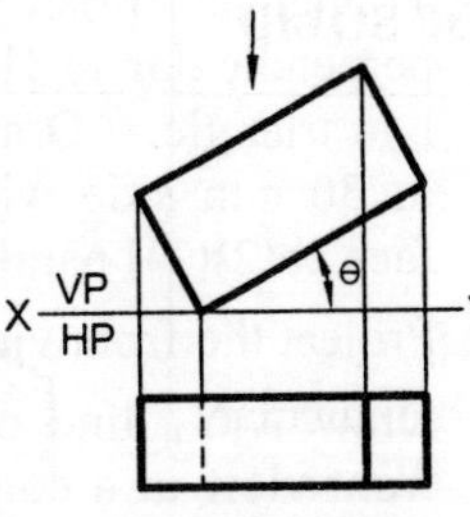

Second Position</td>
</tr>
<tr>
<td>Axis parallel to HP and inclined to VP</td>
<td>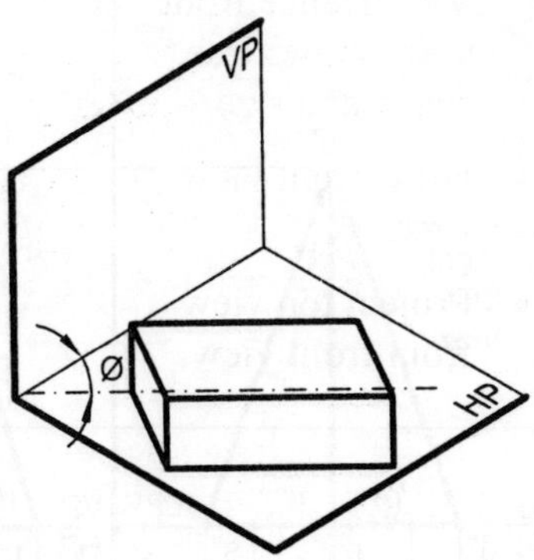
</td>
<td>Simple Position
<ul><li>Draw the front and top views in simple position with axis perpendicular to VP.</li></ul>
Second Position
<ul><li>Draw second top view in the tilted position.</li>
<li>Project the second front view from the second top view.</li></ul></td>
<td>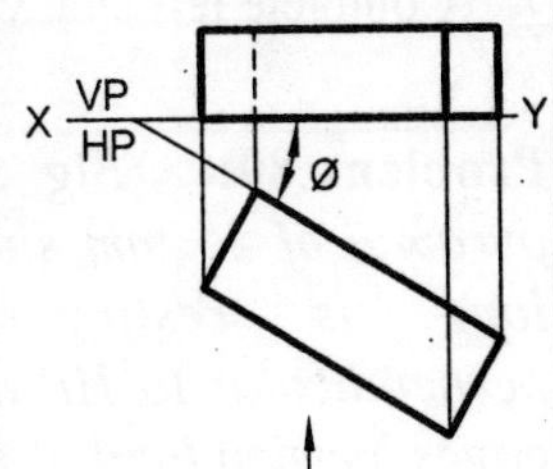

Second Position</td>
</tr>
</table>

7. SIDE VIEWS OF SOLIDS

7.1 AXIS OF SOLID PERPENDICULAR TO HP

Problem 79: *(Fig. 58) A triangular prism of 30 mm side and altitude 50 mm rests with its axis perpendicular to HP, having one of its rectangular faces parallel to VP and nearer to it. Draw the front, top and left side views.*

1. Triangular prism rests with its axis perpendicular to HP. Hence, its top view is a triangle. Draw an equilateral triangle of 30 mm side with one side [rectangular face cb(2)(3)] parallel to XY & nearer to it.
2. Project the front view.
3. Draw X_1Y_1 line on the right side of the front view at a convenient distance from it to represent the Right Profile Plane (R.P.P).
4. Draw the Mitre Line at 45° at the intersection of XY and X_1Y_1.
5. Draw the horizontal projectors from the corners of the prism in the top view to intersect the Mitre Line.
6. From these points of intersection, draw vertical projectors to cut the horizontal projectors drawn from the front view.
7. Complete left side view as shown.

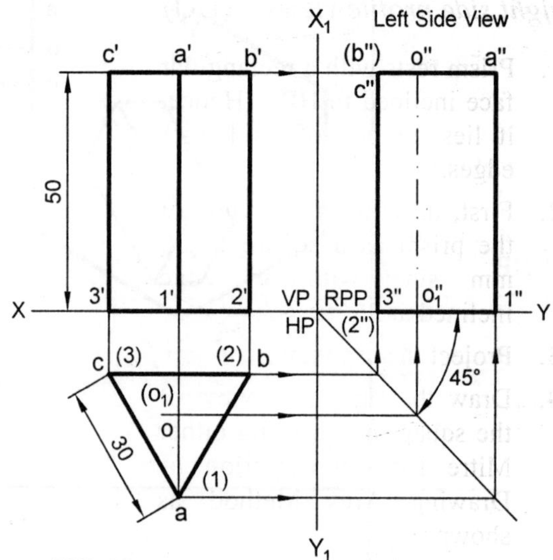

FIG. 58

Problem 80: *(Fig. 59) A pentagonal pyramid of 25 mm side and axis 55 mm long is resting with its axis perpendicular to HP and one of the base edges perpendicular to VP lying on its left.*

Draw its front, top and right side views. (UQ)

1. Draw the top view as a pentagon with a base edge perpendicular to XY lying on left side.
2. Project the front view.
3. Project the right side view using the Mitre Line construction as shown.

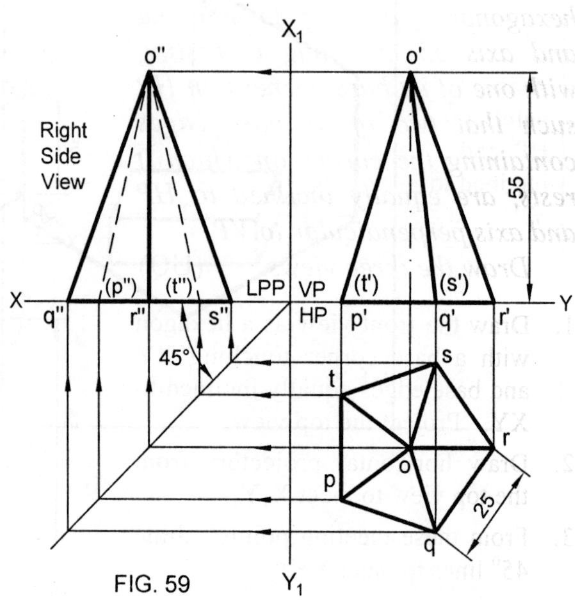

FIG. 59

7.2 AXIS OF SOLID PERPENDICULAR TO VP

Problem 81: *(Fig. 60) A square prism of 30 mm side and altitude 50 mm is placed with its axis perpendicular to VP on one of its longer edges on HP with a rectangular face containing the longer edge on which it rests inclined at 25° to HP.*

Draw the top, front and right side profile views. (UQ)

1. Prism rests with a rectangular face inclined to HP. Hence, it lies on one of its longer edges.
2. First, draw the front view of the prism as a square of 30 mm side with an edge inclined at 25° to XY.
3. Project the top view.
4. Draw the right side view of the square prism using either Mitre Line construction *or* Drawing Arcs Method as shown.

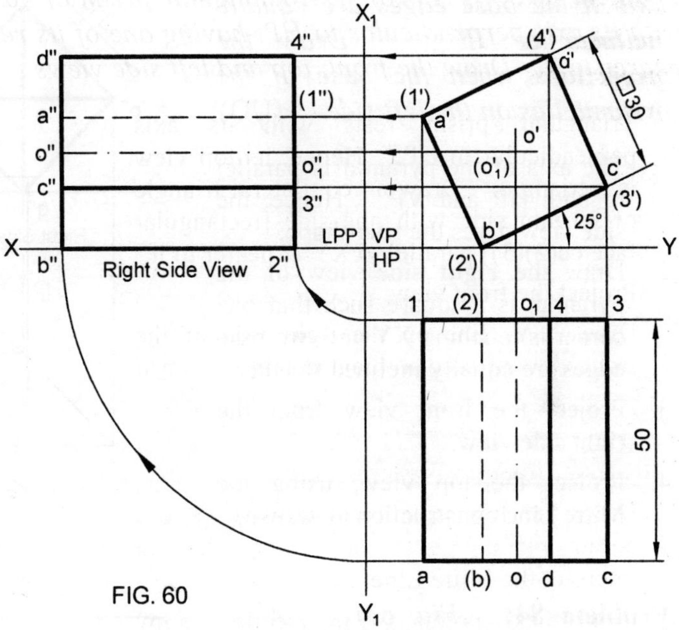

FIG. 60

Problem 82: *(Fig. 61) A hexagonal pyramid of 25 mm side and axis 55 mm long is resting with one of its base corners on HP such that two of its base edges containing the corner on which it rests, are equally inclined to HP and axis perpendicular to VP.*

Draw the three views. (UQ)

1. Draw the front view as a hexagon with a base corner touching XY and base edges equally inclined to XY. Project the top view.
2. Draw horizontal projectors from the top view to meet X_1Y_1.
3. From these meeting points, draw 45° lines to meet XY.
4. Complete the left side view.

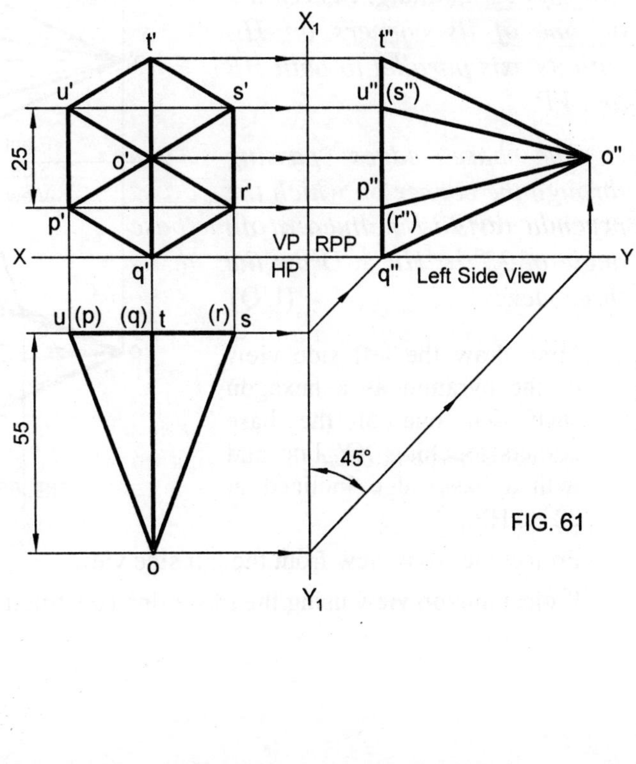

FIG. 61

7.3 AXIS OF SOLID PARALLEL TO BOTH HP AND VP

Problem 83: *(Fig. 62) A square pyramid of 35 mm side and altitude 55 mm rests on one of its corners on HP with its axis parallel to both HP and VP. Two of the base edges are equally inclined to HP. Draw the projections when the base of the pyramid lies on the **left** side.* (UQ)

1. The axis of the pyramid is parallel to both HP and VP. Hence, the side view gives the true shape.
2. Draw the **right** side view of the pyramid as a square such that one corner is touching XY and two base edges are equally inclined to HP.
3. Project the front view from the right side view.
4. Project the top view, using the Mitre Line construction as shown.

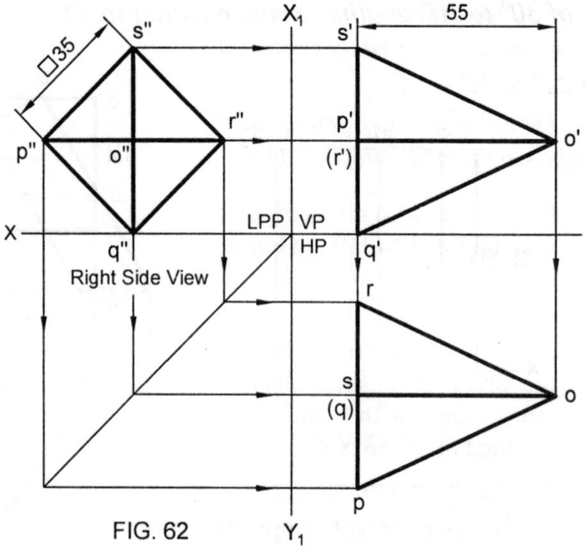

FIG. 62

Problem 84: *(Fig. 63) A hexagonal pyramid of 25 mm side and 55 mm height is resting on one of its corners on HP with its axis parallel to both HP and VP.*

The base edges passing through the corner on which the pyramid rests is inclined at an angle of 22° to HP. Draw the three views. (UQ)

1. First, draw the left side view of the pyramid as a hexagon such that one of the base corners touching XY line and with a base edge inclined at 22° to HP.
2. Project the front view from the left side view.
3. Project the top view using the Mitre line construction as shown.

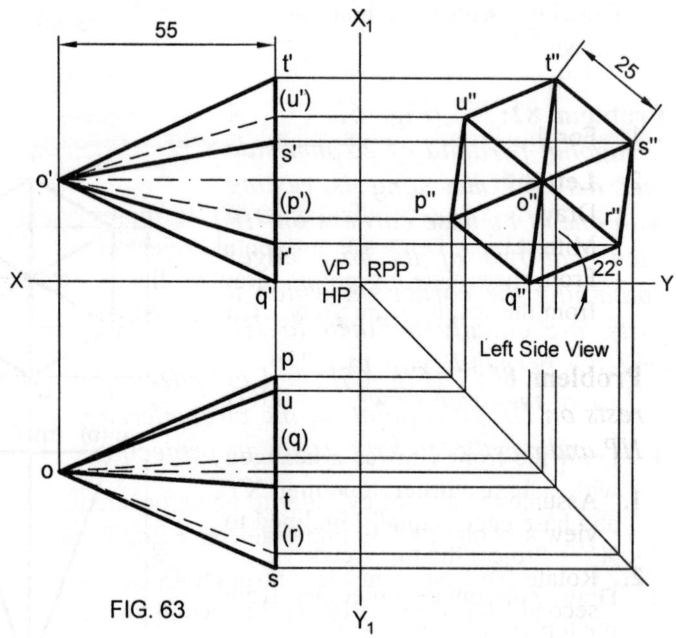

FIG. 63

7.4 AXIS OF SOLID INCLINED TO HP

Problem 85: *(Fig. 64) A pentagonal prism, side of base 25 mm and axis 50 mm long, rests with one of its edges on HP such that the base containing that edge makes an angle of 30° to HP and its axis is parallel to VP. Draw its top, front and profile views.* (UQ)

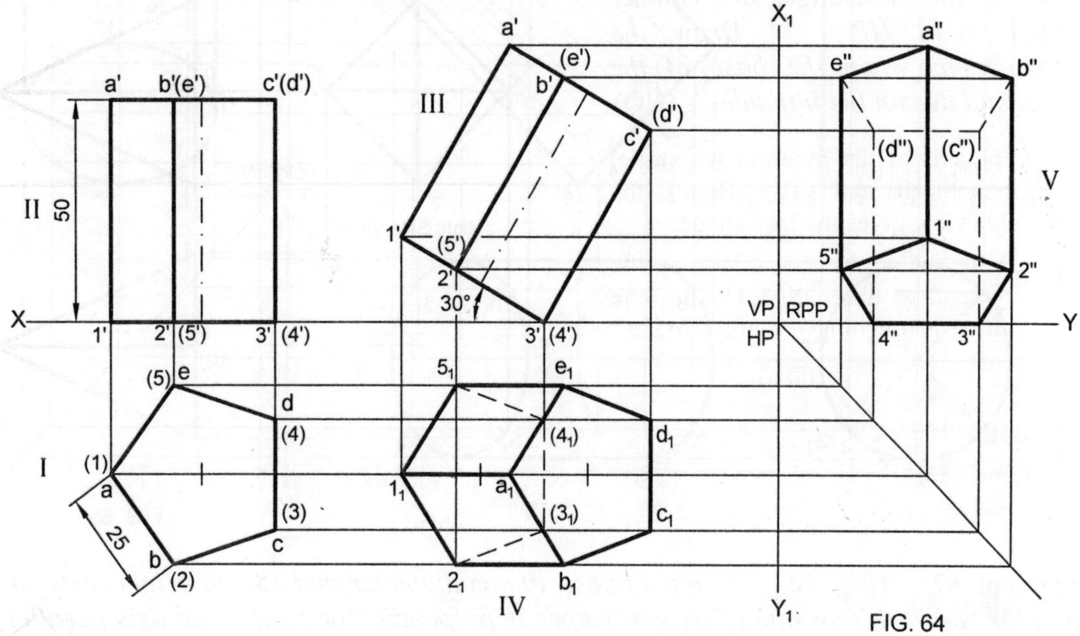

FIG. 64

1. For this Problem, required top & front views are already drawn in Change of Position Method.
2. **Left Side View**
   Draw X_1Y_1 perpendicular to XY to represent the Right Profile Plane (R.P.P). Draw the Mitre Line. Draw the horizontal projectors from the second top view to meet the Mitre Line. From these meeting points, draw vertical projectors to meet the horizontal projectors drawn from the second front view. Complete the left side view as shown.

Problem 86: *(Fig. 65) A cylinder of base diameter 40 mm and axis 55 mm long rests on HP on a point on the circumference of the base with its axis inclined at 45° to HP and parallel to VP. Draw its projections.* (UQ)

1. Assume the axis of cylinder to be perpendicular to HP and parallel to VP. Draw the first top view as a circle of 40 mm diameter. Project the first front view.
2. Rotate the first front view through 45° to HP such that one point touching XY to obtain the second front view. Project the second top view as shown.
3. **Left Side View:** Draw X_1Y_1 perpendicular to XY to represent the Right Profile Plane (R.P.P). Draw the Mitre Line. Draw the horizontal projectors from the second top view to meet the Mitre Line. From these meeting points, draw vertical projectors to meet the horizontal projectors drawn from the second front View. Complete left side view as shown.

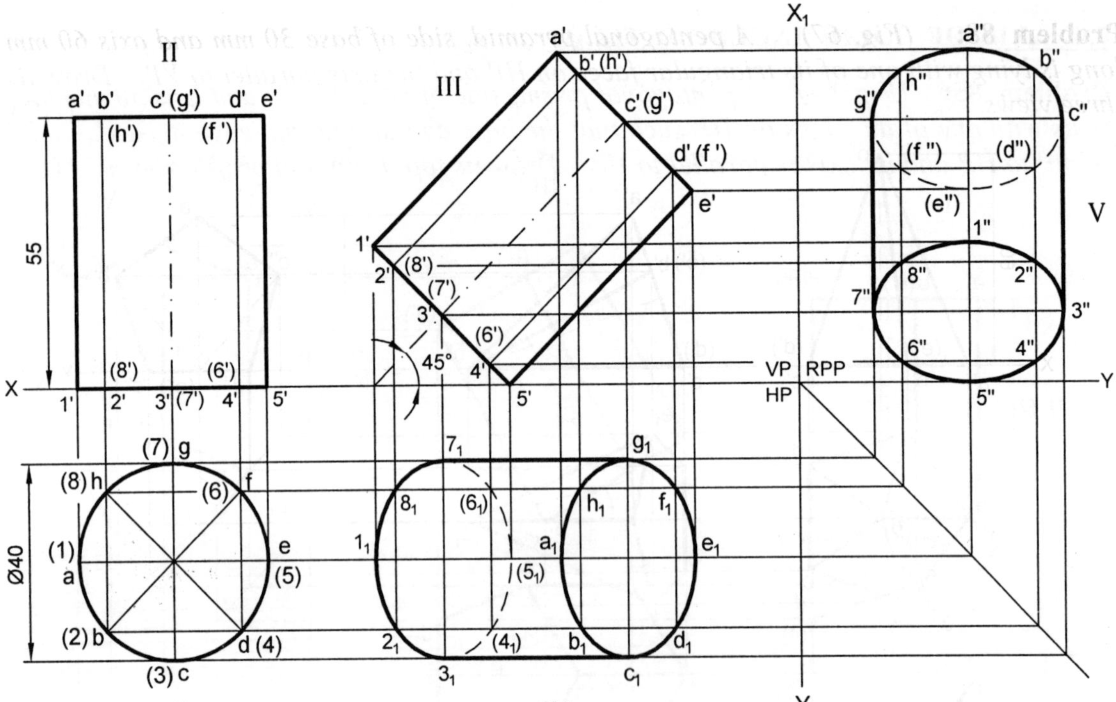

FIG. 65

Problem 87: *(Fig. 66) A cone of base 40 mm diameter and 55 mm height rests on its circular rim such that one of its generators is perpendicular to HP and axis parallel to VP. Draw its three views.* (UQ)

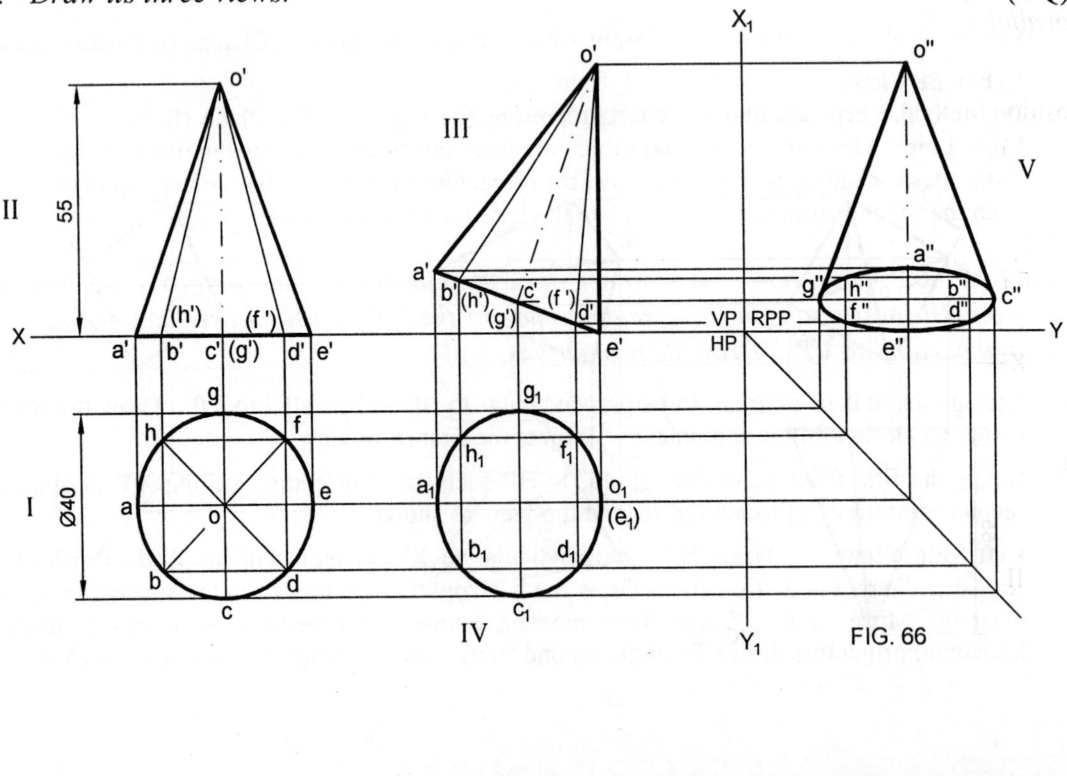

FIG. 66

Problem 88: *(Fig. 67) A pentagonal pyramid, side of base 30 mm and axis 60 mm long is lying with one of its triangular faces on HP and the axis parallel to VP. Draw its three views.* (UQ)

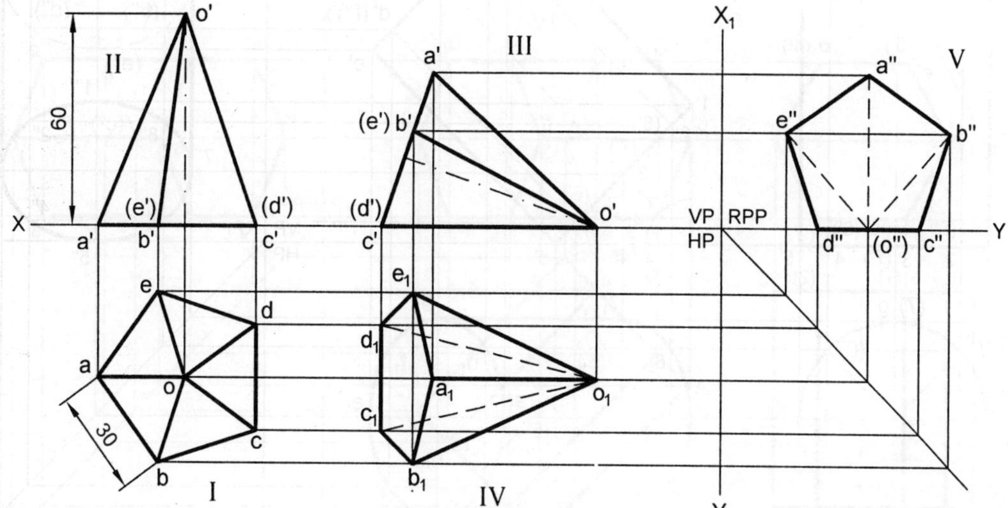

FIG. 67

7.5 AXIS OF SOLID INCLINED TO VP

Problem 89: *(Fig. 68) Draw the three views of a* ***square pyramid*** *of 32 mm side of base and axis 55 mm. It is resting on HP on one of its base corners with a base side containing the corner making 30° with HP. The axis is inclined at 30° to VP and is parallel to HP. The vertex is away from VP.* (UQ)

For this Problem, the required front and top views are already drawn in Change of Position Method. See Fig. 68 for the left profile (side) view.

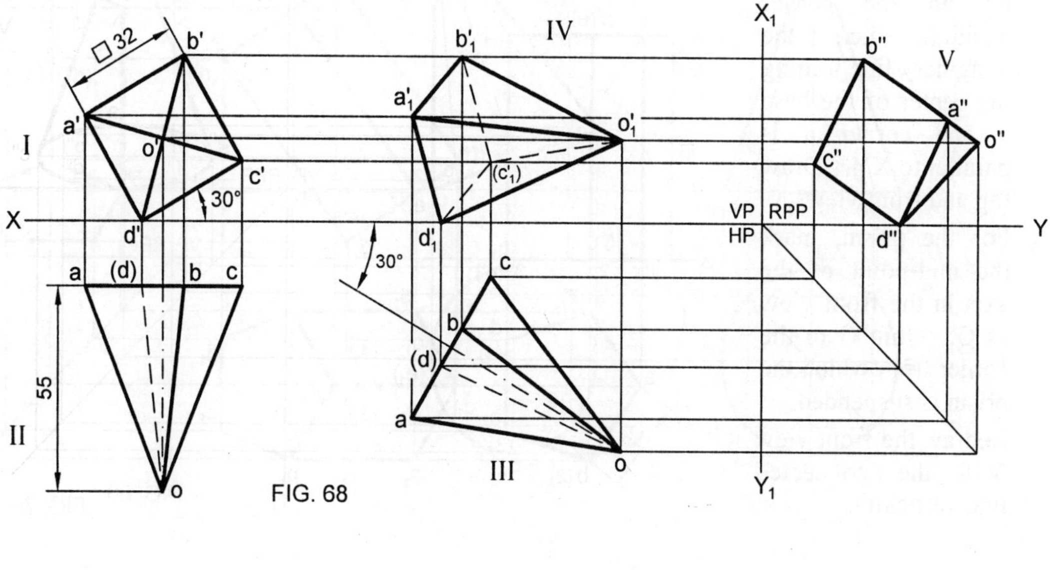

FIG. 68

Problem 90: *(Fig. 69) A* ***hexagonal prism****, side of base 25 mm and axis 60 mm long, lies with one of its rectangular faces on HP, such that the axis is inclined at 45° to VP. Draw its front, top and profile views.* (UQ)

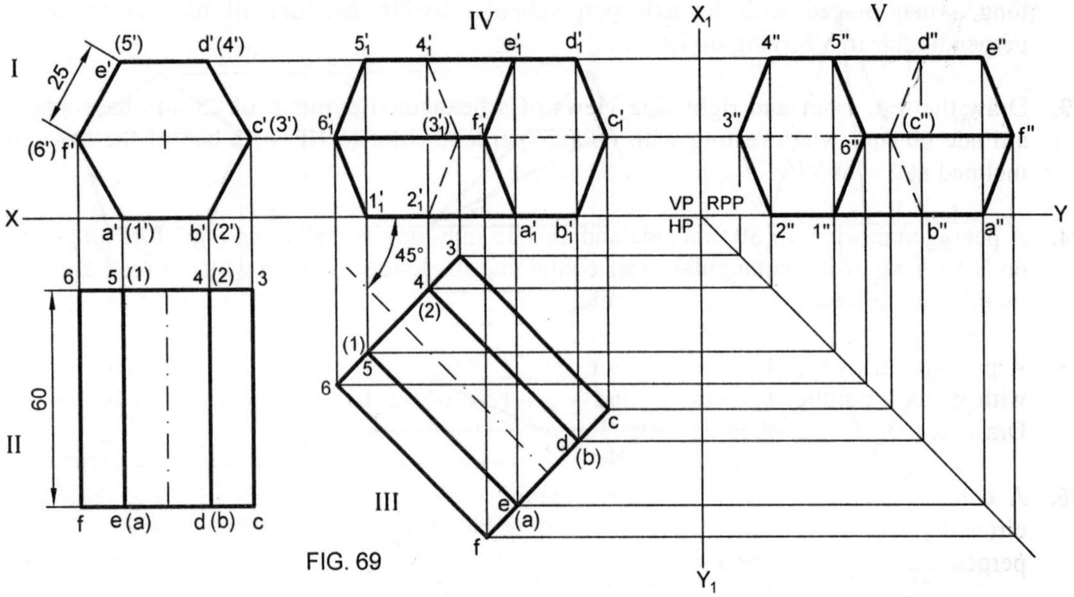

FIG. 69

7.6 FREELY SUSPENDED SOLID

Problem 91: *(Fig. 70) A pentagonal prism of 25 mm side of base and height 55 mm is suspended freely from a corner such that axis is parallel to VP. Draw the three views.*

1. Initially, place the prism with its base on HP in the corner position, i.e., the imaginary line joining the center of the base o and corner a is parallel to XY. Draw top and front views.
2. For the prism, mark the mid-point of the axis in the front view as G. Join G to the corner from which the prism is suspended.
3. Redraw the front view with the connected line vertically.

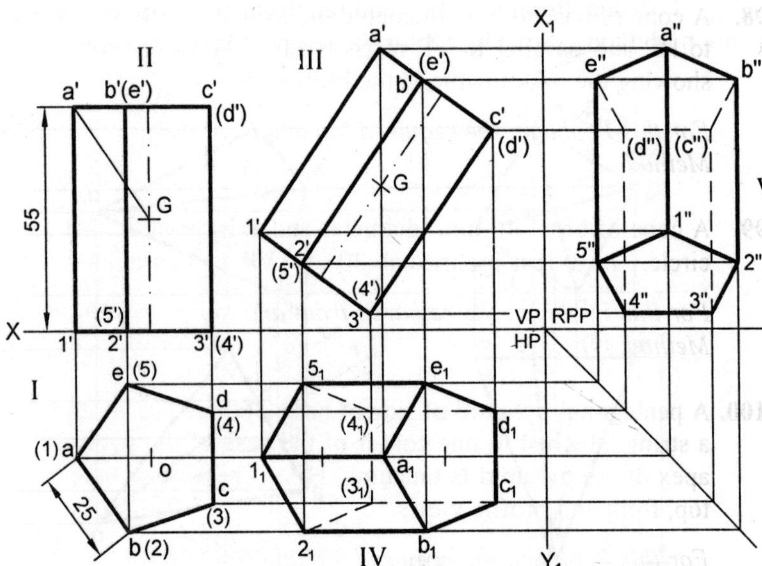

FIG. 70

8. ADDITIONAL PROBLEMS – II

92. Draw the top, front and left side views of a pentagonal prism of 25 mm side and axis 50 mm long, when placed with its axis perpendicular to HP and one of its rectangular faces perpendicular to VP lying on its left.

93. Draw the top, front and right side views of a hexagonal pyramid of 25 mm base edge and altitude 60 mm when resting with its axis perpendicular to HP with one of the base edges inclined at 25° to VP.

94. A pentagonal prism of 30 mm side and axis 55 mm long is resting on one of its longer edges on HP. Two of its rectangular faces containing the longer edge on which it rests are equally inclined to HP and axis is perpendicular to VP. Draw the top, front and left side views.

95. A pentagonal pyramid of 26 mm side and 55 mm height rests on one of its corners on HP with its axis parallel to both HP and VP. One of the base edges is perpendicular to HP. Draw its projections when its vertex lies on the right side. (UQ)

96. A cone of 40 mm base diameter and height 60 mm is placed on HP on a point on the circumference of the base. Draw its profile view, front and top views when the base is perpendicular to both HP and VP. (UQ)

Hint: When the base of a cone is perpendicular to both HP and VP, its axis in this position is parallel to both HP and VP.

97. A cube of 40 mm side is resting on HP on one of its ends with its vertical faces equally inclined to VP. Draw the top, front and right side views.

98. A cone of base 40 mm diameter and axis 60 mm long lies on HP with its axis inclined at 45° to HP and parallel to VP respectively. Draw the top, front and profile views of the cone showing the base visible in the profile view. (UQ)

For this Problem, the required top and front views are already drawn in Change of Position Method.

99. A cone of base 40 mm diameter and axis 50 mm long touches VP on a point of its base circle. Its axis is inclined at 30° to VP and parallel to HP. Draw its three views. (UQ)

For this Problem, the required front and top views are already drawn in Change of Position Method.

100. A pentagonal pyramid of side of base 30 mm and altitude 68 mm is suspended freely from a string attached to one corner of the base of the pyramid. The string is held such that the apex of the pyramid is touching HP. The axis of the pyramid is parallel to VP. Draw its top, front and profile views. (UQ)

For this Problem, the required top and front views are already drawn in Change of Position Method.

9. AXIS OF SOLID INCLINED TO BOTH HP AND VP

Projections of a solid with axis inclined to both HP and VP are drawn in ***three stages*** as:

Stage I: Keep the axis of solid in Simple Position (i.e., $\perp^r$ to one of the reference planes), suitable to reach the final required position. Then draw the first set of views.

Stage II: Tilt the solid so as to bring the axis inclined to one of the reference planes and parallel to the other. Then draw the second set of views.

Stage III: Tilt the solid further so as to bring the axis inclined to both the reference planes. Then draw the third and final set of views.

Problem 101: *(Fig. 71) A **square prism**, side of base 30 mm and axis 50 mm long, has its axis inclined at 60° to HP. It has an edge of its base in the HP and inclined at 45° to VP. Draw its projections.* (UQ)

Simple Position : *See Fig. (i).*

1. When the prism has to rest with an edge of its base on H.P., first assume it in its simple position, i.e., resting with its base on H.P. such that an edge of its base is perpendicular to V.P. This is to enable to tilt the prism about the edge 23 of its base on H.P. For this position, draw the top view first. Then project the corresponding front view.
2. Here in the top view, (2)(3) is perpendicular to XY and on the right side. In the front view, the same edge 2′ (3′) appears as a point.

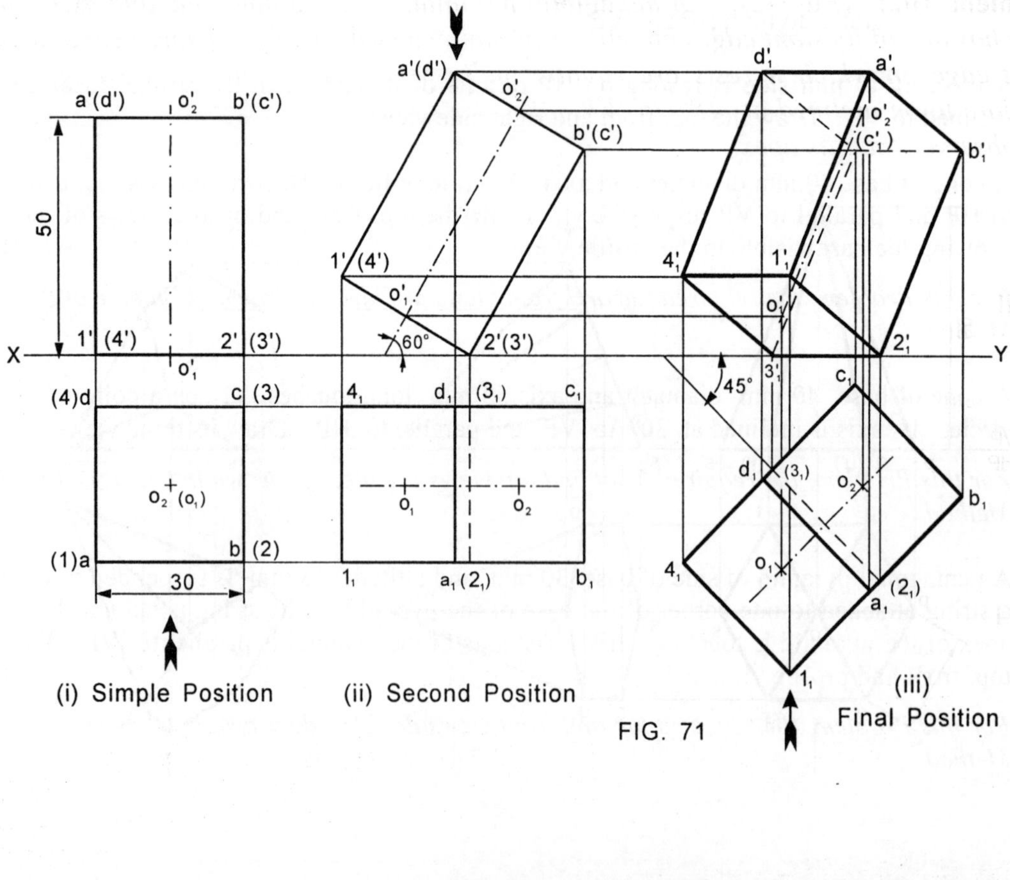

(i) Simple Position (ii) Second Position (iii) Final Position

FIG. 71

Second position : *See Fig. (ii).*

3. The prism has to rest with an edge of its base on H.P. such that its axis is inclined at $\theta = 60^\circ$ to H.P. So tilt the first front view such that $2'$ $(3')$ is on XY and its base is inclined at 30° (i.e., $90^\circ - 60^\circ$) to XY. Thus obtain the second front view.
4. Project the corresponding second top view in which $(2_1)(3_1)$ is perpendicular to XY.

Final position : *See Fig. (iii).*

5. As per the given condition, again the edge $(2_1)(3_1)$ has to make 45° with VP, keeping the inclination of the axis with HP constant. Hence, the shape and size of the second top view will remain the same, but its position alone will change. So, tilt second top view such that (2_1) (3_1) is inclined at 45° to and below XY. Thus obtain the final top view.
6. In the final front view, height of all the corners from XY will remain the same as in the second front view. Therefore draw projectors from the final top view and horizontals from the second front view and obtain the final front view.
7. **Visible and Invisible edges in final front view:** First draw the boundary lines of the final front view by *thick* lines. Look at the final top view in the direction of arrow. $1_1\,a_1$, $1_1\,(2_1)$ and $1_1\,4_1$ are visible. Hence, draw $1_1'\,a_1'$, $1_1'\,2_1'$ and $1_1'\,4_1'$ by *thick* lines.

   c_1 is invisible. Hence, mark it as (c_1') in the final front view. $c_1(3_1)$, c_1b_1 and c_1d_1 are invisible. Hence, draw $(c_1')\,3_1'$, $(c_1')\,b_1'$ and $(c_1')\,d_1'$ by *dashed* lines.

Problem 102: *(Fig. 72)* *A* ***hexagonal pyramid****, base 25 mm side and axis 60 mm long has one of its slant edges on HP such that two of its triangular faces containing the slant edge on which it rests are equally inclined to HP. The top view of the axis appears to be inclined at 45° to VP. Draw its projections when its base is nearer to the observer than its apex.* (UQ)

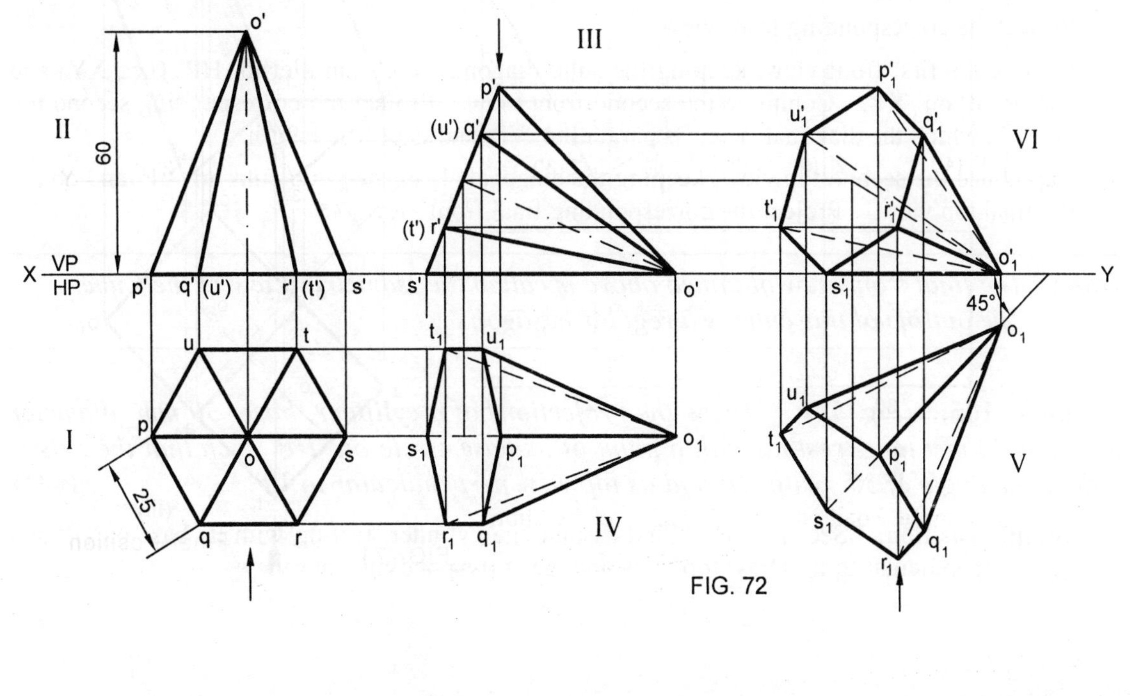

FIG. 72

Problem 103: *(Exercise) Draw the projections of a pentagonal pyramid of 30 mm side and height 60 mm, when it is resting on one of its base edges. The axis is making 30° with HP and the top view of the axis is making 45° with VP.* (UQ)

Problem 104: *(Fig. 73) A* ***hexahedron*** *of edge 35 mm rests on HP on one of its corners with a solid diagonal perpendicular to VP. Draw its projections.* (UQ)

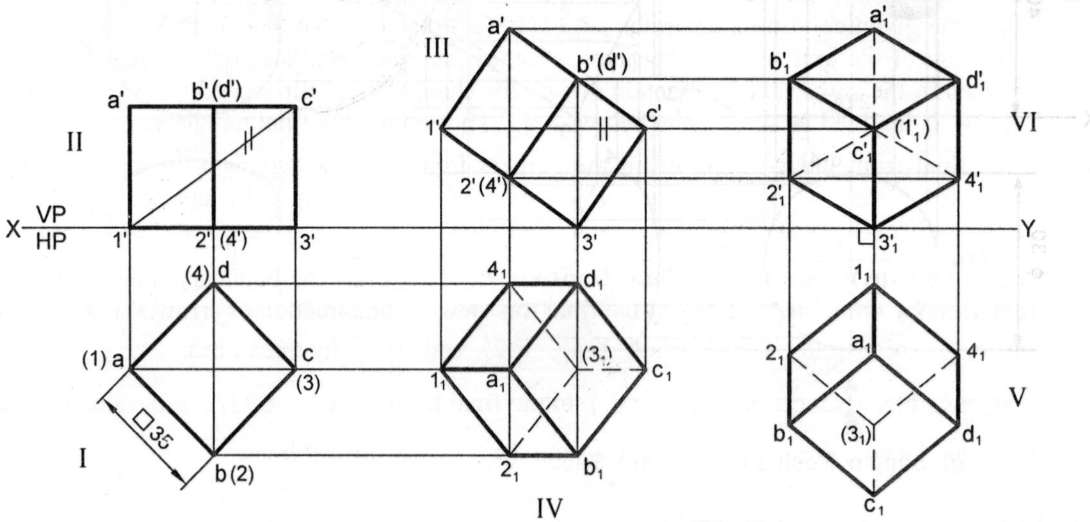

FIG. 73

1. Keep the hexahedron (cube) in *simple position* such that one of the solid diagonals (1)c is parallel to VP one of the square faces is on HP. For this position, draw the top view as a square having edges equally inclined to XY. Thus, diagonal (1)c is parallel to XY.
2. Project the corresponding front view.
3. Redraw the first front view, keeping the solid diagonal 1′c′ parallel to HP (i.e., XY) and corner 3′ on XY. Complete the second front view. Project the corresponding second top view, in which the diagonal $1_1 c_1$ is parallel to VP and is of true length.
4. Reproduce the second top view, keeping the diagonal $1_1 c_1$ perpendicular to VP and obtain the final top view. Project the corresponding final front view.

Note: *The final front view obtained above is called the Isometric View of the Cube. The outline of this cube is a regular hexagon.*

Problem 105: *(Fig. 74) Draw the projections of a* ***cylinder****, base 30 mm diameter and axis 40 mm long, resting with a point of its base circle on HP such that the axis is making an angle of 30° with HP and its top view perpendicular to VP.* (UQ)

1. **Simple position:** See *Fig. (i)*. First assume the cylinder, resting with its base on HP and axis perpendicular to it. Draw top view and the corresponding front view.

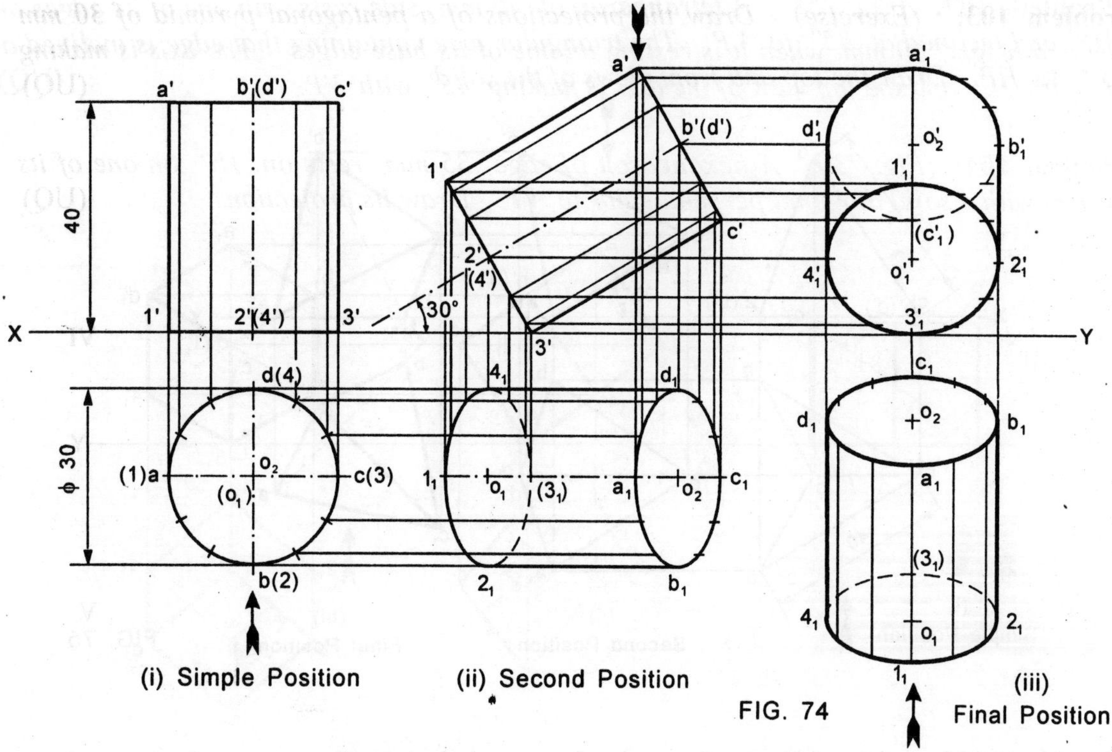

(i) Simple Position (ii) Second Position (iii) Final Position

FIG. 74

2. **Second position:** See *Fig. (ii)*. The cylinder has to rest with a point of its base circle on HP such that its axis makes $\theta = 30^\circ$ with HP. Hence tilt the front view such that the point 3′ is on XY and its base is inclined at 60° (*i.e.*, $90^\circ - \theta$) to XY. Thus obtain the second front view. Project second top view, in which axis is parallel to XY.

3. **Final position:** As per the given condition the cylinder is to be tilted such that top view of the axis is perpendicular to XY, keeping the inclination of the axis with HP constant. Hence, the shape and size of second top view remain the same, but its position alone will change. So, tilt the second top view such that $o_1 o_2$, the top view of the axis is perpendicular to XY and obtain the final top view. Then project the final front view as in *Fig. (iii)*.

4. Look at the top view in the direction of arrow shown. $1_1\ 2_1\ (3_1)\ 4_1$ is nearer to the observer. Hence, draw the curve $1'_1\ 2'_1\ 3'_1\ 4'_1$ by *thick line*.

   $b_1\ c_1\ d_1$, the portion of the top face is invisible in final top view. Hence, draw the curve $b'_1(c'_1)\,d'_1$ by *dashed line* in the final front view.

Problem 106: *(Exercise) A cylinder of diameter 30 mm and axis height 60 mm lies on the ground on a point of its base circle such that the axis is inclined at 45° to the HP. The plane containing the axis makes an angle of 30° with the VP. Draw the projections of the cylinder.*

Problem 107: *(Fig. 75) A **tetrahedron** of 40 mm side rests with one of its edges on HP and inclined at 45° to VP. The triangular face containing that edge is inclined at 30° to HP. Draw the top and front views of the solid.* (UQ)

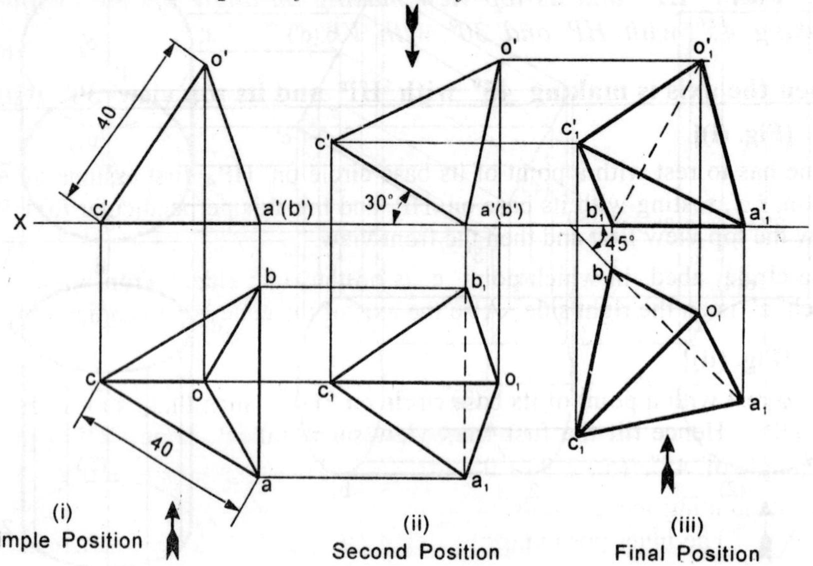

FIG. 75

Problem 108: *(Fig. 76) Draw the front and top views of an **octahedron** of edges 30 mm when one face is in HP and one edge of that face makes 45° with VP.* (UQ)

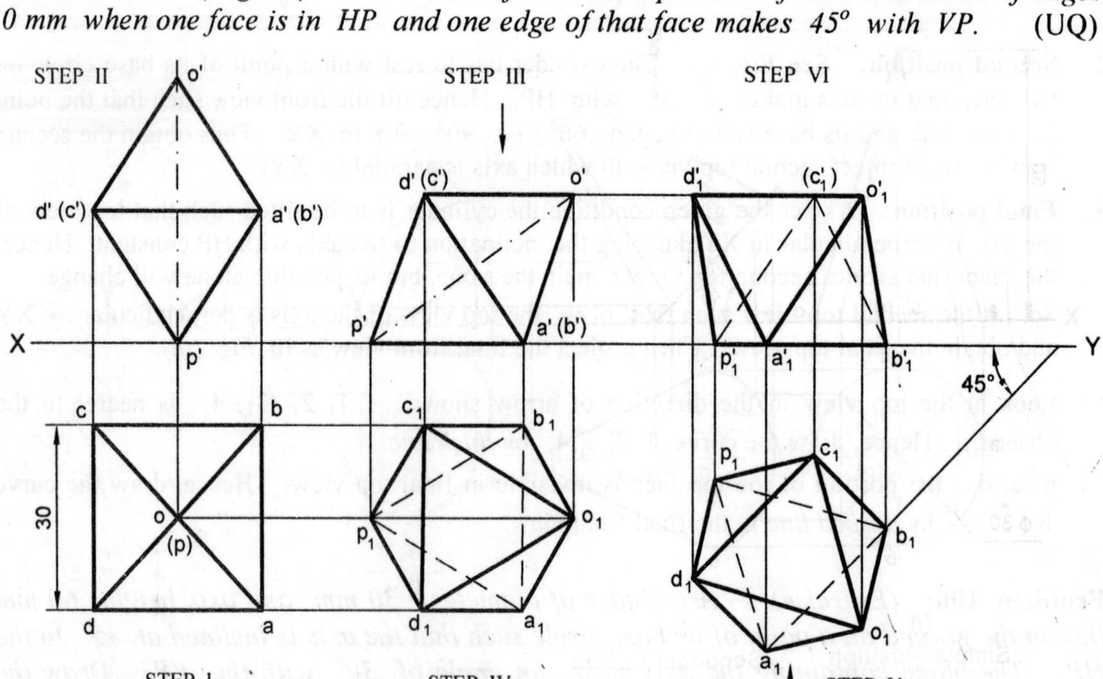

FIG. 76

Problem 109: *(Fig. 77) Draw the projections of a* **cone**, *base 30 mm diameter and axis 50 mm long, resting on HP on a point of its base circle with (A) the axis making an angle of 45° with HP and its top view making an angle of 30° with VP and (B) the axis making 45° with HP and 30° with VP.* (UQ)

Case (A): When the axis is making 45° with HP and its top view 30° with VP

Simple Position [Fig. (i)]

1. When the cone has to rest with a point of its base circle on HP, first assume the cone in its simple position, *i.e.*, resting with its base on HP and the axis perpendicular to HP. For this position, draw the top view first and then the front view.
2. Top view is a circle abcd in which point c is on the right side. Front view is a triangle a'o'c' in which c' is on the right side. Also the axis of the cone is perpendicular to XY.

Second Position [Fig. (ii)]

3. The cone has to rest with a point of its base circle on HP, such that its axis has to make θ = 45° with HP. Hence tilt the first front view such that c' is on XY and its base is inclined at an angle of 45°, (*i.e.*, 90 – θ) to XY. Thus obtain the second front view.
4. Project the corresponding second top view in which the base is seen as an ellipse and the axis is parallel to XY. The inner portion of the ellipse b_1 (c_1) d_1 between the tangents is shown as *dashed line*.

Note: *These tangents need not be any of the particular generators that have been chosen, i.e., need not be* $o_1 b_1$ *or* $o_1 d_1$.

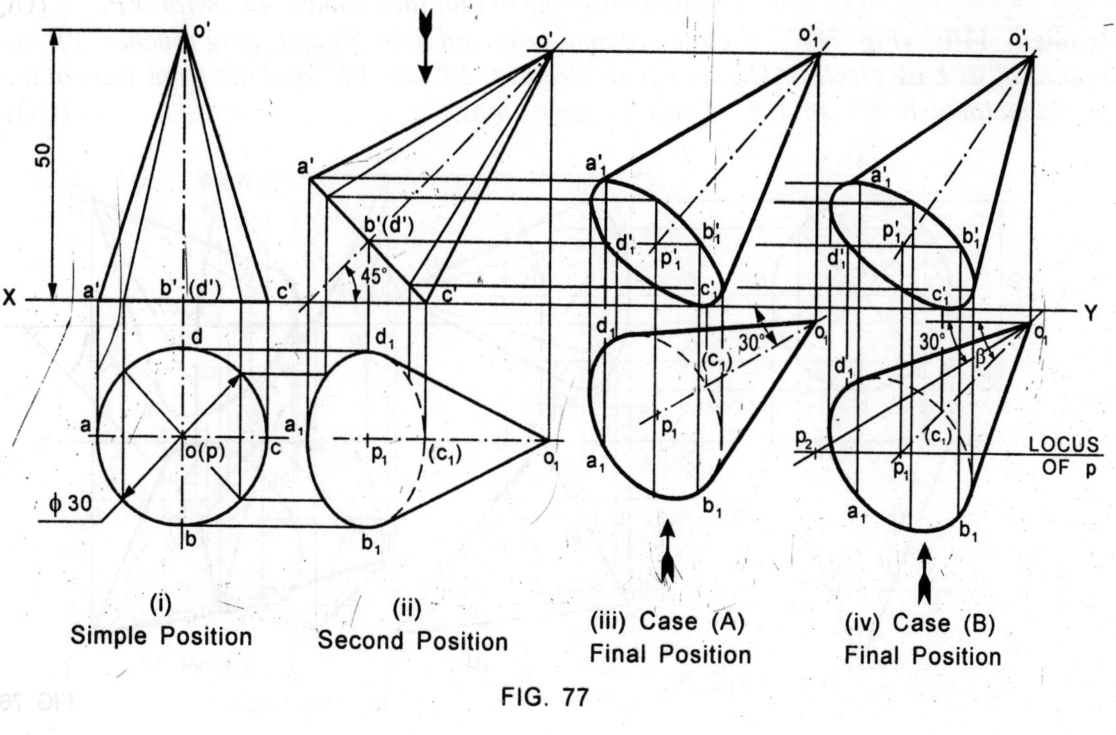

FIG. 77

Final position [Fig. (iii)]

5. As per the given data, the top view of the axis has to make 30° with VP, keeping the inclination of the axis with HP constant. So the shape and size of the second top view will remain the same, but its position alone will change.

   Therefore tilt the second top view such that $o_1 p_1$, the top view of the axis is inclined at 30° to XY and obtain the final top view. Then project the corresponding final front view.

6. Look at the final top view in the direction of arrow. Base is closer than the apex. So draw the base completely by *thick line* in the final front view. Also from o_1' draw two tangents to the ellipse.

Case (B): When the axis is making 45° with HP and 30° with VP

1. Repeat steps 1 to 4 in Case (a) and obtain the second top view.
2. **Final position:** Since the axis makes 45° with HP and 30° with VP, its top view will make an apparent angle β greater than 30° with XY (as per the concept of the projections of straight line inclined to both HP and VP)
3. **To determine β:** From o_1 draw a line inclined at 30° to XY and mark $o_1 p_2$ on it such that it is equal to the true length of the axis = 50 mm.

   o_1 as center and $o_1 p_1$ (length of the axis in the top view) as radius draw an arc to cut the locus of p at p_1. Now the angle made by $p_1 o_1$ with XY is β (> 30°).
4. Redraw the second top view with respect to $p_1 o_1$ which makes an angle of β with XY. Project the corresponding final front view. [*Fig. (iv)*]

Problem 110: *(Fig. 78) A cone 40 mm base and axis 50 mm long touches VP on a point of its base circle. The axis is inclined at 30° to VP and the front view of the axis is inclined at 45° to HP. Draw its projections.* (UQ)

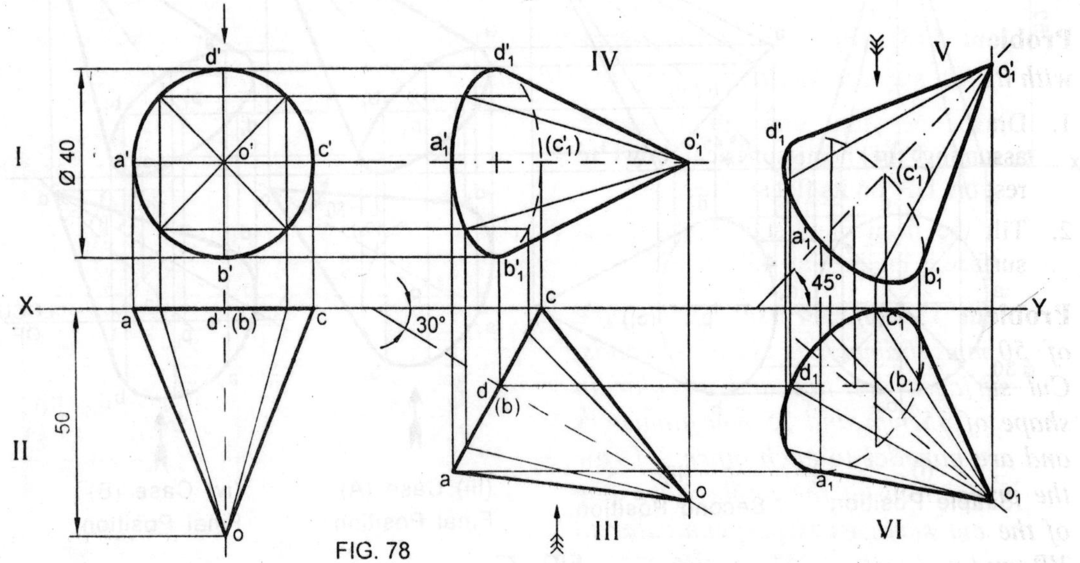

FIG. 78

10. PROJECTIONS OF SPHERES

Projection of a sphere on any plane is always a *circle.* When a hemi-sphere rests on its flat surface on HP, its top view is a *circle.* Front view is a *semi-circle.* When the flat surface of a hemi-sphere is inclined to HP and ⊥r to VP, its top view is an *ellipse.* Periphery of the hemi-sphere is an arc of a circle, the radius of which is equal to the radius of the sphere.

Contacting Spheres: When two or more spheres of *equal radii* resting on HP make contact wiih each other, then the line joining their centers will be *parallel* to XY. When the contacting spheres are of *different radii*, their line of centers will be *inclined* to XY.

Problem 111: *(Fig. 79) One hockey ball of radius 26 mm and a cricket ball of radius 13 mm are kept on HP such that both are touching each other. Draw the projections, when the line of centers in the top view is parallel to VP.* (UQ)

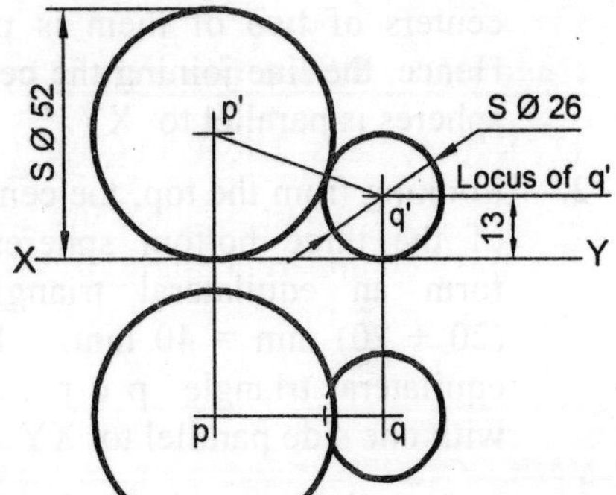

FIG. 79

1. Draw the front view of the hockey ball (ϕ 52 mm).
2. To draw the cricket ball, draw a locus line 13 mm above HP for its center. Cut an arc on it with p' as center and 39 (= 26 + 13) mm radius. q' as center draw the front view of the cricket ball.
3. Since the line joining the centers of spheres is parallel to VP, draw a line for it in top view. Draw projectors to obtain p and q. Complete the top view.

Problem 112: *(Exercise) Two spheres of radii 24 mm and 36 mm are lying on the HP touching each other, such that the line joining the centers makes an angle of 45° with VP. Draw the projections.*

Problem 113: *(Fig. 80) A hemi-spherical bowl of 48 mm diameter is resting on HP with its flat surface inclined at 45° to HP and ⊥r to VP. Draw the projections.* (UQ)

1. Draw the top and front views, assuming the hemi-spherical bowl to rest on HP on its flat surface.
2. Tilt the front view such that the flat surface is inclined at 45° to XY.

Problem 114: *(Exercise) An apple of 50 mm diameter is cut at two ends. Cut surfaces are flat and circular in shape of 35 mm and 25 mm diameters and are parallel to each other. Draw the projections of the apple when one of the cut surfaces is perpendicular to VP and inclined at 35° to HP.*

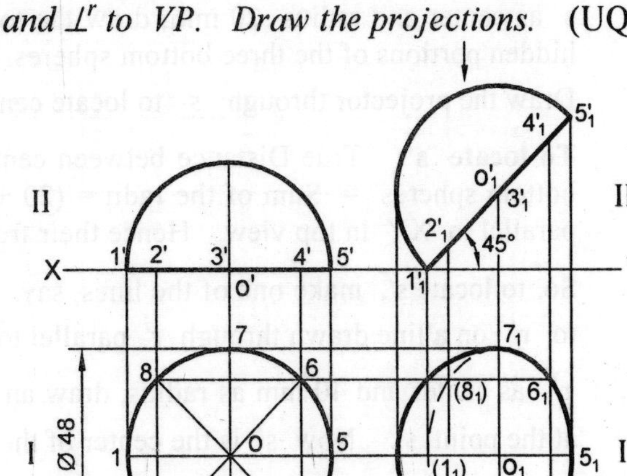

FIG. 80

Problem 115: *(Fig. 81) Four equal spheres of diameters 40 mm are arranged in a pyramidal form such that three spheres are resting on HP. The line joining the centers of two of them is parallel to VP. Each sphere is in contact with the other three. Draw the projections of sphere assembly.* (UQ)

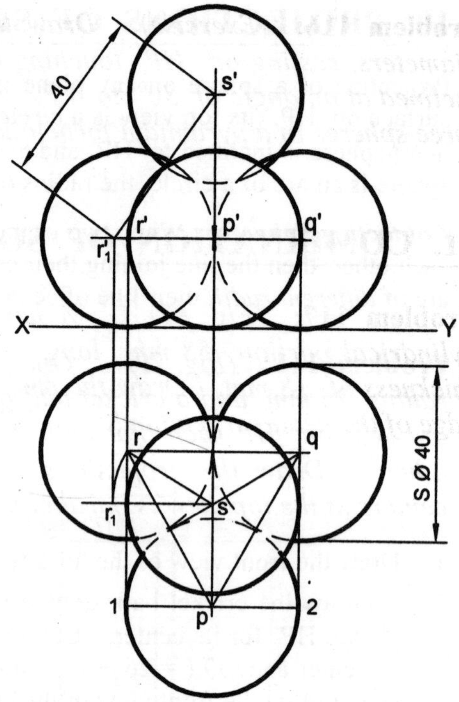

FIG. 81

1. *Given:* Three equal spheres rest on HP (bottom spheres) and the line joining the centers of two of them is parallel to VP. Hence, the line joining the centers of the two spheres is parallel to XY.

2. Looking from the top, the centers p, q and r of the three bottom spheres lying on HP form an equilateral triangle of sides = (20 + 20) mm = 40 mm. Hence, draw an equilateral triangle p q r of side 40 mm with one side parallel to XY in the top view.

3. p, q and r as centers, draw three circles of 40 mm diameter to represent the three bottom spheres. Draw vertical projectors from the top view to obtain the front view.

4. In the front view mark the centers p′, q′ and r′ such that they lie on a line parallel to and 20 mm above XY. Draw the front view as three circles with p′, q′ and r′ as center and 20 mm radius.

5. **Fourth (Top) Sphere:** Now place fourth sphere on top of the 3 spheres. In top view, the center **s** of top sphere is nothing but the center of the equilateral triangle p q r.

   s as center and radius 20 mm, draw the top view of the top sphere as a circle. Mark the hidden portions of the three bottom spheres, hidden by the top sphere.

   Draw the projector through s to locate center s′ of the top sphere in the front view.

   To locate s′: True Distance between center of the top sphere and that of any one of the bottom spheres = Sum of the radii = (20 + 20) mm. But, lines sp, sq and sr are not parallel to XY in top view. Hence their front views will not show their true lengths.

6. So, to locate s′, make one of the lines, say, sr is turned to sr_1 parallel to XY. Project r_1 to r_1' on a line drawn through r′ parallel to XY (which is the locus of r′).

7. r_1' as center and 40 mm as radius, draw an arc to intersect the vertical projector through s at the point s′. Now s′ is the center of the top sphere in the front view.

8. s′ as center and 20 mm as radius, draw a circle to represent top sphere. In the front view, a portion of upper circle is hidden. Also, circles with centers r′ and q′ are partly hidden. The bottom sphere P nearer to the observer is seen fully in the front view.

Problem 116: *(Exercise) Draw the projections of three equal spheres of 40 mm diameters, resting on HP touching each other. The line of centers of two of them is inclined at an angle of 30° to VP. A fourth sphere of diameter 44 mm is kept over the three spheres in a pyramidal form touching the bottom three spheres.*

11. COMBINATION OF SOLIDS

Problem 117: *(Fig. 82) A hexagonal headed bolt of 27 mm diameter has its cylindrical portion 58 mm long. The hexagonal head is of side 27 mm, having a thickness of 18 mm. Draw the projections of the bolt, when it is lying on HP with an edge of the head perpendicular to VP and axis inclined at 60° to HP.* (UQ)

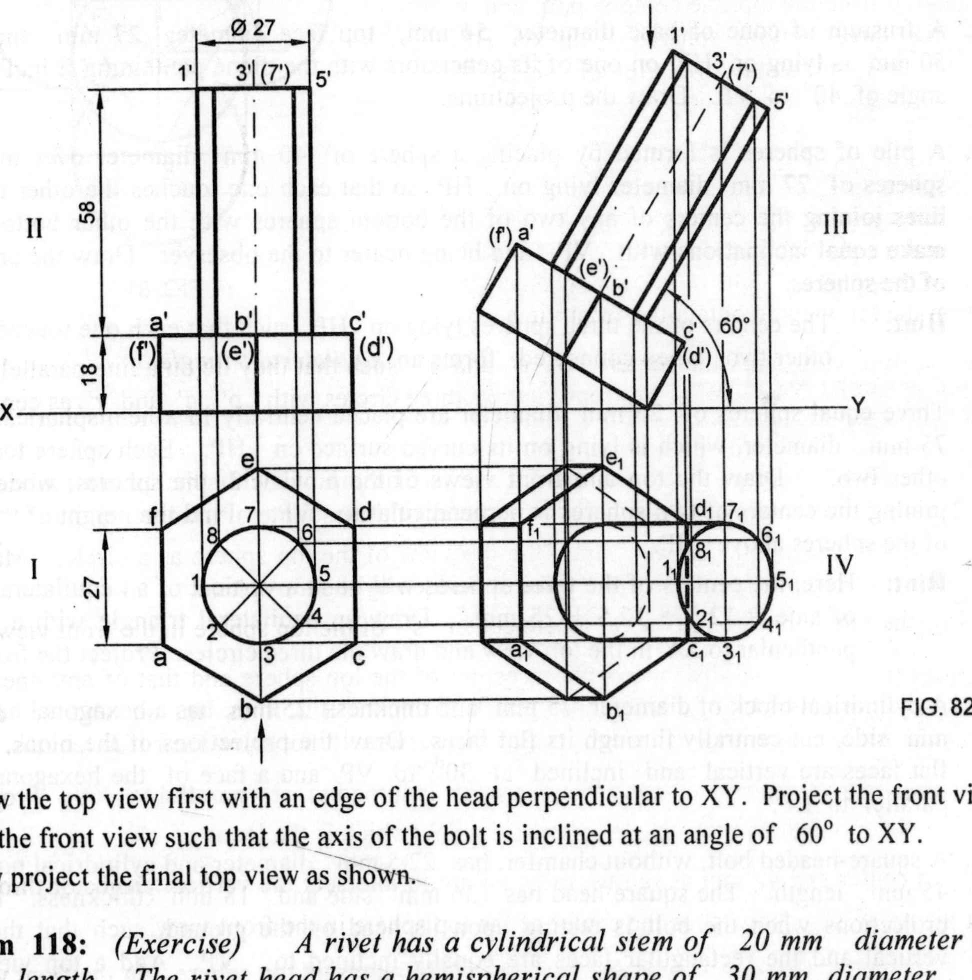

FIG. 82

1. Draw the top view first with an edge of the head perpendicular to XY. Project the front view.
2. Tilt the front view such that the axis of the bolt is inclined at an angle of 60° to XY.
3. Now project the final top view as shown.

Problem 118: *(Exercise) A rivet has a cylindrical stem of 20 mm diameter and 50 mm length. The rivet head has a hemispherical shape of 30 mm diameter. The rivet is lying on HP with its head touching VP and the axis making 45° with VP. Draw its projections.*

12. ADDITIONAL PROBLEMS – III

119. A pentagonal prism, side of base 30 mm and altitude 60 mm is resting on one of the corners of its base on the HP. The longer edge containing that corner is inclined at an angle of 45° to HP and the vertical plane containing that edge and the axis are inclined at 30° to VP. Draw the projections.

120. A hexagonal pyramid of base edge 25 mm and height 55 mm rests on one of its base edges on HP such that the slant face containing the resting edge is perpendicular to HP. The resting edge is inclined at 45° to VP. Draw the projections.

121. A cone of base 50 mm diameter and height 70 mm is lying with one of its generators on HP and the axis appears to be inclined to VP at 40° in the top view. Draw its top and front views. (UQ)

122. A frustum of cone of base diameter 54 mm, top face diameter 27 mm and altitude 50 mm is lying on HP on one of its generators with the plane containing it inclined at an angle of 40° to VP. Draw the projections. (UQ)

123. A pile of spheres is formed by placing a sphere of 40 mm diameter over three other spheres of 27 mm diameter lying on HP so that each one touches the other two. The lines joining the centers of any two of the bottom spheres with the other bottom sphere make equal inclinations with VP and being nearer to the observer. Draw the projections of the spheres. (UQ)

Hint: The centers of the three spheres lying on HP, such that each one touches the other two, when connected, forms an *equilateral triangle.*

124. Three equal spheres of 25 mm diameter are placed centrally in a hemispherical bowl of 75 mm diameter, which is lying on its curved surface on HP. Each sphere touches the other two. Draw the top and front views of the bowl with the spheres, when the line joining the centers of two spheres is perpendicular to VP. Find the height of the centers of the spheres above HP. (UQ)

Hint: Here, the centers of the three spheres will be at a vertical of an equilateral triangle of side = 12.5 + 12.5 + 25 mm. Draw an equilateral triangle with a side perpendicular to XY in the top view and draw the three circles. Project the front view.

125. A cylindrical block of diameter 75 mm and thickness 25 mm, has a hexagonal hole of 25 mm side, cut centrally through its flat faces. Draw the projections of the block, when its flat faces are vertical and inclined at 30° to VP and a face of the hexagonal hole is parallel to HP.

126. A square-headed bolt, without chamfer, has 22.5 mm diameter and cylindrical portion has 45 mm length. The square head has 36 mm side and 18 mm thickness. Draw the projections when the bolt is resting upon its head on the ground, such that the axis is vertical and the rectangular faces are equally inclined to VP. Add a top view on an Auxiliary Inclined Plane inclined at 60° to HP. (UQ)

Note: The square-headed bolt is a combination of a square prism and a cylinder.

13. COLOUR CHART-PAPER MODELS FOR SOLIDS

Students are advised to make their own Padding Sheet (Colour Chart Paper) models for various solids. The models will facilitate developing a good concept of Projections of Solids in various positions with respect to HP and VP.

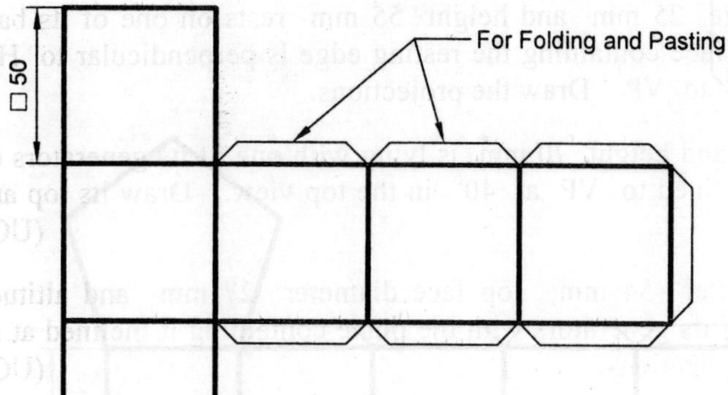

CUBE

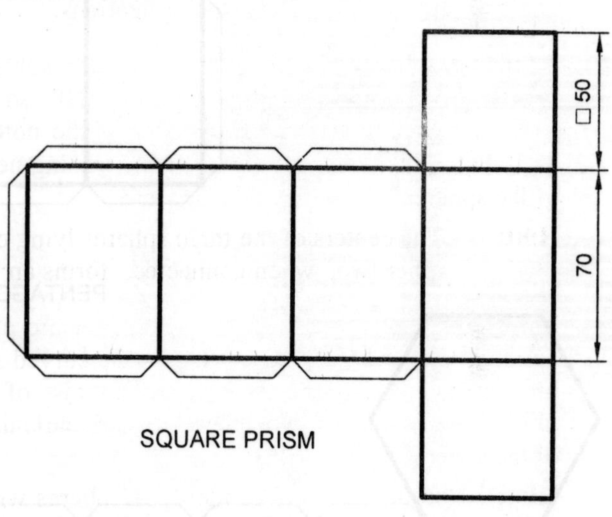

SQUARE PRISM

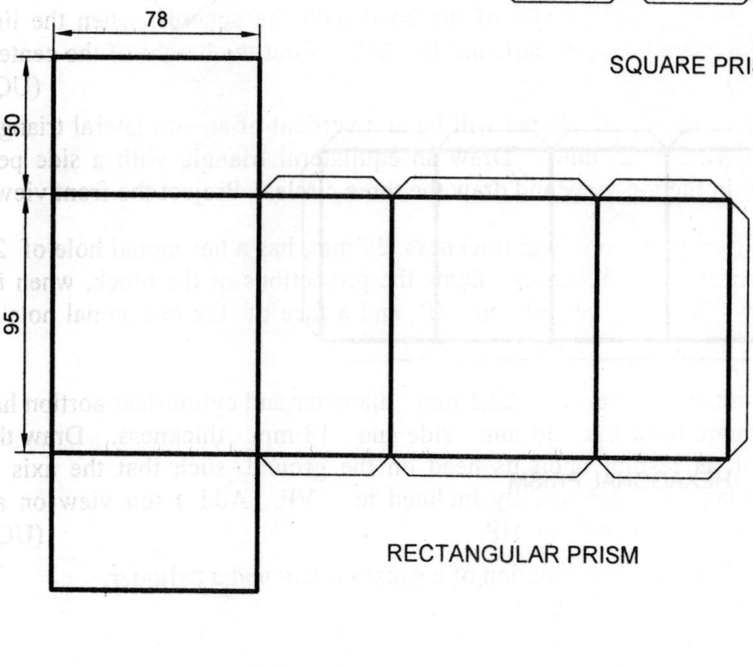

RECTANGULAR PRISM

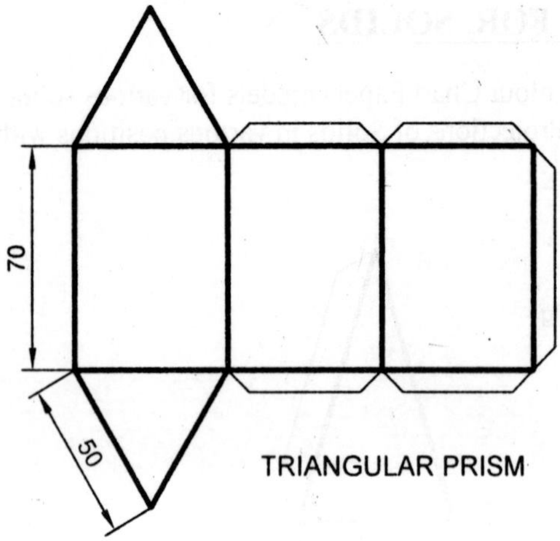

TRIANGULAR PRISM

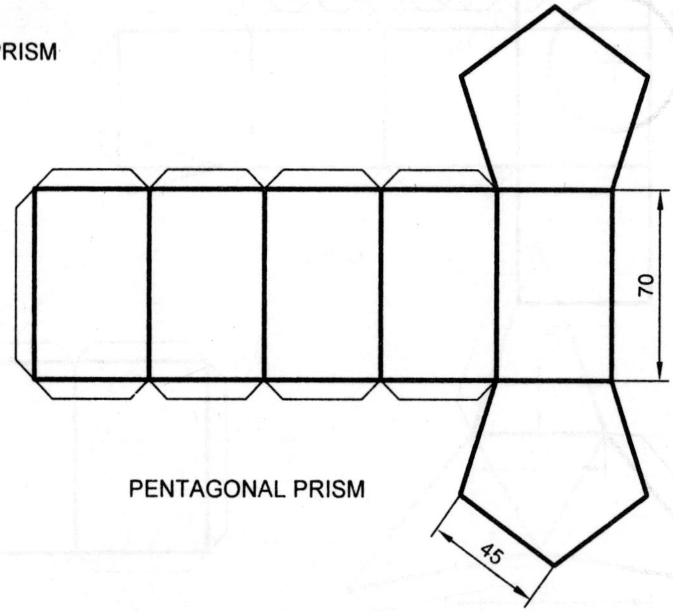

PENTAGONAL PRISM

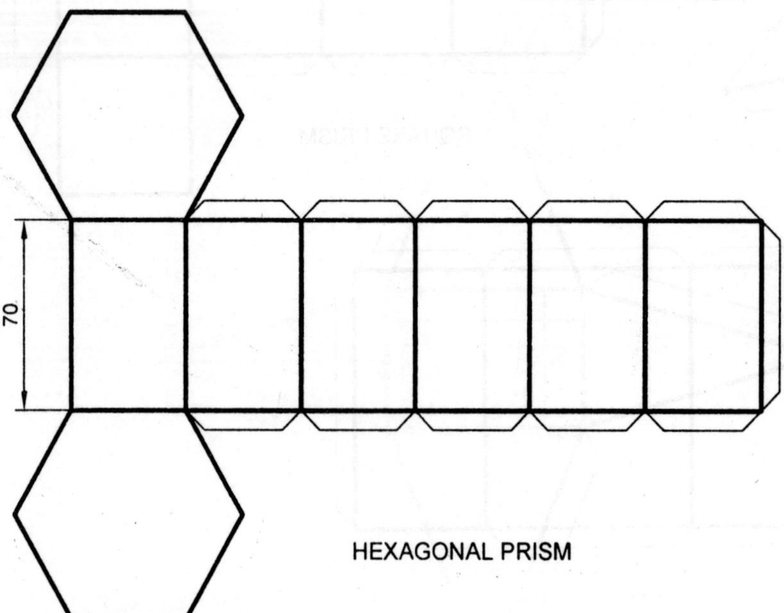

HEXAGONAL PRISM

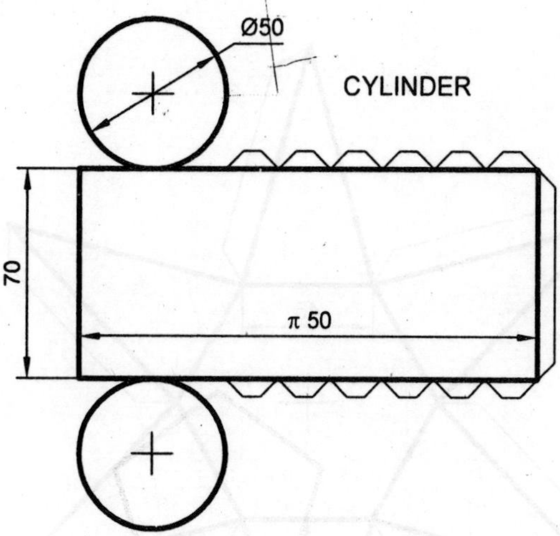

CYLINDER

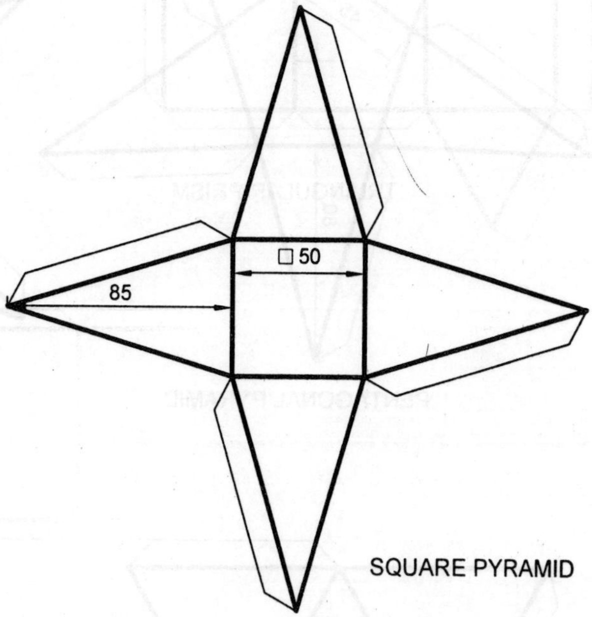

SQUARE PYRAMID

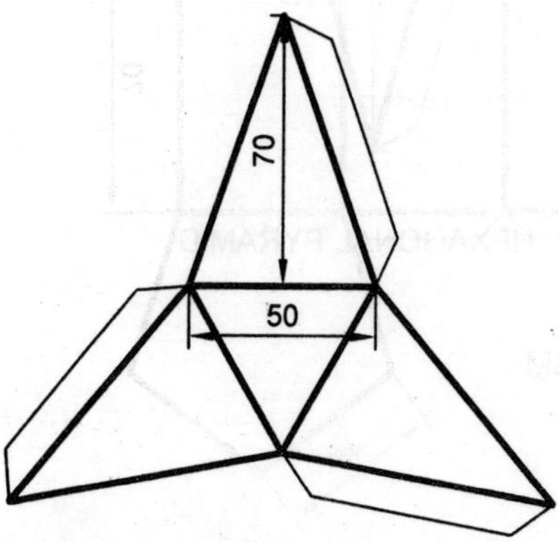

TRIANGULAR PYRAMID

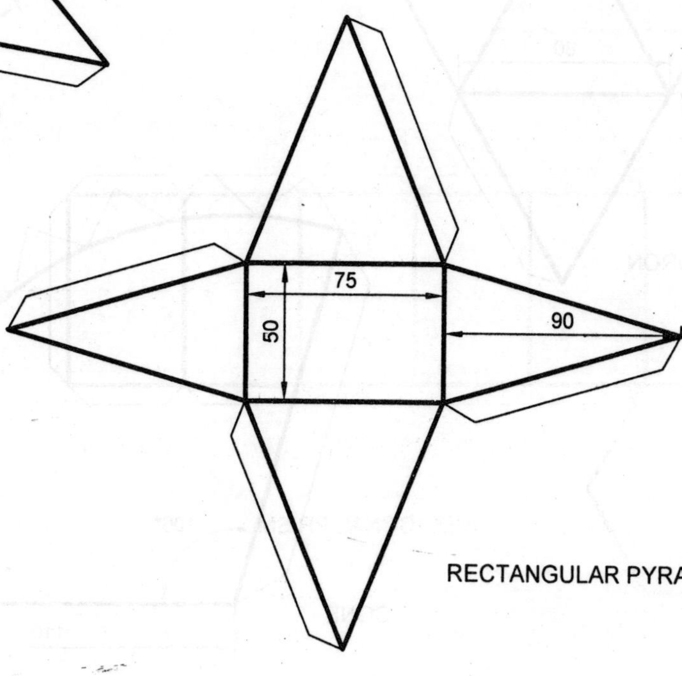

RECTANGULAR PYRAMID

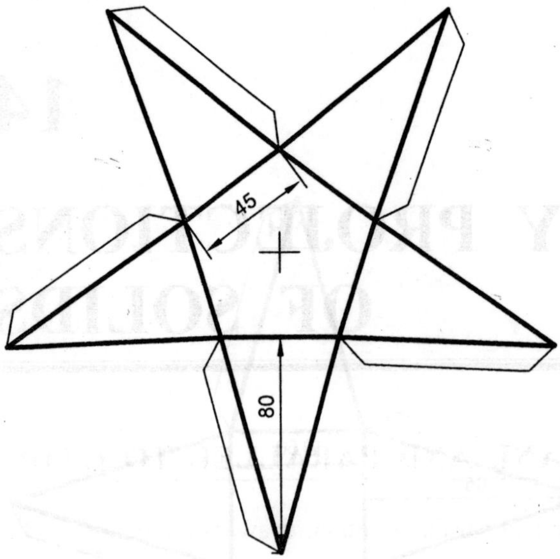

PENTAGONAL PYRAMID

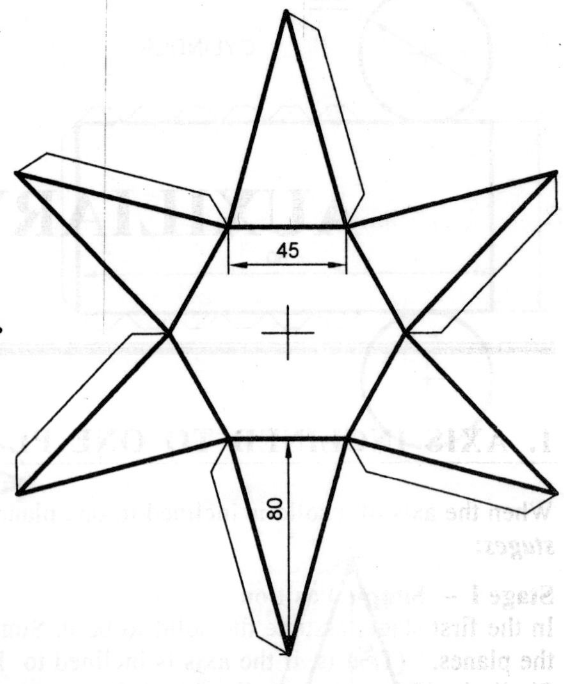

HEXAGONAL PYRAMID

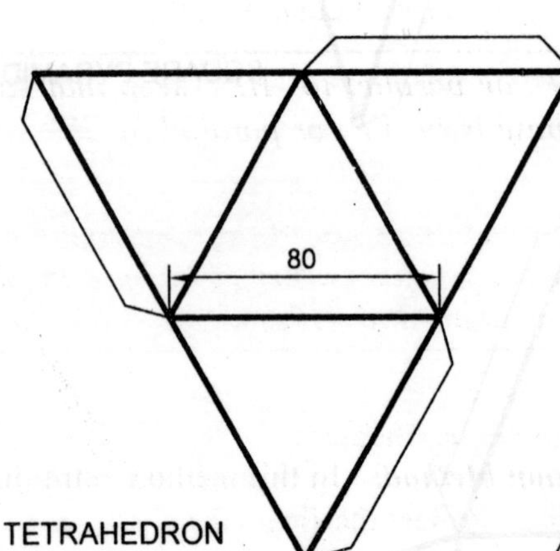

TETRAHEDRON

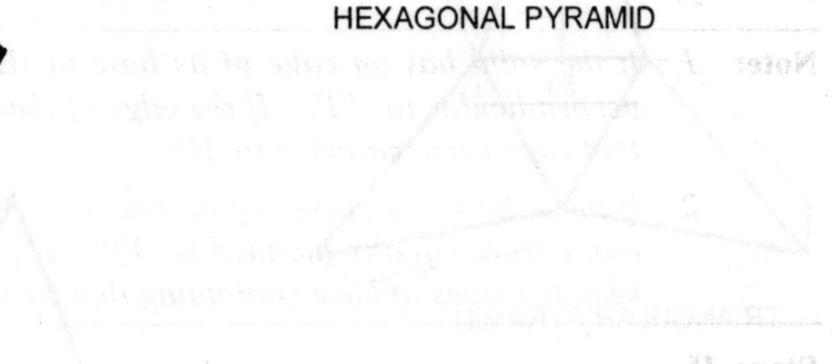

CONE

14

AUXILIARY PROJECTIONS OF SOLIDS

1. AXIS INCLINED TO ONE PLANE AND PARALLEL TO OTHER

When the axis of a solid is inclined to one plane and parallel to other, draw its projections in ***two stages***:

Stage I – Simple Position

In the first stage, assume the solid to be in Simple Position, *i.e.*, its axis perpendicular to one of the planes. (That is, if the axis is inclined to HP, assume the solid to be perpendicular to HP. Similarly, if the axis is inclined to VP, assume the solid to be perpendicular to VP.)

Note: *1. If the solid has an edge of its base in HP or parallel to HP, keep that edge perpendicular to VP. If the edge of the base is in VP or parallel to VP, keep that edge perpendicular to HP.*

2. If the solid has a corner of its base on HP, keep the sides of the base containing that corner equally inclined to VP. If the solid has a corner of its base in VP, keep the sides of base containing that corner inclined to HP.

Stage II

- ***Change of Position Method:*** Discussed in the previous Chapter.
- ***Change of Reference Line*** or ***Auxiliary Projection Method:*** In this method, intro-duce an auxiliary plane according to required conditions. Project the final views.

Problem 1: *(Fig. 1) A hexagonal prism, side of base 20 mm and axis 48 mm long, rests with its base on HP such that an edge of the base is parallel to VP. Draw the projection of the prism on an Auxiliary Inclined Plane (A.I.P.) which makes an angle of 60° with the HP.* (UQ)

1. **Simple Position:** Prism rests with its base on HP and an edge of the base is parallel to VP. For this position, draw top and front views.

Auxiliary Top View on A.I.P.

2. A.I.P. is inclined at 60° to HP. Hence draw a new reference line X_1Y_1 at an angle of 60° to XY to project Auxiliary Top View.
3. From all corners of the prism in the front view, draw projectors perpendicular to X_1Y_1.

4. Measure the distances of all the corners from XY in the first top view and transfer the same with reference to X_1Y_1 on the corresponding projectors.

   Thus locate all the points in the Auxiliary Top View.

To mark visible and invisible edges in the Auxiliary Top View

5. Look at the front view in the direction of arrow shown.
6. End face nearer to the observer is visible. So draw $a_1b_1c_1d_1e_1f_1$ by *continuous thick lines* in Auxiliary Top View.
7. Draw the boundary lines of the above view by *thick* lines.
8. a'1' is visible and hence draw $a_1 1_1$ by *thick* line.
9. 4' is invisible in the front view. Hence draw the edges $(4_1)\ 3_1$, $(4_1)\ 5_1$ and $(4_1)\ d_1$ as dashed lines in the Auxiliary Top View.

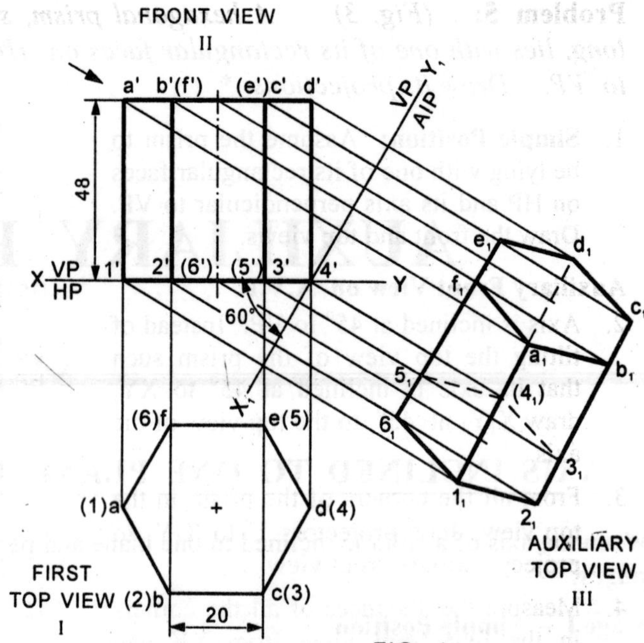

FIG. 1

Problem 2: *(Exercise) A pentagonal prism, side of base 25 mm and axis 50 mm long, rests with one of its shorter edges on HP such that the base containing that edge makes an angle of 30° to HP and its axis is parallel to VP. Draw the projections.*

Problem 3: *(Fig. 2) Draw the projections of a cube of 30 mm side, resting in the HP on one of its corners with a solid diagonal perpendicular to the HP.*

See Fig. 2 for Solution.

Problem 4: *(Exercise) Draw the projections of a cube of 35 mm side, when it rests on HP on one of its corners and a face containing that corner is inclined at an angle of 30° to HP and perpendicular to VP.*

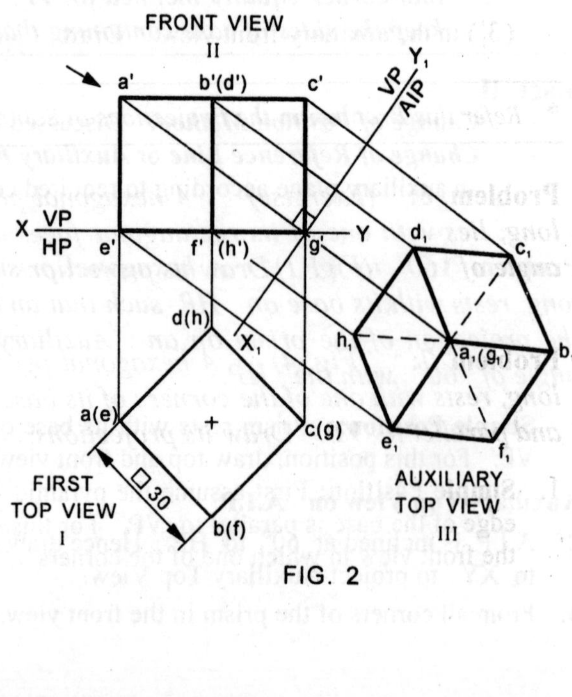

FIG. 2

Problem 5: *(Fig. 3) A hexagonal prism, side of base 25 mm and axis 60 mm long, lies with one of its rectangular faces on HP, such that the axis is inclined at 45° to VP. Draw its projections.* * (UQ)

1. **Simple Position:** Assume the prism to be lying with one of its rectangular faces on HP and its axis perpendicular to VP. Draw the front and top views.

Auxiliary Front View on A.V.P.

2. Axis is inclined at 45° to VP. Instead of tilting the top view of the prism such that its axis is inclined at 45° to XY, draw X_1Y_1 at 45° to the top view of the axis.
3. From all the corners of the prism in the top view, draw projectors $\perp^r$ to X_1Y_1 to project auxiliary front view.
4. Measure the distances of all the corners in the first front view from XY and transfer the same with reference to X_1Y_1 on the corresponding projectors.
5. Look at the top view in the direction of arrow. The end face nearer to the observer is visible and hence draw a_1' b_1' c_1' d_1' e_1' f_1' by *thick lines*.

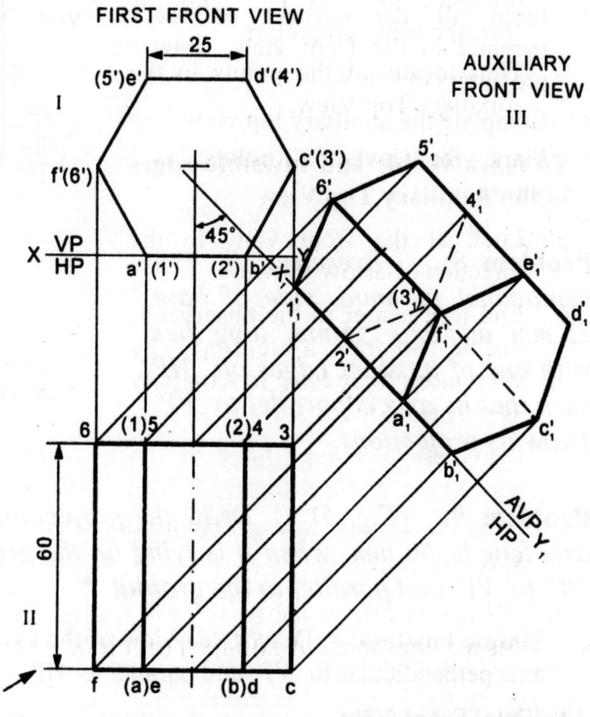

FIG. 3

6. f6 is visible. Draw f_1' $6_1'$ by *thick* line in auxiliary front view. 3 is invisible. Mark it as $(3_1')$ in the auxiliary front view. Draw $(3_1')$ $2_1'$, $(3_1')$ $4_1'$ and $(3_1')$ c_1' as *dashed lines.*

---

* *Refer this Problem in the Projections of Solids Chapter by the Change of Position Method.*

Problem 6: *(Exercise) A hexagonal prism, side of base 22 mm and axis 58 mm long, lies with one of its rectangular faces on HP, such that the axis is inclined at an angle of 60° to VP. Draw its projections.*

Problem 7: *(Fig. 4) A hexagonal pyramid, side of base 25 mm and axis 50 mm long, rests with one of the corners of its base on HP. The axis is inclined at 30° to HP and parallel to VP. Draw its projections.* (UQ)

1. **Simple Position:** First assume the pyramid to be resting with its base on HP such that an edge of the base is parallel to VP. For this simple position, draw the top view first. Project the front view in which one of the corners a' is on the right side.

Auxiliary Top View

2. Draw X_1Y_1 through a′ at 30° to the front view of the axis.
3. From all the corners of the pyramid in the front view, draw projectors to X_1Y_1.
4. Complete the auxiliary top view.
5. Mark the invisible points as shown.

Problem 8: *(Exercise) A pentagonal pyramid, side of base 25 mm and axis 55 mm long, lies with one of its slant edges on HP such that its axis is parallel to VP. Draw its projections.*

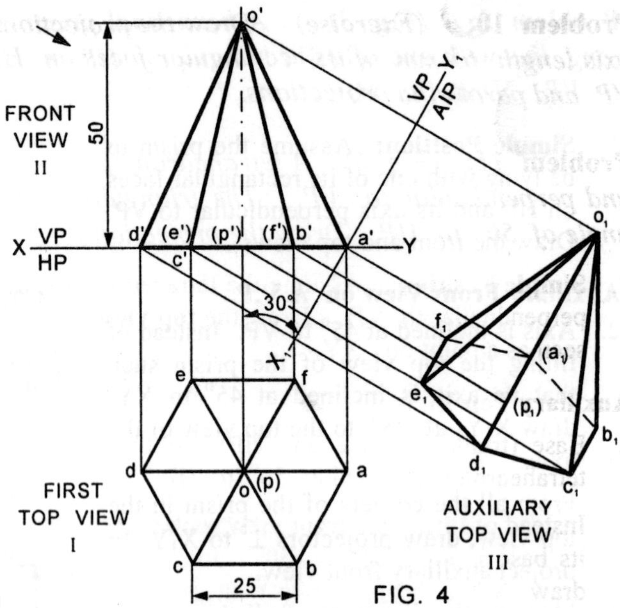

FIG. 4

Problem 9: *(Fig. 5) Draw the projections of a cylinder of diameter 45 mm and axis length 60 mm when it is lying on the* ***ground*** *with its axis inclined at an angle of 30° to VP and parallel to the* ***ground***. * (Ground means HP.) (UQ)

1. **Simple Position:** Draw front view of the cylinder as a circle of 45 mm diameter with its axis perpendicular to VP and parallel to HP. Project the corresponding top view.

Auxiliary Front View

2. Introduce an AVP inclined at 30° to the axis. Draw X_1Y_1 inclined at 30° to the axis.
3. Draw projectors from the different points in the top view perpendicular to X_1Y_1.
4. Mark points on the above projectors at distances from X_1Y_1 equal to their respective distances from XY in the front view.
5. Join the ellipses for the bases by common tangents.
6. Look at the top view in the direction of arrow. Half of the ellipse for the hidden base of cylinder is invisible.

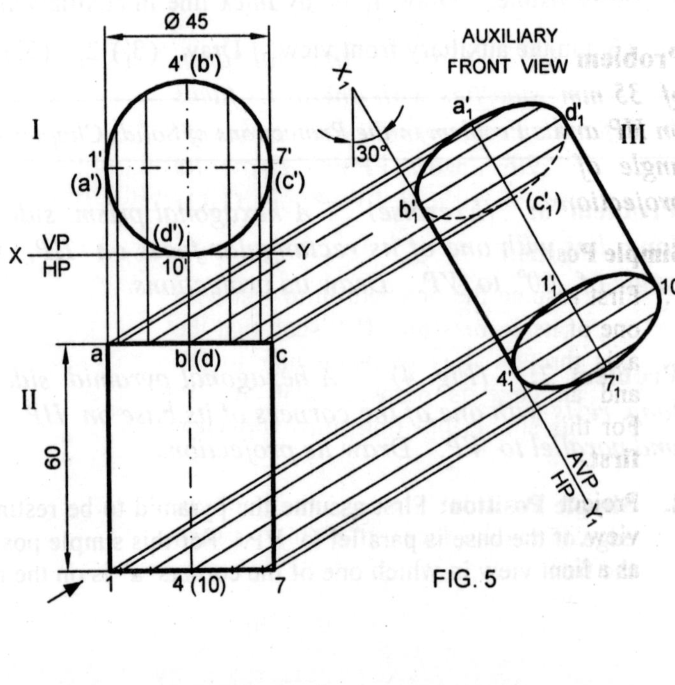

FIG. 5

---

* *Refer this Problem in Projections of Solids Chapter by the Change of Position Method.*

Problem 10: *(Exercise) Draw the projections of a cylinder of diameter 50 mm and axis length 65 mm when it is lying on HP with its axis inclined at an angle of 45° to VP and parallel to HP.*

Problem 11: *(Fig. 6) A tetrahedron of 40 mm side rests with one of its edges on HP and perpendicular to VP. The triangular face containing that edge is inclined at an angle of 50° to HP. Draw the projections.* (UQ)

1. **Simple Position:** Assume the tetrahedron resting with its base on HP such that an edge is perpendicular to VP. Draw the **top view first**. Project the front view. The edge a′(b′) is seen as a point on XY.

Auxiliary Top View

2. Base (triangular face on HP) of the tetrahedron is inclined at 50° to HP.

   Instead of tilting the front view such that its base is inclined at 50° to XY, draw X_1Y_1 through a′(b′) at 50° to the base in the front view.

3. From all corners in the front view draw projectors perpendicular to X_1Y_1 to project the auxiliary top view.
4. Measure the distances of all the corners in first top view from XY and transfer the same with reference to X_1Y_1 on the corresponding projectors. Complete the Auxiliary Top View.

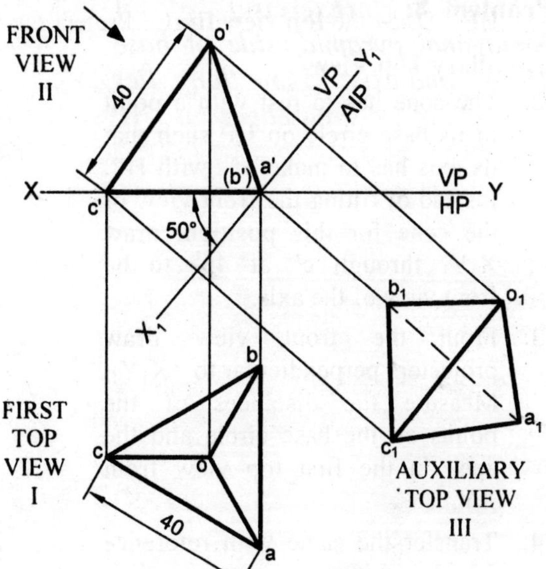

FIG. 6

Problem 12: *(Fig. 7) An octahedron of 35 mm side lies with one of its faces on HP and an edge of that face makes an angle of 90° with VP. Draw its projections.* * (UQ)

Simple Position

1. First assume the octahedron to rest with one of its corners on HP such that the axis through it is perpendicular to HP and an edge is perpendicular to VP. For this simple position draw **top view first.**
2. Project the front view. In the front view, triangular face p′ a′ (b′) is seen as a line.

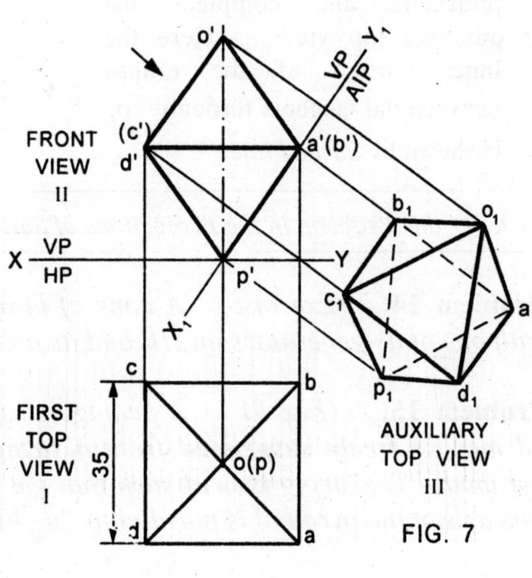

FIG. 7

3. **Auxiliary Top View:** Draw X_1Y_1 so as to contain p′ a′ (b′) in the front view. Draw the projectors $\perp^r$ to X_1Y_1 passing through all corners in front view. Transfer corresponding distances from the first top view. Thus, complete the auxiliary top view.

---

* *Refer this Problem in the Projections of Solids Chapter by the Change of Position Method.*

Problem 13: *(Fig. 8) Draw the projections of a cone, base 30 mm diameter and axis 50 mm long, when it is resting on HP on a point of its base circle with the axis making 45° with HP and parallel to VP.* * (UQ)

1. **Simple Position:** Assume the cone to rest with its base on HP such that its axis is $\perp^r$ to HP. Draw the **top view first**. Project the front view.

Auxiliary Top View

2. The cone has to rest with a point of its base circle on HP such that its axis has to make 45° with HP. Instead of tilting the front view of the cone for this position, draw X_1Y_1 through c′ at 45° to the front view of the axis.

3. From the front view draw projectors perpendicular to X_1Y_1. Measure the distances of the points on the base circle and the apex in the first top view from XY.

4. Transfer the same with reference to X_1Y_1 on the corresponding projectors and complete the auxiliary top view. Here the inner portion of the ellipse between the tangents through o_1' is shown by *dashed line*.

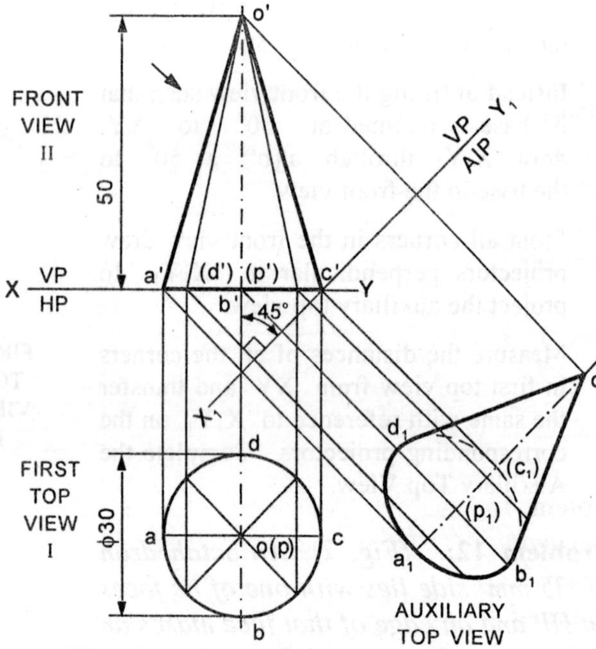

FIG. 8

---

* *Refer this Problem in the Projections of Solids Chapter by the Change of Position Method.*

Problem 14: *(Exercise) A cone of base 50 mm diameter and axis 65 mm long, lies with one of its generators on HP and its axis is parallel to VP. Draw its projections.*

Problem 15: *(Fig. 9) A pentagonal pyramid of side of base 20 mm and altitude 48 mm is **freely suspended** from a string attached to one corner of the base of the pyramid. The string is held such that the apex of the pyramid is just touching the HP. The axis of the pyramid is parallel to the VP. Draw the projections.* (UQ)

Simple Position

1. Top view is a b c d e. **c** is the corner of the base of the pyramid from which the pyramid is suspended.
2. Draw the top view such that the line joining **c** and the center of gravity (g) of the pyramid is parallel to XY.
3. Project the front view as shown.

Auxiliary Top View

4. Mark the front view **g′** of the center of gravity and the axis at 12 mm (= 1/4 th height) from the base.
5. Draw X_1Y_1 perpendicular to c′ g′. Also, the apex of the pyramid is just touching the HP (Given).

   Thus o′ should lie on X_1Y_1 line. Hence draw X_1Y_1 perpendicular to c′ g′ so as to pass through o′.
6. Project the Auxiliary Top View.

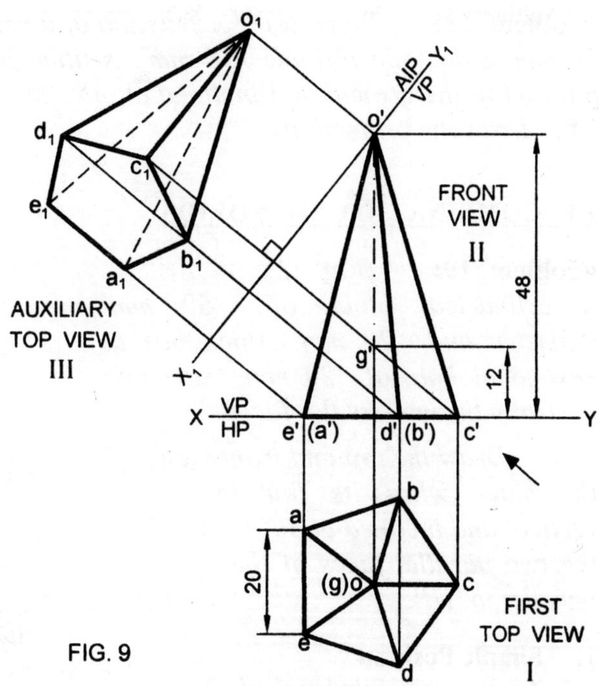

FIG. 9

Note: 1. *If the solid suspended is cone,* **g′** *also lies at one-fourth height from base.*
2. X_1Y_1 *need not pass through* **o′**, *if the apex is not touching the HP, but still should be perpendicular to* **c′g′**.

Problem 16: *(Exercise)*
A hexagonal pyramid of side of base 25 mm and height 55 mm is ***suspended freely*** *from a corner such that the axis is parallel to VP. Draw its projections.*

Problem 17: *(Fig. 10)*
A ***frustum*** *of a square pyramid, base side 40 mm, top side 22 mm and altitude 53 mm has its bases equally inclined to VP. Its axis is parallel to VP. It is tilted such that its axis is inclined at 30° to HP. Draw the projections.* (UQ)

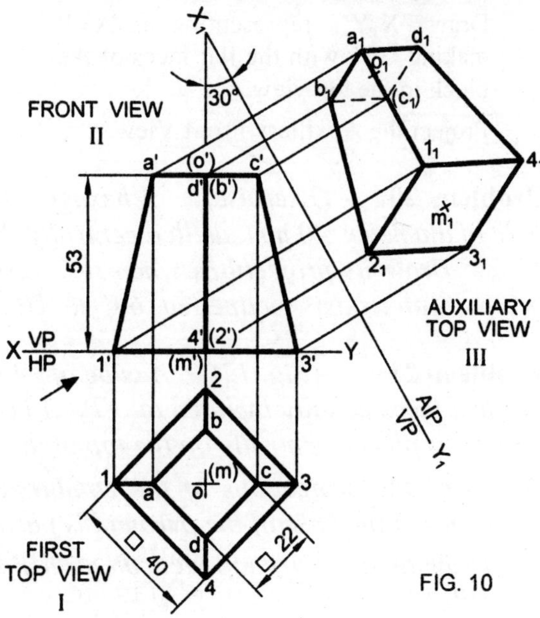

FIG. 10

Problem 18: *(Exercise) A* ***frustum*** *of a pentagonal pyramid, base 32 mm side, top 26 mm side and altitude 58 mm is tilted about a corner of the base till its axis is parallel to the ground and inclined at 45° to VP. Top view of the frustum is nearer to VP. Draw the projections.* (UQ)

1.1 COMBINATION OF SOLIDS

Problem 19: *(Fig. 11)*
A cylindrical block of 60 mm diameter and 20 mm thick has a hexagonal hole of 20 mm side cut centrally through its flat faces.

Draw the top and front views of the block when its flat faces are vertical and inclined at 30° to VP and the two parallel faces of the hole are parallel to HP. (UQ)

1. **Simple Position**
   Assume the cylindrical block to rest on HP with its axis perpendicular to VP and an edge of the hexagonal hole is parallel to HP.
   Draw the front and top views.

2. **Second Position**
   Draw X_1Y_1 representing an AVP making 30° with the flat faces of the block in the top view.
   Project the Auxiliary Front View.

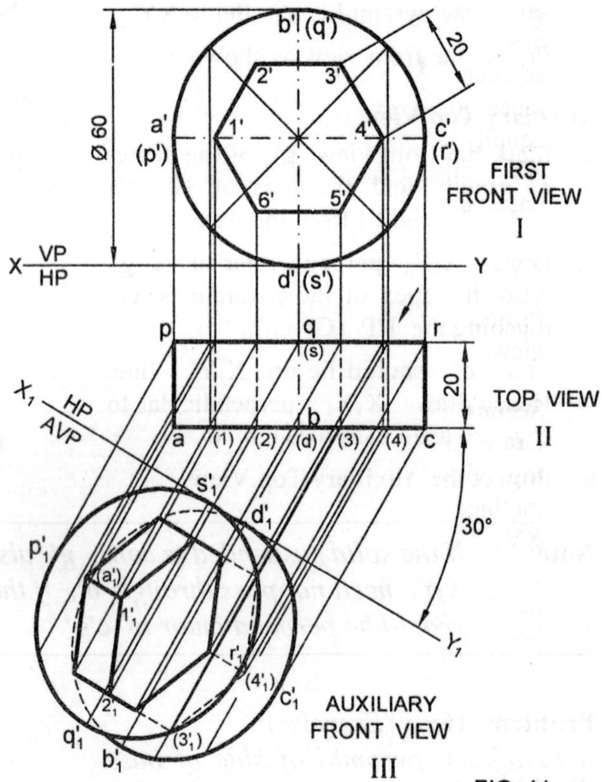

FIG. 11

Problem 20: *(Exercise) A hexagonal nut of side 40 mm and height 40 mm has a hole of diameter 40 mm drilled centrally through its ends.*

Draw its projections when it is resting on one of its corners of the base on the ground with its axis inclined at 60° to HP and two of its faces parallel to VP.

Problem 21: *(Fig. 12) A cube of 40 mm side is resting on HP with its adjacent vertical faces equally inclined to VP. A hexagonal pyramid of 16 mm base and altitude 48 mm is placed centrally on the top of the cube.*

i) *Draw the projections of the* ***combination****, if the axes are collinear and a pair of edges of the base of the pyramid is parallel to VP.*
ii) *Draw an auxiliary top view on an AIP perpendicular to VP and inclined at 45° to HP.* (UQ)

1. **Simple Position**
   Draw the first top view (*thick lines*) such that the axes of the cube and hexagonal pyramid are collinear (i.e., axes coincide), adjacent vertical faces of the cube equally inclined to XY and a pair of base edges of the pyramid parallel to XY.

   Project the front view.

2. **Second Position**
   Draw the auxiliary top view on X_1Y_1 inclined at 45° to XY.

   See Fig. 12 for Solution

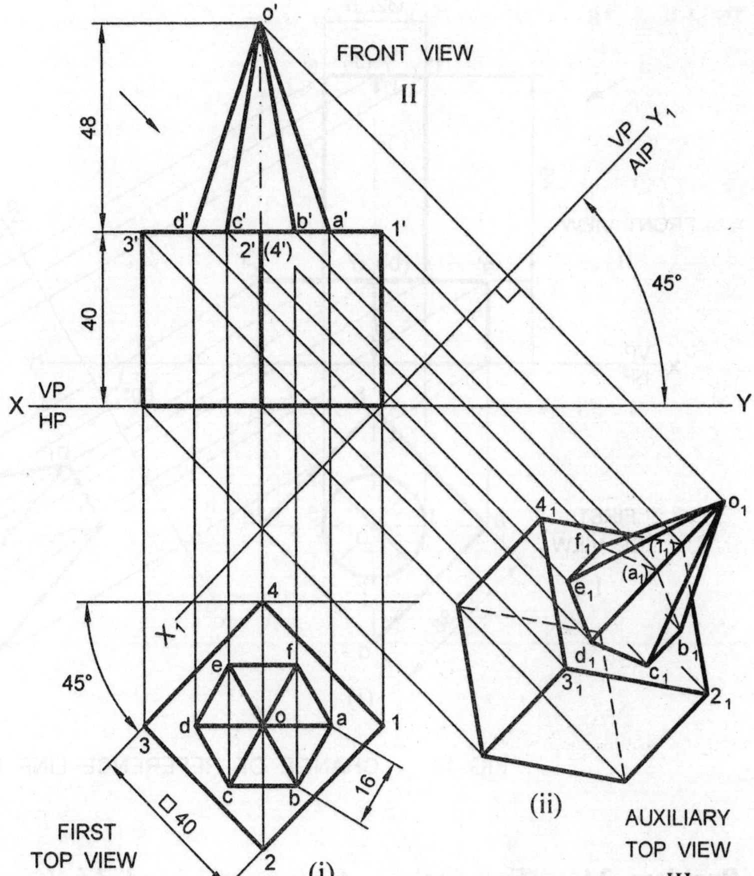

FIG. 12

Problem 22: (*Exercise*) *A right regular hexagonal prism of side of base 20 mm and axis 24 mm long is surmounted coaxially by a hexagonal pyramid of side of base 20 mm and axis 40 mm long.*

Draw the projections of the ***combination*** *when resting on HP on one of the edges of the base of the prism perpendicular to VP with the axis inclined at 30° to HP.* (UQ)

Problem 23: (*Fig. 13*) *A square-headed bolt, without chamfer, has 22.5 mm diameter and cylindrical portion has 45 mm length. The square head has 36 mm side and 18 mm thickness.*

(i) *Draw the projections when the bolt is resting upon its head on the ground, such that the axis is vertical and the rectangular faces are equally inclined to VP.*

(ii) *Add a top view on an Auxiliary Inclined Plane inclined at 60° to HP.* (UQ)

Note: *The square-headed bolt is a combination of a square prism and a cylinder.*

See Fig. (i) for Problem 23(i).
See Fig. (ii) for Problem 23(ii).

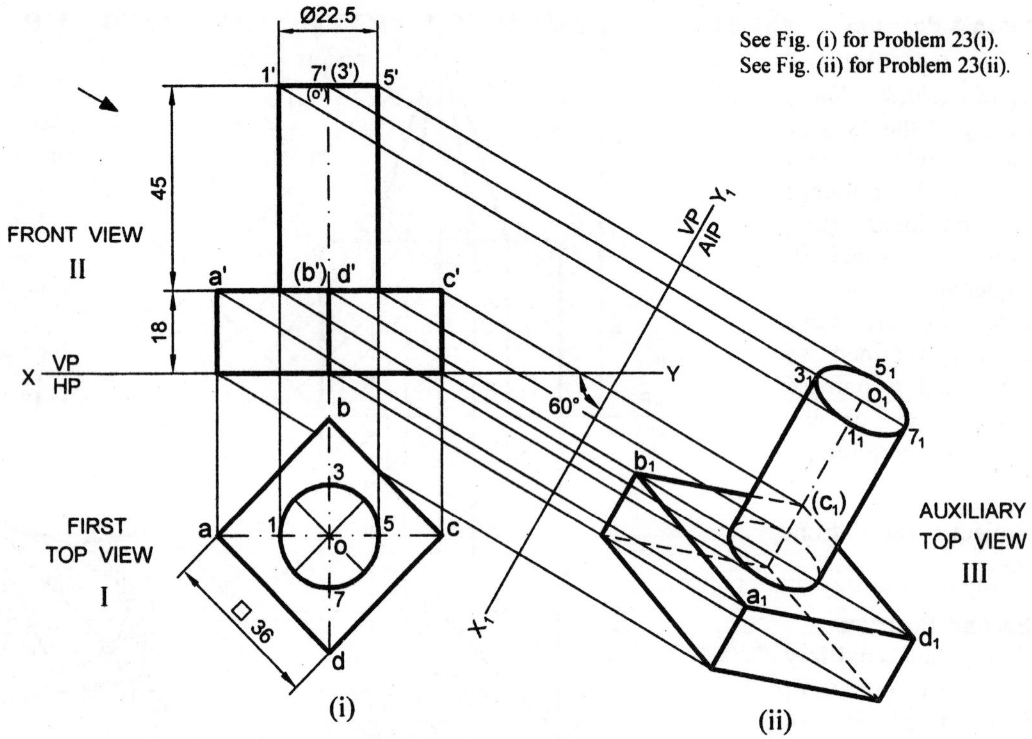

FIG. 13 CHANGE OF REFERENCE LINE METHOD

Problem 24: *(Exercise) A hexagonal-headed bolt, without chamfer, is 27 mm in diameter and the cylindrical portion is 58 mm long. The side of the hexagonal head is 27 mm having a thickness of 18 mm. The bolt is placed with a horizontal edge of the head on HP such that it is perpendicular to VP and the axis of the bolt is inclined at 60° to HP.*

Draw the projections of the bolt, neglecting its threads. (UQ)

Note: Refer Projections of Solids Chapter for Change of Position Method.

2. ADDITIONAL PROBLEMS – I

Problems 25 to 69: (Exercises)

*Refer to the Problems from **34 to 78** involving "Axis of Solid Inclined to One Plane and Parallel to the Other" by the Change of Position Method in the previous Chapter: Projections of Solids.*

Students are advised to solve the above Problems by the Change of Reference Line Method (Auxiliary Projection Method). Check the auxiliary views with the previous method.

3. AXIS OF SOLID INCLINED TO BOTH HP AND VP

Problem 70: *(Fig. 14) A square prism, side of base 30 mm and axis 50 mm long, has its axis inclined at 60° to HP. It has an edge of its base in HP and inclined at 45° to VP. Draw its projections.* (UQ)

1. **Simple Position:** When the prism has to rest with an edge of its base on HP, first assume it in its simple position, *i.e.,* resting with its base on HP, such that an edge of its base is perpendicular to VP. For this position draw the top view first.
2. **Auxiliary Top View:** The axis is inclined at 60° to HP. Instead of tilting the front view of the prism such that its axis is inclined at 60° to XY, draw X_1Y_1 through 2′(3′) at 60° to the front view of the axis.

Note: *Here X_1Y_1 represents the intersection of VP and A.I.P.*

From all the corners of the prism in the first front view draw projectors perpendicular to X_1Y_1 to project the auxiliary top view.

Measure the distances of all the corners in the first top view from XY. Transfer the same with reference to X_1Y_1 on the corresponding projectors.

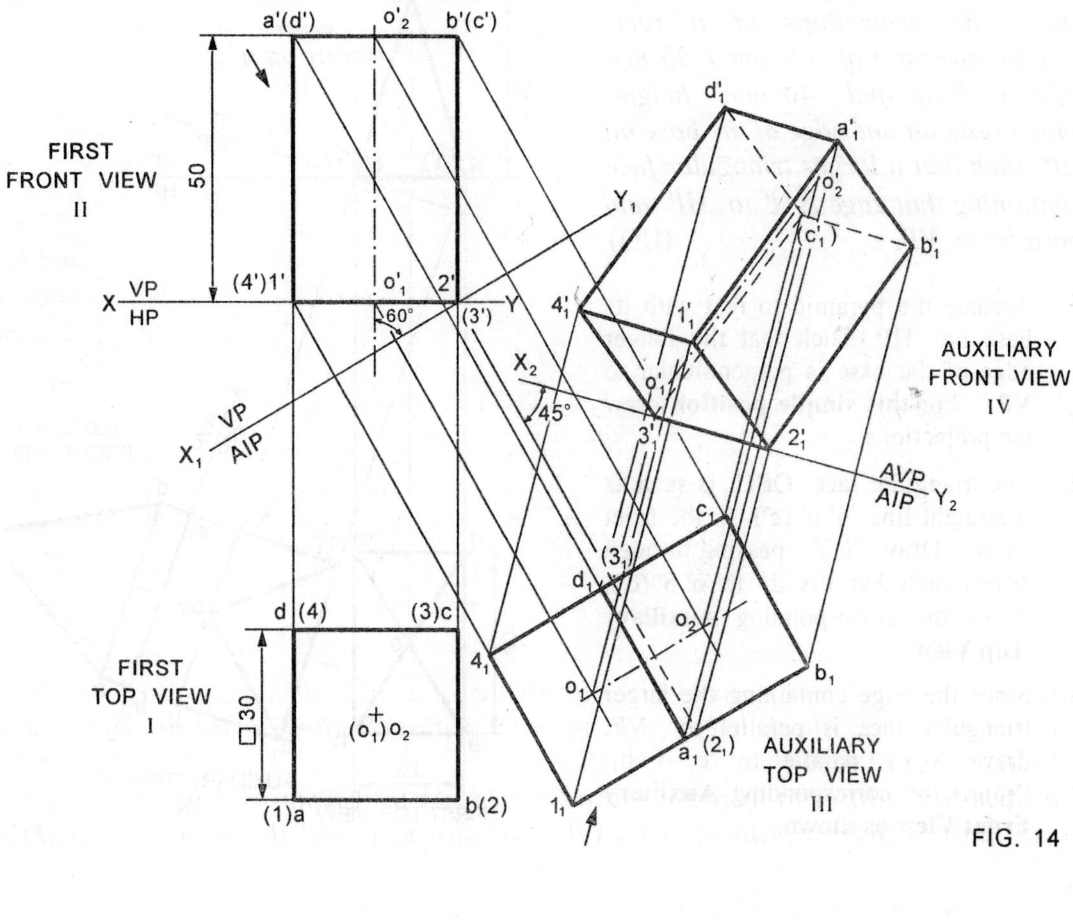

FIG. 14

3. **Visible and Invisible Edges:** Look at the first front view in the direction of arrow shown. The edge $2'\,(3')$ is invisible. Hence draw $(2_1)\,(3_1)$ by *dashed line* in the auxiliary top view. Here this edge is perpendicular to VP.

4. **Auxiliary Front View:** The edge 23 is inclined at 45° to VP (given). Draw X_2Y_2 at an angle of 45° to $(2_1)\,(3_1)$ to project the auxiliary front view.

Note: *Here X_2Y_2 represents the intersection of A.I.P. and A.V.P.*

From all the corners in auxiliary top view draw projectors perpendicular to X_2Y_2.

Measure the distances of all the corners in the first front view from X_1Y_1 and transfer the same with reference to X_2Y_2 on the corresponding projectors.

5. **Visible and Invisible Edges:** Look at the auxiliary top view in the direction of arrow shown. The rectangular face $1_1\, a_1\, b_1\, (2_1)$ is visible. Hence draw $1_1'\, a_1'\, b_1'\, 2_1'$ as continuous *thick* lines in the auxiliary front view.

   Draw the boundary lines as *thick* lines. $1_1\, 4_1$ is nearer to the observer and hence draw $1_1'\, 4_1'$ as thick line. Since c_1 is invisible draw $(c_1')\, b_1'$, $(c_1')\, d_1'$, $(c_1')\, 3_1'$ as *dashed lines.*

Problem 71: *(Fig. 15)*
Draw the projections of a rectangular pyramid of 40 mm x 25 mm sides of base and 40 mm height, which rests on one edge of the base on HP such that a larger triangular face containing that edge is $\perp^r$ to HP and parallel to VP. (UQ)

1. Assume the pyramid to rest with its base on HP such that the longer edge of the base is perpendicular to VP. For this **simple position** draw the projections.
2. The triangular face OBC is seen as a straight line $o'\, b'\, (c')$ in the front view. Draw X_1Y_1 passing through $b'\,(c')$ such that it is $\perp^r$ to $o'\, b'\, (c')$. Draw the corresponding **Auxiliary Top View.**
3. Since the edge containing the larger triangular face is parallel to VP, draw X_2Y_2 parallel to $c_1\, o_1\, b_1$. Project the corresponding **Auxiliary Front View** as shown.

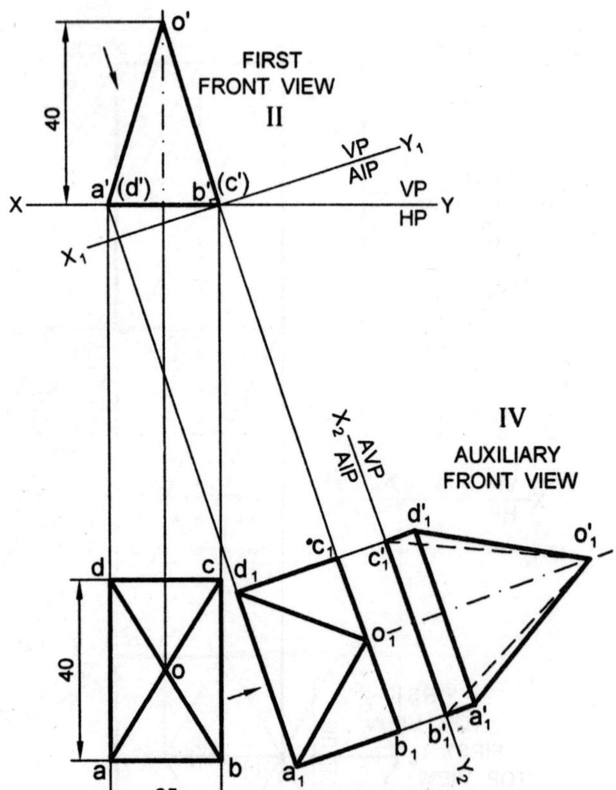

FIG. 15

Problem 72: *(Exercise) A square pyramid of base edges 35 mm and height 50 mm has its base on HP with one of the base edges inclined at 30° to VP. Draw the top and front views. Obtain an auxiliary top view from the front view on an auxiliary plane inclined at 45° to HP.*

Problem 73: *(Fig. 16) A hexagonal pyramid, side of base 25 mm and axis 50 mm long, rests with one of the corners of its base on HP. The axis is inclined at 30° to HP and 45° to VP. Draw its projections.* (UQ)

1. **Simple Position:** First assume the pyramid to be resting with its base on HP such that an edge of the base is parallel to VP. Draw the projections.
2. **Auxiliary Top View:** Draw X_1Y_1 through a′ at 30° to the front view of the axis. From all the corners of the pyramid in the front view, draw projectors to X_1Y_1 and complete the auxiliary top view.
3. **Auxiliary Front View:** Since the axis is inclined at an angle of 30° to HP and 45° to VP, its top view will make an apparent angle β (> 45°) with X_2Y_2. Find β as shown in the auxiliary top view. Draw X_2Y_2 at an angle of β to the auxiliary top view of the axis and complete the auxiliary front view as shown.

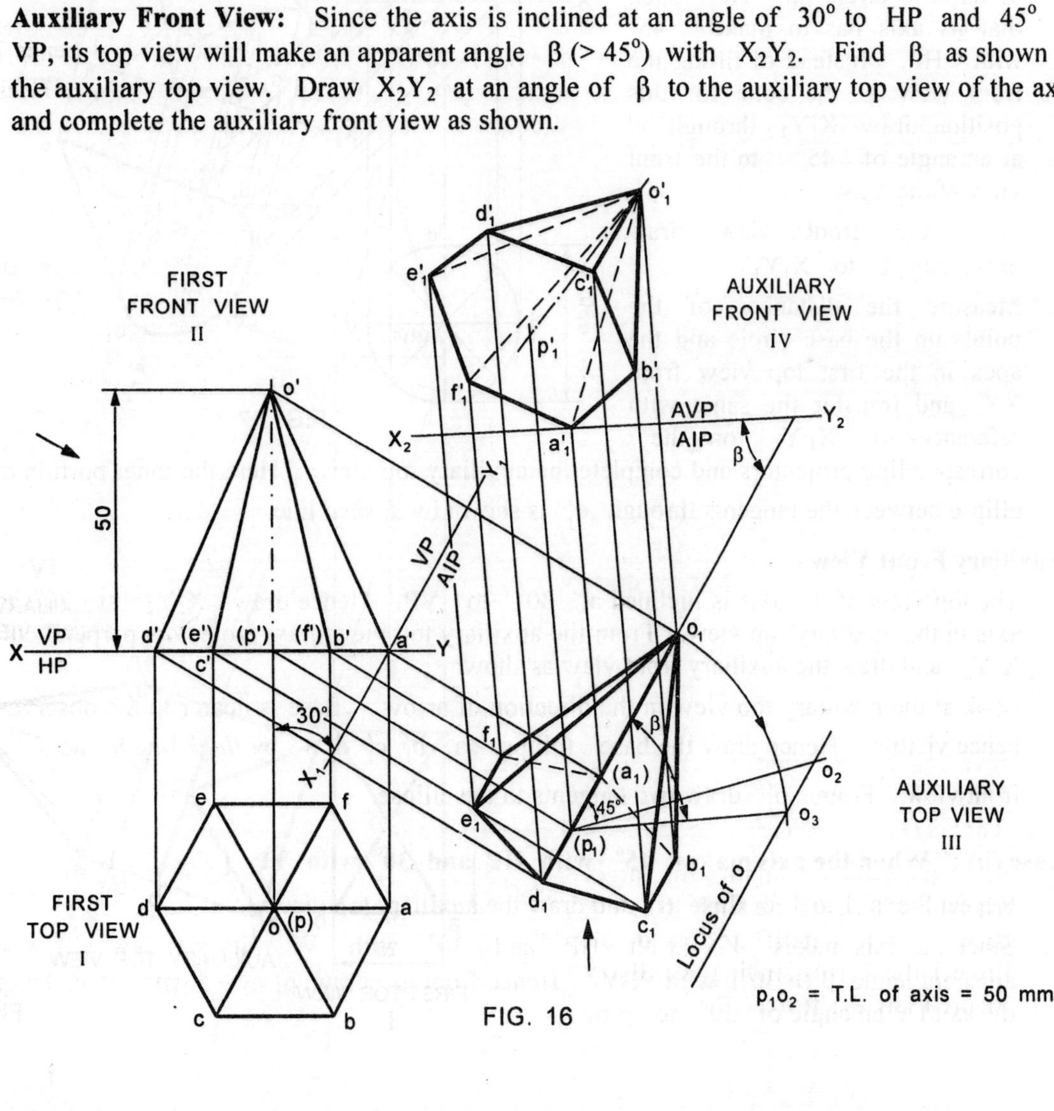

FIG. 16

Problem 74: *(Fig. 17) Draw the projections of a cone, base 30 mm diameter and axis 50 mm long, when it is resting on HP on a point of its base circle with (i) the axis making an angle of 45° with HP and its top view making 30° with VP (ii) the axis making an angle of 45° with HP and 30° with VP.* (UQ)

Case (i) : When the axis makes 45° with HP and its top view 30° with VP [*Fig. (i)*]

1. **Simple Position:** First assume the cone resting with its base on HP such that its axis is perpendicular to HP. Draw the top view first and then front view.

Auxiliary Top View

2. The cone has to rest with a point of its base circle on HP such that its axis has to make 45° with HP. Instead of tilting the front view of the cone for this position, draw X_1Y_1 through c′ at an angle of 45° to the front view of the axis.
3. From the front view draw projectors $\perp^r$ to X_1Y_1.

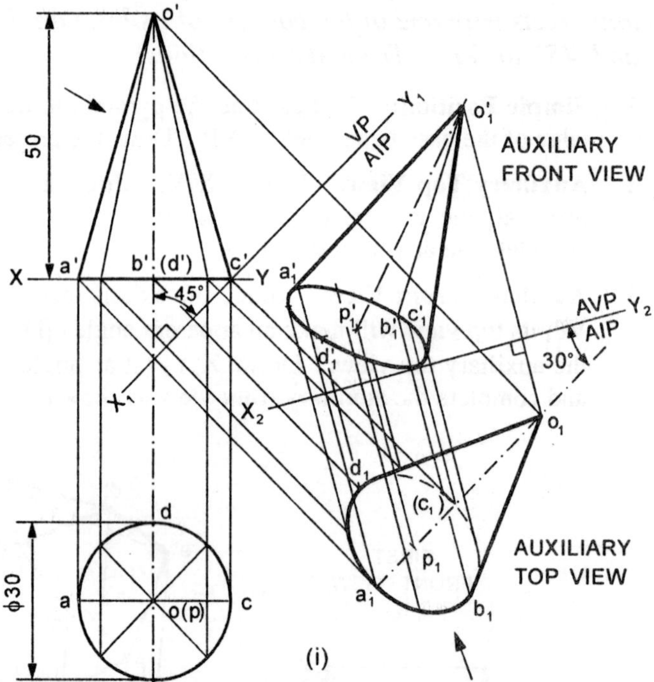

(i)

FIG. 17

4. Measure the distances of the points on the base circle and the apex in the first top view from XY and transfer the same with reference to X_1Y_1 on the corresponding projectors and complete the auxiliary top view. Here the inner portion of the ellipse between the tangents through o_1' is shown by *dashed* line.

Auxiliary Front View

5. The top view of the axis is inclined at 30° to VP. Hence draw X_2Y_2 at 30° to the axis in the auxiliary top view. From the auxiliary top view draw projectors perpendicular to X_2Y_2 and draw the auxiliary front view as shown.
6. Look at the auxiliary top view in the direction of arrow. Base is nearer to the observer and hence visible. Hence draw the base (ellipse a_1' b_1' c_1' d_1') by *thick* line in the auxiliary front view. From o_1' draw two tangents to the ellipse.

Case (ii) : When the axis makes 45° with HP and 30° with VP [*Fig. (ii)*]

1. Repeat Steps 1 to 4 in **Case (i)** and draw the auxiliary top view.
2. Since the axis makes 45° with HP and 30° with VP, its top view will make an apparent angle β (> 30°) with X_2Y_2. Hence from p_1 draw $p_1 o_2$ = 50 mm (true length of the axis) at an angle of 30° to p_1o_1.

3. Through o_2 draw the locus of o parallel to p_1o_1. p_1 as center and p_1 o_1 (length of the axis in the auxiliary top view) as radius draw an arc to cut the locus of o at o_3.
4. Join p_1 o_3 and draw X_2Y_2 parallel to it. Now X_2Y_2 makes an angle of β (> 30°) with respect to the axis in auxiliary top view.
5. Draw projectors perpendicular to X_2Y_2 from the auxiliary top view and complete the auxiliary front view.

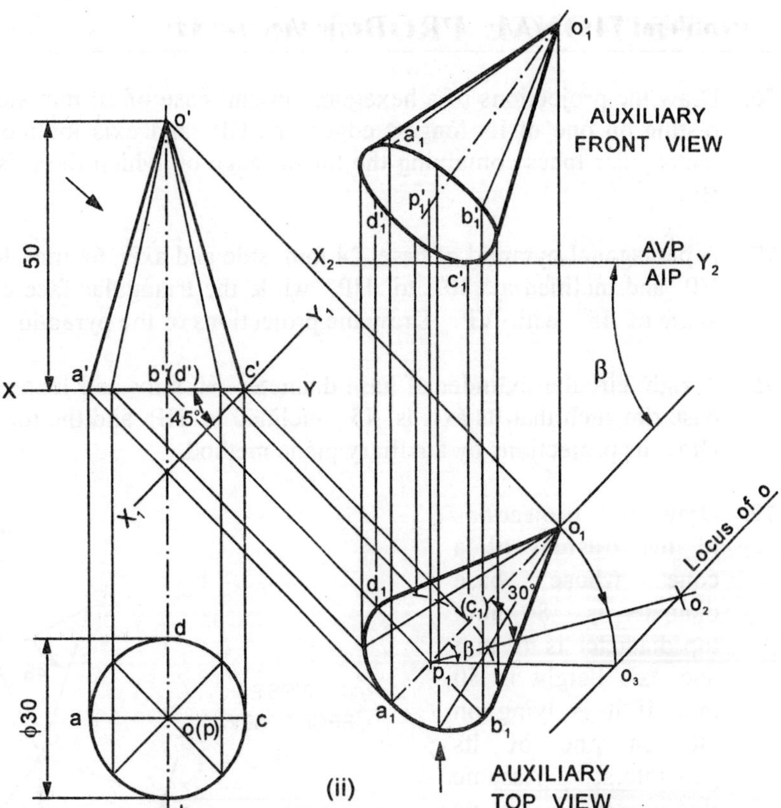

(ii)
FIG. 17

Note: *Above Problem is solved by Change of Position Method in the previous Chapter.*

Problem 75: *(Fig. 18) A pentagonal pyramid of 26 mm side of base and 70 mm altitude rests on an edge of its base on HP so that the highest point in the base is 26 mm above HP. Draw its projection when the axis is parallel to VP.*

Draw another front view on a reference line inclined at 30° to the edge on which it is resting so that the base is visible. (UQ)

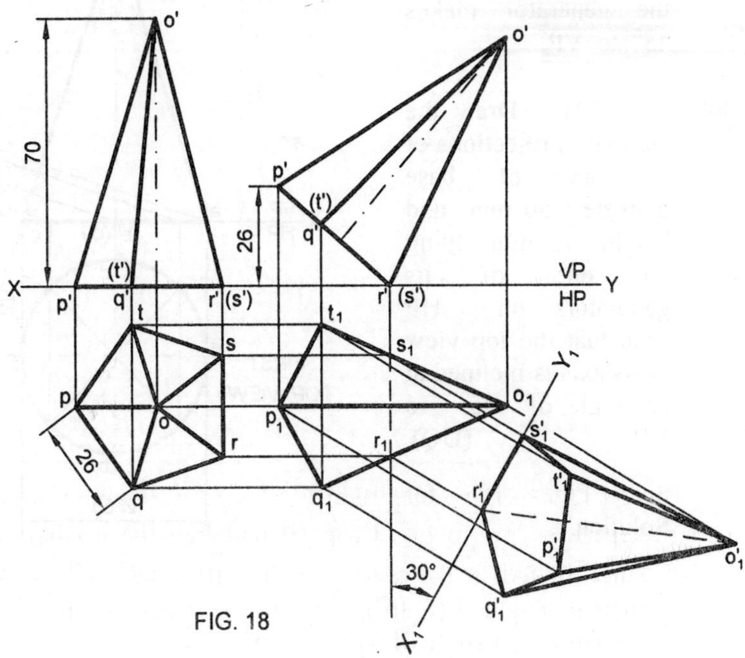

FIG. 18

4. ADDITIONAL PROBLEMS – II

76. Draw the projections of a hexagonal prism, base of 30 mm side and axis 50 mm long when resting on one of its longest edges on HP and axis inclined at 40° to VP. One of the rectangular faces containing the longer edge on which the prism rests is inclined at 30° to HP.

77. A pentagonal pyramid of base 24 mm side and axis 64 mm long has an edge of the base in VP and inclined at 30° to HP, while the triangular face containing that edge makes an angle of 45° with VP. Draw the projections of the pyramid.

78. A right circular cylinder of base diameter 40 mm and length of axis 50 mm, rests on its base rim such that its axis is 45° inclined to HP and the top view is 60° inclined to VP. Draw its projections by auxiliary plane method.

79. Draw the projections of the frustum of a cone whose base diameter is 80 mm, top diameter is 40 mm and axis height is 50 mm, if it is lying on HP on one of its generators. Assume that the plane containing the axis and the generator makes 45° to VP.

80. (Fig. 19) Draw the auxiliary projections of a cone of base diameter 50 mm and height 75 mm lying on one of its generators on HP such that the top view of its axis is inclined at an angle of 45° to VP. (UQ)

See Fig. 19 for Solution.

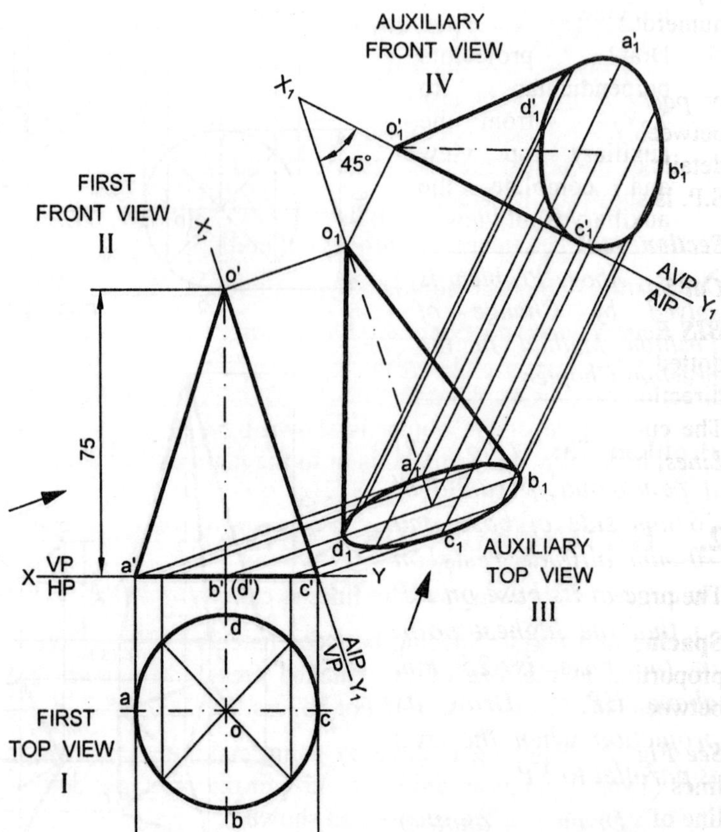

FIG. 19

*

15

SECTIONS OF SOLIDS

1. WHY IS SECTIONING NECESSARY?

Section Plane (S.P.) ***or*** **Cutting Plane**

In the orthographic views, the interior hollow portions and invisible outer edges / faces are represented by *dashed lines*. When the interior of the object is complex, these dashed lines are numerous. Therefore, the drawing is difficult to interpret.

To overcome this difficulty, it is imagined that the object is being cut through *completely* or *partially* by an *imaginary plane* called *Cutting* or *Section Plane.* Portion of the object between the S.P. and the observer is *assumed to be removed* to show the invisible and interior details of the object clearly. Show the S.P. in the view in which it is seen as a straight line, i.e., S.P. is shown by its Trace.

Sectioning: The imaginary process of cutting or sectioning the object is called *Sectioning*.

Cut Surface: Surface obtained by cutting an object by S.P. is called *Cut Surface* or *Section*.

BIS Representation of Section Plane Line: S.P. is conventionally represented as a long-dashed dotted *narrow* line and is drawn *wide* at the ends, bends and the changes of direction. The direction of viewing is perpendicular to the S.P. It is shown by arrows resting on the *thick* ends. The cut surface of the object is shown by a series of *narrow* lines (Type 01.1) called *Section Lines*, inclined at 45° to the axis or to the main outline of section.

2. HATCHING – BIS CONVENTIONS

The process of drawing section lines is called *Hatching*.

Spacing of Lines: Spacing between hatching lines should be in proportion to the size of the hatched areas. Minimum spacing between hatching lines should not be less than 0.7 mm.

See Fig. 1. Hatching is done by using continuous narrow (*thin*) lines (Type 01.1) at an angle of 45° to the principle outlines or line of symmetry of the sections as shown.

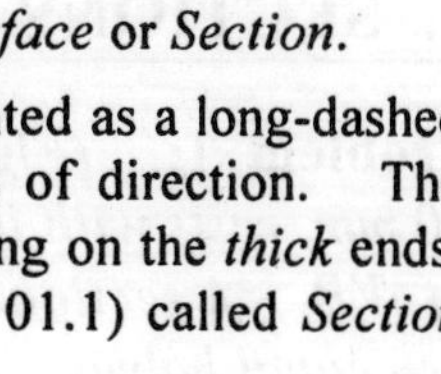

FIG. 1

See Fig. 2. Separate areas of a section of the same component shall be hatched in an identical manner. The hatching of adjacent components shall be carried out with different directions or spacings as shown.

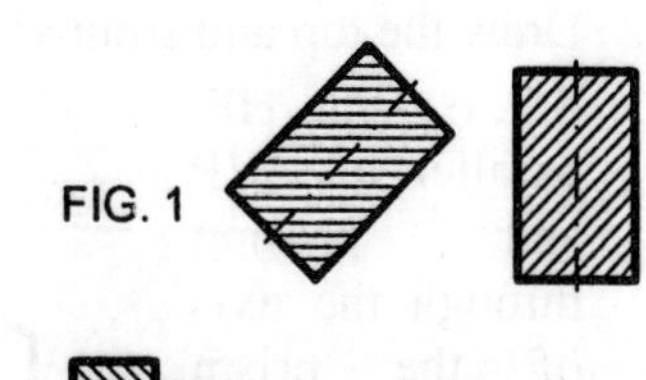

FIG. 2

See Fig. 3. Hatching is interrupted when it is not possible to place inscriptions, say, S outside the hatched area as shown.

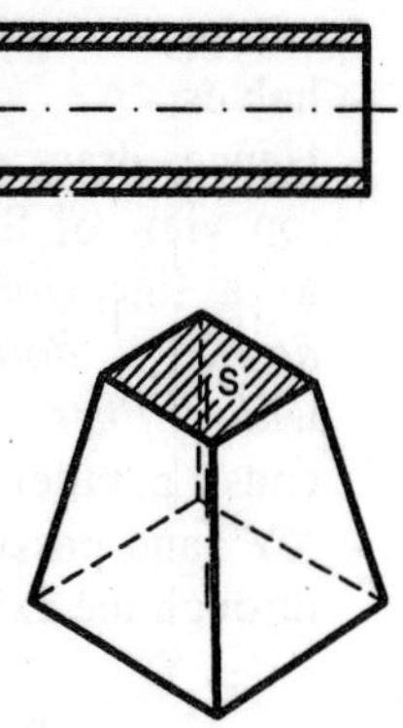

FIG. 3

Note: 1. *If the boundary of the object itself is inclined at 45° to the horizontal, section lines can be drawn at 45° to the main axis of the object.*

2. *Section lines should be equally spaced. Depending on the area to be hatched, spacing normally varies from 1.0 to 3.0 mm.*

Sectional View: It is the drawing showing that part of the object which is between the plane of projection and the S.P. It may be either a Sectional Front View or Sectional Top View.

True Shape of Section: If a solid is sectioned by a S.P. and it is projected to a plane parallel to it, the shape of the section obtained will be exactly the section exposed by the S.P. This shape is called *True Shape of Section.*

Apparent Shape of Section: If the S.P. is inclined to the plane of projection, the shape obtained will not be the true shape. Such a shape is called *Apparent Shape of Section.*

2.2 TYPES OF SECTION PLANES

(i) S.P. Perpendicular to HP and Parallel to VP
(ii) S.P. Perpendicular to VP and Parallel to HP
(iii) S.P. Perpendicular to VP and Inclined to HP
(iv) S.P. Perpendicular to HP and Inclined to VP
(v) S.P. Perpendicular to both HP and VP

3. SECTIONS OF PRISMS AND CUBES

Problem 1: *(Fig. 4)* ***A rectangular prism, side of base 40 mm x 25 mm and height 60 mm, rests with its base on HP such that one of its larger rectangular faces is parallel to VP. A section plane perpendicular to HP and parallel to VP cuts the prism into two equal halves. Draw its top view and sectional front view.***

1. Draw the top and front views of the prism by *thin* lines. Name all the corners.
2. S.P. is $\perp^r$ to HP, parallel to VP and passing through the axis of the prism cutting it into 2 halves. Hence, draw the top view of S.P. as a line (*long-dashed dotted wide lines* at ends) parallel to XY and passing through the axis.

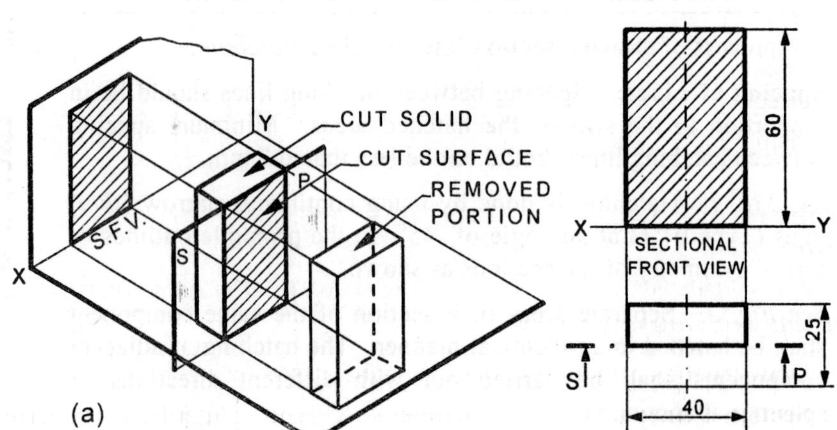

FIG. 4 (b) TOP VIEW

3. Assume that the portion of the prism between the S.P. and the observer is removed. The remaining portion is then shown by *thick* (wide) lines in the top view.

Note: *(i)* *Within the solid portion, the S.P. coincides with the cut surface which is to be shown by full wide line necessarily.*

(ii) *When S.P. is parallel to HP or VP, the Sectional Front View or Sectional Top View of the cut surface itself shows the* ***True Shape of Section****.*

---

4. Then draw the sectional front view for the remaining portion of the prism by *thick* lines (H). In this view the cut surface is shown by hatching (*thin*) lines using 2H.

Problem 2: *(Fig. 5) A* ***hexagonal prism****, side of base 30 mm and axis 60 mm long, rests with its base on HP such that one of its rectangular faces is parallel to VP. A section plane perpendicular to HP and parallel to VP cuts the prism at a distance of 10 mm from its axis. Draw its top and sectional front views.*

1. Draw top and front views of the prism by *thin lines* and name all the corners.
2. Draw SP parallel to XY and 10 mm from the axis towards the observer in the top view.

   Note: *Always major portion of the solid should be retained for projections.*

3. Name the section points 1 and 2 where the S.P. cuts the visible edges ab and cd respectively. Name the points (3) and (4) where the S.P. cuts the invisible edges (d_1) (c_1) and (b_1) (a_1) respectively. Show remaining portion of the prism as *thick* lines.
4. **Cut surface:** Project the above section points on the corresponding edges in the front view. From 1(4) draw a projector to cut a′ b′ at 1′ and a'_1 b'_1 at 4′. Similarly mark 2′ and 3′ in the front view.

   Join 1′ 2′ 3′ 4′ by *thick* lines and hatch this area to represent the front view of the cut surface.
5. Look at the remaining portion of prism in top view in the direction of arrows. 1 a (a_1) (4) and 2d (d_1) (3) are visible. Show 1′ a′ a'_1 4′ and 2′ d′ d'_1 3′ as *thick* lines.

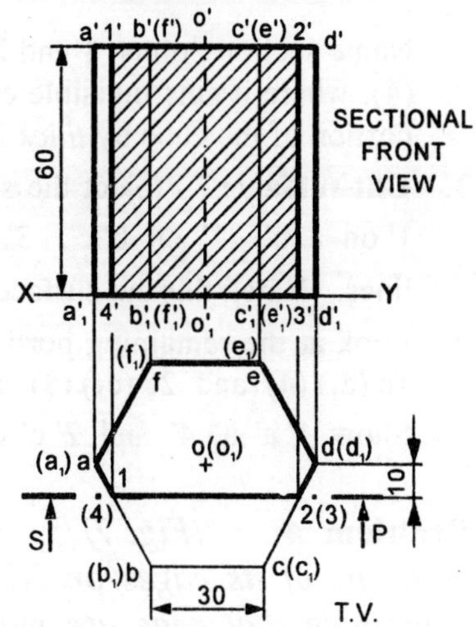

FIG. 5

Problem 3: *(Fig. 6) A* ***cube*** *of 45 mm side rests with a face on HP such that one of its vertical faces is inclined at 30° to VP. A section plane, parallel to VP cuts the cube at a distance of 15 mm from the vertical edge nearer to the observer. Draw its top and sectional front views.*

1. Draw the top and front views of the cube.
2. **Section Plane:** Draw S.P. parallel to XY and 15 mm from the vertical edge $b(b_1)$ nearer to the observer in the top view.

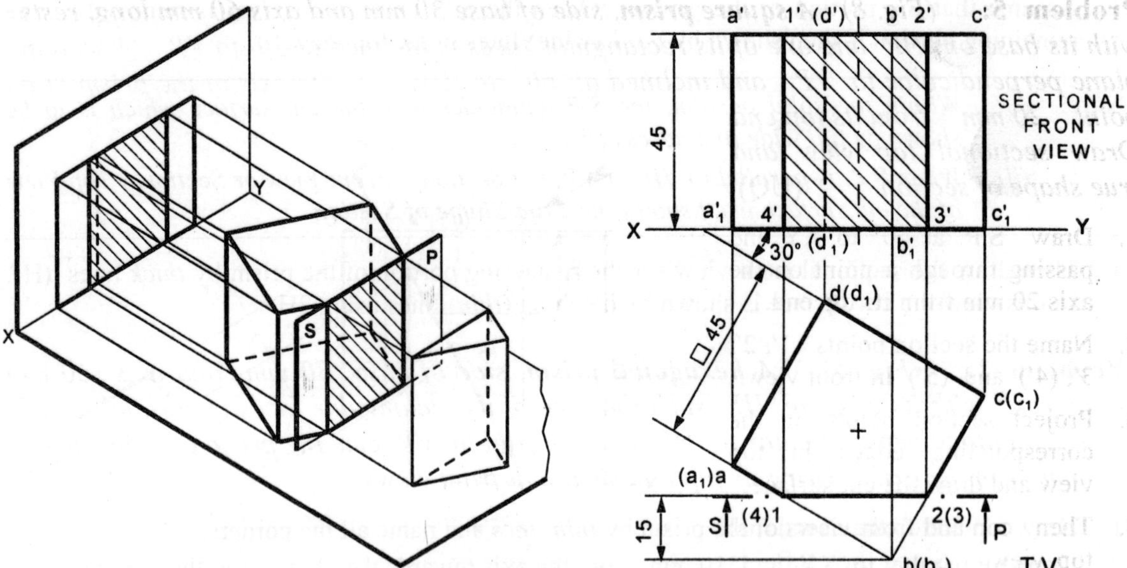

FIG. 6

Name section points 1 and 2 where S.P. cuts the visible edges ab and bc. Name (3) and (4) where it cuts invisible edges (c_1) (b_1) and (b_1) (a_1) respectively. Show the remaining portion of the cube by *thick* lines in the top view.

3. **Cut surface:** Project the section points on the corresponding edges in the front view, *i.e.*, 1′ on a′ b′, 2′ on b′ c′, 3′ on c'_1 b'_1 and 4′ on b'_1 a'_1 respectively. Join 1′2′3′4′ by *thick* lines. Hatch this cut surface.
4. Look at the remaining portion of the cube in the top view in the direction of arrows shown. 1a (a_1) (4) and 2c (c_1) (3) are visible.
   Show 1′ a′ a'_1 4′ and 2′ c′ c'_1 3′ as *thick* lines.

Problem 4: *(Fig. 7) A **cube** of 50 mm side rests with one of its edges on HP such that the square faces containing that edge are making equal inclinations with HP. A horizontal section plane cuts the cube at a distance of 18 mm below the horizontal edge nearer to the observer. Obtain front and sectional top views.*

1. **Section Plane:** Draw S.P. parallel to XY and 18 mm below d′ (d'_1). Mark 1′, 2′, (3′) and (4′) as shown in front view. Show the remaining portion by *thick* lines.
2. **Cut Surface:** Project the section points on the corresponding edges in the top view. Join 1 2 3 4 and hatch this area. Complete the sectional top view as shown.

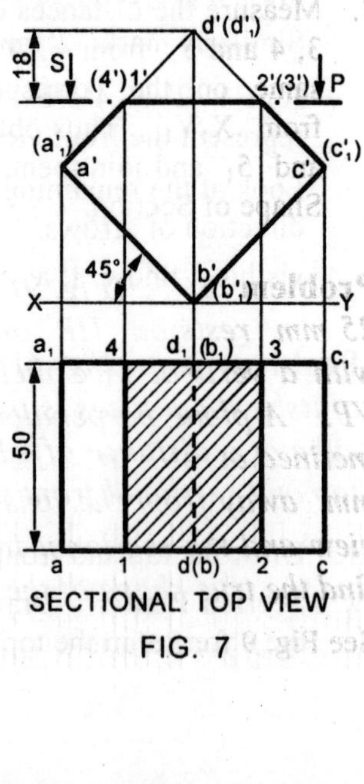

FIG. 7

Problem 5: *(Fig. 8) A* ***square prism****, side of base 30 mm and axis 60 mm long, rests with its base on HP and one of its rectangular faces is inclined at 30° to VP. A section plane perpendicular to VP and inclined at 60° to HP cuts the axis of the prism at a point 20 mm from its top end. Draw sectional top view and true shape of section.* (UQ)

1. Draw S.P. at 60° to XY and passing through a point on the axis 20 mm from its top end.
2. Name the section points 1′, 2′, 3′, (4′) and (5′) in front view.
3. Project section points on the corresponding edges in top view and draw the cut surface.
4. Then complete the sectional top view. Since the S.P. is inclined to HP, cut surface obtained does not appear in true shape and is called **Apparent Shape of Section**.

True Shape of Section

5. Draw X_1Y_1 parallel to S.P. & at any convenient distance from it.
6. Draw projectors perpendicular to X_1Y_1 from 1′, 2′, 3′, (4′) and (5′).
7. Measure the distances of the points 1, 2, 3, 4 and 5 from XY and transfer the same on the corresponding projectors from X_1Y_1. Thus obtain 1_1, 2_1, 3_1, 4_1 and 5_1 and join them, which is the True Shape of Section.

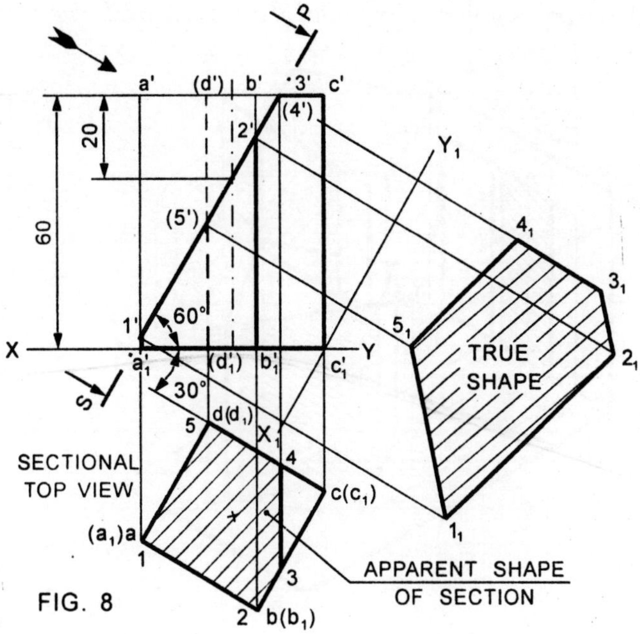

FIG. 8

Problem 6: *(Fig. 9) A* ***cube*** *of side 25 mm rests on HP on one of its faces with a vertical face inclined at 35° to VP. A plane perpendicular to HP and inclined at 50° to VP cuts the cube, 3 mm away from the axis. Draw the top view and the sectional front view. Also, find the true shape of the section.* (UQ)

See Fig. 9 for Solution.

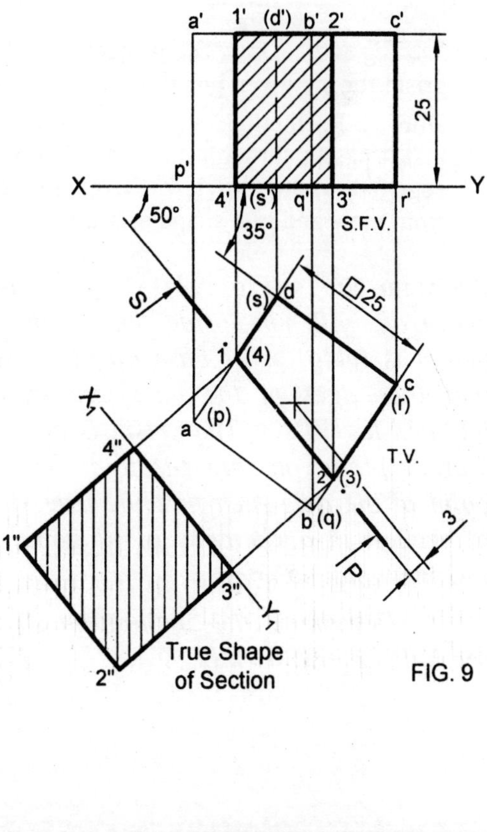

FIG. 9

Problem 7: *(Fig. 10) A* ***square prism****, side of base 40 mm and axis 60 mm long, rests with its base on HP such that one of its rectangular faces is inclined at 30° to VP. A section plane perpendicular to HP and inclined at 60° to VP passes through the prism such that a rectangular face which is making 60° with VP is cut into two halves. Draw the top view, sectional front view and true shape of section.* (UQ)

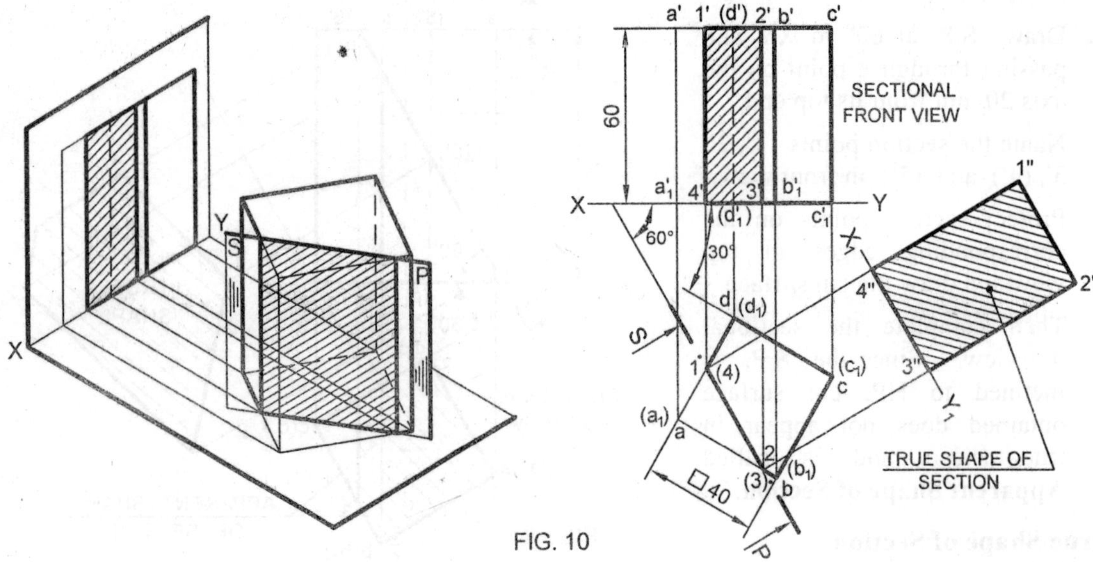

FIG. 10

1. *See Fig. 10.* Draw the projections of the prism for the given position.
2. Draw S.P. inclined at 60° to XY such that it passes through the mid-point of the edge ad in the top view. Draw sectional front view and true shape of section.

Problem 8: *(Fig. 11) A* ***square prism*** *of 32 mm side and 100 mm height is lying on its base on HP such that the edges of the base are equally inclined to VP. The prism is cut by a section plane passing through the mid-point of the axis such that the true shape of section is a rhombus of diagonals of 102 mm and 45 mm.*

Determine the inclination of the section plane with HP. (UQ)

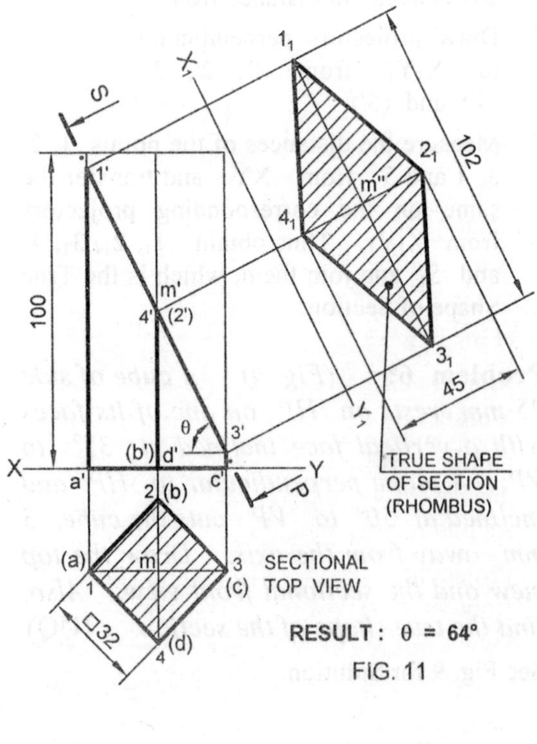

RESULT : θ = 64°

FIG. 11

Given the True Shape of Section, to draw the S.P. **(*Fig. 11*)**

1. Draw the top and front views. Mark the mid-point of the axis m′ in the front view.
2. m′ as center and radius = Half of the longer diagonal (= 51 mm), draw 2 arcs cutting 2 opposite vertical sides of the prism in front view. Thus mark Section Plane position SP.
3. Project the points from the S.P. on X_1Y_1 and draw the true shape of section, *i.e.*, rhombus. Measure the inclination θ of the S.P. with HP. Result: **θ = 64°.**

Problem 9: *(Fig. 12) A cube of 70 mm long edges has its vertical faces equally inclined to the VP. It is cut by an auxiliary inclined plane in such a way that the true shape of the cut part is a regular hexagon. Determine the inclination of the cutting plane with HP. Draw front view, sectional top view and true shape of the section.* (UQ)

Note: As the true shape of the section is a hexagon, the S.P. must cut the cube at 6 points, i.e., the S.P. will cut 2 edges of the top two edges of the base and two *vertical edges*. If the true shape is a *regular* hexagon, S.P. should cut these 6 edges at *mid-points*.

The points of intersection at the 2 *vertical edges* will be farthest from each other. These points represent 2 opposite corners of the hexagon. Distance between them is equal to b d.

Location of S.P.: If the 6 edges of the cube should be cut by a S.P., it has to be placed with one of its square faces on HP and 2 of the vertical faces making equal inclination with VP.

1. First, draw the top view of the cube such that the cube is placed with one of its square faces on HP and two of the vertical faces making equal inclination with VP. Project the corresponding front view.
2. Join bd. Draw a circle of diameter bd.
3. Inscribe a hexagon 123456 in the circle as shown. Measure the distance between the lines 16 and 34 = MN.
4. In the front view, locate 2′ at the mid-point of q′ b′.
5. With 2′ as center and radius equal to ½ MN, draw arcs to cut a′b′ and q′r′ at the points 1′ and 3′ respectively.
6. Join 1′3′, which is the *required S.P.* Measure the angle **θ = 55°**.
7. Draw X_1Y_1 parallel to 1′3′. Redraw the hexagon 123456 as $1_12_13_14_15_16_1$ such that mn is parallel to X_1Y_1.
8. Project all the corners of the hexagon in front view. Project 1′, 2′, ..., 6′ to top view. Obtain the sectional top view.

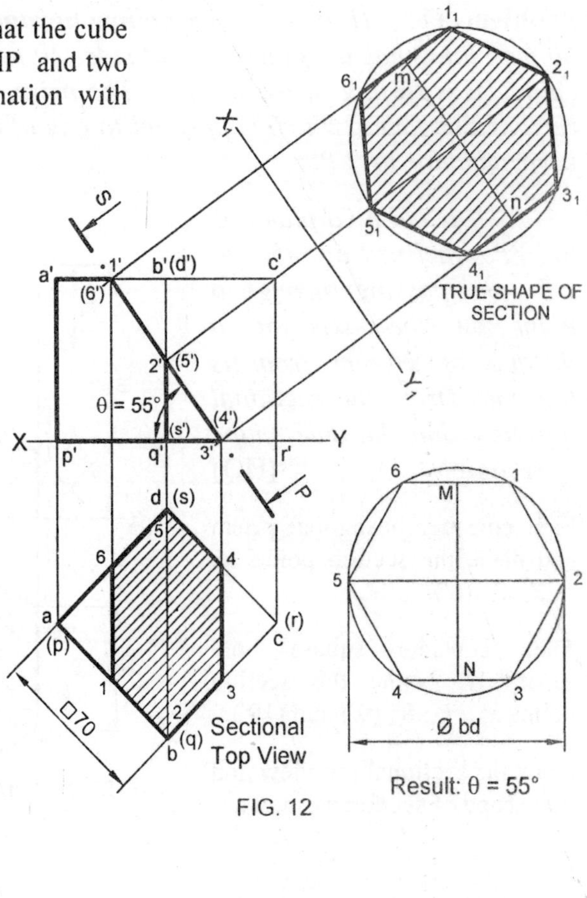

FIG. 12

Problem 10: *(Fig. 13) A* ***triangular prism*** *with a base side of 45 mm and height 60 mm rests on its base on HP with a rectangular face perpendicular to VP. It is cut by an Auxiliary Inclined Plane such that the true shape of the section is an isosceles triangle of maximum size.* (UQ)

Note: *For the maximum size of isosceles triangle, the S.P. must pass through a top edge of the prism and the base corner opposite to that edge.*

1. Draw the top view and front view as shown.
2. Mark S.P. through the points a′(b′) and 3′.
3. Project a′, (b′) and 3′ on the S.P. and obtain the auxiliary view a_1, b_1 and 3_1. Thus, the true shape of section is drawn as an isosceles triangle of maximum size.

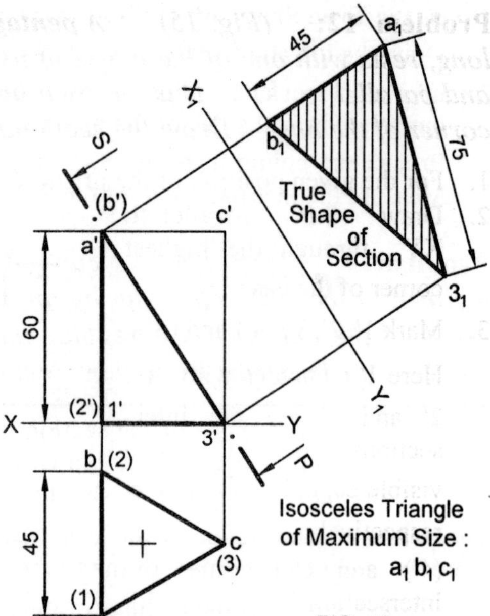

FIG. 13

Problem 11: *(Fig. 14) A* ***hollow hexagonal prism****, side of base 30 mm and axis 60 mm long, has a* ***square hole*** *of side 20 mm through it such that the axis of the hole coincides with that of the prism. The prism rests with its base on HP and one of its rectangular faces (which is parallel to one of the rectangular faces of the square hole) is perpendicular to VP.*

It is cut by a plane ⊥r to VP, inclined at 45° to HP and passing through a point on the axis at a distance of 20 mm from its top end. Draw the sectional top view and the true shape of section. (UQ)

First consider hexagonal prism and name the section points as 1′, 2′, ..., (6′).

Then consider square hole separately. Name the section points as 7′, 8′, (9′) and (10′).

Draw the sectional top view and true shape of section as shown.

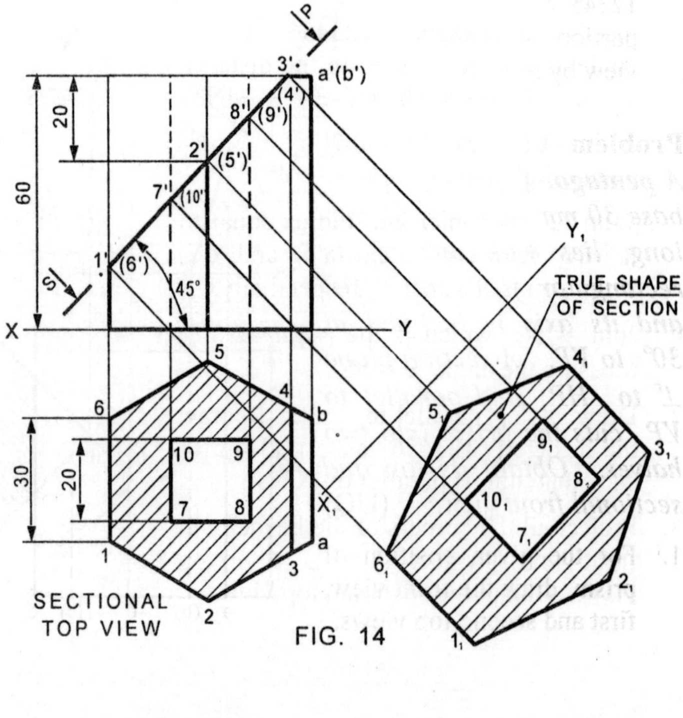

FIG. 14

Problem 12: *(Fig. 15)* ***A pentagonal prism,** side of base 25 mm and axis 60 mm long, rests with one of the edges of its base on HP. Its axis is inclined at 30° to HP and parallel to VP. It is cut by a horizontal section plane passing through the highest corner of the base. Draw the sectional top view.* (UQ)

1. For the given position of the prism, draw the top view, first and second front views as shown.
2. Draw S.P. parallel to XY through the highest corner of the base a_1'.
3. Mark 1′, 2′, 3′, (4′) and (5′).

   Here 1′ coincides with a_1'.

   2′ and 3′ are the intersections of S.P. with visible edges $b'b_1'$ and b′c′ respectively.

   (4′) and (5′) are the intersection of S.P. with invisible edges (d′)(e′) and (e′)(e_1') respectively.

   Show remaining portion of the prism by *thick* lines.
4. Project the section points on the corresponding edges in the II top view. Join 12345. Show the remaining portion in the sectional top view by *thick* lines.

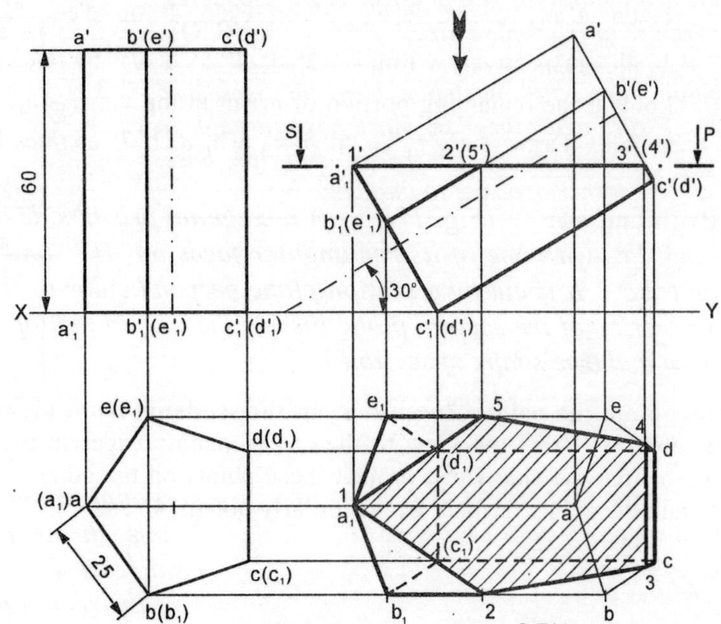

FIG. 15

Problem 13: *(Fig. 16)* ***A pentagonal prism,** side of base 30 mm and axis 60 mm long, lies with one of its rectangular faces on HP and its axis is inclined at 30° to VP. A section plane ⊥r to HP and parallel to VP cuts the prism into two halves. Obtain its top and sectional front views.* (UQ)

1. For the given position of prism, draw the front view, first and second top views.

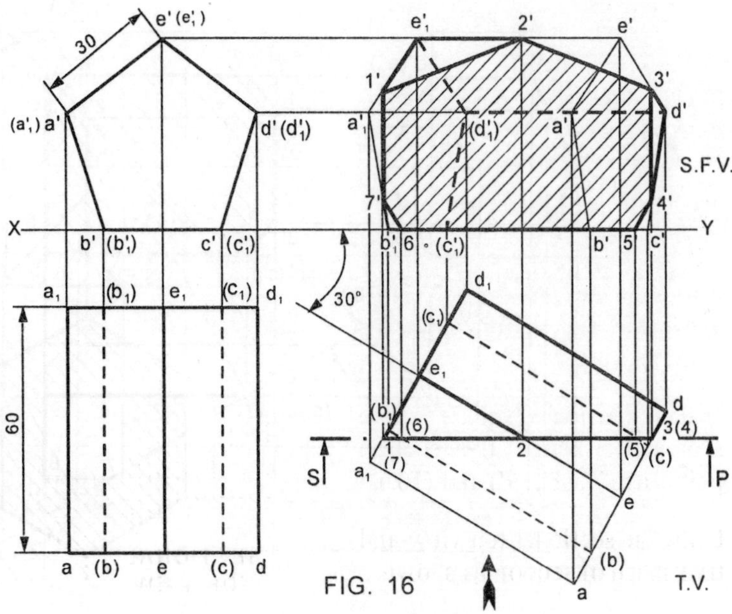

FIG. 16

2. Draw S.P. parallel to XY and passing through the mid-point of the axis in final top view.
3. Name the points 1, 2 & 3 where S.P. cuts visible edges a_1e_1, e_1e & ed respectively.
4. Similarly mark (4), (5), (6) and (7) where S.P. cuts the invisible edges d (c), (c) (c_1), (b)(b_1) and a_1(b_1) respectively. Show the remaining portion of the prism by *thick* lines.
5. **Cut surface:** Project the section points on the corresponding edges in final front view, *i.e.*, 1′ on $a_1' e_1'$, 2′ on $e_1'e'$, 3′ on e′d′, 4′ on d′c′, 5′ on c′(c_1'), 6′ on $b'b_1'$ and 7′ on $b_1' a_1'$ respectively. Join 1′ 2′ 3′ 4′ 5′ 6′ 7′ by *thick* lines and hatch this area.
6. Look at the remaining portion of prism in top view. $1e_1$, e_12, 3d, d(4), (6)(b_1) & (b_1)(7) are visible. Draw $1'e_1'$, $e_1'2'$, 3′d′, d′4′, $6'b_1'$ & $b_1'7'$ as *thick* lines in Sectional F.V.

Problem 14: *(Fig. 17) A* ***hexagonal prism****, side of base 25 mm and axis 65 mm long, lies with one of its rectangular faces on HP and its axis is parallel to both HP and VP. It is cut by a section plane perpendicular to HP and inclined at 30° to VP. The H.T. of the section plane bisects the axis in the top view. Draw the sectional front view and true shape of section.* (UQ)

To obtain the points 1′ and 8′ in the sectional front view: *See Fig. 17.* Project the points 1 and (8) from top view to the corresponding edges in the side view and mark 1″ and 8″. Then from the side view, project these points on the corresponding edges in the front view and obtain the points 1′ and 8′. Similarly obtain 4′ and 5′.

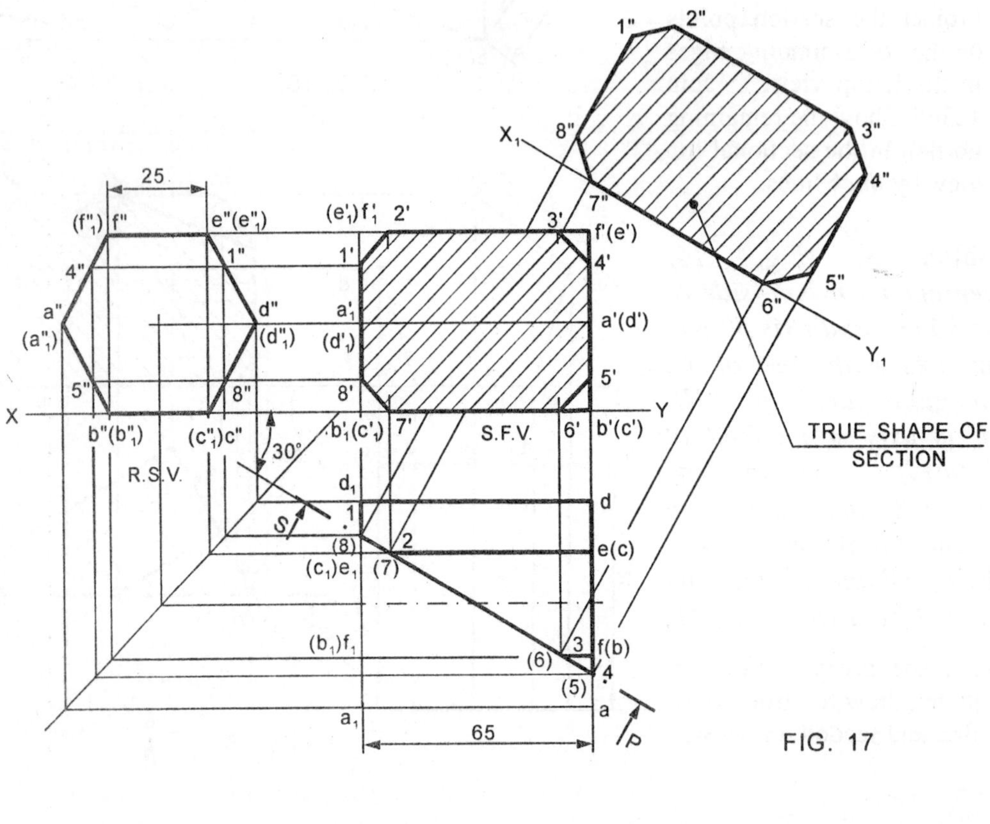

FIG. 17

Problem 15: *(Fig. 18)* ***A triangular prism*** *of 35 mm side of base and axis 56 mm long rests on its base on HP with a side of the base perpendicular to VP. It is cut by an Auxiliary Inclined Plane such that the true shape of section is a trapezium of parallel sides 28 mm and 8 mm. Draw the sectional top and front views and an auxiliary view showing the true shape of section. Determine the angle made by S.P. with HP.* (UQ)

1. Draw the top and front views of the triangular prism.
2. In the top view, mark the points a and b such that ab = 28 mm. Also, mark the points c and d such that cd = 8 mm. Complete the sectional top view.
3. Project a, b, c and d to front view and mark a′, (b′), (c′) and d′ on the corresponding edges.
4. Draw the S.P. through the points a′(b′) and (c′)d′.
5. Measure the angle made by the S.P. with XY = θ = 73°. Thus, complete the sectional front view.
6. Draw X_1Y_1 parallel to S.P. Project front view on X_1Y_1 and complete auxiliary view showing true shape of section.

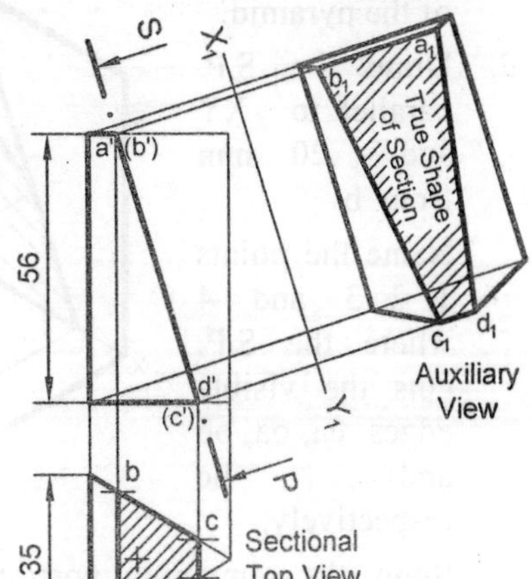

θ = 73°

FIG. 18

Problem 16: *(Fig. 19)* ***A hexagonal prism*** *of 35 mm side and 65 mm height has a concentric hexagonal hole of 18 mm side. It rests with its base on HP such that two rectangular faces make equal inclination to VP. The prism is cut by a section plane whose vertical trace is inclined at 60° to XY. The section plane bisects the axis of the prism. Draw the sectional top and profile views. Add the true shape of section.* (UQ)

1. Draw top and front views of prism.
2. Draw the hexagonal hole in top view and project it to front view.
3. In front view, draw the S.P. at 60° to XY and bisecting the axis.
4. Draw the sectional top view. Also, draw the sectional left side view using the mitre line.
5. Draw X_1Y_1 parallel to S.P. Project the front view on X_1Y_1 and draw the true shape of section.

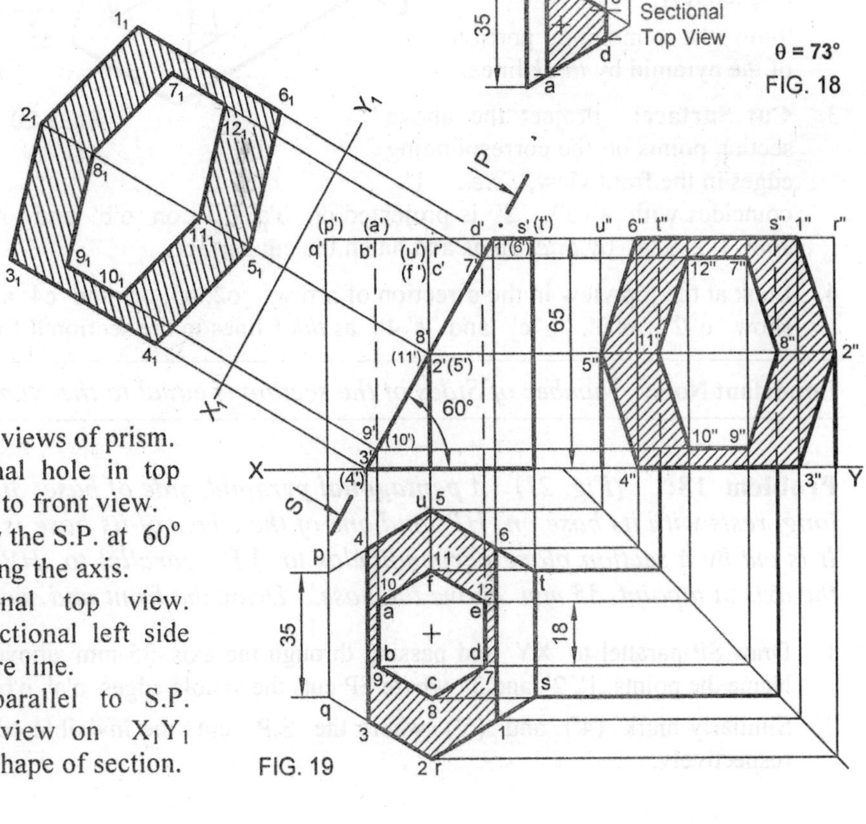

FIG. 19

4. SECTIONS OF PYRAMIDS AND TETRAHEDRONS

Problem 17: *(Fig. 20)* ***A pentagonal pyramid****, side of base 35 mm and axis 60 mm long, rests with its base on HP such that one of the edges of the base is perpendicular to VP. A section plane perpendicular to HP and parallel to VP cuts the pyramid at a distance of 20 mm from the corner of the base nearer to the observer. Draw its top and sectional front views.* (UQ)

1. Draw the top and front views of the pyramid.
2. Draw the S.P. parallel to XY and 20 mm from b.

   Name the points 1, 2, 3 and 4 where the S.P. cuts the visible edges ea, oa, ob and bc respectively.

   Show the remaining portion of the pyramid by *thick* lines.

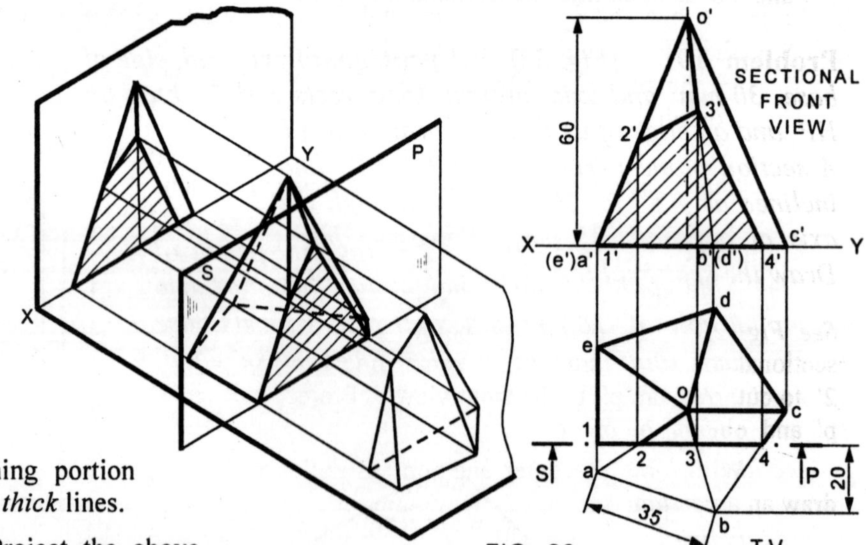

FIG. 20

3. **Cut Surface:** Project the above section points on the corresponding edges in the front view, *i.e.*, 1′ coincides with a′(e′), 2′ is projected on o′a′, 3′ on o′b′ and 4′ on b′c′ respectively. Join 1′ 2′ 3′ 4′ by *thick* lines and hatch the cut surface.
4. Look at the top view in the direction of arrows. o2, o3, oc and c4 are visible edges. Hence show o′ 2′, o′ 3′, o′ c′ and c′ 4′ as *thick* lines in the sectional front view.

Important Note: *Number of Sides of the section is equal to the Number of Edges cut.*

Problem 18: *(Fig. 21)* ***A pentagonal pyramid****, side of base 30 mm and axis 60 mm long, rests with its base on HP and one of the edges of its base is perpendicular to VP. It is cut by a section plane perpendicular to VP, parallel to HP and passing through the axis at a point 35 mm above the base. Draw the front and sectional top views.*

1. Draw SP parallel to XY and passing through the axis 35 mm above the base in front view. Name the points 1′, 2′ and 3′ where SP cuts the visible edges o′a′, o′b′ and o′c′ respectively.
2. Similarly mark (4′) and (5′) where the S.P. cuts the invisible edges o′ (d′) and o′ (e′) respectively.

3. Show the remaining portion of the pyramid a′ b′ c′ (d′) (e′) and the slant edges 1′ a′, 2′ b′, 3′ c′, (4′)(d′) and (5′)(e′) as *thick* lines in the front view.
4. **Cut surface:** Project the section points on the corresponding edges in the top view, *i.e.*, mark 1 on oa, 2 on ob and so on. Join 1 2 3 4 5 by *thick* lines and hatch this area.
5. Look at the remaining portion of the pyramid in the front view in the direction of arrows. Show abcde, 1a, 2b, 3c, 4d and 5e as *thick* lines in sectional top view.

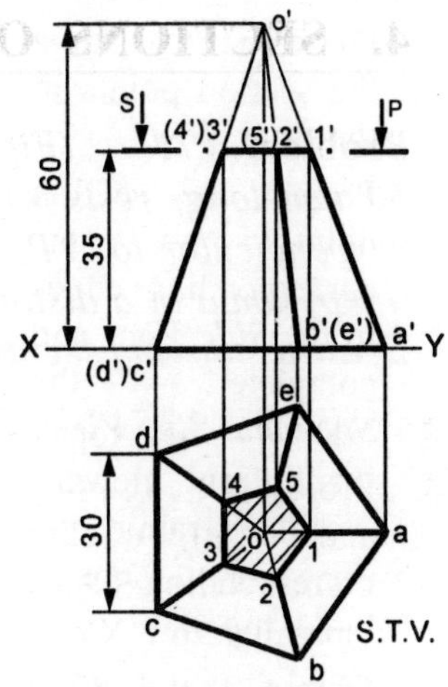

FIG. 21

Problem 19: *(Fig. 22) A **pentagonal pyramid**, side of base 30 mm and axis 60 mm long, rests with its base on HP and an edge of its base is parallel to VP. A section plane perpendicular to VP and inclined at 45° to HP passes through the axis at a point 35 mm above the base. Draw the sectional top view.* (UQ)

See Fig. 22. To obtain the point 2 in the sectional top view, draw a horizontal line from 2′ to cut o′ c′ at p′ in the front view. Project p′ and obtain p on oc.

With o as center and op as radius draw an arc to cut ob at 2.

Problem 20: *(Exercise) A **hexagonal pyramid**, side of base 30 mm and height 65 mm is resting on its base on HP with two edges parallel to VP. It is cut by a plane perpendicular to VP, inclined at 45° to HP and intersecting the axis at a point 25 mm above the base. Draw the sectional top view and true shape of section.* (UQ)

Hint: In this Problem, the base is also cut. Therefore the true shape of section obtained is a seven-sided irregular polygon called *Heptagon*.

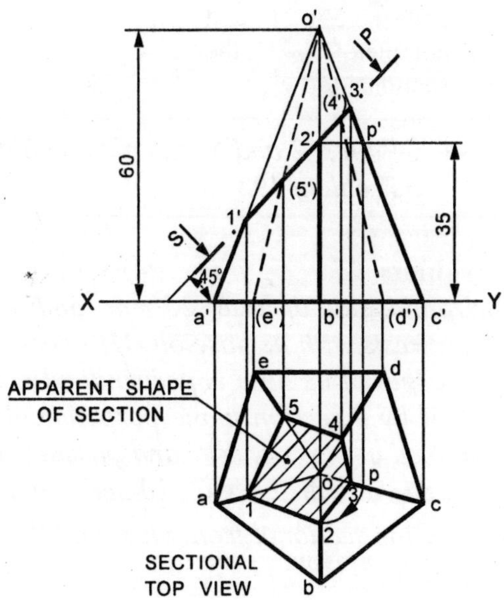

FIG. 22

Problem 21: *(Fig. 23) A **hexagonal pyramid**, side of base 25 mm and axis 50 mm long, rests with its base on HP and an edge of its base is perpendicular to VP. It is cut by a section plane perpendicular to VP, inclined at 30° to HP and passing through a point on the axis 20 mm below the apex. Draw the sectional side view and sectional top view.* (UQ)

1. Draw the top, front and left side views.

2. Draw S.P. Name the section points in the front view.
3. Project the section points on the corresponding edges in the ***Side View*** and complete the ***Sectional Side View.***
4. Project 1′, 3′, (4′) and (6′) on the corresponding edges in the top view.
5. Obtain 2 and 5 by projecting them from the sectional side view (*i.e.*, from 2″ and 5″) and complete the sectional top view.

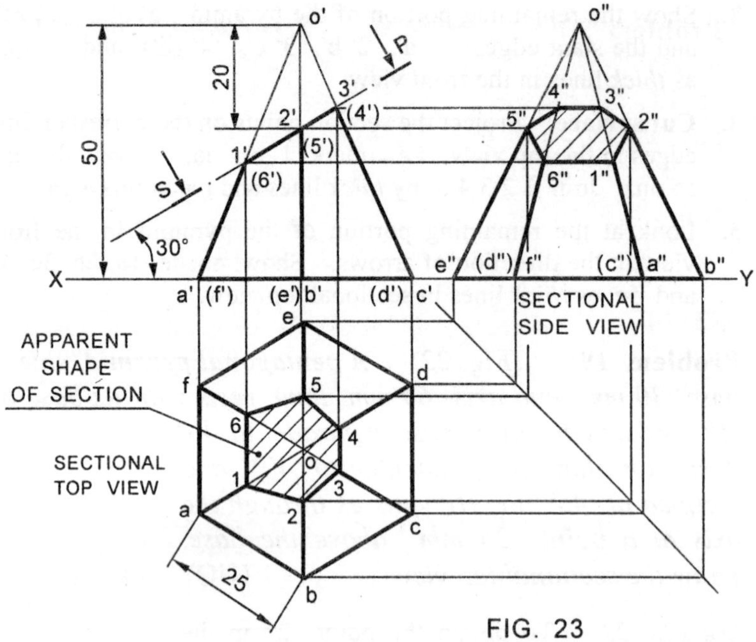

FIG. 23

Note: *Points 2 and 5 can also be obtained directly from the Front View as explained in* ***Problem 19.***

Problem 22: (*Fig. 24*) A ***hexagonal pyramid,*** *side of base 25 mm and axis 55 mm long, rests with its base on HP such that one of the edges of its base is perpendicular to VP. It is cut by a section plane perpendicular to HP, inclined at 45° to VP and passing through the pyramid at a distance of 10 mm from the axis. Draw the sectional front view and the true shape of section.* (UQ)

1. Draw the projections of the pyramid for the given position. Draw S.P. and name the section points.
2. Point 3 is on ob. To locate this point in the front view, with o as center and o3 as radius draw an arc to intersect oc at p.
3. Project p and obtain p′ on o′ c′. From p′ draw a horizontal line to intersect o′b′ at 3′.
4. Complete the sectional front view and the true shape of section as shown.

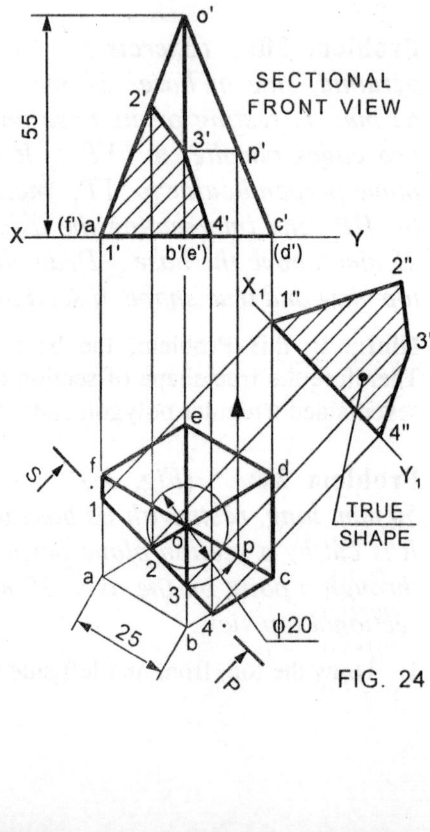

FIG. 24

Problem 23: *(Exercise)* *A **pentagonal pyramid**, side of base 30 mm and axis 60 mm long, rests with its base on HP such that one of the edges of its base is parallel to VP. It is cut by a section plane perpendicular to HP, inclined at 45° to VP and passing through the pyramid at a distance of 8 mm from its axis. Draw the sectional front view and the true shape of section.* (UQ)

Hint: To draw the S.P., first draw a circle with the top view of the axis as center and 8 mm as radius. Then draw a tangent to the circle at 45° to XY.

Problem 24: *(Fig. 25)* *A **pentagonal pyramid**, side of base 26 mm and 52 mm height, rests with its base on HP. One of the edges of its base is perpendicular to VP. A section plane perpendicular to VP and inclined at 45° to HP bisects the axis. Draw the sectional top view and project another top view on an A.I.P. parallel to the section plane showing true shape of section.* (UQ)

See Fig. 25. Draw the sectional top view. Project the remaining portion of the pyramid along with cut surface on an AIP parallel to S.P. Area $1_1 2_1 3_1 4_1 5_1$ is the true shape of section.

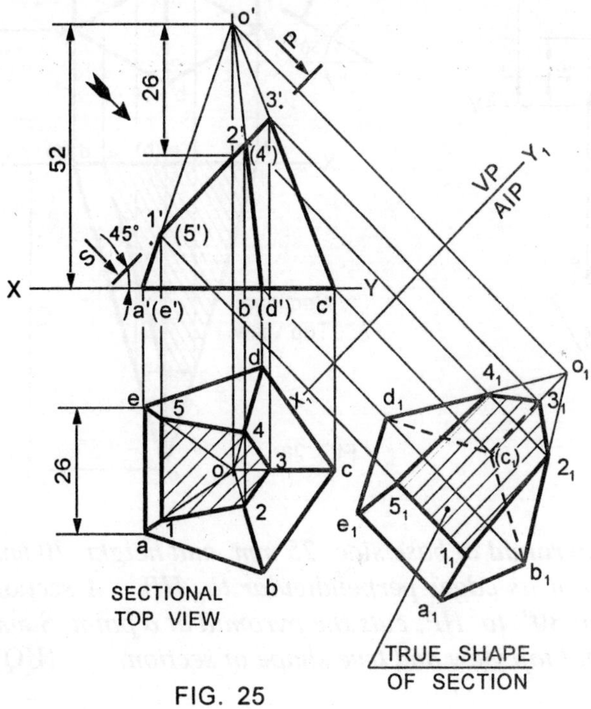

FIG. 25

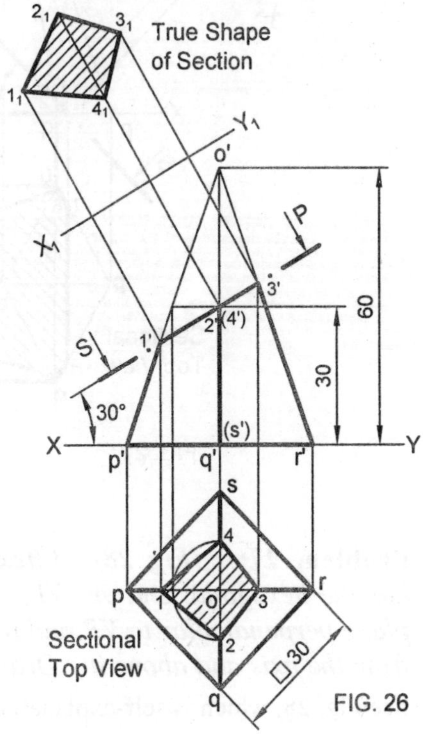

FIG. 26

Problem 25: *(Fig. 26)* *A **square pyramid** of base 30 mm and axis 60 mm long is standing on HP with its base edges equally inclined to VP. It is cut by a section plane perpendicular to VP and inclined at 30° to HP, bisecting the axis. Draw the sectional top view and the true shape of the section, if the upper portion is removed.* (UQ)

See Fig. 26 for Solution.

Problem 26: *(Fig. 27) A* ***hexagonal pyramid****, side of base 25 mm and altitude 70 mm long, rests with its base on HP with two of its base sides parallel to VP. It is cut by a section plane perpendicular to VP and inclined at 45° to HP and passing through a point 15 mm above the base and located on the axis. Draw the sectional top view and the true shape of the section.* (UQ)

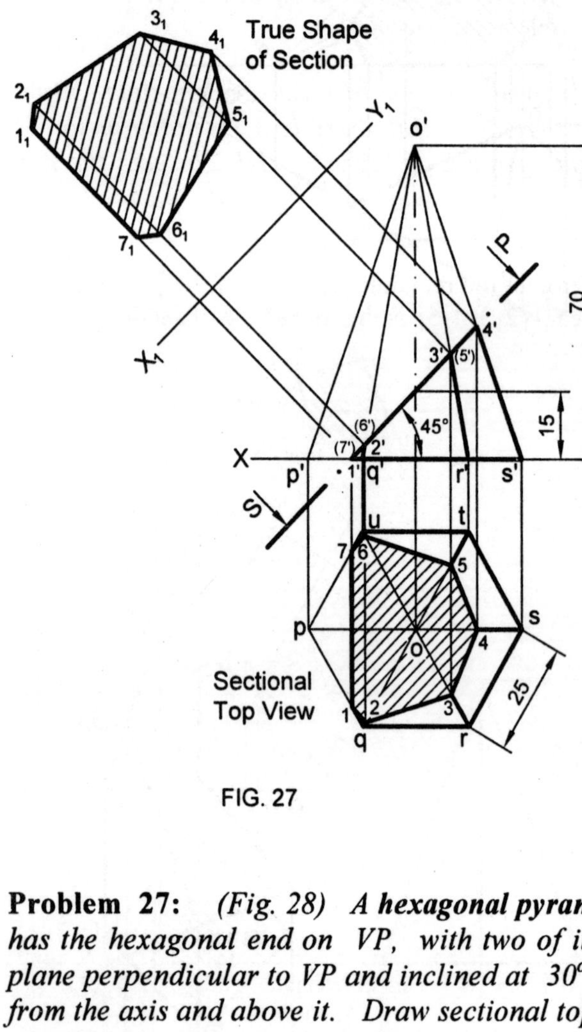

FIG. 27

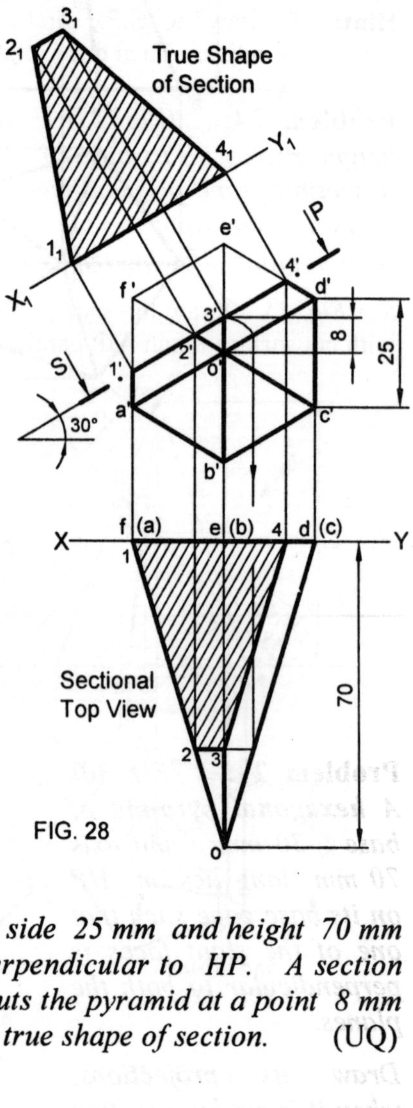

FIG. 28

Problem 27: *(Fig. 28) A* ***hexagonal pyramid*** *of base side 25 mm and height 70 mm has the hexagonal end on VP, with two of its edges perpendicular to HP. A section plane perpendicular to VP and inclined at 30° to HP, cuts the pyramid at a point 8 mm from the axis and above it. Draw sectional top view and true shape of section.* (UQ)

See Fig. 28, which is self-explanatory.

Problem 28: *(Fig. 29) A* ***pentagonal pyramid*** *side of base 25 mm and length of axis 50 mm, lies on one of its triangular faces on HP with its axis parallel to VP. A section plane perpendicular to HP and inclined to VP at 30° cuts the pyramid bisecting its axis. Draw its top view, sectional front view and true shape of section.* (UQ)

See Fig. 29, which is self-explanatory.

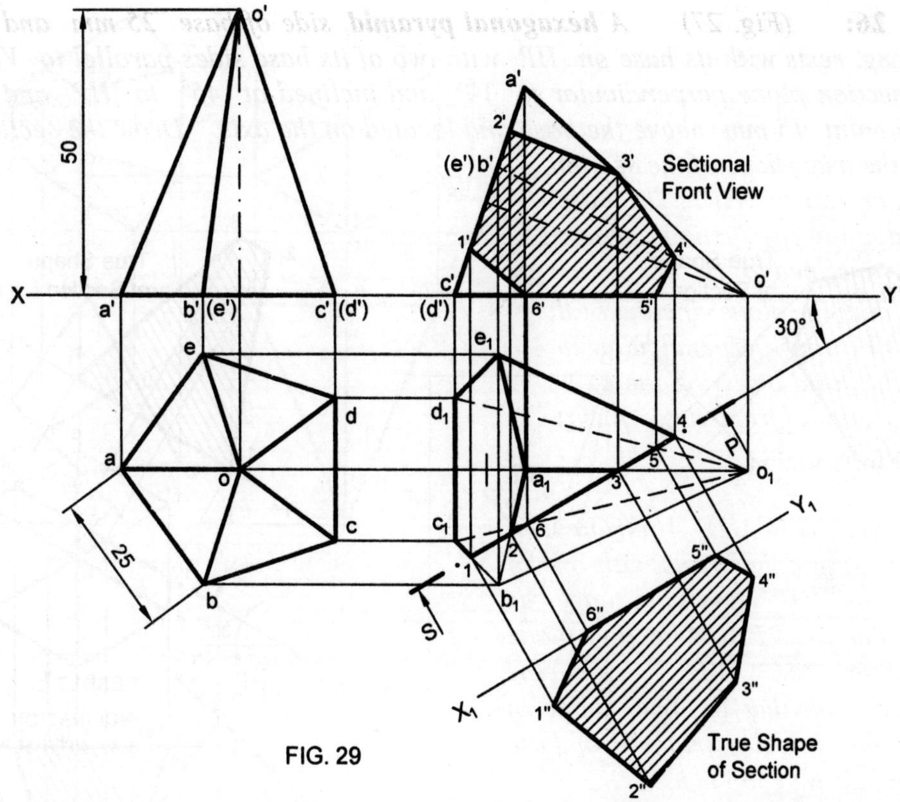

FIG. 29

Problem 29: *(Fig. 30)* ***A hexagonal pyramid*** *of base 30 mm and axis 70 mm long lies on HP on its base edge such that one of the slant faces is perpendicular to both the planes.*

Draw its projections, when it is cut by a section plane parallel to HP and at a distance of 15 mm from apex point. (UQ)

See Fig. 30 for Solution.

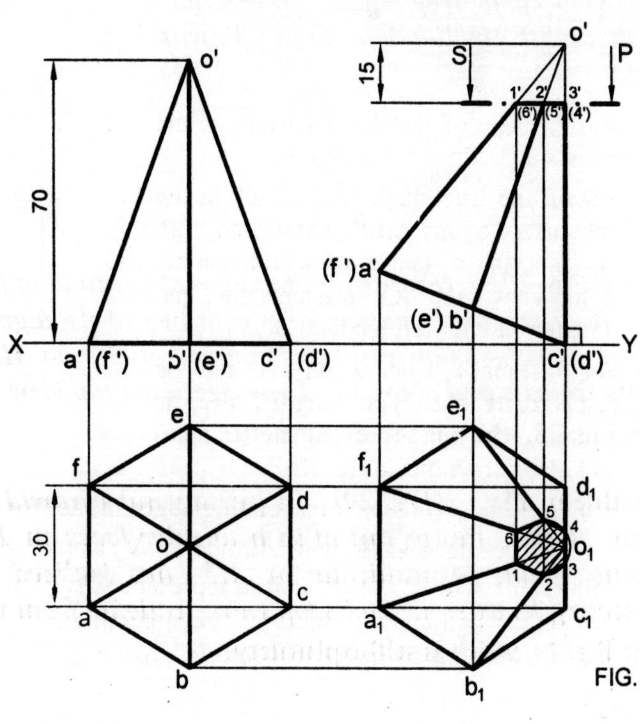

FIG. 30

Problem 30: *(Fig. 31)*
*A **hexagonal pyramid** of base side 30 mm and altitude 60 mm rests on one corner of the base on HP such that the axis is inclined at 30° to VP and parallel to HP. The base edges containing the resting corners are equally inclined to HP.*

A cutting plane inclined at 45° to VP and perpendicular to HP cuts the solid bisecting the axis. Draw the sectional front view. (UQ)

See Fig. 31 for Solution.

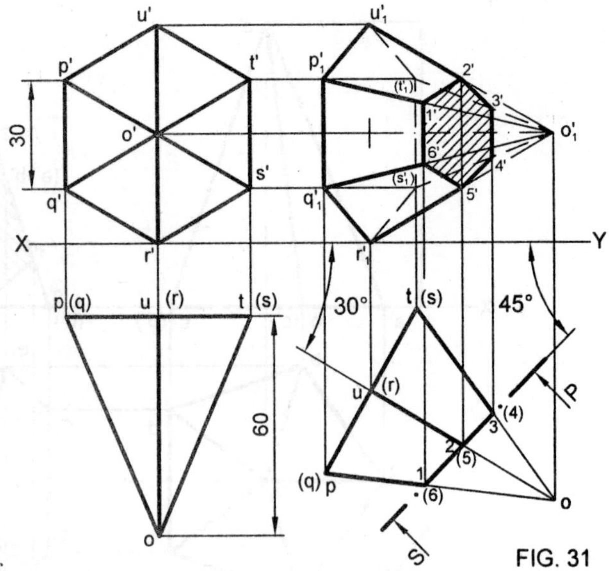

FIG. 31

Problem 31: *(Fig. 32) A **tetrahedron** of 60 mm long edges rests with one of its faces on HP and an edge is perpendicular to VP. A section plane perpendicular to VP cuts the tetrahedron such that the true shape of section is an isosceles triangle of base 50 mm and altitude 36 mm.*

Draw the front view, sectional top view and true shape of section. Also, find the inclination of the S.P. with HP. (UQ)

If the true shape of section is a triangle, the S.P. must cut 3 edges of the tetrahedron.

1. Obtain the true shape by projecting the cut surface on an A.I.P. parallel to S.P. For this, the distances of section points in top view from XY are measured and transferred with reference to X_1Y_1.

   For this, mark 1 and 3 on oa and oc respectively such that length 1-3 is equal to 50 mm (base of the triangle) and also perpendicular to XY.
2. Project 1 and 3 on the corresponding edges in the front view.
3. 1′ (3′) as center and 36 mm (altitude of Δ) as radius draw an arc to cut o′ b′ at 2′.
4. Join 1′(3′) with 2′. Extend it to represent S.P. Angle θ between S.P. and XY is **42°**.
5. Project 2′ and mark 2 on ob. Then complete sectional top view. Draw the true shape of section, which is an ***isosceles triangle***. *See Fig. 32.*

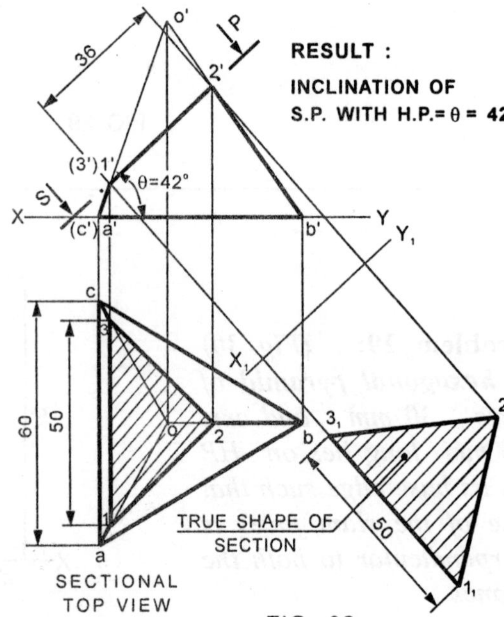

FIG. 32

Problem 32: *(Fig. 33) A* ***pentagonal pyramid*** *of side of base 25 mm and altitude 50 mm is resting on its base on HP with one side of the base perpendicular to VP.*

It is cut by a plane perpendicular to VP such that the true shape of section is an isosceles triangle of maximum possible base and maximum height.

Draw the front view, sectional top view and true shape of section. Measure the inclination of the cutting plane with HP. (UQ)

Results: *See Fig. 33.*

Inclination of S.P. with HP = θ = 83°

Maximum base of isosceles triangle = 40.5 mm

Maximum height of isosceles triangle = 50.5 mm

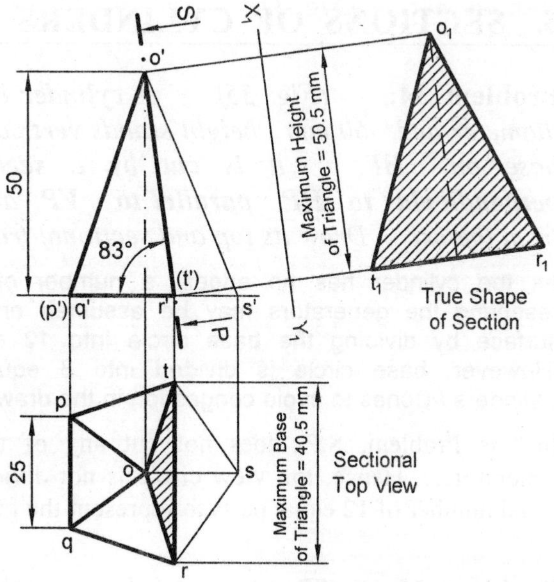

FIG. 33

Problem 33: *(Fig. 34) A* ***square pyramid*** *of 45 mm side of base and altitude 60 mm rests on its base on the ground. A side of base is perpendicular to VP. It is cut by two section planes. One is parallel to its extreme left face and 8 mm away from it, while the other is parallel to the extreme right face and 8 mm away from it. Both the planes intersect each other on the axis of the pyramid.*

Draw the sectional top and front views and Sectional Right Side View. (UQ)

1. Draw the top and front views of the square pyramid.
2. Draw the section planes S_1P_1 and S_2P_2 in the front view as shown. Both the planes intersect at points 2′ and 3′. Mark the other cutting points as shown.
3. Project the corresponding sectional top view. Points 2′ and 3′ are transferred on the slant edge o'r' and then projected to the top view as shown. Draw the hatching lines in the opposite directions for the left and right faces.
4. Draw the Sectional Right Side View of the remaining portion of the square pyramid as shown.

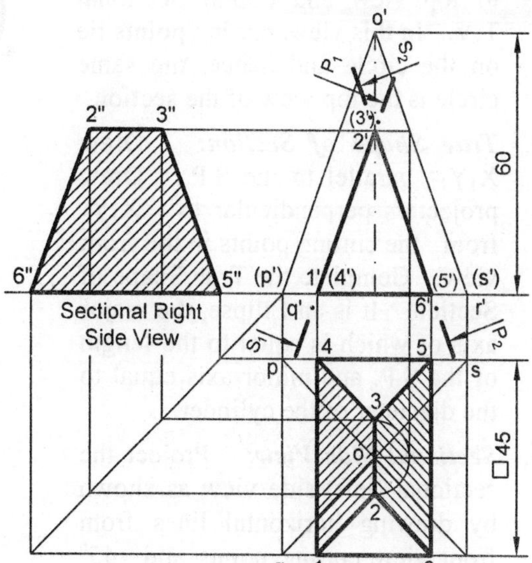

FIG. 34

5. SECTIONS OF CYLINDERS

Problem 34: *(Fig. 35) A cylinder of 50 mm diameter and 60 mm height stands vertically with its base on HP. It is cut by a* ***section plane perpendicular to HP, parallel to VP*** *and 15 mm from the axis. Draw its top and sectional front views.*

As the cylinder has no edges, a number of lines representing the generators may be assumed on its curved surface by dividing the base circle into 12 equal parts. [However, base circle is divided into 8 equal parts in Cylinders / Cones to avoid congestion in the drawings.]

In this Problem, S.P. does not cut any of the cylinder generators. Hence, top view circle is not divided into the usual number of 12 equal parts to represent the generators.

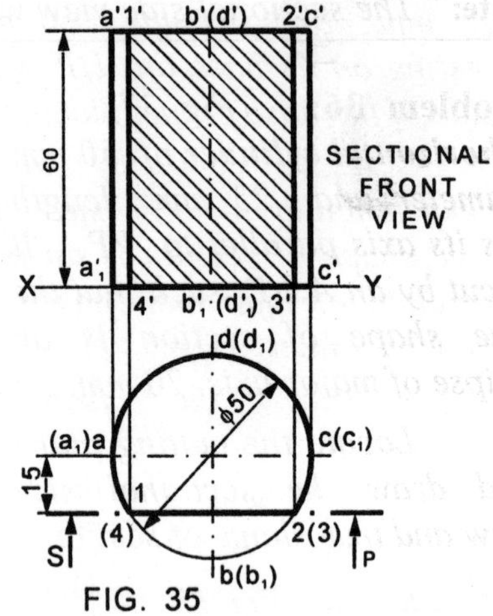

FIG. 35

Problem 35: *(Fig. 36) A cylinder of diameter 40 mm and height 60 mm is having its axis vertical. It is cut by a* ***plane perpendicular to VP*** *and* ***inclined at 30° to HP****. The plane bisects the axis of the cylinder. Draw its front view, sectional top view, sectional side view and true shape of section.* **(UQ)**

1. Draw the top and front views of the cylinder. In front view, draw a S.P. at 30° and bisecting the axis of the cylinder.
2. Name the cutting points in the front view. Project these cutting points to top view and obtain Sectional T.V. In this view, cutting points lie on the circle and hence, the same circle is the top view of the section.
3. ***True Shape of Section:*** Draw X_1Y_1 parallel to the S.P. Draw projectors perpendicular to X_1Y_1 from the cutting points in the front view. Complete the True Shape of Section. It is an Ellipse, the major axis of which is equal to the length of the S.P. and minor axis equal to the diameter of the cylinder.
4. ***Sectional Side View:*** Project the sectional (left) side view as shown by drawing horizontal lines from front view cutting points and 45° projectors from the top view cutting points.

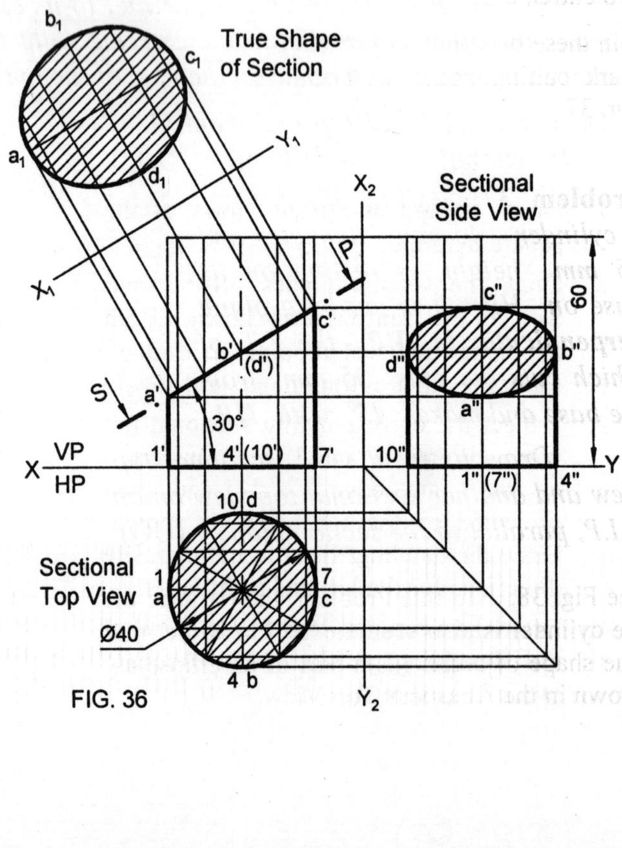

FIG. 36

Section is seen as an ellipse in side view. Major axis is equal to the diameter of the cylinder.

Note: *The sectional side view will be a circle, if the S.P. is inclined at 45° with HP.*

Problem 36: *(Fig. 37)*
A ***horizontal cylinder*** *of 50 mm diameter and 75 mm length has its axis parallel to VP. It is cut by an* ***A.I.P.*** *such that the true shape of section is an ellipse of major axis 70 mm.*

Locate the cutting plane and draw the sectional side view and true shape of section. (UQ)

Length of S.P. between two extreme generators of the cylinder should be equal to 70 mm.

With the midpoint of the axis as center and 35 mm (=70/2) as radius, draw two arcs to cut the two extreme generators as shown.

Join these two points to obtain S.P. Mark cutting points as shown in Fig. 37.

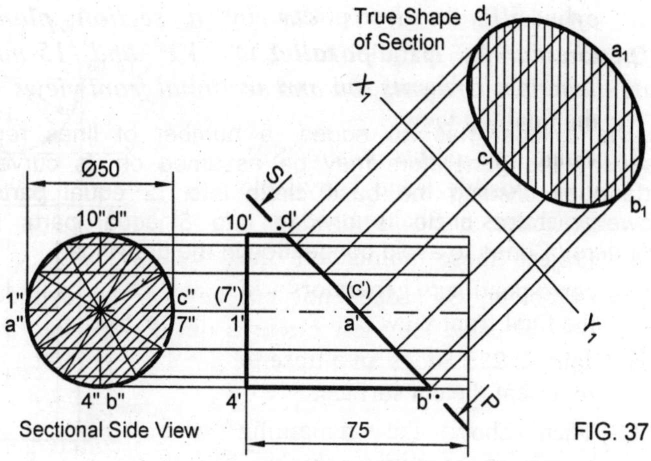

FIG. 37

Problem 37: *(Fig. 38)*
A cylinder 45 mm diameter and 55 mm height, is resting on its base on HP. It is cut by a ***plane perpendicular to VP****, the V.T. of which cuts the axis 35 mm from the base and makes* ***45° with HP****.*

Draw its front view, sectional top view and another sectional top view on an A.I.P. parallel to the section plane. (UQ)

See Fig. 38. In this Problem, the top face of the cylinder is also sectioned. Therefore, the true shape of section is a part of an ellipse as shown in the Auxiliary Top View.

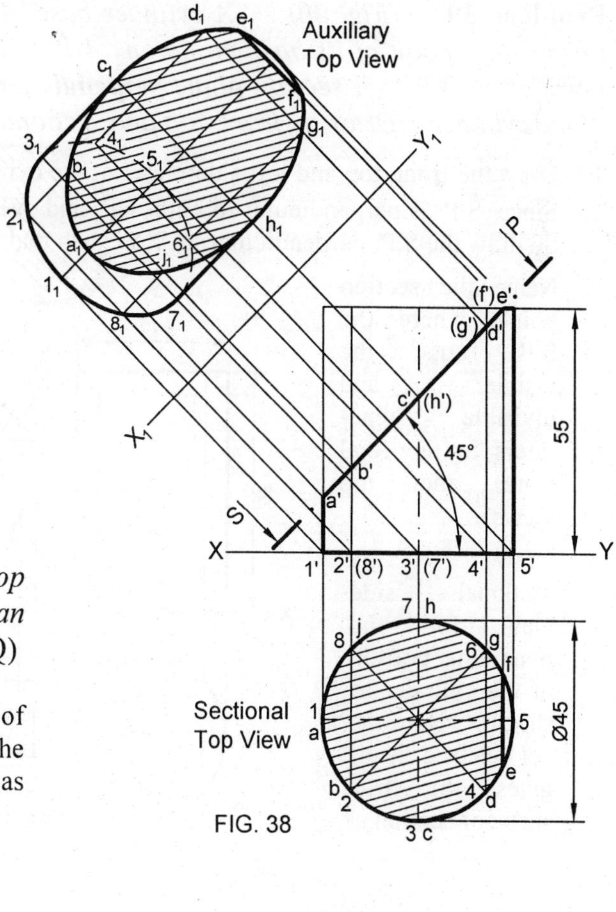

FIG. 38

Problem 38: *(Fig. 39) A cylinder of 50 mm diameter and axis 80 mm long, lies with one of its generators on HP such that its axis is inclined at 45° to VP.* ***A section plane parallel to VP*** *bisects axis of the cylinder. Draw the top and sectional front views.* (UQ)

1. Draw the projections of the cylinder for the given position.
2. Draw S.P. parallel to XY such that it bisects the axis in the final top view.
3. Name the points where S.P. cuts visible and invisible generators in final top view.
4. Project these points on the corresponding generators in the final front view.
5. Join 1′ 2′ 3′ 4′ as an ellipse to represent the cut surface.
6. Then show the remaining portion of the cylinder *thick*.

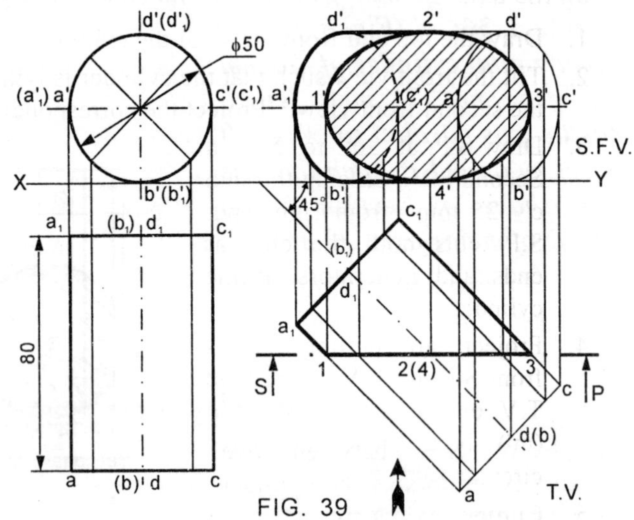

FIG. 39

Problem 39: *(Fig. 40) A cylinder base 40 mm diameter and axis 58 mm long, rests with a point of its base circle on HP. Its axis is inclined at 45° to HP and parallel to VP.* ***A section plane perpendicular to both HP and VP*** *bisects the axis of the cylinder. Draw its front, top and sectional side views.* (UQ)

1. Draw the front, top and left side views of the cylinder for the given position.
2. Since S.P. is perpendicular to both HP and VP, it appears as a straight line in both views. So draw the S.P. perpendicular to XY such that it bisects the axis in both the views.
3. Name the section points where the S.P. cuts the visible and invisible generators in the final front and top views.
4. Draw the sectional side view for the remaining portion of the cylinder as shown. Here the cut surface itself gives the true shape of section.

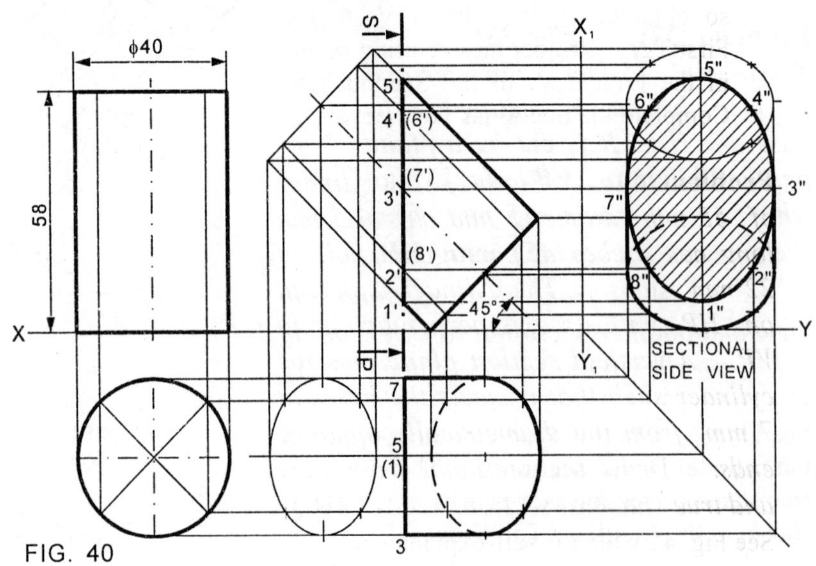

FIG. 40

Problem 40: *(Fig. 41) A cylinder of 45 mm diameter and 60 mm long rests with one of its generators on HP such that its axis is parallel to both HP and VP.* ***A section plane perpendicular to VP*** *and* ***inclined at 30° to HP*** *passes through a point on the axis 25 mm from one of its ends. Draw the sectional top view.* (UQ)

1. Draw the top and front views of the cylinder in simple position.
2. Tilt the front view such that the cylinder is lying with one of its generators on HP and obtain the second front view. Project the corresponding Second Top View.
3. Draw S.P. at 30° to XY in Second F.V., passing through c′ 25 mm from right end. S.P. cuts both the circular ends and generators of the cylinder.
4. Project all cutting points from Second F.V. to Second T.V. except a′, f′, (g′) and (m′) lying on left and right circular ends.
5. Cuttings points a′, f′, (g′) and (m′) cannot be projected directly because the two circular ends of the cylinder appear ⊥ʳ to XY in both Second top and front views. Mark these points in the top view as follows: Transfer the distances f′ 1′ and a′ p′ from the Second F.V. to First F.V. Project the cutting points so obtained in the First F.V. to the First T.V. Project these cutting points in the First T.V. to the Second T.V. Complete the Sectional Top View.

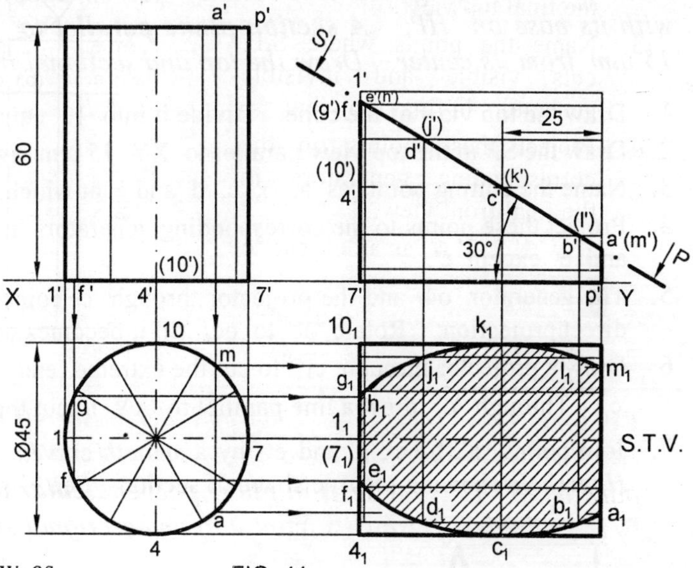

FIG. 41

Problem 41: *(Fig. 42) A cylinder of 55 mm diameter and axis 65 mm long has a coaxial pentagonal hole of 22 mm sides. The cylinder is placed on HP with its axis perpendicular to VP. A* ***vertical section plane*** *cuts the cylinder such that it meets the bases at 7 mm from the diametrically opposite ends. Draw the sectional front view and true shape of section.* (UQ)

See Fig. 42 which is self-explanatory.

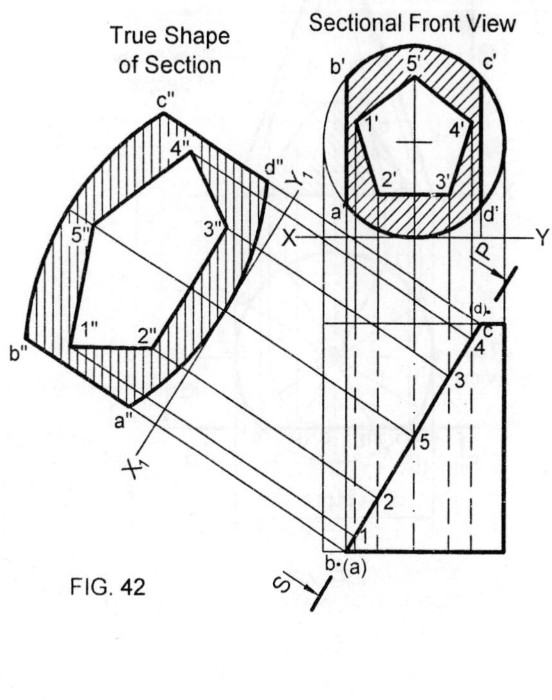

FIG. 42

6. SECTIONS OF CONES

Sectional views of cones are drawn by using any one of the *two* methods, viz., ***Circle Method*** and ***Generator Method.*** Circle Method is more time-consuming and is rather more difficult when the cone is in inclined position. Generator Method is more suitable, particularly when the cone is inclined. Therefore, Generator Method is followed in the following problems.

Problem 42: *(Fig. 43) A cone of 45 mm diameter and altitude 60 mm is resting with its base on HP.* ***A section plane parallel to VP*** *cuts the cone at a distance of 15 mm from its center. Draw the top and sectional front views.*

1. Draw the top view of the cone. Divide it into 12 equal parts. Project the front view.
2. Draw the SP in the top view parallel to XY 15 mm away from the center o.
3. Name the cutting points as a, b, c, d and e at which generators are cut in the top view.
4. Project these points to the corresponding generators in front view. Mark them as a′, b′, d′ and e′ except c′.
5. The generator o4 and the projector through c coincide. Hence, c′ cannot be located by direct projection. Rotate oc to oc_1 till it becomes parallel to XY.
6. Draw a projector through c_1 to cut the extreme (end) generator in the front view at the point c_1'. Through c_1' draw a line parallel to XY to cut the generator o′4′ at c′.
7. Join points a′, b′, c′, d′ and e′ by a *smooth* curve. The curve obtained is a ***Rectangular Hyperbola*** showing the true shape of section. See Fig. 43.

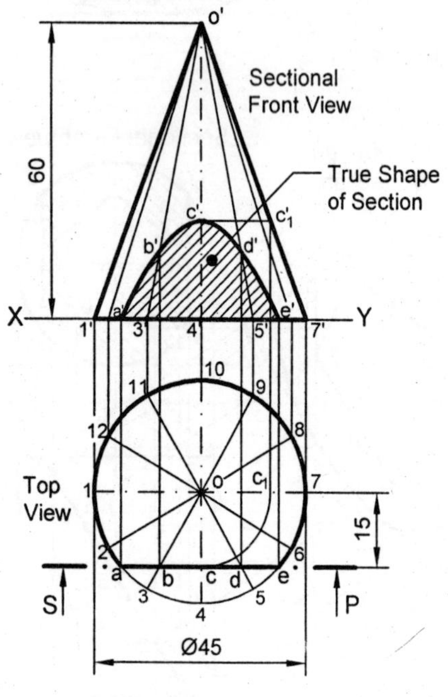

FIG. 43

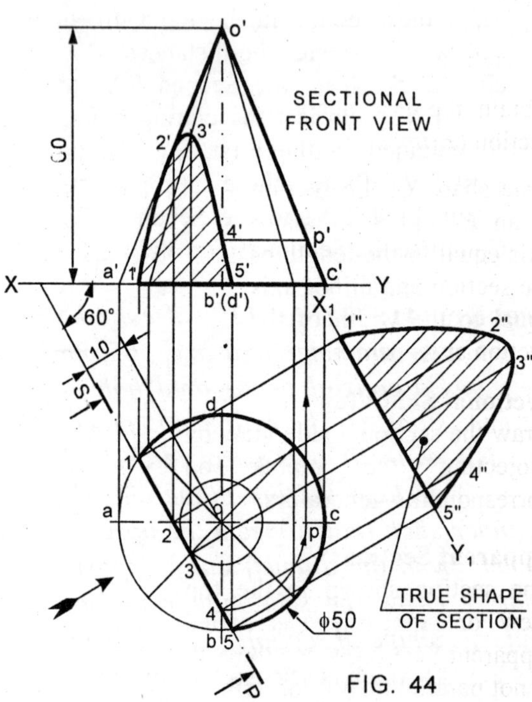

FIG. 44

Problem 43: *(Fig. 44) A* ***cone****, base 50 mm diameter and axis 60 mm long, rests with its base on HP. It is cut by a section plane perpendicular to HP, inclined at 60° to VP and at a distance of 10 mm from its axis. Draw the sectional front view and the true shape of section.* (UQ)

See Fig. 44. Section point 4 is on ob. How to locate 4′ in the sectional front view? For this, with o as center and o4 as radius draw an arc to cut oc at p. Project p and obtain p′ on o′c′. From p′ draw a horizontal to cut o′ b′ at 4′.

Problem 44: *(Fig. 45) A* ***cone****, base 50 mm diameter and axis 65 mm long, rests with its base on HP. It is cut by a section plane perpendicular to VP, inclined at 45° to HP and passing through a point on the axis 35 mm above the base. Draw the sectional top view, sectional side view and true shape of section.* (UQ)

In this Problem, S.P. is inclined to the base of the cone at an angle smaller than the angle of inclination of the generators with the base.

Note: *Points 3′ and (7′) cannot be projected directly. Therefore, the same method as in the case of pyramids should be used to find the positions of 3 and 7 in top view.*

To obtain the points 3 and 7, draw horizontal from 3′ (7′) to cut o′e′ at p′. Project p′ on oe. With o as center and op as radius draw a semi-circle to cut oc and og at 3 and 7 respectively.

True Shape of Section
Draw X_1Y_1 parallel to S.P. Draw projectors through 1′, 2′, 3′, etc.

Obtain the true shape of section (***Ellipse***).

True shape of the section is an ***Ellipse*** with major axis equal to the length of the section and minor axis equal to the width of the section at its center.

Sectional Side View
Draw the sectional side view by projecting the points on corresponding generators.

Apparent Section
The section shown in the top view or front view is called Apparent Section, since the S.P. is not parallel to VP or HP.

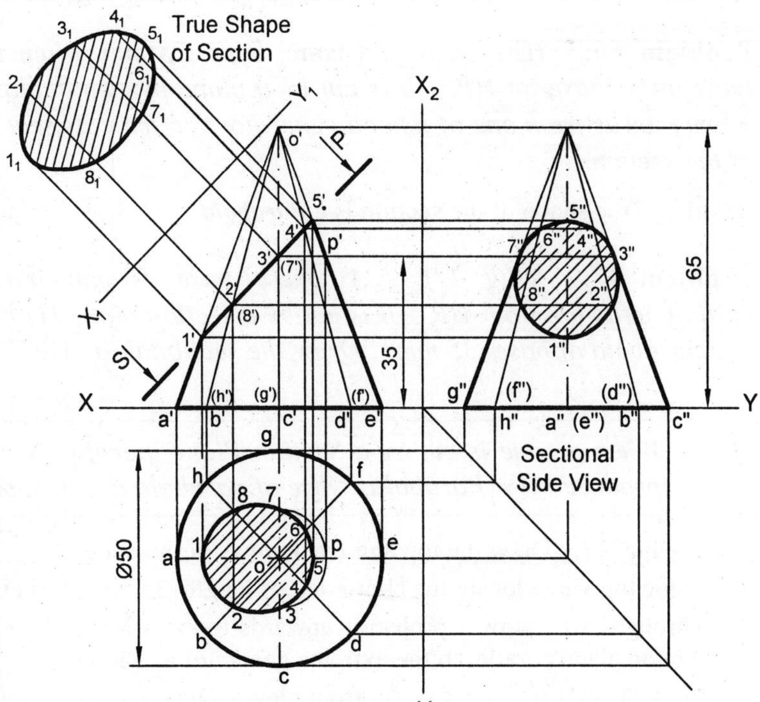

Sectional Top View (Apparent Section)

FIG. 45

Problem 45: *(Fig. 46) A cone, base 40 mm diameter and axis 60 mm long, rests with its base on HP. It is cut by a section plane perpendicular to VP, parallel to one of the end generators and passing through a point on the axis 25 mm from the apex. Draw sectional top view and true shape of the section.* (UQ)

Here, the S.P. is parallel to an end/extreme generator of the cone.

In this Problem, S.P. cuts not only the generators, but also the base of the cone at 1′(7′) in the front view. Hence project these points to the top view.

Obtain other section points in sectional top view as explained in previous Problem. Draw X_1Y_1 parallel to S.P. and obtain true shape of section (***Parabola***).

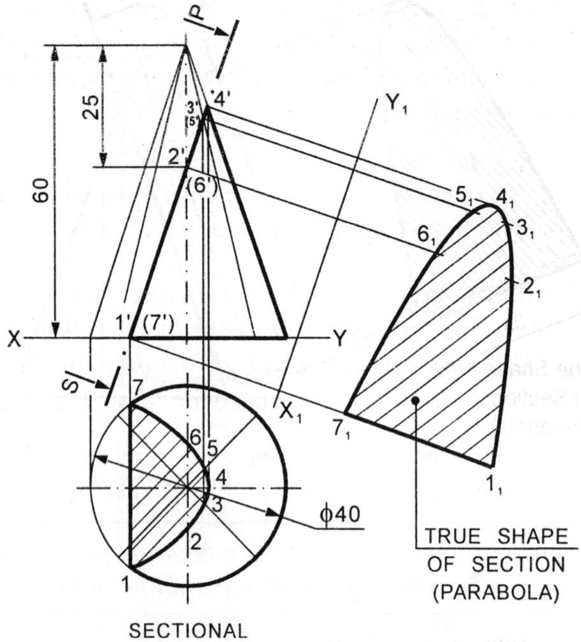

FIG. 46

Note: *For a section to be a Parabola, the S.P. must be parallel to the End Generator.*

Problem 46: *(Exercise) A* ***cone****, base 75 mm diameter and axis 80 mm long is lying on its base on HP. It is cut by a plane perpendicular to VP and parallel to and 12 mm away from one of its end generators. Draw its sectional top view and true shape of the section.* (UQ)

Result: True shape of the section is a ***Parabola***.

Problem 47: *(Fig. 47) A cone of base 50 mm diameter and 65 mm height is resting on its base on HP. It is cut by a section plane such that the true shape produced is a parabola of base 35 mm. Draw the sectional top view. Also, find its true shape.* (UQ)

Note: *When a cone is cut by a Section Plane, parallel to an end generator, the section produced is a Parabola. Size of parabola depends on the position of S.P.*

1. See Fig. 47. Draw the top and front views of the cone.
2. In the top view, locate the chord a j of length 35 mm perpendicular to XY.
3. Through a j, draw a projector upwards to meet XY at a′ (j′). Through a′ (j′), draw the section plane parallel to the extreme generator as shown.
4. Mark the cutting points in the front view. Draw the corresponding sectional top view.
5. Mark X_1Y_1 parallel to the S.P. and draw the true shape of section as a parabola.

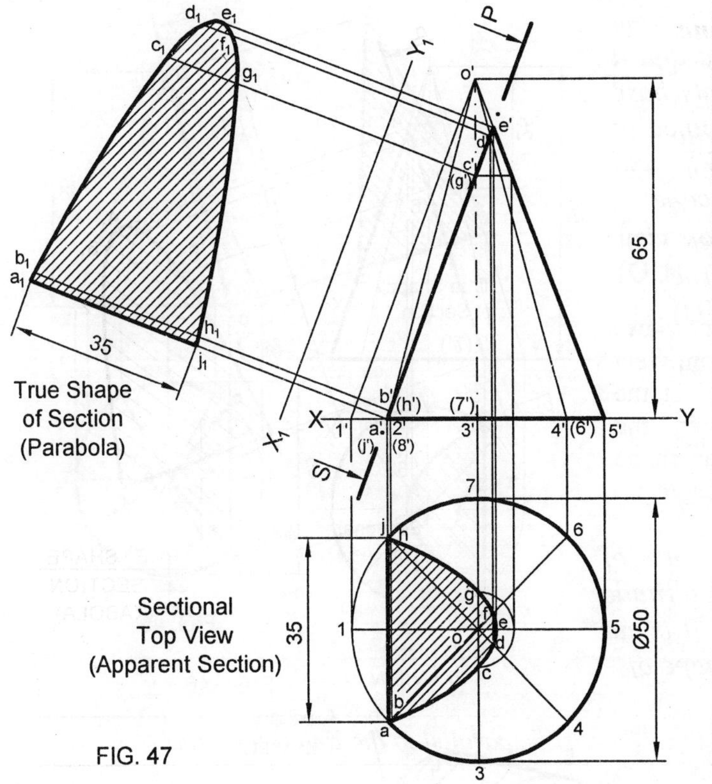

FIG. 47

Problem 48: *(Exercise)* *A* ***cone*** *of base 70 mm diameter, standing upright is cut by a section plane such that, the true shape is a parabola of maximum double ordinate 50 mm and the vertex of the parabola is 70 mm away from this ordinate.*

Draw the front view, top view and the true shape of section. (UQ)

Hint: First find the height of the cone.

For this, draw the base circle and mark the double ordinate 50 mm of the parabola equally about the horizontal center line.

Problem 49: *(Fig. 48)* *A* ***cone*** *of base 50 mm diameter and axis 65 mm long, lies on HP on one of its generators with its axis parallel to VP. A horizontal section plane bisects the axis of the cone.*

Draw the front view and sectional top view. (UQ)

1. Draw S.P. parallel to XY such that it bisects the axis in the final front view.
2. Name the section points at which the base and the generators are cut by the S.P. in correct sequence as shown.
3. Project the section points to the final top view and draw the cut surface.
4. Then complete the sectional top view.

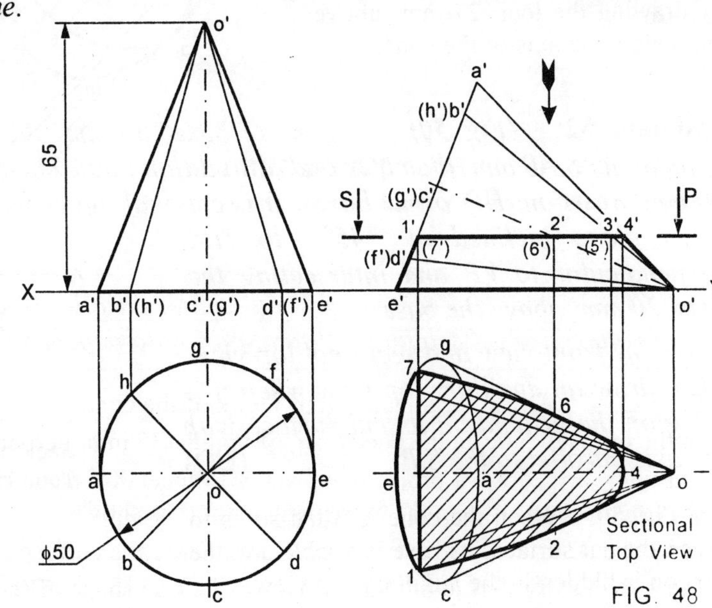

FIG. 48

Problem 50: *(Fig. 49) A cone with a 45 mm diameter of base and a 55 mm length of axis, rests on its base on HP. It is cut by a plane inclined at 75° to HP, perpendicular to VP and passing through the vertex of the cone. Draw the front view, sectional top view and true shape of section.* (UQ)

See Fig. 49. Draw the top and front views of the cone. Mark the S.P. in the front view and name the cutting points. Project the same to the top view and obtain the Sectional Top View. Draw the true shape of section, which is an *Isosceles Triangle*.

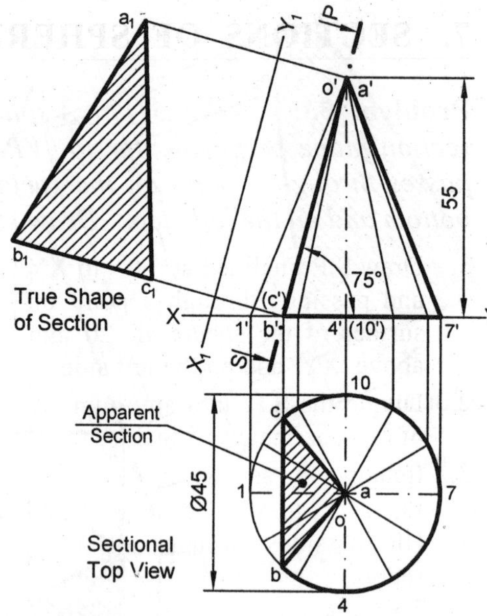

FIG. 49

Problem 51: *(Exercise) A cone of base diameter 70 mm and altitude 90 mm rests on its base on HP. It is cut by an A.I.P. such that the true shape of section is an isosceles triangle with a 54 mm base. Draw the S.P., sectional top view and true shape of section.*

Hint: In the top view, locate base of the triangle 54 mm in top view, by drawing the loci 27 mm above and below the axis of the cone.

Problem 52: *(Fig. 50)*
A cone, base 40 mm diameter and altitude 50 mm, rests on HP on its base. It is cut by a plane inclined at 45° to HP, perpendicular to VP and intersecting the axis 20 mm above the base.

Draw the front view and sectional top view. Also, draw an auxiliary top view when it is lying on the ground on its cut surface with the axis parallel to the S.P. (UQ)

See Fig. 50. When the cone is tilted so as to lie on the cut surface, its base is visible and the section is hidden in the auxiliary top view.

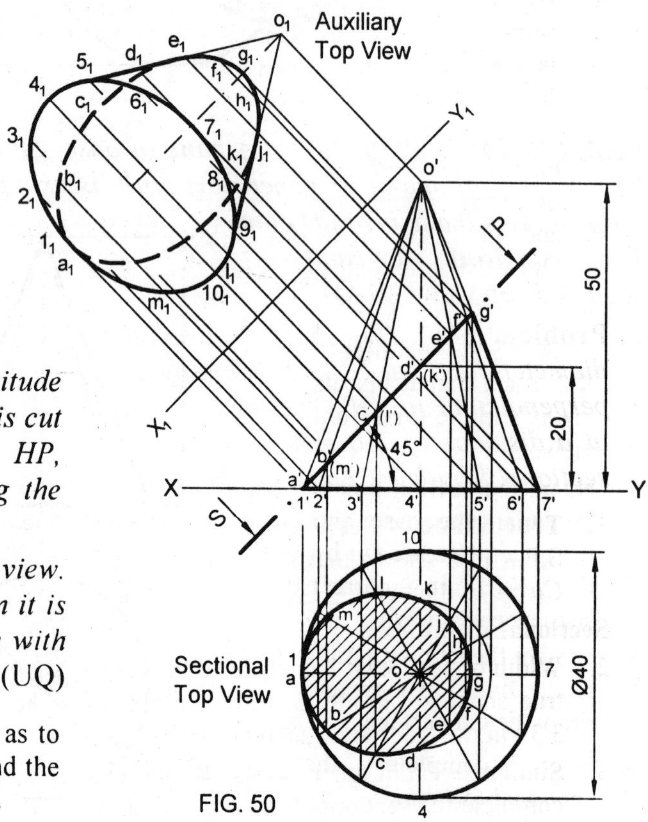

FIG. 50

7. SECTIONS OF SPHERES

Problem 53: *(Fig. 51) A* ***sphere*** *of 60 mm diameter rests on HP. It is cut by a section plane perpendicular to VP and inclined at 60° to HP. The section plane passes through a point on the surface of the sphere at a distance of 20 mm from its bottom and on the left side. Draw sectional top view and true shape of section.* (UQ)

1. Draw S.P. inclined at 60° to XY and passing through a point on surface of the sphere at 20 mm above XY and on the left side.
2. Divide the S.P. into any number of (approximately) equal parts.
3. Imagine a series of Section Planes ⊥^{r} to VP and parallel to HP passing through all the section points. Let AA be the S.P. passing through 6′(8′), cutting the sphere in a circle of radius aa.
4. Top view of the section is a circle of radius aa. Hence, o as center and aa as radius draw a circle in the top view.
5. From 6′ (8′) draw a projector to cut this circle at 6 & 8. Similarly obtain other points and complete Sectional T.V. Draw the True Shape (*Circle*).

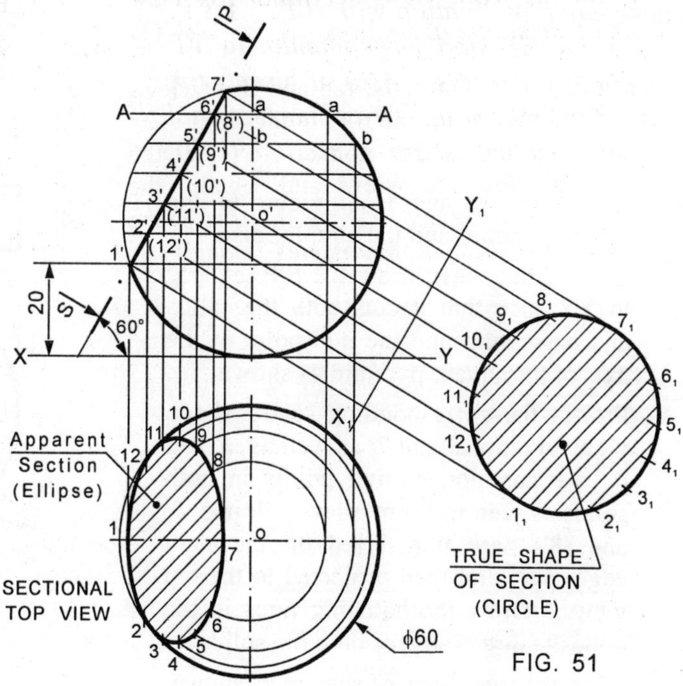

FIG. 51

Problem 54: *(Fig. 52) A* ***sphere*** *of 60 mm diameter rests on HP. It is cut by a section plane perpendicular to HP, inclined at 45° to VP and at a distance of 10 mm from its center. Draw sectional front view and true shape of section.* (UQ)

1. **True Shape of Section:** For the given position of S.P., it is known that the true shape of section is a *Circle* of diameter = 1-9. Hence draw the circle.

Sectional Front View

2. Width of section at 3 is equal to the chord 3″3″ (in true shape). From 3 draw a projector and mark 3′3′ in the front view such that 3′3′ = 3″3″.
3. Similarly obtain other points 1′, 5′5′, 7′7′ & 9′ and complete the sectional front view as shown.

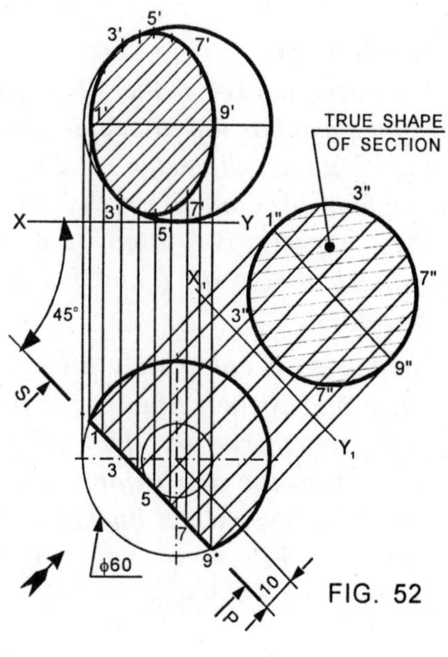

FIG. 52

8. SECTIONS OF COMBINATION OF SOLIDS

Problem 55: *(Fig. 53) A **square pyramid** of 45 mm side of base and 55 mm altitude rests centrally on a **circular disc** of 80 mm diameter and 20 mm thickness. Two of the base edges of the pyramid make equal inclination with VP. A S.P. at 45° to HP and perpendicular to VP cuts both the solids. Also, it bisects the axis of the pyramid. Draw the sectional top view and true shape of section.* (UQ)

1. Draw the top and front views of the square pyramid and the circular disc.
2. In the front view, draw the S.P. at 45° to XY such that it cuts both the solids and passes through the mid-point of the axis of the square pyramid as shown.
3. Project the corresponding sectional top view. Points **3** and **7** as well as **a** and **b** are common cutting points in both solids as seen in Front View. Points 2′ and (8′) are transferred on the slant edge p′o′ and then projected to the top view. Draw the hatching lines in the opposite directions for both the solids.
4. Draw the true shape of section as shown.

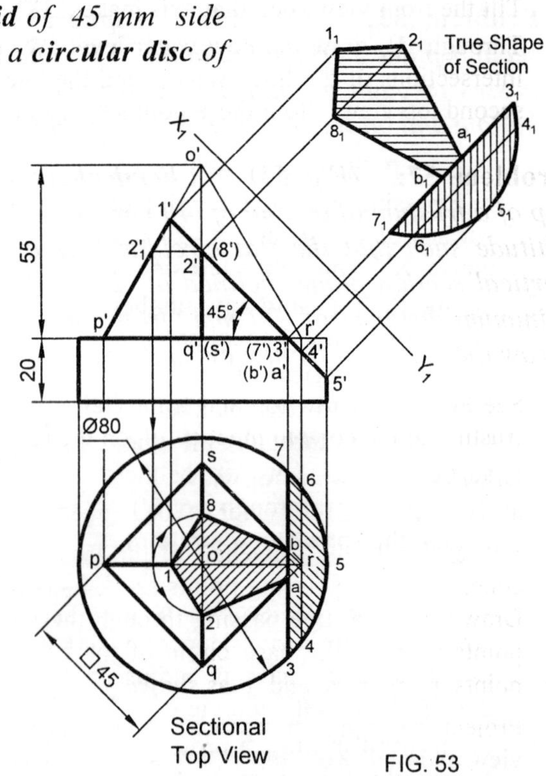

FIG. 53

Problem 56: *(Fig. 54)*
A conical bucket made up of thin metal sheet has top diameter 36 cm and bottom diameter 20 cm. Height of the conical part is 30 cm. The bucket is fitted with a cylindrical ring of 6 cm height at the bottom.

The bucket is completely filled with water and is then tilted on a point of its bottom rim on HP through 38°. Draw the projections showing the water level when bucket axis is parallel to VP. (UQ)

See Fig. 54 for Solution.

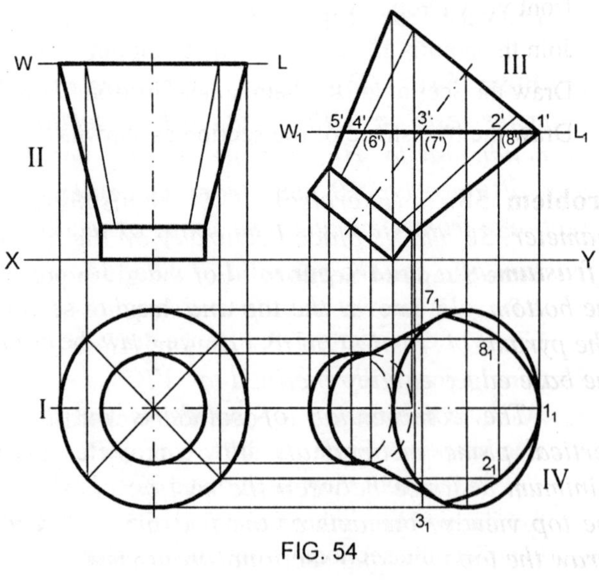

FIG. 54

1. Draw top and front views of the bucket. Divide outer circle in top view into 8/12 equal parts. Project these division points to the front view. W-L indicates the water level in the bucket.
2. Tilt the front view such that axis makes 38° with vertical. Project the corresponding top view.
3. Through 1′ draw the new water level W_1-L_1 parallel to XY. Mark 2′, 3′, etc., at the intersections of the line W_1-L_1 and the lateral lines. Project 1′, 2′, etc., to 1_1, 2_1, etc., in second top view. Join these points by a smooth curve to represent the water level as shown.

Problem 57: *(Fig. 55) A **hemisphere** of 50 mm diameter is placed centrally on the top of a frustum of a **cone** of 40 mm top diameter, 70 mm bottom diameter and 45 mm altitude such that the flat circular face of the hemisphere will be at the bottom. A vertical section plane inclined at 25° to VP passes such that the section plane is at a minimum distance of 20 mm from the axis of the cone and center of the hemisphere. Draw the sectional front view and true shape of section of the combination of solids.*(UQ)

1. See Fig. 55. Draw top and front views of the solids. Project the two end generators of the frustum of the cone to meet at o′ on the axis.
2. Draw the S.P. in the top view at 25° to XY and at 20 mm from the center of the hemisphere.
3. S.P. cuts only the front half of the frustum of the cone. Mark the cutting points as 1, 2, 3, 4 and 5. Draw the radial line passing through these cutting points to meet the base circle of the cone at the points p, q, r, s and t as shown.
4. Project the points p, q, r, s and t to the front view and mark p′, q′, r′, s′ and t′. Join these points with the apex o′.
5. Project the cutting points 1, 2, 3, ..., 8 to the front view to obtain 1′, 2′, 3′, ..., 8′.
6. Join the points 1′, 2′, ..., 5′ by a smooth curve.
7. Draw the section of the hemisphere in front view.
8. Draw the true shape of the section as shown.

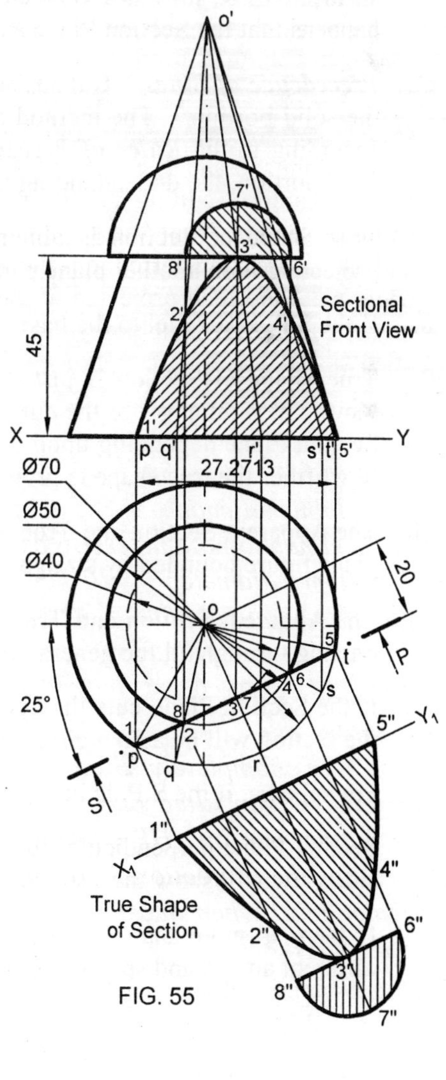

FIG. 55

Problem 58: *(Exercise) A **hemisphere** of diameter 30 mm is placed centrally on the top of a **frustum of a square pyramid** of side 36 mm at the bottom, 18 mm at the top and height 48 mm. The pyramid is resting on its base on HP with all the base edges equally inclined to VP.*

The combination of solids is cut by a vertical plane inclined at 40° to VP. The minimum distance between the section plane and the top view of the axis of the frustum is 6 mm. Draw the top view and sectional front view.

9. POINTS TO REMEMBER

1. Generally, Section Planes are chosen so as to be perpendicular to one of the reference planes and either perpendicular, parallel or inclined to the other plane. Projection of a Section Plane on the plane to which it is perpendicular is always a straight line. This line will be parallel, perpendicular or inclined to XY, depending upon the S.P. being parallel, perpendicular or inclined respectively to the other reference plane.
2. Section Planes are represented by their Traces and marked as S.P.
3. Sectioning of a solid is generally such that a smaller portion of the solid is cut and removed and the projections of bigger portion of the solid with the cut surface are drawn.
4. In the sectional views, the lines corresponding to the removed portion are drawn *thin*.
5. ***Cut Surface:*** When a solid is cut by a Section Plane and the cut portion removed, the cut surface will be seen as a visible edge in the view in which the Section Plane is shown. It so happens that the Section Plane and the cut surface coincide within the solid portion.
6. ***Precedence of Lines:*** Cut surface seen as visible edge should be drawn as a *full line* within the solid portion. The method of showing the S.P. even in the solid portion is not correct according to the ***Rules of Precedence of Lines***. If the Section Plane is shown within the solid portion, the understanding of the presence of a visible edge will be spoiled.
7. True Shape of Section is obtained by viewing the object normal to the cut surface and projecting it on another plane parallel to the Section Plane.
8. If the S.P. is parallel to the base of a solid, then the Apparent Section itself is in true shape.
9. True Shape of Section is just a mirror image of the cut surface showing its real shape as viewed at right angle to the cut surface. True shape may be positioned on either side of the Section Plane depending upon the convenience of space. X_1 and Y_1 are placed according to whether the true shape is projected from H.T. in the top view or V.T. in the front view.
10. The Apparent Section and True Shape of the Section of a prism or a pyramid is a polygon of plane figure bounded by straight line segments.
11. The Apparent Section and True Shape of the Section of a cylinder or cone, cut by a plane passing through all the generators, will be a plane figure bounded by a smooth curve.
12. If the Section Plane cuts the cylinder or cone and passes through the base, then that part of the section will be a *Straight Line*.
13. For spheres, if the S.P. is inclined to HP / VP, then the Apparent Section will be an *Ellipse*.
14. Section Plane perpendicular to both HP and VP will have both H.T. and V.T. and the side view will show the True Shape of Section.
15. If there is more than one Section Plane, then the sectional area should be hatched with different angles and spacing to differentiate.

10. ADDITIONAL PROBLEMS

59. (Fig. 56) A **cube** of side 40 mm is resting on the ground with one of its faces. Two of its vertical faces are equally inclined to VP. It is cut by a plane perpendicular to VP and inclined at 30° to HP and passing through a point on the axis at a distance of 30 mm from the base.

Draw the front view, sectional top view and true shape of section. (UQ)

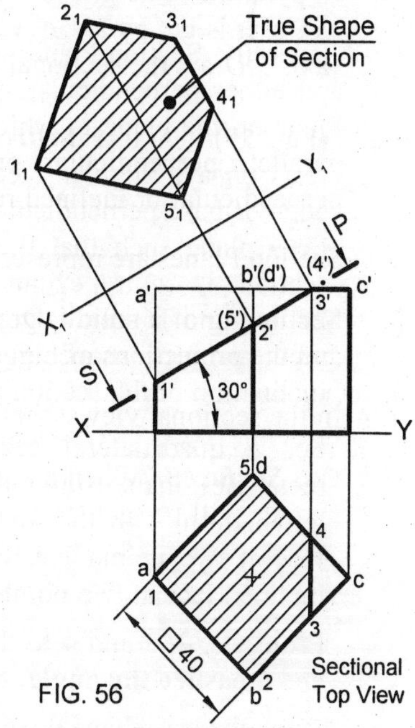

FIG. 56

60. A **square prism**, having base of side 30 mm, is cut by a section plane such that the true shape is a hexagon having two opposite sides 25 mm long. The remaining four sides are 40 mm long. Draw the top view, front view and the true shape of section. Also, determine the height of the prism. (UQ)

Draw the top and front views of the prism for any height, keeping the side faces equally inclined to VP. Measure the diagonal of the square in the top view. Using this diagonal, draw the true shape diagram as a hexagon of two sides = 25 mm and the other four sides = 40 mm. Complete the apparent shape of section on the top view by marking 25 mm sides. Height of the prism = **76 mm**.

61. A **pentagonal prism** of side of base 33 mm and height 77 mm has a coaxial hole of 33 mm diameter and is placed with its axis vertical. It is cut by a plane inclined at 40° to HP and bisecting the axis. Draw the true shape of section. Name the shape of the hollow portion in true shape. (UQ) [***Result:*** True shape of hollow portion is ***Ellipse***.]

62. A **hexagonal pyramid** of base 30 mm and axis 70 mm long rests on its slant edge of the face on HP. A cutting plane perpendicular to VP and inclined to HP passes through the highest corner of the base and intersecting the axis at a distance of 25 mm from the base. Draw the projections of the solid and find the inclination of section plane with HP. (UQ)

***Hint*:** Draw the top and front views with one of the sides of the base parallel to XY. The inclination of the section plane with **HP** = **75°**.

63. A **square pyramid** of side of base 40 mm and altitude 65 mm is lying on HP with a base edge parallel to VP. It is cut by a plane perpendicular to VP, inclined to HP such that the true shape is a trapezium of sides 36 mm, 16 mm and height 40 mm. Draw the projections. Also, determine the angle of the section plane with HP. (UQ)

Result: Angle of inclination of the Section Plane with HP = **50.5°**.

64. A **pentagonal pyramid** of side of base 30 mm and height 70 mm rests on HP on its base with a side of base parallel to VP. It is cut by an A.I.P. inclined at 30° to HP. It is also cut by a horizontal cutting plane. Both the cutting planes pass through a point on the axis 40 mm from the vertex. Draw the sectional top view and true shape of section.

65. A **frustum of hexagonal pyramid** of side of base 35 mm and height 50 mm, is cut from a pyramid of height 85 mm. It is standing upright with base on HP and axis parallel to VP. It is cut by an A.V.P. making 55° with VP and remaining 20 mm away from the axis. Draw the sectional front view, sectional side view and true shape of section. (UQ)

66. (Fig. 57) A **tetrahedron** of edge 55 mm rests on a triangular face on HP with an edge of base perpendicular to VP. It is cut by plane such that the true shape of the section is a trapezium of parallel sides 28 mm and 16 mm. Draw the sectional top view and true shape of section. Find the inclination of the section plane. (UQ)

For a quadrilateral section (trapezium / square / rectangle), the section plane must cut four edges of the tetrahedron. In the top view, the S.P. cuts the base face at two points 1 & 2, 16 mm apart and the other inclined face at two points 3 & 4, 28 mm apart.

Project 1, 2, 3 and 4 to the front view. Mark the S.P. and **measure the angle 46°**.

Draw the true shape of section as shown.

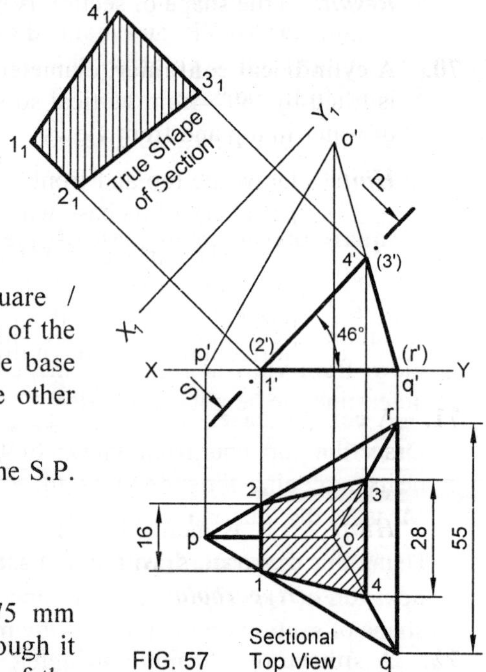

FIG. 57

67. A **cylinder**, base 50 mm diameter and axis 75 mm long, has a square hole of 25 mm side cut through it so that the axis of the hole coincides with that of the cylinder. The cylinder is resting on HP with the axis perpendicular to VP and the faces of the hole equally inclined to HP. A vertical section plane inclined at 60° to VP cuts the cylinder in two equal halves. Draw the front view of the cylinder on an A.V.P. parallel to the section plane. (UQ)

Hint: Draw the auxiliary front view of the cylinder as a whole. Project the points at which the generators of the cylinder and the edges of the hole are cut. The section of the cylinder will be a part of an ellipse. Join the points at which the edges of the hole are cut. The rear edges of the hole will be visible within the section.

68. A **cylinder**, 52 mm diameter and 63 mm long, has its axis parallel to both the HP and the VP. It is cut by a vertical section plane inclined at 30° to the VP, so that the axis is cut at a point 24 mm from one of its ends and both the bases of the cylinder are partly cut. Draw its sectional front view and true shape of the section. (UQ)

Hint: Draw the top and front views of the cylinder. Draw a line SP to represent the S.P. Project the points at which the bases of the cylinder and the lines are cut.

Note that the points on the bases cannot be projected directly. Hence, project them to the first top view, then to the first front view and finally transfer them to second front view. Draw the true shape of section on a new XY or symmetrically around the center line.

69. A **cylinder** 33 mm diameter and 88 mm long rests on HP with its axis parallel to HP and inclined at 30° to VP. It is cut by a plane inclined at 40° to HP, perpendicular to VP and passing through the mid-point of the axis of the cylinder. Obtain the dimensions of the true shape of section. (UQ)

Result: True shape of section is Ellipse of major axis **58.5 mm** and minor axis **33 mm**.

70. A **cylindrical container**, diameter of base 45 mm and height 60 mm, is full of water. It is tilted by 60° from vertical so that some water from it is drained out. Draw the surface of water in top and front views. (UQ)

Hint: Draw the top and front views of the cylinder, when it rests on HP on the point of periphery of its base with axis inclined at 60° with vertical and parallel to VP.

Since the surface of water always remains parallel to HP and as water is allowed to drain from top of the cylinder, in the given position water level line will be parallel to XY and passing through the lowest point of the top face. Draw a horizontal line through this lowest point in front view.

71. A **cone** of base 55 mm diameter and axis 65 mm long, rests with its base on HP. A section plane perpendicular to both HP and VP cuts the cone at a distance of 8 mm from its axis. Draw its top view, front view and sectional side view. (UQ)

Hint: S.P. is inclined to the base at an angle greater than the angle of inclination of the generators with the base. Side view shows the true shape of section, *viz.*, ***Hyperbola.***

72. A **sphere** of 70 mm diameter is cut by a section plane perpendicular to the VP and inclined at 55° to HP in such a way that the true shape of the section is a circle of 45 mm diameter. Draw its front view and sectional top view. (UQ)

Hint: Draw a circle of diameter 70 mm below XY. Project the front view. True shape of section is a circle of diameter 45 mm. So, draw a chord of 45 mm length in the front view. The perpendicular drawn to this chord gives inner circle radius. Fix the drafter at 55°. Draw a line tangential to the inner circle to represent S.P. in front view. Draw the Apparent Section of sphere.

73. A 60 mm diameter **sphere** is cut by a section plane inclined at 45° to HP, perpendicular to VP and 10 mm from its center. Draw sectional top view and true shape of section, after removing smaller part of sphere. (UQ)

Hint: True Shape of Section of a sphere is always a *circle* whatever be the position of the S.P. Therefore, draw the true shape of the section and then project back to obtain the Apparent Shape of Section.

74. A **cone** of 40 mm base diameter and 50 mm altitude is placed coaxially on a **circular disc** of 75 mm diameter and 25 mm thickness. An Auxiliary Inclined Plane bisects the cone axis and passes tangential to the base circle of the cone. Draw the projections. Obtain the true shape of section of the solids. (UQ)

11. SECTIONAL VIEWS OF ENGINEERING COMPONENTS

Sectional Views of engineering components are drawn by taking section along their axes.

11.1 TYPES OF SECTIONS AND SELECTION OF SECTIONAL VIEWS

1. **Full Section:** When a Section Plane passes through an object so as to cut it completely, the sectional view is called *Full Sectional View* or simply *Sectional View*. It provides all inner details of the object, better than the un-sectioned view with dashed lines for inner details.

   Note: Hidden lines should be omitted in sectional views to improve clarity of drawing.

2. **Half Section:** For a symmetrical object, the drawing of its full sectional view involves repetition of hatching on both sides of its axis. Therefore, only one half of the object may be shown in section, while its other half need not be shown as sectioned. Such a view is called *Half-Sectional View*. It is used to indicate both interior and exterior details in the same view.

   To obtain Half-Sectional View, the object is cut by two *imaginary* section planes perpendicular to each other. ¼ th of the object is *imagined* to be removed. See Fig. 65(i) and Fig. 66(ii). Orthographic view of the remaining ¾ th portion of the object is called Half-Sectional View.

   Note: In Half-sectional view, axis is preferred than the object line. Because, the removed quarter portion of the object is only *imagined* to be removed and not actually cut out.

3. **Rib in Section:** Ribs, webs, shafts and the like have no internal details. Therefore, they are not cut in the longitudinal sections and should not be hatched. Only visible outline is shown. Another practice is to show the rib by means of dashed lines and to draw section lines in it, omitting alternate lines of hatching.

4. **Rotated or Revolved Section:** (Fig. 58) The cross-sectional view, if rotated through 90° is known as *Rotated* or *Revolved Section*. A revolved section is obtained by passing a S.P. through the part, perpendicular to the axis.

   The true shape of external surfaces of certain objects such as spokes of wheels, ribs, arms, webs, etc., with negligible thickness may be shown without an additional view by means of revolved sections.

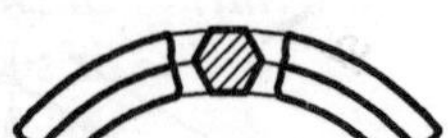

FIG. 58

REVOLVED SECTION

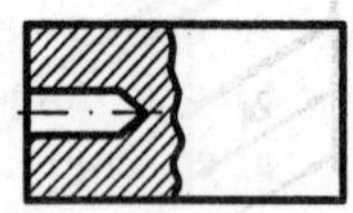

FIG. 59

BROKEN-OUT SECTION

5. **Partial or Broken-out Section:** (Fig. 59) A small portion of a part is broken to show the individual interior details of the part. Irregular wavy freehand line is used to represent the partial section.

   Inner details of certain local portions such as drilled hole, key on a shaft, etc., which cannot be sectioned, may be shown as broken-out section on the object. Full or half section of the complete object need not be shown.

6. **Removed Section:** (Fig. 60) Removed section shows the construction of certain parts of machine components outside the view. It is similar to the revolved section except that the

sectional view is removed from the actual view and drawn separately either, near to and connected with the view or in a different position.

Removed sections are preferred to revolved sections in cases where revolved or superimposed cross section is not desirable.

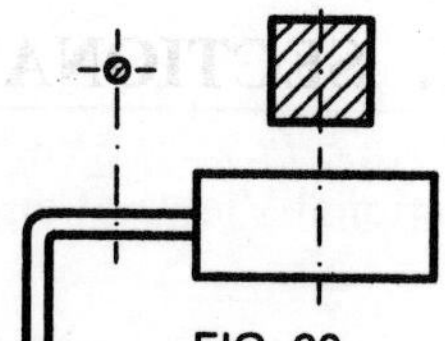

FIG. 60
REMOVED SECTION

7. **Offset Section**

The section planes are assumed to pass through the axis of the symmetry of the object.

When features like hole, recess, slot, cut-out, etc., of the machined parts do not lie on the axis of symmetry of the object, the S.P. is offset to include the axes of the different features.

See Fig. 61. The cutting planes are aligned into a single plane.

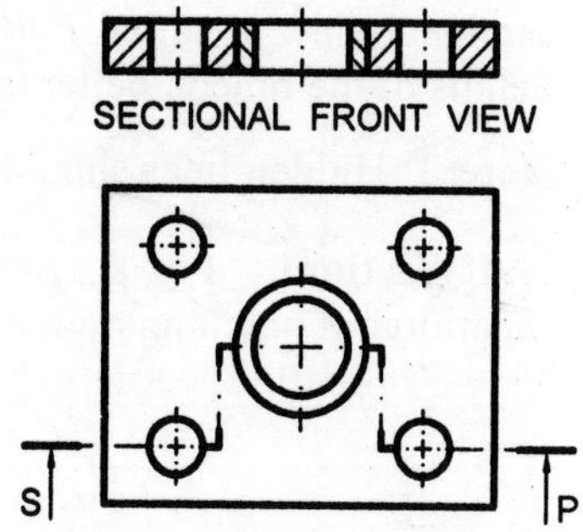

FIG. 61 OFFSET SECTION

11.2 SECTIONAL VIEWS OF ENGINEERING COMPONENTS

Problem 75: *(Fig. 62) Fig. 62(i) shows the pictorial view of STOP BLOCK. Draw the Sectional Front view (Central Cutting Plane), Top View and Right Side View.* (G.U.)

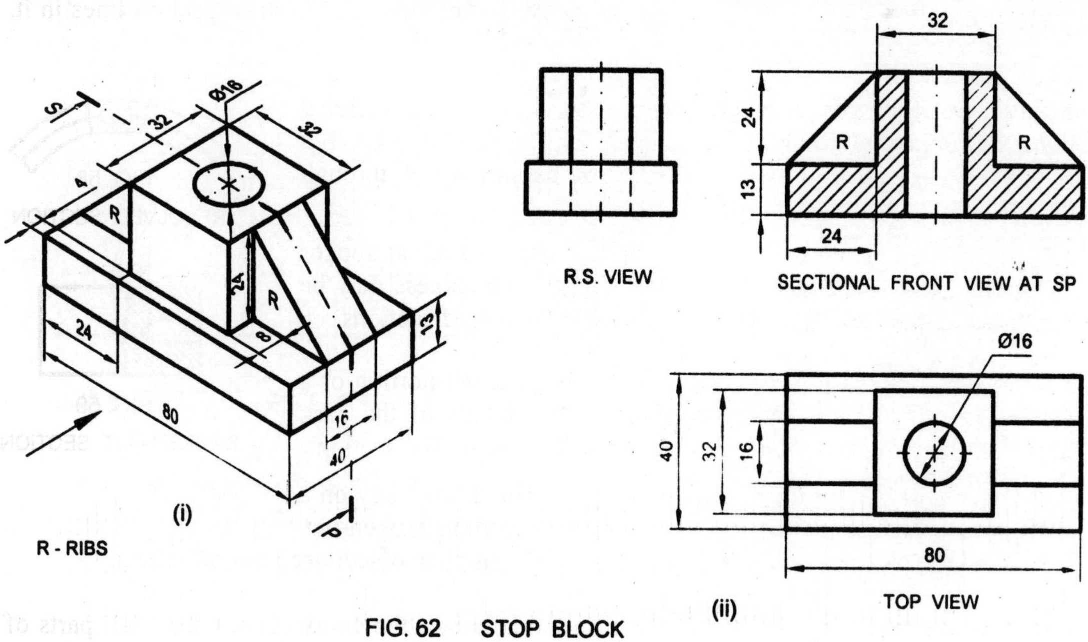

FIG. 62 STOP BLOCK

Problem 76: *(Fig. 63) Fig. 63(i) shows the pictorial view of OPEN BEARING. Draw the Sectional Front View, Sectional Left Side View and Top View.* (K.U.)

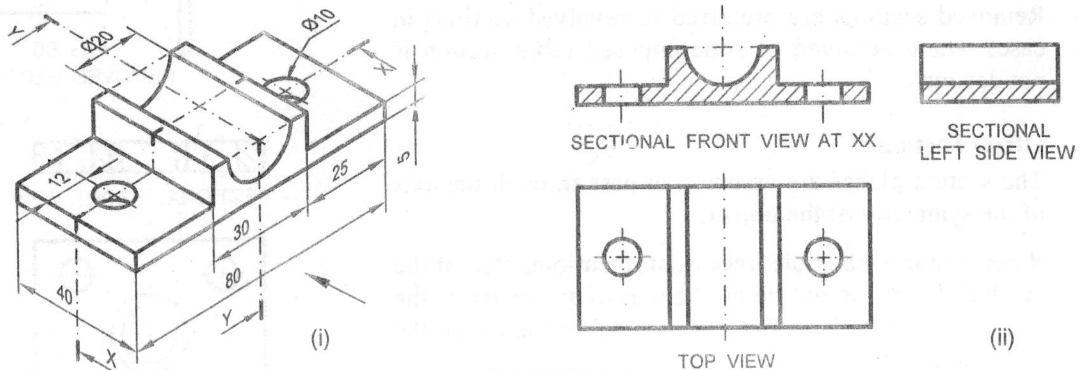

FIG. 63 OPEN BEARING

Problem 77: *(Fig. 64) Fig. 64(i) shows the pictorial view of CYLINDRICAL SUPPORT. Draw the Sectional Front View, Top View and Sectional Left Side View.*

Imagine ½ object after cutting the full object by a Section Plane CC perpendicular to the direction of sight, passing through the plane of symmetry as shown in Fig. (ii). Similarly, draw Fig. (iii).

Draw full sectional front view, full sectional left side view and top view as shown in Fig. (iv).

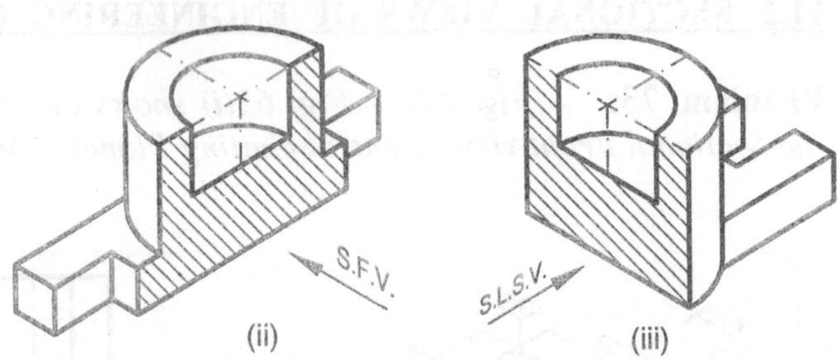

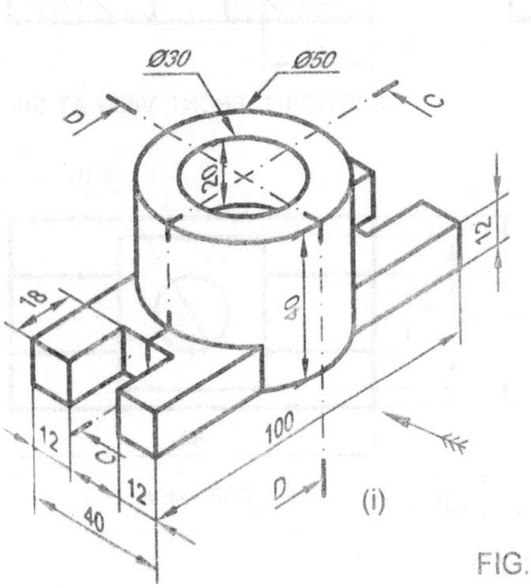

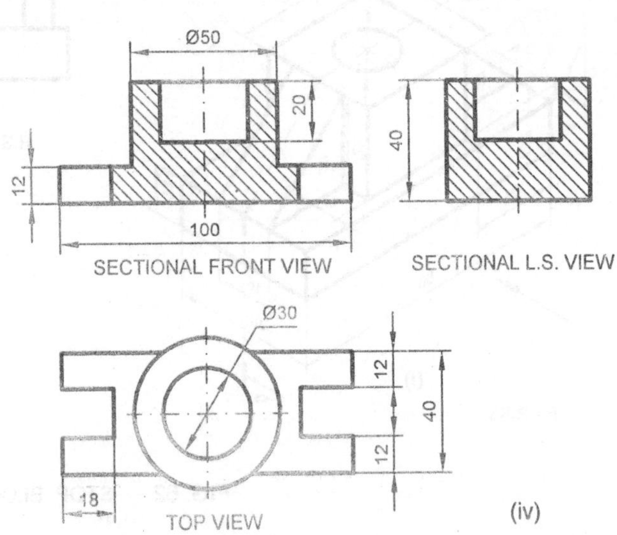

FIG. 64

Problem 78: *(Fig. 65) Fig. 64(i) shows the pictorial view of CYLINDRICAL SUPPORT. Draw the Half-Sectional Front View and Half-Sectional Left Side View.*

To draw half-sectional front and left side views, cut the full object by two mutually perpendicular Cutting Planes CC and DD passing through the planes of symmetry. After cutting, ¾ th object will be seen as in Fig. 65(i). Draw half-sectional F.V. and half-sectional L.S.V. as in Fig. 65(ii).

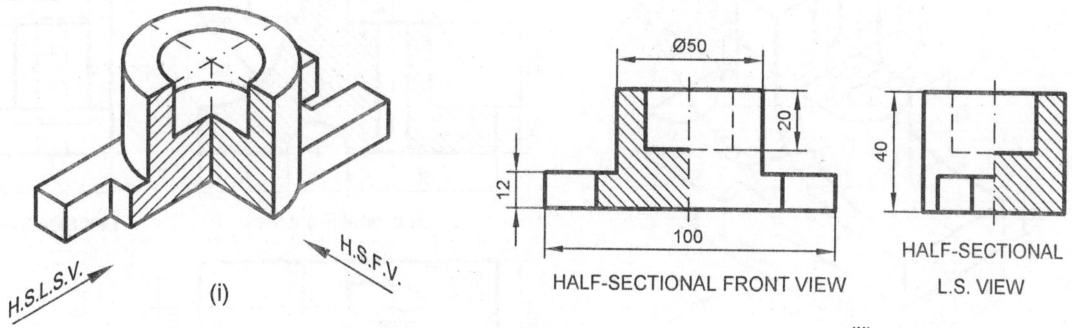

FIG. 65

Problem 79: *(Fig. 66) Fig. 66(i) shows the pictorial view of a machine component. Draw the Half Sectional Front View, Half Sectional Left Side View and Top View.* (G.U.)

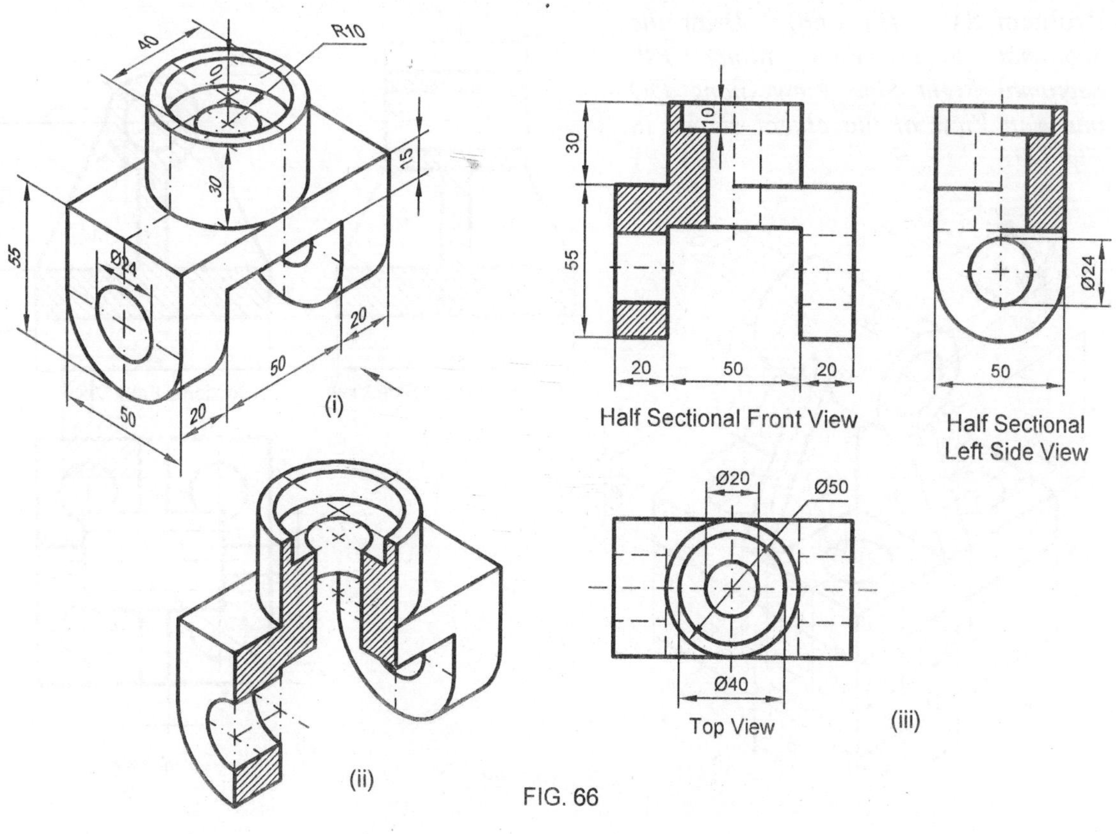

FIG. 66

Problem 80: *(Fig. 67) Draw the Sectional Front View along AA, Left Side View and Top View of the object shown in Fig. 67(i).* (K.U.)

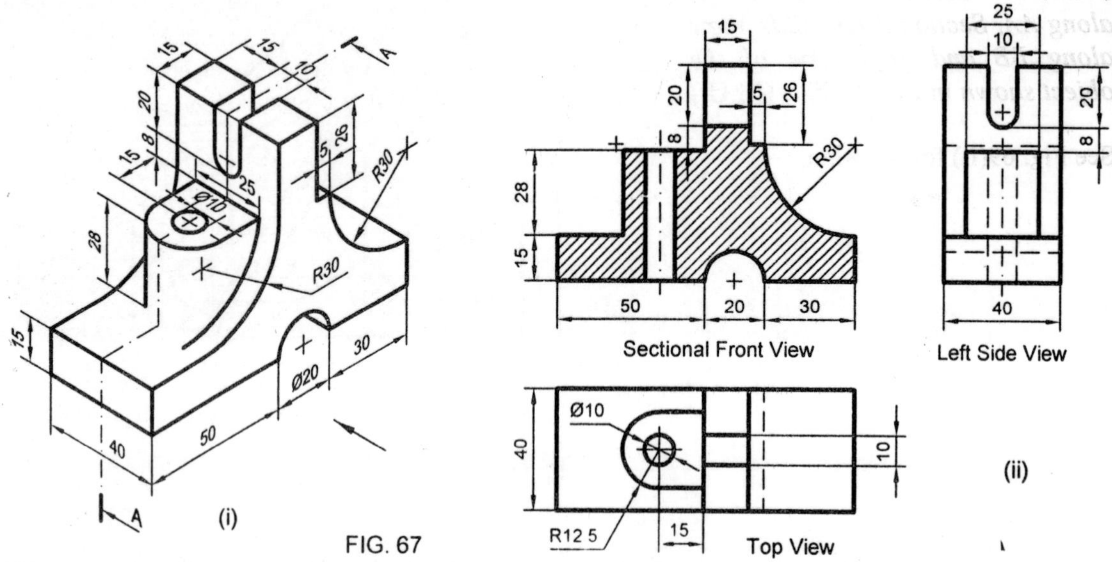

FIG. 67

Problem 81: *(Fig. 68) Draw the Sectional Front View along PP, Sectional Right Side View along QQ and Top View of the object shown in Fig. 68(i).* (G.U.)

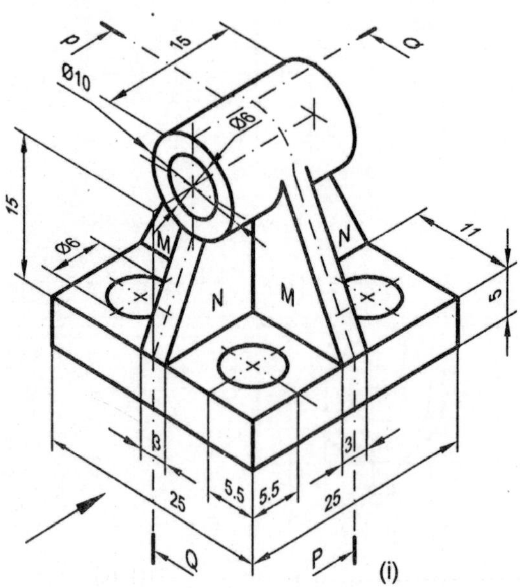

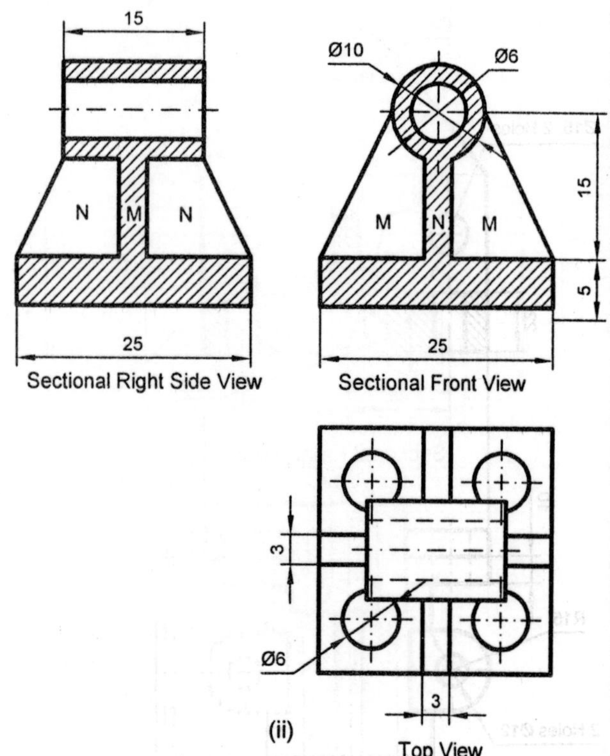

FIG. 68

Problem 82: *(Fig. 69)*
Draw the Sectional Front View along AA, Sectional Left Side View along BB and Top View of the object shown in Fig. 69(i). (M.U.)

See Fig. 69(ii) for Solution.

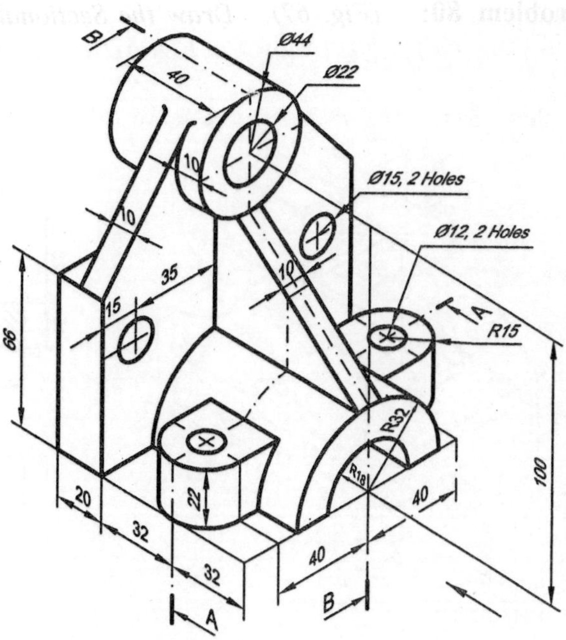

(i)

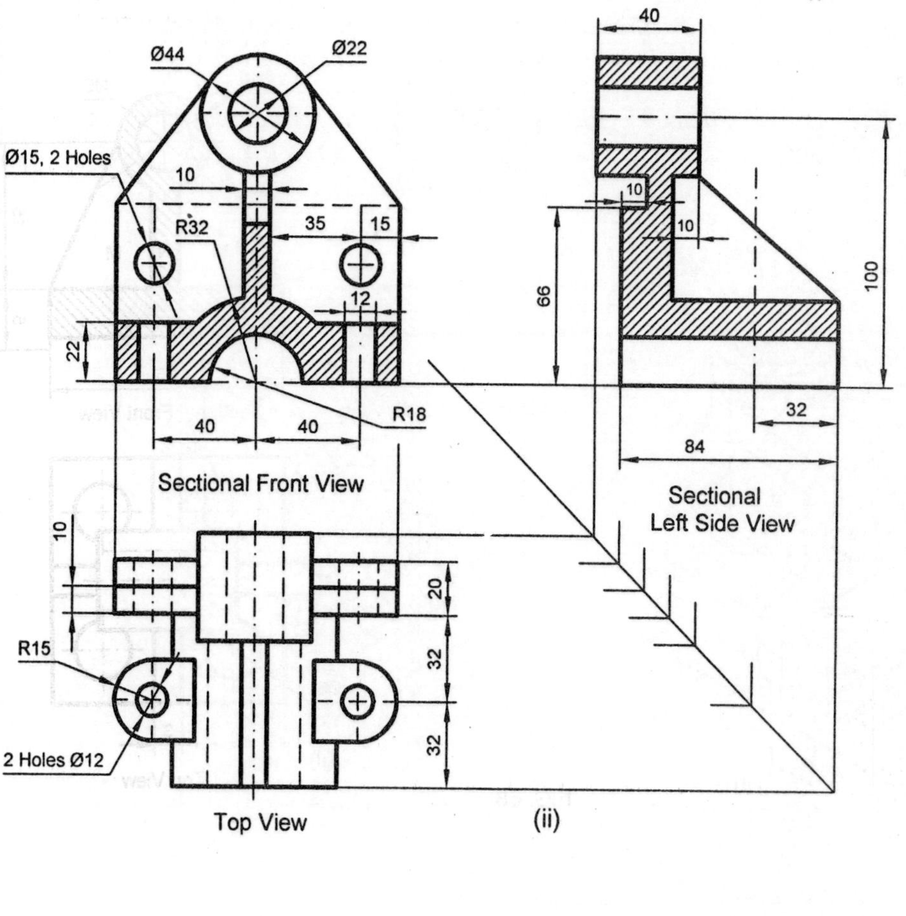

(ii)

FIG. 69

Problem 83: *(Exercise)* *Draw the following views of the TAPER BLOCK shown in Fig. 70: i) Sectional Front view ii) Sectional Left Side View and iii) Top View*

Problem 84: *(Exercise)* *Draw the following views of the V-BLOCK shown in Fig. 71: i) Half-Sectional Front view ii) Half-Sectional Left Side View and iii) Top View.*

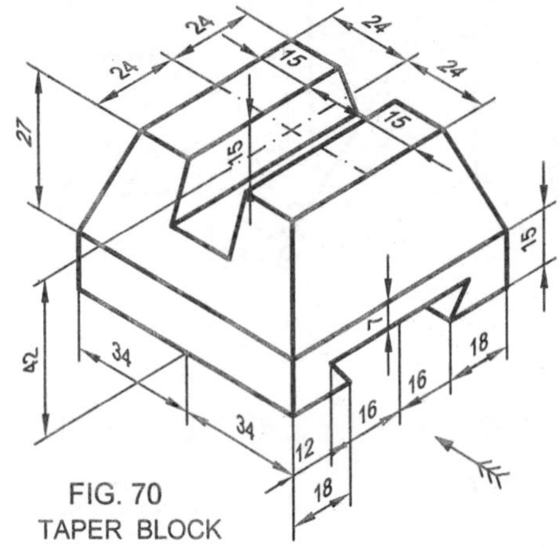

FIG. 70
TAPER BLOCK

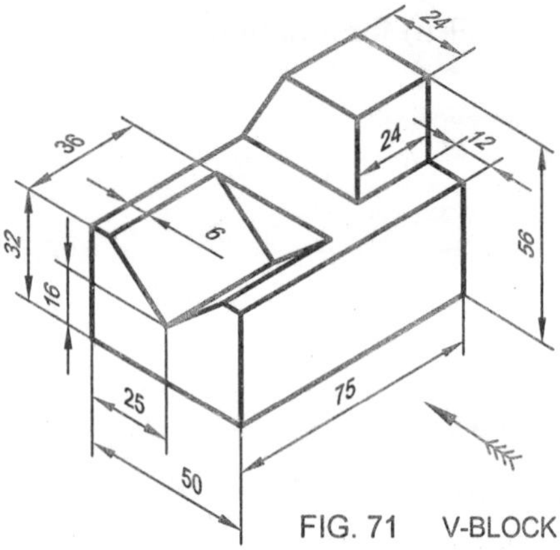

FIG. 71 V-BLOCK

Problem 85: *(Exercise)* *Draw the following views of the object shown in Fig. 72: i) Sectional Front view ii) Sectional Left Side View and iii) Top View.*

Problem 86: *(Exercise)* *Draw the following views of the object shown in Fig. 73: i) Half-Sectional Front view ii) Half-Sectional Left Side View and iii) Top View.*

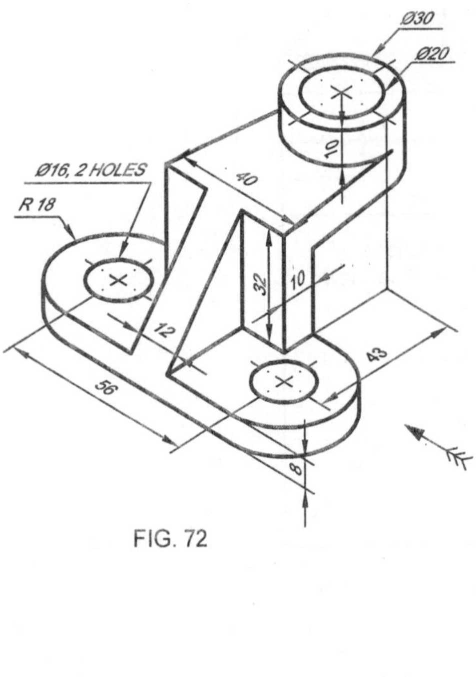

FIG. 72

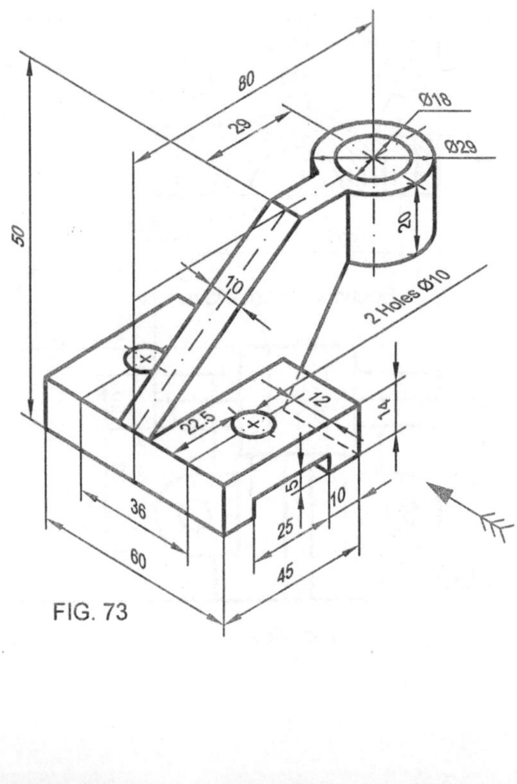

FIG. 73

*

16

DEVELOPMENT OF SURFACES

In industrial world, an engineer is frequently confronted with problems where the development of surfaces of an object has to be made in the fabrication of furnaces, chimneys, boilers, automobile body building, ship building, aircraft, etc. Development of surface of an object means the unrolling or unfolding of all surfaces of the object on a plane. A typical example is your tooth paste carton of the shape of a square / rectangular prism. Open its two bases and cut the carton along a longer edge. Place it on a planė surface to obtain the development of the carton.

Layout of only the sides of the object without its top or bottom is called *Lateral Surface.*

Principle of Development: Every line on the development should show the *true length* of the corresponding line on the surface which is developed. Surface generated by moving a straight line is called *Plane Surface* and that generated by a curve is called *Curved Surface.*

Solids bounded by Plane Surfaces : Prism, Pyramid, Tetrahedron
Solids bounded by Single-curved Surfaces : Cylinder, Cone
Solids bounded by Double-curved Surfaces : Sphere, Torus, Paraboloid

Sheet Metal Work: To fabricate a sheet metal product, first cut the metal sheet to the shape formed by laying all the lateral surfaces one adjacent to other, bend, fold and then join by welding or riveting. It is called *Sheet Metal Work.*

1. METHODS OF DEVELOPMENT OF LATERAL SURFACES

(i) Parallel Line Development
See Figs. 1(a) and (b). This method is used to develop *cubes, prisms* and *cylinders,* in which all the edges / generators of lateral surfaces are *parallel* to each other. Prism and cylinder roll out into rectangular stretch-outs by this method.

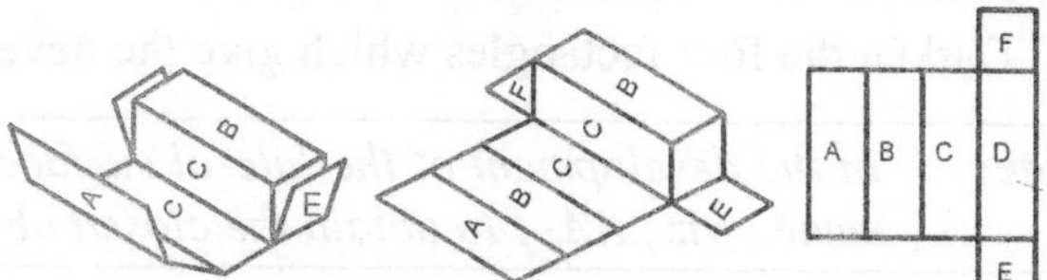

FIG. 1(a) PRISM

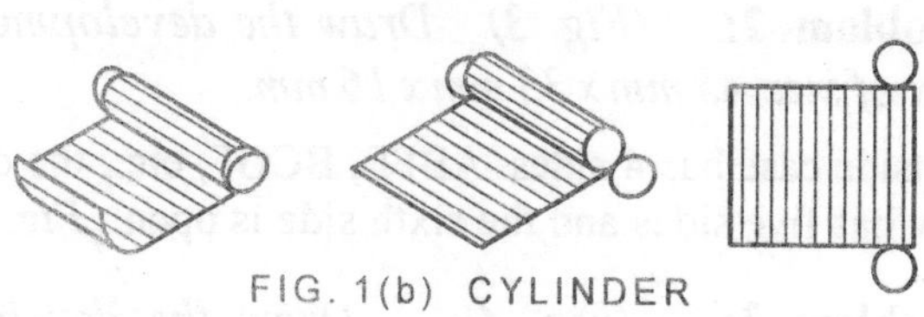

FIG. 1(b) CYLINDER

(ii) Radial Line Development
See Figs. 1(c) and (d). Radial This method is employed for *pyramids* and *cones* in which the apex is taken as center and the true length of the slant edge or generator as radius for its development.

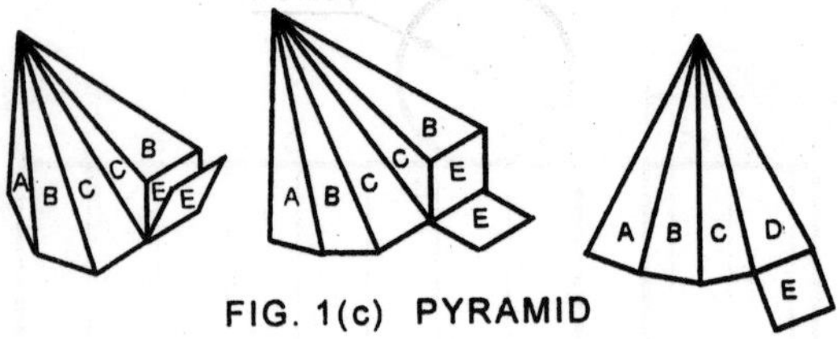

FIG. 1(c) PYRAMID

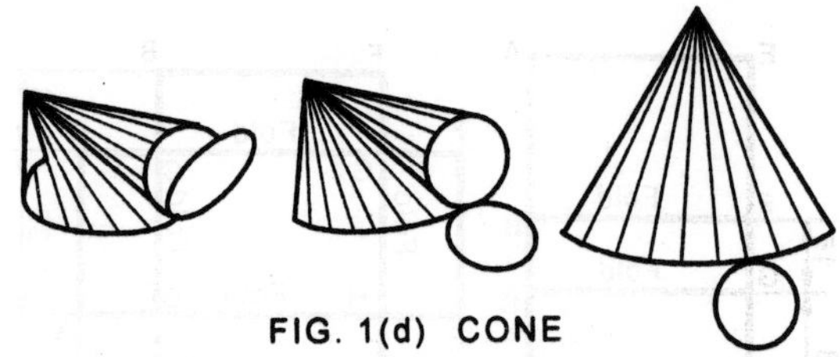

FIG. 1(d) CONE

Pyramids and cones roll out into isosceles triangles and sectors of a circle respectively.

(iii) Approximate Development: This method is employed for double-curved surfaces like spheres, as they are theoretically un-developable in a single plane.

(iv) Triangulation Development: This method is used for developing transition pieces.

2. PRISMS AND CYLINDERS – Parallel Line Development

Problem 1: *(Fig. 2) Draw the development of the lateral surfaces of a right* ***square prism*** *of edge of base 30 mm and axis 50 mm long.*

1. Draw the top and front views.
2. Development consists of four equal rectangles of size 50 mm x 30 mm in contact and in sequence. Hence draw a rectangle $A_1 A_1 AA$ such that $A_1 A_1$ = Perimeter of the base of the prism = Length of the stretch-out line = AA. AA_1 = its Height.
3. On A_1A_1, mark four equal divisions A_1B_1, B_1C_1, etc., each equal to the side of base of the prism 30 mm. Erect perpendiculars at B_1, C_1 and D_1.
4. Darken the four rectangles which give the development of the lateral surface of the prism.

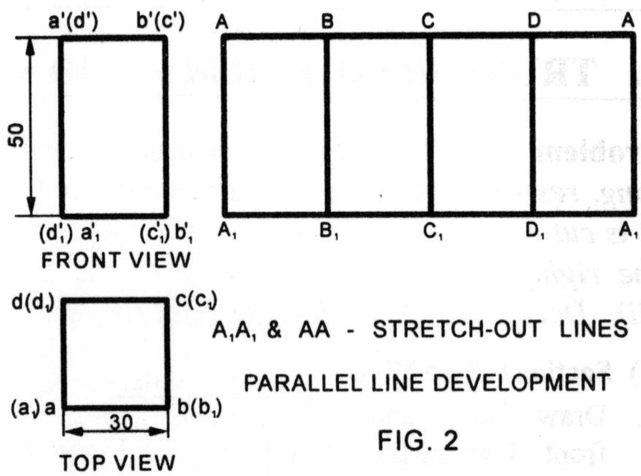

FIG. 2

Note: *In the development of the lateral surfaces, the starting and closing edges should be the same (viz., AA_1) to obtain the closed object.*

Problem 2: *(Fig. 3) Draw the development of the outside case and tray of a match-box of size 45 mm x 33 mm x 16 mm.*

Outside case has 4 sides ABFE, BCGF, etc., the development of which is shown in *Fig. (i)*. The tray has five sides and the sixth side is open. *Fig. (ii)* shows the development of the tray.

Problem 3: *(Fig. 4) Draw the development of the* ***complete surface*** *of a G.I.* ***cylindrical drum*** *with lid. Diameter is 30 cm and height is 1.6 times the diameter.*

Since the drum has bottom base and lid, the complete development is as shown.

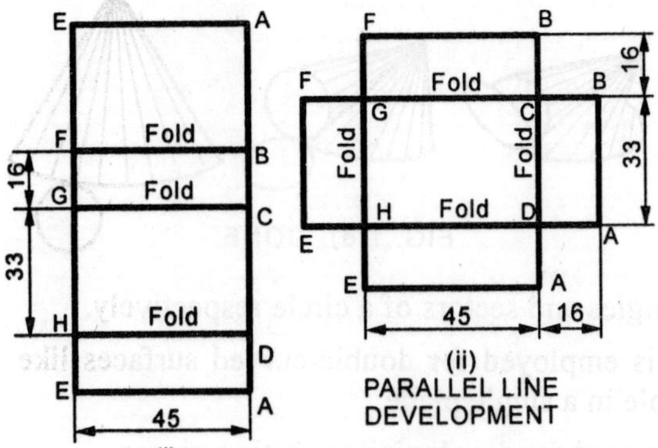

(i)

(ii) PARALLEL LINE DEVELOPMENT

FIG. 3

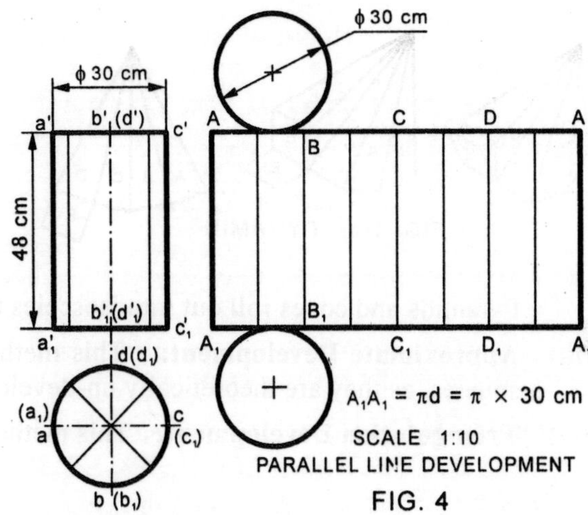

$A_1A_1 = \pi d = \pi \times 30$ cm

SCALE 1:10

PARALLEL LINE DEVELOPMENT

FIG. 4

Note: *1.* *It is advisable to draw the top and bottom circles at the* ***extreme end*** *as shown, to facilitate* ***economical way of cutting*** *the sheet metal.*

2. *Usually the latest surfaces of solids are developed and the ends or bases are omitted from the developments. They can easily be added whenever required.*

3. TRUNCATED PRISMS AND CYLINDERS

Problem 4: *(Fig. 5) A hexagonal prism, edge of base 20 mm and axis 50 mm long, rests with its base on HP such that one of its rectangular faces is parallel to VP. It is cut by a plane perpendicular to VP, inclined at 45° to HP and passing through the right corner of the top face of the prism. (i) Draw the* ***sectional*** *top view. (ii) Develop the lateral surfaces of the* ***truncated prism.*** (UQ)

(i) Sectional Top View

1. Draw top and front views of the prism for the given position.
2. Draw the S.P. in front view. Mark section points.
3. Draw sectional top view and hatch it.

(ii) To draw the development of the Lateral Surfaces of the Truncated Hexagonal Prism

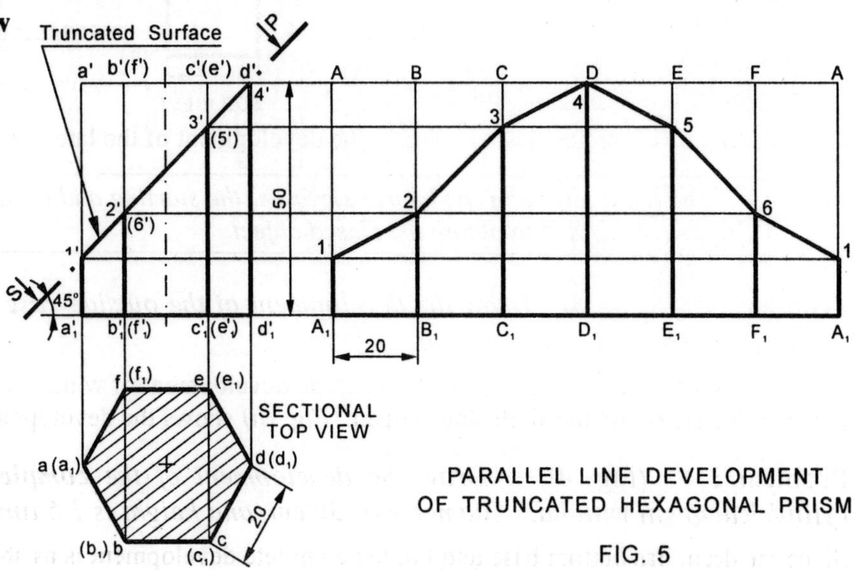

PARALLEL LINE DEVELOPMENT OF TRUNCATED HEXAGONAL PRISM

FIG. 5

4. Draw two stretch-out lines AA and A_1A_1 each equal to perimeter of the base (120 mm).
5. Divide A_1A_1 into six equal parts. Draw six equal rectangles to represent the development of the **lateral surfaces of the hexagonal prism** in *thin* lines.
6. From the section point 1 draw a horizontal line and mark 1 on A A_1.
7. Similarly obtain the other points 2, 3, ..., 6 in the development.
8. Join 12, 23, 34, ... 61 as straight lines and complete the development.

Rule: For economical development of lateral surfaces of the truncated solids (prisms / pyramids / cylinders), draw the development starting with the shortest edge / generator to facilitate minimum welding or riveting at the joint.

Problem 5: *(Exercise) A rectangular prism (60 x 40) and height 60 mm stands on its base on HP with a longer edge parallel to VP. It is cut by a plane normal to VP, inclined at 60° to HP and passing through left corner of the top face of the prism. Develop the lateral surfaces of the truncated prism.*

Problem 6: *(Fig. 6) A pentagonal prism, side of base 25 mm and altitude 50 mm, rests on its base on the HP such that an edge of the base is parallel to VP and nearer to the observer. It is cut by a plane inclined at 45° to HP, perpendicular to VP and passing through the center of the axis. (i) Draw the true shape of section. (ii) Draw the development of the complete surfaces of the truncated prism.* (UQ)

1. Draw the development of lateral surfaces of truncated pentagonal prism.
2. At the line TP, draw the base of the pentagon to represent the development of the prism base.
3. Draw the true shape of section on X_1Y_1 parallel to the S.P. It is an **irregular** pentagon. Transfer the same in the development.

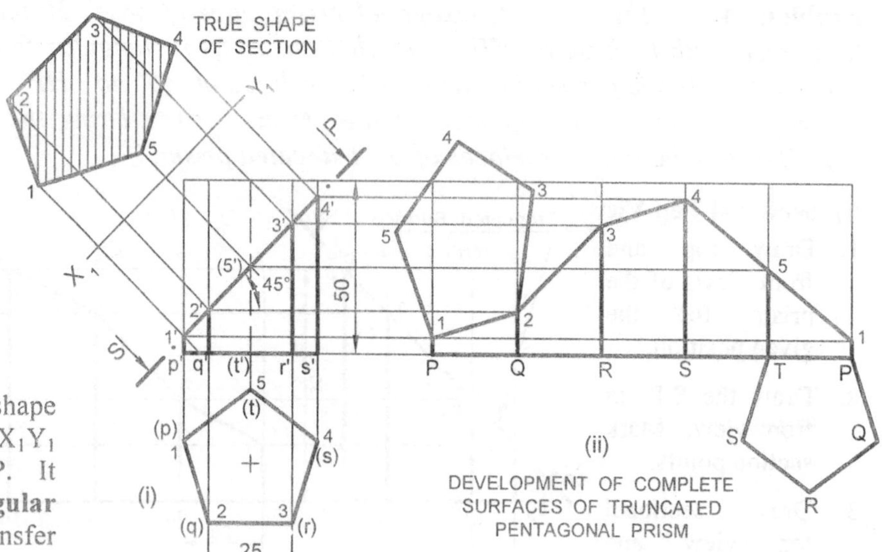

FIG. 6

4. *Fig. 6* shows the development of the complete surfaces of the truncated prism.

Problem 7: *(Exercise) A pentagonal prism of side of base 30 mm and altitude 60 mm stands on its base on HP such that a vertical face is parallel to VP and away* *from observer. It is cut by a plane perpendicular to VP, inclined at an angle of 50° to HP and passing through the axis 35 mm above the base. Draw the development of the lower portion of the prism.*

Problem 8: *(Fig. 7) A cube of side 33 mm rests on its base on HP with a vertical face inclined at 30° to VP. It is cut by a cutting plane perpendicular to VP and inclined at 50° to HP. Cutting plane bisects the axis of the cube. Develop the complete surfaces of the right portion of the cut cube.* (UQ)

1. Draw top view of the cube resting on its base with a vertical face inclined at 30° to VP.
2. Project the front view.
3. Draw S.P. at 50° to HP and bisecting axis of cube.
4. Mark section points in front view. Project these points to the top view.
5. Develop the lateral surface of the complete cube in thin lines.
6. Project horizontally from the section points of the front view to the development of the cube. Join all the intersecting points in the correct sequence.
7. Complete the development of the surfaces of the right portion of the cut cube as shown.

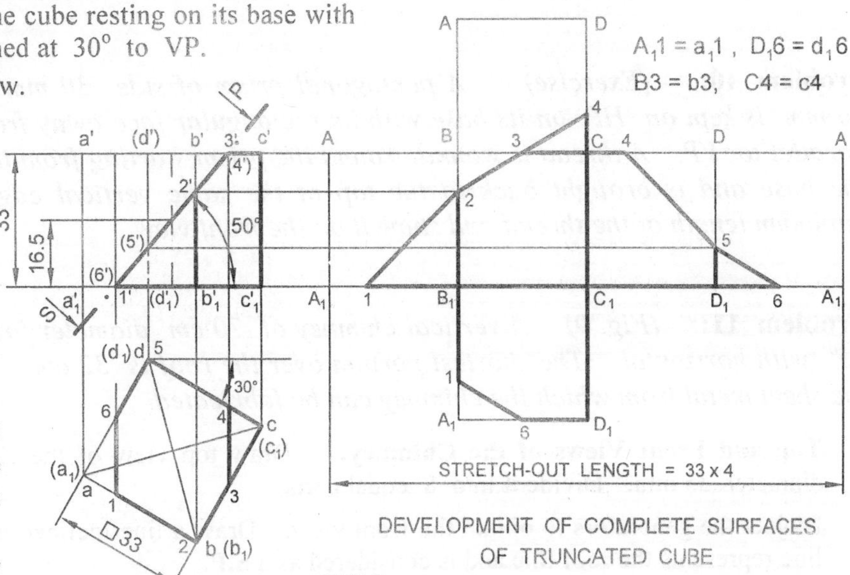

FIG. 7

Problem 9: *(Fig. 8) Draw the development of the lateral surface of a hexagonal prism of 24 mm base edge and 68 mm height. An insect moves on its surface from a corner on the base to the diametrically opposite corner of the top face by the* ***shortest route.*** *Trace graphically the* ***path of the insect*** *in front view.* (UQ)

1. Draw the top view of the given hexagonal prism. Project the corresponding front view.

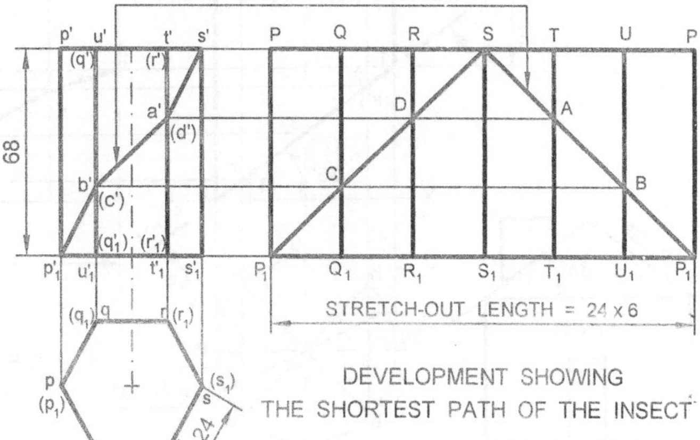

FIG. 8

2. Draw the stretch-out length = 24 mm x 6 = 144 mm. Divide it into 6 equal parts.
3. Join P_1S which represents the shortest path of the insect moving from one corner at the **base** of the prism to its diametrically opposite corner at the **top** of the prism.
4. Similarly join SP_1 to represent the path of the insect from the top S to the base P_1 cutting the edges TT_1, UU_1 at the points A and B respectively.
5. From A, draw a horizontal line to meet the vertical drawn through t′ at the point a′.
6. Similarly, mark the point b′ on the vertical line through u′.
7. Join p_1' b′ a′ s′ which represents the **shortest path of the insect** in the front view.

Problem 10: *(Exercise) A pentagonal prism of side 30 mm and length of axis 60 mm is kept on HP on its base with its rectangular face away from the observer and parallel to VP. A thread is wound around the prism starting from the nearest corner of the base and is brought back to the top of the same vertical edge. Determine the minimum length of the thread and show it on the front view.* (UQ)

Problem 11: *(Fig. 9) A vertical chimney of 70 cm diameter joins a roof sloping at 35° with horizontal. The shortest portion over the roof is 32 cm. Obtain the shape of the sheet metal from which the chimney can be fabricated.* (UQ)

1. **Top and Front Views of the Chimney:** Draw top view of the chimney as a circle of diameter 35 mm. Divide it into 8 equal parts.

   Project the generators to obtain the front view. Draw a line inclined at 35° to XY. This line represents the roof line and is considered as a **S.P.**

   Mark a height of 16 mm on the *left extreme generator* (*shortest height of the chimney*) and complete the front view of the chimney.

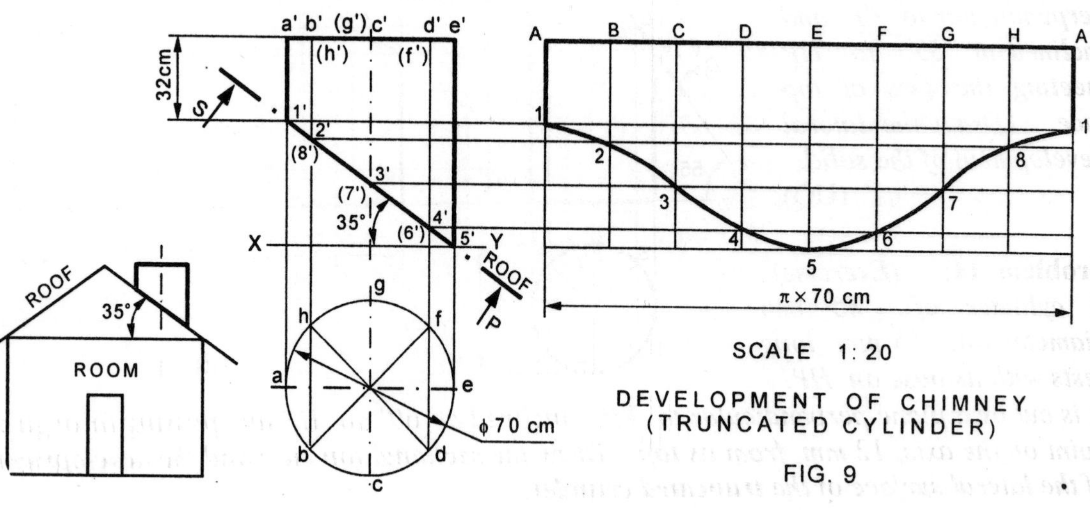

FIG. 9

2. **Shape of the Sheet Metal to Fabricate the Chimney:** Draw a stretch-out line of length AA = π x 35 mm. Divide it into 8 equal parts. Show the generators in the development. Project the section points on the corresponding generators in the development.

Note: ***1. Stretch-out Length of the Cylinder:*** *Stretch-out line AA of length π x 70 can be divided into 8 equal parts by Geometrical Construction. Instead, the chord length ab in the top view can be had on the small bow divider and 8 equal divisions can be marked on the line AA. This method results in reduction of the circumference by about 1% only, which is within limits.*

2. The stretching-out can be started from either of the extreme ends, whatever be the solid. Both the forms of development with different shapes yield the same object after proper rolling or folding and hence are accepted.

Problem 12: *(Fig. 10)* *Draw the development of the lateral surface of the lower portion of a cylinder of diameter 50 mm and axis 70 mm when sectioned by a plane inclined at 40° to HP and ⊥r to VP and bisecting the axis.* (UQ)

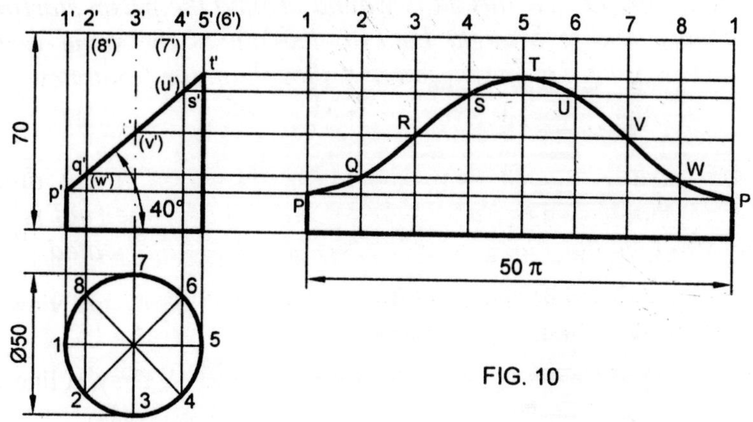

FIG. 10

Problem 13: *(Fig. 11)* *A cylinder of diameter 40 mm, height 75 mm is cut by a plane perpendicular to VP and inclined at 55° to HP meeting the axis at top face. Draw the lateral development of the solid.* (UQ)

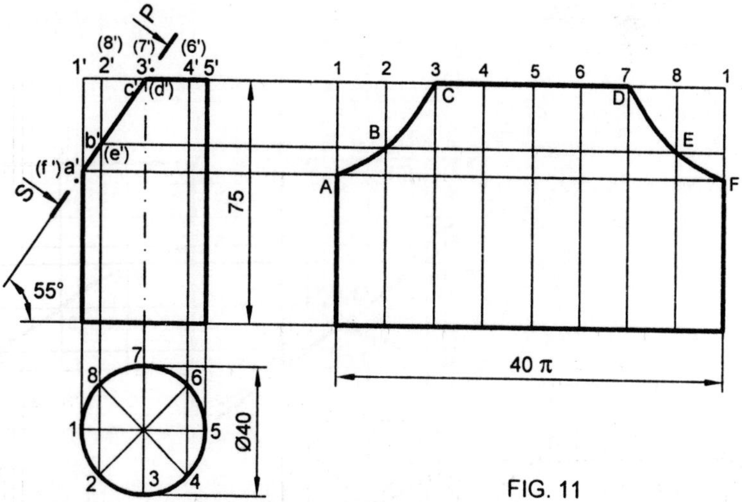

FIG. 11

Problem 14: *(Exercise)* *A cylinder of 45 mm diameter and 55 mm long rests with its base on HP. It is cut by a plane perpendicular to VP, inclined at 60° to HP and passing through a point on the axis, 12 mm from its top. Draw the sectional top view and the development of the lateral surface of the* ***truncated cylinder.***

Problem 15: *(Fig. 12) Draw the development of the lateral surface of the Part I of the cylinder shown in Fig. (i).* (UQ)

1. Draw the development of the cylinder.
2. Consider the 30° line as a S.P. perpendicular to VP and inclined at 30° to HP.
3. Project section points on the corresponding generators in the development and complete the bottom curve.
4. Then consider the portion of a circle of radius 24 mm as another curved S.P. Mark section points.

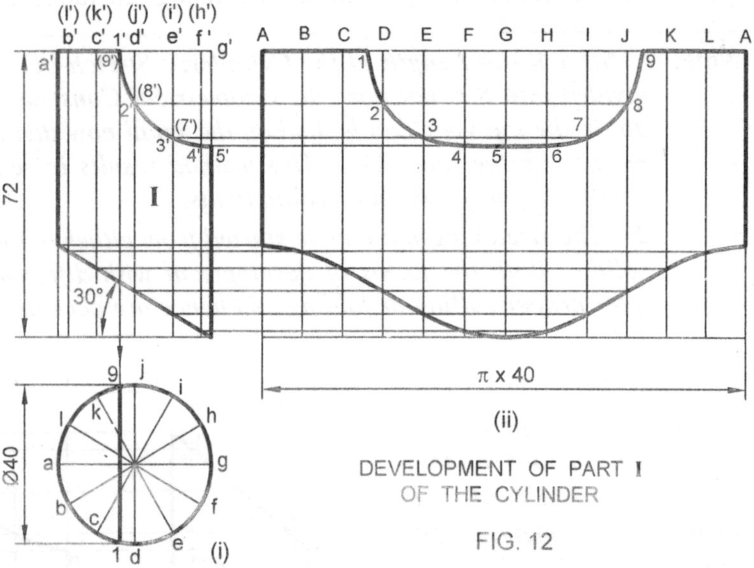

DEVELOPMENT OF PART I OF THE CYLINDER

FIG. 12

5. Project all section points on corresponding generators in the development except 1′ & (9′).
6. To obtain 1 and 9 in the development, project 1′ and (9′) to the top view.
7. Mark 1 in the development such that C1 = c1.
8. Similarly, obtain point 9 and complete the development as shown in *Fig. (ii)*.

Problem 16: *(Fig. 13) Draw the development of the truncated cylinder shown in Fig. (i).*

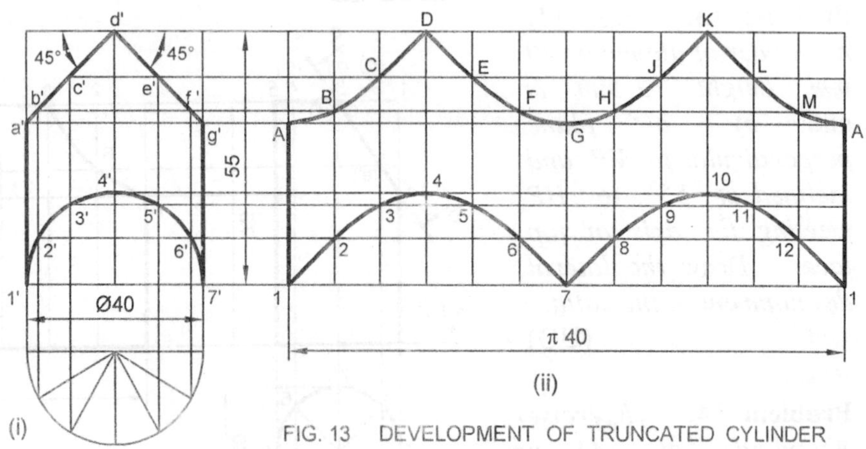

FIG. 13 DEVELOPMENT OF TRUNCATED CYLINDER

Problem 17: *(Exercise) A cylinder of 40 mm diameter and 75 mm altitude is cut by a plane perpendicular to VP, inclined at 55° to HP and meeting the axis at the top face. Another plane of 25 mm radius with right bottom corner as center cuts the cylinder. Develop the lateral surface of truncated cylinder.* (UQ)

Problem 18: *(Fig. 14)* ***Draw the development of a two-piece Cylindrical 90° Elbow shown in Fig. (i).***

1. Consider the given elbow as two truncated cylinders I and II. Draw the development of horizontal cylinder as a rectangle of size π 50 mm x 85 mm.
2. Draw the development of the truncated horizontal cylinder I in the above rectangle. Similarly, draw development of truncated vertical cylinder II.

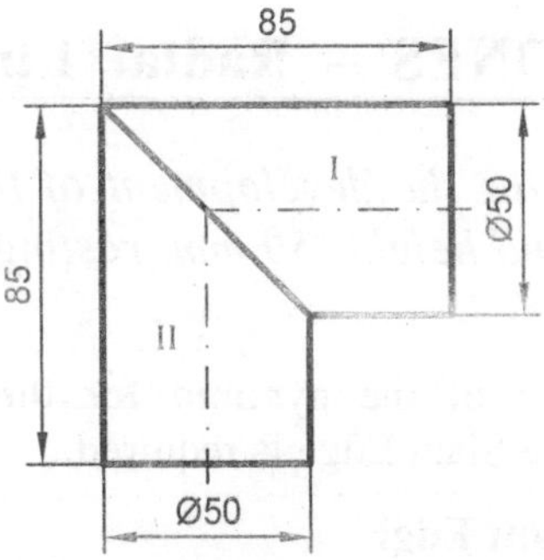

(i) TWO-PIECE CYLINDRICAL 90° ELBOW

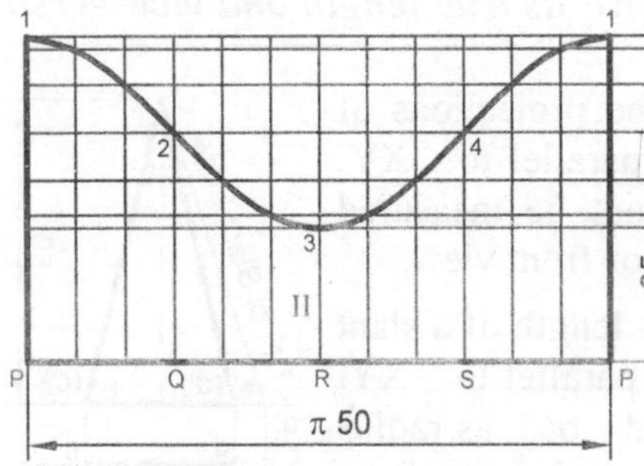

(ii) DEVELOPMENT OF TWO-PIECE CYLINDRICAL 90° ELBOW

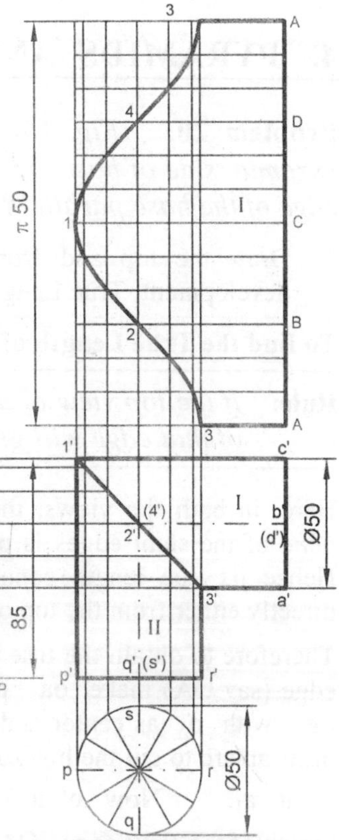

FIG. 14

Problem 19: *(Fig. 15)* *The development of a cylinder is a rectangle of size 110 mm x 55 mm. A semi-circle of 55 mm radius is inscribed in the rectangle. Draw the projections of the cylinder showing the semi-circle on them.* (UQ)

1. Draw development of cylinder as a rectangle 110 x 55. *[Fig. (i)].* Inscribe the given semi-circle of 55 mm radius in the above rectangle.

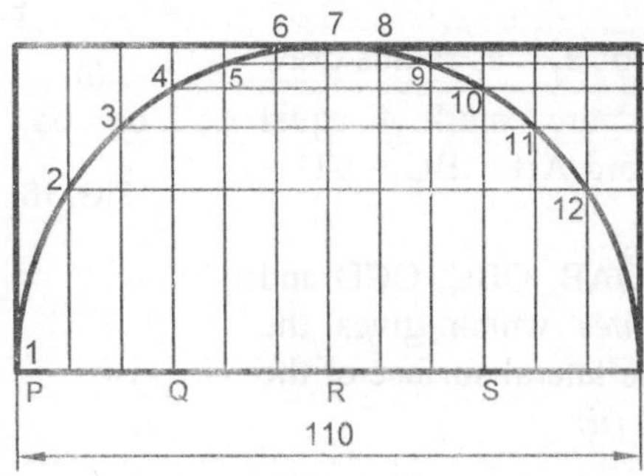

(i) DEVELOPMENT OF GIVEN CYLINDER

FIG. 15

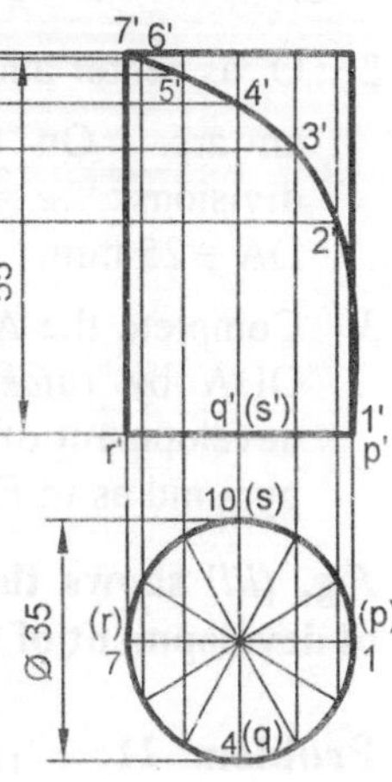

(ii) PROJECTIONS OF CYLINDER

Projections of Cylinder – Reverse Procedure

2. Diameter of cylinder = 110/π = 35 mm.
3. Height of cylinder = Height of given rectangle = 55 mm.
4. Draw front view of the cylinder as a rectangle 35 x 55. *(Fig. (ii)*
5. Draw the corresponding top view as a circle of 35 mm diameter.
6. Divide the above circle into 12 equal parts. Name them and project to the front view.
7. Divide the rectangle into 12 equal parts. Name the points of intersection of the semi-circle with the above divisions.
8. Project above points to front view to meet vertical projectors drawn from top view. Draw a smooth curve to represent the semi-circle in front view.

4. PYRAMIDS AND CONES – Radial Line Development

Problem 20: *(Fig. 16) Draw the development of the lateral surfaces of a square pyramid, side of base 25 mm and height 50 mm, resting with its base on HP and an edge of the base parallel to VP.*

1. Draw the top and front views of the pyramid for the given position. To draw its development, True Length of the Slant Edge is required.

To find the True Length of the Slant Edge

Rule: *If the top view of a slant edge of a pyramid is parallel to XY, then the front view of that edge will give its true length and vice-versa.*

Here, in both the views, the projections of *none* of the slant edges is parallel to XY. Hence its true length cannot be measured directly either from the top or front view.

Therefore to obtain the true length of a slant edge (say OA) make oa parallel to XY, *i.e.*, with o as center and oa as radius draw an arc to cut the horizontal drawn from o at a_1. Now $o'\,a_1'$ will be the true length of the slant edge OA.

2. O as center and $o'\,a_1'$ as radius draw an arc. On this arc, mark 4 equal divisions, *i.e.*, chord AB = BC = CD = DA = 25 mm.
3. Complete the Δs OAB, OBC, OCD and ODA by *thick lines* which gives the development of the lateral surface of the pyramid as in *Fig. (ii)*.

***Fig. (iii)* shows the *uneconomical method* of development of the pyramid.**

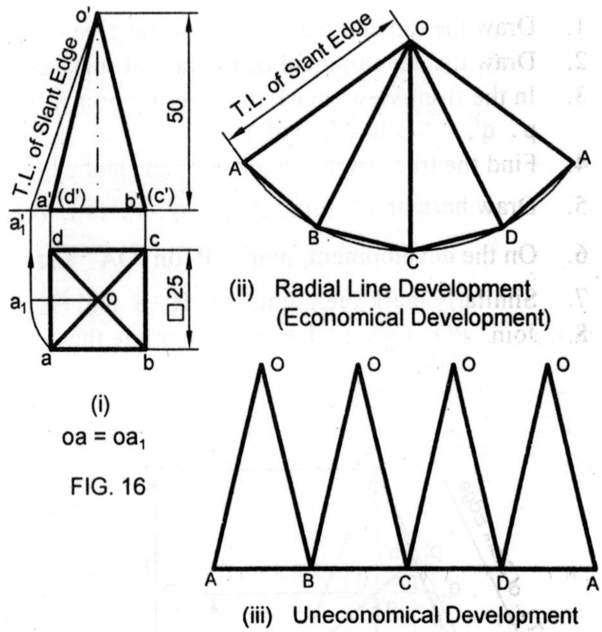

(ii) Radial Line Development (Economical Development)

(iii) Uneconomical Development

(i) oa = oa_1

FIG. 16

Problem 21: *(Fig. 17) Draw the development of the lateral surface of a* ***cone*** *of base diameter 48 mm and altitude 55 mm.*

1. Draw the top view and front view of the cone. In top view, oa is parallel to XY. Hence, o′a′ = L = True Length of the Slant Generator / End Generator.

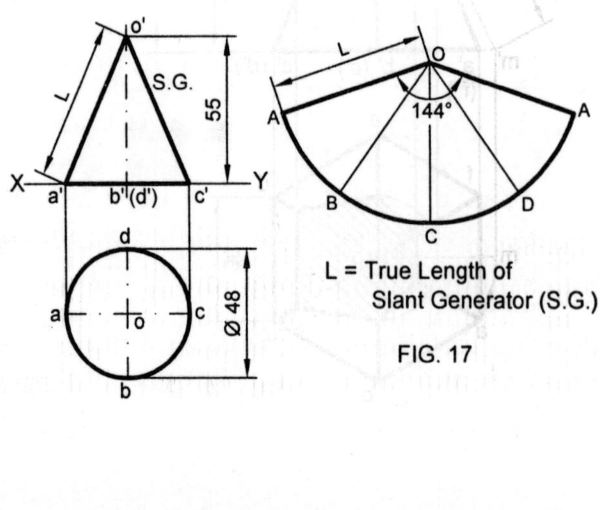

L = True Length of Slant Generator (S.G.)

FIG. 17

2. Development of the lateral surface of the cone is a sector of a circle of radius = L. Length of the arc of the sector = AA = Circumference of the base circle = $2\pi r$. Angle subtended by the arc at the center is θ. Thus, $L\theta = 2\pi r = 360^\circ \times r$. So, $\theta = 360^\circ \times r/L$.
3. O as center and L as radius, draw an arc subtending angle θ at O.

5. FRUSTUM AND TRUNCATED PYRAMIDS AND CONES

Problem 22: *(Fig. 18) A regular hexagonal pyramid of side of base 30 mm and height 60 mm is resting vertically on its base on HP such that two of the sides of the base are perpendicular to VP. It is cut by a plane inclined at 40° to HP and perpendicular to VP. The cutting plane bisects the axis of the pyramid. Obtain the development of the lateral surface of the* ***truncated pyramid****.* (UQ)

1. Draw the top and front views of the given pyramid. Mark the cut surface as shown.
2. Draw the development of the lateral surfaces of the pyramid.
3. In the front view, mark the meeting points of the cut surface with the edges of the pyramid as p′, q′, r′, (s′), (t′) and (u′).
4. Find the true length o′ m′ of the slant edge graphically as shown.
5. Draw horizontals through p′, q′, r′, (s′), (t′) and (u′) to cut $o' m_1'$ at p_1', q_1', etc.
6. On the development, mark P on OA and U on OF such that OP = $o' p_1'$ = OU.
7. Similarly mark the points Q, R, S and T.
8. Join PQ, QR, ..., UP and complete the development of the truncated pyramid.

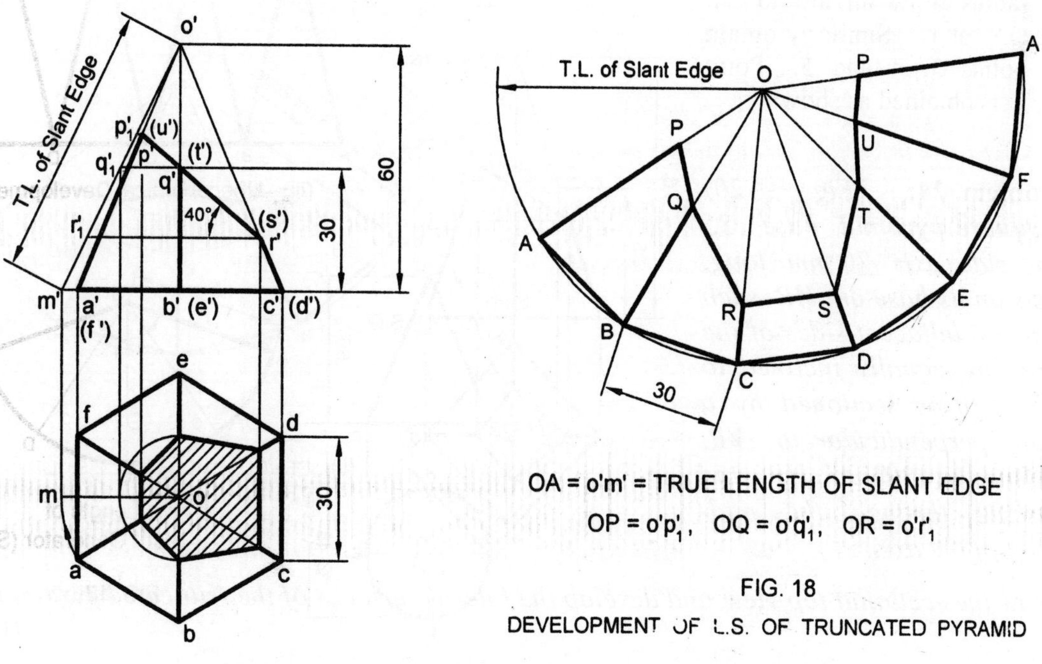

FIG. 18
DEVELOPMENT OF L.S. OF TRUNCATED PYRAMID

Problem 23: *(Fig. 19) A pentagonal pyramid, side of base 30 mm and height 52 mm, stands with its base on HP and an edge of the base is parallel to VP and nearer to it. It is cut by a plane perpendicular to VP, inclined at 40° to HP and passing through a point on the axis, 32 mm above the base. Draw the* ***sectional top*** *view. Develop the lateral surface of the* ***truncated pyramid****.* (UQ)

1. Draw top and front views of the pyramid. Find the true length of slant edge.
2. Draw the development of lateral surface of pyramid.
3. Draw S.P. in front view. Mark the section points. Draw sectional top view and hatch the cut surface.
4. **True Length of remaining portion of Slant Edges:** Draw horizontals through section points 1′, 2′, etc., to cut $o'a_1'$ (=True Length of Slant Edge OA) at $1_1'$, $2_1'$, etc.
5. O as center and $o'1_1'$ as radius draw an arc to cut OA at 1. Similarly obtain points 3, 4 and 5. Point 2 is obtained as shown.

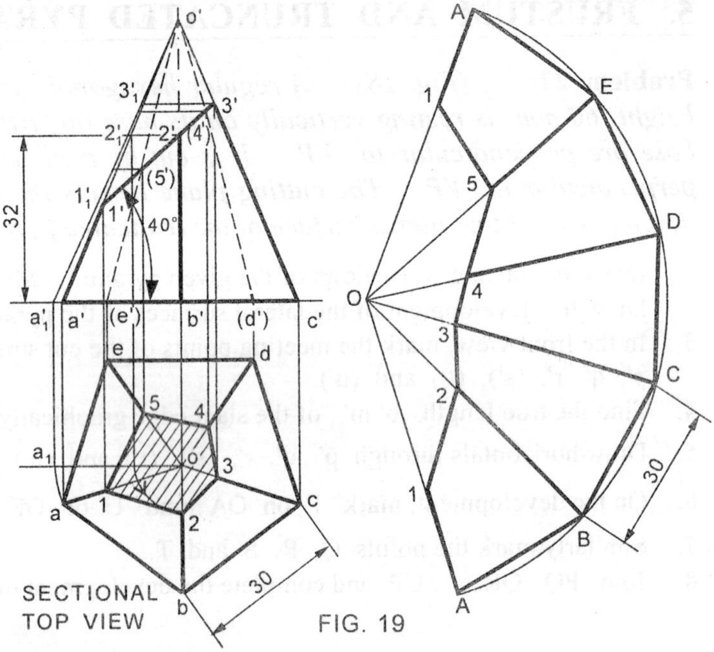

FIG. 19

Problem 24: *(Fig. 20) A* ***square pyramid****, base 35 mm side, axis 70 mm long, rests on its base on HP such that two adjacent sides of the base are equally inclined to VP. It is sectioned by a plane perpendicular to VP, inclined at 30° to HP and passing through the mid-point of the axis.*

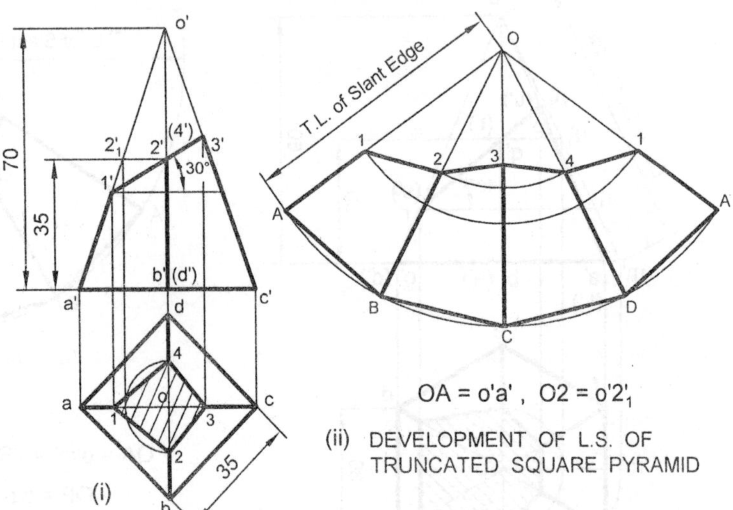

FIG. 20

Draw the ***sectional top view*** *and develop the lateral surfaces of the* ***truncated pyramid****.*

(UQ)

Problem 25: *(Fig. 21)* *A square pyramid of base side 30 mm and altitude 65 mm is resting on HP on its base with a side of base inclined at 25° to VP. It is cut by a plane inclined at 35° to HP, perpendicular to VP and bisecting the axis.*

Draw the development of remaining lower portion of the pyramid. (UQ)

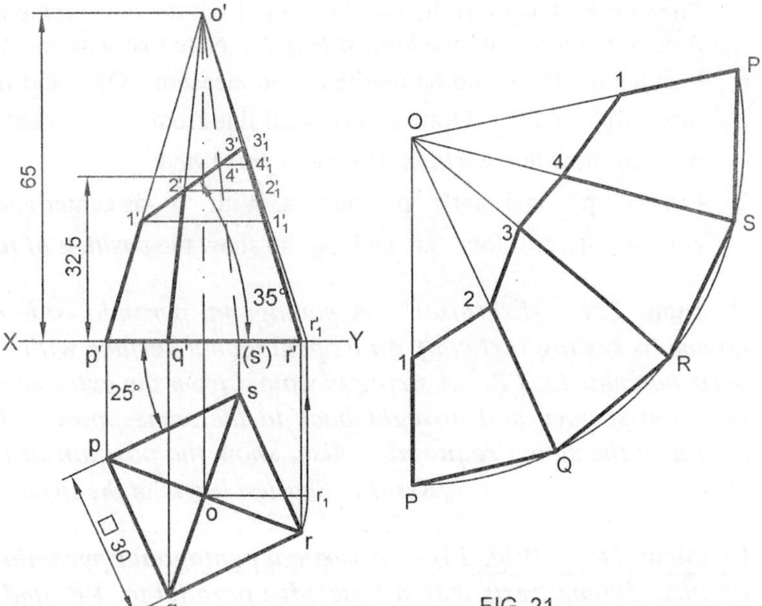

FIG. 21

Problem 26: *(Fig. 22)* *A **monument** is in the form of frustum of a **square pyramid** of base 1.2 m side, top 0.5 m side and height 1.0 m. An electrical connection is to be made along the surface of this monument between one corner of the base and the diagonally opposite corner on the top. Find the shortest length of the wire required and show the position of the wire in the top and front views.* (UQ)

1. To obtain the true length of the slant edge directly in the front view, imagine the edges of the base making equal inclinations with VP. For this position draw the top view.
2. Project the front view of the frustum of pyramid. Extend a'e' and c'g' to fix o'.
3. Since oa is parallel to XY in the top view, o'a' will directly give True Length of that Slant Edge. Also e'a' is true length of EA (slant edge of the frustum). So O as center and o'a' as radius draw an arc and ***complete the development of the pyramid.***
4. With O as center and o'e' as radius draw an arc and obtain E, F, G and H on the respective edges and ***complete the development of the frustum.***
5. Let electrical connection be given from A (one corner of the base) to G (diagonally opposite corner on the top face).

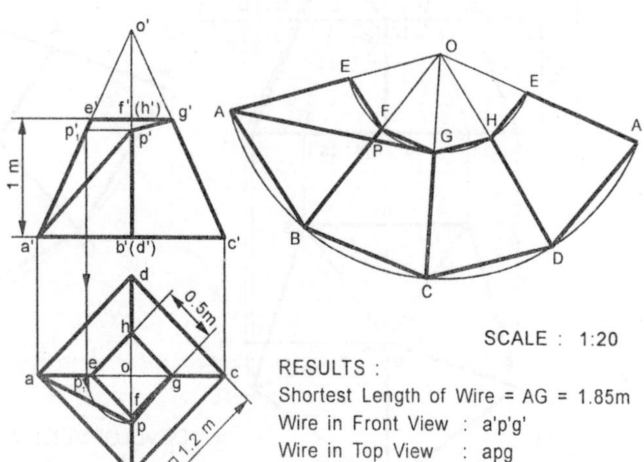

FIG. 22

The shortest length of the wire is found from the development drawn in true lengths. So join A & G to represent the ***shortest length***. Note that wire meets slant edge FB at P.

6. Position of P should be marked. So measure OP and transfer the same on o'a' such that $o'p_1' = OP$. Draw a horizontal line from p_1' to cut o' b' at p'. Join a'p' and p'g' to show the ***position of wire in front view.***
7. Project p_1' and mark p_1 on oa. With o as center and op_1 as radius draw an arc to cut ob at p. Join ap and pg to show the ***position of the wire in top view.***

Problem 27: *(Exercise) A pentagonal pyramid, side of base 28 mm and altitude 66 mm is resting vertically on its pentagonal surface with one of the sides of that surface perpendicular to VP. A string is wound from the extreme right point in the base around the slant surface and brought back to the same point. Find graphically the **shortest length** of the string required. Also, show the position of the string in the top and front views.* *[**Result:** Shortest length of the string = 118 mm.]*

Problem 28: *(Fig. 23) A vertical **pentagonal pyramid** of side of base 27 mm and altitude 50 mm rests with a base edge parallel to VP and nearer to it. It is cut by two planes perpendicular to VP. One is horizontal and cuts the portion of the pyramid on the left of the axis at a height of 18 mm above the base of the pyramid. The other plane inclined at 45° to HP cuts the portion of the pyramid to the right of the axis passing through a point on it 18 mm above the base and leans upwards.*

*Draw the development of the lateral surfaces of the **truncated pyramid**.* (UQ)

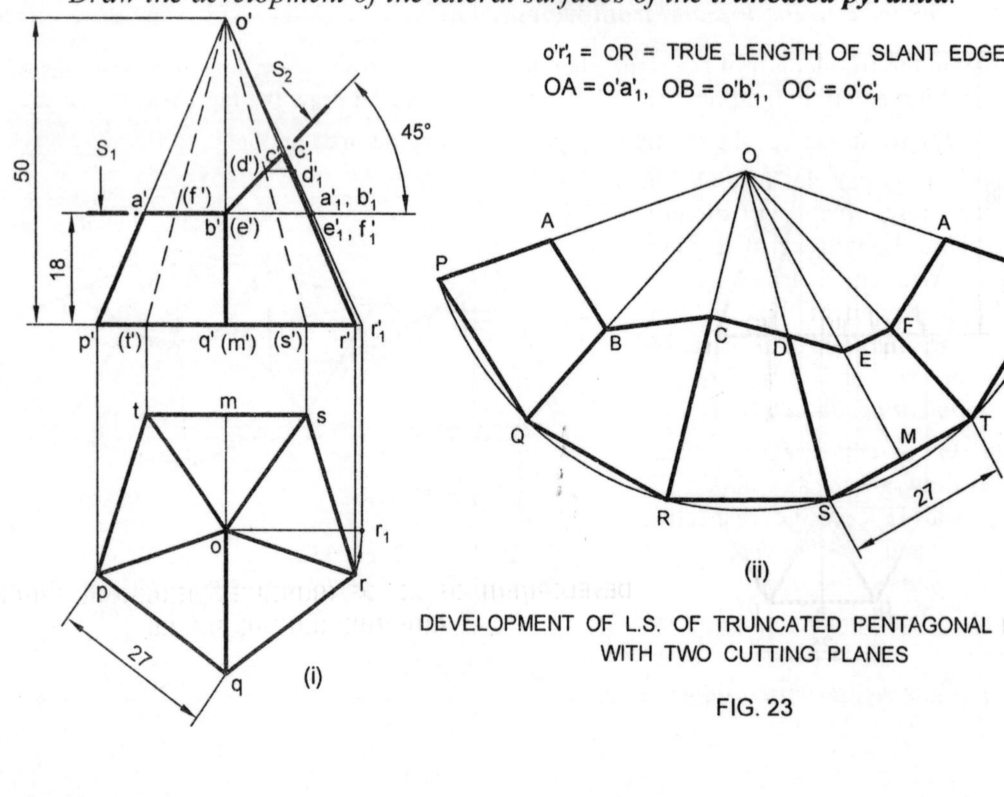

DEVELOPMENT OF L.S. OF TRUNCATED PENTAGONAL PYRAMID WITH TWO CUTTING PLANES

FIG. 23

1. *See Fig. (i).* Draw the top view of the pentagonal pyramid such that the base edge s t is parallel and nearer to XY. Project the corresponding front view of the pyramid.
2. Note that in the front view, none of the slant edges is in true length. Find the True Length of the slant edge $o'\ r_1'$ as shown.
3. Draw the development of the lateral surface of the full pyramid.
4. Mark the two given cutting planes S_1 and S_2 in the front view.
5. Mark the cutting points in the front view as shown. Mark the cutting point (e′) on an imaginary line o′ (m′) which is lying on the rear side triangular face o′ (t′) (s′).
6. Complete the development of the lateral surfaces of the truncated pyramid as in *Fig. (ii)*.

Problem 29: *(Fig. 24) A hexagonal pyramid of side of base 25 mm and altitude 65 mm rests on its base on HP such that a base edge is parallel to VP. It is sectioned by a plane normal to VP, inclined at 30° to HP and passing through a point on the axis 22 mm from the vertex leaning downwards. It is cut by another plane normal to VP, passing through the same point and perpendicular to the previous section plane. The portion containing the vertex is removed.*

(i) Draw the sectional top view of the cut solid.

(ii) Develop the lateral surfaces of the remaining solid such that the developed surface is symmetrical. (UQ)

See Fig. 24 for Solution.

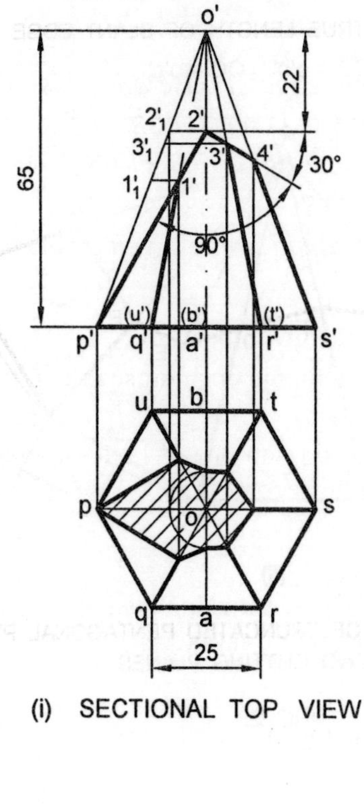

(i) SECTIONAL TOP VIEW

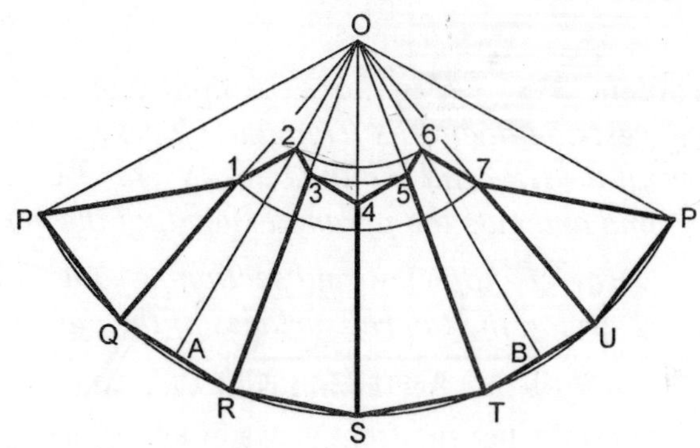

OP = o'p' = TRUE LENGTH OF SLANT EDGE

$O1 = o'1_1'$, $O2 = o'2_1'$, $O3 = o'3_1$

(ii)

DEVELOPMENT OF L.S. OF TRUNCATED HEXAGONAL PYRAMID WITH TWO CUTTING PLANES

(DEVELOPED SURFACE IS SYMMETRICAL)

FIG. 24

Problem 30: *(Fig. 25) A hexagonal pyramid of side of base 25 mm and altitude 70 mm rests on its base on HP such that a base edge is parallel to VP. It is cut by two cutting planes, perpendicular to VP. One of the planes is inclined at 30° to HP and meeting the axis at a point 40 mm from the base. The other plane is curved of 25 mm radius with the right corner of the base as center.*

Develop the lateral surfaces of the remaining portion of the pyramid. (UQ)

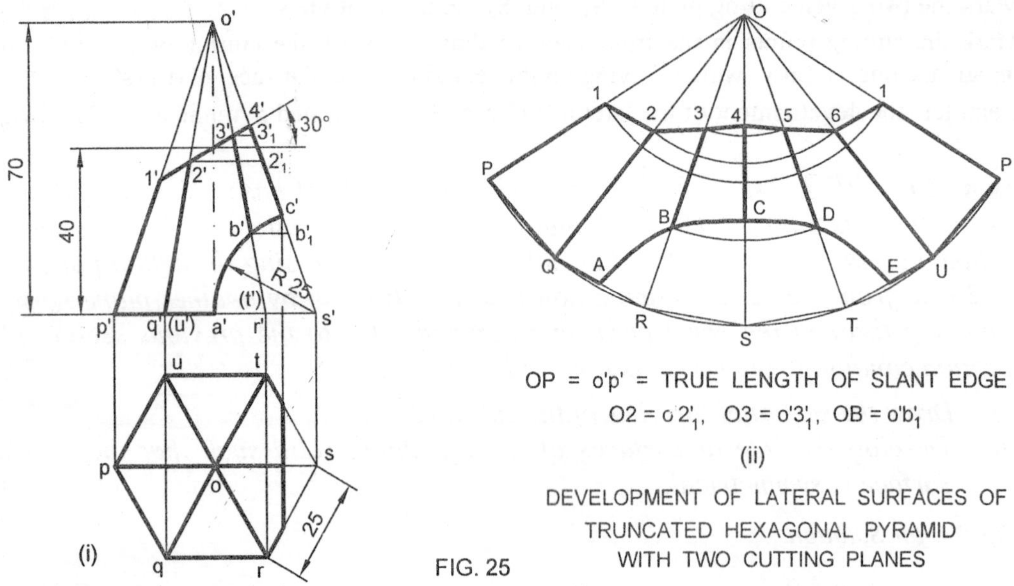

(i)

OP = o'p' = TRUE LENGTH OF SLANT EDGE

O2 = o'$2'_1$, O3 = o'$3'_1$, OB = o'b'_1

(ii)

DEVELOPMENT OF LATERAL SURFACES OF TRUNCATED HEXAGONAL PYRAMID WITH TWO CUTTING PLANES

FIG. 25

Problem 31: *(Fig. 26) A square pyramid of 35 mm edge of base and axis 60 mm long rests on one of its triangular faces on HP and the axis is parallel to VP. A plane normal to HP and* ***inclined at 45° to VP*** *passes through the center of the base of the pyramid and cuts the pyramid such that the vertex is retained.*

(i) Draw the top view and sectional front view of the pyramid in the given position.
(ii) Develop the lateral surfaces of the cut solid. (UQ)

(i) Top View and Sectional Front View of the Pyramid in the given position

1. Draw the top and front views of the square pyramid in the simple position such that the axis of the pyramid is vertical.
2. The axis of the pyramid is parallel to VP. Also, it is resting on one of its triangular faces on HP and hence the axis is inclined to HP. Draw the second front view such that the pyramid is lying on one of its triangular faces on HP as shown.
3. Project the corresponding second top view.
4. Draw the cutting plane as a line in the second top view such that it is inclined at 45° to XY and passing through the center of the base. Draw the sectional front view.

(ii) Development of lateral surfaces of the cut solid

5. Find the true length of the slant edge of the pyramid in the top and front views.

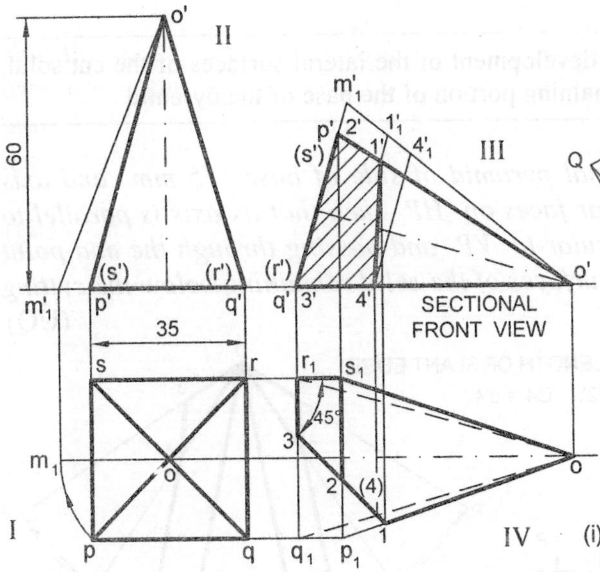

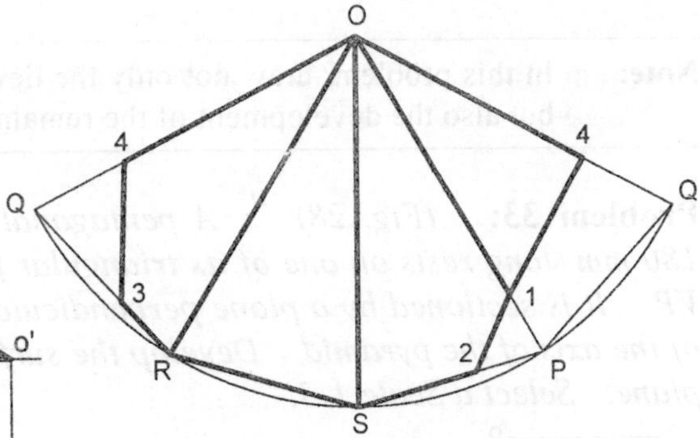

OP = $o'm'_1$ = TRUE LENGTH OF SLANT EDGE

O1 = $o'1'_1$, O4 = $o'4'_1$

(ii)

DEVELOPMENT OF LATERAL SURFACES OF CUT SQUARE PYRAMID

FIG. 26

6. Transfer the true length of the slant edge to the second front view.
7. Draw the development of the square pyramid using the true length.
8. Mark the development of the lateral surfaces of the cut solid as shown.

Problem 32: *(Fig. 27) A hexagonal pyramid, edge of base 25 mm and axis 60 mm long is lying on one of its triangular faces on HP. It is sectioned by a horizontal plane 15 mm from the highest point of the pyramid. (i) Draw the projections of the solid. (ii) Develop the surfaces of the solid remaining below the section plane.* (UQ)

(i)

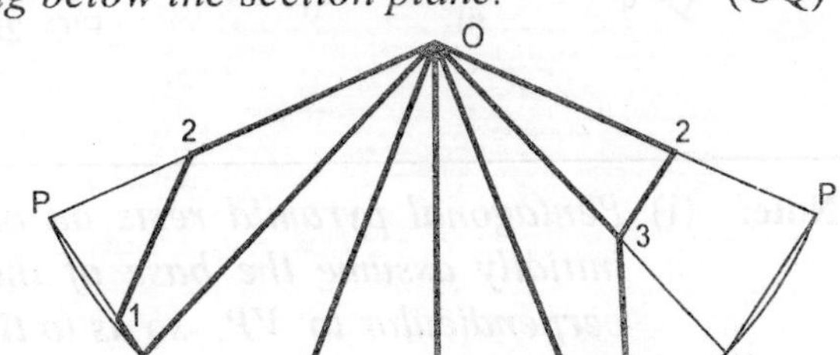

OP = $o'm'_1$ = TRUE LENGTH OF SLANT EDGE

O2 = $o'2'_1$, O3 = $o'3'_1$

(ii) DEVELOPMENT OF CUT HEXAGONAL PYRAMID

FIG. 27

Note: In this problem, draw not only the development of the lateral surfaces of the cut solid, but also the development of the remaining portion of the base of the pyramid.

Problem 33: *(Fig. 28) A pentagonal pyramid of side of base 75 mm and axis 180 mm long rests on one of its triangular faces on HP such that its axis is parallel to VP. It is sectioned by a plane perpendicular to VP and passing through the mid-point of the axis of the pyramid. Develop the surfaces of the solid remaining below the cutting plane. Select a Scale 1:3.* (UQ)

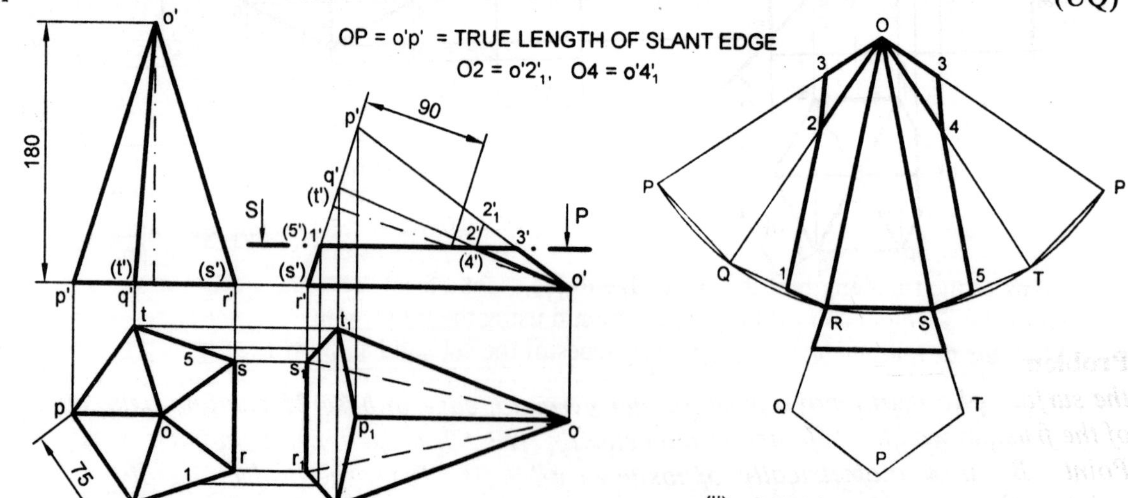

(i) FIG. 28 (ii) DEVELOPMENT OF CUT PENTAGONAL PYRAMID

Note: (i) *Pentagonal pyramid rests on one of its triangular faces on HP. Hence, initially assume the base of the pyramid such that an edge of the base is perpendicular to VP, so as to tilt the pyramid to the required given position.*

(ii) *Refer the top view in the simple position. The slant edge op is parallel to XY. Hence o'p' is equal to the true length of the slant edge OP.*

Problem 34: *(Fig. 29) A* ***hexagonal pyramid*** *of side of base 25 mm and axis 60 mm long rests on a* ***hexagonal prism*** *of side of base 25 mm and height 30 mm, symmetrically such that the corners of the pyramid coincide with the corners of the prism. One of the base edges of the both the solids is parallel to VP. The* ***combined solids*** *are cut by an inclined plane making 45° to the horizontal and passing through the center of the hexagonal faces of the contact of the solids.*

Develop the lateral surfaces of the solids on one side of the cutting plane, where the apex of the pyramid is existing. (UQ)

See Fig. 29 for Solution.

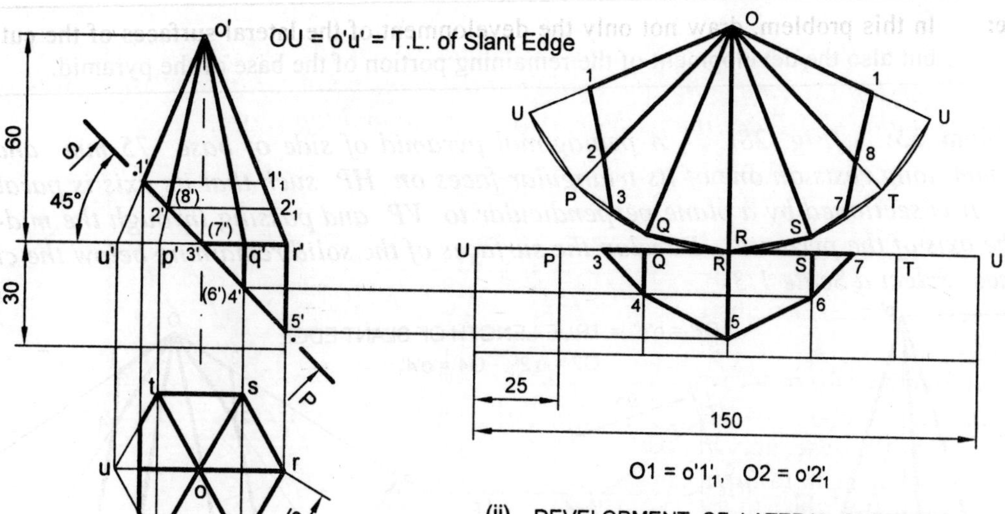

(ii) DEVELOPMENT OF LATERAL SURFACES OF TRUNCATED COMBINATION OF SOLIDS

FIG. 29

Problem 35: *(Fig. 30) Determine graphically the shortest length measured along the surface of a frustum of a cone, between two points A and B. Point A is on the base of the frustum which is 60 mm in diameter. Point B is a diametrically opposite point on the top surface which is 30 mm in diameter. Height of the frustum is 40 mm.* (UQ)

1. Draw the top and front views of the frustum of the cone.
2. Extend the two extreme generators to meet at o′.
3. L = True Length of Slant Generator. Find the subtended angle $\theta = (r/L) \times 360^\circ = (30/85.4) \times 360^\circ = 126.4^\circ$.
4. Draw the development of the frustum of the cone as shown.
5. Mark the points a′ and b′ in the front view as they are diametrically opposite. Mark a and b in top view.

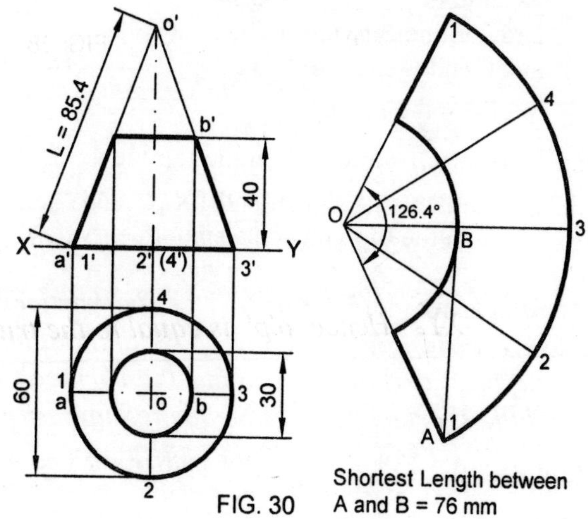

FIG. 30

Problem 36: *(Exercise) The inside of the **hopper** of a flour mill is lined with tin sheet. Top and bottom of the hopper are **circles** of 50 cm and 30 cm diameters respectively. Height of the hopper is 48 cm. Draw the shape of the sheet to which it is to be cut to fit-in the hopper.* (UQ)

Problem 37: *(Fig. 31) A cone of base diameter 50 mm and height 70 mm rests on its base on the ground. A string is wound round the curved surface of the cone starting from left extreme point and ending at the same point. Find the shortest length of the string required. Trace the path of string in front and top views.* (UQ)

1. Draw the top and front views of the cone.
2. Draw the development of the cone. The line AA gives the shortest length of the string.

Path of String in Front View

3. Transfer points A, B, C, ..., H of the generators in the development to front view as a′, b′, c′, ..., h′.
4. Join a′ b′ c′ ... h′ a′ representing the path of string by a smooth curve.

Path of String in Top View

5. Project the points a′, b′, d′, e′, f′, h′ (except c′ and g′) on to the corresponding generators in top view. Draw a line through c′ to meet end generator o′5′ to obtain the point c_1'.

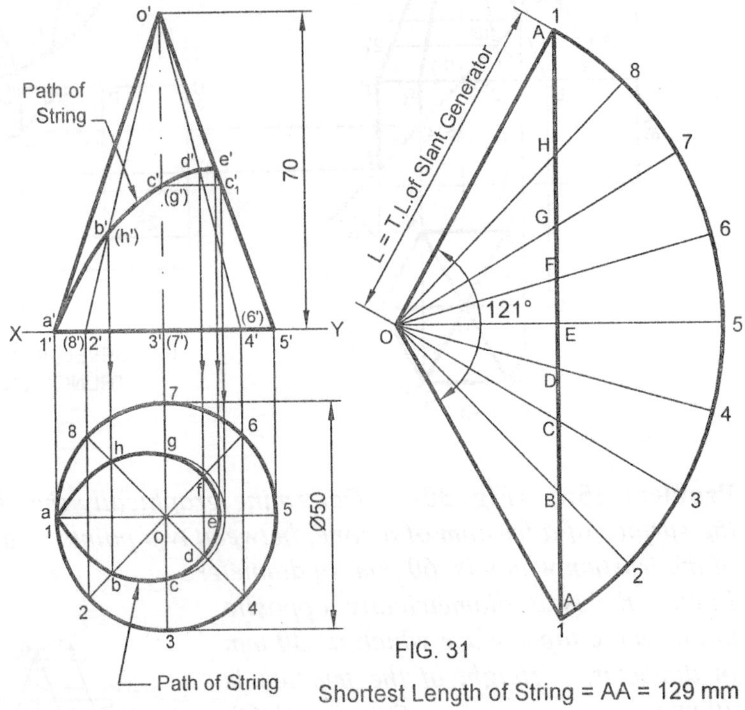

FIG. 31

Shortest Length of String = AA = 129 mm

6. Project c_1' on to o5 in the top view. Transfer the same on to the 3rd and 7th generators by drawing arcs as shown and fix c and g. Join abcdefgha representing the path of the string in the top view by a smooth curve as shown.

Problem 38: *(Exercise) The vertical section of a* **cone** *through the axis is an* ***isosceles triangle*** *of 50 mm base and 60 mm height. A bee sits on the extreme left end of the base and moves around the surface of the cone and returns to the starting point. Find the shortest path of the bee. Show the path in front and top views.* (UQ)

Results: $\theta = 138.5^\circ$; Shortest path of the bee = 121.5 mm.

Problem 39: *(Fig. 32) A cone of diameter of base 50 mm and altitude 65 mm is lying on its base on HP. It is sectioned by a plane normal to both HP and VP at a distance of 10 mm to the right of the axis of the cone. Draw the development of the lateral surface of the left side remaining portion.* (UQ)

Note: *Cutting plane normal to HP and VP is vertical.*

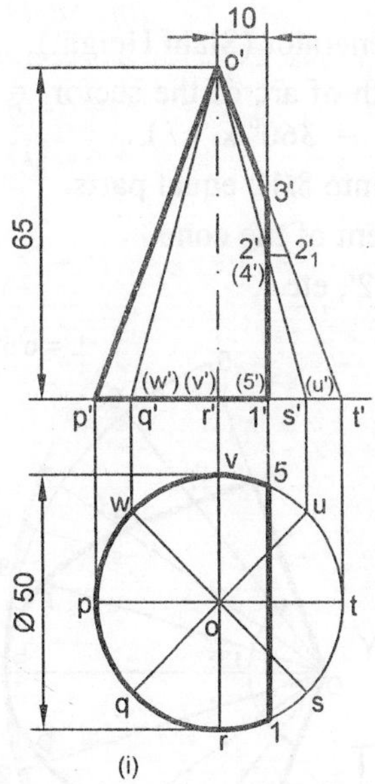

(i)

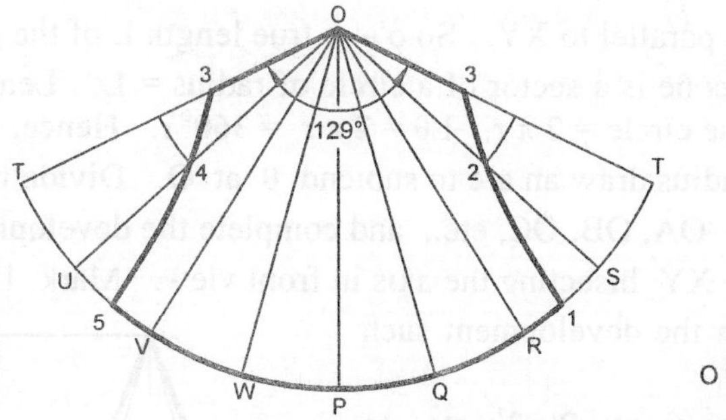

(ii) Development along OT

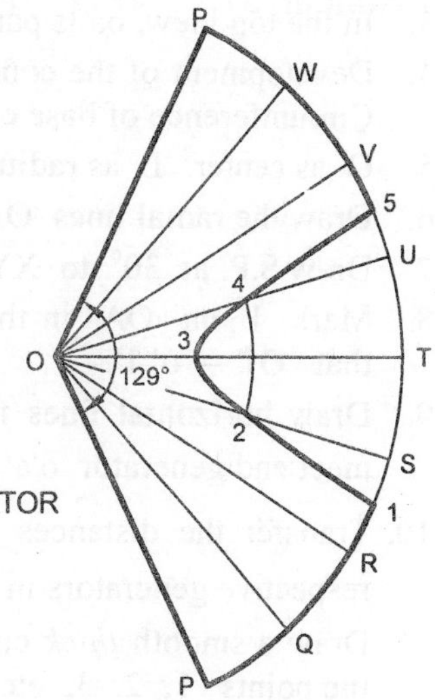

L = OP = o'p' = TRUE LENGTH OF SLANT GENERATOR

$O2 = o'2'_1$, $O4 = o'4'_1$

DEVELOPMENT OF LATERAL SURFACE OF CUT CONE

FIG. 32

(iii) Development along OP

1. Draw the top and front views of the cone as in Fig. (i).
2. Mark the vertical S.P. 10 mm to the right of the axis in front view. Project it to top view.
3. Draw the development of the lateral surface of cone. $\theta = (25/69.6) \times 360^\circ = 129^\circ$.
4. Transfer the cutting points from the top view to the development of the cone.
5. Complete the development of lateral surface of left side remaining portion of the cone.

Note: *The required development may be as shown in Fig. (ii) or in Fig. (iii).*

Problem 40: *(Fig. 33)*
A cone of base 50 mm diameter and height 65 mm rests with its base on HP. A section plane ⊥r to VP and inclined at 30° to HP bisects the axis of the cone. Draw the development of its lateral surface. (UQ)

1. Draw the top and front views of cone.
2. Divide the base circle into 8 / 12 equal parts. Show the generators in both views.

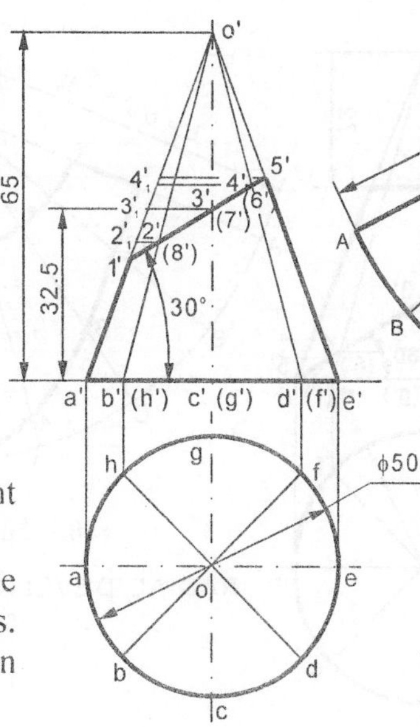

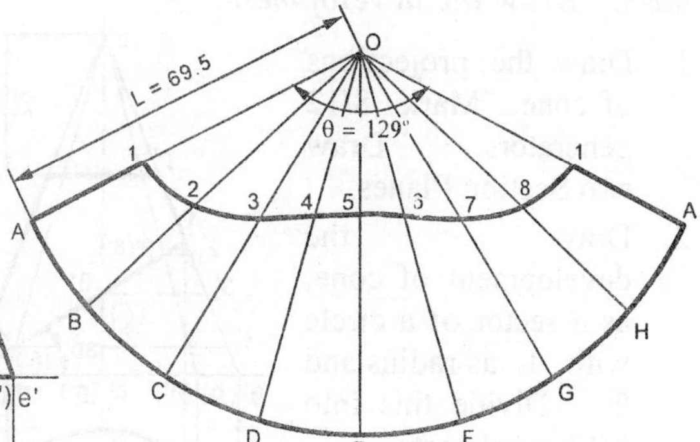

$\theta = 360^\circ \times (r / L) = 129^\circ$

L = o'a' = OA
= True Length of Slant Generator

RADIAL DEVELOPMENT OF TRUNCATED CONE

FIG. 33

3. In the top view, oa is parallel to XY. So o'a' = true length L of the generator (Slant Height).
4. Development of the cone is a sector of a circle of radius = L; Length of arc of the sector = Circumference of base circle = $2\pi r$. $L\theta = 2\pi r = 360^\circ r$. Hence, $\theta = 360^\circ \times r / L$.
5. O as center, L as radius draw an arc to subtend θ at O. Divide it into 8/12 equal parts.
6. Draw the radial lines OA, OB, OC, etc., and complete the development of the cone.
7. Draw S.P. at 30° to XY bisecting the axis in front view. Mark 1', 2', etc.
8. Mark 1 on OA in the development such that O1 = o' 1'.
9. Draw horizontal lines from 2', 3', etc., to meet end generator o'a' at $2'_1$, $3'_1$, etc.
10. Transfer the distances $o'2'_1$, $o'3'_1$, etc., on respective generators in the development.
11. Draw a smooth *thick* curve passing through the points 1, 2, 3, etc.

Problem 41: *(Fig. 34) A lamp shade is formed by cutting a cone of base 140 mm diameter and 170 mm height by a horizontal plane inclined at 30° to HP, passing through one extremity of the base. Draw the development.* (UQ)

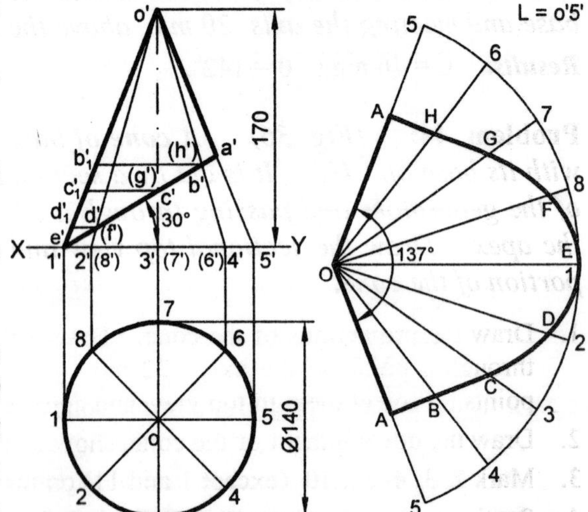

FIG. 34

Problem 42: *(Fig. 35) A* ***lamp shade*** *is formed by cutting a* ***cone*** *of base 144 mm diameter and 174 mm height by a horizontal plane at a distance of 72 mm from the apex and another plane inclined at 30° to HP, passing through one extremity of the base. Draw the development.* (UQ)

1. Draw the projections of cone. Mark 8/12 generators. Draw two Section Planes.
2. Draw the development of cone, as a sector of a circle with L as radius and θ. Divide this into 8/12 equal parts.
3. Horizontal S.P. gives frustum of the cone. To develop this, with O as center and o' p' as radius, draw an arc cutting OA at P.

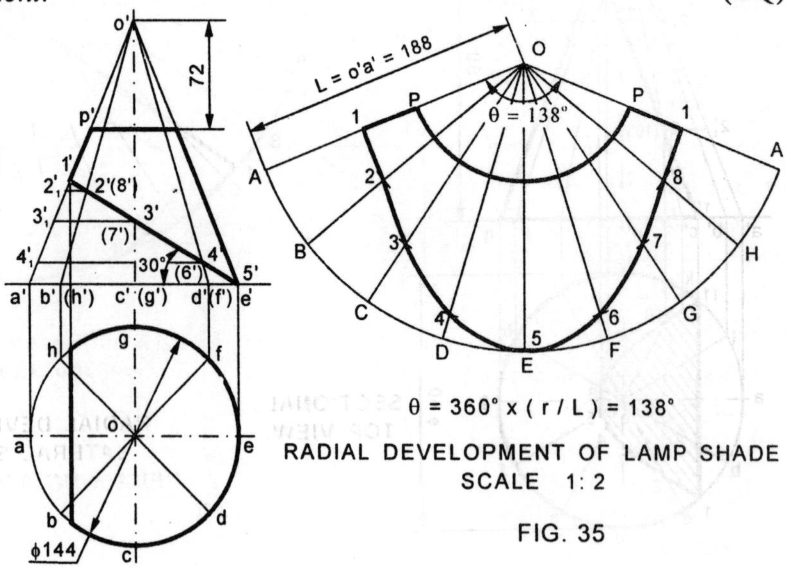

θ = 360° x (r / L) = 138°

RADIAL DEVELOPMENT OF LAMP SHADE

SCALE 1: 2

FIG. 35

4. *Inclined S.P.:* In the development mark 1 on OA such that O1 = o′ 1′.
5. Draw horizontal line from 2′ to cut o′a′ at $2'_1$. Mark 2 on OB such that O2 = o′ $2'_1$.
6. Similarly obtain 3, 4, ..., 8 in the development. Join all the points by a smooth curve.

Problem 43: *(Exercise) A cone of base diameter 60 mm and height 70 mm is resting on its base on HP. It is cut by a plane perpendicular to VP and parallel to HP at a distance of 20 mm from the vertex. It is also cut by a plane inclined at 40° to the base and meeting the axis 20 mm above the base. Draw the development.* (UQ)

Results: L = 76 mm; θ = 142°.

Problem 44: *(Fig. 36) A cone of base 50 mm diameter and 60 mm height, rests with its base on HP. It is cut by a section plane perpendicular to VP, parallel to one of the generators and passing through a point on the axis at a distance of 22 mm from the apex. Draw the sectional top view and develop the lateral surface of the remaining portion of the cone.* (UQ)

1. Draw the projections of the cone. Draw SP parallel to the extreme generator and passing through a point on the axis, 22 mm from the apex in the front view. Mark the section points. Project them to top view and complete the Sectional Top View.
2. Draw the development of the cone, showing 12 generators.
3. Mark 2, 3, 4, ..., 10 (except 1 and 11) on the corresponding generators in the development.
4. Section point 1′ is on the portion of the base b′ c′ in the front view. b′ c′ does not represent the true length of that portion of the base. Hence 1 cannot be marked in the development directly from the front view.

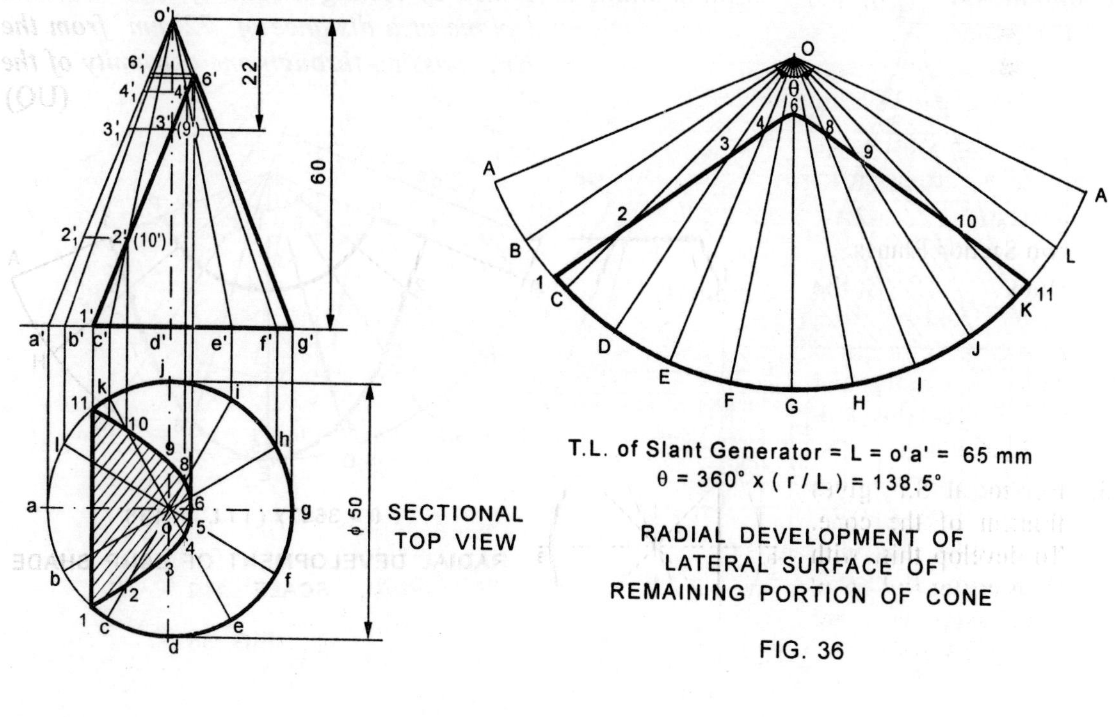

FIG. 36

5. For this, project 1′ to the top view and obtain 1 on bc which is in true length. So mark 1 in the development such that B1 = b1.
6. Similarly mark 11 such that K-11 = k-11. Complete the development as shown.

Problem 45: *(Exercise) A cone, base 50 mm diameter and axis 60 mm long, rests with its base on HP. A section plane perpendicular to VP and inclined at 45° to HP bisects the axis of the cone. Draw the sectional top view and the development of the lateral surface of the remaining portion of the cone.*

Problem 46: *(Fig. 37) A cone, base 54 mm diameter and height 72 mm, rests with its base on HP. A section plane perpendicular to HP and inclined at 25° to VP cuts the cone at a distance of 13.5 mm from the axis. Draw the sectional front view and develop the lateral surface of the remaining portion of the cone.* (UQ)

1. Draw the projections of the cone. Draw the SP in the top view and project the sectional front view. Draw the development of the cone.
2. In the top view, base is in true size. Section points 1 and 6 lie on the base. Hence mark 1 and 6 in the development directly from top view such that B1 = b1 and F6 = f6.
3. In the top view oc, od, oe and of are *not in true size*.
   Hence 2, 3, 4 and 5 cannot be directly marked in the development.
   Now project these points to the front view and obtain $2'_1$, $3'_1$, $4'_1$ and $5'_1$.
4. Mark 2 in the development such that O2 = o′ $2'_1$.
5. Similarly obtain other points 3, 4 and 5. Complete the development as shown.

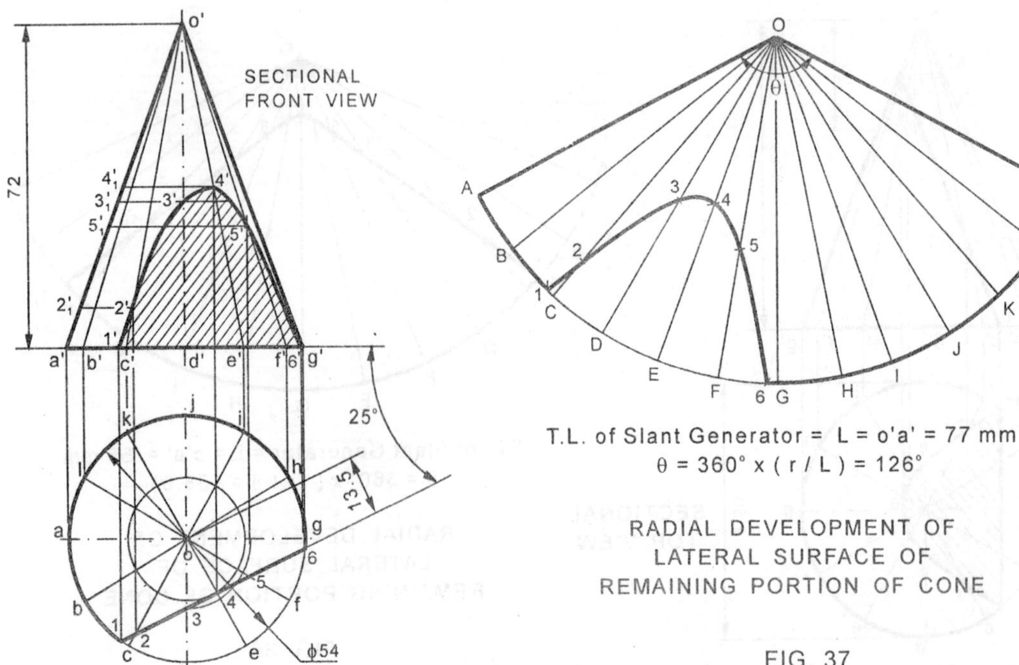

RADIAL DEVELOPMENT OF LATERAL SURFACE OF REMAINING PORTION OF CONE

FIG. 37

Problem 47: *(Fig. 38) A cone of 50 mm diameter of base and 65 mm altitude is lying on one of its generators on HP with its axis parallel to VP. It is sectioned by a plane parallel to VP and 5 mm in front of the axis of the cone. (i) Draw the sectional front view and (ii) Draw the development of the lateral surface of the remaining portion of the cone.* (UQ)

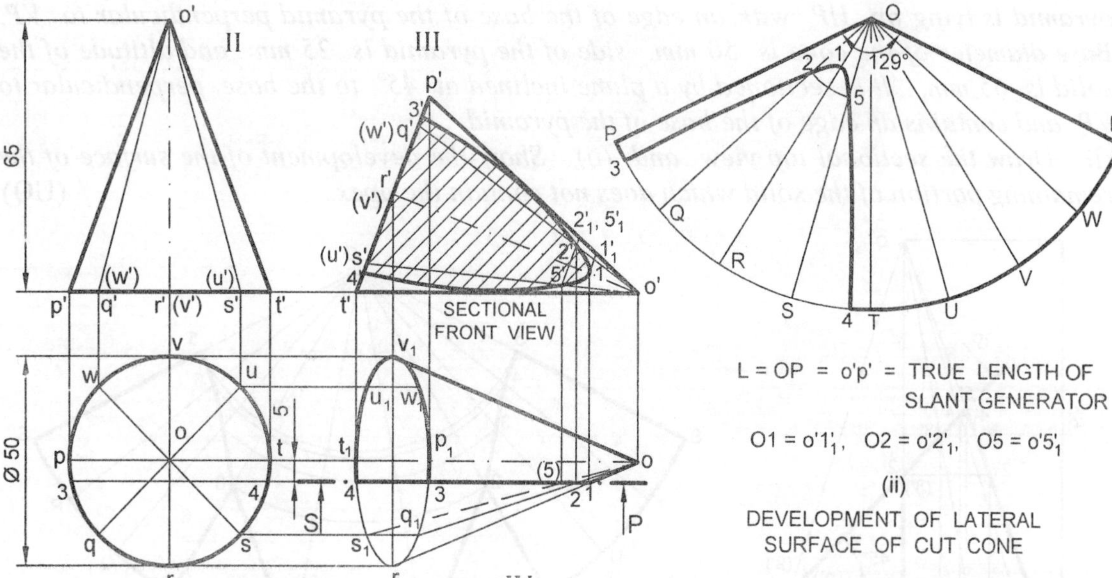

FIG. 38

1. Draw the top and front views of the cone in the simple position.
2. Draw the second front view and second top view of the cone in the tilted position.
3. S.P. is parallel to VP (given). Hence it will be perpendicular to HP. Mark the S.P. in the second top view 5 mm below the axis. Project the cutting points from the second top view to the sectional front view. Complete the sectional front view as shown.
4. Draw the development of the lateral surface of the cut cone.

Problem 48: *(Exercise) A cone of 50 mm diameter of base and 60 mm height is lying on HP on one of its generators with its axis parallel to VP. It is sectioned by a plane perpendicular to VP, parallel to and 30 mm above HP. Draw the development of the lateral surface of the cut cone.* (UQ)

Problem 49: *(Exercise) The development of the conical surface is a sector of the semi-circular plane. Radius of the sector is 90 mm. Angle subtended by the arc at the center is 120°. Construct the cone and draw its projections when the vertex is 42 mm above HP and its axis is parallel to VP.* (UQ)

Hint: Draw the top view of the cone as a circle of 30 mm radius. [Arc Length = A = $R\theta$ = $2\pi r$ and R = 90 mm, $\theta = (\pi \times 120) / 180$. So r = 30 mm.] Draw the front view of height 90 mm. Tilt it so that vertex is 42 mm above XY. Complete the projection.

Problem 50: *(Exercise) The development of a right circular cone is a semi-circle of 110 mm diameter. A square having a diagonal of length 110 mm is removed. Draw the front and top views of the cone showing the views of the portion removed.*

Results: $\theta = 180^\circ$; $r = 55$ mm.

Problem 51: *(Fig. 39) A solid is composed of half-cone and a half-hexagonal pyramid is lying on HP with an edge of the base of the pyramid perpendicular to VP. Base diameter of the cone is 50 mm, side of the pyramid is 25 mm and altitude of the solid is 65 mm. It is sectioned by a plane inclined at 45° to the base, perpendicular to VP and contains an edge of the base of the pyramid.*

(i) Draw the sectional top view and (ii) Show the development of the surface of the remaining portion of the solid which does not contain the apex. (UQ)

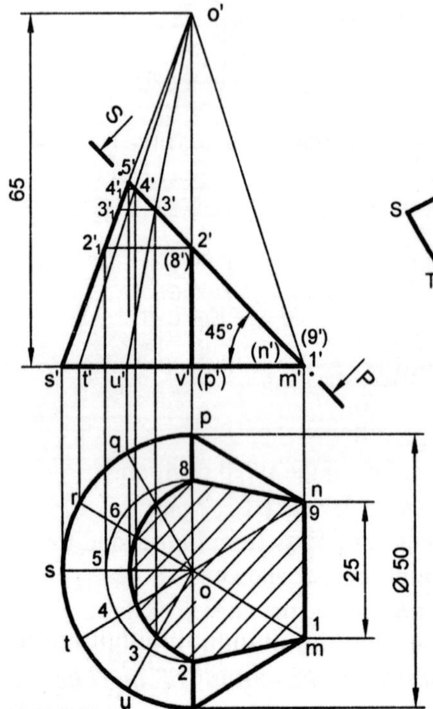

SECTIONAL TOP VIEW

(i)

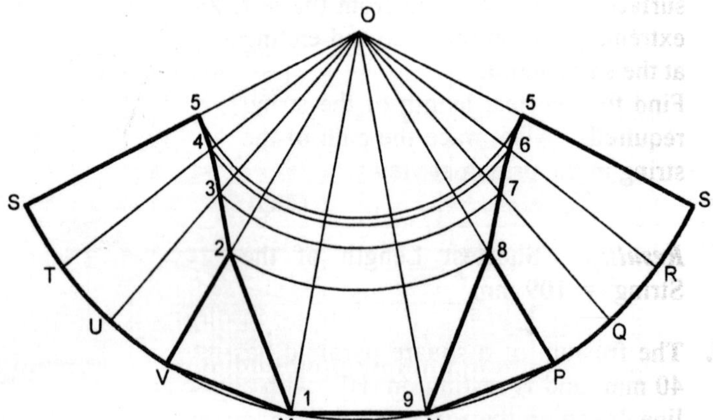

L = OS = o's' = TRUE LENGTH OF SLANT GENERATOR

O2 = o'2'$_1$, O3 = o'3'$_1$, O4 = o'4'$_1$

(ii)

DEVELOPMENT OF LATERAL SURFACES OF HALF-CONE AND HALF-HEXAGONAL PYRAMID

FIG. 39

Problem 52: *(Exercise) A solid is composed of half-cone and a half-hexagonal pyramid is lying on HP with an edge of the base of the pyramid perpendicular to VP. Base diameter of the cone is 50 mm, side of the pyramid is 25 mm and altitude of the solid is 65 mm. It is sectioned by a plane inclined at 60° to the base, perpendicular to VP and contains an edge of the base of the cone.*

(i) Draw the sectional top view and (ii) Show the development of the surface of the remaining portion of the solid which does not contain the apex.

6. ADDITIONAL PROBLEMS

53. **A vertical square prism, side of base 30 mm and axis 60 mm long has one of its rectangular faces leaning to the right at 30° to VP. It is cut by a plane perpendicular to VP and inclined at 60° to its axis. The cutting plane passes through the axis at the mid-height of the prism. Develop the lower portion of lateral surfaces of the prism. (UQ)**

54. **(Fig. 40) A regular hexagonal pyramid of side of base 25 mm and axis 50 mm long is standing vertically on HP with one of the base sides parallel to VP.**
A string is wound around the surfaces of the pyramid from the left extreme point on the base and ending at the same point.
Find the shortest length of the string required. Also, trace the path of the string in the principal views.
(UQ)

Result: **Shortest Length of the String = 109 mm**

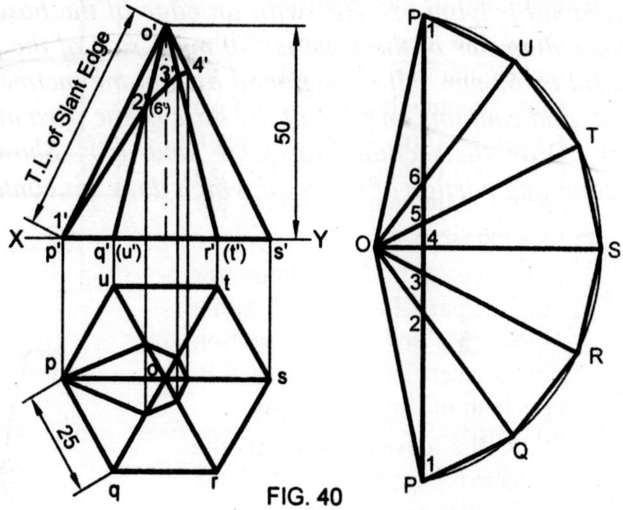

FIG. 40

55. **The frustum of a square pyramid has its base 40 mm side, top 20 mm side and height 40 mm and is resting on HP with a base edge 30° inclined to VP. Find the length of a line drawn on the surface from one top corner to the diametrically opposite bottom corner of the pyramid through the shortest path. Also mark the line in the top and front views. (UQ)**

(***Result*** : *Length = 70 mm*)

56. **Draw the development of a cone of base diameter 60 mm and height 64 mm resting with its base on HP. An insect moves from a point on the base edge to the diametrically opposite point on the same edge through the shortest path along the curved surface. Mark the shortest path in the front and top views of the cone. (UQ)**

(*Results:* θ = 152.5° ; Length of the shortest path = 86 mm.)

57. **A cone of base diameter 150 mm and altitude 210 mm is resting with its base on HP. A cutting plane perpendicular to VP and inclined at 35° to HP bisects the axis of the cone. Draw the development of the truncated cone.** *($\theta = 121^o$)*

58. **A square pyramid of side of base 35 mm and axis 60 mm long is resting on a square prism of side of base 35 mm and height 40 mm, symmetrically such that the corners of the pyramid coincide with the corners of the prism. Two of the base edges of the solids are equally inclined to VP. The combined solids are cut by an inclined plane making 45° to the horizontal and passing through the center of the square faces of the contact of the solids. Develop the lateral surfaces of the solids on one side of the cutting plane, where the apex of the pyramid is existing. (UQ)**

59. *(Fig. 41)* *Draw the development of the* ***duct*** *shown in Fig. (i).*

1. The duct consists of two parts I and II, i.e., two truncated rectangular hollow prisms. Draw its front view.

Development of Part I

2. Draw the development of the rectangular prism (Part I) as four rectangles. Two rectangles are 75 x 15 each. Other two rectangles are 75 x 35 each. Alternate rectangles are of same size.
3. The edge 1′2′ is common for both the parts I and II. Assume 1′, 2′, (3′) and (4′) as section points where an imaginary S.P. cuts both the hollow prisms.
4. Draw the development of the truncated portion of the rectangular prism (Part I).

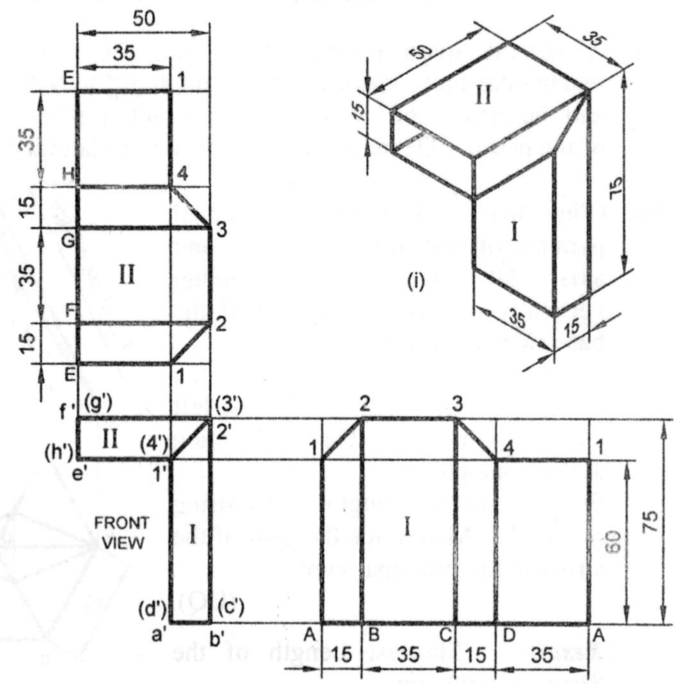

FIG. 41 (ii) PARALLEL LINE DEVELOPMENT OF DUCT

Development of Part II

5. Draw the development of Part II as four rectangles. Two rectangles are of size 50 mm x 15 mm. Other two are of size 50 mm x 35 mm. Alternate rectangles are of same size.
6. Again the edge 1′2′ is common for both the parts I and II.
7. Draw the development of the truncated portion of the rectangular prism (Part II) as shown.

60. *(Fig. 42)* *Draw the development of the Y-piece formed by three cylindrical pipes of 30 mm diameter shown in Fig. (i).*

To draw the given Y piece

1. *See Fig. (i).* Draw the vertical cylinder of 30 mm diameter and 45 mm height.
2. From o′ draw two lines inclined at 30° to the vertical on both sides to represent the axes of the two inclined pipes. On these lines mark o_1' and o_2' such that $o'o_1' = o'o_2' = 60$ mm to represent the length of their axes. Through o_1' and o_2' draw perpendiculars to their respective axes. On these lines represent their bases of diameter 30 mm.
3. Draw the end generators of inclined cylindrical pipes parallel to their respective axes.
4. Mark three intersection points 1′, 5′ and m′ where end generators of the three cylindrical pipes meet. Join them with o′ by *straight lines* which are the intersection curves.

To develop the Y-piece

5. *See Fig. (ii).* Consider each intersection line as common cutting plane for each pair of cylindrical pipes. First draw a rectangle of size 45 mm x $\pi D = \pi \times 30$ mm.

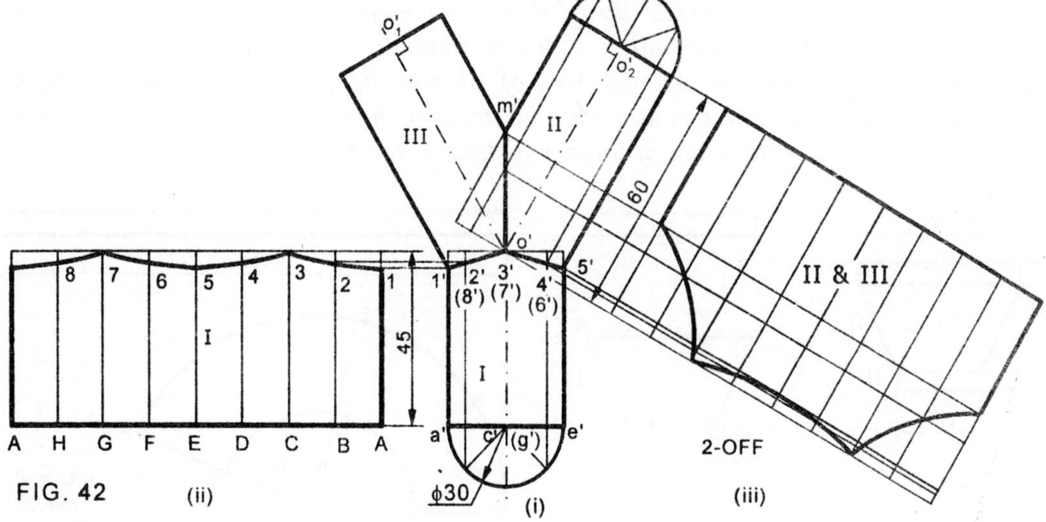

FIG. 42 (ii) (i) (iii)

6. Draw its generators and the development of the vertical cylindrical truncated pipe I.
7. *See Fig. (iii).* Then draw the developments of the truncated pipe II separately.
8. As the truncated cylindrical pipes II and III are identical in all respects including the inclination of both the cutting planes, the developments of both will be identical. Hence mark as 2-OFF as shown in *Fig. (iii).*

61. ***(Fig. 43) Draw the development of the Three Piece Elbow shown in Fig. 43(i).***

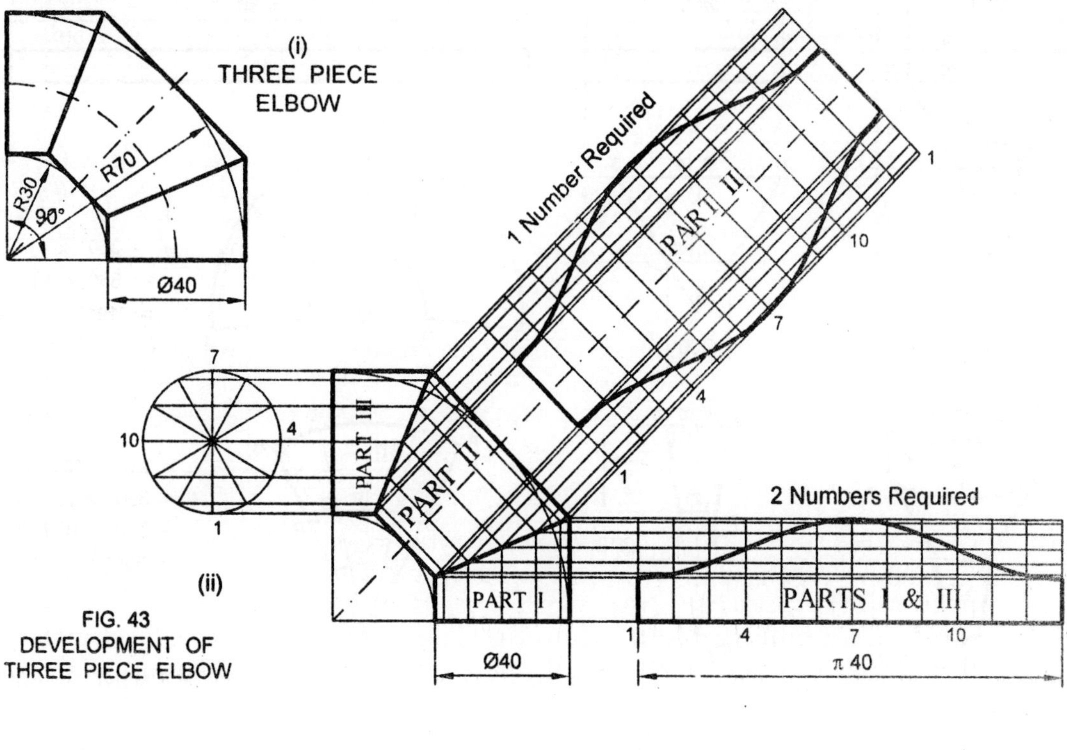

FIG. 43
DEVELOPMENT OF
THREE PIECE ELBOW

62. *(Fig. 44) A hopper made up of thin sheet metal is in the form of an inverted frustum of a cone. Its top end is 450 mm diameter, bottom end 230 mm diameter and height 230 mm. A circular ring of 50 mm height is welded to the frustum at the bottom and another circular ring of 60 mm height is welded to the frustum at top. Draw the development of lateral surface of the hopper.* (UQ)

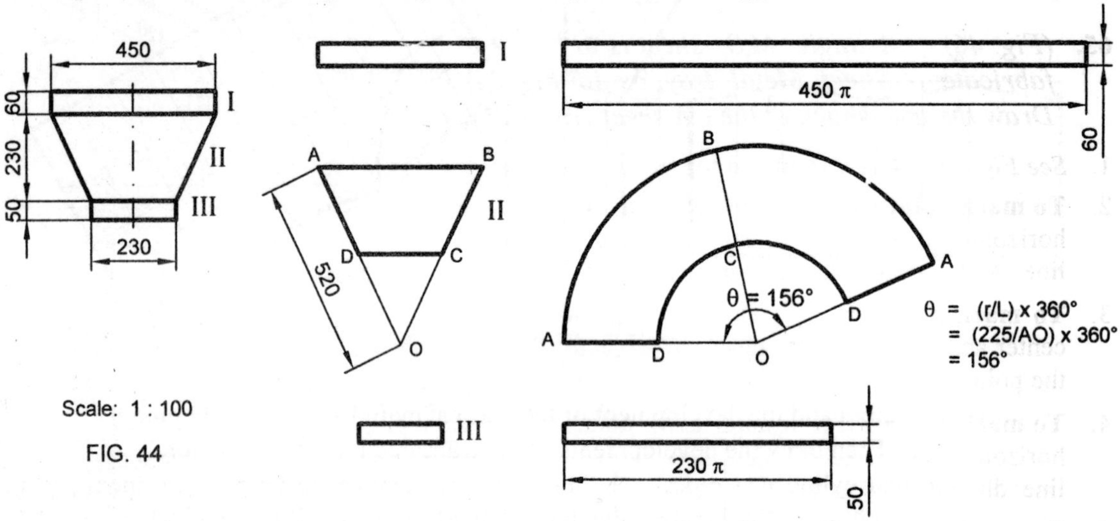

Scale: 1 : 100

FIG. 44

63. *(Fig. 45) Draw the development of the* ***funnel*** *shown in Fig. 45(i).*

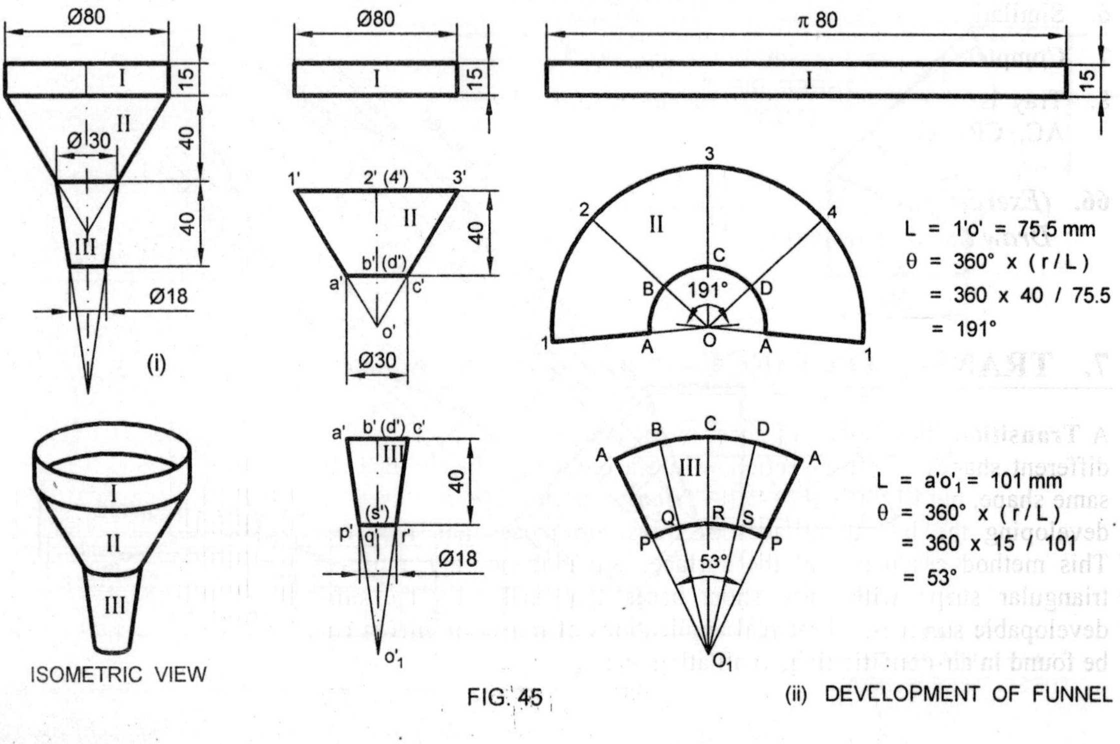

ISOMETRIC VIEW

FIG. 45

(ii) DEVELOPMENT OF FUNNEL

64. A **funnel** is to be made of sheet metal. It tapers from 60 mm diameter to 30 mm diameter to a height of 25 mm and from 30 mm diameter to 20 mm diameter to a height of 50 mm. The bottom of the funnel is beveled-off to a plane inclined at 45° to the axis. Draw its development. (UQ)

65. *(Fig. 46) A single M.S. sheet is to be cut to fabricate a **Sheet Metal tray** as in Fig. (i). Draw the true shape of the cut sheet.* (UQ)

1. *See Fig. (ii).* Draw front and top views of the tray.
2. **To mark point Q :** Rotate a′ b′ about a′ to the horizontal position. Project downwards to cut the line 4-1 extended at the point Q.
3. **To mark point Q_1 :** Draw an arc with P as center and radius PQ to meet the projector b′b at the point Q_1 so that PQ = PQ_1.
4. **To mark corner B :** Rotate a′ b′ about a′ to the horizontal position. Project downwards to cut the line db extended at B.
5. **To mark corner B_1:** Draw an arc with A as center and radius AB to meet the projector b′b at the point B_1 so that AB = AB_1.
6. Similarly, mark the remaining corners S, D, S_1 & D_1.
7. Complete the development of the tray as shown.
8. Tray is obtained by folding the metal sheet along AC, CR, RP and PA.

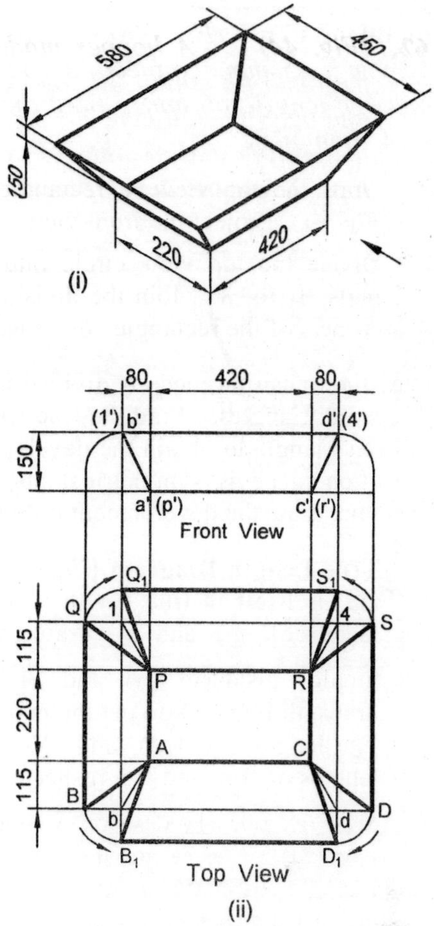

FIG. 46 DEVELOPMENT OF TRAY

66. *(Exercise) **Fig. 47** gives the top and front views of a tray. Draw the development.*

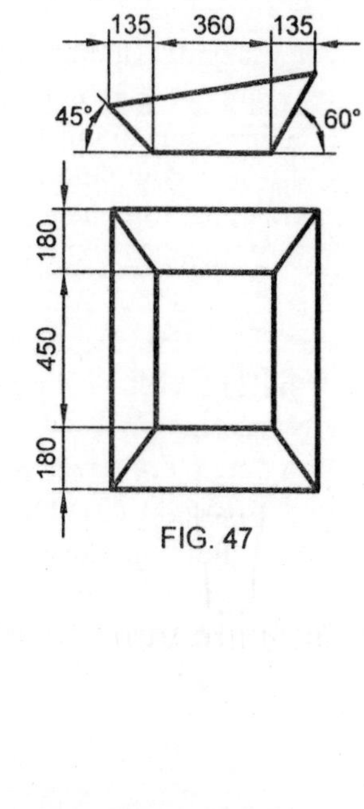

FIG. 47

7. TRANSITION PIECE – Triangulation Development

A **Transition Piece** made of sheet metal is used to connect pipes of different shapes of cross-section, while a reducer connects pipes of same shape, but of different sizes. ***Triangulation Method*** is used for developing the lateral surfaces of transition pieces and reducers. This method assumes that the surfaces are made of a number of triangular strips with very short bases that make up the non-developable surfaces. Practical applications of transition pieces can be found in air-conditioning, ventilating, etc.

Problem 67: *(Fig. 48)* *An **air conditioning duct** of a rectangular cross-section 60 mm x 45 mm connects a circular pipe of 37.5 mm diameter through a transition piece of length 60 mm. Both the sections are co-axial. Develop the lateral surface of transition piece.* **(UQ)**

1. Draw the top view (a rectangle with a co-axial circle) of the transition piece shown in *Fig. (i)*. Project the front view.

2. Divide the top view circle into (say) 8 equal parts 1 to 8. Join the division points with the corners of the rectangle. *See Fig. (ii).*

Now the transition piece is divided into 4 plane and 8 curved triangles. Find the true lengths of the sides of each triangle to obtain the development. Since the transition piece is symmetrical about the horizontal axis mn, draw the development only for one-half of it.

3. **True Length Diagram** [*Fig. (iii)*]: In Δ m1a, ma is itself in true length. To obtain the true lengths of m1 and a1, draw a vertical line VO.

   Step-off distances m1 and a1 on the horizontal drawn through O to get the points 1M′ and 1A′. The lengths V-1M′ and V-1A′ give the true lengths of 1m and 1a respectively.

4. *See Fig. (iv).* To develop Δ 1ma, draw a line 1M = V-1M′. 1 as center and TL_1 as radius draw an arc. M as center and radius ma, draw another arc to cut the previous arc at A. Join 1A and MA. True length of arc 1-2 is approximately = chord length 1-2 (= $\pi D/8$ = 14.7 mm).

5. 1 as center and chord length 1-2 as radius, draw an arc. A as center and TL_2 as radius, draw another arc to cut the previous arc at 2. Repeat this and complete the development of one-half of the transition piece as shown in *Fig. (iv)*.

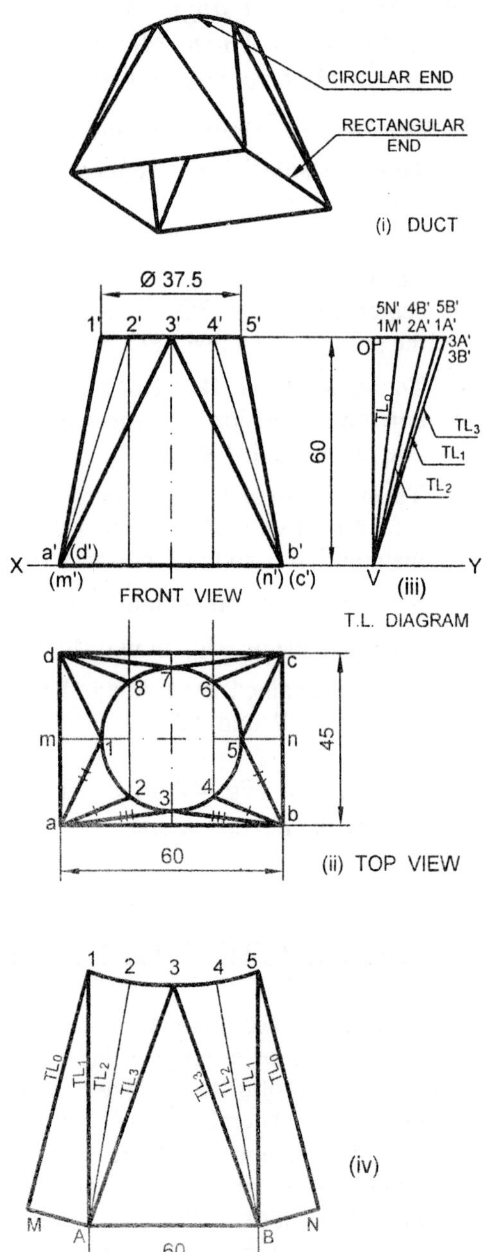

FIG. 48

8. SPHERE – Approximate Development

Sphere is a solid of revolution like right circular cylinder or right circular cone. As the surface of the sphere is a double-curved surface, its development can be produced by some *Approximate Methods*, namely, **(i) Zone Method** and **(ii) Lune or Gore Method**.

8.1 ZONE METHOD : *(Fig. 49)*

When the sphere is cut by a series of cutting planes, the cut section will be circles. The portion of the sphere between two cutting planes perpendicular to the axis is called a *Zone*.

See Fig. 49. Developing the sphere by dividing it into a number of *conical* segments is known as *Poly-Conic / Zone Method.* Each zone is considered as frustum of a cone except the upper most and lower most zones. These are cones of small height.

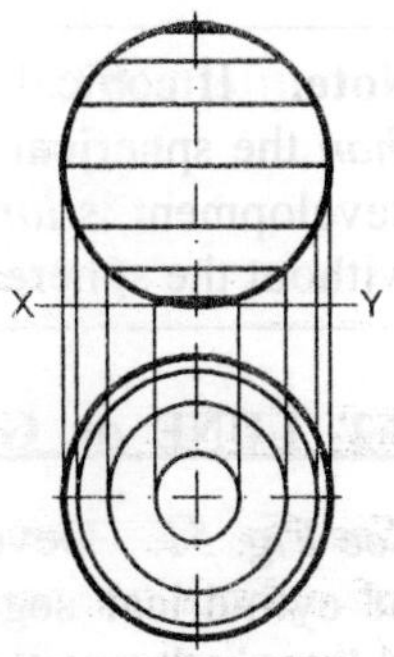

FIG. 49 8 ZONES

Problem 68: *(Fig. 50) Draw the development of a sphere of 50 mm diameter by Zone Method.*

1. Draw the front view of the sphere as a circle (Sϕ 50 mm). Divide it into 12 equal parts as 1′, 2′, ..., 12′. Through these division points draw horizontal lines representing section planes. The sphere is thus divided into 6 zones of equal surface chord width. Name the zones as A, B, C, etc. Consider Zone A as a cone and develop it by Radial Line Method.
2. With O_A as center and $o'_A 1'$ as radius draw an arc to subtend an angle θ_A at the center O_A, where $\theta_A = (r_1 / L_1) \times 360^\circ$; $L_1 = o'_A 1'$ and $2r_1 = 11'1'$.
3. Zone B is approximated as frustum of a cone having its apex at o'_B. O_B as center and $o'_B 2'$ and $o'_B 1'$ as radii draw two arcs subtending an angle θ_B at the center O_B where $\theta_B = (r_2/L_2) \times 360^\circ$; $L_2 = o'_B 2'$ and $2r_2 = 10'2'$. Thus, obtain the development of Zone B.
4. Similarly, draw the development of Zone C. Thus, the development of the top half of the sphere is completed. *See Figs. (i) and (ii).* The three Zones below the horizontal axis are symmetrical in *Fig. (i)* and hence, similar to Zones A, B and C.

   Fig. (iii) shows the applications of the above in an Atomic Reactor.

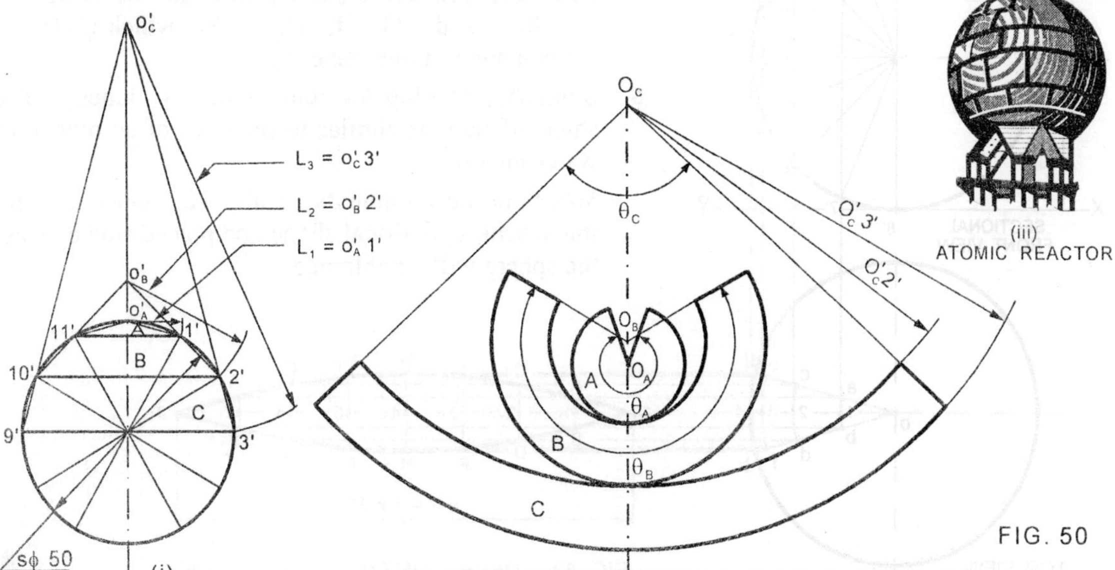

(i)

(ii) DEVELOPMENT OF ZONES A, B AND C, EACH 2-OFF

(iii) ATOMIC REACTOR

FIG. 50

Note: If conical surfaces are inscribed within the sphere, the resulting development is *smaller than* the spherical surface. If conical surfaces are circumscribed about the sphere, the resulting development is *larger than* the spherical surface. If conical surfaces are partly within and partly without the sphere, the resulting development *very closely approximates* the spherical surface.

8.2 LUNE or GORE METHOD

See Fig. 51. Developing the surface of the sphere by dividing it into a number of cylindrical segments is known as *Poly-Cylindric / Lune / Gore Method.* A lune is the portion between two planes containing the axis of sphere.

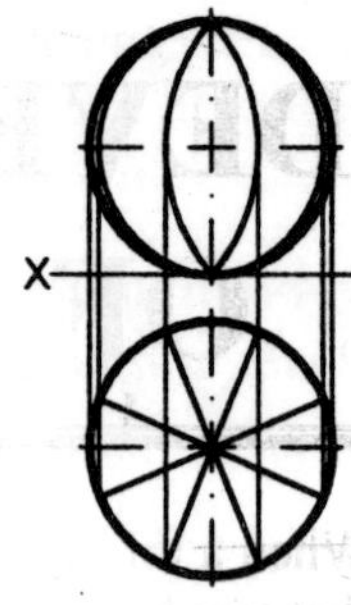

FIG. 51 8 LUNES

Problem 69: *(Fig. 52) Develop a* ***sphere*** *of 62 mm diameter by Lune Method.* (UQ)

1. Draw the top view of the sphere. Divide it into 12 equal lunes.
2. Assume the sphere to be cut by a section plane perpendicular to HP and parallel to VP, passing through the center of the sphere.
3. Draw the sectional front view of the sphere as a *circle.* Consider *one lune* ogh. Its sectional front view is the semi-circle o′8′. Divide this into 8 equal parts at the points o′, 1′, 2′, etc.
4. Project the above points to the top view and mark them as o, 1, 2, etc. With center o and radius o1, o2, etc., draw concentric arcs to cut og and oh at a, b, c, d, etc.
5. The lune is semi-cylindrical. Hence its length in the development is half the periphery of the cylinder. Draw $O_18_1 = \pi R$. Divide it into 8 equal parts by geometrical construction. Draw verticals at these division points 1_1, 2_1, 3_1, etc.

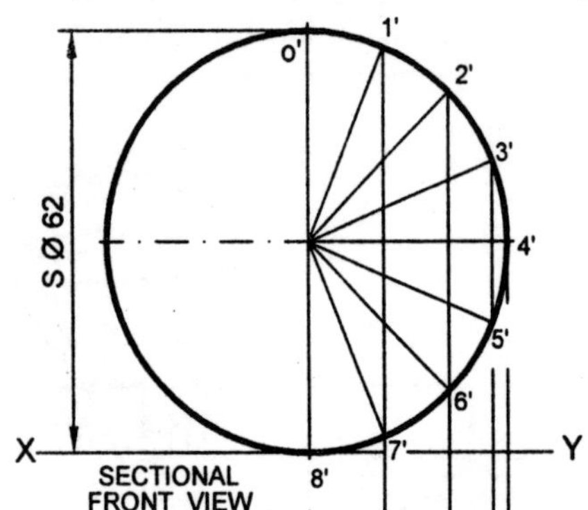

6. Draw horizontals through a, b, c, d, etc., to cut the above verticals at A, B, C, D, etc.
7. Draw a smooth curve passing through O, A, C, ..., A_1, 8_1 and O, B, D, ..., 8_1 which gives the development of one lune ogh.
8. Similarly, develop the remaining 11 lunes. The shape of Lune is similar to the slice of an orange or water-melon.

When the developments of all 12 lunes are bent into a semi-cylindrical shape and placed side by side, the sphere will be obtained.

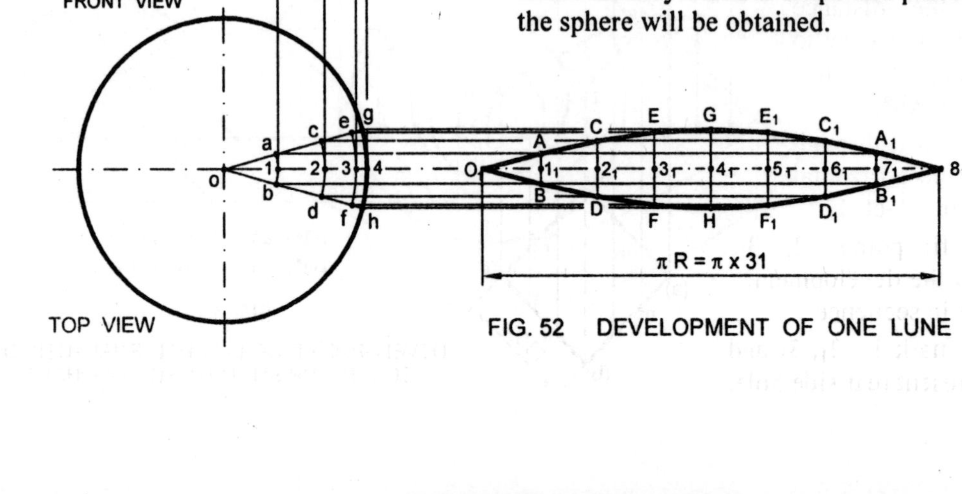

FIG. 52 DEVELOPMENT OF ONE LUNE

*

17

DEVELOPMENT OF SURFACES OF SOLIDS WITH CUTOUTS

When a through cutout (hole or slot) of shapes square / cylindrical, etc., is cut in a solid, it means that two holes are produced on its lateral faces. One hole lies on front face and other on rear face.

1. PRISMS AND CYLINDERS WITH SQUARE CUTOUTS

Problem 1: *(Fig. 1) A* ***square prism*** *of 36 mm edge of base and 64 mm height stands on HP with two of its base edges equally inclined to VP. It has a* ***square hole*** *of 24 mm side centrally cut right through the prism such that its faces are equally inclined to HP. Axis of the hole is parallel to HP and perpendicular to VP. Draw the development of the lateral surfaces of the prism showing the true shape of the square cutout formed on it.* (UQ)

1. Draw the top view of the square prism equally inclined to XY. Project its front view. Draw the development of the lateral surface of the prism.
2. Draw a square hole 1′ 2′ 3′ 4′ of side 24 mm fixing its center at the mid-point of the axis of the prism and two of its faces are equally inclined to HP in the front view.
3. Project the square hole to the top view as dashed lines.
4. Project the corners of the square hole in the front view to the development of prism.
5. Transfer the distance **am** in the top view to the development such that **am** = **AM**.
6. Draw a vertical through M to cut the horizontal drawn from 1′ at 1.
7. Similarly fix points 2, 3 and 4 in the development. Join them in sequence.
8. Similarly mark 1_1, 2_1, 3_1 and 4_1 to represent rear side hole.

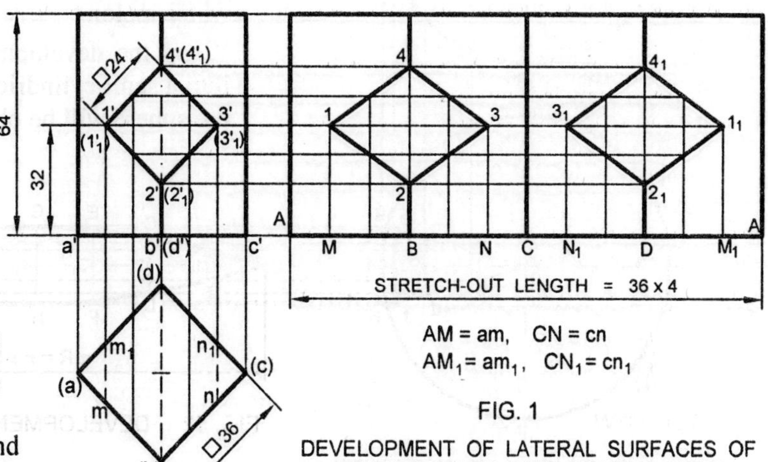

FIG. 1

DEVELOPMENT OF LATERAL SURFACES OF SQUARE PRISM WITH SQUARE HOLE

Problem 2: *(Fig. 2) A* ***hexagonal prism*** *of side of base 24 mm and axis 64 mm is on HP on one of its ends with a base edge parallel to VP. A* ***square hole of side*** *26 mm is drilled such that the axis of the hole is perpendicular to VP and bisects the axis of the prism with all the faces equally inclined to HP. Develop lateral surfaces.* (UQ)

1. Draw the development of the lateral surface of the hexagonal prism.
2. Draw the square hole in front view, fixing its center at the mid-point of the axis and all faces equally inclined to HP.
3. Draw horizontals from the cutting points of the hole in front view to meet the corresponding lines in the development of the prism.
4. Join the above intersecting points. Hole is a square. Shape obtained has line segments.

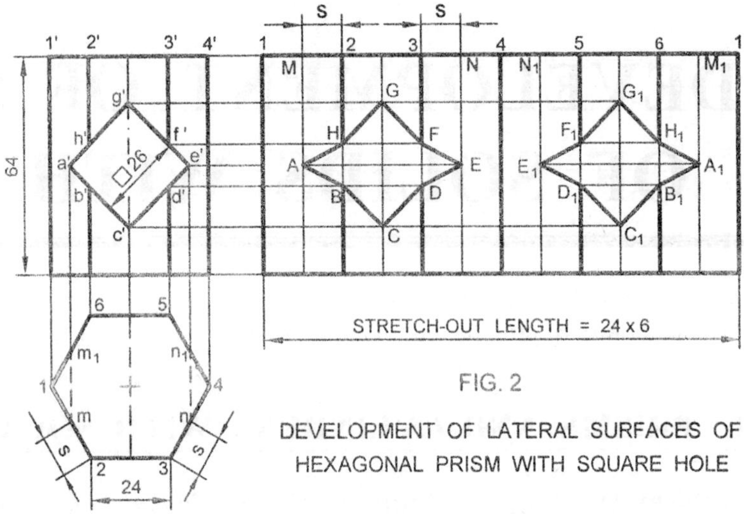

FIG. 2

DEVELOPMENT OF LATERAL SURFACES OF HEXAGONAL PRISM WITH SQUARE HOLE

Problem 3: *(Fig. 3) A* ***pentagonal prism*** *of side of base 25 mm and axis 60 mm is on HP on one of its ends with a base edge parallel to VP and nearer to it. A* ***square hole*** *of side 25 mm is drilled such that axis of the hole is perpendicular to VP and bisects axis of the prism with all the faces equally inclined to HP. Draw the development of lateral surfaces of the prism showing true shape of the hole on it.* (UQ)

1. Draw the top and front views.
2. Draw the square hole in front view, fixing its center at the mid-point of the axis and all faces equally inclined to HP.
3. Draw the full development of the lateral surface of the prism.
4. Draw horizontals from the cutting points of the hole in the front view to meet the corresponding lines in the development of the prism.
5. Join all the above points in sequence and complete.

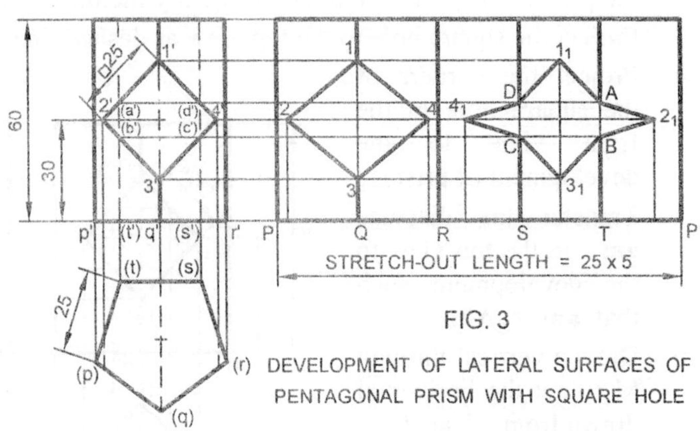

FIG. 3

DEVELOPMENT OF LATERAL SURFACES OF PENTAGONAL PRISM WITH SQUARE HOLE

Problem 4: *(Fig. 4) A* ***cylinder*** *of 48 mm base diameter and 72 mm height rests with its base on HP. A* ***square cutout*** *of 24 mm side is made through it such that the axis of the hole is parallel to HP and perpendicular to VP.*

The faces of the hole are equally inclined to HP and its axis bisects the axis of the cylinder. Develop lateral surfaces.

Note: In the development of square hole in a cylindrical (curved) surface, the edges of square hole are slightly curved. (UQ)

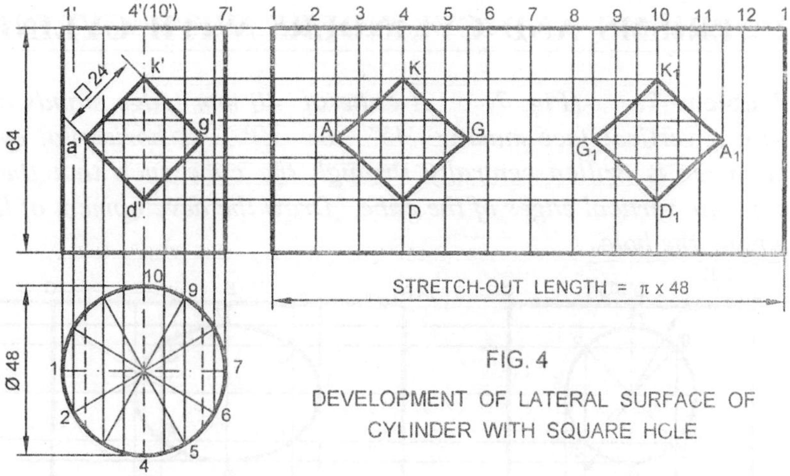

FIG. 4

DEVELOPMENT OF LATERAL SURFACE OF CYLINDER WITH SQUARE HOLE

Problem 5: *(Fig. 5)*
*A **cylinder** of 60 mm diameter and axis 80 mm long rests on its base on HP. A horizontal **square hole** of 35 mm side is cut through the cylinder such that the axis of the hole is parallel to HP and bisects the axis of the cylinder. A side face of the hole is inclined at 20° to HP. Develop lateral surfaces.* (UQ)

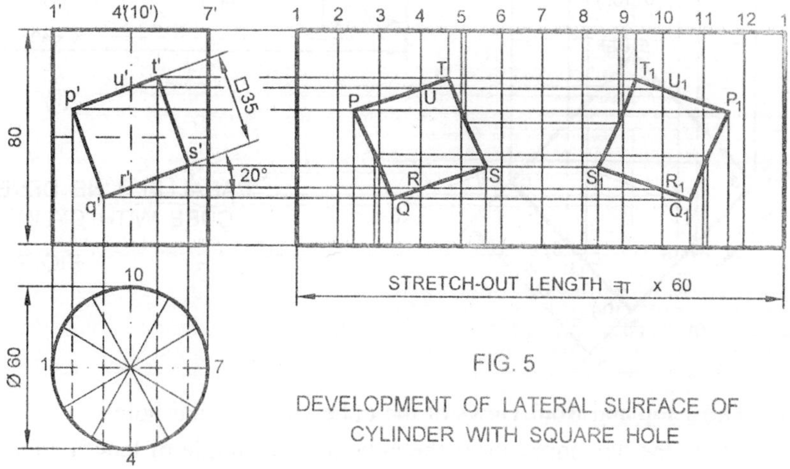

FIG. 5

DEVELOPMENT OF LATERAL SURFACE OF CYLINDER WITH SQUARE HOLE

Problem 6: *(Fig. 6)*
*A **cylinder** of 50 mm base diameter and axis 70 mm long rests on its base on HP. A **square cutout** of 35 mm side is drilled through the cylinder such that axis of cutout is ⊥r to the axis of the cylinder. The center of the cutout is 35 mm above HP and 15 mm away from the axis of cylinder. Two faces of the cutout are equally inclined to HP. Develop lateral surfaces.* (UQ)

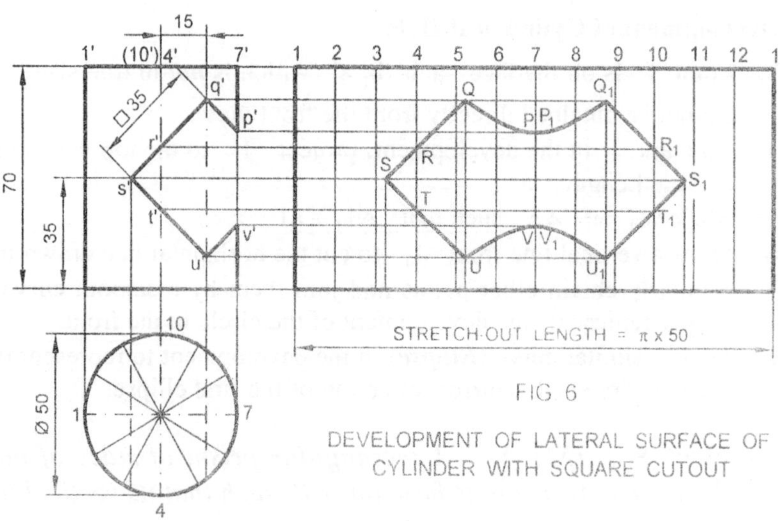

FIG 6

DEVELOPMENT OF LATERAL SURFACE OF CYLINDER WITH SQUARE CUTOUT

2. PRISMS AND CYLINDERS WITH CYLINDRICAL CUTOUTS

Problem 7: *(Fig. 7) A **cube** of 40 mm edge stands on one of its faces on HP with a vertical face making 45° to VP. A horizontal **cylindrical hole** of 30 mm diameter is drilled centrally through the cube such that the hole passes through the opposite vertical edges of the cube. Draw the development of lateral surfaces of the cube having the hole.* (UQ)

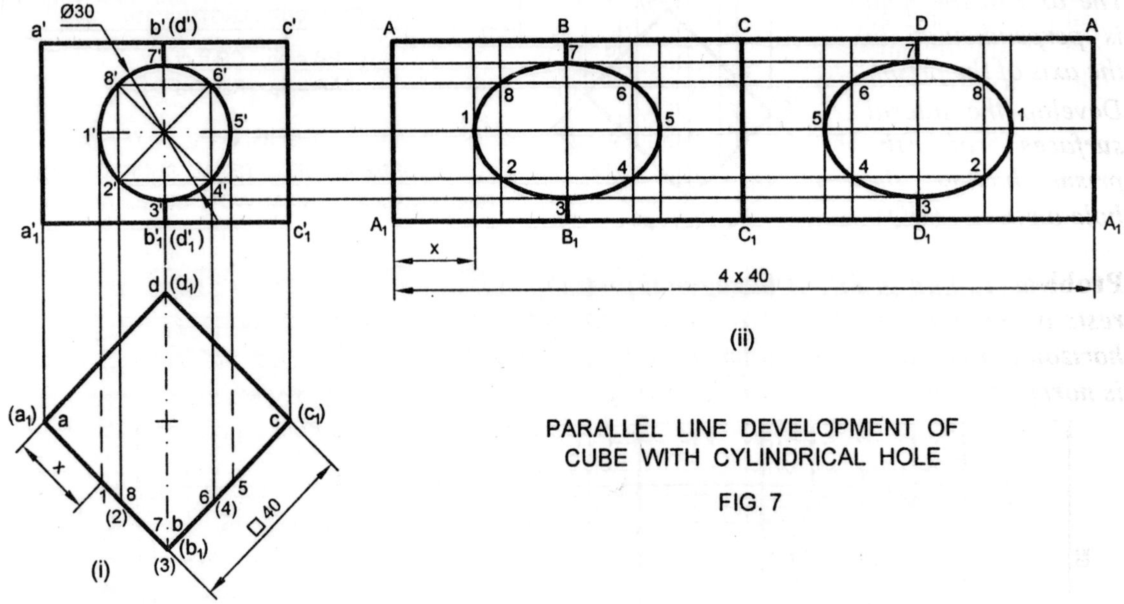

PARALLEL LINE DEVELOPMENT OF CUBE WITH CYLINDRICAL HOLE

FIG. 7

1. Draw top and front views of the cube with circular hole.
2. Draw the development of the cube as a rectangle of length equal to the perimeter of the base and breadth equal to the side of the cube and show the vertical edges.

Development of Cylindrical Hole

3. Point 1′ is on the face $a' \, b' \, b_1' \, a_1'$ which is *not* in true size. Hence 1 in the development cannot be obtained directly from the front view.
4. To mark 1 in the development, project 1′ to the top view and obtain 1 on ab which is in True Length.
5. Mark 1_1 on AA such that $A1_1 = a1 = x$.
6. Draw a vertical line from 1_1 to cut the horizontal line drawn from 1′ at 1.
7. Similarly obtain other points and join them by a smooth curve (*Ellipse* in this case). This ellipse represents the development of the circle in the front.
8. Draw a similar curve (***Ellipse***) in the development to represent the hole in the rear side. This second curve is the mirror reflection of the first ellipse.

Problem 8: *(Fig. 8) A **rectangular prism** of sides of base 50 mm x 40 mm and height 68 mm rests on its base on HP such that its vertical faces are equally inclined to*

VP. It has a horizontal ***hole*** *of 40 mm diameter drilled through the geometrical center of the prism. The axis of the hole is perpendicular to the axis of the prism. Develop the lateral surfaces of the prism having the hole on it.* (UQ)

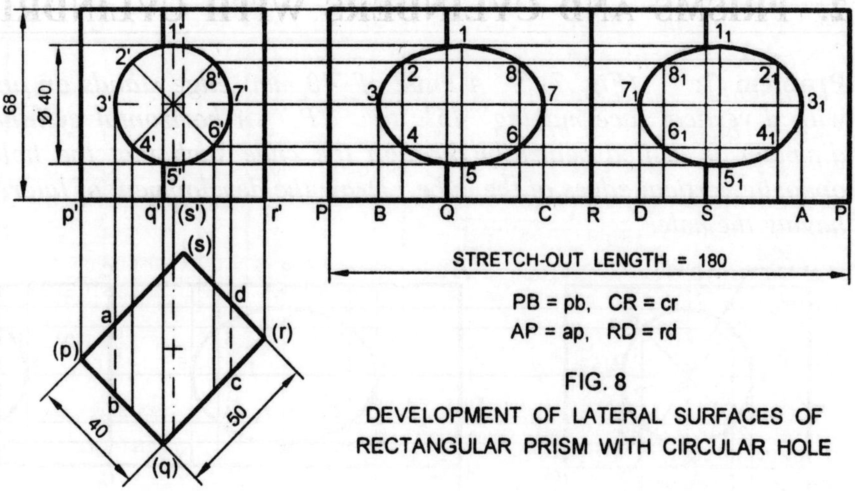

FIG. 8

DEVELOPMENT OF LATERAL SURFACES OF RECTANGULAR PRISM WITH CIRCULAR HOLE

Problem 9: *(Fig. 9) A* ***hexagonal prism*** *of side of base 25 mm and altitude 65 mm rests on its base on HP, having a rectangular face of the prism parallel to VP. A horizontal* ***hole*** *of 35 mm diameter is centrally drilled in it, such that the axis of the hole is normal to VP. Develop the lateral surfaces of the prism with the shape of hole.* (UQ)

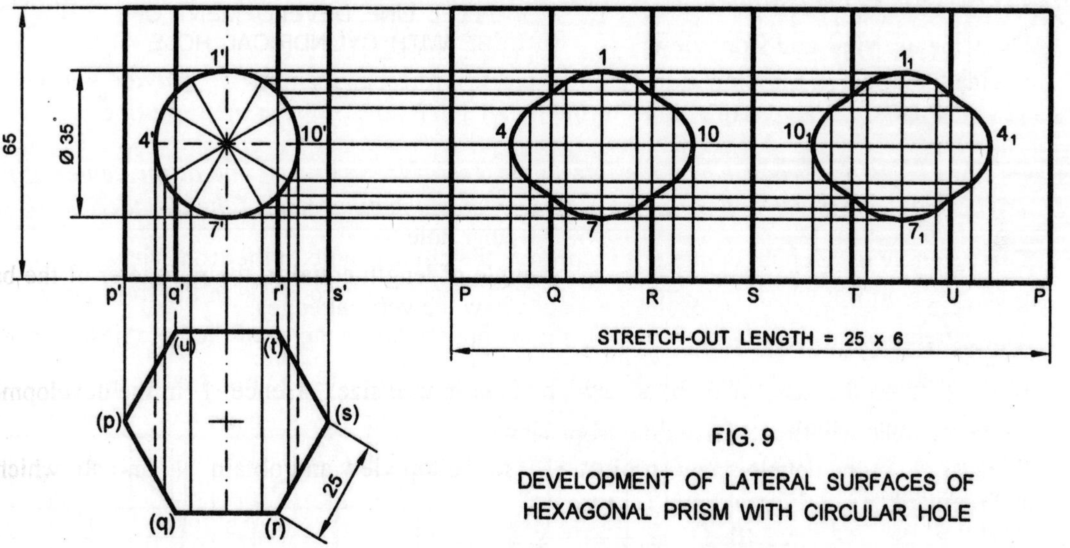

FIG. 9

DEVELOPMENT OF LATERAL SURFACES OF HEXAGONAL PRISM WITH CIRCULAR HOLE

1. Draw the projections of given hexagonal prism. Draw a circle of 35 mm diameter in front view, 32.5 mm above the base. Divide the circle into 12 equal parts. Project it to top view.
2. Draw the development of the lateral surfaces of the hexagonal prism.
3. Draw horizontal lines from the generators of the hole in front view to the development of the prism. Transfer the distances of generators in top view to the development. Join the intersecting points in proper sequence and complete the development.

Problem 10: *(Fig. 10) A right regular* ***pentagonal prism,*** *base edge 28 mm and altitude 62 mm, stands on its base on HP such that one of its base edges is parallel to VP and* *nearer to the observer. Through the center of the rectangular face containing the base edge, a* **hole** *of 32 mm diameter is drilled through, such that the hole axis is horizontal. Develop lateral surfaces.* (UQ)

Assume the circular hole ⊥ʳ to VP. True shapes of the front and rear holes in the pentagonal prism differ in the development.

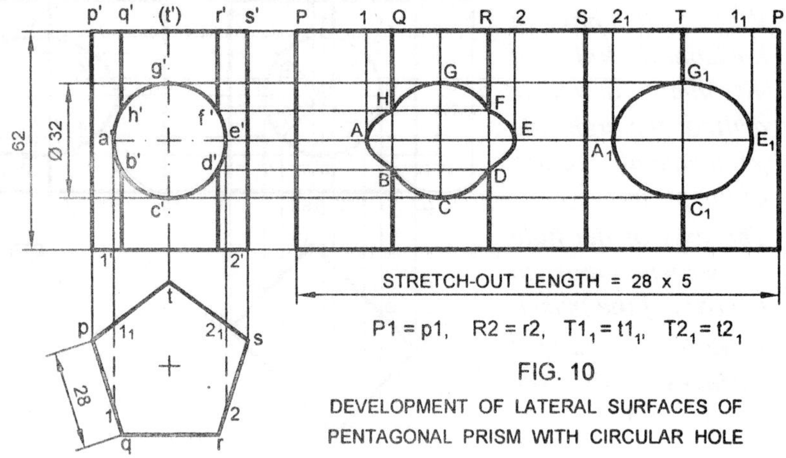

$P1 = p1,\ R2 = r2,\ T1_1 = t1_1,\ T2_1 = t2_1$

FIG. 10

DEVELOPMENT OF LATERAL SURFACES OF PENTAGONAL PRISM WITH CIRCULAR HOLE

Problem 11: *(Fig. 11)* ***A cylinder** of base diameter 64 mm and axis length 80 mm rests on HP on its base. A **hole** of 60 mm diameter is drilled on the surface of the cylinder. The axis of the hole intersects with the axis of the cylinder at right angles and bisects the axis of the cylinder. Draw the development.* (UQ)

1. Draw the top view and front view.
2. Divide the circle in top view into 12 equal parts. Project to front view. Divide line 1-1 into 12 equal parts. Draw vertical lines through division points of the stretch-out line.

Note: *Two holes are produced on the lateral surface of the cylinder, one at the center of front half of the cylinder and the other at the center of the other half of the cylinder.*

---

3. Draw a circle of 60 mm diameter to represent the circular hole in the front view.
4. Mark the points at which the generators of cylinder cut the hole circle in the front view.
5. The extreme points $a'(a_1')$ and $g'(g_1')$ on the horizontal diameter of the hole lies in between the two generators. Project $a'(a_1')$ and $g'(g_1')$ to the top view to obtain m, m_1 and n, n_1. Transfer the above points to the development by marking 1M = 1m.
6. Draw a vertical line through M. Draw a horizontal line through a' (a_1') to meet the above vertical line at A. Similarly obtain the other points and draw the smooth curves.

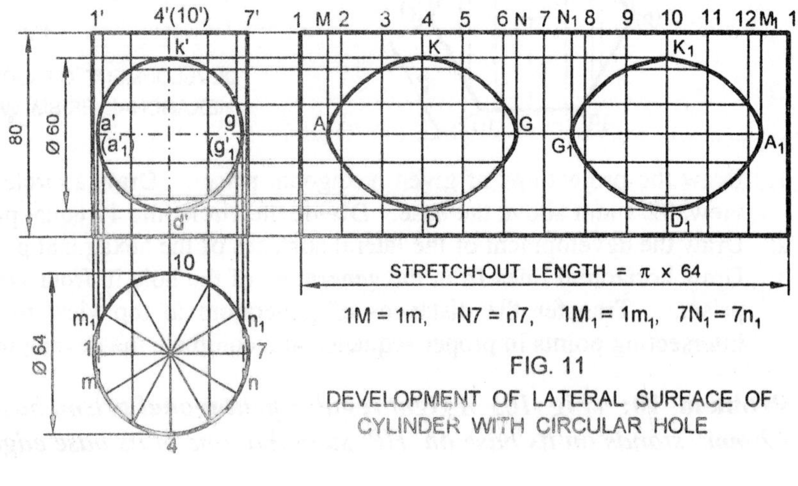

$1M = 1m,\ N7 = n7,\ 1M_1 = 1m_1,\ 7N_1 = 7n_1$

FIG. 11

DEVELOPMENT OF LATERAL SURFACE OF CYLINDER WITH CIRCULAR HOLE

Problem 12: *(Fig. 12) A **square prism** of side of base 45 mm and axis 72 mm long stands on its base on HP with a vertical face inclined at 45° to VP. A **circular cutout** of 40 mm diameter is drilled through the prism. The center of the cutout is 40 mm above the base and is 15 mm off-set from the axis of the prism. Draw the development of the lateral surfaces of the solid.* (UQ)

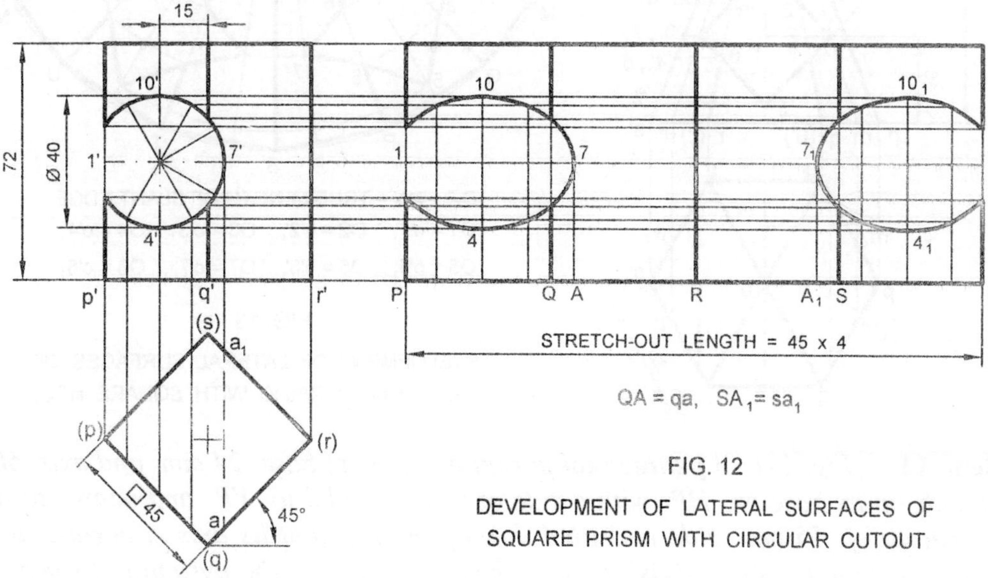

FIG. 12

DEVELOPMENT OF LATERAL SURFACES OF SQUARE PRISM WITH CIRCULAR CUTOUT

3. PYRAMIDS & CONES WITH SQUARE / TRIANGULAR CUTOUTS

Problem 13: *(Fig. 13) A regular **hexagonal pyramid**, 22 mm side of base and height 55 mm, stands on its base with an edge of base parallel to VP. A **square hole** of 18 mm side is cut through it such that the axis of the hole is perpendicular to VP and intersects the axis of the pyramid 16 mm above the base. The side faces of the hole are equally inclined to HP. Develop the solid with the square hole.* (UQ)

1. Draw the top and front views of the hexagonal pyramid with square hole. Draw the development of the lateral surfaces of the hexagonal pyramid.
2. Name the cutting points of hole with the surface of pyramid in front view as 1′, 2′, ..., 8′.
3. Join the apex to the extreme points of the square hole to meet the base, i.e., join o′ 3′. Extend it to meet the base of the pyramid at m′. Similarly mark n′.
4. Project m′ and n′ to the top view and mark m, m_1 and n, n_1. Transfer m, m_1 and n, n_1 to the development of the prism as M, M_1 and N, N_1.
5. Draw horizontals from the cutting points to meet slant edge o′s′. Transfer the corresponding distances to the development. Join the points representing the hole in sequence.

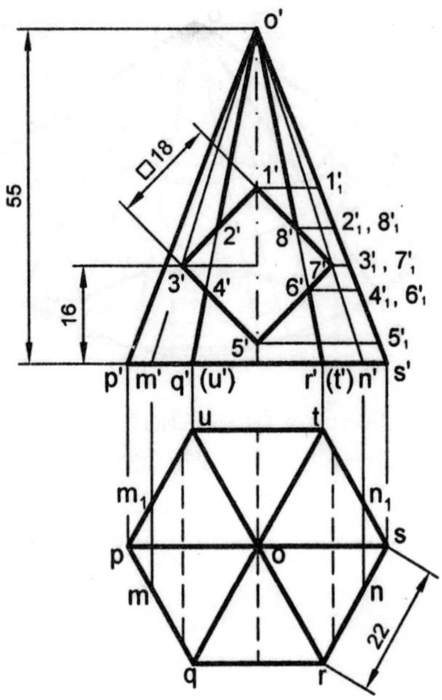

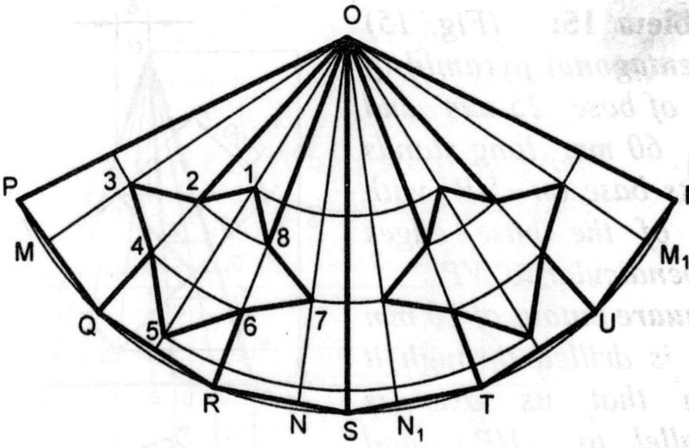

OP = o'p' = TRUE LENGTH OF SLANT EDGE

O1 = o'1'$_1$, O2 = o'2'$_1$, O3 = o'3'$_1$, O4 = o'4'$_1$

O5 = o'5'$_1$, O6 = o'6'$_1$, O7 = o'7'$_1$, O8 = o'8'$_1$

FIG. 13

DEVELOPMENT OF LATERAL SURFACES OF HEXAGONAL PYRAMID WITH SQUARE HOLE

Problem 14: *(Fig. 14) A* ***pentagonal pyramid*** *of side of base 24 mm and axis* ***60 mm*** *long stands on its base on HP with a side of base parallel to VP and nearer to it. A* ***square cutout*** *of 15 mm side is drilled through it such that its axis is parallel to HP and perpendicular to VP. Axis of cutout meets the axis of the pyramid 15 mm from base. Faces of the cutout are equally inclined to HP. Develop the lateral surfaces.* **(UQ)**

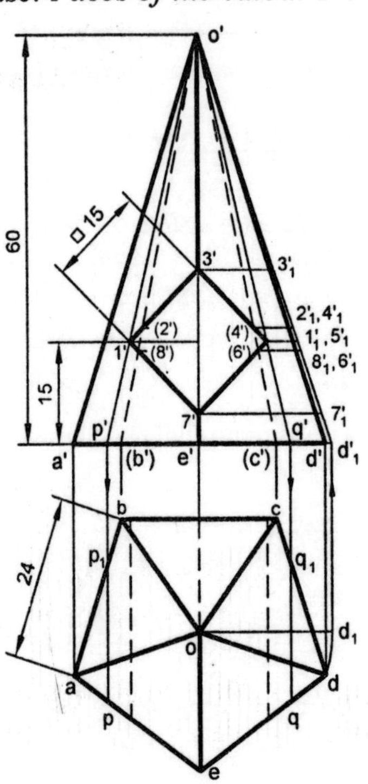

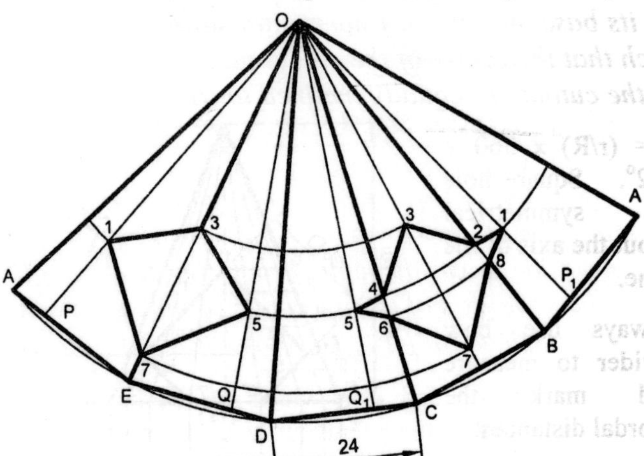

OA = o'd'$_1$ = TRUE LENGTH OF SLANT EDGE

O1 = o'1'$_1$, O2 = o'2'$_1$, O3 = o'3'$_1$, O4 = o'4'$_1$

O5 = o'5'$_1$, O6 = o'6'$_1$, O7 = o'7'$_1$, O8 = o'8'$_1$

FIG. 14 DEVELOPMENT OF LATERAL SURFACES OF PENTAGONAL PYRAMID WITH SQUARE HOLE

Problem 15: *(Fig. 15) A **pentagonal pyramid** of side of base 25 mm and axis 60 mm long stands on its base on HP with one of the base edges perpendicular to VP. A **square cutout** of 20 mm side is drilled through it such that its axis is parallel to HP and perpendicular to VP. The center of the cutout is 20 mm above the base of pyramid and 6 mm away from its axis. The faces of the cutout are equally inclined to HP. Develop the lateral surfaces of the pyramid with the cutout.* (UQ)

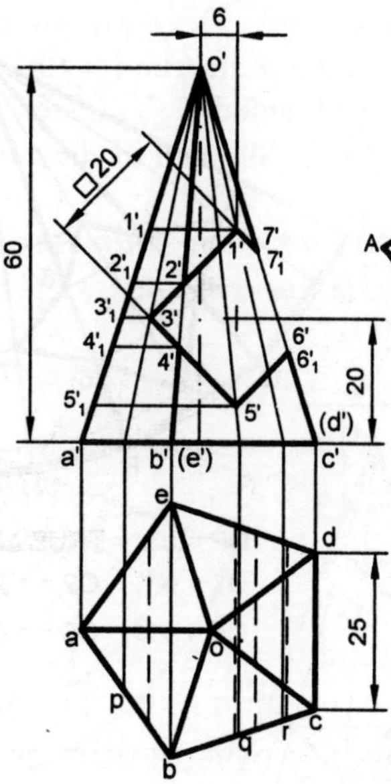

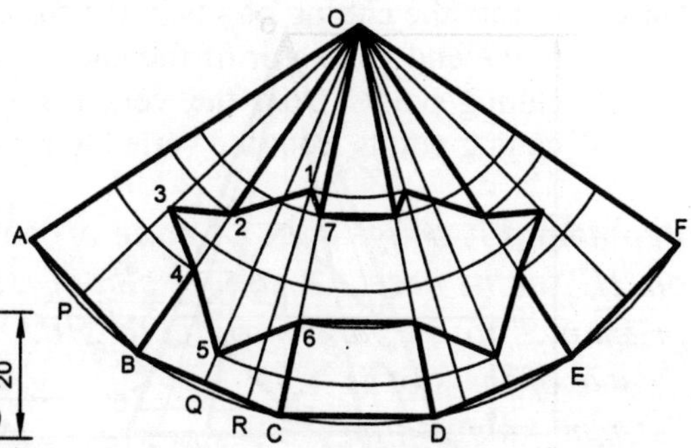

OA = o'a' = TRUE LENGTH OF SLANT EDGE

O1 = o'1'₁, O2 = o'2'₁, O3 = o'3'₁, O4 = o'4'₁

O5 = o'5'₁, O6 = o'6'₁, O7 = o'7'₁

FIG. 15

DEVELOPMENT OF LATERAL SURFACES OF PENTAGONAL PYRAMID WITH SQUARE CUTOUT

Note: When a base edge is perpendicular to VP, one of the slant edges is parallel to XY in the top view. This slant edge gives the true length in the front view.

Problem 16: *(Fig. 16) A **cone** of 60 mm base diameter and axis 70 mm long stands on its base on HP. A horizontal **square cutout** of 30 mm side is cut through the cone such that the center of the cutout meets the axis of the cone 12 mm from the base. Sides of the cutout are equally inclined to HP. Develop the lateral surface of the solid.* (UQ)

$\theta = (r/R) \times 360 = 142^\circ$. Square hole is symmetrical about the axis of the cone.

Always use bow divider to measure and mark the chordal distances.

This kind of marking of small distances may result a shortening of approximately 1% in length. This error is negligible.

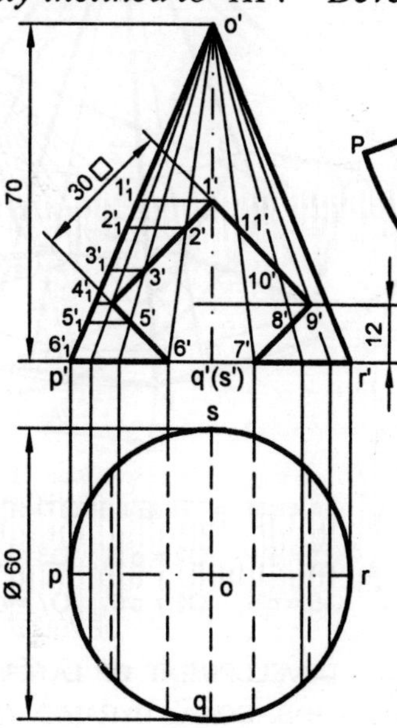

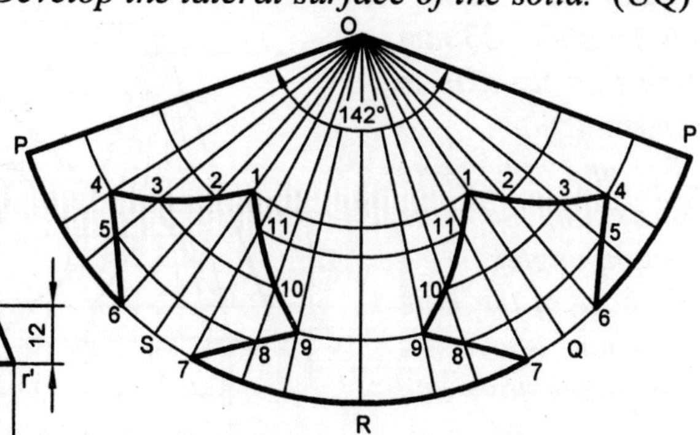

OP = o'p' = TRUE LENGTH OF GENERATOR

O1 = o'1'₁, O2 = o'2'₁, O3 = o'3'₁

O4 = o'4'₁, O5 = o'5'₁, O6 = o'6'₁

FIG. 16

DEVELOPMENT OF LATERAL SURFACE OF CONE WITH SQUARE CUTOUT

Note: Locate the cutting points in the front view. Draw horizontal lines from the cutting points to the end generator of the cone (which is in true length). Thus the true lengths of the cutting points from the vertex are obtained. Transfer the above true lengths of the cutting points from the vertex to the development of the cone.

Problem 17: *(Fig. 17) A cone of 55 mm base diameter and altitude 65 mm stands on HP on its base. A square slot of 30 mm side passes through it such that axis of the prism is ⊥ʳ to VP and parallel to HP. Center of the slot is 24 mm above the base and 6 mm away from the axis of cone. Faces of the slot are equally inclined to HP. Develop the lateral surface.* (UQ)

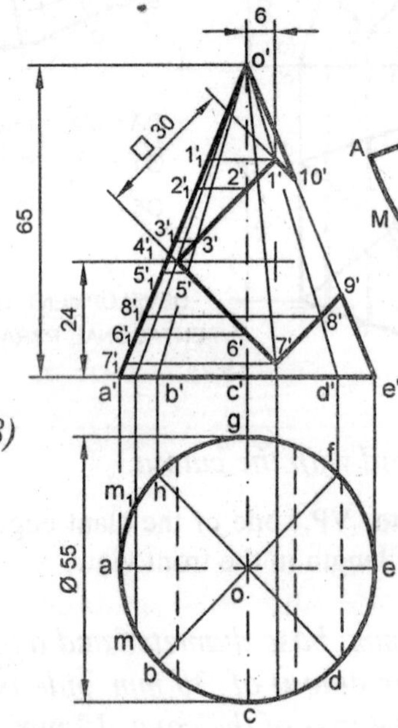

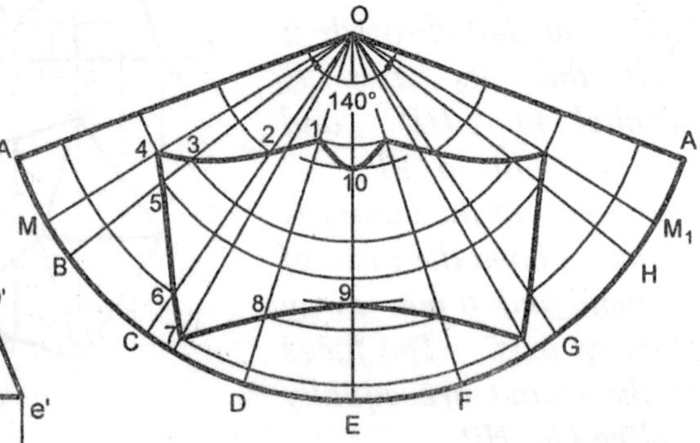

OA = o'a' = TRUE LENGTH OF GENERATOR

O1 = o'1'₁, O2 = o'2'₁, O3 = o'3', O4 = o'4'₁

O5 = o'5'₁, O6 = o'6'₁, O7 = o'7', O8 = o'8'₁

FIG. 17

DEVELOPMENT OF LATERAL SURFACE OF CONE WITH SQUARE SLOT

Problem 18: *(Fig. 18) A vertical cone of base diameter 60 mm and height 75 mm rests on its base on HP. It is cut with a through triangular hole of base 33 mm and height 23 mm such that its axis is perpendicular to VP and intersects the axis of the cone at right angles. The base of the hole is 8 mm above the base of the cone and parallel to it. Draw the development of the cone showing the true shape of hole on it.* (UQ)

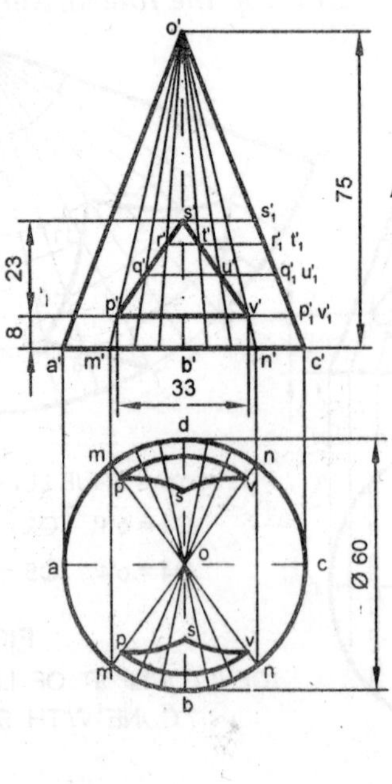

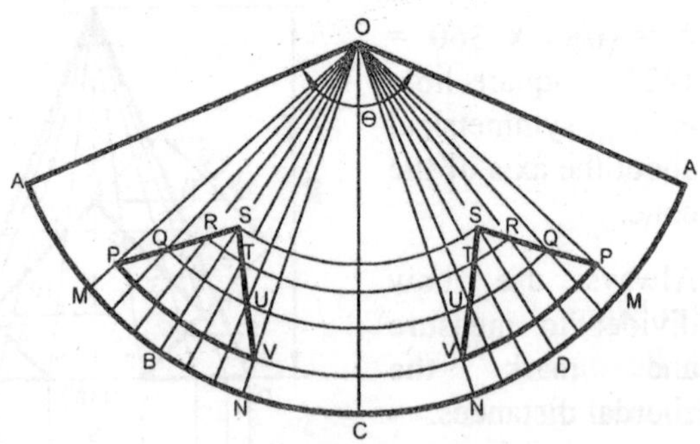

TRUE LENGTH OF SLANT GENERATOR = o' a' = OA

FIG. 18

DEVELOPMENT OF CONE WITH TRIANGULAR HOLE

4. PYRAMIDS AND CONES WITH CYLINDRICAL CUTOUTS

Problem 19: *(Fig. 19) A **hexagonal pyramid** of side of base 24 mm and altitude 56 mm rests with its base on HP with an edge of the base parallel to VP. A through **circular hole** of 24 mm diameter is drilled in the pyramid such that the axis of the hole is perpendicular to VP and cuts the axis of the pyramid 16 mm above the base. Develop the lateral surface of the pyramid with the true shape of the hole cut in it.* (UQ)

1. Draw top and front views of the pyramid. Draw the circular hole in the front view. Draw two tangents o′m′ and o′n′ from the vertex o′ to the circle to meet the circle at a′ and e′.
2. Draw horizontals from a′ and e′ to meet the true slant edge o′ p′ at the points a_1' and e_1'.
3. Project m′ to top view on pq and pu to obtain m and m_1.
4. Similarly, project n′ to top view on rs and st to obtain n and n_1.
5. Draw the development of pyramid. Mark point M on PQ such that PM = pm.
6. Similarly, mark N on RS such that RN = rn. Mark K on QR such that QK = qk.
7. On OM mark A such that OA = $o'a_1'$. On ON mark E such that OE = $o'e_1'$.
8. Similarly mark B, C, D, F, G and H.
9. Complete the development with the true shape of the hole.

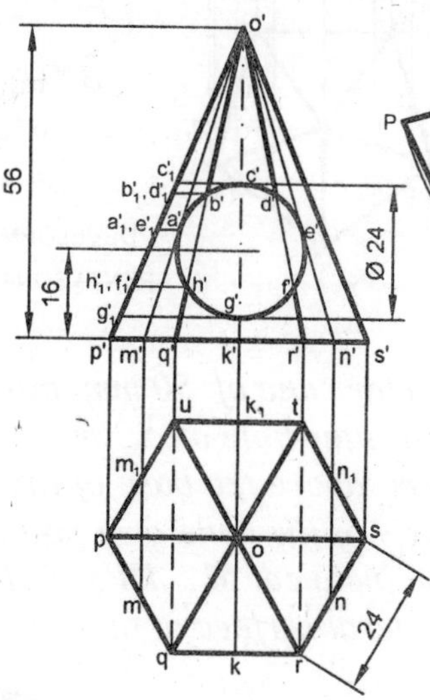

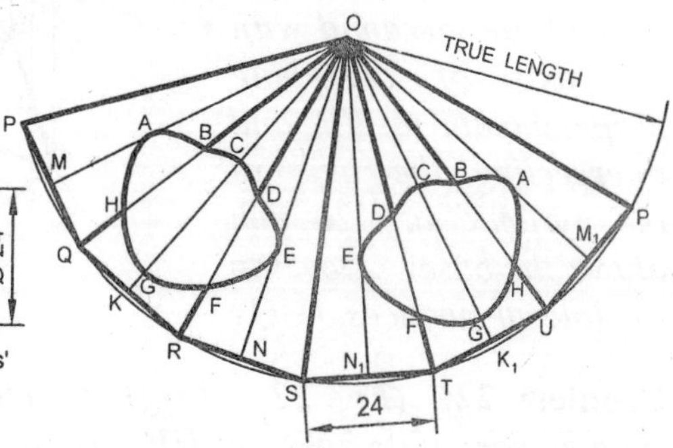

OP = o'p' = TRUE LENGTH OF SLANT EDGE

OA = $o'a_1'$, OB = $o'b_1'$, OC = $o'c_1'$, OD = $o'd_1'$

OE = $o'e_1'$, OF = $o'f_1'$, OG = $o'g_1'$, OH = $o'h_1'$

FIG. 19

DEVELOPMENT OF LATERAL SURFACES OF HEXAGONAL PYRAMID WITH CIRCULAR HOLE

Problem 20: *(Fig. 20) A **hexagonal pyramid** of 25 mm side of base and axis 60 mm long stands vertically with a base edge parallel to VP. A **cylindrical slot** of 32 mm diameter is drilled through the pyramid in such a way that the axis*

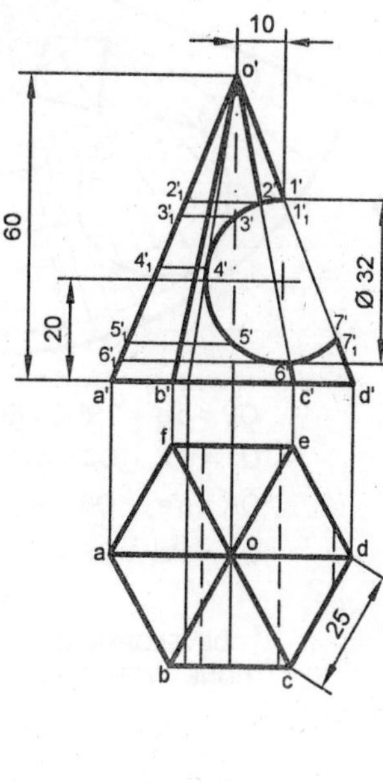

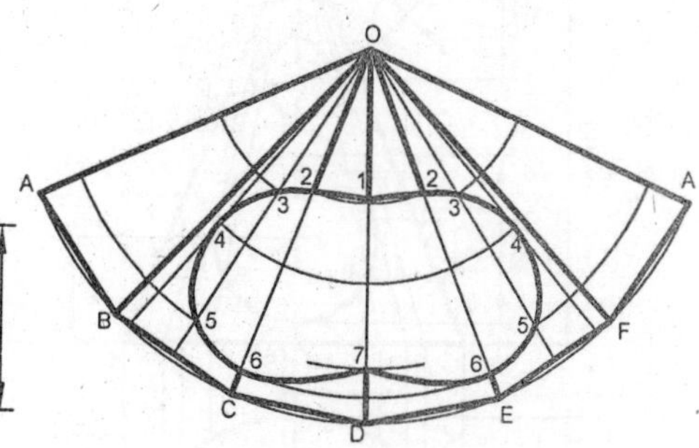

OA = o'a' = TRUE LENGTH OF SLANT EDGE

O1 = $o'1_1'$, O2 = $o'2_1'$, O3 = $o'3_1'$, O4 = $o'4_1'$

O5 = $o'5_1'$, O6 = $o'6_1'$, O7 = $o'7_1'$

FIG. 20

DEVELOPMENT OF LATERAL SURFACES OF HEXAGONAL PYRAMID WITH CYLINDRICAL SLOT

of the slot is perpendicular to VP. The center of the slot is 20 mm above the base and 10 mm away from the axis of the pyramid. Develop the lateral surfaces. (UQ)

Problem 21: *(Fig. 21) A **pentagonal pyramid** of base edge 24 mm and axis 60 mm long rests on its base on HP with a base edge parallel to VP and away from observer.*

*A **hole** of 20 mm diameter is drilled right through the pyramid with the axis of the hole perpendicular to VP and intersecting the axis of the pyramid at 20 mm above the base. Develop the lateral surfaces.* (UQ)

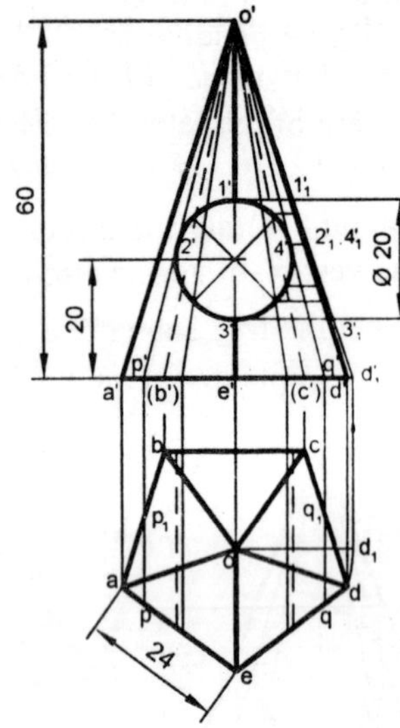

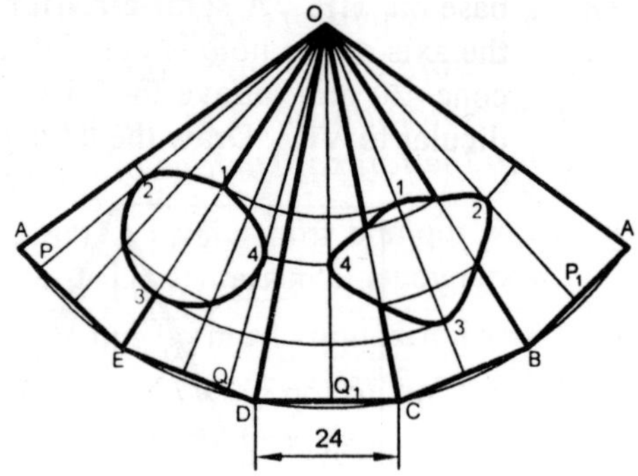

OA = o'd'$_1$ = TRUE LENGTH OF SLANT EDGE

O1 = o'1'$_1$, O2 = o'2'$_1$, O3 = o'3'$_1$, O4 = o'4'$_1$

FIG. 21

DEVELOPMENT OF LATERAL SURFACES OF PENTAGONAL PYRAMID WITH CIRCULAR HOLE

Problem 22: *(Fig. 22) A right regular **cone** of 50 mm base diameter and axis 60 mm long stands on its base on HP. A **circular hole** of 12 mm radius is drilled through the axis of the cone at a height of 15 mm above the base of the cone. Also, a **square hole** of 10 mm side is cut through the axis of the cone at a point 22 mm below the vertex. The sides of the hole are equally inclined to HP. The axes of the holes are perpendicular to VP. Develop the lateral surface of the cone with the holes in it.* (UQ)

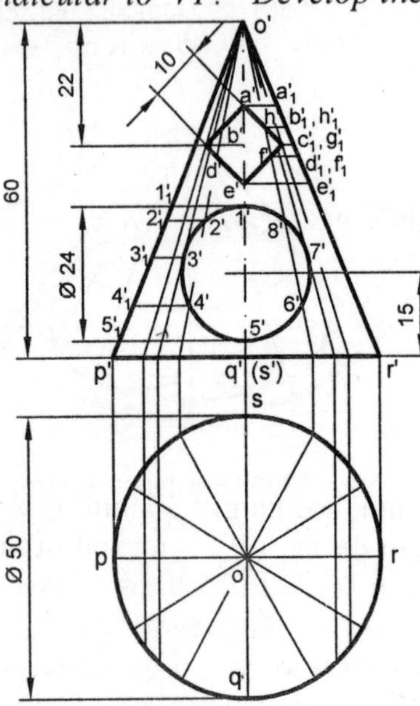

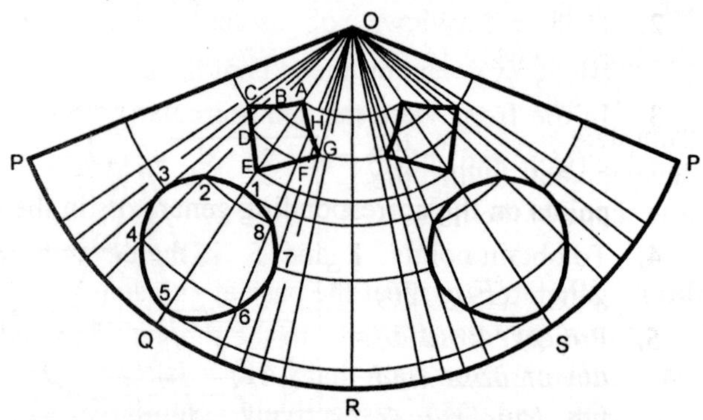

OP = o'p' = TRUE LENGTH OF GENERATOR

O1 = o'1'$_1$, O2 = o'2'$_1$, O3 = o'3'$_1$, O4 = o'4'$_1$

OA = o'a'$_1$, OB = o'b'$_1$, OC = o'c'$_1$, OD = o'd'$_1$

OE = o'e'$_1$, OF = o'f'$_1$, OG = o'g'$_1$, OH = o'h'$_1$

FIG. 22

DEVELOPMENT OF LATERAL SURFACE OF CONE WITH CIRCULAR AND SQUARE HOLES

5. ADDITIONAL PROBLEMS

23. (Fig. 23) A **cone** of base 60 mm diameter and 60 mm height stands vertically with its base on HP. A **semi-circular hole** of 36 mm diameter is cut through the cone such that the axis of the hole is parallel to HP, perpendicular to VP and intersecting the axis of the cone 20 mm above the base. The flat surface of the hole is parallel to HP and perpendicular to VP. Draw the development of the lateral surface of cone with the hole. (UQ)

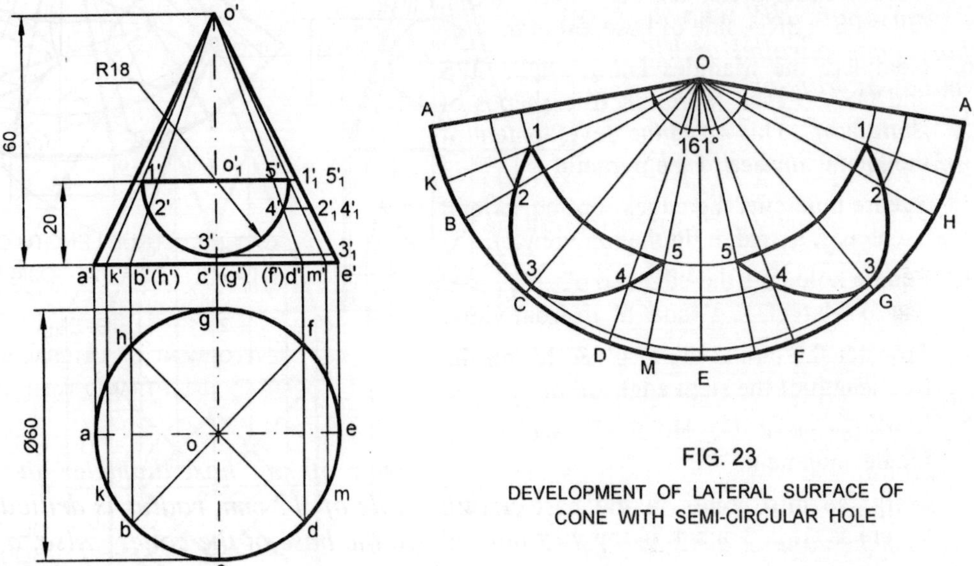

FIG. 23

DEVELOPMENT OF LATERAL SURFACE OF CONE WITH SEMI-CIRCULAR HOLE

1. Draw the projections of the cone and its development as shown.
2. In the front view, o_1' as center and 36 mm as diameter draw a semi-circle at 20 mm above XY and on the axis of the cone.
3. In the front view the semi-circular hole cuts 6 generators o′b′, o′c′, o′d′, o′(f′), o′(g′) and o′(h′) at 2′, 3′, 4′, ($4_1'$), ($3_1'$) and ($2_1'$) respectively. Obtain these points on the corresponding generators in the development as usual.
4. To obtain points 1 and 5 in the development, through 1′ and 5′ draw 2 additional generators to meet the base at k′ and m′.
5. Project k′ and m′ to the top view and obtain k and m. Mark K and M in the development such that AK = ak and DM = dm. Then mark 1 and 5 as usual on OK and OM respectively. Similarly obtain 1_1 and 5_1 on the rear side of the hole.

24. *(Fig. 24)* A **square pyramid**, side of base 50 mm and height 60 mm is resting on HP on its base with two of its sides of the base parallel to VP. An axial **square hole** of 25 mm side is cut through the pyramid so that the vertical faces of the hole are equally inclined to VP. Draw the projections of the pyramid. Show the development of the lateral surface of the solid. (UQ)

1. Draw top and front views of the pyramid.
2. Mark the square hole in the top view.
3. Project the hole to the front view.
4. Obtain the true length of the slant edge of the square pyramid as o′m′.
5. O as center and true length of slant edge o′m′ as radius, draw an arc. On this arc, mark 4 equal divisions, i.e., chord RQ = QP = PS = SR = side of base 50 mm.
6. Complete the triangles ORQ, OQP, OPS and OSR, which represent true shapes of slant face. This gives the development of the lateral surface of the pyramid.
7. Square hole cuts the edges op, oq, or and os at e, f, g and h in the top view.
8. Square hole cuts the edges o′p′, o′q′, o′r′ and o′s′ at e′, f′, g′ and h′ in front view.
9. Transfer the points e′, f′, g′ & h′ on the true length of the slant edge o′ m′.
10. Mark the point G, H, E, F and G in the development.
11. Mark the mid points of the base of the square as 1, 2, 3 and 4 in top view.
12. Join o1, o2, o3 & o4 which pass through key points a, b, c & d in top view.

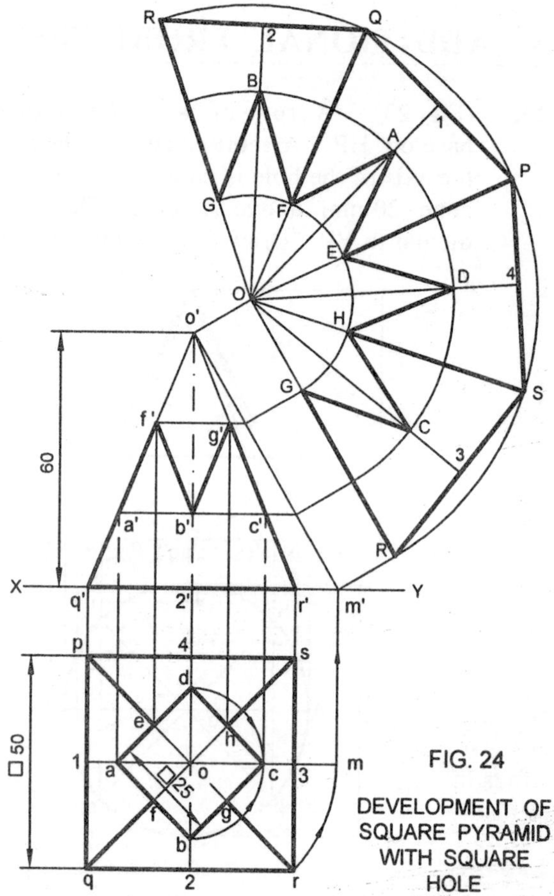

FIG. 24

DEVELOPMENT OF SQUARE PYRAMID WITH SQUARE HOLE

13. Mark the mid-points of the chords PQ, QR, RS and SP as 1, 2, 3 and 4 respectively in the development. Join O1, O2, O3 and O4 in the development.
14. On O1, O2, O3 and O4 transfer the true positions of a′, b′, c′ and d′ as A. B, C and D in the development. Join GC, CH, HD, DE, EA, AF, FB and BG.
15. Complete the development of the pyramid with hole in thick lines as shown.

25. A **cone** of 36 mm base diameter and axis 60 mm long rests on its base on HP. A **square slot** of diagonal 20 mm is drilled right through the cone with one diagonal coinciding with the axis of the cone. Axis of the slot is perpendicular to VP and meets the axis of the cone 18 mm above its base. Develop the lateral surface of the solid. (UQ)

26. A **cone** of 40 mm base diameter and altitude 50 mm rests on its base on HP. A **cylindrical hole** of 20 mm diameter is drilled right through the cone. The axis of the hole is perpendicular to VP and meets the axis of the cone 15 mm above the base of the cone. Develop the lateral surface. (UQ)

*

18

INTERSECTION OF SURFACES

1. INTERSECTION OF SURFACES OF SOLIDS

When one solid enters or penetrates into the other of the same or different cross-section, the surfaces of both the solids come in contact with one another. Thus the outline of the penetrating solid is seen. This is known as ***Interpenetration of Solids.***

In engineering practice, when two solids are joined to make a single object, for example, a dome fitted on a boiler, their surfaces will intersect each other. The line of intersection of both the surfaces will be common.

When two solids of the same or different cross-section come in contact with each other, the surfaces of both the solids intersect one another. This sort of surface contact is termed as ***Intersection of Surfaces of Solids.***

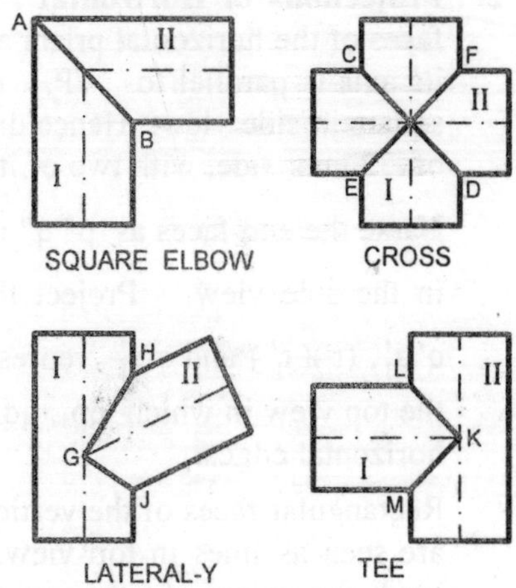

LINE OF INTERSECTION: AB, CD, EF, GH, GH, KL and KM

FIG. 1 PIPE FITTINGS

Engineering Applications of Intersection of Surfaces: *See Fig. 1.* Pipe fittings like elbow, T-joint, Y-joint, etc., Fitting of steam dome of a cylindrical boiler, Fabrication of chimney / hopper, Wings of airplane, Automobile body building, Sheet metal works, Ship building, etc.

Classifications of Intersection of Surfaces of Solids

1. Intersection of two plane surfaces (Prism Vs Prism)
2. Intersection between a plane surface and a curved surface (Prism Vs Cylinder)
3. Intersection of two curved surfaces (Cylinder Vs Cylinder, Cylinder Vs Cone)

Line or Curve of Intersection: Line of intersection of two surfaces of solids is a line common to both. When two solids bounded by plane surfaces such as prism and pyramid penetrate each other, we obtain ***straight lines*** as their lines of intersection. Line of intersection between a plane surface and a curved surface of two solids such as prism and cylinder is a ***curve.*** The line of intersection between two curved surfaces of solids such as cylinder and cone is a ***curve.***

1.1 METHODS OF DETERMINING THE LINE / CURVE OF INTERSECTION

1. Line / Generator Method / Piercing Point Method / Edgewise View Method: Used for Prisms, Cylinders, etc.
2. Section Plane Method: Used for Pyramids, Cones, Cylinders, Spheres, etc.

2. INTERSECTION OF PRISM Vs PRISM

Problem 1: *(Fig. 2) A square prism, side of base 40 mm and height 75 mm stands with its base on HP and two of its rectangular faces are equally inclined to VP. It is completely penetrated by a horizontal square prism, side of base 32 mm and axis 75 mm long, such that the axes of the two prisms intersect each other at right angles. The two rectangular faces of the horizontal prism are equally inclined to HP and its axis is parallel to VP. Draw the projection of the prisms showing the lines of intersection.*

1. **Projections of Vertical Prism:** Draw the top view of vertical prism as a square of 40 mm side such that two of its sides are equally inclined to XY. Name the top and bottom square faces as abcd and (a_1) (b_1) (c_1) (d_1) respectively. Project its corresponding front view in which $a'\,a_1'$, $b'\,b_1'$, $c'\,c_1'$ and $(d')\,(d_1')$ represent vertical edges. Draw Left Side View in which $d''\,d_1''$, $a''\,a_1''$, $b''\,b_1''$ and $(c'')(c_1'')$ represent vertical edges.

2. **Projections of Horizontal Prism:** Since two rectangular faces of the horizontal prism are equally inclined to HP and its axis is parallel to VP, the end faces will be seen as a square in side view. Hence draw Left Side View as a square of 32 mm side, with two of its sides equally inclined to XY.

   Name the end faces as $p''\,q''\,r''\,s''$ and $(p_1'')\,(q_1'')\,(r_1'')\,(s_1'')$ in the side view. Project the front view in which $p'\,p_1'$, $q'\,q_1'$, $(r')(r_1')$ and $s'\,s_1'$ represent horizontal edges. Project the top view in which pp_1, qq_1, rr_1 and $(s)(s_1)$ represent the horizontal edges.

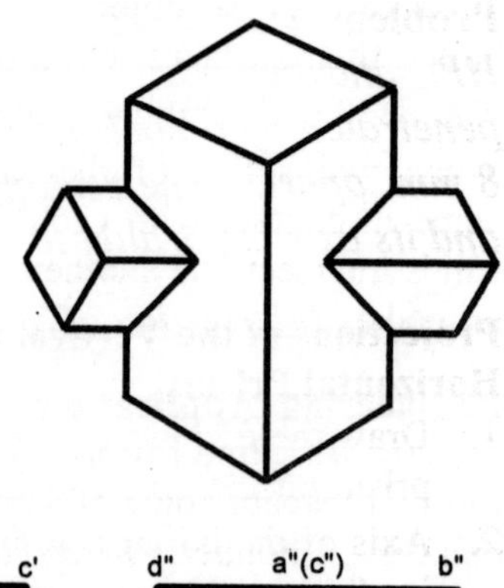

   Rectangular faces of the vertical prism are seen as lines in top view. Hence mark the points of intersection in top view first.

3. **Intersection Points in Top View:** In the top view the horizontal edges pp_1 pierces rectangular face $ab(b_1)(a_1)$ at 1, qq_1 intersects the edge $a(a_1)$ at 2, rr_1 pierces the rectangular face $ad(d_1)(a_1)$ at 3 and $(s)(s_1)$ intersects the edges $a(a_1)$ at (4).

   Similarly mark 5, 6, 7 and (8) in the top view.

4. **Intersection Points in Front View:** From 1 and 2 draw projectors to intersect $p'\,p_1'$ and $q'\,q_1'$ at 1′ and 2′ respectively.

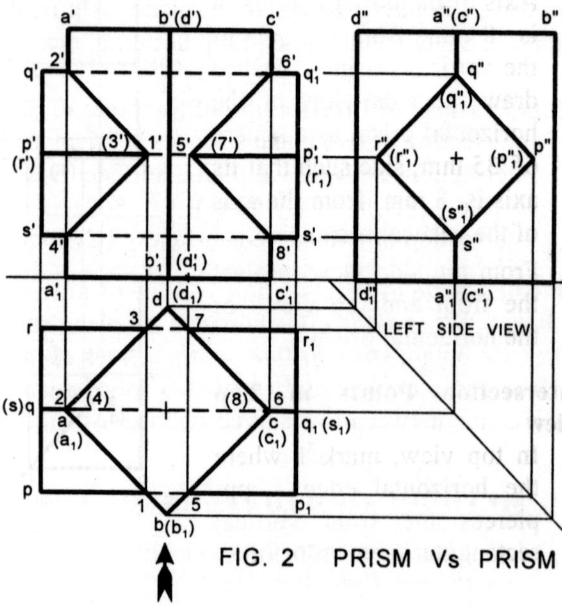

FIG. 2 PRISM Vs PRISM

Also project (3′) and 4′ on (r′)(r_1') and s′s_1' respectively. Similarly obtain 5′, 6′, (7′) and 8′ in the front view.

5. **Visible and Invisible Lines of Intersection in Front View:** Look at the top view in the direction of arrow. Lines of intersection 1-2, 1-(4), 5-6 and 5-(8) are visible. Hence draw 1′2′, 1′4′, 5′6′ and 5′8′ as thick lines in the front view.
   Lines of intersection 2-3, 3-(4), 6-7, 7-(8) are hidden by 1-2, 1-(4), 5-6 and 5-(8) respectively in the top view. Hence show the lines 2′(3′), (3′) 4′, 6′(7′), (7′) 8′ to coincide with 1′2′, 1′4′, 5′6′ and 5′8′ in the front view.

Rules of Visibility

- If a point lies on a visible edge or faces of both the intersecting surfaces, it is visible.
- Line of intersection connecting two visible points is visible.
- Line of intersection connecting one visible and one invisible point is invisible.
- Line of intersection changes from visible to invisible or vice-versa, only on crossing the visible edge or face of one of the intersecting surfaces.

Problem 2: *(Fig. 3) A square prism, side of base 45 mm stands with its base on HP and two of its rectangular faces are equally inclined to VP. It is completely penetrated by a horizontal square prism of side of base 35 mm such that their axes are 8 mm apart. The rectangular faces of the horizontal prism are equally inclined to VP and its axis is parallel to VP. Draw the lines of intersection.* (UQ)

Projections of the Vertical and Horizontal Prisms

1. Draw three views of vertical prism for the given position.
2. Axis of the horizontal prism is 8 mm from the axis of the vertical prism. Hence draw the side view of the horizontal prism as a square of 35 mm side such that its axis is 8 mm from the axis of the vertical prism.
3. From the side view, project the front and top views of the horizontal prism.

Intersection Points in Top View

4. In top view, mark 1 where the horizontal edge pp_1 pierces the front vertical rectangular face $ab(b_1)(a_1)$.

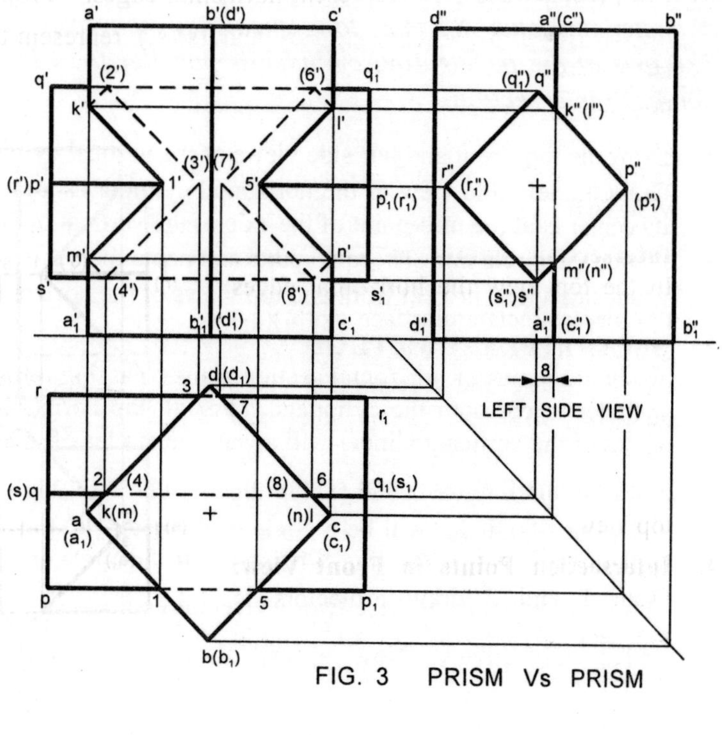

FIG. 3 PRISM Vs PRISM

5. Mark 2, 3 and (4) where the horizontal edges qq_1, rr_1 and (s) (s_1) intersect the rear vertical rectangular face ad (d_1) (a_1) respectively. Similarly mark 5,6,7 and (8) in top view.
6. In the side view, the rectangular face p'' q'' (q_1'') (p_1'') intersects the vertical edges a'' a_1'' and (c'')(c_1'') at k'' and ($1''$) respectively. Similarly the rectangular face $s''p''$ (p_1'') (s_1'') intersects the vertical edges $a''a_1''$ and (c'')(c_1'') at m'' and (n'') respectively. Hence project these intersection points to the top view and obtain k (m) and 1 (n) as shown.

Intersection Points in Front View

7. From 1, 2, 3 and (4) draw projectors to intersect the corresponding edges of the horizontal prism in the front view at $1'$, ($2'$), ($3'$) and ($4'$) respectively.
8. Draw horizontal lines from k'' and m'' from side view to intersect the projector drawn from k (m) at k' and m' respectively.

Visible and Invisible Lines of Intersection in Front View

9. In top view 1k and 1 (m) are visible. Hence draw $1'k'$ and $1'm'$ in front view.
10. Draw $k'(2')$, $(2')(3')$, $(3')(4')$ and $(4')m'$ as dashed lines. Thus complete the lines of intersection on left side. Similarly draw the lines of intersection on right side in front view.

3. INTERSECTION OF CYLINDER Vs CYLINDER

Problem 3: *(Fig. 4) A cylinder of 60 mm diameter and axis 80 mm long stands with its base on HP. It is completely penetrated by a horizontal cylinder of 40 mm diameter and axis 80 mm long such that their axes bisect each other at right angles. The axis of the penetrating cylinder is parallel to VP. Draw the projections showing curves of intersection.* (UQ)

1. Draw the top, front and left side views of the vertical cylinder.
2. Draw the left side view of the horizontal cylinder as a circle of diameter 40 mm such that its center is at the mid-point of the axis of the vertical cylinder.
3. Divide the circle into 12 equal parts and name them.
4. From side view, project front and top views of the horizontal cylinder.

I. SECTION PLANE METHOD

5. Assume a series of horizontal section planes, passing through the generators of the horizontal cylinder to cut both the cylinders. For all the horizontal section planes, the sectional top views of the vertical cylinder will always be circles of 60 mm diameter.
6. First assume the horizontal section plane to pass through $1''$($1_1''$). The sectional top view of the vertical cylinder will be a circle of diameter 60 mm and that of the horizontal cylinder will be a line 1-1_1. The line 1-1_1 intersects the circle at p_1 and q_1 in the top view.
7. Draw projectors from these two intersection points (*i.e.*, from p_1 and q_1) to cut $1'1_1'$ at p_1' and q_1' respectively. Now p_1' and q_1' are the intersection points in front view.

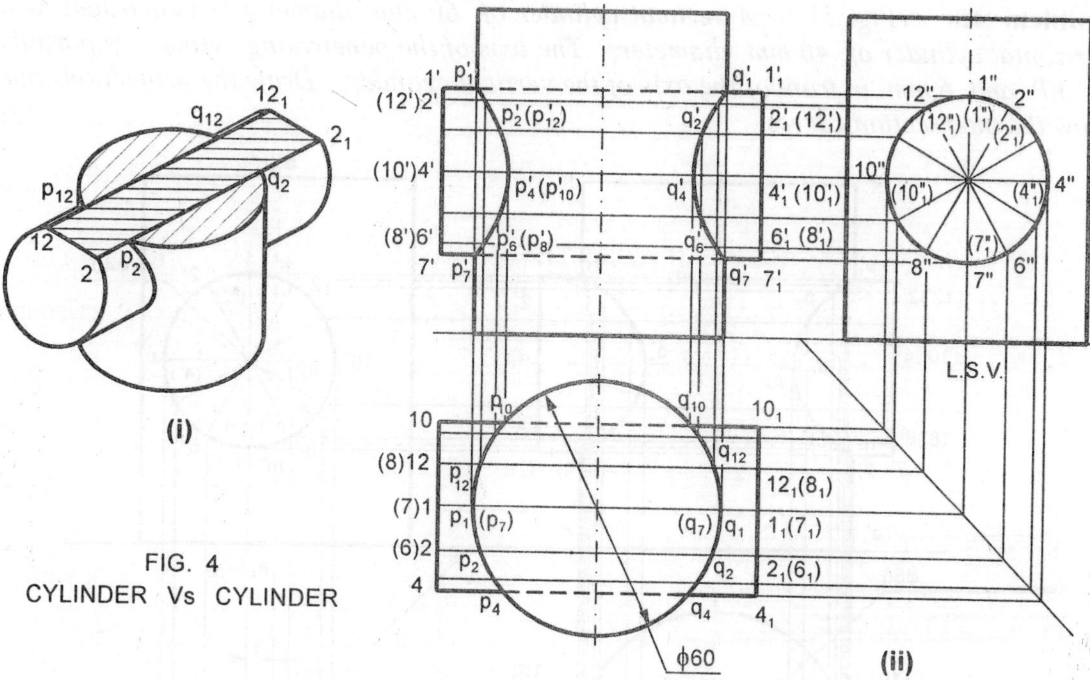

FIG. 4
CYLINDER Vs CYLINDER

8. Assume second horizontal section plane to pass through $2''$ $(2_1'')$ and $12''$ $(12_1'')$.
9. Sectional top view of the vertical cylinder is again a circle of same diameter (60 mm).
10. Sectional top view of the horizontal cylinder will be a rectangle of length $2\text{-}2_1$ and breadth 2-12 (*Fig. i*). This rectangle intersects the circle at p_2, q_2, q_{12} and p_{12} in top view.
11. Project these four intersection points to the corresponding generators in the front view and obtain p_2', (p_{12}'), (q_{12}') and q_2'.
12. Similarly assume series of section planes passing through other generators and obtain the intersection points both in the top and front views.

Visible and Invisible Curves of Intersection in Front View

13. In top view the portion of intersection curve p_1p_2 ..... (p_7) is nearer to the observer and visible. So show the curve p_1', p_2', ..., p_7' as continuous thick line in front view.
14. The rear portion of the curve is hidden and coincides with the visible portion of the curve.
15. Similarly obtain the intersection curve on the right side.

II. GENERATOR METHOD

1. Mark p_1, p_2, etc., where the generators $1\text{-}1_1$, $2\text{-}2_1$, etc., cut the circle in top view.
2. Project the above points. Obtain p_1', p_2', etc., on the corresponding generators in front view.
3. Join these points in correct sequence. Obtain the intersection curve on the left side.
4. Similarly mark q_1, q_2, etc., in the top view.
5. Project these points to get q_1', q_2' in front view. Complete the right side intersection curve.

Problem 4: *(Fig. 5) A vertical cylinder of 60 mm diameter is penetrated by a horizontal cylinder of 40 mm diameter. The axis of the penetrating cylinder is parallel to VP and 6 mm in front of the axis of the vertical cylinder. Draw the projections and show the intersection curves.* (UQ)

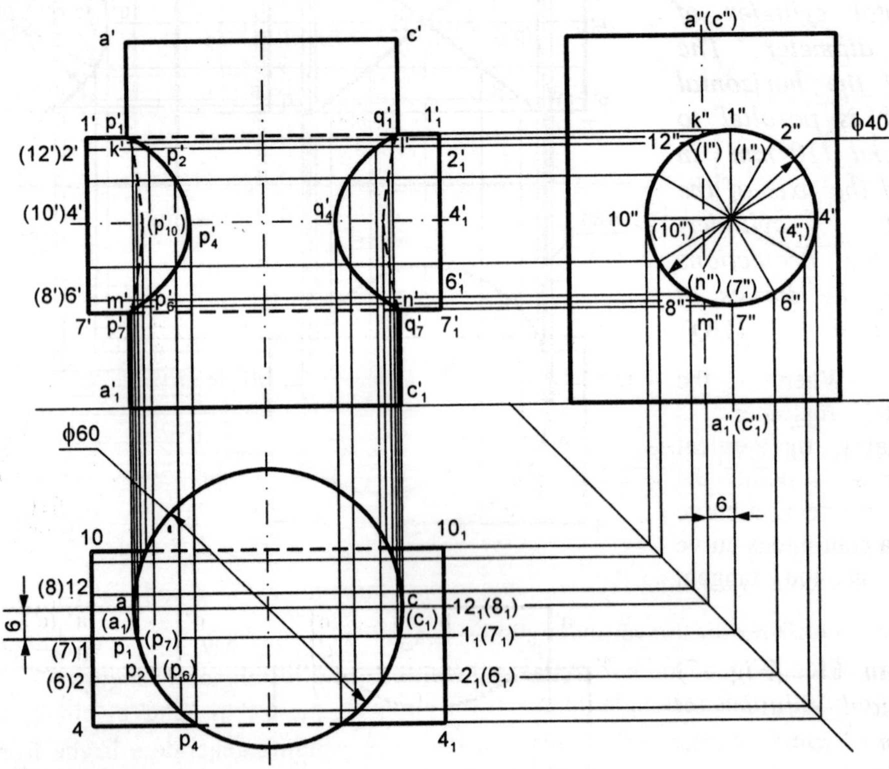

FIG. 5

1. Draw the three views of the vertical cylinder.
2. Axis of the horizontal cylinder is parallel to VP and 6 mm in front of the axis of the vertical cylinder. Hence draw side view of horizontal cylinder as a circle of 40 mm diameter such that its center is 6 mm to right side of the axis of vertical cylinder.
3. Project its front and top views and name the generators as explained in Problem 3.
4. Obtain 12 (twelve) intersection points $p_1', p_2', ..., p_{12}'$ on the left side and another 12 intersection points $q_1', q_2', ..., q_{12}'$ on the right side.

To obtain Critical Intersection Points: These are the intersection points at which the intersection curve changes its direction.

5. The circle in the side view intersects the generator $a'' a_1''$ at k'' and m'' and $(c'')\,(c_1'')$ at (l'') and (n'') respectively. Project these points to the front view.
6. Draw the curve through $p_1', p_2', ..., p_7'$ as thick since this portion is visible.
7. Draw the curve through $p_1', k', (p_{12}'), (p_{10}'), (p_8'), m', p_7'$ as dashed since this portion is hidden.

Problem 5: *(Fig. 6) A vertical cylinder of 60 mm diameter is penetrated by a horizontal cylinder of 40 mm diameter. The axis of the horizontal cylinder is parallel to VP and 10 mm in front of the axis of the vertical cylinder. Draw the projections showing the curves of intersection.* (UQ)

Note: When the horizontal cylinder is tangential to the vertical cylinder at a point, the form of intersection curve will be a continuous curve crossed at the tangent point.

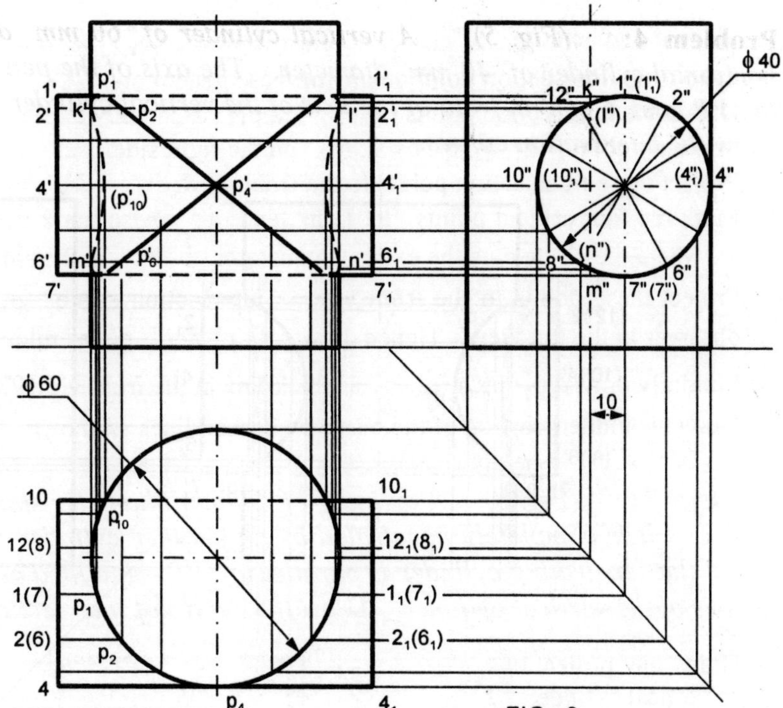

FIG. 6

Problem 6: *(Fig. 7) A vertical cylinder of 60 mm diameter is penetrated by a horizontal cylinder of 50 mm diameter. Axis of horizontal cylinder is parallel to VP and 10 mm in front of the axis of vertical cylinder. Draw the intersection curves.*

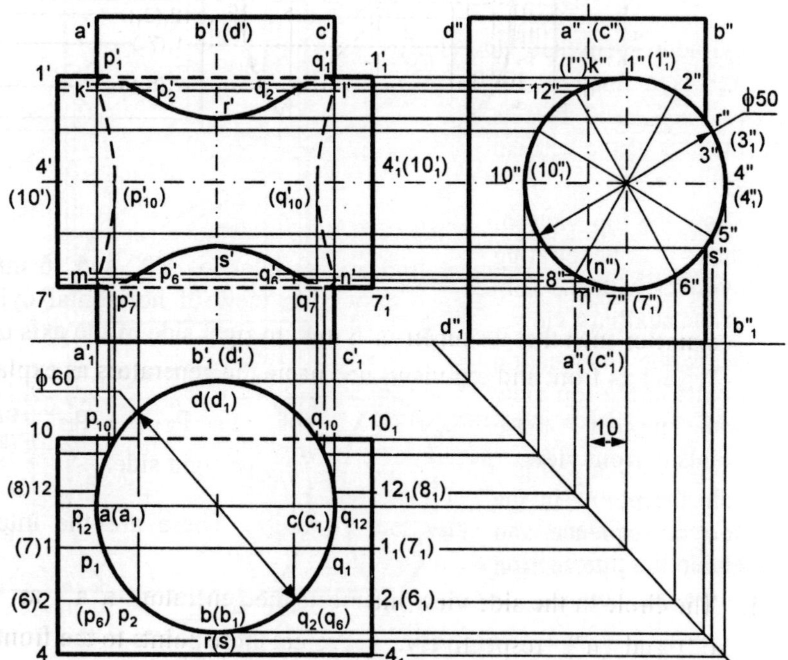

FIG. 7

1. Draw the projections of both cylinders for the given position.
2. Name the generators of the penetrating (horizontal) cylinder in all the three views.
3. Mark p_1, p_2 where the generators $1\text{-}1_1$ and $2\text{-}2_1$ intersect the circle in the top view.

4. Since the generators $3''\ (3_1'')$, $4''\ (4_1'')$ and $5''\ (5_1'')$ do not intersect the vertical cylinder there is no intersection point on any of these generators.
5. Obtain the intersection points $p_6, p_7, \ldots, p_{12}$ as already explained.
6. Similarly obtain $q_1, q_2, q_6, q_7, \ldots q_{12}$ on the right side.
7. Project these intersection points to the front view.
8. Mark the four critical points k', l', m' and n'. Also mark two more critical points r'' and s'' where the circle cuts the extreme generator $b''\ b_1''$ in the side view.
9. Project these points to the front view. Intersection curve $p_1\ p_2\ r\ q_2\ q_1$ is nearer to the observer in the top view. Hence draw $p_1'\ p_2'\ r'\ q_2'\ q_1'$ as thick line in the front view.
10. Similarly draw $p_7'\ p_6'\ s'\ q_6'\ q_7'$ as thick line in the front view.
11. Draw the hidden portion of the curve by dashed lines as shown.

Problem 7: *(Fig. 8) A cylinder of 50 mm diameter stands vertically with its base on HP. It is completely penetrated by an another cylinder of 35 mm diameter. The axis of the penetrating cylinder is parallel to VP, inclined at 30° to HP and bisects the axis of the vertical cylinder. Draw the curves of intersection.* (UQ)

1. Draw the projections of vertical cylinder.
2. Draw the front view of penetrating cylinder such that its axis is inclined at 30° to XY and bisects the axis of the vertical cylinder.
3. Draw the top view of the penetrating cylinder with the help of the auxiliary view.
4. Mark the intersection points in the top view and project these points to front view.
5. Join the points in the correct sequence and obtain the intersection curves as shown.

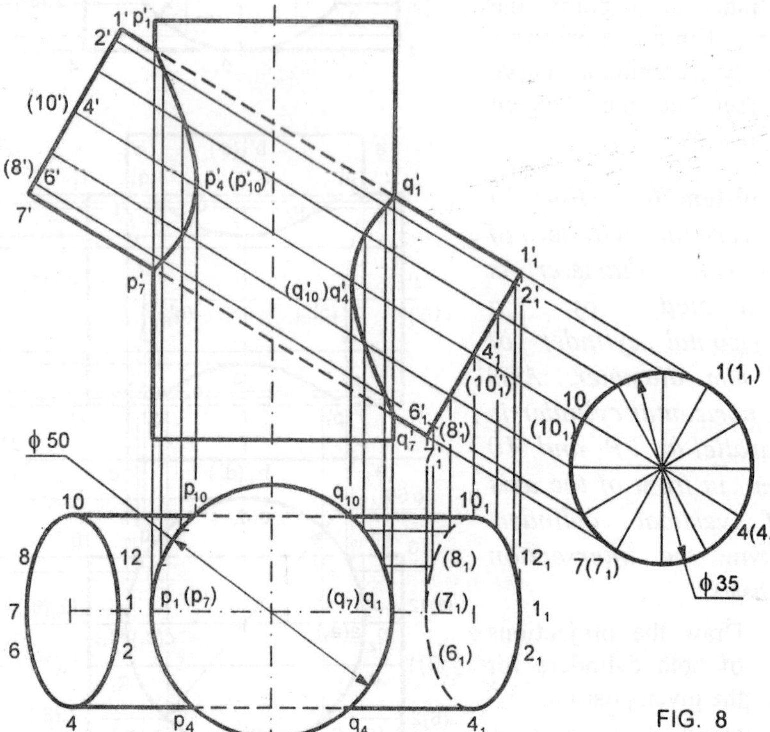

FIG. 8

Problem 8: *(Exercise) A pipe of 25 mm diameter is joined to an another vertical pipe of 50 mm diameter at an angle of 45°. Their axes are 12.5 mm apart. Draw the projections of the pipes showing curves of intersection.*

4. INTERSECTION OF PRISM Vs CYLINDER

Problem 9: *(Fig. 9) A cylinder of 60 mm diameter and axis 80 mm long is standing vertically on its base on HP. It is pierced by a square prism of 30 mm side and 100 mm length, the axis of which is parallel to both the reference planes and the faces equally inclined to HP. The axes of the solids intersect at right angles. The height of the axis of the prism above HP is 40 mm. Draw the projections of the solids showing the curves of intersection in the front view.* (UQ)

1. Draw the top view of the cylinder and square prism in the given position.
2. Project the front view.
3. Draw the corresponding left side view of the intersecting cylinder and prism. From the side view, it is clear that the cylinder envelops the square prism.
4. Apply the Horizontal Section Plane Method for solving this problem.
5. Obtain the curves of intersection in the front view as shown.

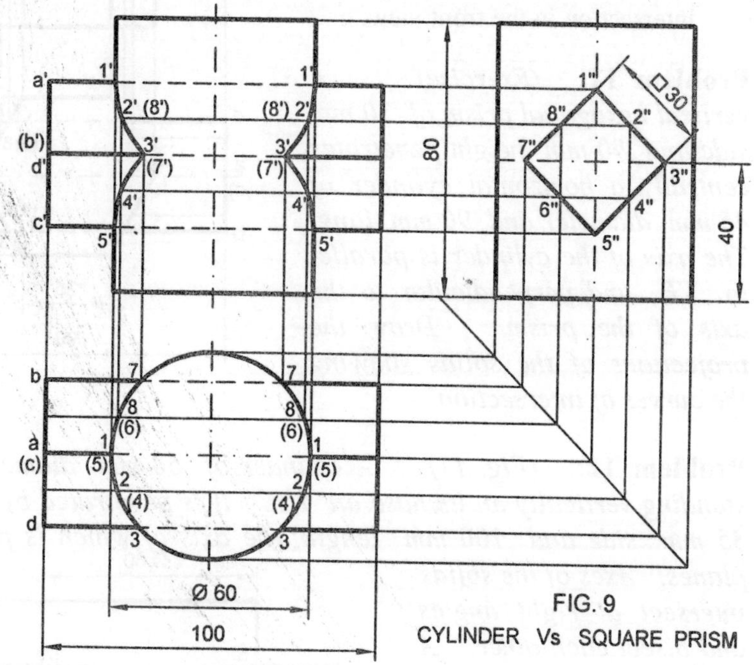

FIG. 9
CYLINDER Vs SQUARE PRISM

Problem 10: *(Fig. 10) A horizontal cylinder of 30 mm diameter and 90 mm long penetrates centrally a vertical hexagonal prism of 25 mm side and 80 mm height. The axis of the cylinder is parallel to VP and perpendicular to the axis of the prism. Draw the projections of the solids showing the curves of intersection in the front view.* (UQ)

1. Draw the top view of the given intersecting cylinder and hexagonal prism.
2. Project the front view as shown.
3. Draw the left side view of the intersecting cylinder and hexagonal prism.
4. Divide the cylinder circle in the side view into 8 or 12 equal parts.
5. Project the generators to the top and front views.

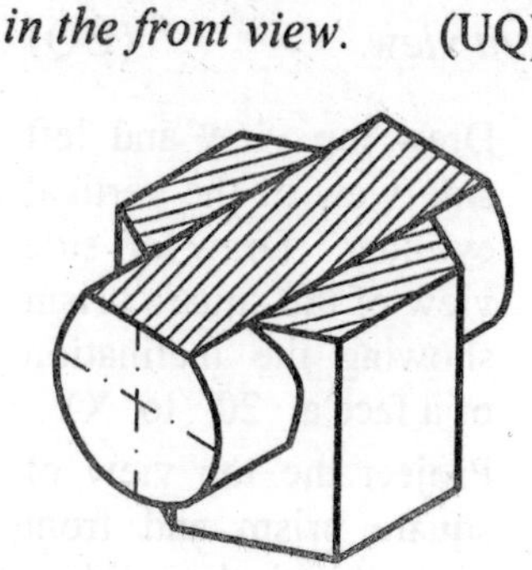

6. Obtain the intersecting points in the top view at the points at which the generators of the cylinder intersect the rectangular faces of the hexagonal prism.
7. Project the intersection points from the top view to front view.
8. Draw smooth curves of intersection in the front view.

Problem 11: *(Exercise) A vertical hexagonal prism of 30 mm side and 90 mm height penetrates centrally a horizontal cylinder of 45 mm diameter and 90 mm long. The axis of the cylinder is parallel to VP and perpendicular to the axis of the prism. Draw the projections of the solids showing the curves of intersection.*

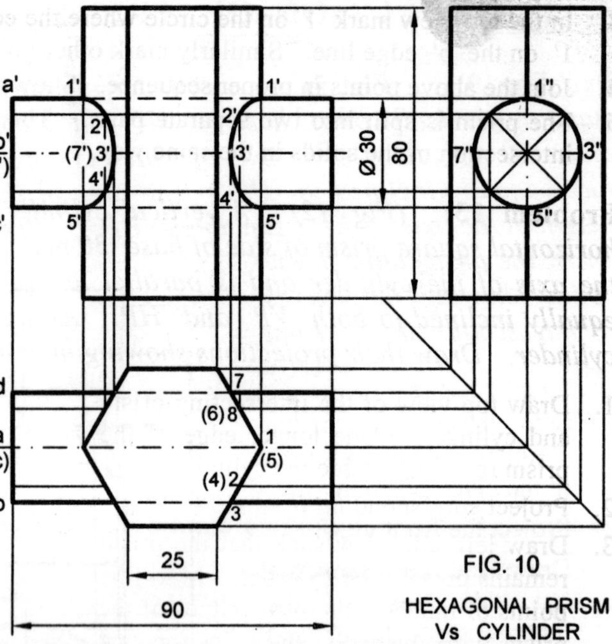

FIG. 10 HEXAGONAL PRISM Vs CYLINDER

Problem 12: *(Fig. 11) A cylinder of 60 mm diameter and axis 80 mm long is standing vertically on its base on HP. It is penetrated by a horizontal square prism of 35 mm side and 100 mm length, the axis of which is parallel to both the reference planes. Axes of the solids intersect at right angles and bisect each other. A lower rectangular face of the horizontal prism is inclined at an angle of 20° to HP. Draw the projections showing the intersection curve in the front view.* (UQ)

1. Draw top view and left side view of the vertical cylinder. Draw left side view of the square prism showing the inclination of a face at 20° to XY.
2. Project the top view of square prism and front view of both the solids.

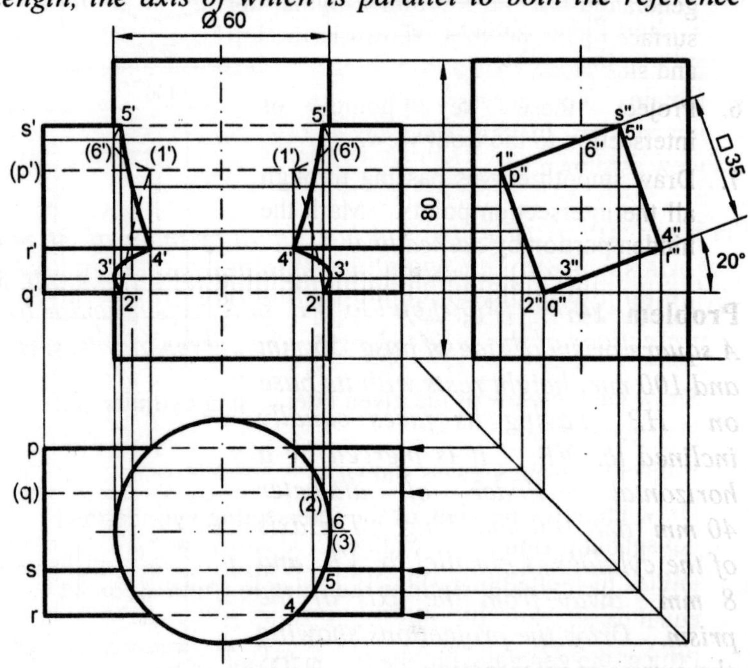

FIG. 11 CYLINDER Vs INCLINED SQUARE PRISM

3. In the top view mark 1 on the circle where the edge line p cuts it. Project it in the front as 1′ on the p′ edge line. Similarly mark other points for curve of intersection.
4. Join the above points in proper sequence. Show the hidden curve in *dashed lines*.
5. The prism is split into two separate parts. Therefore, obtain the curve on the other side of intersection of the solids in the same way.

Problem 13: *(Fig. 12) A vertical cylinder of 45 mm diameter is penetrated by horizontal square prism of side of base 30 mm. Axis of the prism is 7.5 mm away from the axis of the cylinder and is parallel to VP. Rectangular faces of the prism are equally inclined to both VP and HP. Axis of the prism is in front of the axis of the cylinder. Draw their projections showing intersection curves in front view.* (UQ)

1. Draw top view of the intersecting prism and cylinder. One longer edge of the prism remains outside the cylinder.
2. Project corresponding front view.
3. Draw left side view such that the prism remains outside the cylinder. Mark the points of intersection of the edges of the prism with the cylinder in the top and side views.
4. Project these points of intersection to the front view.
5. Also mark the key points of the extreme generators of the cylinder with the surface of the prism as shown in the top and side views.
6. Project these key points of intersection to the front view.
7. Draw smooth curves passing through all the intersection points. Mark the hidden portion by *dashed lines*.

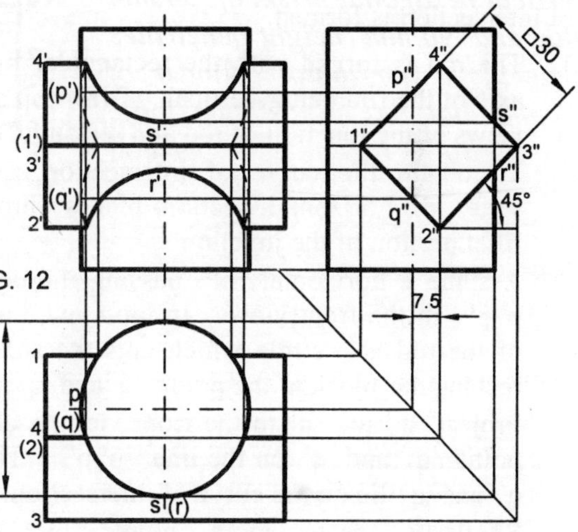

FIG. 12

Problem 14: *(Fig. 13)*
A square prism of side of base 55 mm and 100 mm height rests with its base on HP having its faces equally inclined to VP. It is pierced by a horizontal cylinder of diameter 40 mm and 100 mm long. The axis of the cylinder is parallel to VP and 8 mm away from the axis of the prism. Draw the projections showing the curves of intersection. (UQ)

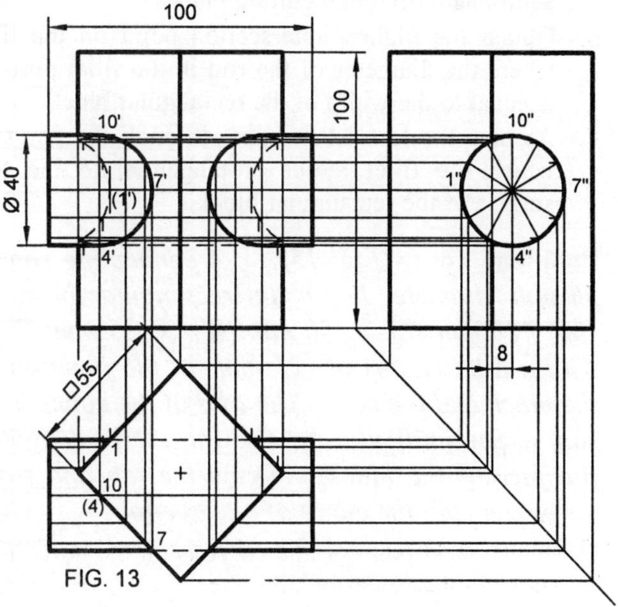

FIG. 13

Problem 15: *(Fig. 14) A connecting rod of 50 mm diameter has a flat rectangular block of size 110 mm (L) x 62 mm (W) x 25 mm thick with a fillet radius of 25 mm at the junction of the block and the rod. The axis of the connecting rod is perpendicular to the largest rectangular flat face of the block. Draw the top and front views showing the curve of intersection.* (UQ)

Connecting rod increases in diameter as it approaches the rectangular block. This forms what is called a **fillet** of 25 mm radius. Note that the fillet gradually increases the diameter of connecting rod. Width of the rectangular block is less than the sum of diameter of the rod and twice the fillet radius. Hence a curve of intersection is formed.

1. The rod is forged with the rectangular block, the axis of the rod being vertical. Draw top and front views of the junction of rod and rectangular block.
2. To obtain the curve of intersection, assume a series of horizontal section planes through the fillet portion of the junction.
3. Assume a horizontal S.P. passing through a line p′-p′ in the front view. In top view, the section of the rod is a circle which cuts the sides of the rectangular block at the points q and q.
4. Project q and q to the front view to obtain the points q′ and q′ on the line p′-p′. The points q′ and q′ lie on the curve of intersection.
5. Similarly locate more points by assuming additional horizontal cutting planes.
6. Obtain the highest intersection point on the line where the diameter of the rod in the fillet portion is equal to the width of the rectangular block.
7. Assume the last S.P. to pass through the point at which the fillet radius is tangential to the top surface of the rectangular block.

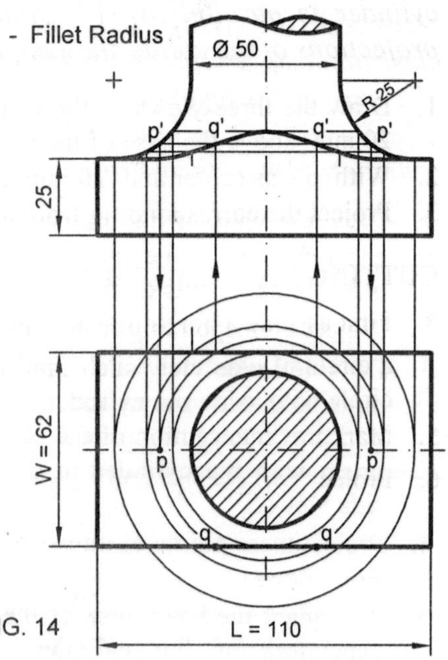

FIG. 14

Problem 16: *(Fig. 15) A connecting rod of* ***50 mm*** *diameter has a flat rectangular block of size 110 mm (L) x* ***50 mm*** *(W) x 25 mm thick with a fillet radius of 25 mm at the junction of the block and the rod. The axis of the connecting rod is perpendicular to the largest rectangular flat face of the block. Draw the top and front views showing the curve of intersection.* (UQ)

Hint: Highest point on the curve is on the horizontal line through the center of fillet arc.

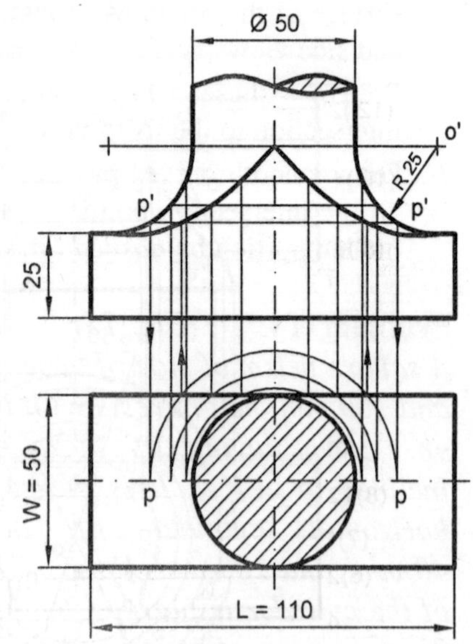

o′ - Center of the Fillet Arc
W = 50 = Ø 50

FIG. 15

5. INTERSECTION OF CYLINDER Vs CONE

Problem 17: *(Fig. 16) A cone of base 60 mm diameter and axis 70 mm long, rests with its base on HP. It is completely penetrated by a horizontal cylinder of 30 mm diameter such that both the axes intersect each other at right angles. The axis of the cylinder is parallel to VP and 20 mm above the base of the cone. Draw the projections of the solids showing the curves of intersection.* (UQ)

1. Draw the three views of the cone. On the axis of the cone in the Side View mark o″ at 20 mm above the base of the cone to represent the axis of the horizontal cylinder.
2. With o″ as center and 30 mm as diameter draw a circle to represent the cylinder.
3. Project the corresponding front and top views of the cylinder and name the generators.

CUTTING PLANE METHOD : *See Fig. (i).*

4. First assume a horizontal S.P. passing through 1″ ($1_1''$). The sectional top view of the cone is a circle of diameter a′a′ and that of the horizontal cylinder is a line $1\text{-}1_1$. Hence draw a circle with o as center and a′a′ as diameter in the top view.
5. Mark the intersection points p_1 and q_1 where the line $1\text{-}1_1$ intersects this circle.
6. Project p_1 and q_1 and obtain p_1' and q_1' on $1'1_1'$ in the front view.

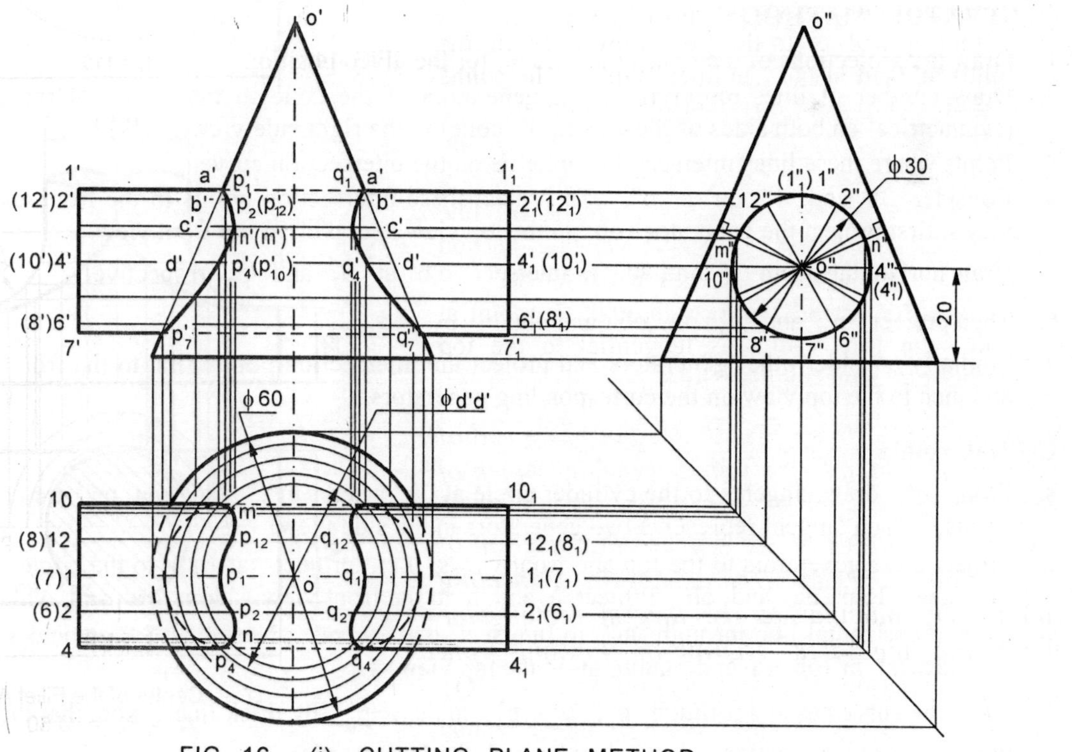

FIG. 16 (i) CUTTING PLANE METHOD

7. Assume second horizontal S.P. passing through 2″ ($2_1''$) and 12″ ($12_1''$). The sectional top view of the cone will be a circle of diameter b′b′ and that of the cylinder will be a rectangle of length $2\text{-}2_1$ and breadth 2-12. The circle and the rectangle intersect at 4 intersection points. Mark them as p_2 q_2 q_{12} and p_{12} in the top view.
8. Project these points on the corresponding generators of the cylinder in the front view and obtain p_2', q_2', (q_{12}') and (p_{12}') respectively.
9. Similarly assume series of horizontal section planes passing through 3″($3_1''$), 4″ ($4_1''$), etc.
10. Obtain the intersection points in top and front views.

Critical Points

11. From o″ draw perpendiculars to the end generators of the cone in the side view. They cut the circle at m″ and n″. Assume a horizontal S.P. passing through m″ and n″.
12. Project these critical points to the top and front views and obtain m, n and m′, n′.

To Mark Visible and Invisible Portions of the Intersection Curves

13. In the front view draw curve passing through p_1', p_2', p_3', n′, p_4', ..., p_7' as *thick* lines.
14. The rear portion of the intersection curve coincides with the front portion. Hence it is not shown as dashed line in this view.
15. In the top view join n, p_2, p_1, p_{12} and m to represent the visible portion of the curve. The hidden portion is shown as *dashed* lines.

GENERATOR METHOD : *See Fig. (ii).*

1. Draw the projections of the cone and cylinder for the given position.
2. Draw number of lines representing the generators of the cone in the region of the circle (symmetrical on both sides of the axis of the cone) in the right side view.
3. Points where these lines intersect the circle lie on the intersection curve.
4. Consider 3″ and 4″ on o″ b″. To project these intersection points in the front and top views, first project the generator ob in the top view and o′b′ in the front view.
5. Draw horizontals from 3″ and 4″ to intersect o′b′ at p_3' and p_4' respectively.
6. Then project p_3' and p_4' on ob and obtain p_3 and p_4.
7. Similarly consider other generators and project the intersection points first to the front view and then to the top view on the corresponding generators.

Critical Points

8. From o″ draw tangents to the cylinder circle at k″ and m″. k″ and m″ are critical points. Each tangent represents two generators in the side view.
9. Project these generators to the top and front views, *i.e.*, project a″ (h″) to the circle in the top view. Join oa and oh. Project a and h to the front view. Join o′a′ and o′h′.
10. Draw a horizontal line through m″ to intersect o′a′ at m′. Note that (k′) coincides with m′. Project m′ on oa and obtain m in the top view.
11. Draw the curve passing through p_1', p_2', p_3', m′, ..., p_6' by thick line. Note that the rear portion of the intersection curve coincides with the front portion.

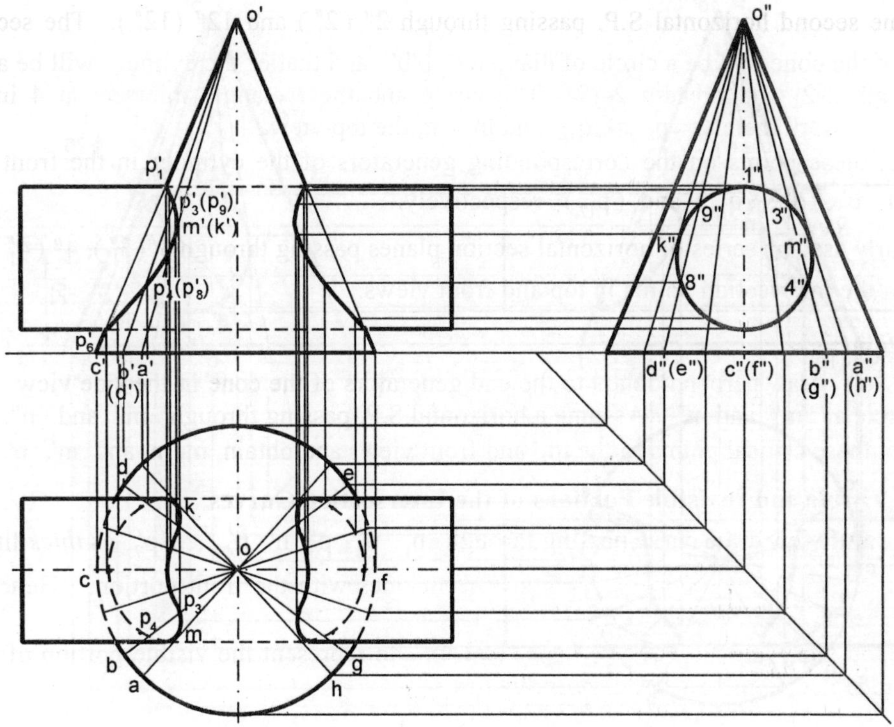

FIG. 16 (ii) GENERATOR METHOD

12. Similarly obtain the intersection curve on the right side in the front view. The hidden portion of the curve is shown dashed in the top view.

Note: *This method is also a type of Cutting Plane Method in which cutting planes pass through the apex and parallel to the axis of the cylinder.*

Problem 18: *(Fig. 17) A cone of base 60 mm diameter and axis 70 mm long, stands vertically with its base on HP. It is penetrated by a horizontal cylinder of 26 mm diameter. The axis of the cylinder is parallel to VP, 20 mm above the base and 5 mm in front of the axis of the cone. Draw the projections of solids showing the curves of intersection.* (UQ)

1. Draw the projections of the cone. Draw side view of the horizontal cylinder as a circle such that its axis is 20 mm above the base and 5 mm to right side of axis of cone.
2. Draw the top and front views of the horizontal cylinder.
3. Draw number of lines, representing the generators of the cone in the region of the circle in left side view. Points where these lines intersect the circle lie on intersection curve.
4. Project these intersection points to front and top views. For example 2″ and 3″ are on o″ b″. Project these points on o′b′ and obtain p_2' and p_3' in front view.

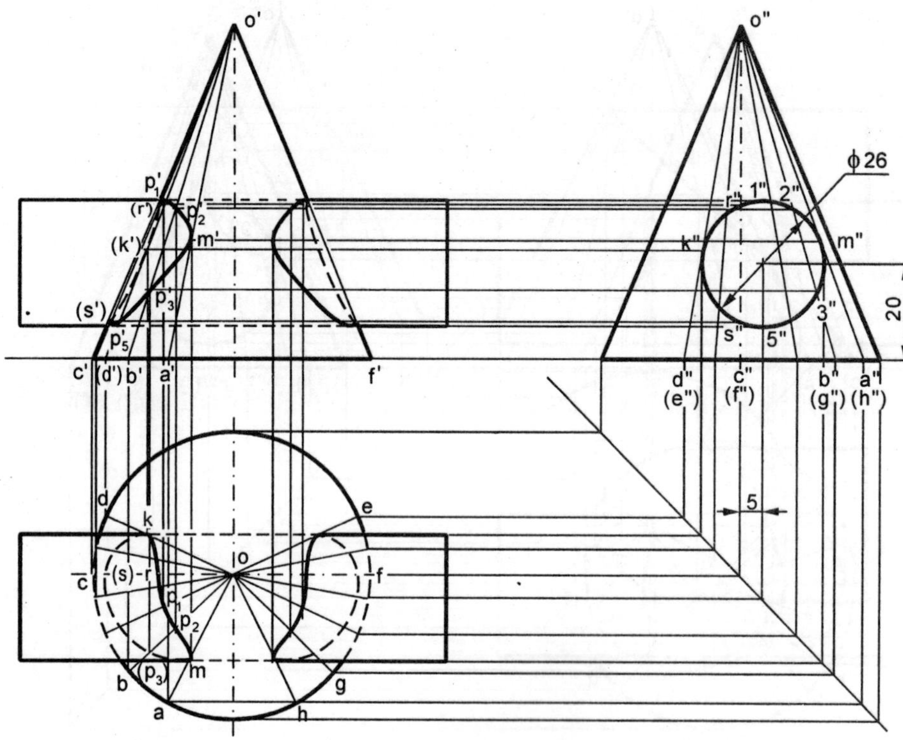

FIG. 17

5. Project these points to the top view and obtain p_2 and (p_3) on ob.
6. Draw two tangents to the circle in the side view and obtain the critical points m″ and k″ on the generators o″ a″ and o″ d″ respectively.
7. Project m″ and k″ on o′a′ and o′ (d′) and obtain m′ and (k′) respectively. Project these points to the top view and obtain m and k on oa and od respectively.
8. Two more critical points r″ and s″ are on o″c″. Project these points to front and top views.
9. Note that 1″ and 5″ are also critical points. To locate these points consider the generators passing through these points in the side view and project them to the other views. Construction is shown to obtain p_1' and p_1 only.
10. Join the intersection points in the correct sequence and obtain the curves on both sides.
11. Show the hidden portion of the curve as *dashed lines* in both views.

Problem 19: *(Fig. 18) A funnel consists of a cylindrical pipe and a conical part, both enveloping an imaginary common sphere of 40 mm diameter, with their axes intersecting at right angles. The diameter of the mouth of the funnel is 80 mm and its total height is 60 mm. The funnel is placed with its mouth on HP and two axes parallel to VP. Draw the curves of intersection.* (UQ)

See Fig. 18 for Solution.

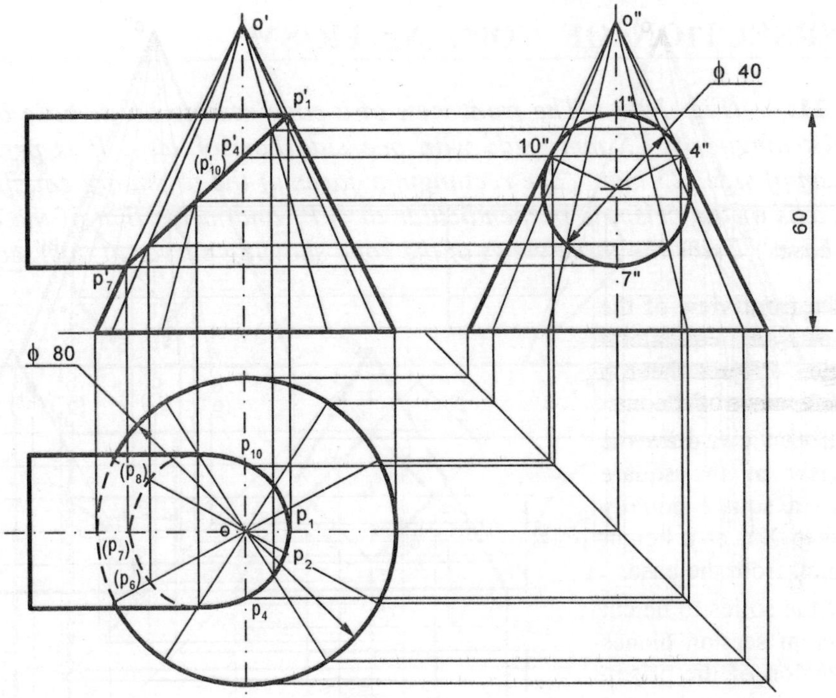

FIG. 18

Problem 20: *(Fig. 19) A vertical cylinder of 50 mm diameter and 80 mm height is penetrated by a cone of base 60 mm diameter and axis 70 mm long, the axes of the two solids bisecting each other at right angles. Draw the projections showing the curves of intersection.* (UQ)

1. Name the intersection points p_1, p_2, ..., p_{12} in top view wherever the corresponding generators of the cone cut the cylinder circle.
2. Project these points and obtain p'_1, p'_2, ..., p'_{12} on the corresponding generators in the front view. Join these points in the correct sequence.
3. Obtain the intersection curve on one side.
4. Similarly obtain the curve on the other side.

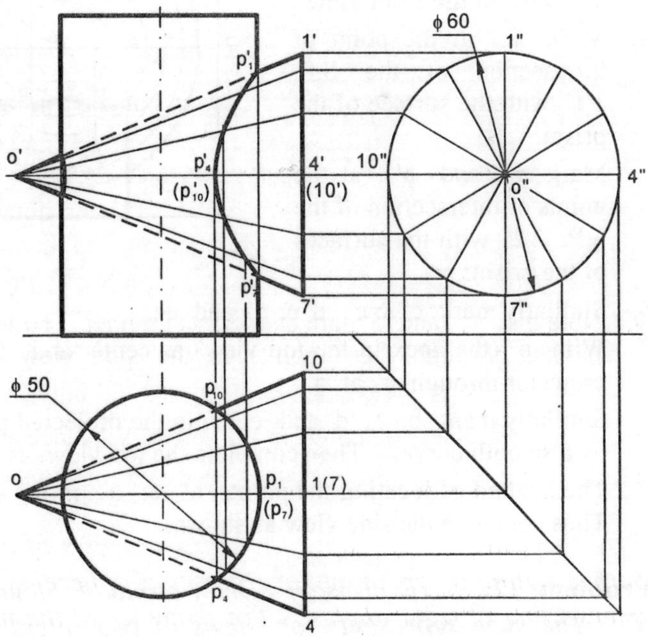

FIG. 19

6. INTERSECTION OF CONE Vs PRISM

Problem 21: *(Fig. 20) The front view of a cone resting on its base on HP is an equilateral triangle of 75 mm side with one side horizontal. It is penetrated by a square prism of side 25 mm. The rectangular faces of the prism are equally inclined to HP. The axis of the prism is perpendicular to VP and meets that of the cone 25 mm above the base. Draw the three views of the cone showing curves of intersection.* (UQ)

1. Draw the front view of the cone as an equilateral triangle. Project the top and side views of the cone.
2. In the front view, draw the front view of the square prism as a square equally inclined to XY at a height of 25 mm from the base.
3. Assume the solids to be cut by different section planes in the region of the prism. Accordingly, mark the section planes as 1′1′, 2′2′, 3′3′, etc., in the front view.
4. Mark a′ at the point of intersection of the S.P. 1′1′ with the surface of the prism.
5. Mark b′ and b′ at the points of intersection of the S.P. 2′2′ with the surfaces of the prism.

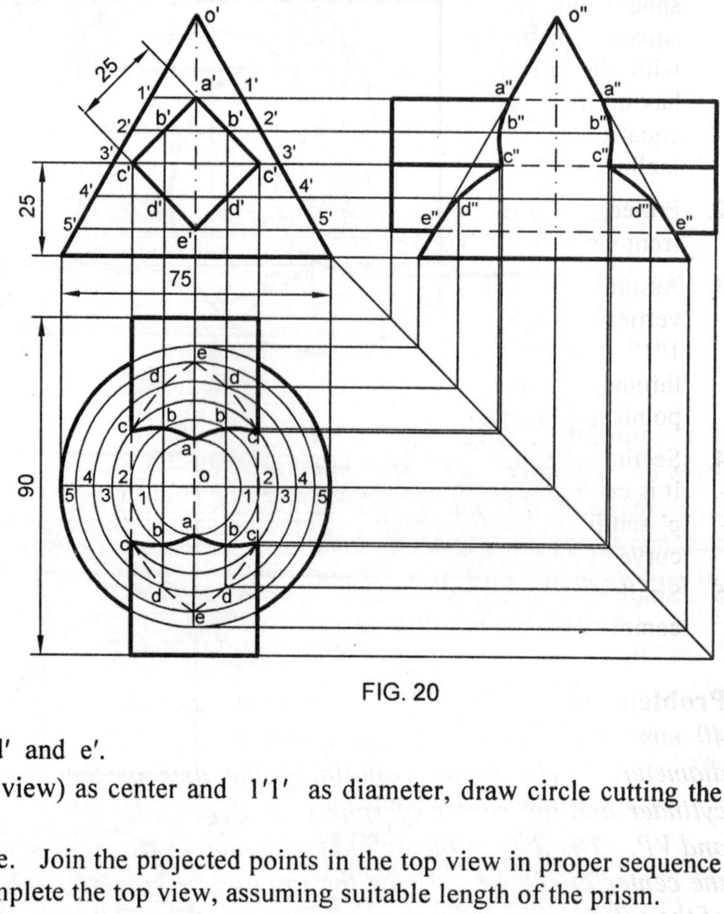

FIG. 20

6. Similarly mark c′ & c′, d′ & d′ and e′.
7. With o (the apex in the top view) as center and 1′1′ as diameter, draw circle cutting the projector through a′ at a.
8. Similarly mark b, c, d and e. Join the projected points in the top view in proper sequence by a smooth curve. Then complete the top view, assuming suitable length of the prism.
9. The method of locating the points a″ & a″ in the **side view** is clearly shown by mitre line. Thus complete the side view as shown.

Problem 22: *(Exercise) Draw circle of 60 mm diameter. Inscribe a pentagon in it. The view represents top views of two intersecting solids, namely, a cone and a pentagonal prism. Both the solids share a common axis length of 60 mm. The base of the cone is on HP. A rectangular face of the prism is parallel and nearer to VP. Draw the front view of the solids and show the curves of intersection.* (UQ)

7. INTERSECTION OF SPHERE Vs PRISM

Problem 23: *(Fig. 21) A sphere of 72 mm diameter is penetrated by a square prism of 40 mm side of base and 100 mm length, the axis of which passes through its center. The axis of the prism is parallel to both HP & VP and the faces are equally inclined to VP. Draw the front view of the solids showing the curves of penetration.* (UQ)

1. Draw the side view of the sphere and the square prism, with the prism having its faces equally inclined to VP.
2. Project the front view.
3. Assume a vertical S.P. 1″1″ passing through the points p″ and p″.
4. Section of sphere will be a circle of diameter 1″1″. It is cut by the section of the prism at the points p′, p′, p′ and p′ in the front view. These points are on the curve of intersection.
5. Similarly, obtain the points q′, r′ & s′ and complete the intersection curve.

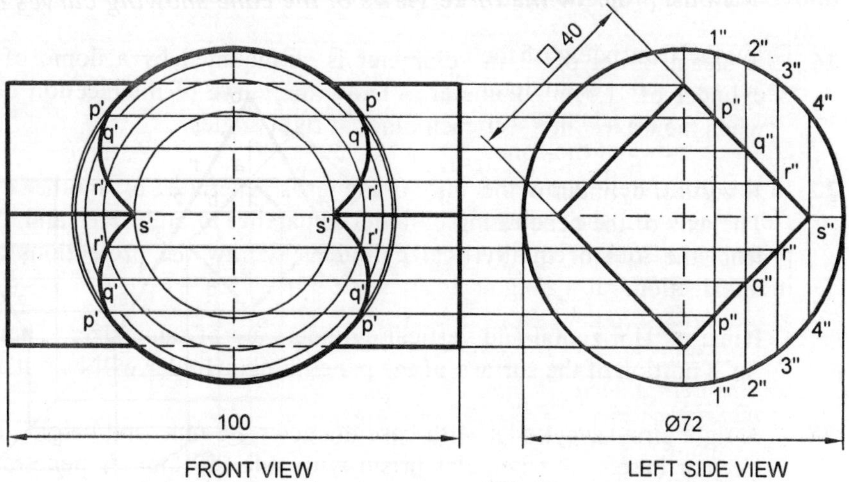

FIG. 21

Problem 24: *(Fig. 22) A vertical cylinder of 40 mm diameter penetrates a sphere of 80 mm diameter. The plane containing the axis of the cylinder and the center of sphere is ⊥ʳ to both HP and VP. The axis of the cylinder 25 mm in front of the center of sphere. Draw the top and front views of the solids. Show the curves of intersection.* (UQ)

1. Draw the front and top views of the sphere.
2. Draw the top view of the vertical cylinder by offsetting its axis by 25 mm from the axis of the sphere and divide it into 12 equal parts. Mark the section planes passing through these division points.
3. Project the cutting points to the front view and obtain the curve of intersection as shown.

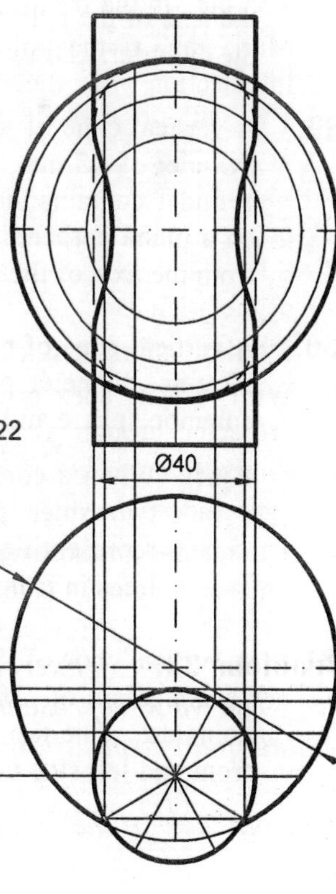

FIG. 22

8. ADDITIONAL PROBLEMS

25. A vertical square prism 60 mm edge of the base is standing on its base with a vertical face making 45° to VP. It is completely penetrated by a horizontal square prism of 45 mm edge of the base. The axis of the horizontal square prism is parallel to VP and 15 mm in front of the axis of the vertical prism, such that a rectangular face makes 30° with HP. Draw the projections showing lines of intersection. (UQ)

26. A steam boiler of 3 m diameter is surmounted by a dome of the shape of a vertical cylinder of 1.5 m diameter. Draw the curve of intersection between the two surfaces when their axes intersect each other at right angles.

27. A vertical cylinder of 70 mm diameter is penetrated by another cylinder of the same size. The axis of the penetrating cylinder is parallel to both HP and VP and is 9 mm away from the axis of the vertical cylinder. Draw the projections and show the curves of intersection. (UQ)

Hint: Horizontal and vertical cylinders are of same size. Their axes are 9 mm apart. So, a portion of the surface of the penetrating cylinder will be outside the vertical cylinder.

28. A right circular cylinder with base diameter 80 mm and height 100 mm is resting on its circular base. A triangular prism with side 80 mm is penetrated through the cylinder such that the axis of the prism is parallel to HP and perpendicular to VP and it bisects the axis of the cylinder. One of the flat faces of the prism is parallel to that of the axis of the cylinder. Draw the projections showing the curves of penetration.

29. A vertical cone of 60 mm base diameter and altitude 75 mm is penetrated by a vertical cylinder of 45 mm base diameter and altitude 75 mm such that the top circular end of the cylinder contains the vertex of the cone. The axis of the cylinder and that of the cone lie on a plane perpendicular to both HP and VP. The axis of the cylinder is 6 mm away from the axis of the cone towards the observer. Draw the curves of intersection. (UQ)

30. A vertical cone of base 60 mm diameter and 54 mm height and a horizontal cylinder of 30 mm diameter penetrate each other. The cylinder and cone envelope an imaginary common sphere and their axes meet at right angles. Draw the curves of intersection. (UQ)

Hint: When a cone and cylinder envelope a common sphere, then in the end view, the two sides of the triangle representing the cone are tangential to the circle representing cylinder. The two lines of intersection appear as crossed straight lines in front view and as ellipse in the top view.

31. A vertical cone of 50 mm diameter of base and height 80 mm is penetrated by a vertical square prism of 30 mm edges and axis 75 mm such that the axes of both the solids coincide. The top square end of the prism contains the apex of the cone. Draw the top, front and left side views of the intersecting solids showing the curves of intersection. (UQ)

*

19

DEVELOPMENT OF LATERAL SURFACES OF TWO INTERSECTING SOLIDS

We have already studied **Development of Lateral Surfaces of Solids** in the **Chapter 15** and **Intersection of Surfaces** in the **Chapter 17**. This Chapter deals with the development of lateral surfaces of two intersecting solids, *viz., Cylinder Vs Cylinder* and *Cylinder Vs Prism.*

1. DEVELOPMENT OF LATERAL SURFACES OF TWO INTERSECTING CYLINDERS

Problem 1: *(Fig. 1) A cylinder of 60 mm diameter and axis 80 mm long stands with its base on HP. It is completely penetrated by a horizontal cylinder of 45 mm diameter and axis 80 mm long such that their axes bisect each other at right angles. The axis of the penetrating cylinder is parallel to VP. Draw the projections showing curves of intersection. Also, draw the development of the lateral surfaces of the two intersecting cylinders.* (UQ)

Note: *First, draw the projections of the two given cylinders with their axes at right angles. Show the curves of intersection. Then draw the development of lateral surfaces.*

1. *See Fig. (i).* Draw the Top, Front and Left Side Views of the vertical cylinder.
2. Draw Left Side View of horizontal cylinder as a circle. Project its front and top views.
3. Divide the circle in the left side view into 12 equal parts and name them as 1″, 2″, 3″, …, 12″.
4. Project the generators of the above circle to the top view circle of the vertical cylinder. Name the intersection points as $p_1, p_2, \ldots\ldots, p_{12}$.
5. Draw vertical projectors from the above points to the front view.
6. Also, project horizontal projectors of the generators of the circle in the side view to the front view. These horizontal projectors intersect the vertical projectors drawn in Step 5.
7. Mark the points of intersection and name them as $p_1', p_2', \ldots \ldots, p_{12}'$. Draw the curves of intersection as shown.

Development of lateral surfaces of intersecting VERTICAL cylinder

8. *See Fig.(ii)*. Draw the development of the vertical cylinder as a rectangle of length = 60 π mm and height = 80 mm.
9. Divide the top view into 12 equal parts and name them as a, b, c, d, ..., l, m and n.
10. Divide the rectangle length also into 12 equal parts. Name them as A, B, ..., N and A.
11. Transfer the division points 1″, 2″, 3″, ..., 12″ in the side view to the height of the rectangle as 1, 2, ..., 12 as shown.
12. Mark x on BC such that Bx = b p_{10}. Draw vertical line through x and mark P_{10}.
13. Mark y on BC such that xy = p_{11} p_{10}. Draw vertical line through y and mark P_{11}.
14. Mark z on CD such that Cz = c p_{12}. Draw vertical line through z and mark P_{12}.
15. Similarly obtain the points P_1, P_2, ..., P_9 in the development as shown.
16. Draw an **ellipse** passing through P_1, P_2, ..., P_{12}.
17. Similarly, draw the ellipse through Q_1, Q_2, ..., Q_{12}.

Development of lateral surfaces of the intersecting HORIZONTAL cylinder

18. *See Fig. (iii)*. Draw the development of horizontal cylinder as a rectangle of size 45 π x 80.
19. Divide the length 45 π mm into 12 equal parts and name them as 1, 2, 12 & 1.

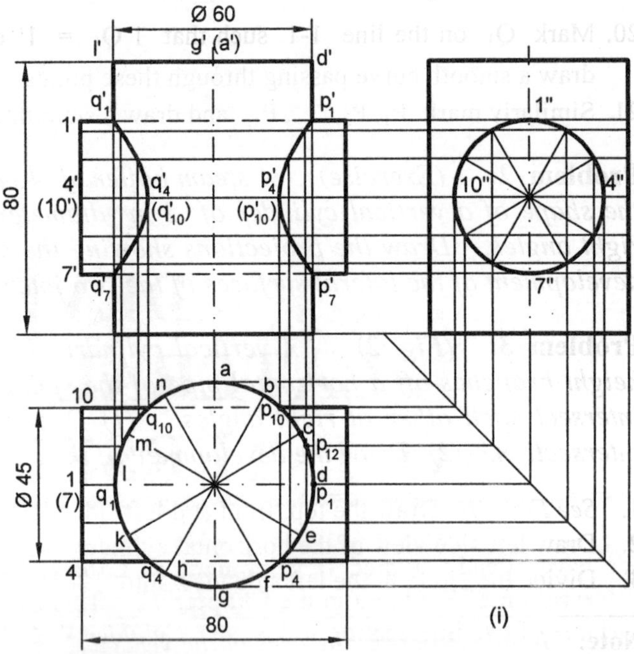

(i)

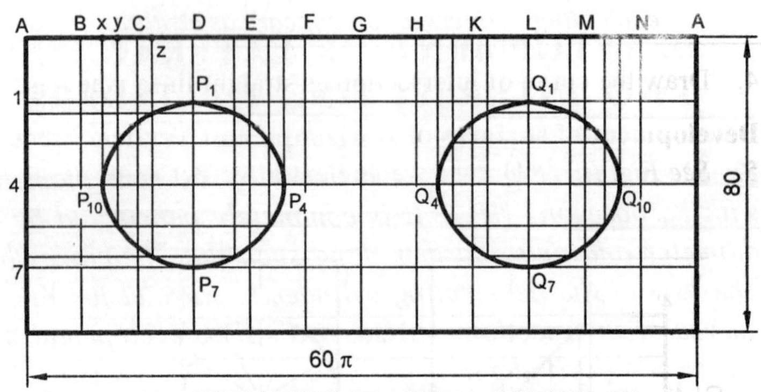

(ii) DEVELOPMENT OF LATERAL SURFACES OF INTERSECTING VERTICAL CYLINDER

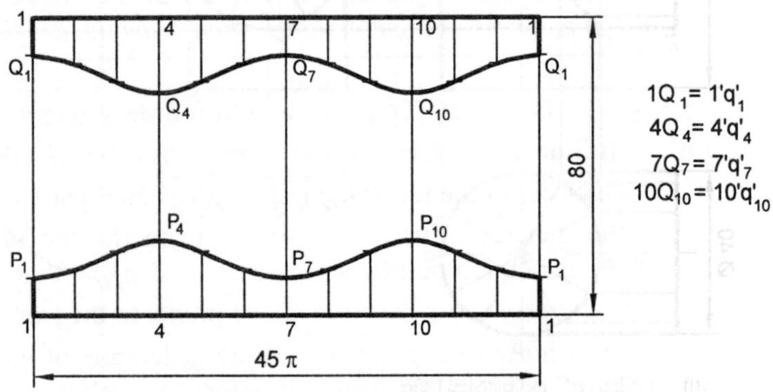

(iii) DEVELOPMENT OF LATERAL SURFACES OF INTERSECTING HORIZONTAL CYLINDER

FIG. 1

20. Mark Q_1 on the line 1-1 such that $1\,Q_1 = 1'\,q_1'$. Similarly mark $Q_2, Q_3, \ldots, Q_{12}$ and draw a smooth curve passing through these points.
21. Similarly mark $P_1, P_2, \ldots, P_{12}$ and draw a smooth curve passing through these points.

Problem 2: *(Exercise) A steam boiler of 4 m diameter is surmounted by a dome of the shape of a vertical cylinder of 2 m diameter. Their axes intersect each other at right angles. Draw the projections showing the curves of intersection. Also, draw the development of the lateral surfaces of the two intersecting cylinders.* (UQ)

Problem 3: *(Fig. 2) A vertical cylindrical pipe of 40 mm diameter and 70 mm height branches-off a horizontal pipe of the* ***same diameter****. The axes of the two pipes intersect each other at right angles. (1) Draw the projections showing the curves of intersection. (2) Draw the development of the surfaces of both pipes.* (UQ)

1. *See Fig. (i).* Draw the top, front and left side views of the vertical cylinder.
2. Draw left side view of the horizontal cylinder as a circle. Project its front and top views.
3. Divide the circle in the left side view into 12 equal parts and draw the projections.

Note: *In this Problem, the axes of the vertical and horizontal cylinders intersect each other at right angles. If the diameters of both the cylinders are* ***same****, both of them are tangent to each other. Intersection appears as* ***straight lines meeting at right angles in front view****.*

---

4. Draw the curve of intersection as straight lines meeting at right angles in the front view.

Development of surfaces of Horizontal and Vertical Pipes

5. *See Fig. (ii).* Draw development of vertical pipe as rectangle π 40 mm x 70 mm.

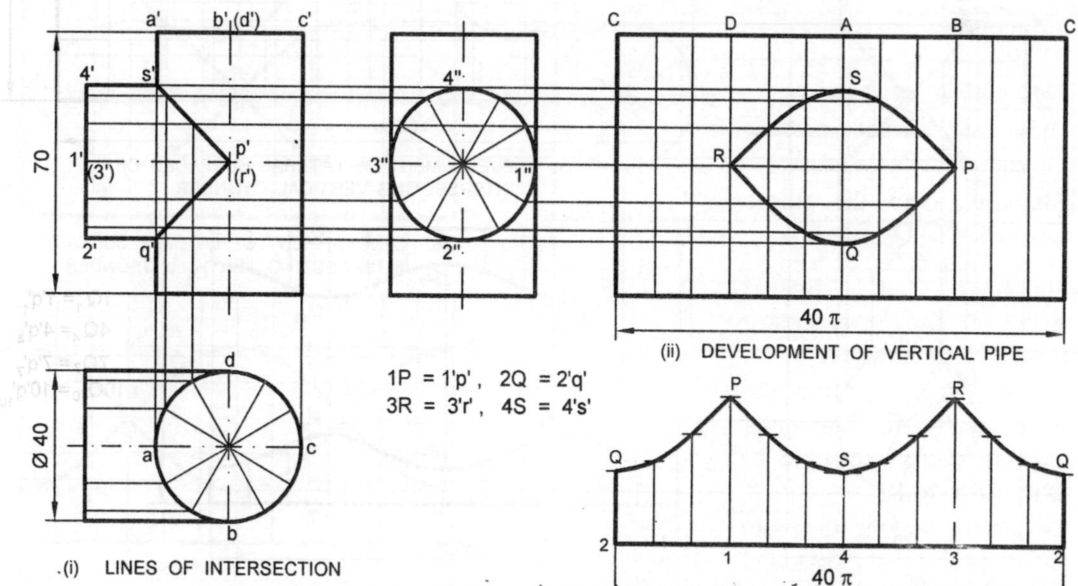

(i) LINES OF INTERSECTION

(ii) DEVELOPMENT OF VERTICAL PIPE

(iii) DEVELOPMENT OF HORIZONTAL PIPE

FIG. 2

6. Draw the curve of intersection in its development by transferring the corresponding distances from the front and left side views as shown.
7. *See Fig. (iii).* Draw the development of the surface of the horizontal pipe by transferring the corresponding distances from the front view.

Note: *As the horizontal pipe is a branch of the vertical one, show the intersection curve on only one side.*

Problem 4: *(Exercise) Draw the curves of intersection and develop a vertical cylindrical pipe of 44 mm diameter branched with a horizontal cylindrical pipe of the same diameter. The axes of both the pipes are parallel to VP and intersect at right angles to each other. Assume suitable lengths for the pipes.*

Problem 5: *(Fig. 3) Two cylinders of equal diameter of 40 mm intersect each other at right angles. Their axes bisect each other and are parallel to VP. Determine the line of intersection of the two cylinders. Also, draw the development of the lateral surfaces of the two intersecting cylinders.* (UQ)

Note: (i) *Assume the lengths of the two intersecting cylinders as 70 mm.*
(ii) *If two cylinders of equal diameter intersect each other at right angles, the curves of intersection are straight lines intersecting at right angle to each other. See Fig. 3, which is self-explanatory.*

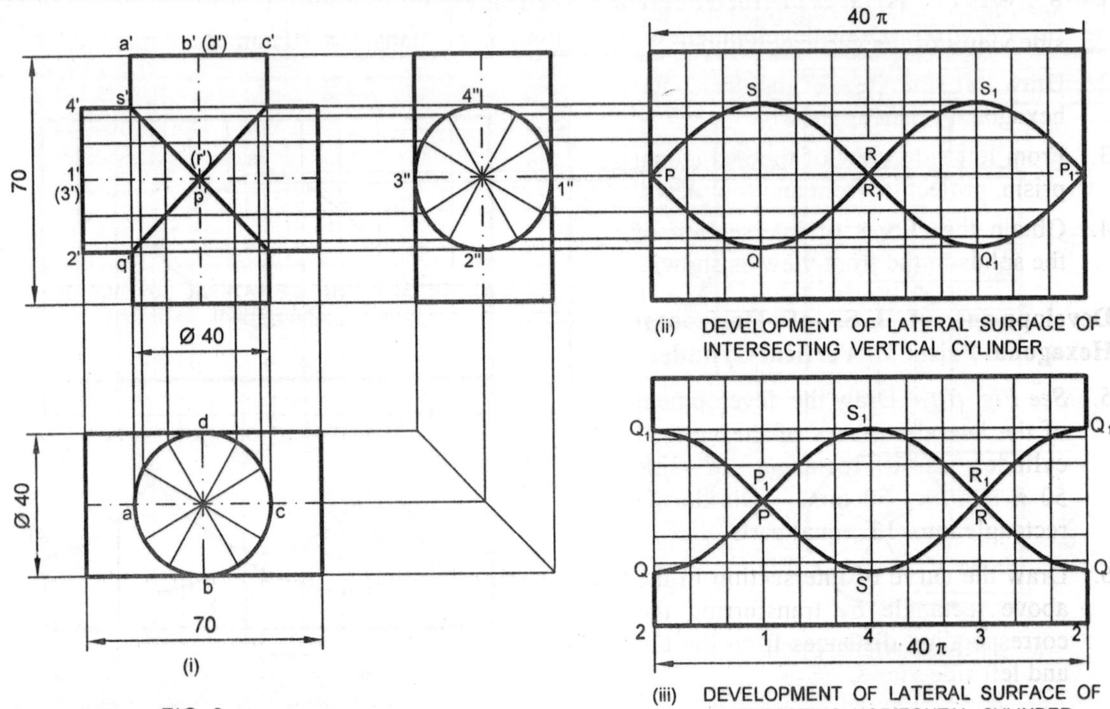

(ii) DEVELOPMENT OF LATERAL SURFACE OF INTERSECTING VERTICAL CYLINDER

(iii) DEVELOPMENT OF LATERAL SURFACE OF INTERSECTING HORIZONTAL CYLINDER

FIG. 3

2. DEVELOPMENT OF LATERAL SURFACES OF INTERSECTING PRISM Vs CYLINDER

If the cylinder is vertical, the intersecting prism will be horizontal without off-set. The other arrangement is that if the cylinder is horizontal, the intersecting prism will be vertical without off-set.

Problem 6: *(Fig. 4) A vertical cylinder of 50 mm diameter and 80 mm height is standing on its base on HP. It is penetrated by a hexagonal prism of 20 mm side and 80 mm long. Axes of the solids intersect at 90° with no off-set.*

(1) Draw the projections of the solids showing curves of intersection.

(2) Draw the development of the lateral surfaces of the intersecting prism and cylinder. (UQ)

1. *See Fig. (i).* Draw top, front and left side views of the vertical cylinder.
2. Draw left side view of the horizontal hexagonal prism as shown.
3. From left side view of the hexagonal prism, project its front and top views.
4. Obtain the curves of intersection of the solids in the front view as shown.

Development of L.S. of Horizontal Hexagonal Prism Vs Vertical Cylinder

5. *See Fig. (ii).* Draw the development of the lateral surfaces of the vertical cylinder as a rectangle of size 50 π mm x 80 mm. Divide the rectangle into 12 equal parts.
6. Draw the curve of intersection in the above rectangle by transferring the corresponding distances from the top and left side views.

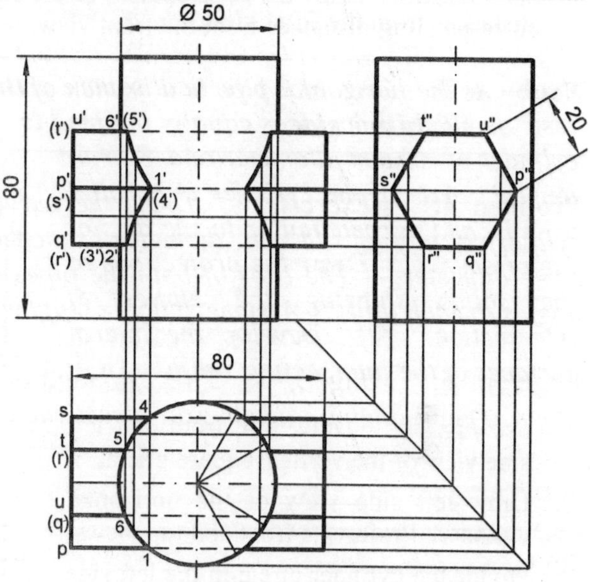

(i) CYLINDER Vs HEXAGONAL PRISM

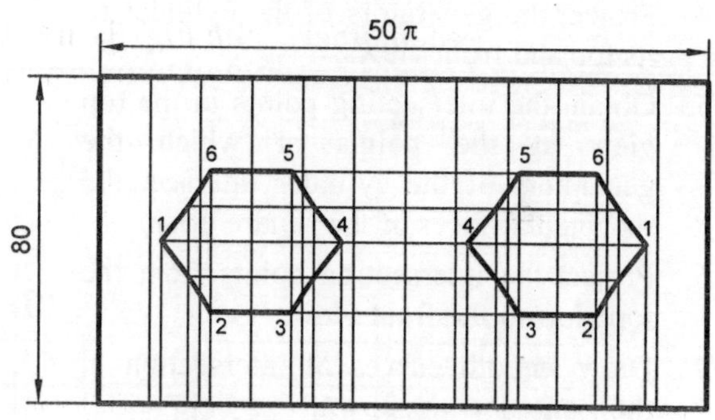

(ii) DEVELOPMENT OF LATERAL SURFACE OF INTERSECTING VERTICAL CYLINDER

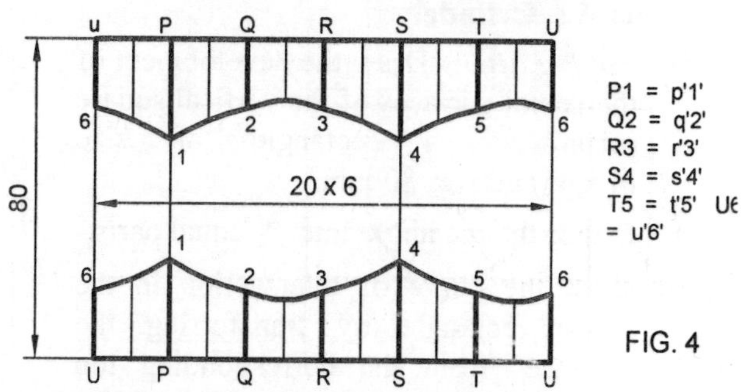

FIG. 4

(iii) DEVELOPMENT OF LATERAL SURFACES OF INTERSECTING HORIZONTAL HEXAGONAL PRISM

7. *See Fig. (iii).* Draw a rectangle of size (6 x 20) mm x 80 mm to represent the development of the horizontal hexagonal prism. Divide the above rectangle into 6 equal parts.
8. Draw the curve of intersection in the above rectangle by transferring the corresponding distances from the front view.

Problem 7: *(Fig. 5) A vertical square prism of side of base 45 mm and height 80 mm rests on HP with its faces equally inclined to VP. It is pierced centrally by a horizontal cylinder of 40 mm diameter and 80 mm length. Axis of the cylinder is parallel to VP and perpendicular to the axis of the prism. (1) Draw the projections of the solids showing the curves of intersection. (2) Develop the lateral surfaces of the intersecting solids.* (UQ)

1. *See Fig. (i).* Draw top, front and left side view of the vertical square prism.
2. Draw left side view of the horizontal cylinder. Project its front and top views.
3. Divide the cylinder circle in the left side view into 8 (or 12) equal parts.
4. Project the generators of the cylinder to its top and front views.
5. Obtain the intersecting points in the top view at the points at which the generators of the cylinder intersect the rectangular faces of the square prism.
6. Project the intersection points from the top view to the front view.
7. Draw smooth curves of intersection in the front view as shown.

Development of lateral surfaces of Square Prism Vs Cylinder

8. *See Fig. (ii).* Draw the development of the lateral surfaces of the vertical square prism as a rectangle of size (4 x 45) mm x 80 mm.
9. Divide the rectangle into 4 equal parts.
10. Draw the curve of intersection in the above rectangle by transferring the distances from the corresponding top and left side views.

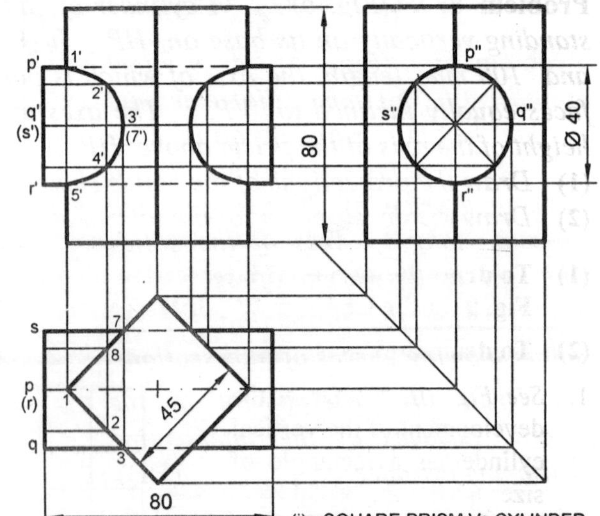

(i) SQUARE PRISM Vs CYLINDER

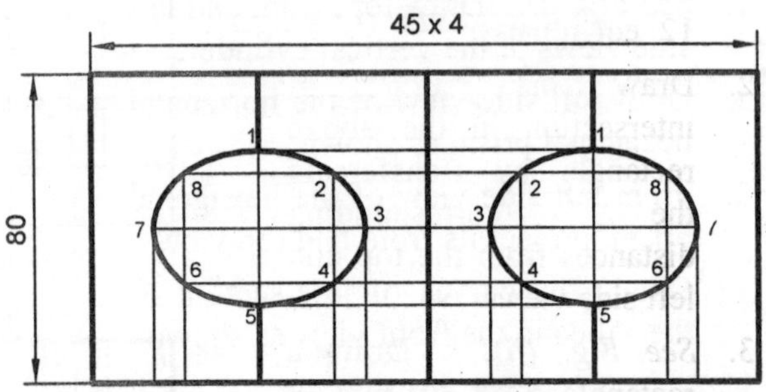

(ii) DEVELOPMENT OF LATERAL SURFACES OF INTERSECTING VERTICAL SQUARE PRISM

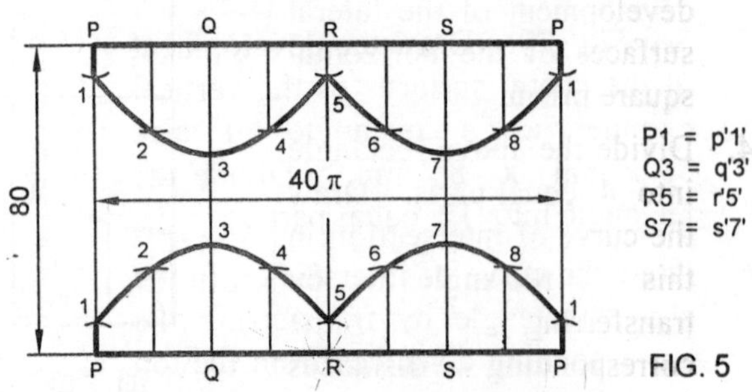

FIG. 5

(iii) DEVELOPMENT OF LATERAL SURFACE OF INTERSECTING HORIZONTAL CYLINDER

11. *See Fig. (iii).* Draw a rectangle of size 40 π mm x 80 mm to represent the development of the horizontal cylinder. Divide this rectangle into 8 (or 12) equal parts.
12. Draw the curve of intersection in the above rectangle by transferring the distances from the corresponding front view.

3. ADDITIONAL PROBLEMS

Problem 8: *(Fig. 6) A cylinder of 60 mm diameter and axis 80 mm long is standing vertically on its base on HP. It is pierced by a square prism of 30 mm side and 100 mm length, the axis of which is parallel to both the reference planes and the faces equally inclined to VP. The axes of the solids intersect at right angles. The height of the axis of the prism above HP is 40 mm.*

(1) *Draw the projections of the solids showing the curves of intersection.*

(2) *Draw the development of the lateral surfaces of the intersecting solids.* (UQ)

(1) To draw the curves of intersection of Square Prism Vs Cylinder, refer Problem 9 and Fig. 9 in the Chapter 17 – Intersection of Surfaces.

(2) To draw the development of lateral surfaces of Square Prism Vs Cylinder

1. *See Fig. (i).* Draw the development of the vertical cylinder as a rectangle of size 60 π mm x 80 mm. Divide the rectangle into 12 equal parts.
2. Draw the curve of intersection in the above rectangle by transferring the corresponding distances from the top and left side views.
3. *See Fig. (ii).* Draw a rectangle (4 x 30) mm x 100 mm to show the development of the lateral surfaces of the horizontal square prism.
4. Divide the above rectangle into 4 equal parts. Draw the curve of intersection in this rectangle by transferring the corresponding distances from the front view.

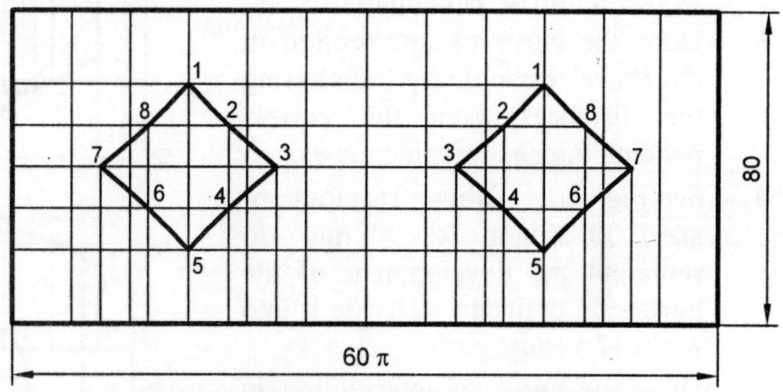

(i) DEVELOPMENT OF LATERAL SURFACE OF INTERSECTING VERTICAL CYLINDER

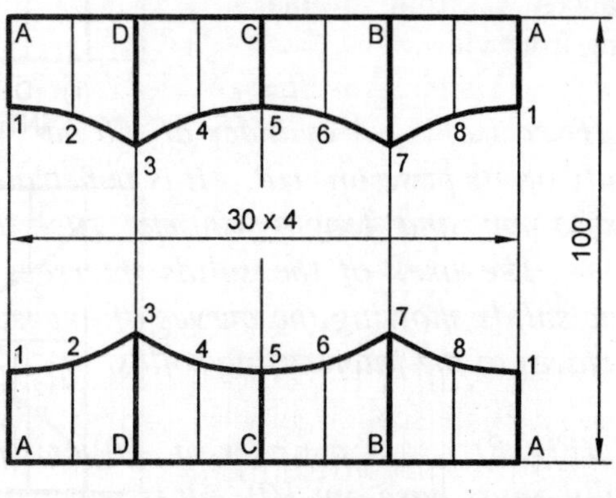

A1 = a'1'
B7 = b'7'
C5 = c'5'
D3 = d'3'

FIG. 6

(ii) DEVELOPMENT OF LATERAL SURFACES OF INTERSECTING HORIZONTAL SQUARE PRISM

Problem 9: *(Fig. 7) A horizontal cylinder of 30 mm diameter and 90 mm long penetrates centrally a vertical hexagonal prism of 25 mm side and 80 mm height. The axis of the cylinder is parallel to VP and perpendicular to the axis of the prism. (1) Draw the projections of the solids showing the curves of intersection. (2) Draw the development of the lateral surfaces of the intersection of the solids.* (UQ)

(1) To draw the projections of the solids showing the curves of intersection, refer Problem 10 and Fig. 10 in the Chapter 17 – Intersection of Surfaces.

(2) To draw the development of lateral surfaces of Hexagonal Prism Vs Cylinder

1. *See Fig. (i).* Draw the development of the lateral surfaces of the vertical hexagonal prism as a rectangle of size (6 x 25) mm x 80 mm.
2. Divide it into 6 equal parts.
3. Draw the curve of intersection in the above rectangle by transferring the distances from the corresponding top and left side views.
4. *See Fig. (ii).* Draw a rectangle of size 30 π mm x 90 mm to represent the development of the horizontal cylinder. Divide it into 8 (or 12) equal parts.
5. Draw the curve of intersection in the above rectangle by transferring the distances from the corresponding front view.

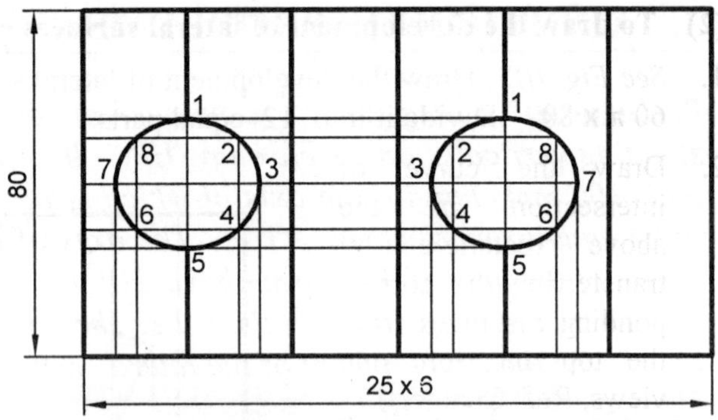

(i) DEVELOPMENT OF LATERAL SURFACES OF INTERSECTING VERTICAL HEXAGONAL PRISM

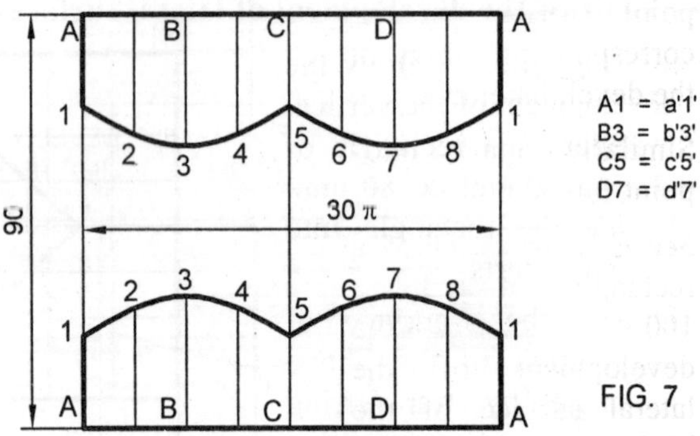

FIG. 7

(ii) DEVELOPMENT OF LATERAL SURFACE OF INTERSECTING HORIZONTAL CYLINDER

Problem 10: *(Exercise) A cylinder of 60 mm diameter and axis 80 mm long is standing vertically on its base on HP. It is penetrated centrally by a rectangular prism of size 30 mm x 25 mm and length 100 mm, such that its longer rectangular faces are parallel to VP. The axes of the solids intersect at right angles. (1) Draw the projections of the solids showing the curves of intersection. (2) Draw the development of the lateral surfaces of the intersecting solids.* (UQ)

Problem 11: *(Fig. 8) A cylinder of 60 mm diameter and axis 80 mm long is standing vertically on its base on HP. It is penetrated by a horizontal square prism of 35 mm side and 100 mm length, the axis of which is parallel to both the reference*

planes. The axes of the solids intersect at right angles and bisect each other. A lower rectangular face of the horizontal prism is inclined at 20° to HP. (1) Draw the projections showing the intersection curve in the front view. (2) Draw the development of the lateral surfaces of the intersecting solids. (UQ)

(1) To draw the curves of intersection, refer Problem 12 and Fig. 11 in Chapter 17.

(2) To draw the development of lateral surfaces of Square Prism Vs Cylinder

1. *See Fig. (i).* Draw the development of lateral surfaces of the vertical cylinder as a rectangle 60 π x 80. Divide it into 12 equal parts.
2. Draw the curve of intersection in the above rectangle by transferring the corresponding distances from the top and left side views. Ref: Chapter 17.
3. For side 1″2″ in side view, mark its mid–point. Transfer the corresponding point in the development.
4. Similarly, mark mid-point for side 4″5″.
5. *See Fig. (ii).* Draw a rectangle (4 x 35) x 100 for the development of the lateral surface of the horizontal square prism. Divide it into 4 equal parts.
6. Draw the curve of intersection in above rectangle by transferring the corresponding distances from the front view.

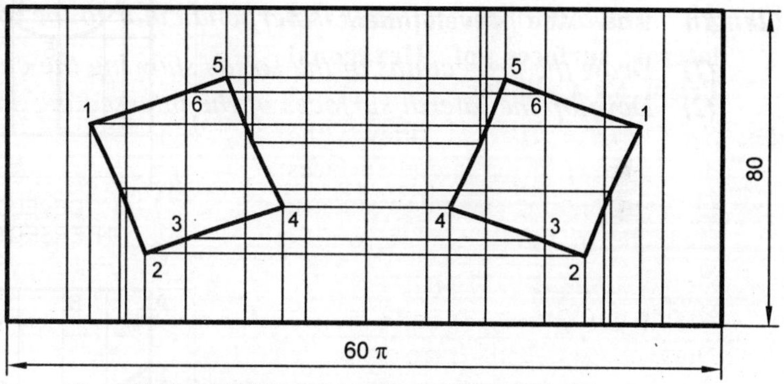

(i) DEVELOPMENT OF LATERAL SURFACE OF INTERSECTING VERTICAL CYLINDER

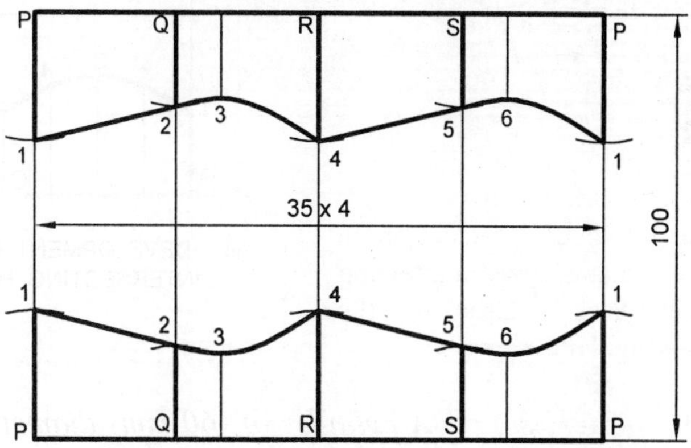

FIG. 8

(ii) DEVELOPMENT OF LATERAL SURFACES OF INTERSECTING HORIZONTAL SQUARE PRISM

Problem 12: *(Exercise) A vertical cylinder of 60 mm diameter is penetrated by a horizontal cylinder of 40 mm diameter. The axis of the penetrating cylinder is parallel to VP and 6 mm in front of the axis of the vertical cylinder. Draw the development of the lateral surfaces of the two intersecting cylinders.*

Problem 13: *(Exercise) A vertical cylinder of 50 mm diameter and 80 mm height is standing on its base on HP. It is penetrated by a pentagonal prism of 24 mm side and 80 mm length such that one of its rectangular faces is parallel to VP and farther away from it. The axes of the solids intersect at right angles with no off-set.*

(1) Draw the projections of the solids showing the curves of intersection.
(2) Draw the development of the lateral surfaces of the intersecting prism and cylinder.

Problem 14: *(Exercise) A rectangular prism of side of base 45 mm x 40 mm and height 80 mm rests on its base on HP with its rectangular faces equally inclined to VP. It is pierced centrally by a horizontal cylinder of 35 mm diameter and 80 mm length. The axis of the cylinder is perpendicular to the axis of the prism.*

(1) Draw the projections of the solids showing the curves of intersection.
(2) Develop the lateral surfaces of the intersecting solids.

*

20

ISOMETRIC PROJECTION

1. PICTORIAL PROJECTIONS

In Orthographic Projection, two or more views are drawn to describe an object completely. To visualize its shape, all the orthographic views have to be read simultaneously and a conclusion is to be drawn *mentally* about its shape. This difficulty is overcome by pictorial projection. It displays all the three dimensions of an object in one and the same view. It is obtained by projecting an object only on one plane of projection. Types of Pictorial Projections are:

1. ***Axonometric Projection:*** In this projection, only one view showing all the three dimensions of an object is drawn on a Plane of Projection. Axonometric Projections are classified into:

 (i) Isometric Projection: It is a pictorial projection in which three dimensions of a solid are not only shown in one view, but also their dimensions can be measured from it.

 (ii) Dimetric Projection and ***(iii) Trimetric Projection***

 [Only isometric projection is dealt with in this Chapter.]

2. ***Oblique Projection:*** Unlike axonometric projection, one of the faces of the object is kept parallel to the plane of projection in Oblique Projection.

3. ***Perspective Projection:*** It shows an object as it appears instead of its true shape and size.

2. PRINCIPLE OF ISOMETRIC PROJECTION

What is Isometric Projection ? *Iso* (in *Greek*) means *equal* ; ***metric*** means ***measure***. Hence ***Isometric Projection*** means ***a system of projection of equality of measure***.

Orthographic Projection of a Cube: *See Fig. 1.* The principle involved in drawing an Isometric Projection can best be explained by drawing *Orthographic Projection of a Cube*, resting on HP on one of its corners with a *solid diagonal* (Solid Diagonal is an imaginary line joining one of the corners at the top and the diametrically opposite corner at the bottom) perpendicular to HP.

Isometric Projection: *Fig. 2* shows the top view of the cube, obtained from orthographic projection in *Fig. 1.* It gives a fair idea of the pictorial form of the cube and is called its *Isometric Projection.*

Rule: To distinguish Isometric Projection from Orthographic Projection, use CAPITAL LETTERS for the corners in the Isometric Projection. *See Fig. 2.* ABCDEFGH is the Isometric Projection of the Cube.

- ***Refer Fig. 2.*** All the edges and faces of the cube are equally inclined to the plane of projection. Hence the square faces are seen as similar and equal rhombuses.
- The three edges / sides of the cube AB, AD and AE are meeting at the point A. These edges are mutually perpendicular to each other in the solid.
- Above three edges form 90° in the Orthographic Projection in *Fig. 1* and are now seen to include 120° with each other in Isometric Projection in *Fig. 2*.
- Edges AB and AD make 30° with the horizontal. Edge AE is vertical.

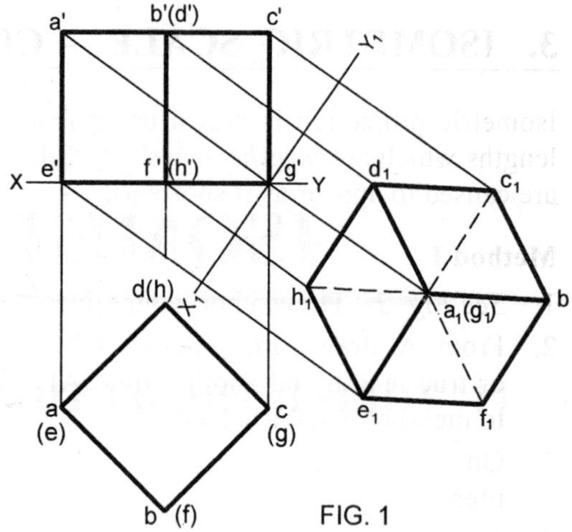

FIG. 1

- The three edges AB, AD and AE meeting at A and forming a solid right angle are also equally inclined to VP and therefore they are *equally foreshortened.*

  All the other edges like CB and BF are parallel to any of the lines AB, AD and AE and are also *equally foreshortened*. These edges being *equally foreshortened*, leads us to the problem of selecting an ***Isometric Scale.***
- BD and AC represent equal diagonals of the cube, but are not equally shortened in Isometric Projection. Diagonal DB of the top face of the cube is parallel to the VP. Hence it retains its true length in Isometric. However, diagonal AC is shortened.

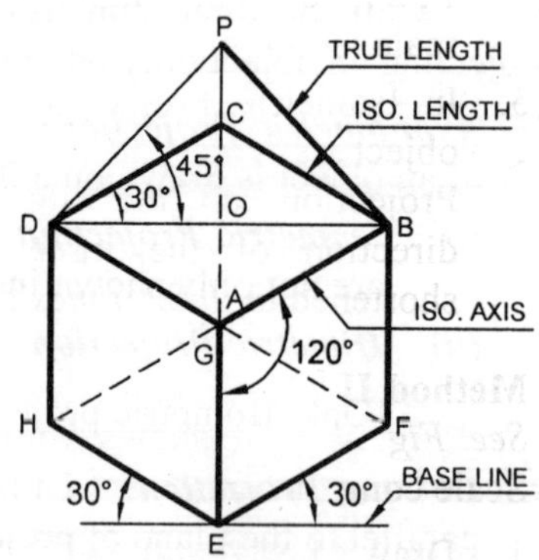

FIG. 2 ISOMETRIC CUBE

2.1 TERMINOLOGY

Isometric Axes: Lines AB, AD and AE meeting at a point A and making an angle of 120° with each other are termed *Isometric Axes.*

Isometric Lines: Lines parallel to the isometric axes are termed *Isometric Lines.* Lines CD, CB, etc., are examples of Isometric Lines.

Non-isometric Lines: Lines which are not parallel to isometric axes are termed *Non-isometric Lines.* Line BD is an example.

Isometric Planes: Planes representing the faces of the cube as well as other planes parallel to these planes are termed *Isometric Planes.*

Note: *There are three Isometric Planes, viz., Left Isometric Plane, Right Isometric Plane and Top Isometric Plane, containing any two isometric axes; Also, planes parallel to these planes are termed Isometric Planes.*

Non-isometric Planes: Planes not parallel to isometric planes are Non-isometric Planes.

3. ISOMETRIC SCALE – CONSTRUCTION

Isometric projection is drawn using *Isometric Scale*, which converts true lengths into isometric lengths which are *foreshortened*. ("Edges of an object are foreshortened" *means that* "the edges are caused to appear as if shortened".)

Method I

1. *See Fig. 3*. Draw a horizontal line AB.
2. From A draw a line AC at 45° to represent actual or true length and another line AD at 30° to AB to measure isometric length.
3. On AC mark the points 0, 1, 2, etc., to represent actual lengths.
4. From these points draw verticals to meet AD at 0′, 1′, 2′, etc. The length A1′ represents the isometric scale length of A1 and so on.
5. In Isometric Projection, actual (true) length of an object is foreshortened *i.e.*, in the Isometric Projection, all the edges of an object along the direction of the three isometric axes are foreshortened to *0.82 times* their actual / true lengths.

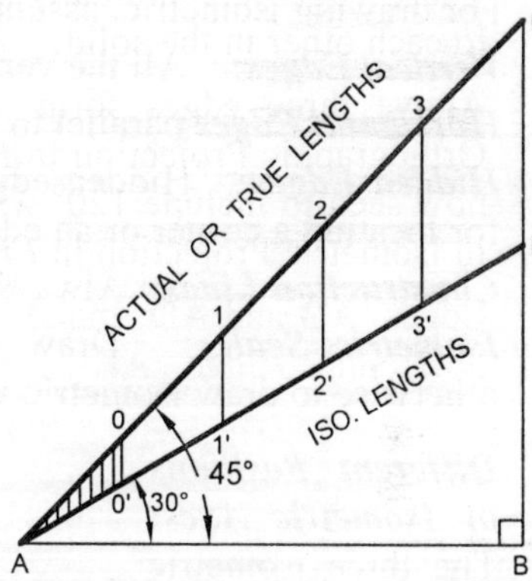

FIG. 3 ISOMETRIC SCALE

Method II

See Fig. 4. It shows **simplified** Isometric Scale construction.

1. Draw a horizontal line AC = true length.
2. Mark the division points of true length on AC as 0, 1, 2 and 3.
3. At the point A, draw a line inclined at an angle of 15° to AC.
4. Also, at the point C draw a line at an angle of 45° to CA. Both meet at D.
5. Draw lines parallel to CD through the points 0, 1, 2 and 3 to meet AD at the points 0′, 1′, 2′ and 3′. The line AD gives the isometric lengths.

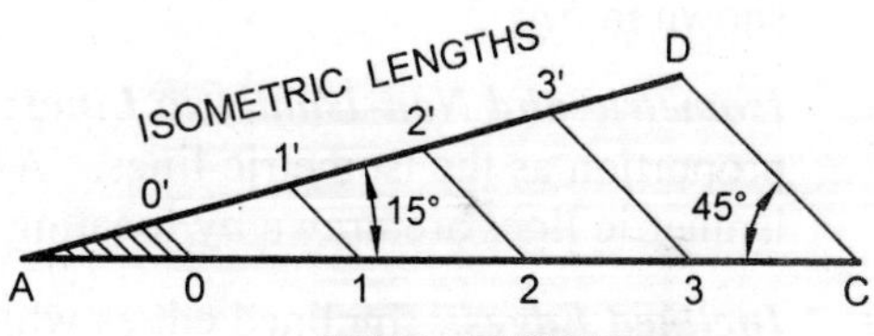

FIG. 4 ISOMETRIC SCALE (PREFERRED)

4. ISOMETRIC VIEW

Isometric View (Isometric Drawing)	Isometric Projection
Drawn to *actual scale*. When lines are drawn parallel to isometric axes, the true lengths are laid off.	Drawn to *isometric scale*. When lines are drawn parallel to isometric axes, the lengths are ***foreshortened to 0.82 times*** the actual lengths.

Isometric Projection is drawn by reducing each dimension to Isometric Scale. It is time-consuming. Distances between any two points cannot be measured with actual scale. Hence, the general practice by the engineers is to draw Isometric View which is exactly similar in shape of Isometric Projection. Direct measurement of actual dimensions is possible.

CONVENTIONS

1. For drawing isometric, assume the object to be placed on HP only.
2. ***Vertical Edges:*** All the vertical edges of the object are drawn vertical in the isometric.
3. ***Horizontal Edges*** parallel to the isometric axes are drawn at 30° to the horizontal.
4. ***Hidden Edges:*** Hidden edges need not be shown in isometric, unless specified otherwise for locating a corner or an edge or a face.
5. ***Construction Lines:*** Always retain the construction lines after completing the isometric.
6. ***Isometric Scale:*** Draw isometric projection using isometric scale, unless specified otherwise to draw isometric view or isometric drawing.
7. ***Different Positions of Isometric Axes:*** The three isometric axes may be arranged in any one of the positions shown in *Fig. 5.*

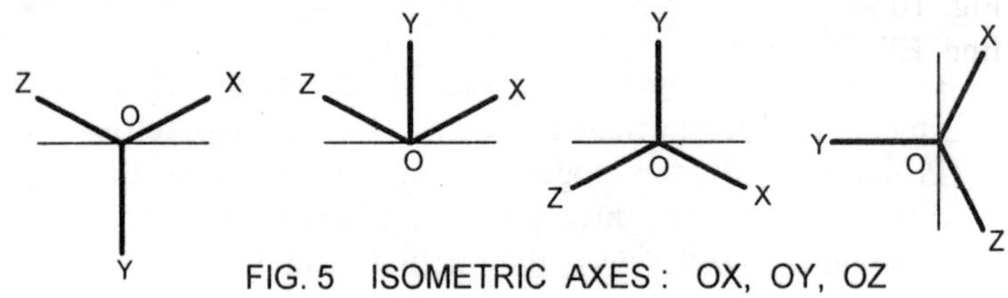

FIG. 5 ISOMETRIC AXES : OX, OY, OZ

8. ***Isometric and Non-Isometric Lines:*** Non-isometric lines are not foreshortened in the same proportion as the isometric lines. Also, they should not be drawn at an angle of 30°. Non-isometric lines are drawn by locating their ends and then joining them.
9. ***Inclined Edges:*** Inclined edges which are not parallel to the isometric axes should never be drawn at the given inclination in the isometric.
   Draw these inclined edges by first locating the end points in isometric and then joining them.
10. ***Angles in Isometric:*** Angles cannot be set-off directly in isometric. For this, it is necessary to locate the end points of the lines which are subtending the angle.
11. ***Isometric and Non-Isometric Planes***
   In Fig. 6, Isometric Planes are marked as 1 and non-isometric Planes as 2.

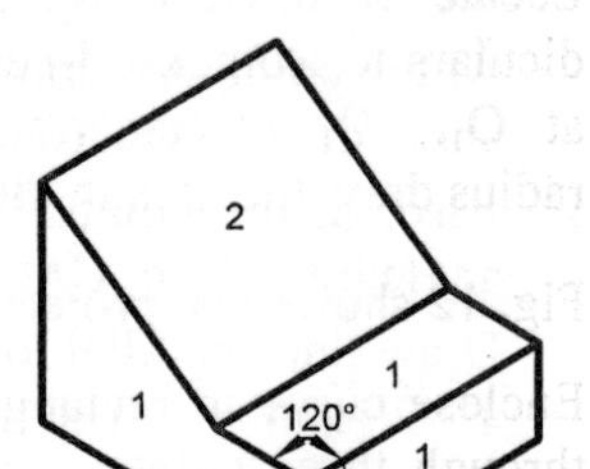

FIG. 6

12. ***Dimensioning in Isometric***
   In *Figs. 7(a) & (b),* all the dimension lines are drawn in line with the vertical or sloping lines at 30°. No other lines are allowed.

   The strokes of the letters and numerals are to be parallel to the isometric axes.

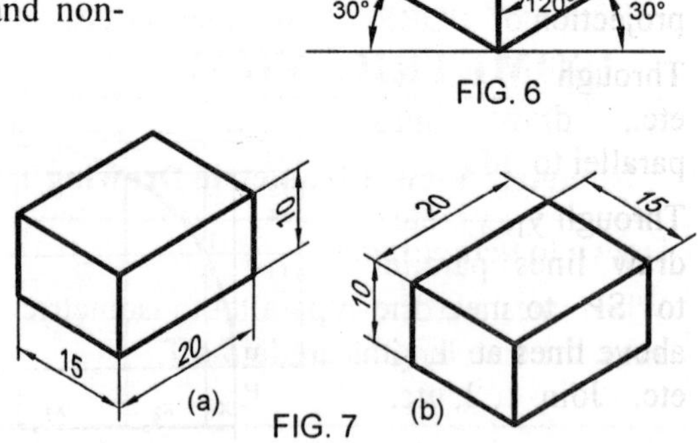

FIG. 7

5. ISOMETRIC PROJECTION OF LINES

Fig. 8 shows the isometric projection and isometric view of a horizontal line AB. Fig. 9 shows the isometric projection and isometric view of a vertical line CD.

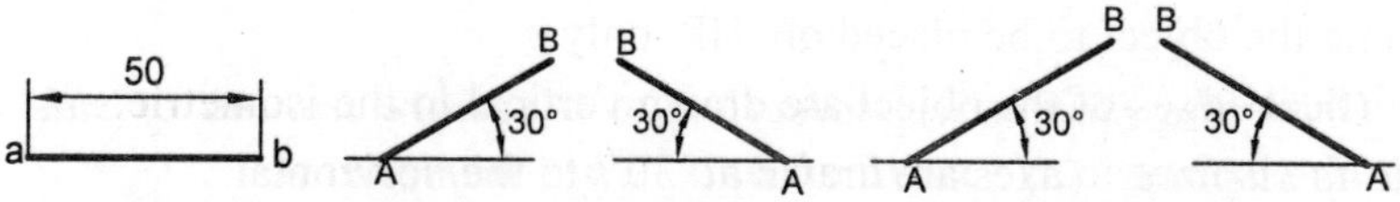

TOP VIEW ISOMETRIC PROJECTION ISOMETRIC VIEW

FIG. 8

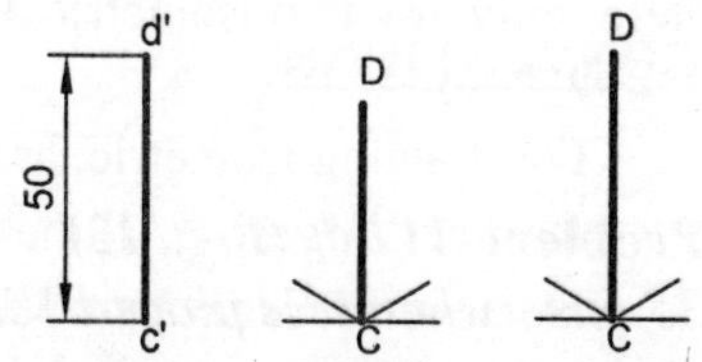

FRONT VIEW ISOMETRIC PROJECTION ISOMETRIC VIEW

FIG. 9

Fig. 10 shows the three possible ways of representing the isometric projection of a non-isometric line EF inclined at 45° to the horizontal.

FIG. 10

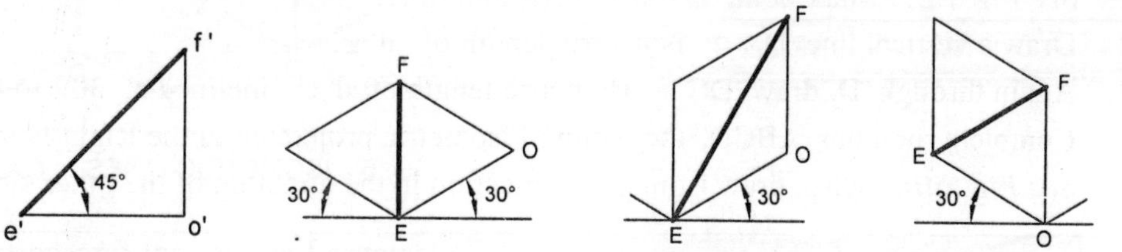

(i) FRONT VIEW (ii) TOP ISO. PLANE (ii) RIGHT ISO. PLANE (iii) LEFT ISO. PLANE

Fig. 11 shows isometric projection of a 90° **arc** of a circle. Arc pq is tangential to pr and qr. Locate R, P and Q. At P and Q draw perpendiculars to isometric lines PR and QR to meet at O_1. O_1 as center and O_1P (= O_1Q) as radius draw the arcs in the isometric projection.

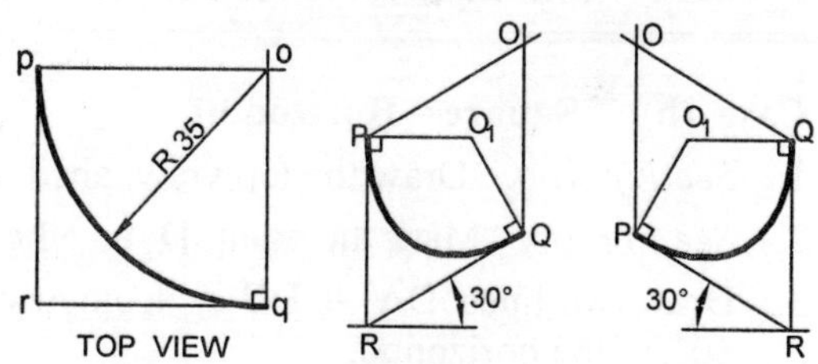

TOP VIEW RIGHT I.P. LEFT I.P.

FIG. 11

Fig. 12 shows iso. projection of **irregular curve** EF.

Enclose curve in rectangle PQRS. Mark 1, 2, etc., on it. Draw horizontals and verticals through these points. Mark their coordinates as (x_1, y_1), (x_2, y_2), etc. Draw the isometric projection of PQRS. Mark distances (x_1, y_1), (x_2, y_2), etc., on it.

Through x_1, x_2, etc., draw lines parallel to SR.

Through y_1, y_2, etc., draw lines parallel to SP to meet the above lines at 1, 2, etc. Join 1, 2, etc.

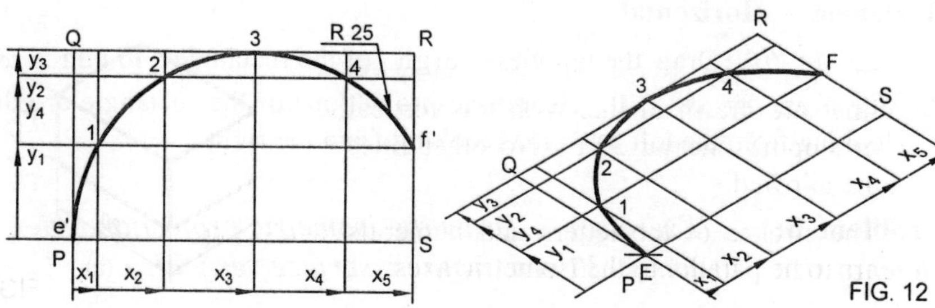

FIG. 12

6. ISOMETRIC PROJECTION OF PLANE FIGURES

Isometric Projection or Isometric View of a Plane Figure / Lamina is generally drawn on any one of the Isometric Planes, *viz.*, Left Isometric Plane, Right Isometric Plane or Top Isometric Plane. Here, only any two isometric axes are considered at a time, because the object (Plane Figure) is a 2-D body.

Problem 1: *(Fig. 13) Draw the isometric projection of a* ***square*** *lamina of side 35 mm when it is placed with its surface (a)* ***vertical*** *and (b)* ***horizontal****.*

Case (a) : Square – Vertical

1. *See Fig. (i).* Draw the front view a′ b′ c′ d′ of the square of side 35 mm.

Note : *In isometric projection / view, draw the vertical lines vertical and the horizontal lines inclined at an angle of 30° to the horizontal.*

---

2. *See Fig. (ii).* Mark point D on a horizontal line.
3. Draw a vertical line DA = isometric length of d′ a′.
4. Again through D, draw DC = isometric length of d′ c′ inclined at 30° to the horizontal.
5. Complete rhombus ABCD, the required isometric projection in the Right Isometric Plane.
6. *See Fig. (iii).* Also, draw isometric projection in the direction of the other sloping axis.

Note : *Arrows shown in Figs. (ii) & (iii) indicate the directions of viewing in Right and Left Isometric Planes respectively, to obtain given front view.*

Case (b) : Square – Horizontal

1. *See Fig. (iv).* Draw the top view abcd of the square of 35 mm side.
2. *See Fig. (v).* Mark the point D on a horizontal line.
3. Draw two lines DA = DC = isometric length of da = isometric length of dc, inclined at 30° to the horizontal.
4. Complete the rhombus ABCD, which is the required isometric projection.

Problem 2: *(Fig. 14) Draw the isometric projection of a* ***thin rectangular sheet*** *of size 40 mm x 30 mm, with its surface* ***horizontal****.*

Rectangle – Horizontal

1. *See Fig. (i).* Draw the top view efgh of the rectangle 40 mm x 30 mm as shown.
2. *See Figs. (ii) & (iii).* Isometric projections of the rectangle are drawn with the side HG looking from the left as well as looking from the right.

Problem 3: *(Exercise) Draw the isometric projection / view of a* ***thin rectangular sheet*** *of size 40 mm x 30 mm, with its surface* ***vertical****.*

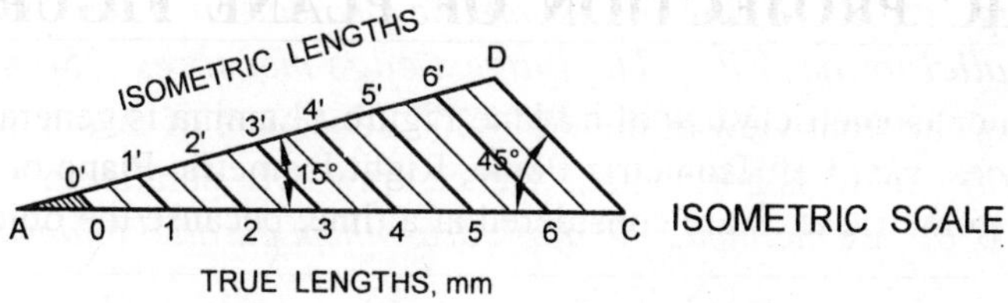

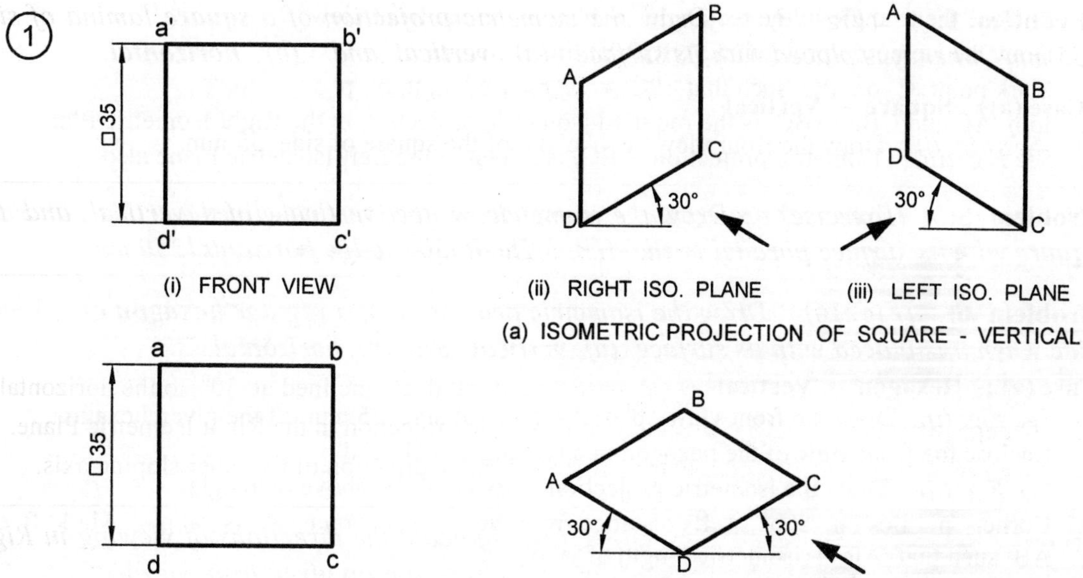

(i) FRONT VIEW (ii) RIGHT ISO. PLANE (iii) LEFT ISO. PLANE

(a) ISOMETRIC PROJECTION OF SQUARE - VERTICAL

(iv) TOP VIEW (v) HORIZONTAL ISO. PLANE

(b) ISOMETRIC PROJECTION OF SQUARE - HORIZONTAL

FIG. 13

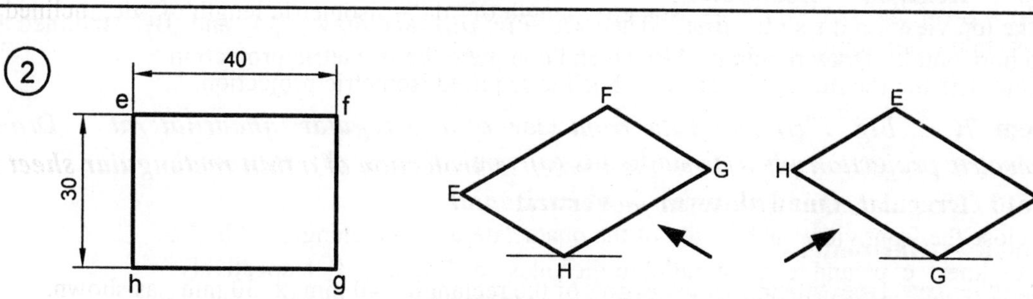

(i) TOP VIEW (ii) RIGHT ISO. PLANE (iii) LEFT ISO. PLANE

FIG. 14 ISOMETRIC PROJECTION OF RECTANGLE - HORIZONTAL

Note : Use True Scale (Actual Scale) for drawing Isometric View / Isometric Drawing.

Problem 4: *(Fig. 15) Draw the isometric projection of a 60° - 30°* ***set-square*** *with its surface parallel to the VP. The longest edge measures 250 mm.*

1. *See Fig. (i).* Draw the front view of the 60° - 30° set-square of 250 mm base.
2. *See Figs. (i) & (ii).* a′ b′ is horizontal. Hence draw AB parallel to a sloping axis.
3. Sides a′ c′ and b′ c′ are *inclined. So don't draw them parallel to any isometric axis.*

Note: *In isometric, draw an angle by fixing positions of its ends on isometric lines.*

4. Enclose the triangle in the rectangle a′ b′ q′ p′ in the front view.
5. Draw the isometric projection ABQP of this rectangle.
6. Mark point C on PQ such that PC = isometric length of p′ c′.
7. Join AC and CB. ABC is the required isometric projection in the Right Isometric Plane.
8. *See Fig. (iii).* Isometric projection ABC is drawn in the Left Isometric Plane also.

Problem 5: *(Exercise) Draw the isometric projection / view of the 60° - 30°* ***set-square*** *with its surface parallel to the HP. The longest edge measures 250 mm.*

Problem 6: *(Fig. 16) Draw the isometric projection of a regular* ***hexagon*** *of 25 mm side when it is placed with its surface (a) vertical and (b) horizontal.*

Case (a) : Hexagon – Vertical

1. *See Fig. (i).* Draw the front view p′ q′ r′ s′ t′ u′ of side 25 mm of the given hexagon.
2. Enclose the front view of the hexagon in a rectangle a′ b′ c′ d′.
3. *See Fig. (ii).* Draw the isometric projection ABCD of the above rectangle.
4. Corner p′ lies on a′ b′. Its distance from the nearest corner a′ is a′ p′. Mark P on AB such that AP = isometric length of a′ p′.
5. Similarly transfer the distances of each of the corners q′, r′, s′, t′ and u′ of the hexagon as measured from the nearest corners b′, c′, d′, a′ respectively.
6. Join P Q R S T U, the isometric projection of hexagon on Left Isometric Plane when vertical.

Exercise: *Draw the isometric projection of the above hexagon on the Right Isometric Plane.*

Case (b) : Hexagon – Horizontal

Draw the top view p q r s t u first. Then see *Fig. (iii)* and draw BA and BC inclined at 30° to horizontal. Draw rhombus ABCD and complete the isometric projection.

Problem 7: *Fig. 17(i) shows the front view of an* ***irregular quadrilateral****. Draw the isometric projection when its surface is (a) vertical and (b) horizontal.*

Case (a) : Irregular Quadrilateral – Vertical

1. Enclose the front view a′ b′ c′ d′ of the quadrilateral in a rectangle a′ b′ f′ e′.
2. Draw lines c′ p′ and c′ q′ parallel to the sides e′ f′ and b′ f′ respectively.
3. *See Fig. (ii).* Draw the isometric projection ABFE of the rectangle a′ b′ f′ e′.
4. Fix the point D in EF as explained in Problem 11.
5. Draw the iso. projection CPFQ of the rectangle c′ p′ f′ q′ as shown to fix the point C.
6. Draw the quadrilateral ABCD, the required isometric projection on Right Isometric Plane.

Case (b) : Irregular Quadrilateral – Horizontal

Draw top view abcd first. Draw the isometric projection in horizontal position [*Fig. (iii)*].

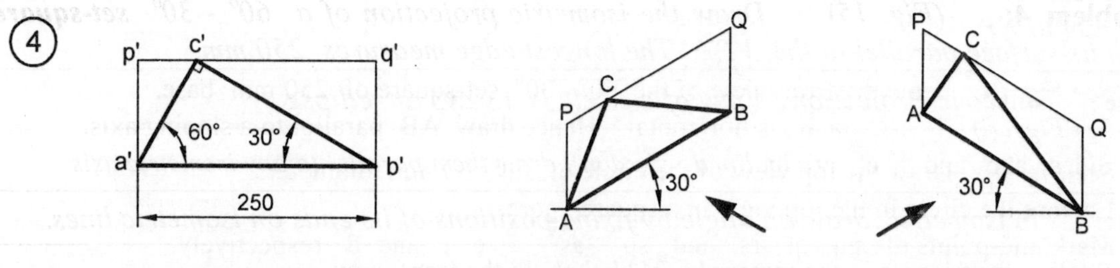

(i) FRONT VIEW (ii) RIGHT ISO. PLANE (iii) LEFT ISO. PLANE

FIG. 15 ISOMETRIC PROJECTION OF SET-SQUARE

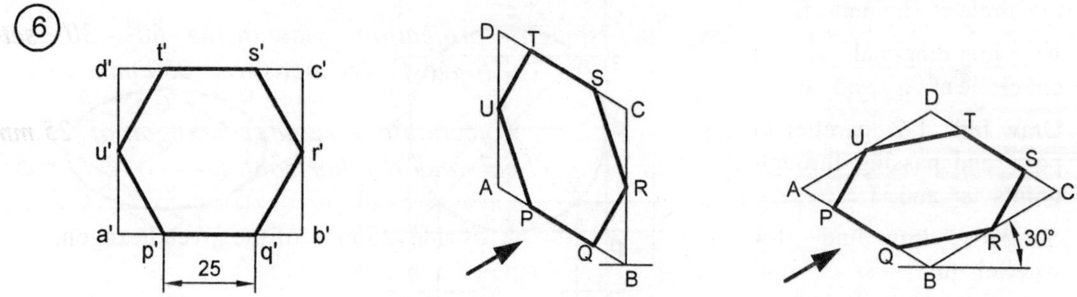

(i) FRONT VIEW (ii) SURFACE VERTICAL ON LIP (iii) SURFACE HORIZONTAL

FIG. 16 ISOMETRIC PROJECTION OF HEXAGON

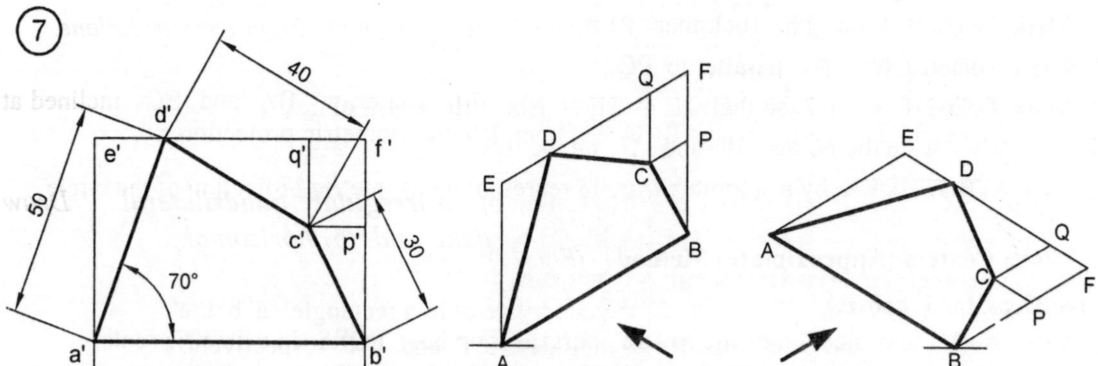

(i) FRONT VIEW (ii) SURFACE VERTICAL (iii) SURFACE HORIZONTAL

FIG. 17 ISOMETRIC PROJECTION OF QUADRILATERAL

Problem 8: *(Figs. 18 & 19) Draw isometric projection of a* ***circle*** *with its surface parallel to HP.*

Note : *Isometric Projection / View of a circle is always an* ***ellipse****.*

---

1. *See Fig. (i).* Draw the top view of the circle of, say, 50 mm diameter.
2. Enclose the circle in the top view in a square pqrs.
3. Mark mid-points of pq, qr, rs and sp as a, b, c and d respectively.

Two Methods of Construction of Ellipse are as follows:

(i) Method of Points (*Fig. 18*)

4. Join diagonal sq to cut the circle at h and f.
5. Also join diagonal rp to cut circle at g and e.
6. Draw line 1-2 parallel to pq and passing through points e and f.
7. Similarly, draw line 3-4 parallel to sr and passing through h and g.
8. *See Fig. (ii).* Draw the isometric projection of the square pqrs as rhombus PQRS in isometric scale.

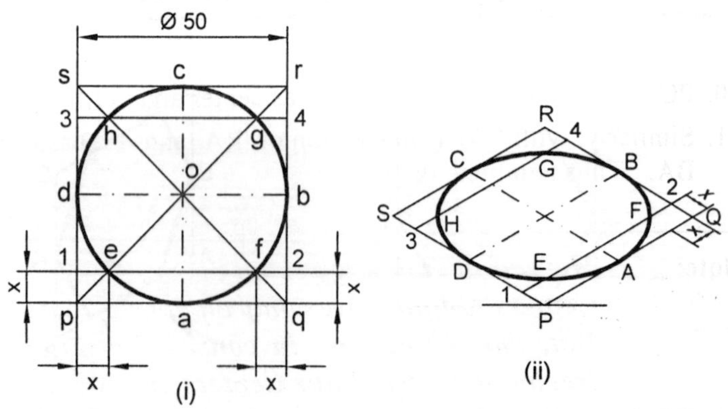

FIG. 18 METHOD OF POINTS

9. Mark the mid-points of PQ, QR, RS and SP as A, B, C and D respectively.
10. Mark the point 1 on PS such that P1 = x.
11. Draw isometric line 1-2 parallel to PQ.
12. Mark E and F on 1-2 so that 1E = 2F = isometric length of x.
13. Similarly, locate the points H and G on the isometric line 3-4.
14. Join AFBGCHDEA by a smooth *ellipse* to represent the isometric projection of the circle.

(ii) Four Centers (Approximate) Method *(Fig. 19)*

Refer Steps 1 – 3 above.

4. Draw isometric projection of the square pqrs as rhombus PQRS in isometric scale.
5. Mark mid-points of PQ, QR, RS and SP as A, B, C and D respectively.
6. Join P with B and C, the mid-points of opposite sides of the rhombus.
7. Join R with A and D.
8. P as center and PC as radius draw an arc CB.
9. R as center and RA as radius draw an arc AD.

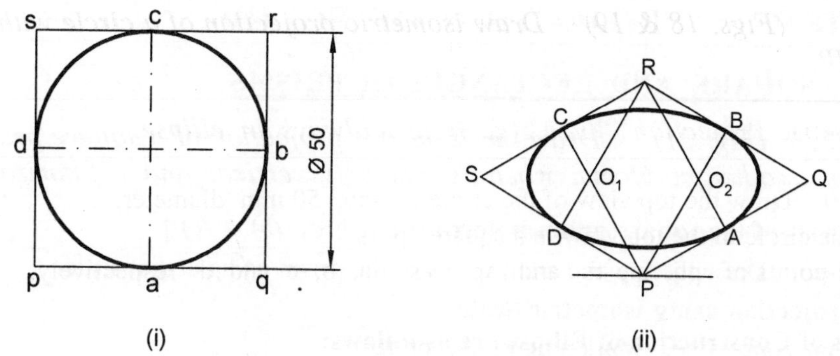

FIG. 19 FOUR CENTERS METHOD

10. PC and RD cut at O_1. O_1 as center and O_1D as radius draw an arc DC.
11. Similarly, with O_2 (intersection of RA and PB) as center and O_2B as radius draw an arc BA. Thus complete the *ellipse*.

Note: ***Method of Points*** *gives an accurate ellipse, but it is time-consuming.* ***Four Centers Method*** *gives only an approximate ellipse.*
But, due to the ease in construction and to avoid the labour of sketching freehand ellipse, Four Centers Method is preferable.

Problem 9: *(Fig. 20) Draw the isometric projection of a* ***circle*** *of 50 mm diameter with its surface parallel to VP.*

Fig. 20 shows the isometric projection of the circle with its surface parallel to VP as an ellipse obtained in the rhombus on Left Isometric Plane.

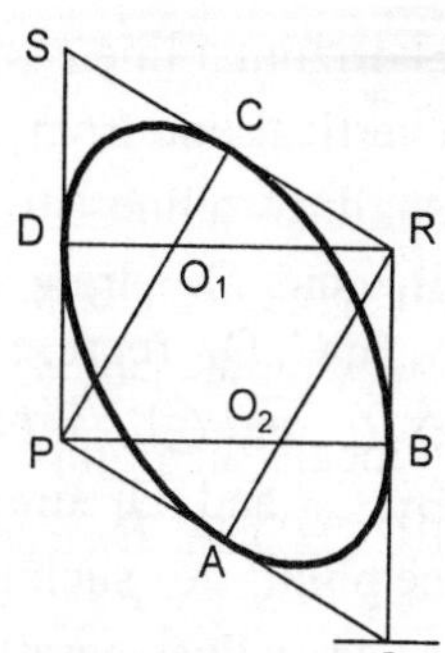

FIG. 20
ELLIPSE ON LIP

Problem 10: *(Exercise) Draw the isometric projection of the above* ***circle*** *with its surface parallel to VP such that the ellipse is on Right Isometric Plane.*

Problem 11: *(Exercise) Draw the isometric projection of a* ***semi-circle*** *of radius of 28 mm when its surface is parallel to HP.*

Hint: First consider the semicircle as a full circle and draw the corresponding ellipse in the isometric projection. Then, retain the required portion of the semicircle and its corresponding semi-ellipse in the projection.

7. ISOMETRIC PROJECTION OF SOLIDS

7.1 CUBE, SQUARE AND RECTANGULAR PRISMS

Problem 12: *(Fig. 21) Draw the isometric projection of a* ***square prism*** *of side of base 35 mm and height 65 mm when its axis is (i) vertical and (ii) horizontal.*

Case (i) : Axis of the square prism is vertical : [*Figs. (b) & (c)*]

1. *See Fig. (a).* Draw the top and front views of the prism using actual scale.

Isometric Projection using Isometric Scale

2. Draw a horizontal line. Mark a point A_1 on it.
3. From A_1 draw two lines making 30° with horizontal, representing two isometric axes.
4. On these isometric axes, mark B_1 and D_1 such that $A_1B_1 = A_1D_1$ = Side of the square, measured from the Isometric Scale using the divider.
5. From B_1 draw a line parallel to A_1D_1. From D_1 draw a line parallel to A_1B_1. Both meet at C_1. *Rhombus* $A_1B_1C_1D_1$ is the *isometric projection of the base.* *[Fig. (b)]*
6. *See Fig. (c).* At A_1 erect vertical to represent the third axis.
7. Mark the point A on it such that A_1A = Height of the prism in isometric length.
8. From B_1, C_1 and D_1 draw vertical lines.
9. From A draw lines parallel to A_1B_1 and A_1D_1 and obtain the points B and D on the respective vertical lines.
10. From B and D draw lines parallel to AD and AB respectively to meet at C.
11. All the edges of top face, vertical edges DD_1, AA_1, BB_1 and base edges D_1A_1, A_1B_1 are visible in the isometric projection. Hence draw them as *thick* lines as shown.

Case (ii) : Axis of the square prism is horizontal : [*Figs. (d), (e), (f) and (g)*]

1. Draw a horizontal line. Mark a point A_1 on it.
2. Draw a vertical line from A_1. Mark D_1 on it such that A_1D_1 = 35 mm in Iso. Scale.
3. From A_1 draw a line 30° to the horizontal. Mark B_1 on it such that $A_1B_1 = A_1D_1$.
4. From B_1 and D_1 draw lines parallel to A_1D_1 and A_1B_1 respectively. Both meet at C_1. Now $A_1\,B_1\,C_1\,D_1$ represents the end face of the prism as shown in *Figs. (d) and (e).*
5. *See Figs. (f) and (g).* From A_1 draw a 30° line on the other side to represent third axis.
6. From B_1, C_1 and D_1 draw lines parallel to the third axis.
7. Mark the point A such that A_1A = Height of the prism in isometric length.
8. From A draw lines parallel to A_1B_1 and A_1D_1 to obtain B and D respectively.
9. From B and D draw lines parallel to AD and AB to obtain C.
10. Draw all the visible edges in the isometric projection as *thick* lines.

Note : In isometric projection / view, hidden lines are not generally shown.

Problem 13: *(Exercise) Draw the isometric projection of a* ***square prism*** *of 40 mm side of base and 60 mm altitude when its axis is (i) vertical and (ii) horizontal.*

Problem 14: *(Exercise) Draw the isometric projection of a* ***cube*** *of side 45 mm.*

12

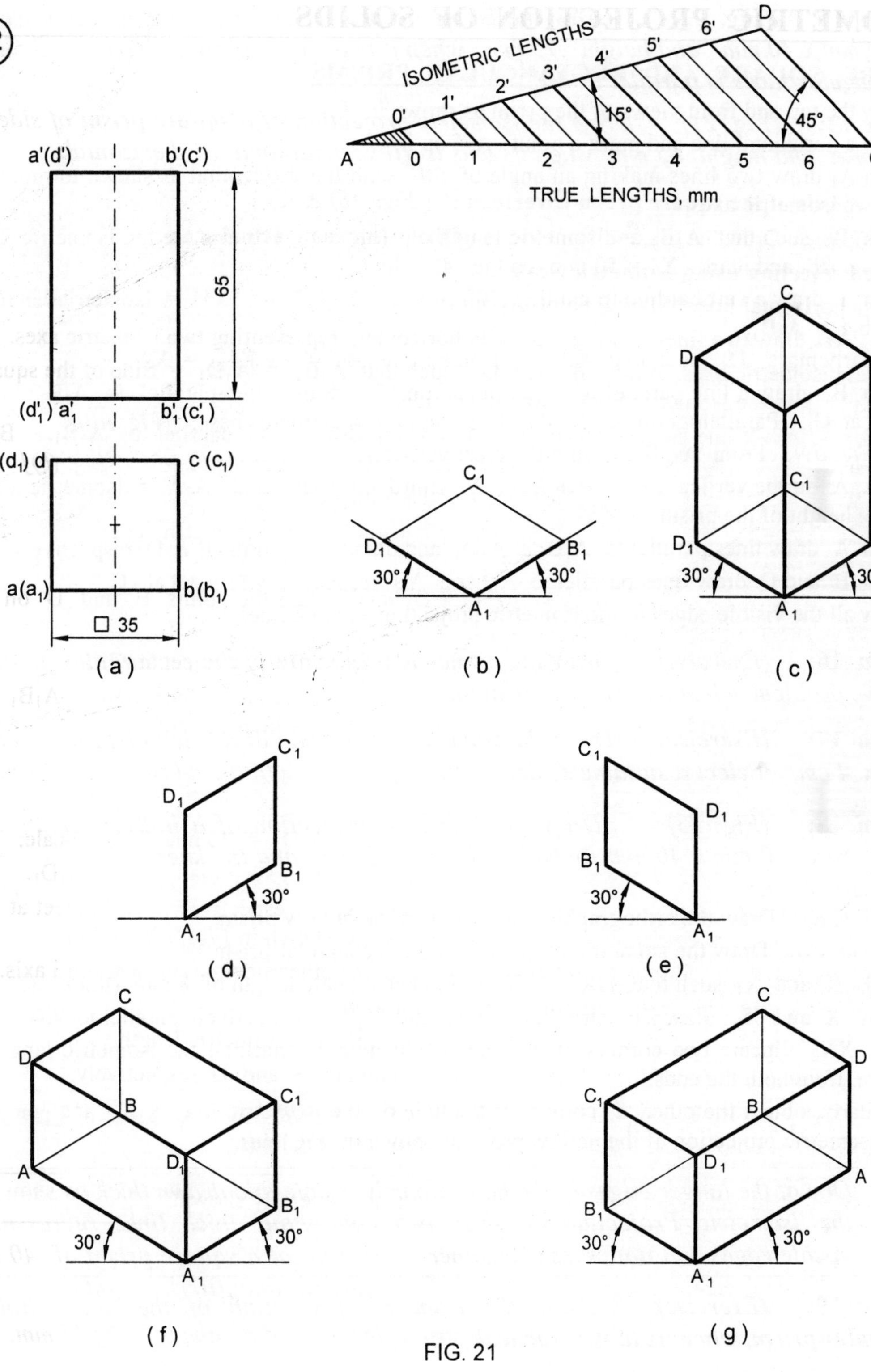
ISOMETRIC LENGTHS
0'
1'
2'
3'
4'
5'
6'
D
15°
45°
A
0
1
2
3
4
5
6
C
TRUE LENGTHS, mm
a'(d')
b'(c')
65
(d'$_1$) a'$_1$
b'$_1$ (c'$_1$)
(d$_1$) d
c (c$_1$)
a(a$_1$)
b(b$_1$)
□ 35
(a)
C$_1$
D$_1$
B$_1$
30°
30°
A$_1$
(b)
C
D
B
A
C$_1$
D$_1$
B$_1$
30°
30°
A$_1$
(c)
C$_1$
D$_1$
B$_1$
30°
A$_1$
(d)
C$_1$
D$_1$
B$_1$
30°
A$_1$
(e)
C
D
B
C$_1$
A
D$_1$
B$_1$
30°
30°
A$_1$
(f)
C
D
C$_1$
B
D$_1$
A
B$_1$
30°
30°
A$_1$
(g)

FIG. 21

Problem 15: *(Fig. 22) Draw the isometric projection of a* ***rectangular prism*** *of base 50 mm x 40 mm and height 75 mm, when it rests with its base on HP and one of its rectangular faces is parallel to VP.*

1. Draw the top and front views of the prism as shown in *Fig. (i)*.
2. Draw a horizontal line and mark a point A_1 on it.
3. From A_1 draw two lines making an angle of 30° with the horizontal as shown to represent the two isometric axes.
4. Mark B_1 such that A_1B_1 = Isometric length of $(a_1)(b_1)$. For this see the Isometric Scale in *Fig. (ii)* and mark X1 = 50 mm on the 45° line.
5. From 1 draw a vertical line to cut the 30° line at 1′. Now X1′ = Isometric length of $(a_1)(b_1)$ = A_1B_1.
6. Similarly mark D_1 such that A_1D_1 = Isometric length of $(a_1)(d_1)$ = X2′.
7. From B_1 draw a line parallel to A_1D_1 and from D_1 draw a line parallel to A_1B_1. Both meet at C_1. Parallelogram $A_1B_1C_1D_1$ is isometric projection of base. [*Fig. (iii)*]
8. *See Fig. (iv)*. From A_1, B_1, C_1 and D_1 erect verticals.
9. Mark A on the vertical line drawn from A_1 (third axis) such that A_1A = Isometric length of the height of the prism = X3′.
10. From A draw lines parallel to A_1B_1 & A_1D_1 and obtain the points B & D respectively.
11. From B and D draw lines parallel to AD and AB respectively to meet at C.
12. Draw all the visible edges in the isometric projection as *thick* lines.

Problem 16: *(Exercise) Draw the isometric projection of the* ***rectangular prism*** *in the above problem, when its axis is horizontal.*

Problem 17: *(Exercise) Draw the isometric projection of a* ***brick*** *of size 22 cm x 11 cm x 4 cm. Select a suitable scale.*

Problem 18: *(Fig. 23) Draw the isometric projection of a* ***hollow rectangular prism*** *of base 50 mm x 40 mm (outside), height 75 mm and thickness 8 mm when its axis is horizontal.* (UQ)

1. *See Fig. (i)*. Draw the orthographic projections of the hollow prism.
2. *See Fig. (ii)*. Draw the isometric projection of the rectangular prism.
3. Mark X and X_1 such that DX = AX_1 = Isometric scale length of 8 mm (thickness).
4. From X and X_1 draw isometric lines XX_2 and X_1X_3 respectively parallel to AB.
5. On XX_2 locate two corners of the rectangular hole by marking the isometric length of 8 mm from both the ends.
6. Similarly, obtain the other 2 corners of the hole on the isometric line X_1X_3 and complete the isometric projection of the hollow prism as shown in *Fig. (iii)*.

Note: *One of the longer edges of the hole is partly visible and drawn thick as shown in the Isometric Projection, because two continuous thick lines representing visible edges will* ***not*** *cross each other.*

Problem 19: *(Exercise) Draw the isometric projection of the above* ***hollow rectangular prism****, when its axis is vertical.*

15

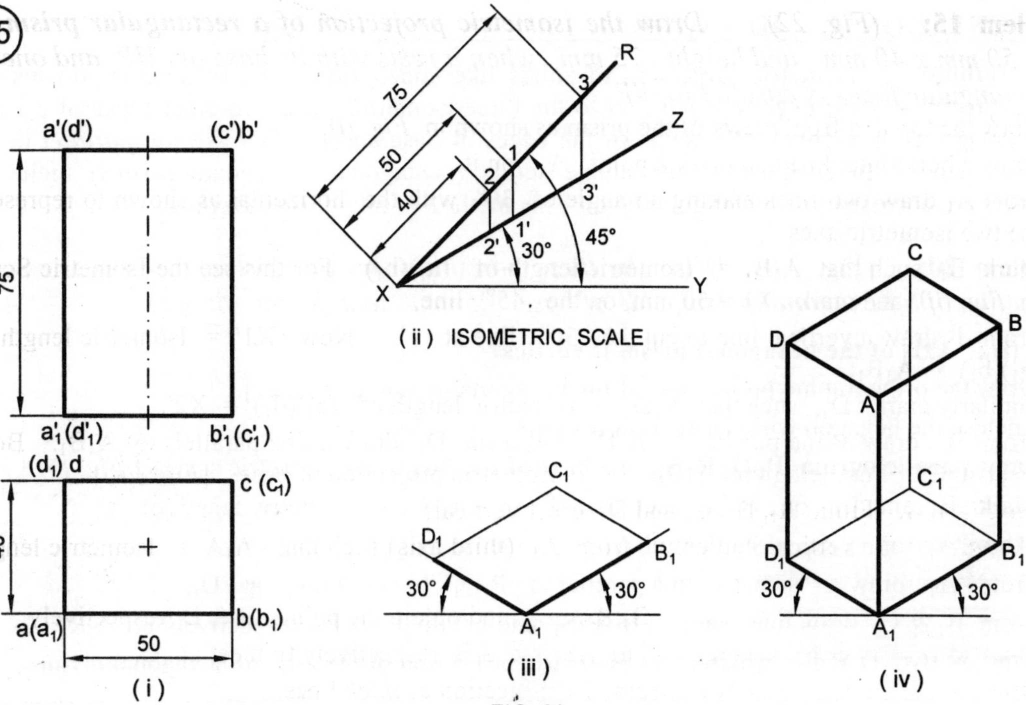

FIG. 22

18

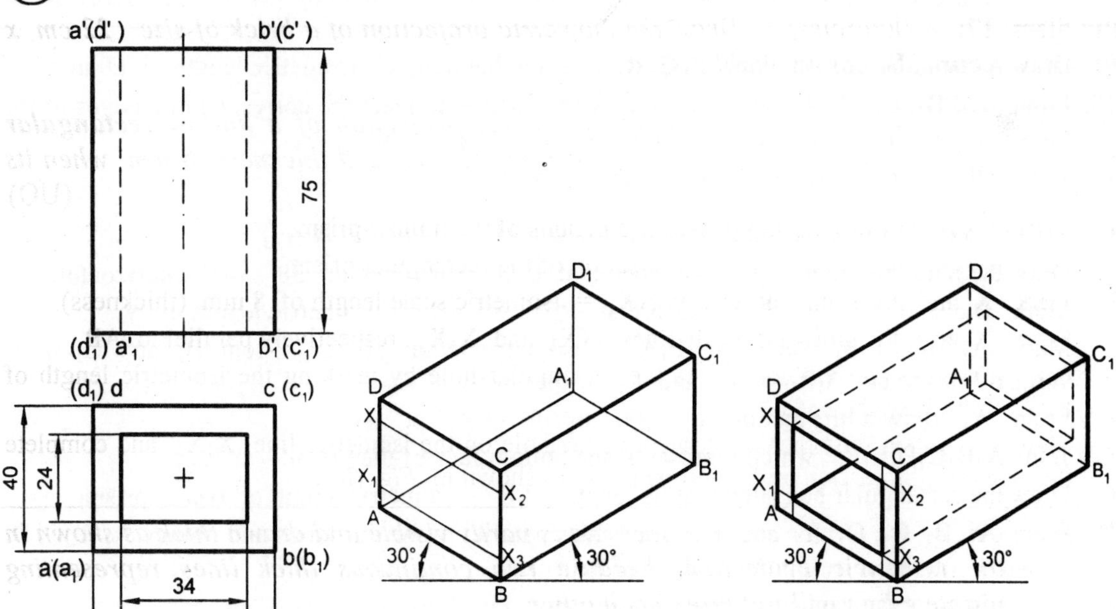

FIG. 23

7.2 PENTAGONAL / HEXAGONAL PRISMS AND CYLINDERS

Box Method: Isometric projection of solids like cube, square and rectangular prisms are drawn directly as their edges are *parallel* to the three isometric axes. In other types of prisms, *viz.*, pentagonal, hexagonal, etc., some of the edges of base and top face are ***not parallel*** to the isometric axes. In such cases, the above solids (pentagonal, hexagonal prisms) including cylinders are enclosed in a rectangular box. This method is called ***Box Method.***

Problem 20: *(Fig. 24) Draw the three possible ways of representing the isometric projection of a **hexagonal prism**, side of base 25 mm and height 60 mm.*

Case (i) : Axis of the hexagonal prism is vertical

1. Draw the orthographic projections of the hexagonal prism [*Fig. (a)*].
2. Enclose the hexagon within a **rectangle** pqrs in the top view.
3. Draw **parallelogram $P_1 Q_1 R_1 S_1$**, the **base of the box** using isometric scale. [*Fig. (b)*]
4. Mark B_1 and C_1 on P_1Q_1 such that $P_1B_1 = Q_1C_1 =$ Isometric length of y.
5. Mark A_1 on P_1S_1 such that $P_1A_1 =$ Isometric length of x.
6. From A_1 draw an isometric line parallel to P_1Q_1 to cut Q_1R_1 at D_1.
7. From B_1 & C_1 draw lines parallel to P_1S_1 and obtain F_1 & E_1 respectively on S_1R_1.
8. Join $A_1B_1C_1D_1E_1F_1$ which is the isometric projection of the *base* of hexagonal prism.

Rule: *Make direct measurements only on Isometric Lines. Lengths of edges cannot be transferred on Non-isometric Lines. (e.g., B_1 should not be marked on P_1Q_1 by cutting an arc on it with A_1 as center and isometric length of $(a_1)(b_1)$ as radius.)*

---

9. Draw rectangular box on plane $P_1 Q_1 R_1 S_1$ such that height = Isometric length of 60 mm.
10. From A_1, B_1, C_1, D_1, E_1 and F_1 draw vertical lines to meet the corresponding edges of the top surface of the box at A, B, C, D, E and F respectively.
11. Join all the visible edges as thick lines as shown in *Fig.(c)*.

Case (ii) : Axis of the hexagonal prism is horizontal : [*Figs.(d) & (e)*]

1. Draw the parallelogram $P_1Q_1R_1S_1$ such that P_1Q_1 is inclined at 30° to the horizontal.
2. Mark B_1 and C_1 on P_1Q_1 such that $P_1B_1 = Q_1C_1 =$ isometric length of y.
3. Mark A_1 on P_1S_1 such that $P_1A_1 =$ isometric length of x. From B_1 and C_1 draw vertical lines to cut S_1R_1 at F_1 and E_1 respectively.
4. From A_1 draw a line parallel to P_1Q_1 to cut Q_1R_1 at D_1.
5. Join $A_1B_1C_1D_1E_1F_1$ which represents one bottom face of the prism.
6. Draw the rectangular box such that its length P_1P = isometric length of axis of prism.
7. From A_1, B_1, C_1, D_1, E_1 and F_1 draw lines parallel to the third axis and obtain A, B, C, D, E and F on the corresponding edges of the box.
8. Then complete the isometric projections as shown in *Figs. (d) & (e)*.

Problem 21: *(Exercise) Draw the three possible ways of representing the isometric projection of a **pentagonal prism**, side of base 30 mm and height 60 mm.*

20

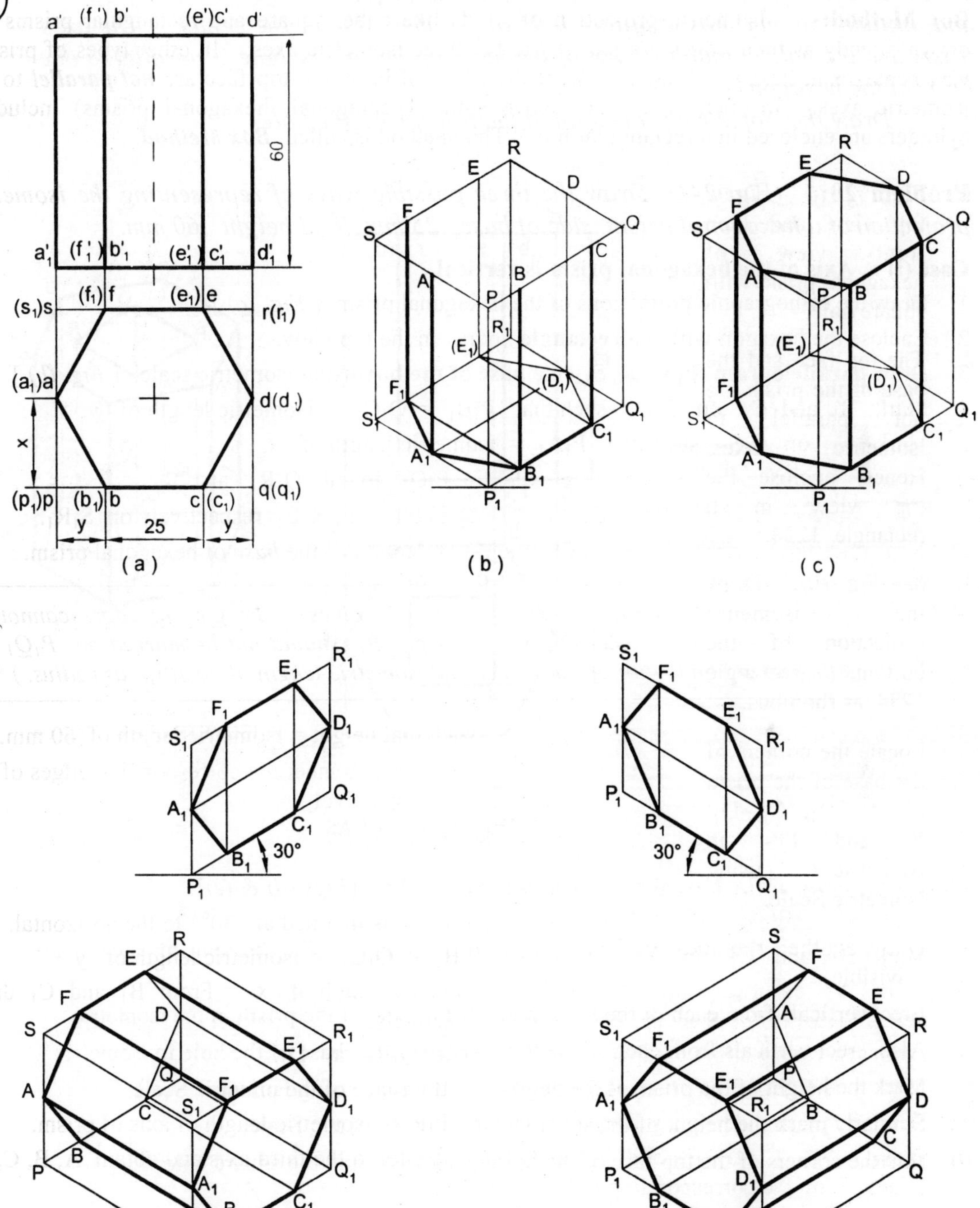

FIG. 24

Problem 22: *(Fig. 25)* *A right* ***regular hexagonal prism*** *of side of base 25 mm and altitude 56 mm has a* ***square hole*** *of 20 mm side at the center. The axes of the prism and the hole coincide. One of the faces of the square hole is parallel to one of the faces of the hexagon.*

Draw the isometric projection of the prism with the hole. (UQ)

1. *See Fig. (i).* Draw top view and then front view of hexagonal prism with square hole.
2. The top face and the base of the prism are not parallel to isometric axes. Hence, enclose the top view in a rectangle 1234.
3. *See Fig. (ii).* Draw the isometric projection of the bottom rectangle 1234 as rhombus.
4. Locate the corners of the base of the prism as A_1, B_1, C_1, D_1, E_1 and F_1 in isometric using Isometric Scale.

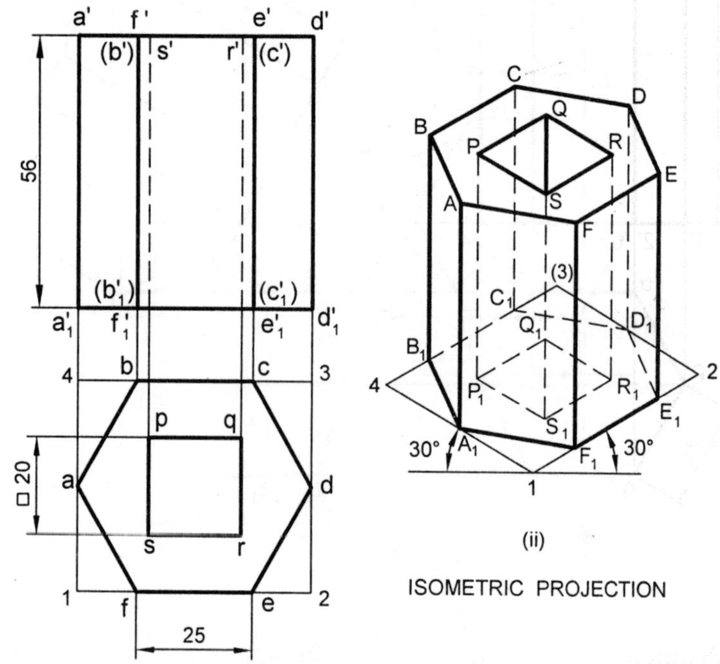

(i) ORTHOGRAPHIC VIEWS

(ii)

ISOMETRIC PROJECTION

FIG. 25

5. Draw another rhombus in isometric corresponding to the bottom of the square hole (invisible).
6. Erect verticals from each of the six corners of the base of the prism in the isometric.
7. Also, erect verticals from each of the four corners of the base of the hole in isometric.
8. Mark the height of the prism at the points A_1, B_1, etc., using Isometric Scale.
9. Similarly mark the height of the square hole using Isometric Scale.
10. Join the corners of the top face of the hexagonal prism and the square hole.

Problem 23: *(Exercise)* *Draw the isometric projection of a vertical* ***pentagonal hollow prism*** *of 25 mm side of base outside, height 60 mm and thickness 8 mm, when resting with one of its rectangular faces parallel to VP and nearer to it.*

Problem 24: *(Fig. 26) A right hexagonal prism of side of base 24 mm and axis 56 mm long is lying on one of the corners of the base. Its axis is inclined at an angle of 30° to HP. Draw the isometric projection of the solid.* (UQ)

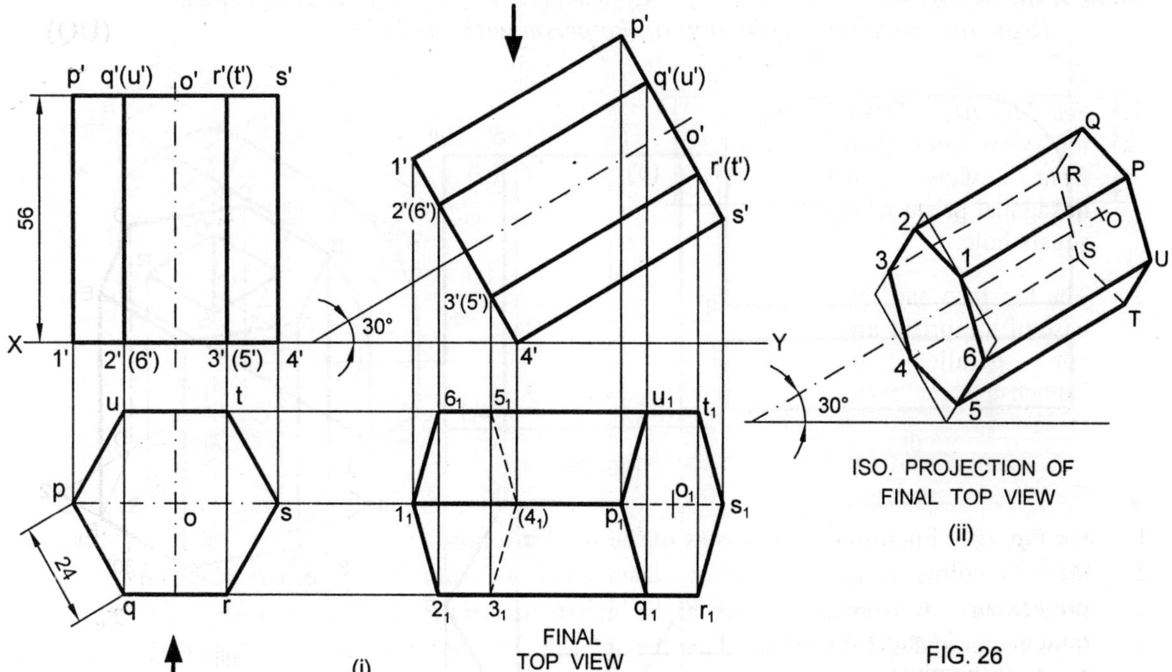

FIG. 26

Problem 25: *(Exercise) A pentagonal prism, side of base 25 mm and axis 50 mm long rests, with one of its shorter edges of HP such that the base containing that edge makes 30° to HP and its axis is parallel to VP. Draw both the orthographic projections and the isometric projection of the solid.*

Angle in Isometric

Problem 26: *(Fig. 27)* Draw the isometric projection of a taper block with 45° inclination.

1. Do not set-off angles directly in isometric.
2. Translate the angle to linear measurements along the isometric axes.
3. Locate the end points of the lines which subtend the angle.

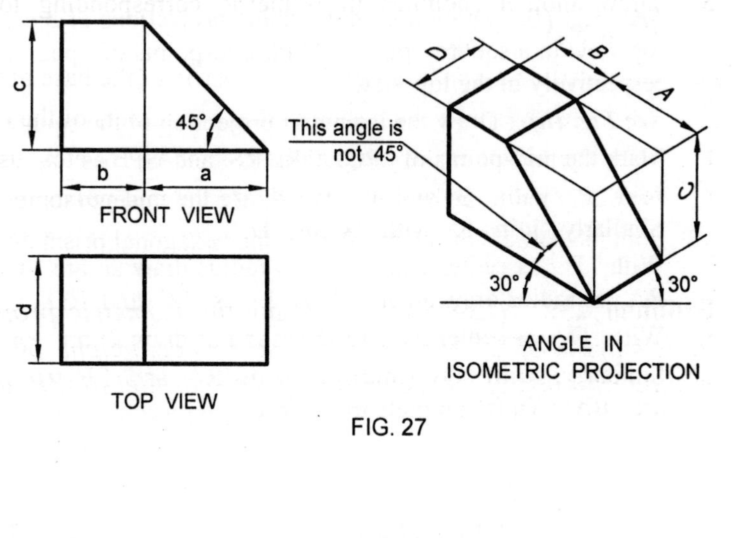

FIG. 27

Curved Solid

Problem 27: *(Fig. 28) Draw the isometric projection of the **Curved Solid** shown.*
(Have you observed the rounded corners of your table? Corners of any machine element should not be sharp, as they may easily chip-off. Rounded corner has fillet radius as shown.)

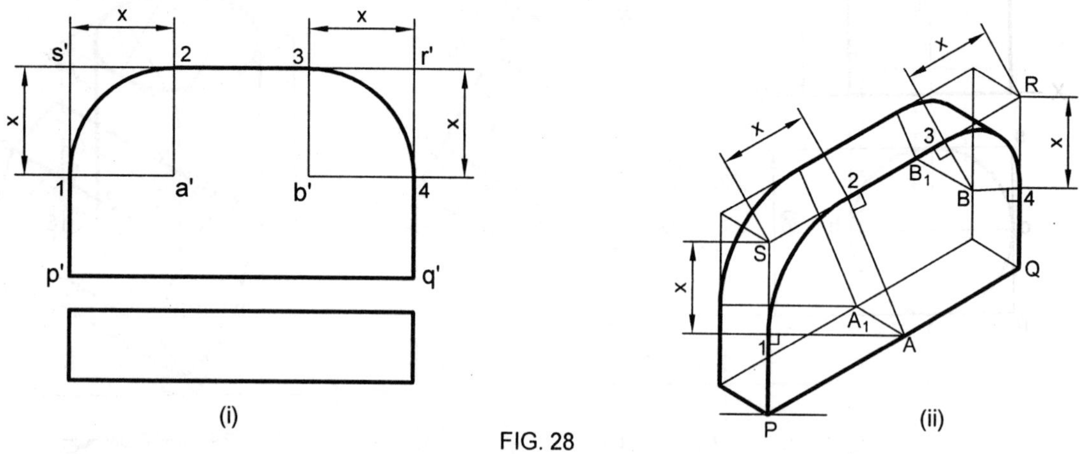

FIG. 28

1. *See Fig. (i).* Enclose the front view of the solid in a box.
2. Mark the points 1, 2, 3 and 4 at a distance of x from the top corners of the box.
3. *See Fig. (ii).* In isometric, fix A at the intersection of the ⊥rs drawn from 1 and 2.
4. A as center and A1 as radius, draw the arc 1-2. Fix point B and draw the arc 3-4.
5. Similarly fix the centers A_1 and B_1 and draw the arcs at the rear side of the solid.

Problem 28: *(Fig. 29) Draw the isometric projection of a **cylinder** of base 50 mm diameter and 70 mm height when it rests with its base on HP.*

Use ***Box Method*** for isometric projection / view of ***cylinder.*** For this, enclose the cylinder in a square prism. [The invisible part of the base of the cylinder need not be shown.] For easy and quick construction of the ***ellipse***, adopt the ***Four Centers (Approximate) Method.***

1. ***See Fig. (i).*** Draw the orthographic projections of the cylinder. Enclose the circle in the top view in a square pqrs. Mark mid-points of pq, qr, rs and sp as a, b, c and d respectively in the top view.
2. ***See Fig. (ii).*** Draw the isometric projection of the square as rhombus PQRS as shown.
3. Mark the mid-points of PQ, QR, RS and SP as A, B, C and D respectively.
4. Join P with B and C which are the mid-points of the opposite sides of the rhombus. Similarly, join R with A and D.
5. With P as center and PC as radius draw an arc CB. Similarly, R as center and RA as radius draw an arc AD. Lines PC and RD intersect at O_1.
6. With O_1 as center and O_1D as radius draw an arc DC.
7. Similarly, with O_2 (intersection of RA and PB) as center and O_2B as radius draw an arc BA. Thus complete the ellipse to represent the base circle. [*Fig. (ii)*].

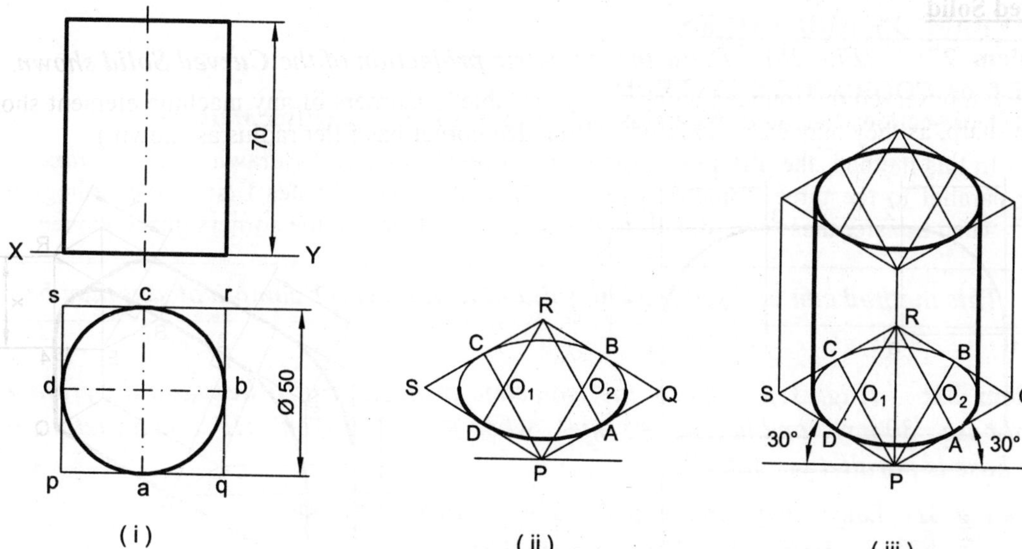

FIG. 29 ISOMETRIC PROJECTION OF CYLINDER (FOUR CENTERS METHOD)

8. *See Fig. (iii).* Draw the isometric projection of the box, taking its height = Height of the cylinder in isometric length. Construct another ellipse on the top surface of the box. Draw two common tangents to the ellipses. Complete the isometric projection.

Problem 29: *(Fig. 30)* ***A cylinder** of base 50 mm diameter and axis 70 mm long is lying on the HP. Draw its isometric projection when the axis is horizontal.*

Fig. 30 shows the isometric projection of the cylinder when the axis is horizontal.

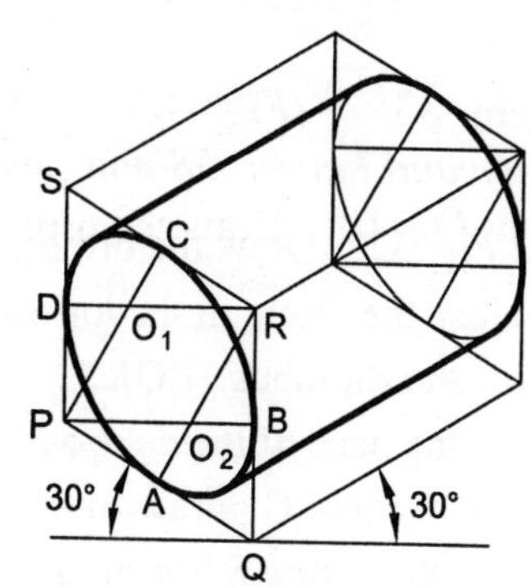

FIG. 30
ISOMETRIC PROJECTION OF CYLINDER WHEN THE AXIS IS HORIZONTAL

Problem 30: *(Exercise)* ***A hollow cylinder** of base 50 mm diameter and axis 70 mm long has a square hole of 20 mm side. The axes of the cylinder and the hole coincide. Draw the three possible ways of representing the isometric projection of the solid.* (UQ)

7.3 PYRAMIDS AND CONES

OFFSET or COORDINATE METHOD

Isometric projections / views of pyramids and cones are drawn by ***Offset Method.***

In this method, the different corners of a solid are located by drawing the *coordinates* or *offsets* parallel to the three isometric axes. Draw these coordinates first in the orthographic views. Then transfer these coordinates to the isometric. Connect the corners thus obtained.

Note: ***This method can be used for any solid/object, however complicated it may be.***

Problem 31: *(Fig. 31) Draw the isometric projection of a* ***hexagonal pyramid*** *of side of base 30 mm and height 75 mm, when it is resting on HP such that an edge of the base is parallel to VP.* (UQ)

1. *See Fig. 31.* Draw the top and front views of the hexagonal pyramid.
2. Enclose the hexagon in a rectangle pqrs in the top view.
3. Draw the isometric projection of the base of the pyramid in the parallelogram PQRS.
4. Note that FC is an isometric line on which O_1 lies. Hence mark O_1 on the **isometric line FC** such that FO_1 = isometric length of y.
5. From O_1 erect a vertical line and mark the apex O such that O_1O = isometric length of the axis.
6. Join O with all the corners of the base of the pyramid and complete the isometric projection.

Problem 32: *(Exercise) Draw the isometric projection of a* ***triangular pyramid*** *of side of base 35 mm and height 75 mm, when it is resting on HP such that an edge of the base is parallel to VP.*

Problem 33: *(Fig. 32) Draw the isometric projection of a* ***cone*** *of base 40 mm diameter and height 58 mm when it rests with its base on HP. (i.e., Axis is vertical)*

1. *See Fig. 32.* Draw the orthographic projections of the cone.
2. Enclose the circle in the top view in a square pqrs.
3. Draw the rhombus PQRS using the Isometric Scale.
4. Draw the base of the cone as an ellipse by ***Four Centers (Approximate) Method.***
5. From B and C draw lines parallel to isometric axes and obtain O_1 (***Offset Method***).
6. From O_1 draw a line parallel to the third axis.
7. On this line mark O such that O_1O = isometric length of the height of the cone.
8. From O draw 2 tangents to the ellipse and complete the isometric projection of the cone.

Problem 34: *(Exercise) Draw the isometric projection of the* ***cone*** *in the previous Problem when its axis is horizontal.*

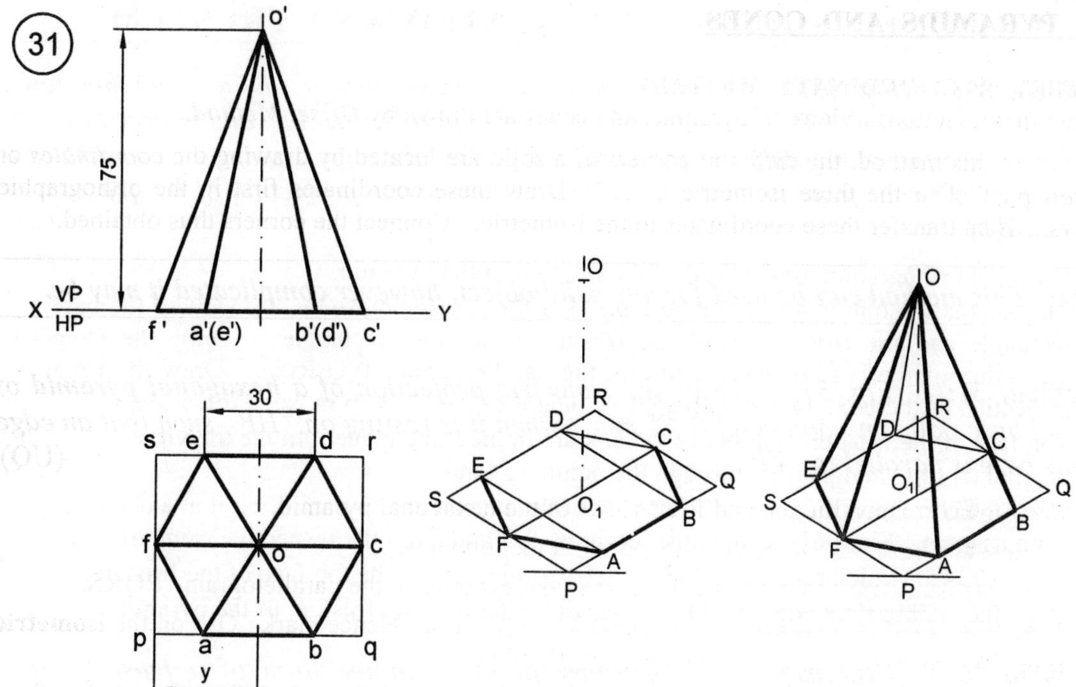

FIG. 31 ISOMETRIC PROJECTION OF HEXAGONAL PYRAMID

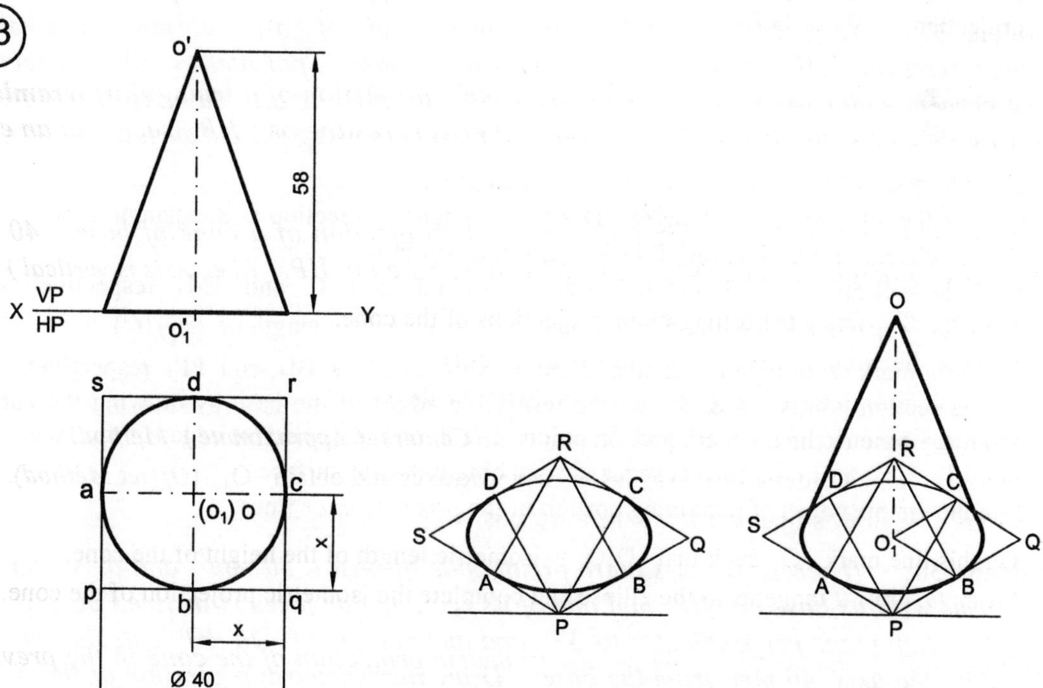

FIG. 32 ISOMETRIC PROJECTION OF CONE - AXIS VERTICAL

8. SECTIONED SOLIDS – TRUNCATED AND FRUSTUM

Problem 35: *(Fig. 33) A pentagonal pyramid, base 30 mm and axis 65 mm long, rests with its base on HP. An edge of the base is parallel to VP and nearer to it. A horizontal section plane cuts the pyramid and passes through a point on the axis at a distance of 25 mm from the apex. Draw the isometric projection of the* ***frustum of the pyramid****.* (UQ)

1. Draw the orthographic projections of the frustum of the pyramid.
2. Enclose the frustum in a rectangular box. For this, first enclose the top view of the base in a rectangle and the front view of the frustum in another rectangle. Draw the isometric projection of the base of the frustum in the parallelogram $P_1Q_1R_1S_1$. Draw the isometric projection of the box, taking its height = height of the frustum in isometric scale.
3. Top face of the frustum and that of the box are in the same plane. In the top view draw a line parallel to pq and passing through the points 5 and 4.
4. Draw the corresponding isometric line. Locate the corners 5 and 4 on it at a distance of x from its ends. Similarly locate other corners 1, 3 and 2 on the respective isometric lines.
5. Join 12345 which represents the isometric projection of the top face of the frustum.
6. Draw the visible slant edges and the edges of the base of the frustum of the pyramid.

Problem 36: *(Exercise) A masonry pillar is in the form of a* ***frustum of a hexagonal pyramid****. The pillar is 2 m high and the sides of its base and top face are 0.5 m and 0.3 m respectively. Draw the isometric projection.* (UQ)

Problem 37: *(Fig. 34) A hexagonal prism, side of base 25 mm and height 50 mm rests on HP and one of the edges of its base is parallel to VP. A section plane perpendicular to VP and inclined at 50° to HP bisects the axis of the prism. Draw the isometric projection of the* ***truncated prism****, showing the cut surface.* (UQ)

1. Draw front and sectional top views of truncated prism.
2. Enclose the top view in a rectangle. Draw its isometric projection as a rectangular box.
3. Draw the isometric projection of the hexagonal prism in it.
4. Mark the section points 3 and 6 on the vertical edges C_1C and E_1E respectively such that C_13 = isometric length of $c_1'3'$ and E_16 = isometric length of $(e_1')(6')$.
5. Similarly mark section points 2 and 7 on the vertical edges BB_1 and FF_1 respectively.
6. Locate section points 4 & 5 on the respective edges of the base by drawing the corresponding isometric line. Mark section points 1 & 8 on the respective edges of top face.
7. Join 1, 2, ..., 8 in the correct sequence and obtain the cut surface.
8. Join the visible edges of remaining portion of the prism as *thick* lines.

Problem 38: *(Exercise) A* ***square prism****, side of base 30 mm and axis 65 mm long, rests with its base on HP and one of its rectangular faces is inclined at 30° to VP. A section plane perpendicular to VP and inclined at 60° to HP cuts the prism at a point on the axis 40 mm from the base. Draw the isometric projection of the prism, showing the cut surface.* (UQ)

(35)

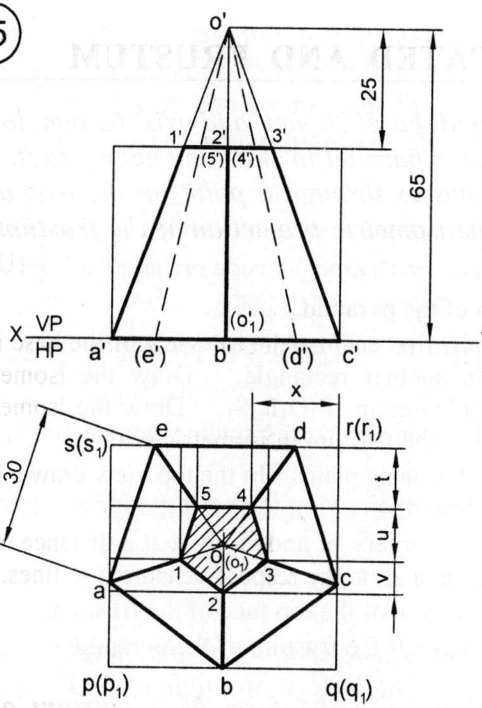

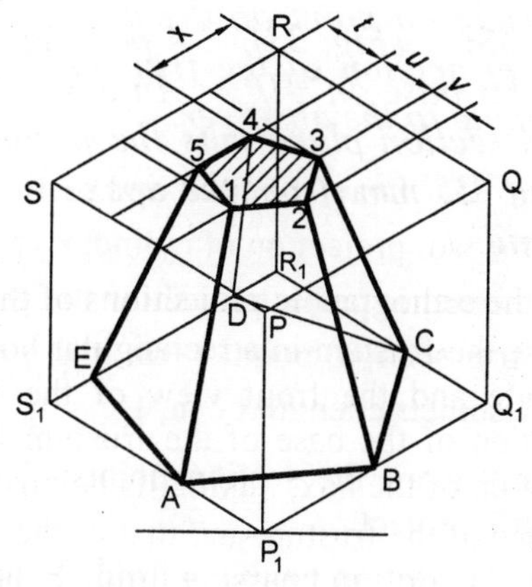

FIG. 33
ISOMETRIC PROJECTION OF
FRUSTUM OF PENTAGONAL PYRAMID

(37)

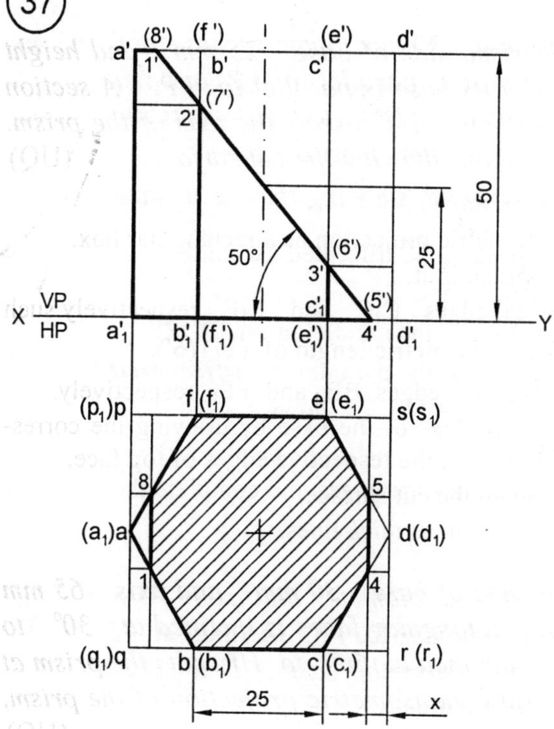

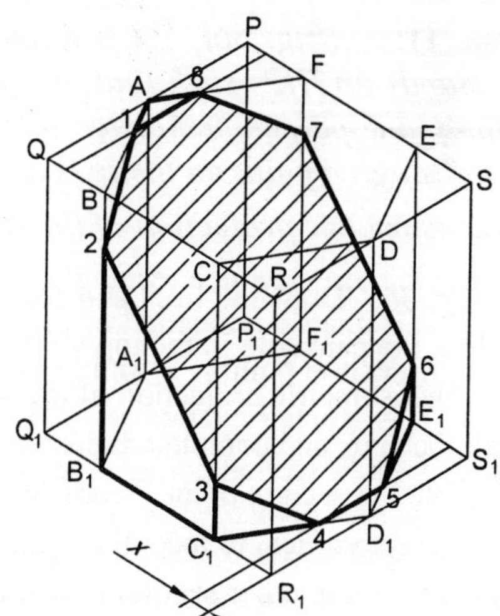

FIG. 34
ISOMETRIC PROJECTION OF
TRUNCATED HEXAGONAL PRISM

Problem 39: *(Fig. 35) A cylinder 50 mm diameter and 60 mm height stands on HP. A section plane perpendicular to VP, inclined at 55° to HP cuts the cylinder and passes through a point on the axis at a height of 45 mm above the base. Draw the isometric projection of the* ***truncated portion of the cylinder****, when the cut surface is clearly visible to the observer.* (UQ)

1. *See Fig. 35.* Draw the front and sectional top views of the truncated cylinder.
2. Draw the iso. projection of cylinder (by ***Four Centers Method***) inside rectangular box.
3. In the isometric projection, show the generators cut by the section plane.
4. Mark the section point 1 on the generator A_1A in the isometric projection such that $A_1 1$ = isometric length of $a_1' 1'$.
5. Similarly mark the section points 2, 3, 6 and 7 on the corresponding generators in the isometric projection.
6. Locate the section points 4 and 5 on the top face by drawing the corresponding isometric line.
7. Join 4321765 by a smooth curve. Also, join 5 and 4 by a straight line.

Problem 40: *(Exercise) A cylinder of 60 mm diameter and 70 mm height, stands on HP. A section plane perpendicular to VP and inclined at an angle of 45° to HP bisects the axis. Draw the isometric projection of the* ***truncated cylinder****, showing the cut surface.* (UQ)

Problem 41: *(Fig. 36) A pentagonal pyramid, 30 mm edge of base and 65 mm height, stands on HP such that an edge of the base is parallel to VP and nearer to it. A section plane perpendicular to VP and inclined at 30° to HP cuts the pyramid passing through a point on the axis at a height of 35 mm from the base.*
Draw the isometric projection of the ***truncated pyramid****, showing the cut surface.* (UQ)

1. *See Fig. 36.* Draw the front and sectional top views of the truncated pyramid.
2. Enclose the pentagon in the top view in a rectangle pqrs.
3. Draw the isometric projection of the pentagonal base in the parallelogram PQRS.
4. Mark the apex and complete the isometric projection of the pyramid (by ***Offset Method***).
5. To locate the section point 1 obtain 1_1 in the plane PQRS as shown.
6. At 1_1 erect vertical to meet the slant edge OA at 1.
7. Similarly obtain other section points on the respective slant edges and complete the isometric projection of the truncated pyramid.

Problem 42: *(Exercise) A cone, base 50 mm diameter and axis 70 mm long, stands on HP. It is cut by a section plane perpendicular to VP, inclined at 45° to HP and passing through a point on the axis 35 mm above the base. Draw the isometric projection of the* ***truncated cone****, showing the cut surface.* (UQ)

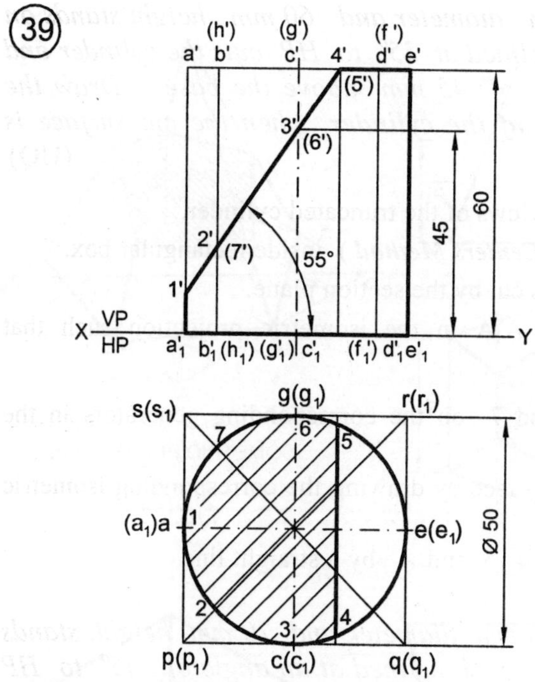

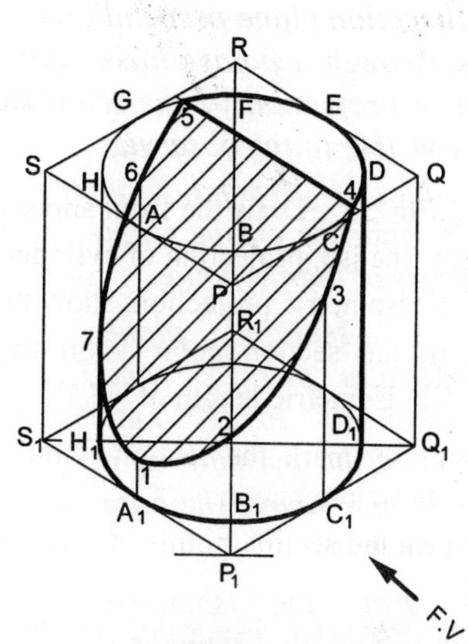

FIG. 35
ISOMETRIC PROJECTION OF
TRUNCATED CYLINDER

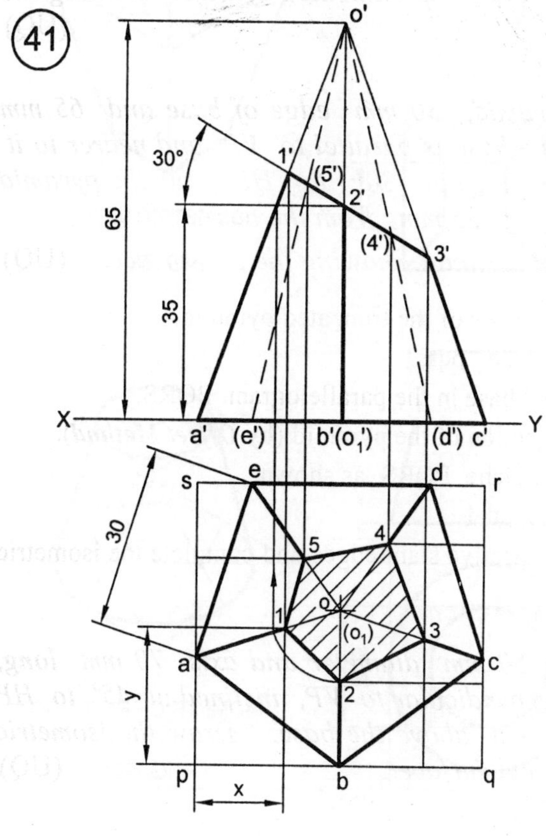

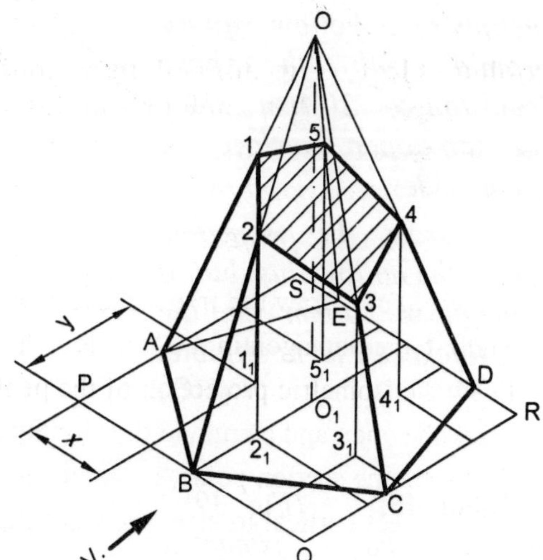

FIG. 36
ISOMETRIC PROJECTION OF
TRUNCATED PENTAGONAL PYRAMID

Problem 43: *(Fig. 37)*
A pentagonal pyramid of 40 mm edge of base and height 70 mm rests with its base on HP. One of the base edges is perpendicular to VP and lies on the right of the axis of the pyramid.

A section plane perpendicular to VP and inclined at 30° to HP cuts the axis of the pyramid at a point 30 mm above the base of the pyramid.

Draw the isometric projection of the truncated pyramid. (UQ)

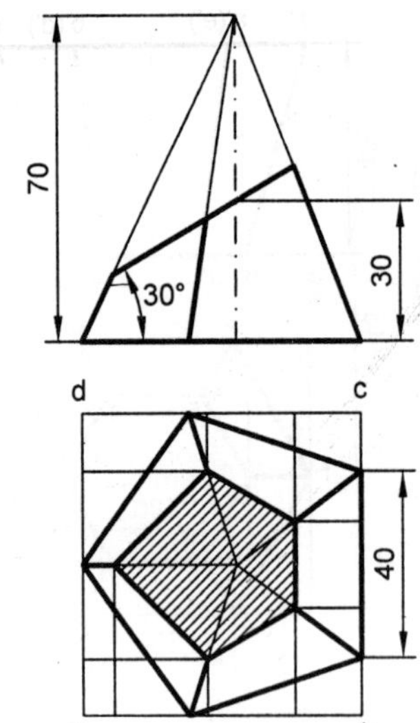

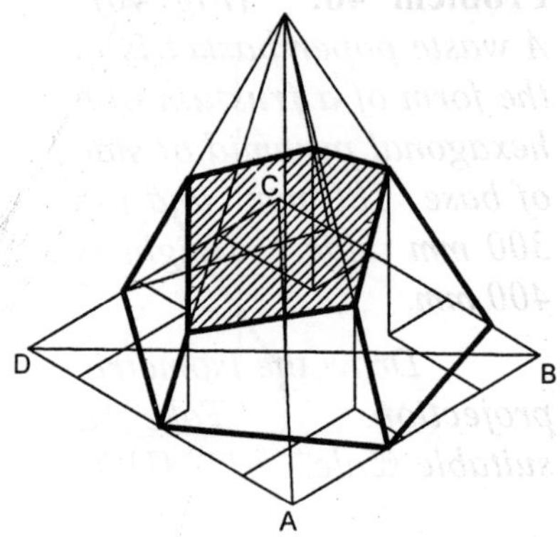

ISOMETRIC PROJECTION

FIG. 37

Problem 44: *(Fig. 38)*
A dust bin is in the form of a ***frustum*** *of a hollow square pyramid with the base dimension of 20 mm sides and top open surface of 45 mm sides.*

Draw the isometric view of the hollow dust bin, if its height is 50 mm and the wall thickness is negligible. (UQ)

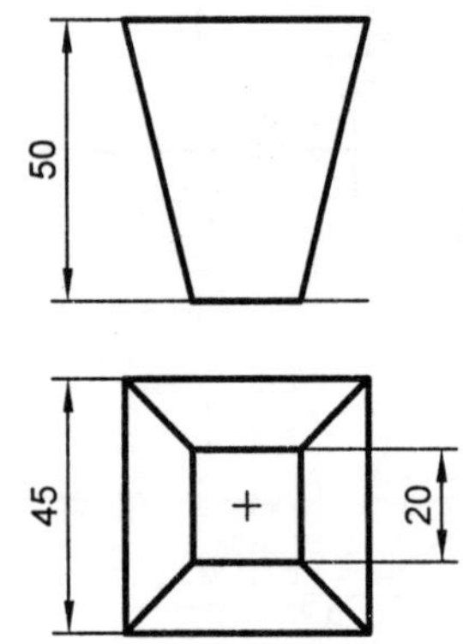

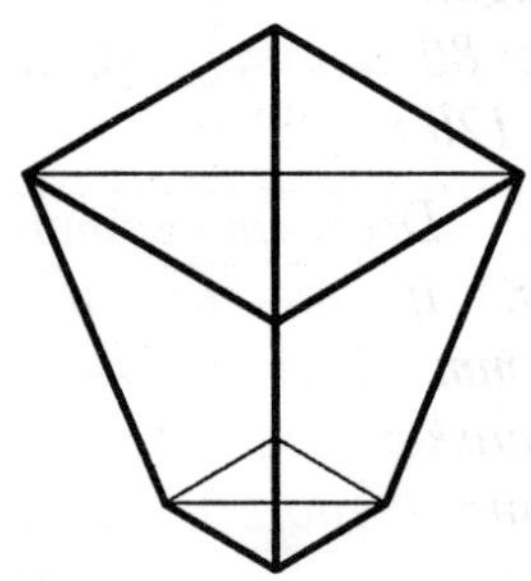

ISOMETRIC VIEW

FIG. 38

Problem 45: *(Fig. 39)*
Draw the isometric projection of the ***frustum*** *of a cone of base diameter 60 mm and top face diameter 35 mm and axis length 50 mm resting on HP on its base.* (UQ)

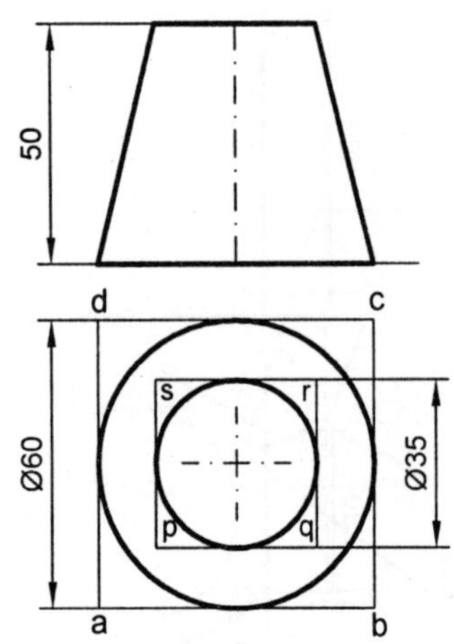

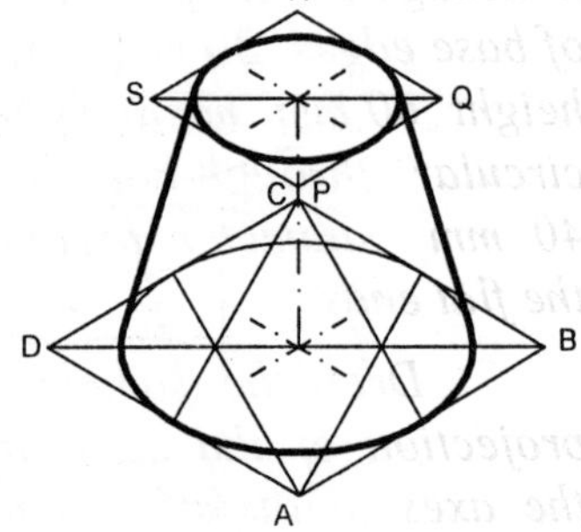

ISOMETRIC PROJECTION

FIG. 39

Problem 46: *(Fig. 40)* *A waste paper basket is in the form of a **frustum** of a hexagonal pyramid of side of base 180 mm and top 300 mm side. Height is 400 mm.*

Draw the isometric projection. Take a suitable scale. (UQ)

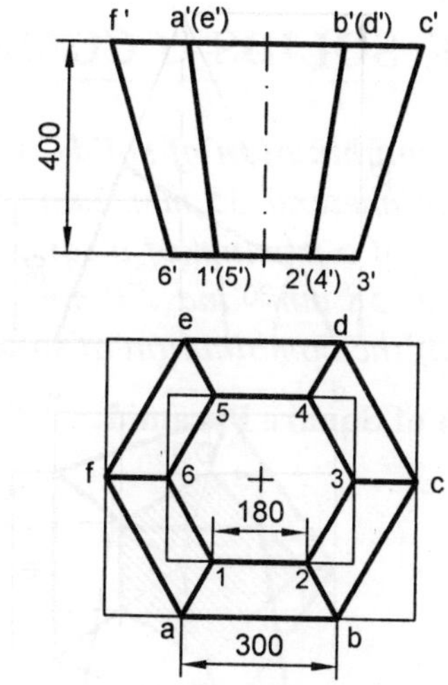

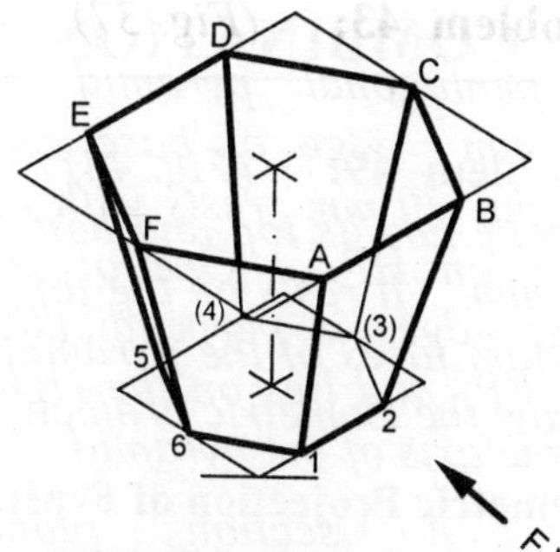

ISOMETRIC PROJECTION OF BASKET (FRUSTUM OF A HEXAGONAL PYRAMID)

FIG. 40

Problem 47: *(Fig. 41)* *A waste paper basket is in the form of a **frustum** of hexagonal pyramid with base 80 mm hexagon and top 120 mm.*

Draw the isometric view, if its height is 200 mm. The thickness of the basket can be taken as 12 mm. Adopt a suitable scale. (UQ)

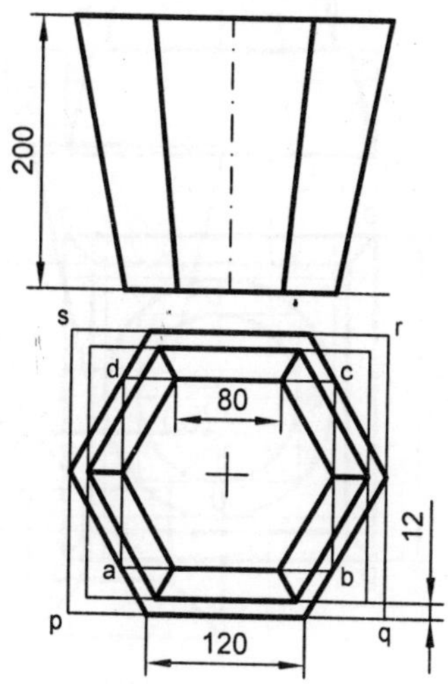

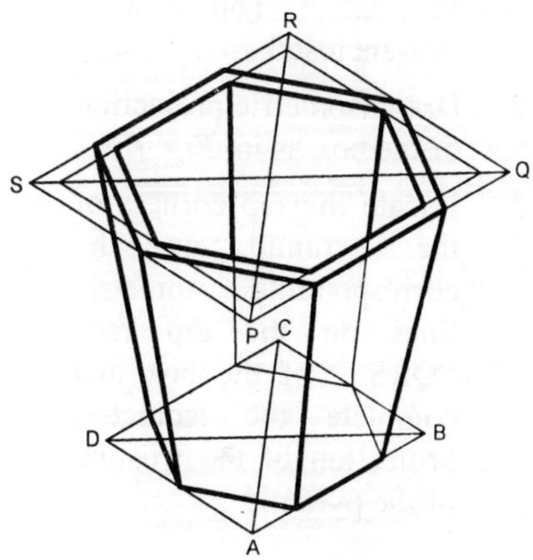

ISOMETRIC VIEW

FIG. 41

Problem 48: *(Fig 42)* *A hexagonal shaped slab of base edges 30 mm and height 60 mm has a right circular through-hole of 40 mm diameter across the flat ends.*

Draw the isometric projection of the slab, if the axes of the hole and the slab coincide. (UQ)

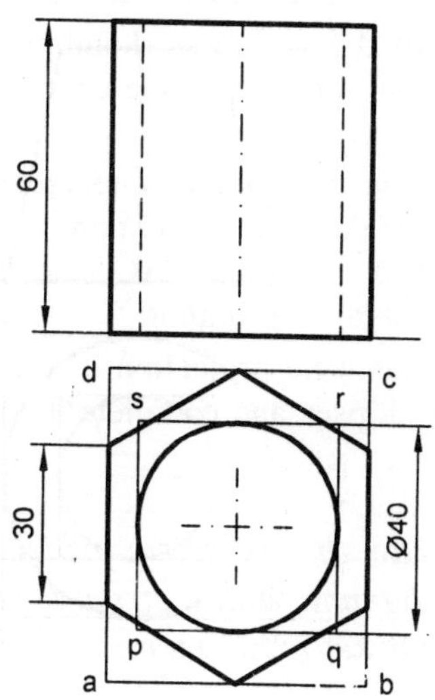

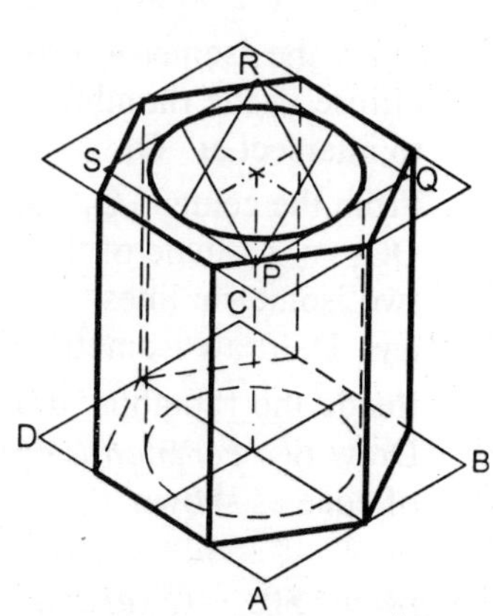

ISOMETRIC PROJECTION

FIG. 42

9. COMBINATION OF SOLIDS (COMPOUND SOLIDS)

Problem 49: *(Fig. 43)* ***A combination of solids** is formed as follows: A frustum of a cone has its top and bottom diameters 35 mm and 50 mm respectively and altitude 53 mm. It rests on the top face of a frustum of a square pyramid. Sides of the top and bottom faces of the pyramid are 58 mm and 70 mm respectively. Height is 22 mm. Draw the isometric projection of the combination of solids.* (UQ)

Isometric Projection of Frustum of Square Pyramid

1. Enclose the frustum of the pyramid in a box. For this, enclose the front view of the frustum in a rectangle as shown in *Fig. (a)*. Top view is self-enclosed.
2. Draw isometric projection of the box as in *Fig. (b)*.
3. Locate the top corners of the pyramid on the corresponding isometric lines on the top face PQRS of the box and complete the isometric projection of the frustum of the pyramid.

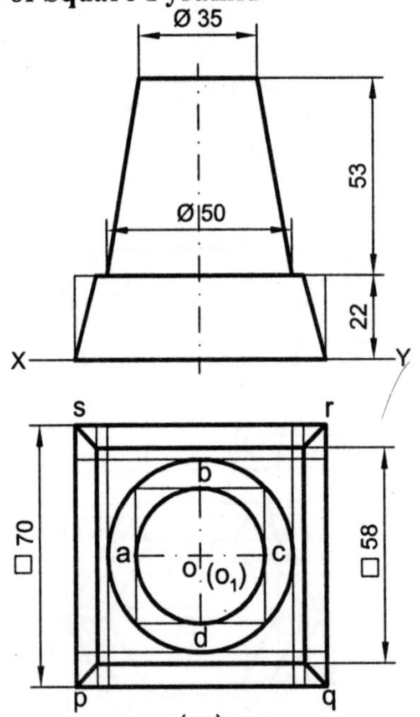

(a)

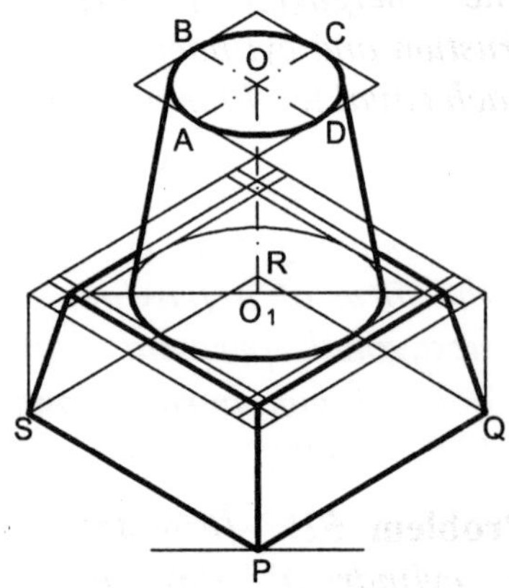

(b)

FIG. 43
ISOMETRIC PROJECTION OF
COMBINATION OF SOLIDS

Isometric projection of Frustum of Cone

4. Enclose the bottom face of the cone in a square in the top view in *Fig. (a)*.
5. Draw the isometric projection of the square as rhombus on top face of the box. Draw an ellipse in the rhombus to represent the base of the cone. Join the diagonals of the rhombus to intersect at O_1.
6. From the center O_1 of the bottom face of the cone, erect a vertical and mark O such that OO_1 = Altitude of the frustum of the cone = 53 mm in isometric scale. Through O draw two isometric lines AC and BD such that AC = ac and BD = bd. Through A, B, C and D, draw isometric lines to obtain a rhombus.
7. Inside the rhombus, draw the isometric projection of the top face of the cone (*top ellipse*). Draw *two common tangents* to ellipses and complete the isometric projection of the frustum of cone as shown.

Problem 50: *(Exercise) Draw the isometric projection of a bucket of top diameter 250 mm and bottom diameter 160 mm with a circular ring of 50 mm width attached at the bottom. Total height of the bucket is 320 mm.* (UQ)

Problem 51: *(Fig. 44) Draw the isometric projection of a funnel consisting of a cylinder and a frustum of a cone.*

The diameter of the cylinder is 20 mm and the top diameter of the funnel is 68 mm. The heights of the frustum and cylinder are each equal to 40 mm. (UQ)

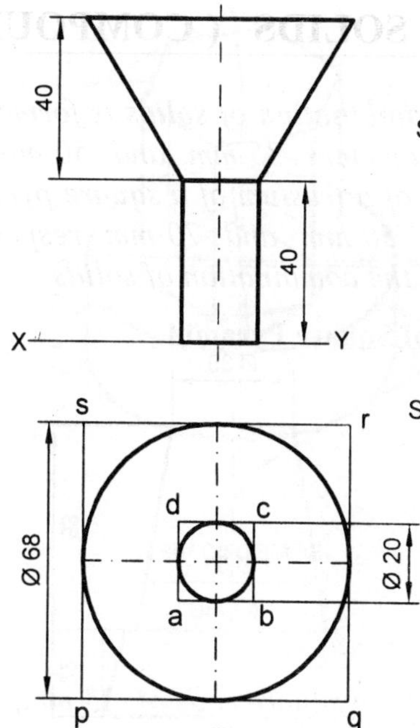

(i)

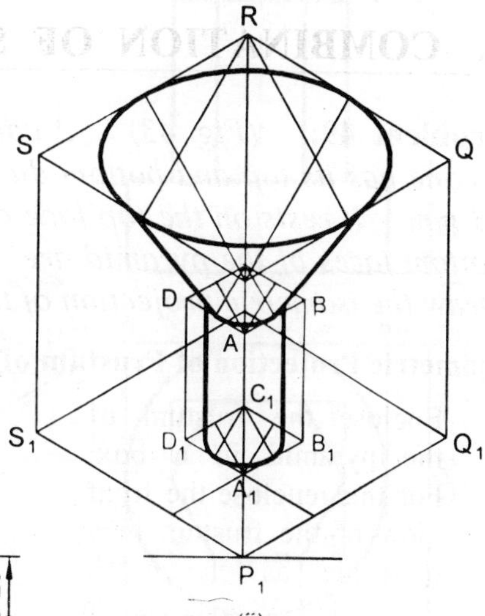

(ii)

FIG. 44
ISOMETRIC PROJECTION OF FUNNEL

Problem 52: *(Fig. 45) A cylindrical slab of 75 mm diameter and 45 mm thick is surmounted by a cube of 38 mm side.*

On the top of the cube rests a square pyramid of altitude of 38 mm and side of base 25 mm. The axes of the solids are in the same straight line.

Draw the isometric view of the setup. (UQ)

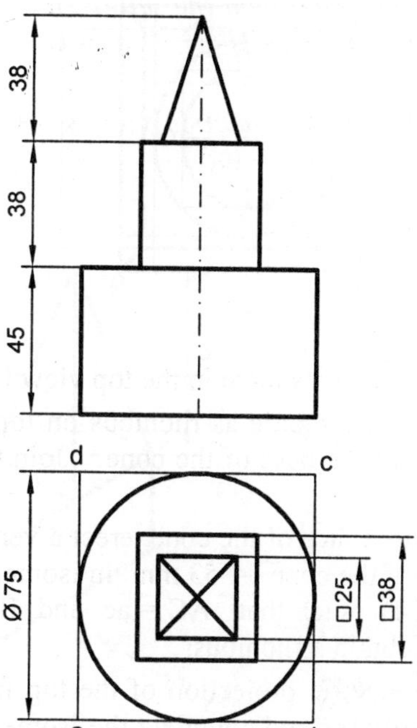

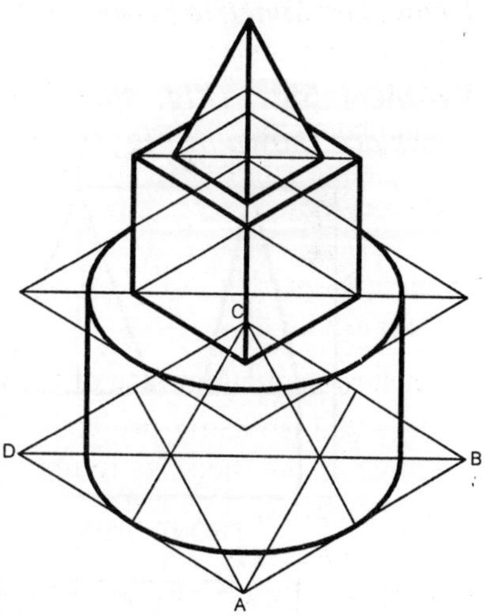

ISOMETRIC VIEW

FIG. 45

Problem 53: *(Fig. 46) A hexagonal shaped solid of base edge 20 mm and height 40 mm lies centrally on a cylinder of 60 mm diameter and 20 mm thick. Draw the isometric view of the solids if their axes lie on the same line.* (UQ)

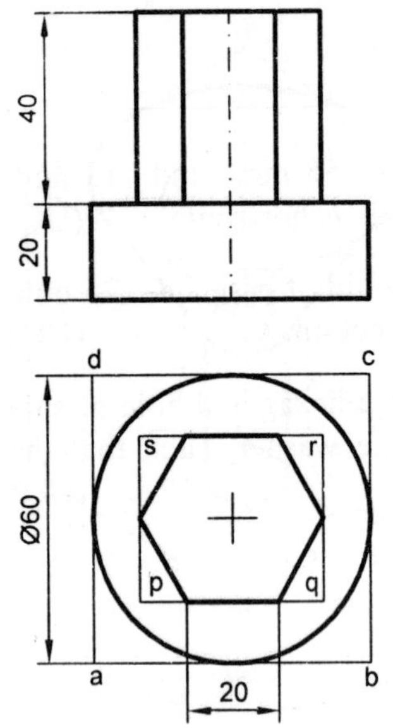

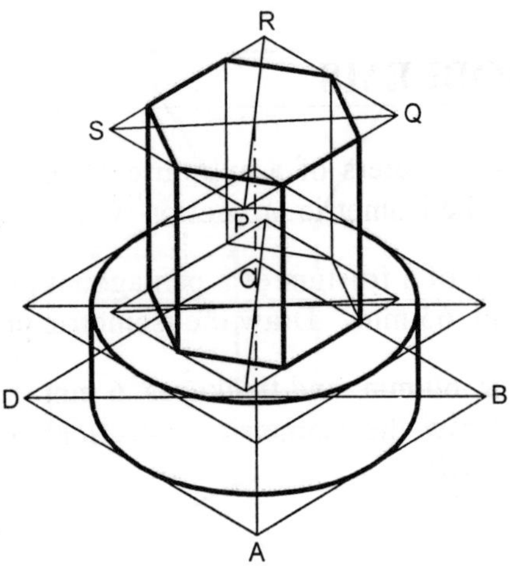

ISOMETRIC VIEW

FIG. 46

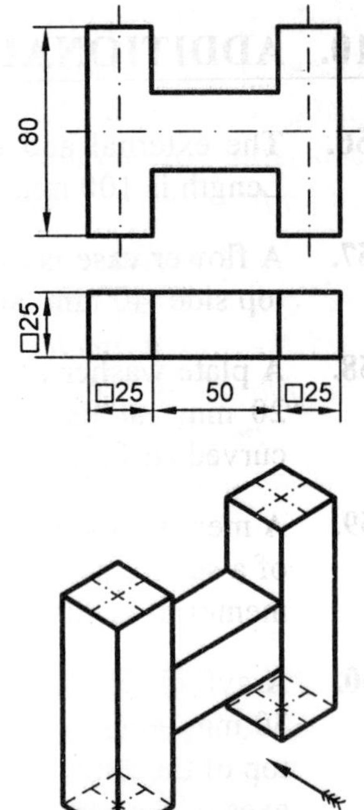

FIG. 47 LETTER - H

Problem 54: *(Fig. 47) Three square rods of 25 mm x 25 mm cross section and lengths 80 mm, 80 mm and 50 mm are so nailed together that they form the letter H. Draw the isometric projection of the letter H.* (UQ)

Problem 55: *(Fig. 48) Draw the isometric view of the object whose front and top views are shown in Fig. (i).* (UQ)

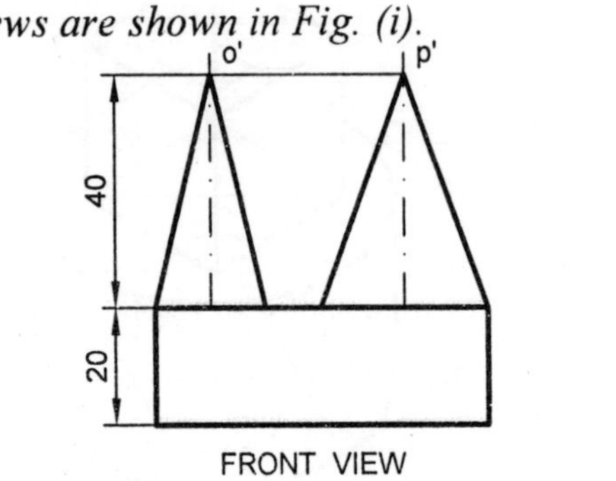

FRONT VIEW

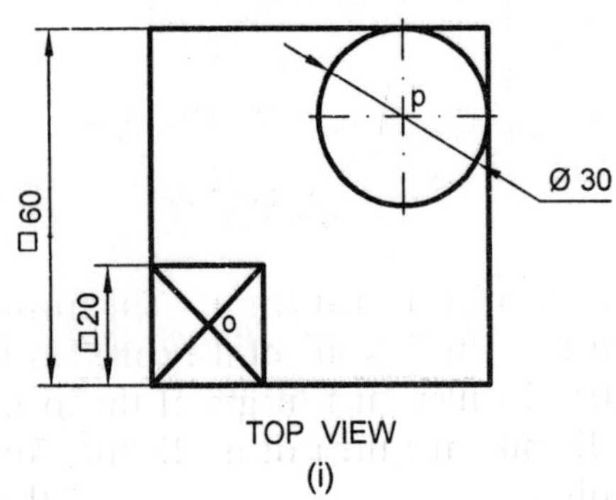

TOP VIEW
(i)

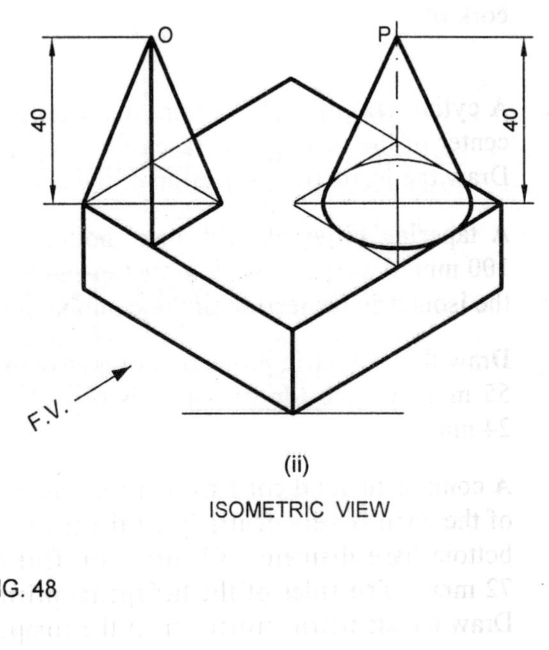

(ii)
ISOMETRIC VIEW

FIG. 48

10. ADDITIONAL PROBLEMS

56. The external and internal diameters of a cylindrical bush are 55 mm and 35 mm. Length is 100 mm. Draw the isometric projection when its axis is horizontal. (UQ)

57. A flower vase is in the form of a frustum of a pentagonal pyramid of base side 25 mm, top side 40 mm and height 65 mm. Draw the isometric projection. (UQ)

58. A plate washer of diameter 60 mm and thickness 6 mm has a hexagonal hole of side 20 mm at its center. Draw the isometric projection of the washer, such that the curved surface is lying on HP. (UQ)

59. A memorial is in the form of a cube of side 50 cm and thereafter tapers into a frustum of a square pyramid whose top surface is a square of 30 cm side. Total height of the memorial is 80 cm. Draw its isometric projection. (UQ)

60. A cylindrical slab 50 mm in diameter and 45 mm thick is surmounted by a cube of 30 mm edge. A square pyramid of altitude 45 mm and side of 25 mm rests on the top of the cube with the side of its base parallel to the sides of the top of the cube. The axes of the solids are collinear. Draw the isometric projection. (UQ)

61. A combination of solids is formed as follows: A hollow cylinder of 50 mm and 35 mm outside and inside diameters and height 60 mm stands vertically on a square prism of 65 mm side and 35 mm height. Draw the isometric projection.

62. A glass bottle consists of a cylinder of 60 mm diameter and 40 mm height with a cylindrical neck of 20 mm diameter and 20 mm height. The bottle is fitted with a cork of 18 mm diameter which projects by 5 mm. Draw its isometric projection. (UQ)

63. A cylindrical disc of 20 mm diameter and altitude 20 mm is placed vertically at the center of the rectangular face of a hexagonal prism of 28 mm sides and axis 60 mm. Draw the isometric projection of the combination. (UQ)

64. A tapering pillar of 100 mm bottom diameter and 50 mm top diameter holds a 100 mm diameter sphere at its top center. The height of the pillar is 100 mm. Draw the isometric projection of the combination of solids. (UQ)

65. Draw the isometric projection of a hexagonal prism of side of base 25 mm and height 55 mm, on the top of which is placed a cone of base diameter 50 mm and height 24 mm.

66. A compound solid consists of a frustum of a cone and a hexagonal prism. The frustum of the cone is kept centrally on the top of the prism. The frustum of the cone has the bottom base diameter 45 mm, top face diameter 27 mm and length of the axis is 72 mm. The sides of the hexagonal prism are 32 mm and the axis is 22 mm long. Draw the isometric projection of the compound solid. (UQ)

11. ISOMETRIC PROJECTION OF SPHERES

The orthographic view of a sphere seen from any direction is a circle of diameter equal to the diameter of the sphere. ***Isometric Projection of a sphere is also a circle of diameter equal to the actual diameter of the sphere.*** We shall prove this concept in the *following Problem*:

Rule: Use Isometric Scale for isometric projections of spheres or having spherical parts.

Problem 67: *(Fig. 49) Draw the isometric projection of a* ***sphere*** *of diameter 50 mm resting centrally on the* ***top of a cube*** *of side 60 mm.*

1. *See Fig. (i).* Draw the front view of the combination of solids.
2. *See Fig. (ii).* Draw the isometric projection of the square block using isometric scale.
3. Locate the center A of the top surface of the block. (A is the point of contact of sphere)
4. Draw a vertical at A. Mark point O on it such that AO = Isometric radius of sphere.
5. O as center and radius equal to the true radius of the sphere, draw a circle which is the isometric projection of the sphere.

Problem 68: *(Fig. 50) A sphere of radius 20 mm is kept on the top face of a square prism of side of base 40 mm and height 20 mm. The latter is placed on the top face of a cylinder of 65 mm diameter and 25 mm height. All the three solids have the common axis. Draw the isometric projection of* ***combination of solids****.* (UQ)

1. Draw the front and top views of the given combination of solids as shown in *Fig. (a)*..
2. **To draw the isometric projection of the cylinder:** Enclose the cylinder in a box. Draw the isometric projection of the box. Draw the isometric projection of the cylinder inside the box by ***Four Centers Method.***
3. **To draw the isometric projection of the square prism:** The top face of the cylinder and bottom face of the prism are on same plane PQRS. Locate 4 corners of the base of the square prism on the corresponding isometric lines on the top face PQRS of the box as in *Fig. (b)*. Then complete the isometric projection of the prism as in *Fig. (c)*.

To draw the isometric projection of the sphere

4. *See Fig. (c).* Join the diagonals of the top face of the prism to intersect at O_1.

Note: *When a sphere is viewed in any direction, its shape will be a circle of radius = Actual radius of the sphere. Therefore, the isometric projection of the sphere will be a circle of radius = Actual radius of the sphere.*

5. However, the center of the sphere in isometric projection is located as follows:
   The imaginary line joining the center of the sphere and the point of contact of the sphere with the center of the top face of the prism will be foreshortened to its isometric length. Therefore from the center O_1 of the top face of the prism, erect vertical and mark the center O of the sphere such that OO_1 = isometric length of radius of the sphere.
6. O as center and the actual radius (20 mm) of the sphere as radius, draw a circle to represent the isometric projection of the sphere as shown.

(67)

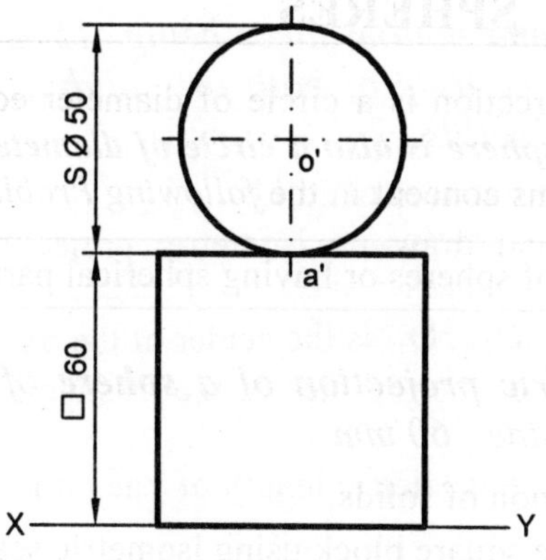

(i) FRONT VIEW

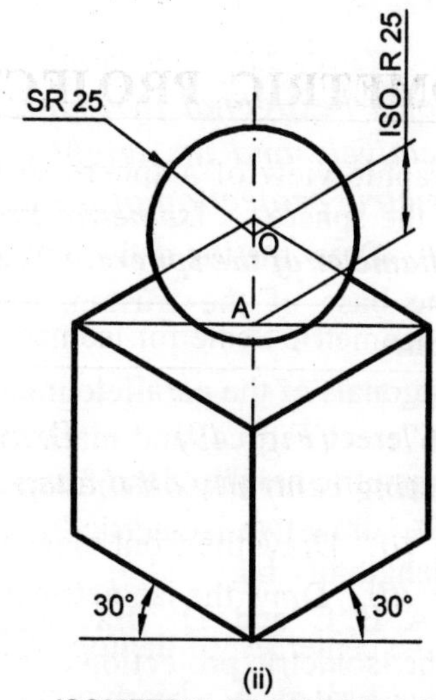

(ii) ISOMETRIC PROJECTION OF SPHERE ON A CUBE

FIG. 49

(68)

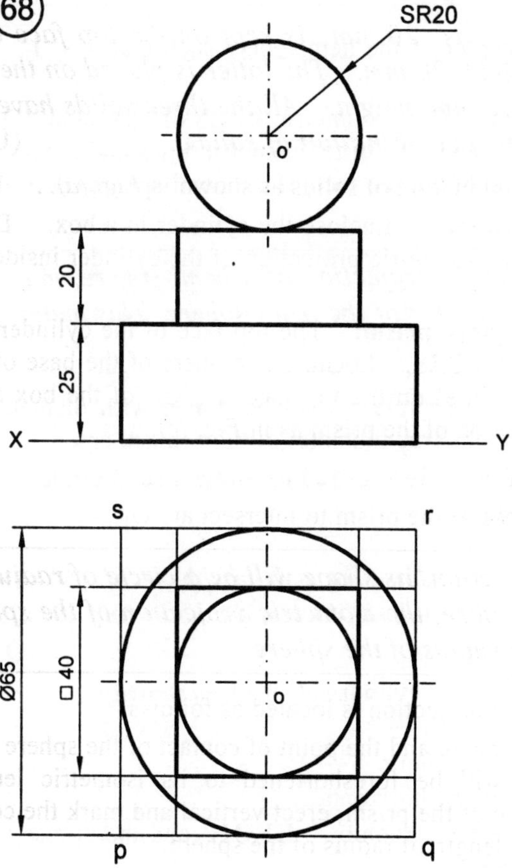

(a)

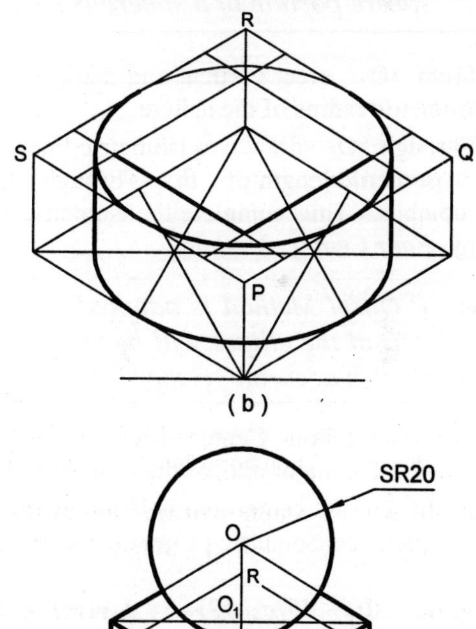

(b)

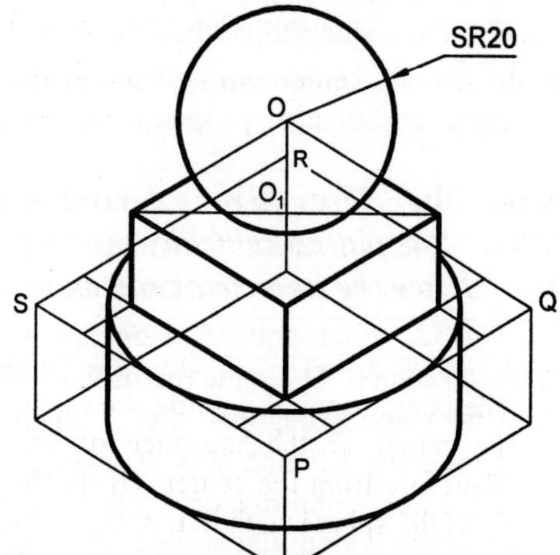

FIG. 50

(c) ISOMETRIC PROJECTION

Problem 69: *(Fig. 51) A hemisphere of 44 mm diameter is nailed on the top face of a frustum of a hexagonal pyramid, sides of top and bottom faces being 15 mm and 30 mm respectively and its height is 50 mm. The axes of both the solids coincide. Draw the isometric projection of the* ***combination of solids****.* (UQ)

1. *See Fig. 51.* Draw the top and front views of the combination of solids.
2. Enclose the base of the frustum in a rectangle and draw its isometric projection as a parallelogram.
3. Join the diagonals of the parallelogram to intersect at O_1. O_1 is the center of the base.
4. From O_1 erect vertical and mark O_2, the center of the top face of the frustum such that O_1O_2 = isometric height of the frustum.
5. Through O_2 draw two isometric lines such that AC = isometric length of ac and BD = Isometric length of bd.
6. Through A, B, C and D draw isometric lines and obtain the parallelogram as shown and complete the isometric projection of the top face.

Note: *Procedure to obtain the isometric projection of the surface of a hemi-sphere/portion of a sphere is similar to that of the sphere.*

---

7. From O_2 erect vertical and mark O_3, the center of the hemisphere such that O_2O_3 = Isometric radius of the sphere.
8. Through O_3 draw two isometric lines such that EG = isometric length of eg and FH = isometric length of fh. Through E, F, G and H draw isometric lines and obtain a rhombus. Thus complete the isometric projection of the top face of the hemi-sphere (ellipse) by ***Four Centers Method***.

Note: *If Offset Method is adopted for the isometric projection of a hemi-sphere, major axis of the ellipse will be equal to true diameter of the hemi-sphere. This method, though accurate, is more time-consuming.*

Here, since Four Centers Method is used, the semi-circle should be drawn with diameter equal to the major axis of the ellipse, and not with the true diameter of the hemisphere.

9. With O_3 as center and half the major axis of the ellipse as radius, draw a semi-circle and complete the isometric projection of the hemisphere.

Problem 70: *(Fig. 52) A* ***rivet head*** *has the shape of a hemisphere of 60 mm diameter. It is placed centrally over a cylindrical shank of 40 mm diameter and 60 mm length. Draw the isometric projection of the rivet.* (UQ)

1. *See Fig. 52.* First, draw the isometric projection of the cylindrical shank as shown.
2. Mark the center O at the top of the cylinder.
3. Draw the vertical diagonal EG = diameter of hemisphere.
4. Draw the sides EF and EH at an angle of 30° to the horizontal.
5. Draw the horizontal diagonal FH and complete the rhombus EFGH and the ellipse to represent the bottom flat surface of the hemisphere.
6. O as center and true radius of the sphere, draw a semi-circle and complete the isometric.

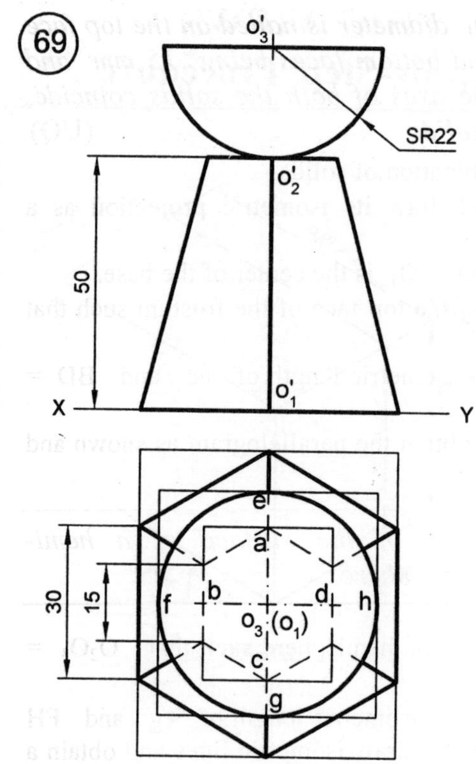

(i)

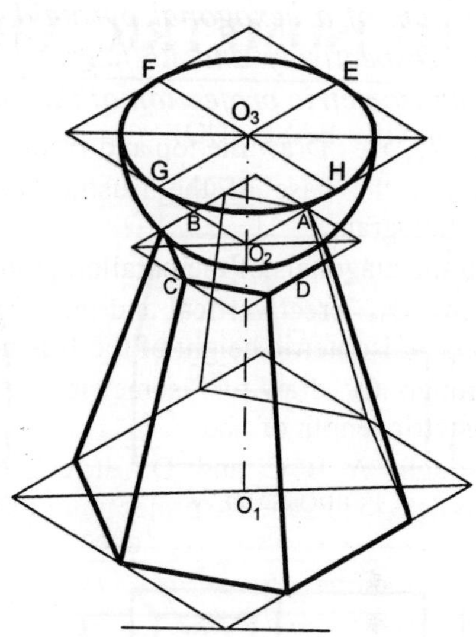

(ii)

ISOMETRIC PROJECTION OF
HEMISPHERE ON HEXAGONAL PYRAMID

FIG. 51

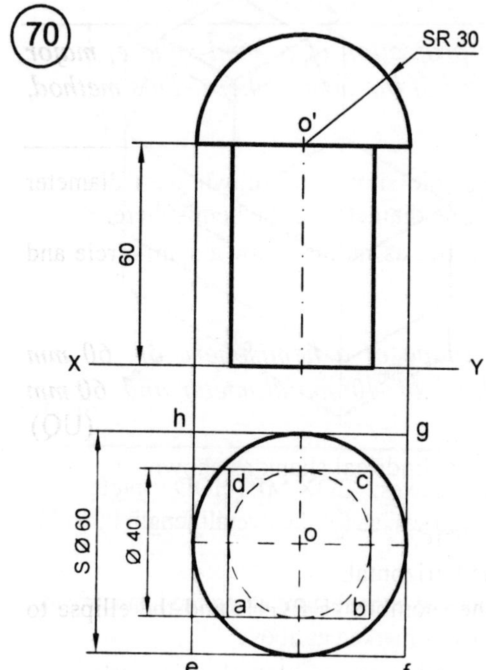

(i)

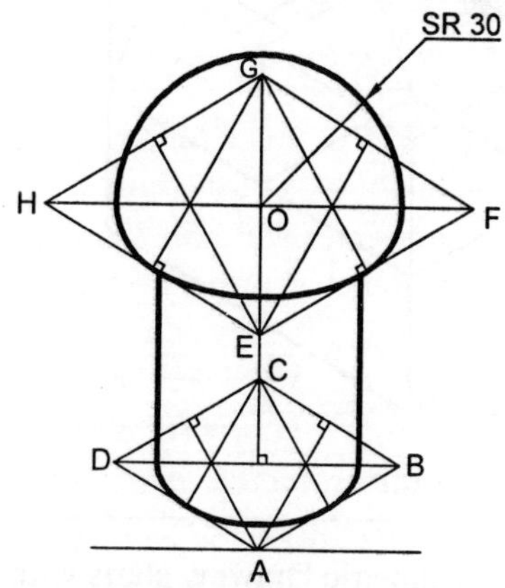

(ii)

ISOMETRIC PROJECTION OF
RIVET HEAD

FIG. 52

12. CONVERSION OF ORTHOGRAPHIC VIEWS TO ISOMETRIC VIEWS - Step-by-Step Procedure

(71) MIXIE - BASE (UQ)

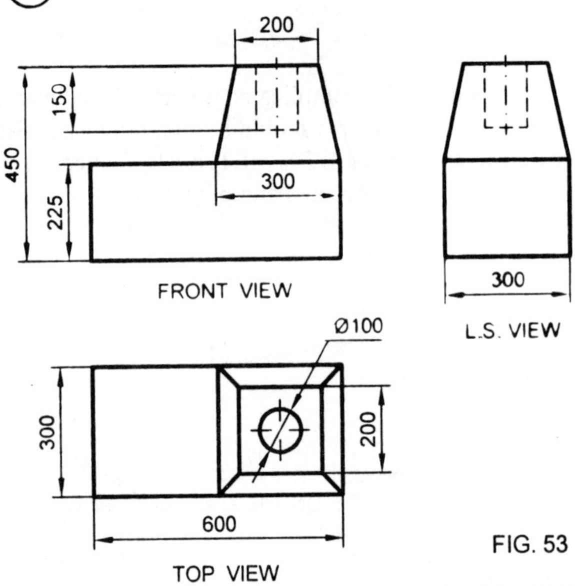

FIG. 53

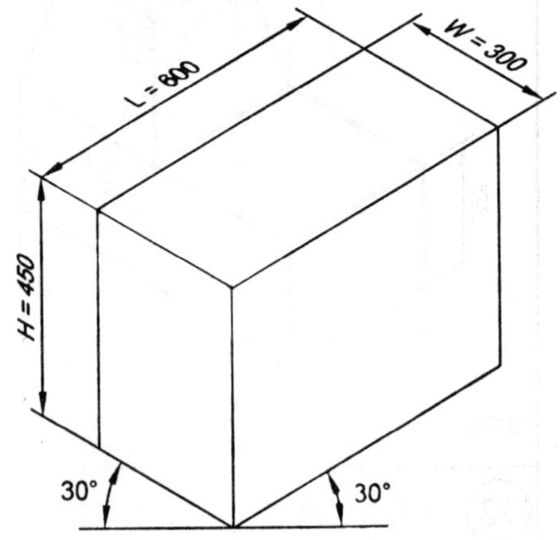

STEP I

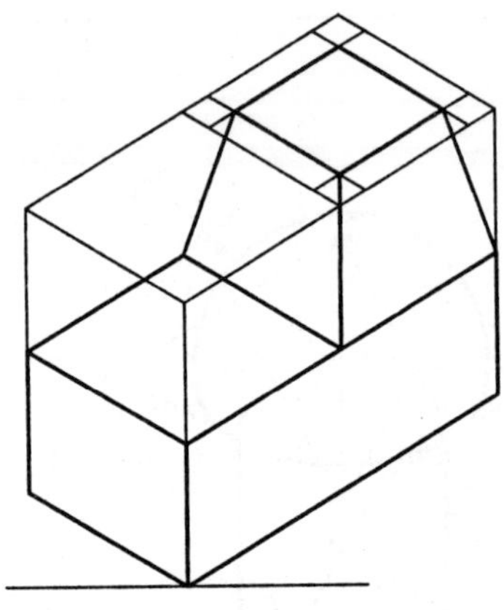

STEP II

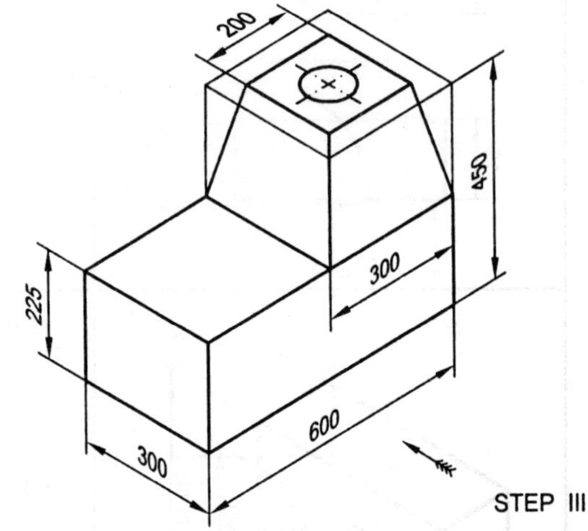

STEP III

STEP I : Isometric Drawing starts with the three isometric axes. BOX METHOD is preferable. Hence, draw a box in thin lines along the isometric axes for the overall length L, width W and height H of the object.

STEP II : Mark the various points at appropriate distances of each feature along the isometric axes. Draw lightly, the lines corresponding to the points marked as above.

STEP III : Draw the ellipse to represent the hole of 100 mm diameter.

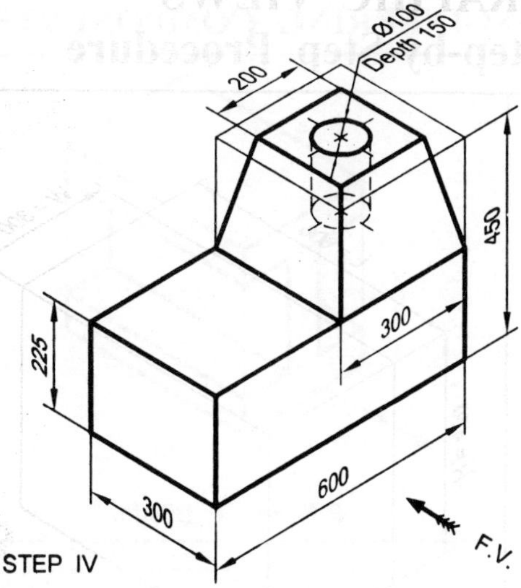

STEP IV

STEP IV : Darken the required visible and invisible lines of all the features in the Isometric View.

Note: For complicated objects, the cylindrical features should be first drawn as prisms using BOX METHOD. Centers of holes are located and blocked in.

(72) SQUARE BLOCK WITH HOLE

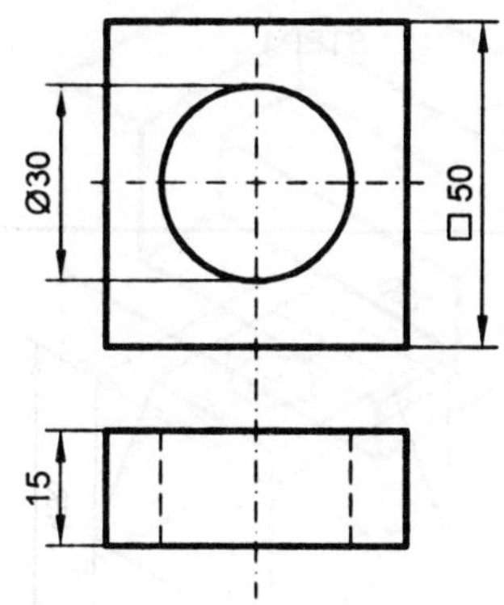

FIG. 54

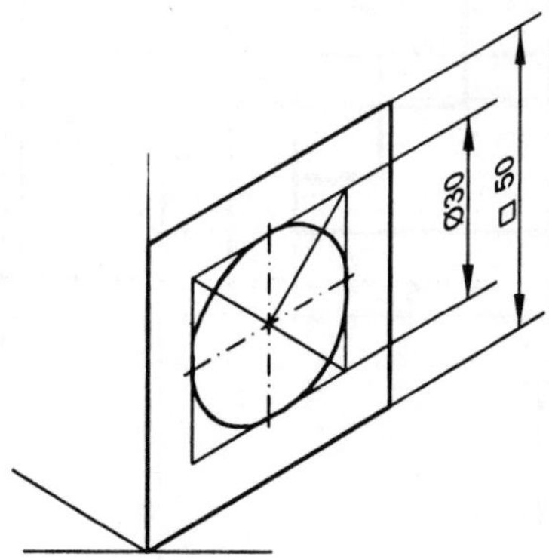

STEP I

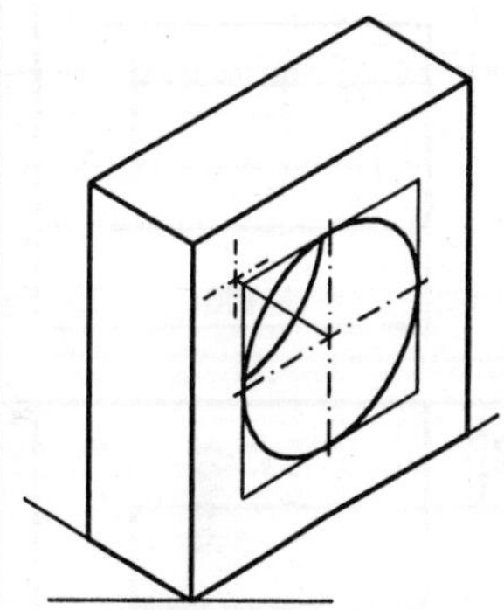

STEP II

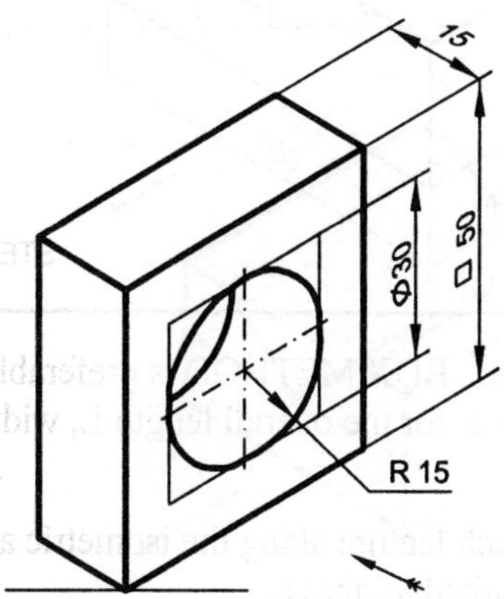

STEP III

Note: In the isometric view, draw the hole as two parallel ellipses corresponding to the front and rear circles of the hole.

STEP I : Draw the rhombus in the isometric to represent the square in the front view. Draw the ellipse nearer to the observer as shown.

STEP II : Mark the various points at appropriate distances of each feature along the isometric axes. Draw lightly, the lines corresponding to the points marked as above.

STEP III : Retain only the visible portion of the rear ellipse.

13. TYPICAL PRACTICAL PROBLEMS - ENGINEERING COMPONENTS

(73)

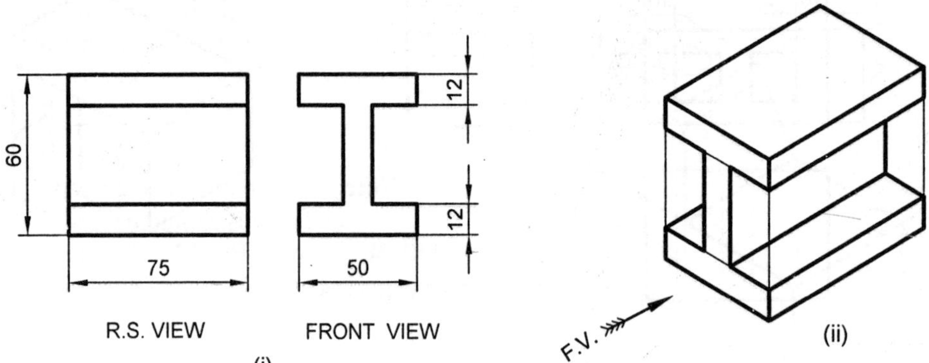

FIG. 55 I - BEAM

(74)

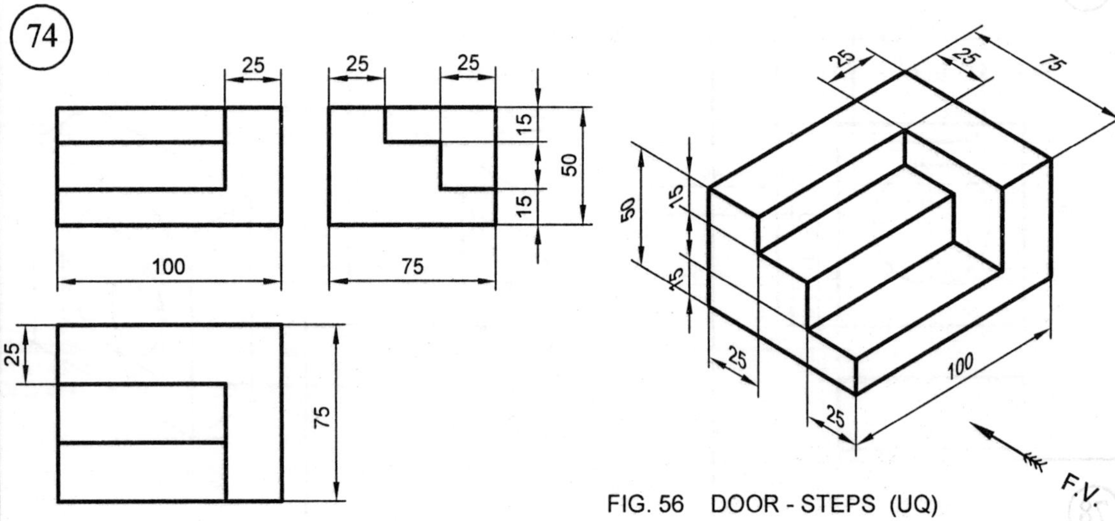

FIG. 56 DOOR - STEPS (UQ)

(75)

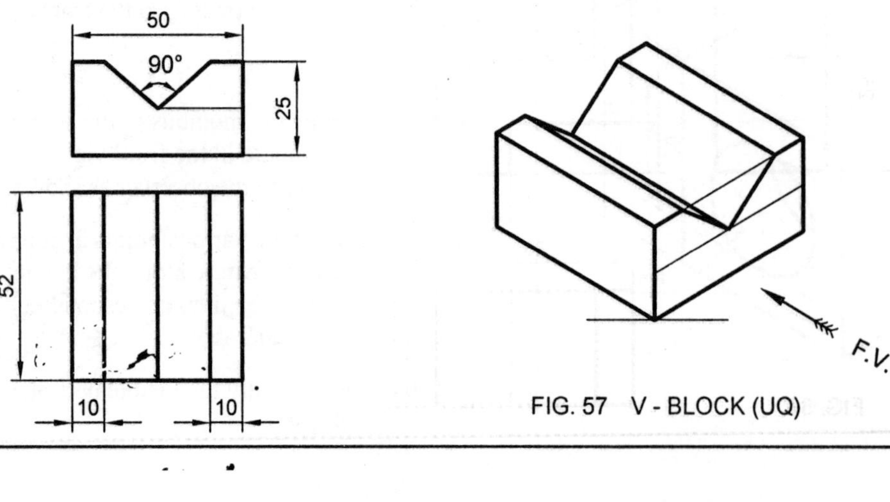

FIG. 57 V - BLOCK (UQ)

(76)

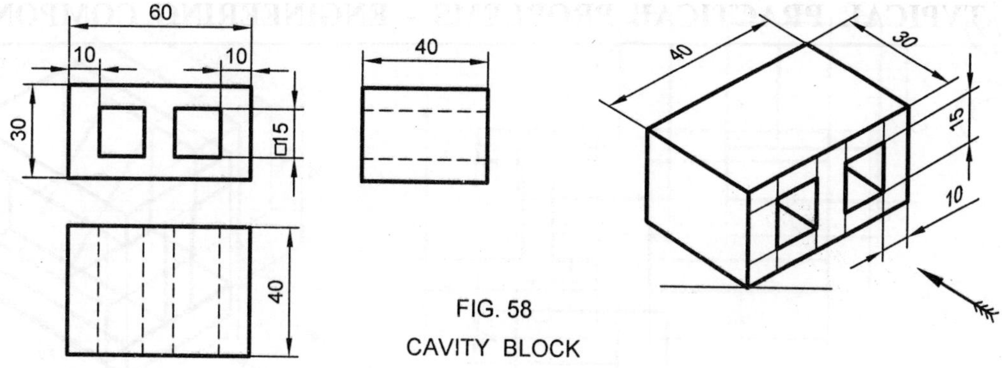

FIG. 58

CAVITY BLOCK

(77)

(UQ)

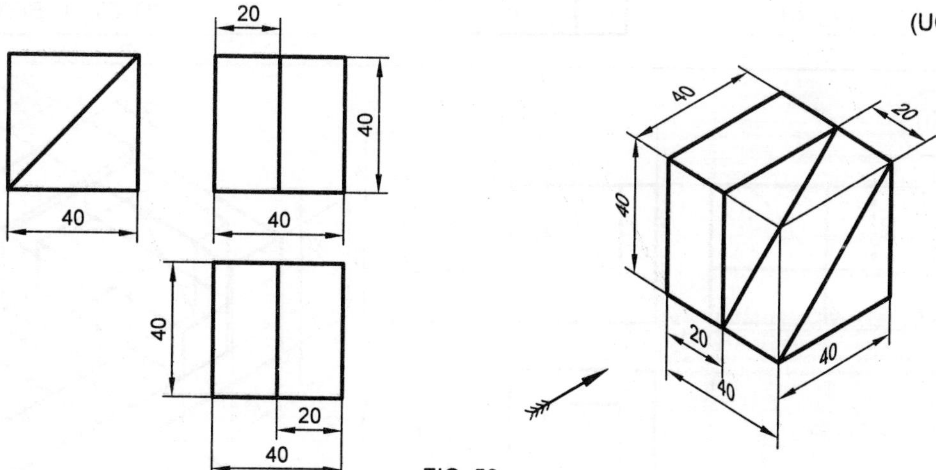

FIG. 59

(78)

(UQ)

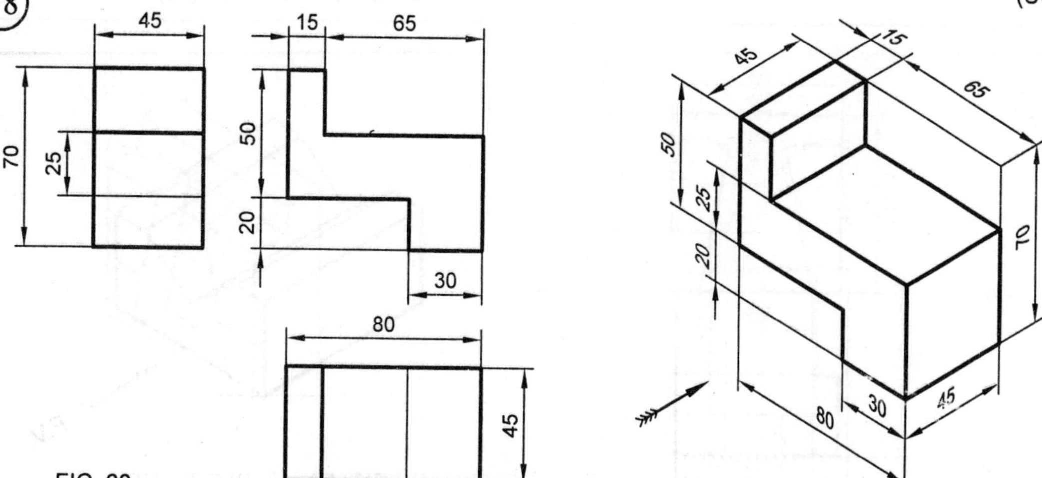

FIG. 60

(79) (UQ)

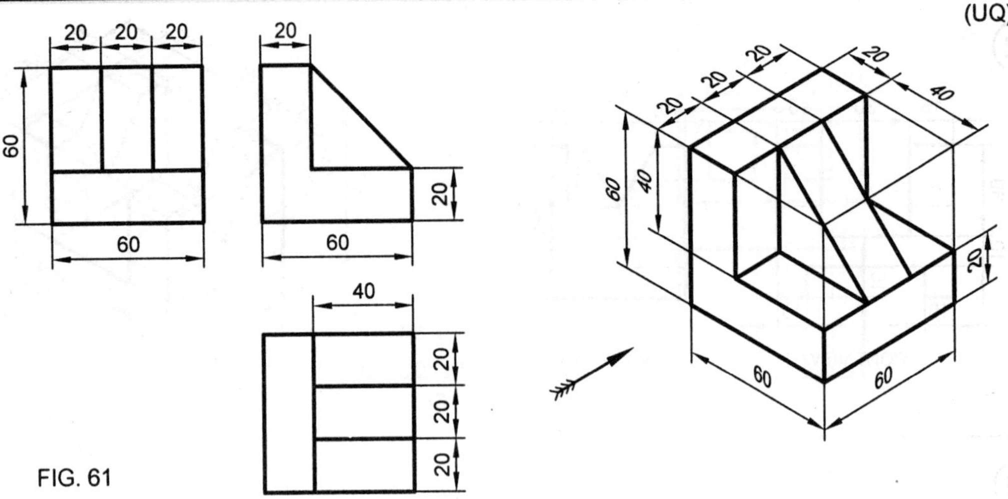

FIG. 61

(80) (UQ)

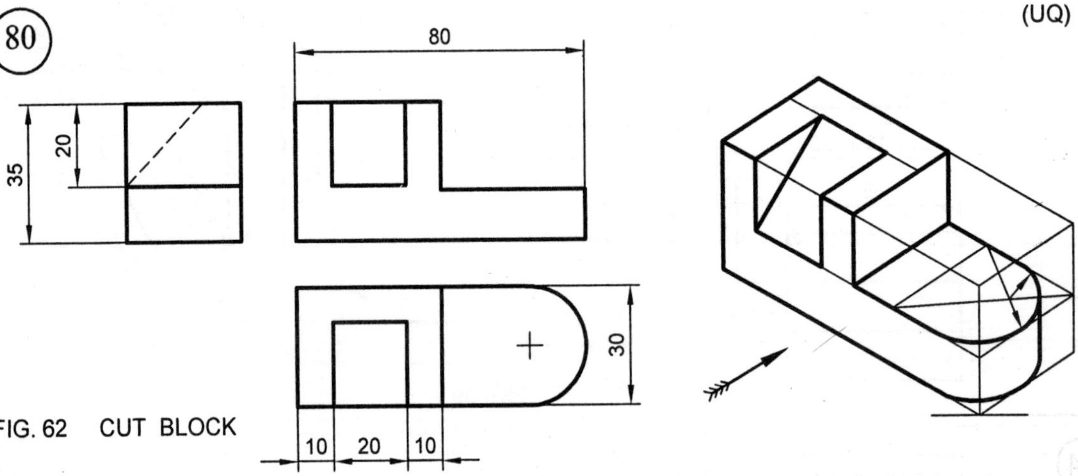

FIG. 62 CUT BLOCK

(81)

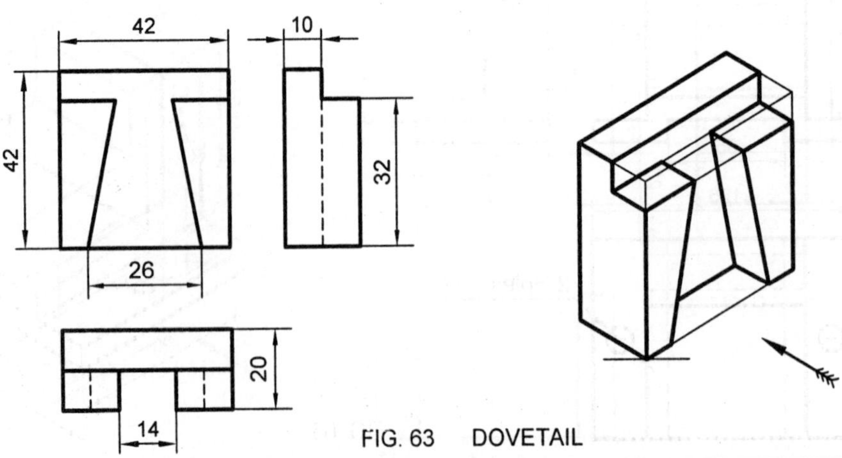

FIG. 63 DOVETAIL

(82) (UQ)

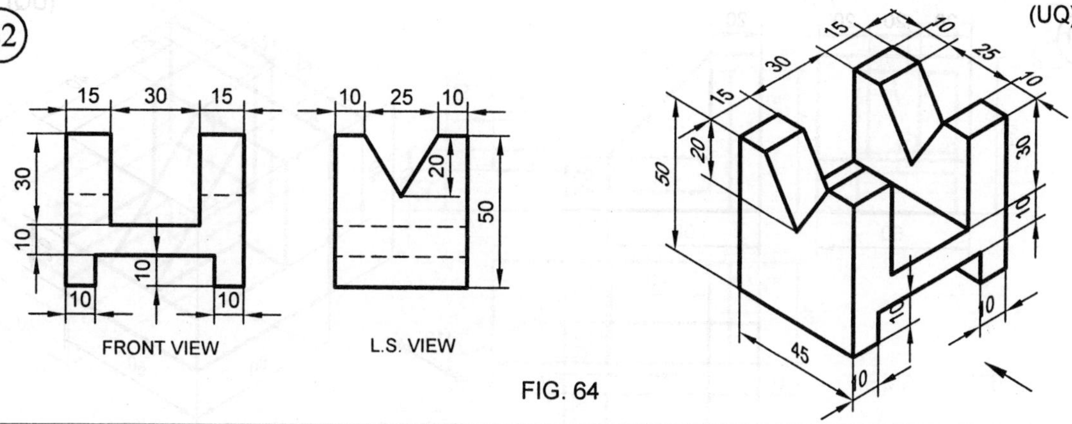

FIG. 64

(83) (UQ)

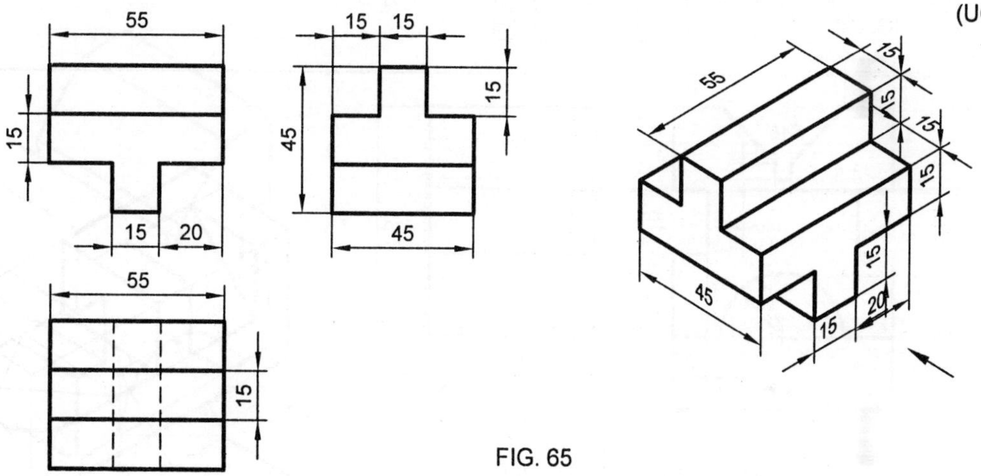

FIG. 65

(84) (UQ)

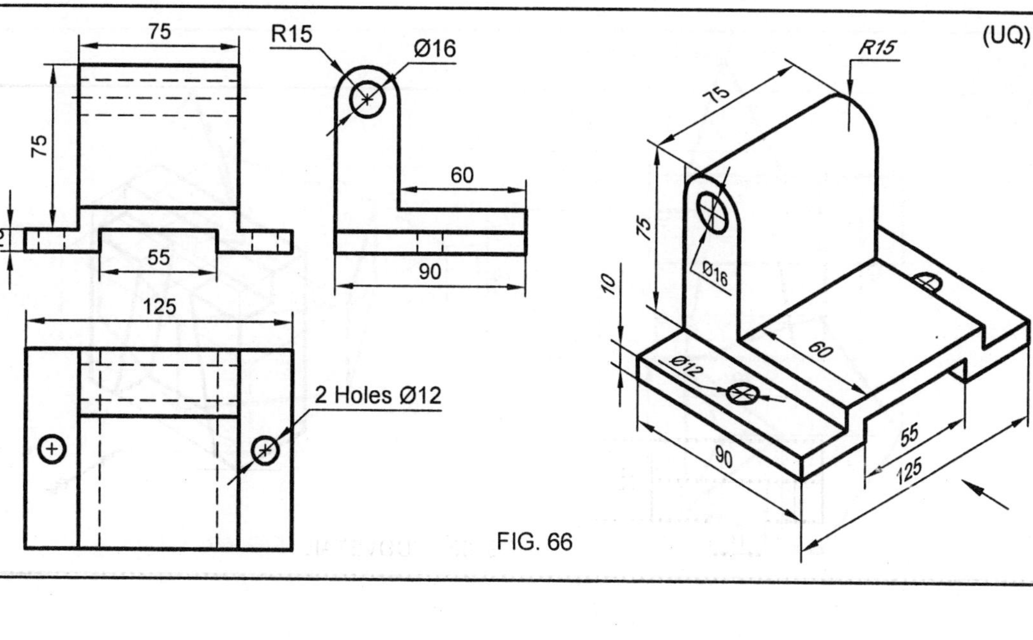

FIG. 66

85

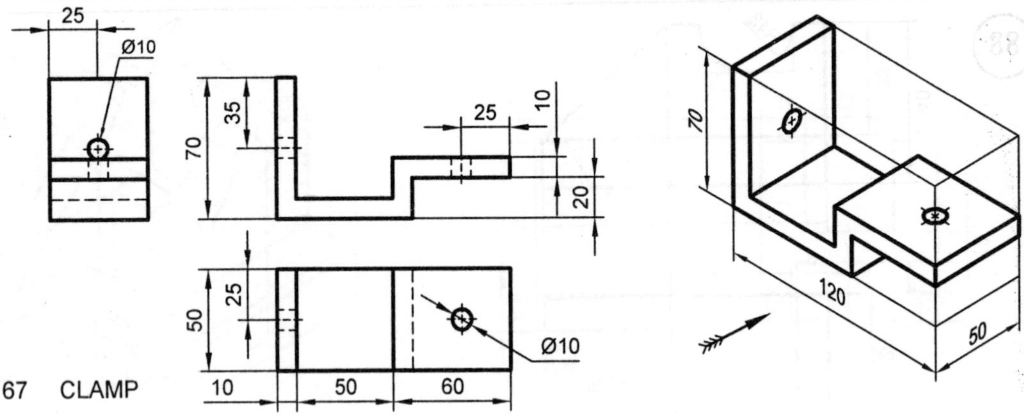

FIG. 67 CLAMP

86

(UQ)

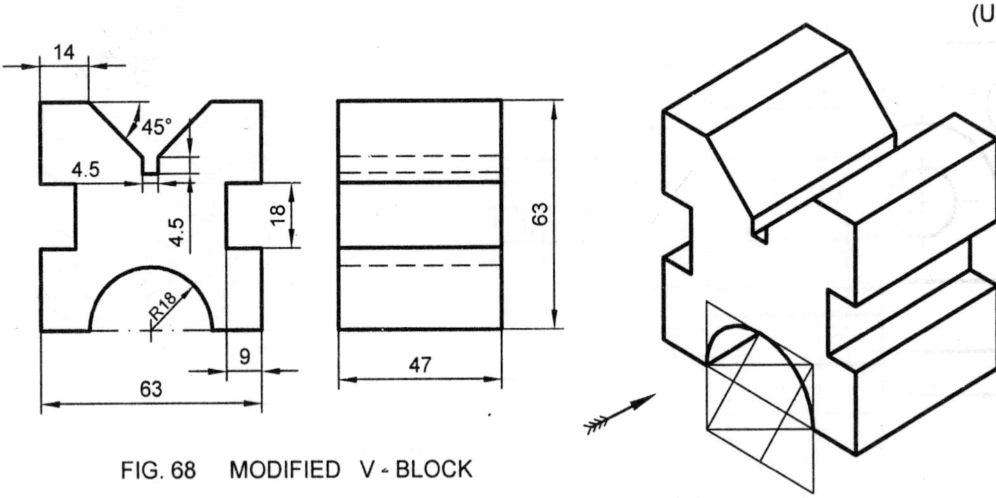

FIG. 68 MODIFIED V - BLOCK

87

(UQ)

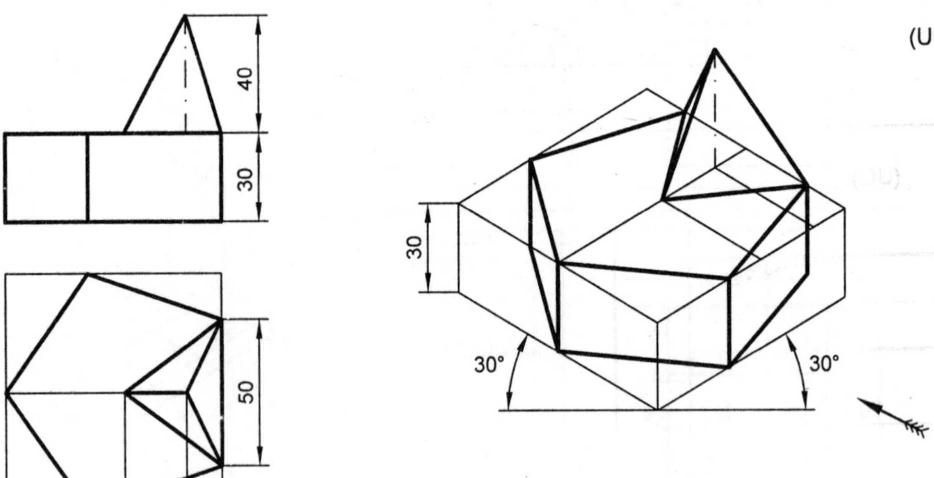

FIG. 69

88

(UQ)

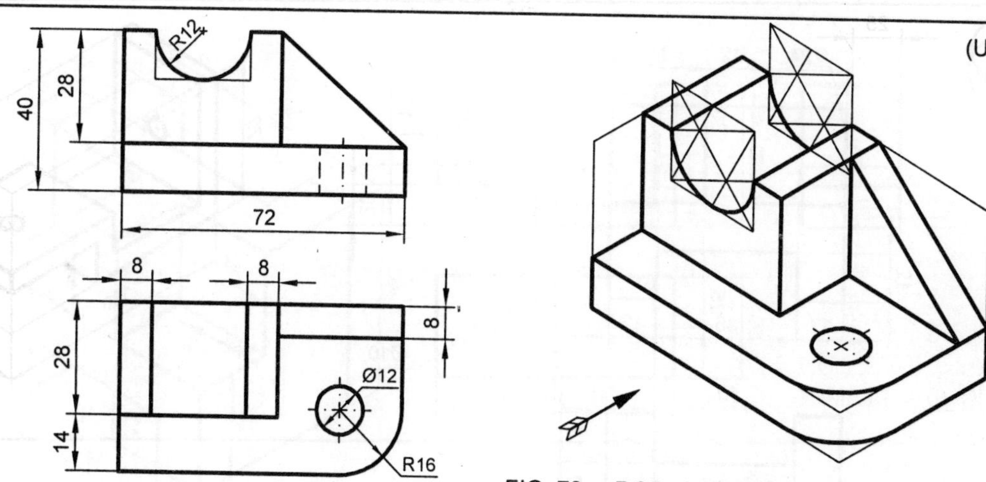

FIG. 70 ROD SUPPORT

89

(UQ)

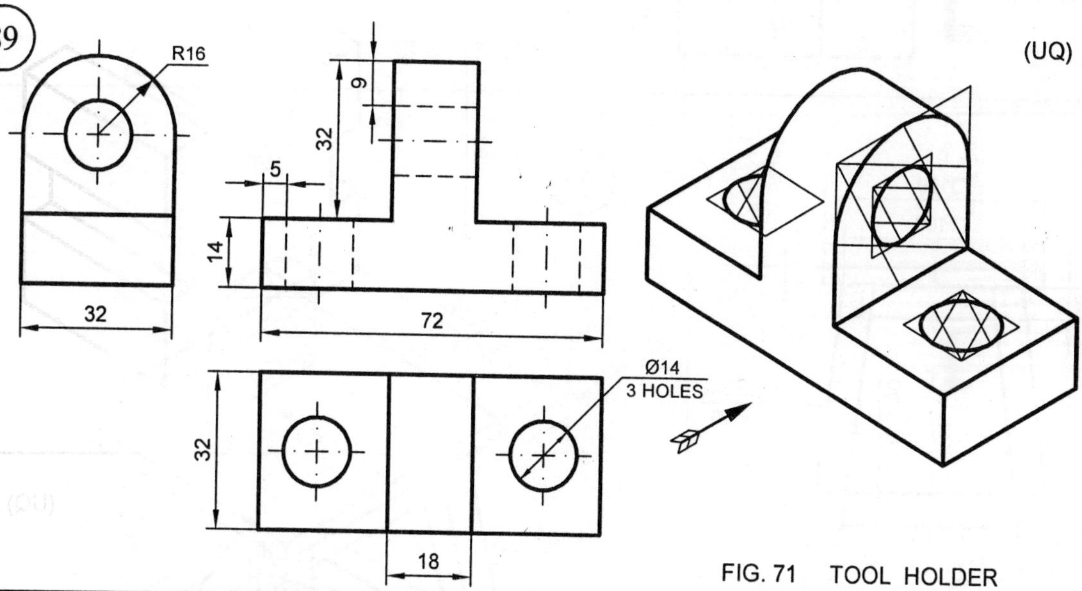

FIG. 71 TOOL HOLDER

90

(UQ)

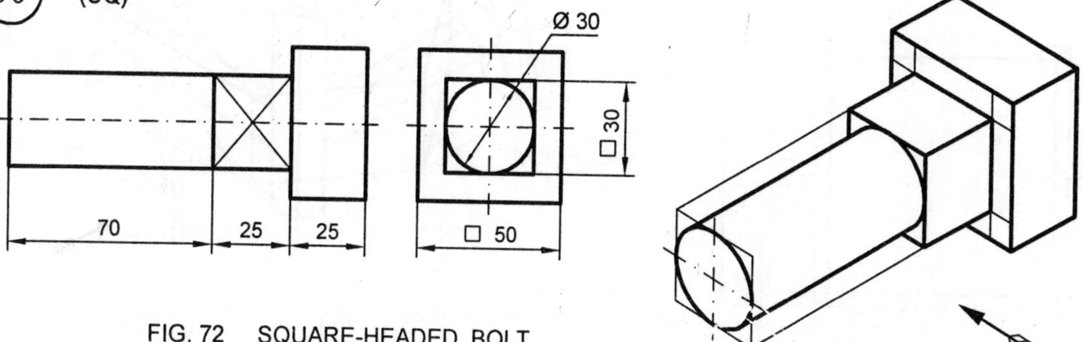

FIG. 72 SQUARE-HEADED BOLT

91

(UQ)

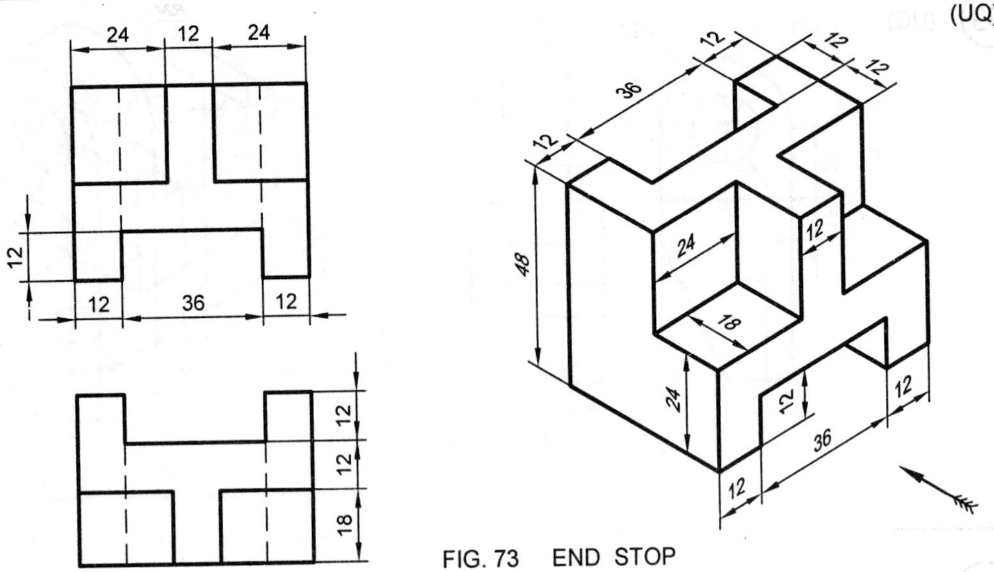

FIG. 73 END STOP

92

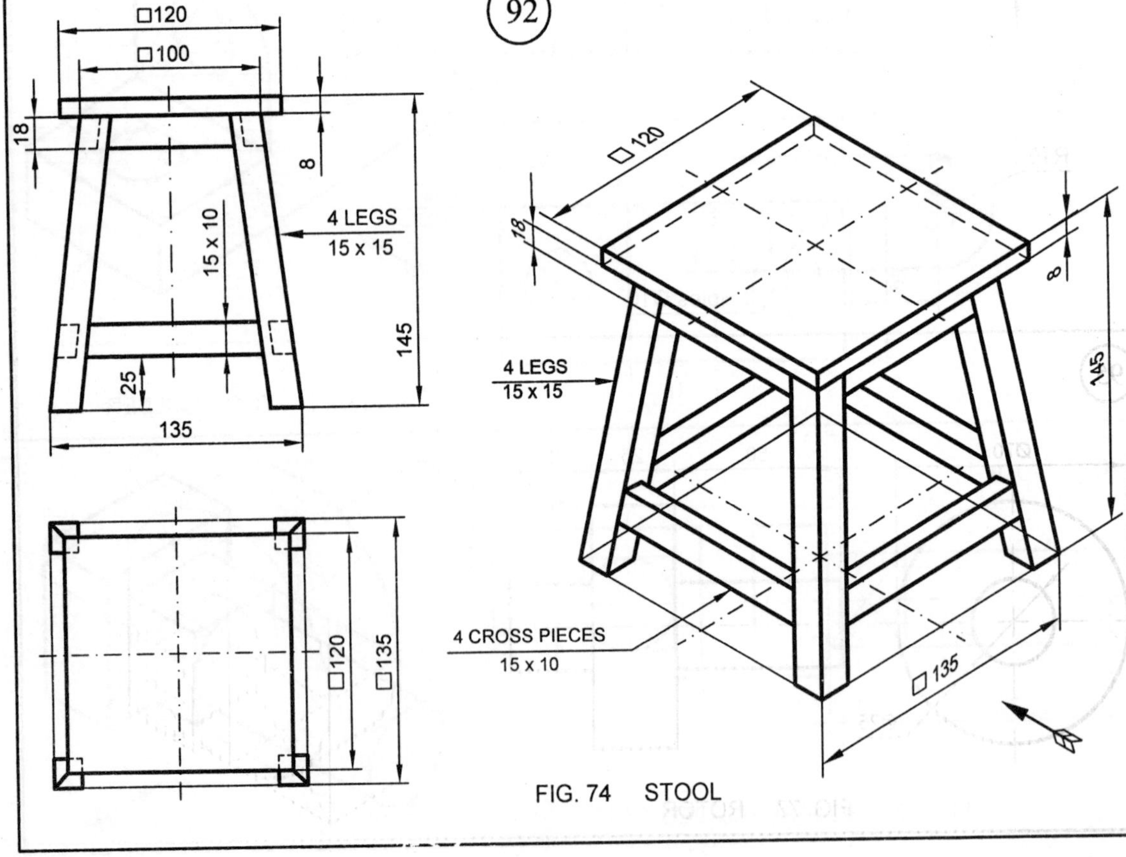

FIG. 74 STOOL

(93) (UQ)

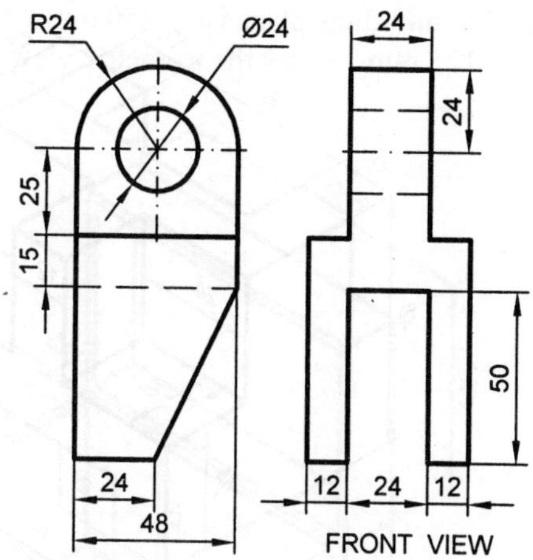

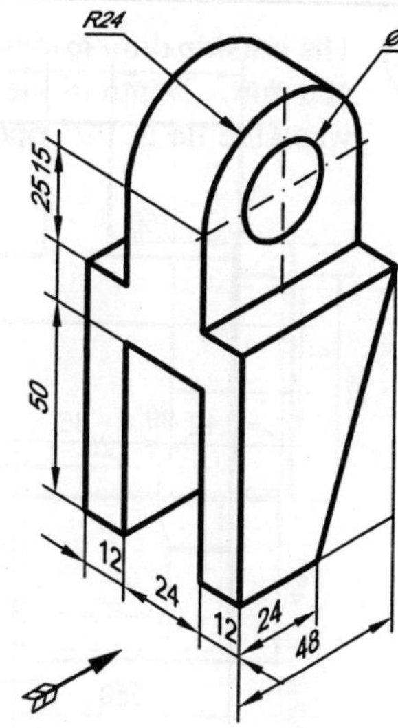

FIG. 75

(94) (UQ)

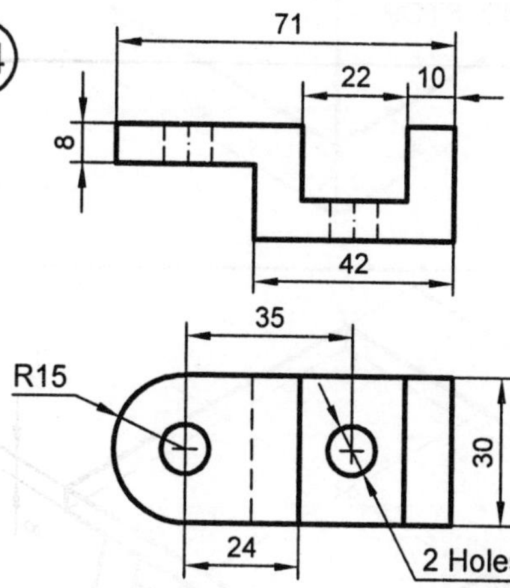

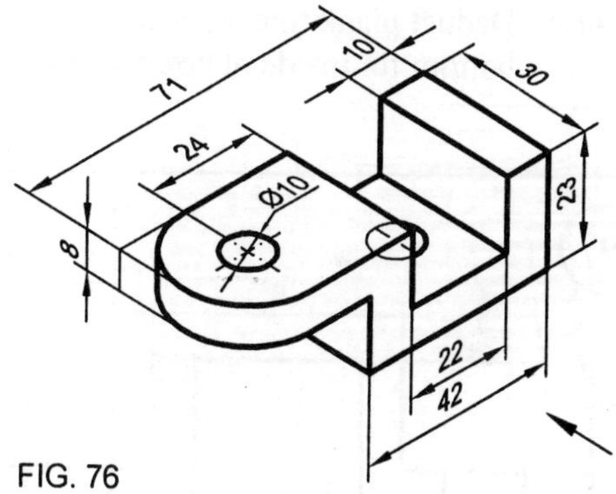

FIG. 76

(95)

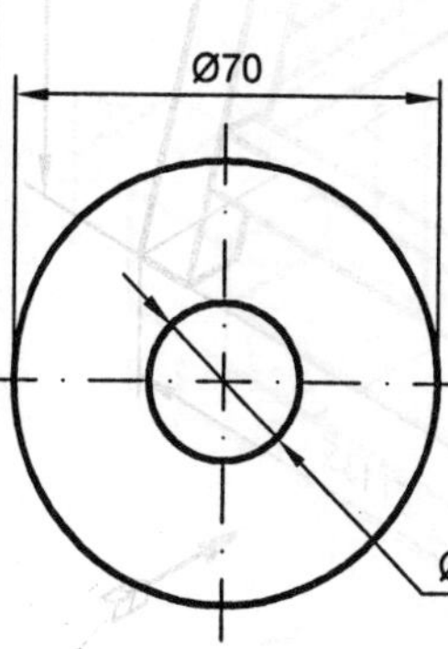

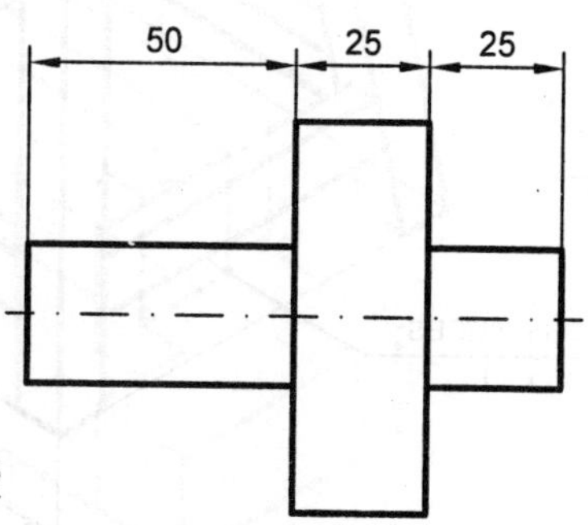

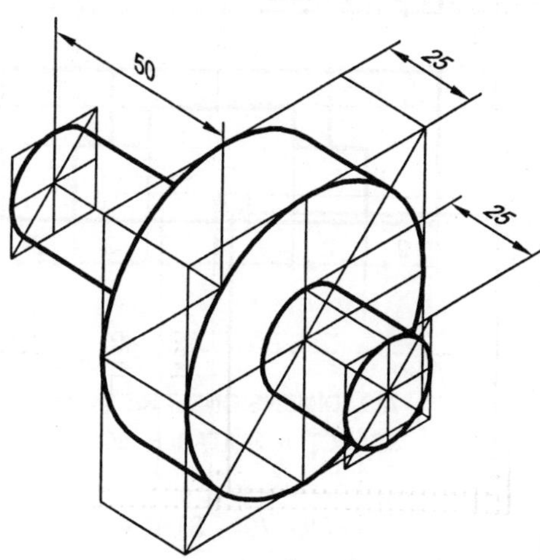

FIG. 77 ROTOR

96 The outside dimensions of a box made of 25 mm thick planks are 850 mm x 550 mm x 400 mm. Depth of the lid on the inside is 125 mm. Draw the isometric view of the box when the lid is 90° open. (UQ)

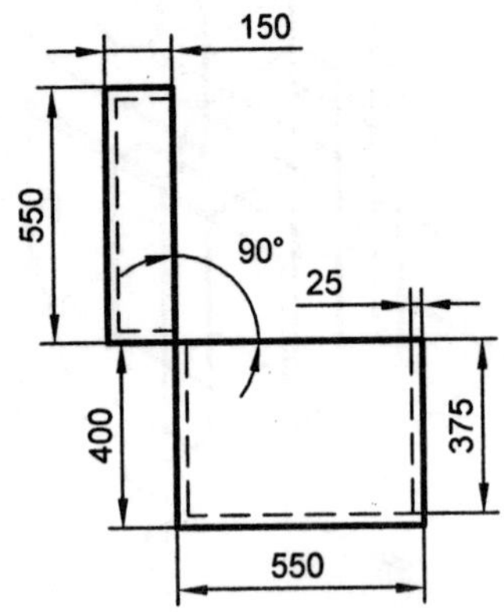

FRONT VIEW

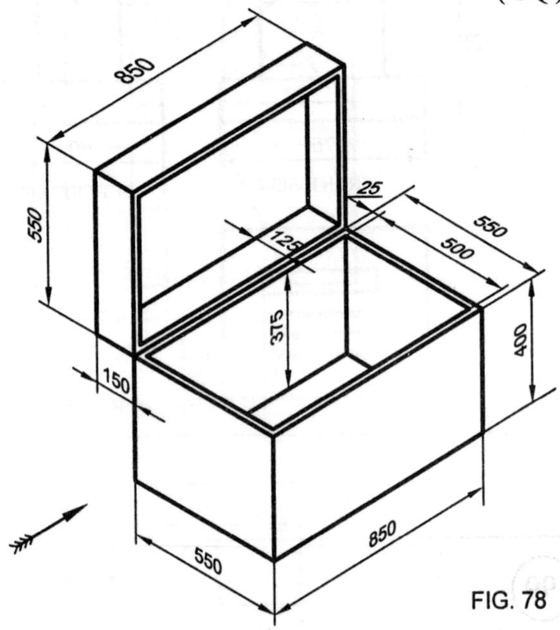

Note: Deduct plank thickness at top and bottom for inside of box and lid.

FIG. 78

97 (UQ)

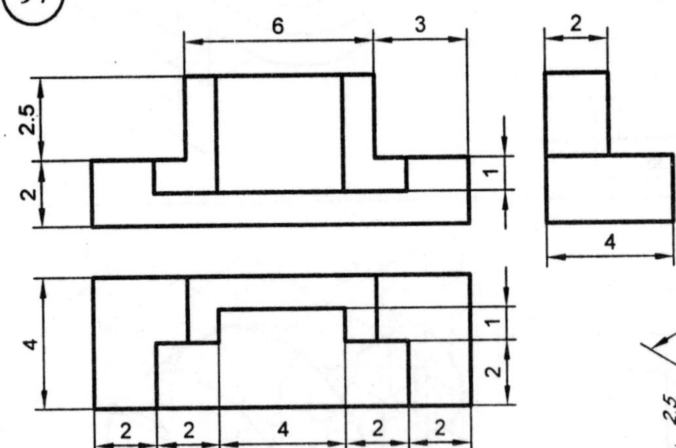

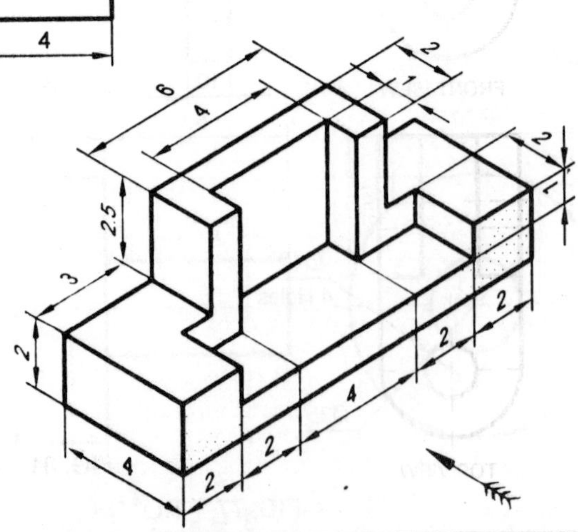

ALL DIMENSIONS ARE IN m.
SELECT A SUITABLE SCALE.

FIG. 79 STAGE

(98) (UQ)

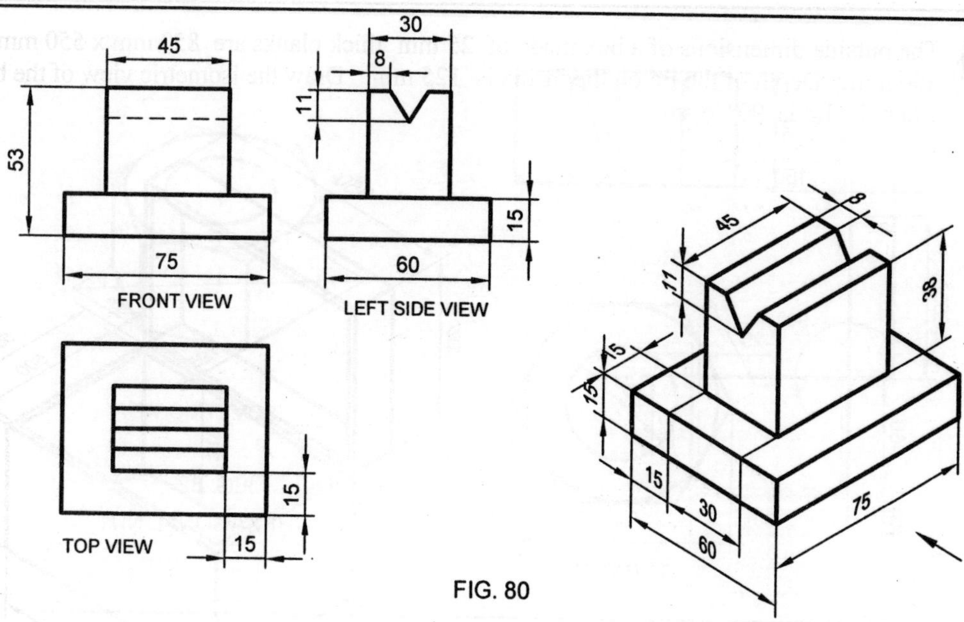

FIG. 80

(99) (UQ)

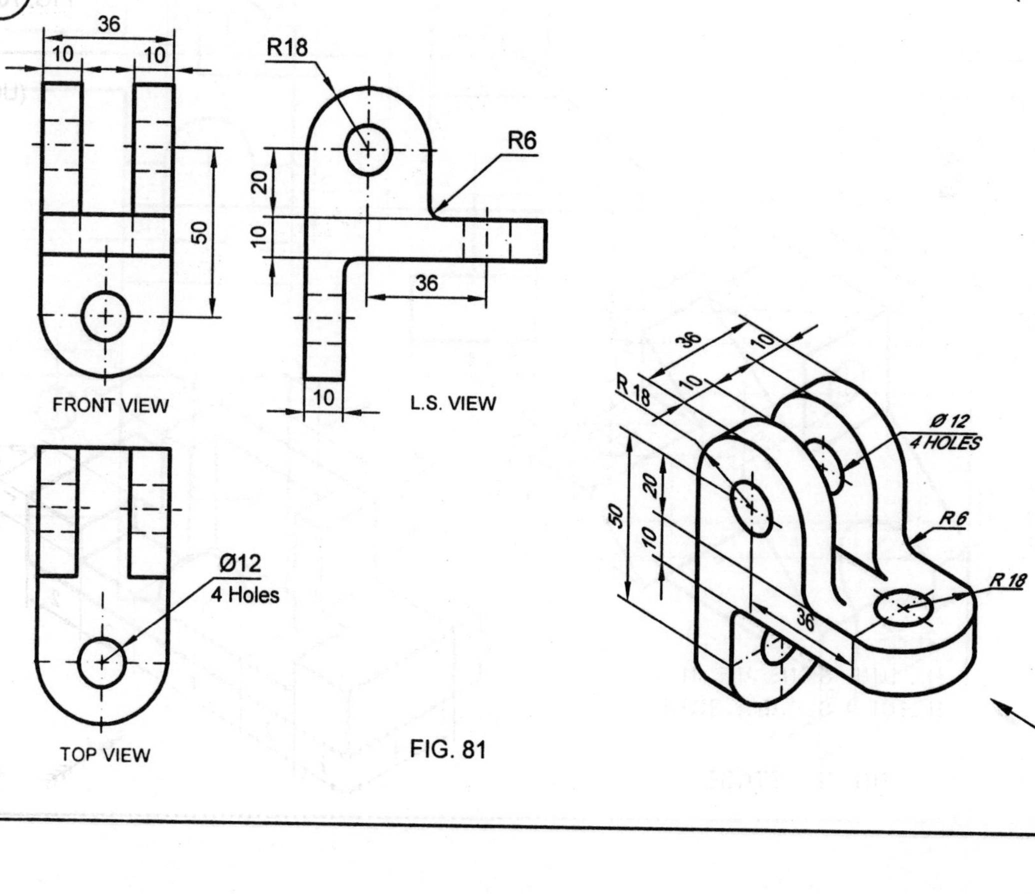

FIG. 81

(100)

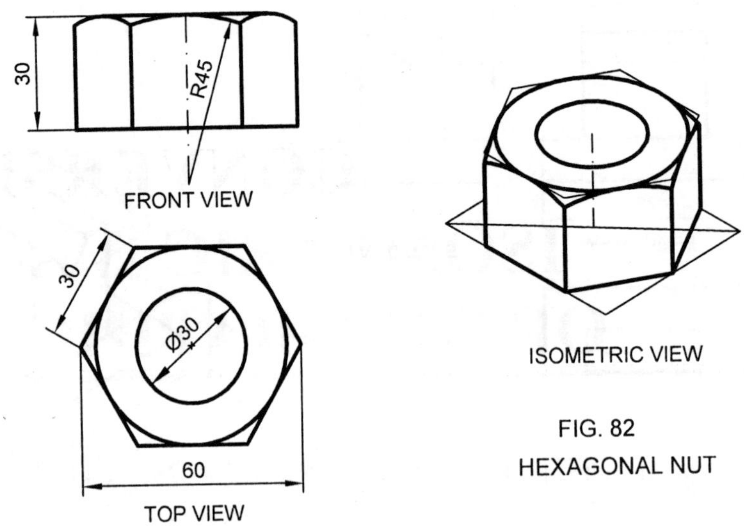

FIG. 82
HEXAGONAL NUT

(101) (UQ)

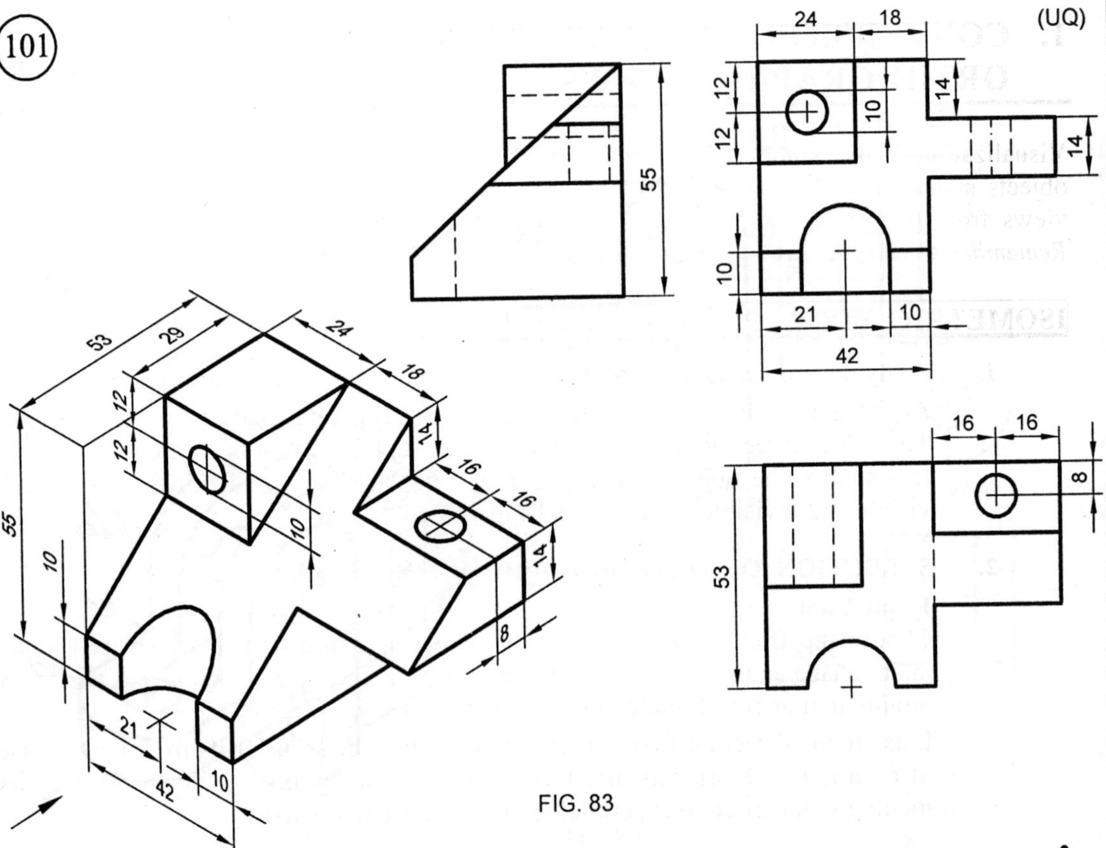

FIG. 83

*

21

CONVERSION OF ISOMETRIC VIEWS TO ORTHOGRAPHIC VIEWS

You are now familiar with the Principles of Orthographic and Isometric (Pictorial) Projections, including BIS Conventions for Hidden Lines and Precedence of Lines.

1. CONVERSION OF ISOMETRIC VIEWS TO ORTHOGRAPHIC VIEWS – Step-by-Step Procedure

Visualization of the isometric (pictorial) views may be started from simple objects. Complex objects may be visualized after sufficient practice. While preparing drawings of orthographic views from pictorial views, follow a systematic **Step-by-Step** procedure as explained below: *Remember !* BIS conventions are also to be observed *in toto.*

ISOMETRIC VIEW (PICTORIAL VIEW)

1. Study the Isometric View of the given object and its details carefully.

   An isometric view shows three mutually perpendicular planes, *i.e.*, Left, Right and Top. The three faces on these three planes and details on them are read one-by-one to visualize the object. Thus, each part of the object is identified, thereby all the visible and invisible surfaces can be identified.

2. **SELECTION OF ORTHOGRAPHIC VIEWS**

   Front View

   Choose the front view by seeing the object in a direction in which its length, the major shape and maximum details are seen. Further, place the object so as to obtain minimum number of hidden lines in the front view.

   Indicate the direction from which the object is to be seen for its front view by means of an arrow. Mark this arrow parallel to a sloping axis. Obtain other views by looking in the directions parallel to other two isometric axes.

   Other Orthographic Views

   Decide the necessary *other views* to be drawn.

ORTHOGRAPHIC VIEWS

3. ***Overall Dimensions:*** Determine the overall dimensions of the required orthographic views, *i.e.*, Length L, Width W and Height H (or Thickness T).

4. ***Proportioning:*** Estimate *proportioning of dimensions* suitably in drawing.

5. ***Block-in Views:*** Draw rectangular blocks (*in thin lines*) for the overall dimensions of each view, keeping sufficient space between the views for legible placing of dimensions, Note, etc. on the views. This is known as *Block-in views*.

6. ***Center Lines:*** Mark centers for circles in the front view. Draw center lines passing through these centers for the axes of symmetry. When a cylindrical part / hole is seen as a rectangle, draw only one center line for its axis; when seen as a circle draw two axes perpendicular to each other.

7. Draw the details of the object (*in thin lines*) ***simultaneously*** in all the views by drawing projectors in the following order:
   (i) Circles and Arcs of circles,
   (ii) Visible lines of the general shape of the object,
   (iii) Minor details of the object such as lines, small fillet curves, etc.

8. Darken all the visible lines and circles / arcs. Mark the invisible lines, if any.

9. Mark the dimensions.

1.1 IDENTIFICATION OF SURFACES

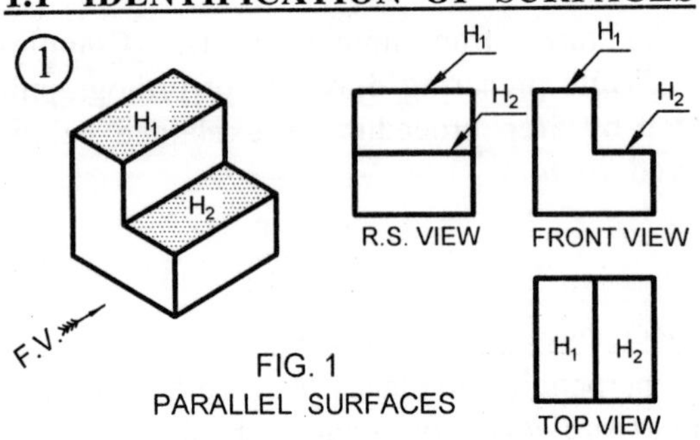

FIG. 1
PARALLEL SURFACES

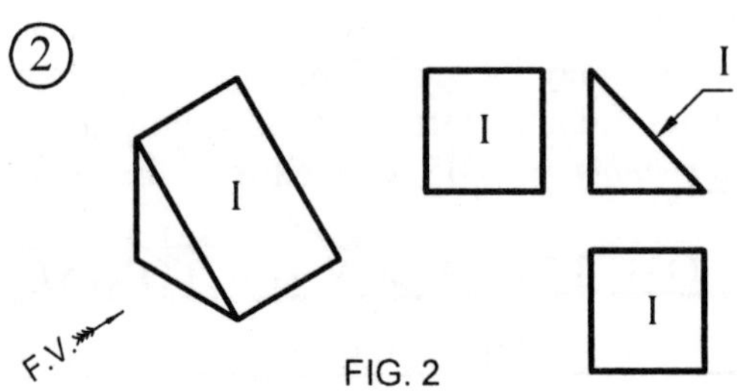

FIG. 2
INCLINED SURFACE

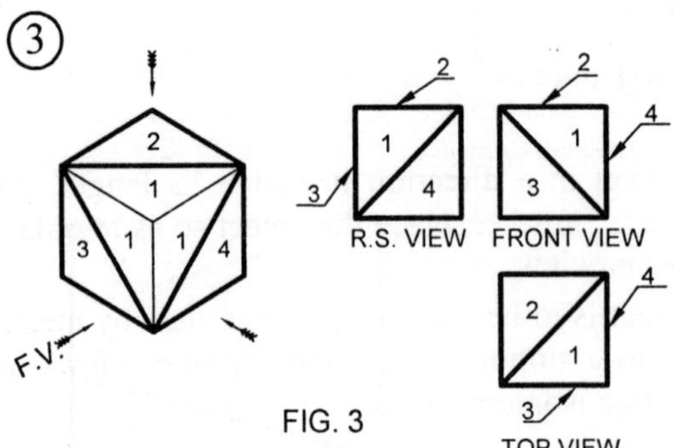

FIG. 3
OBLIQUE SURFACES

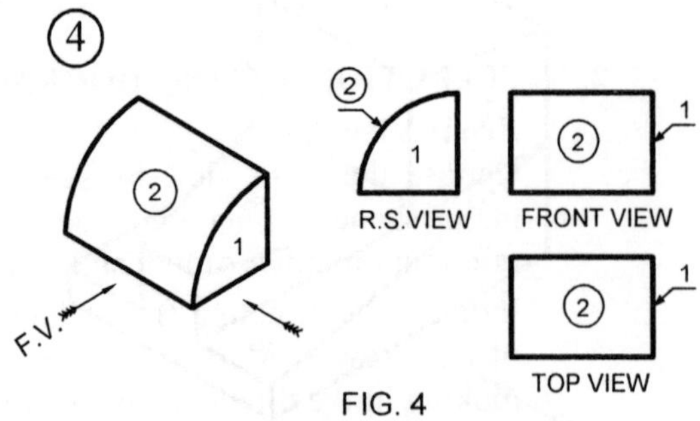

FIG. 4
CURVED SURFACE

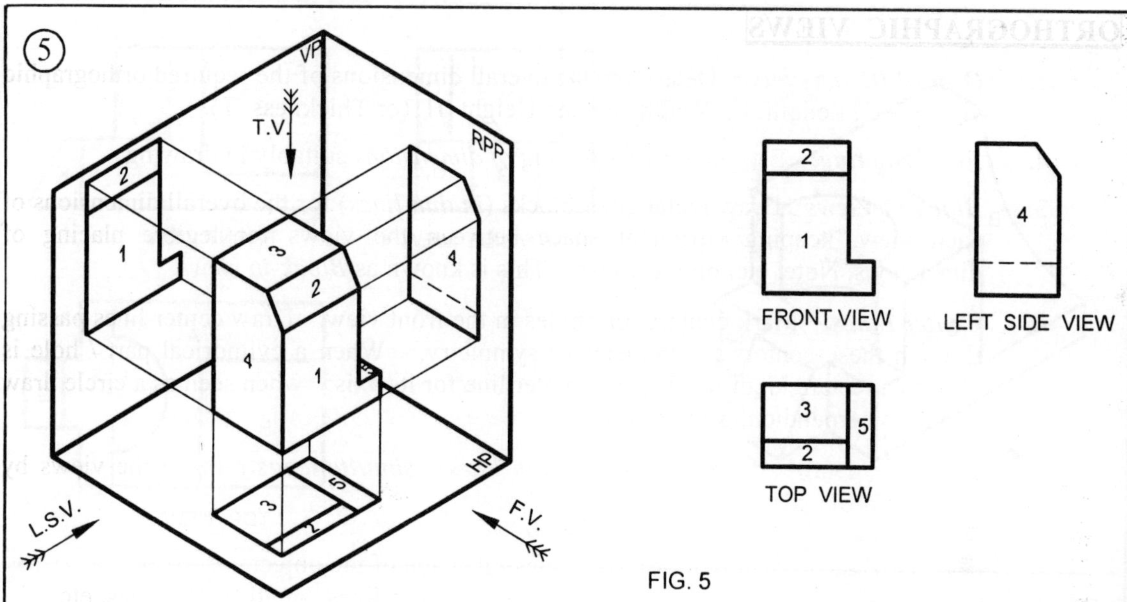

FIG. 5

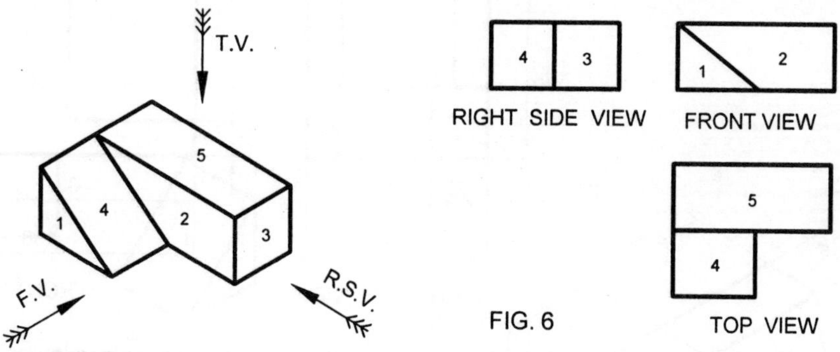

FIG. 6

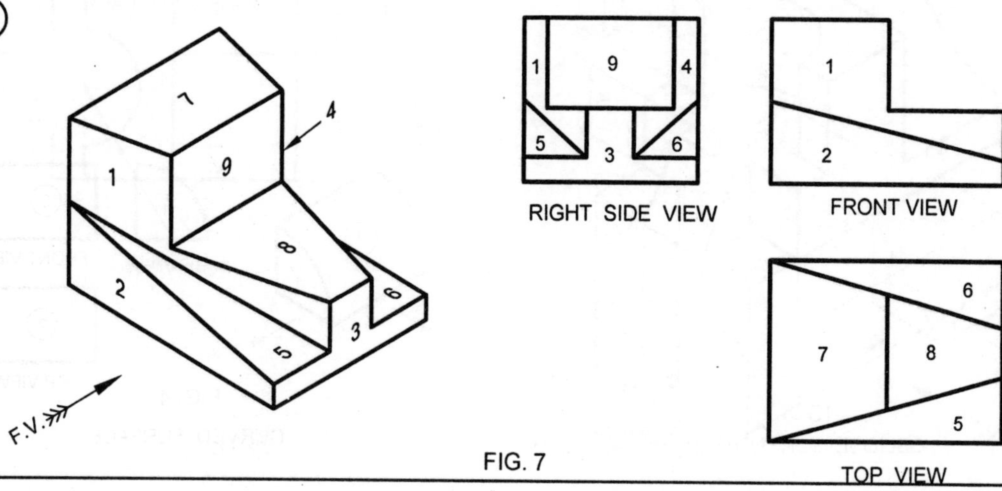

FIG. 7

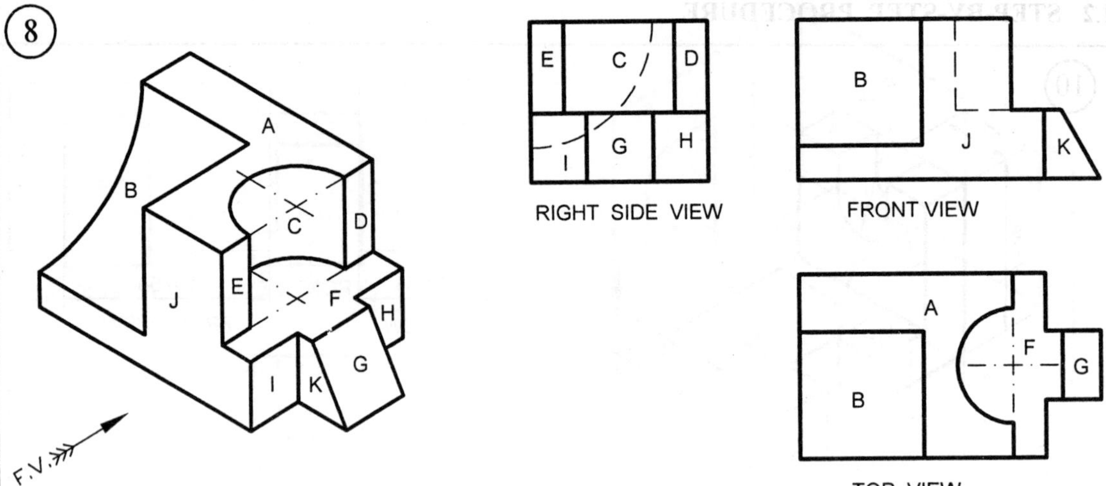
8
A
B
C
D
E
F
G
H
I
J
K
F.V.
RIGHT SIDE VIEW
FRONT VIEW
TOP VIEW

FIG. 8

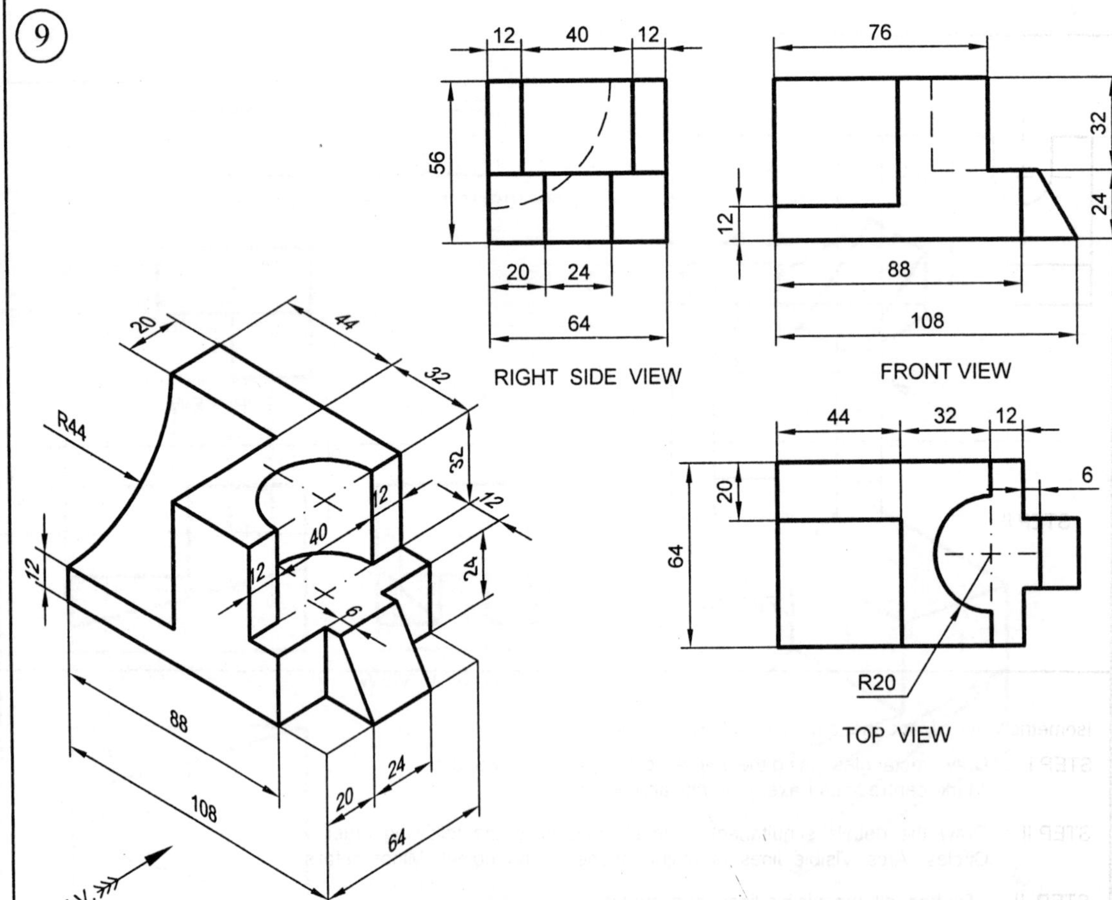
9
RIGHT SIDE VIEW
FRONT VIEW
TOP VIEW
ISOMETRIC VIEW
R44
R20
F.V.

FIG. 9

1.2 STEP-BY-STEP PROCEDURE

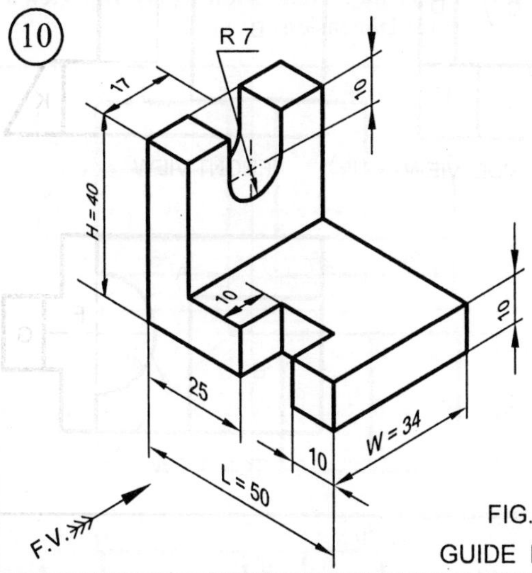

FIG. 10
GUIDE BLOCK

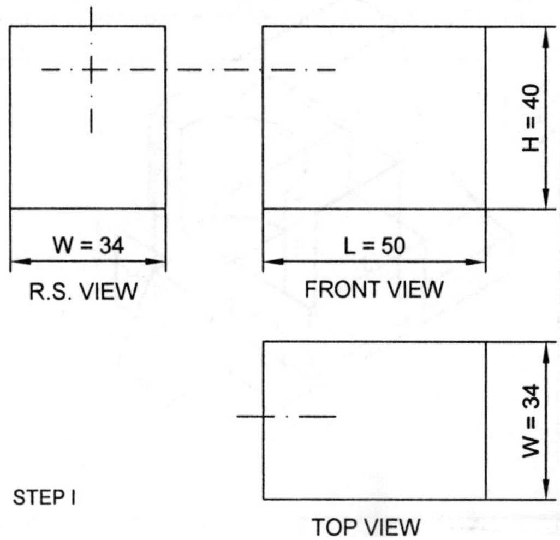

STEP I

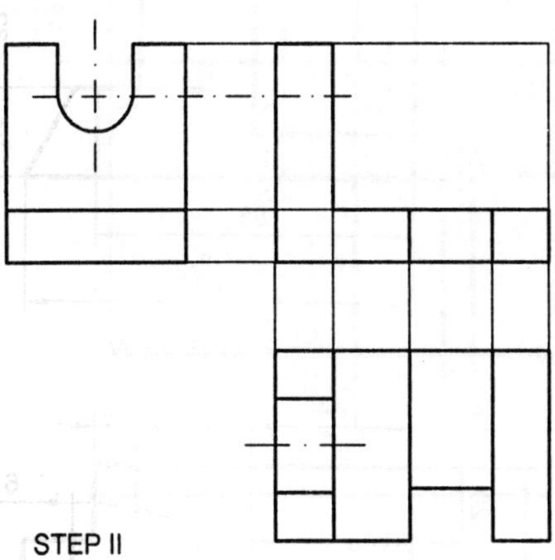

STEP II

FIG. 10

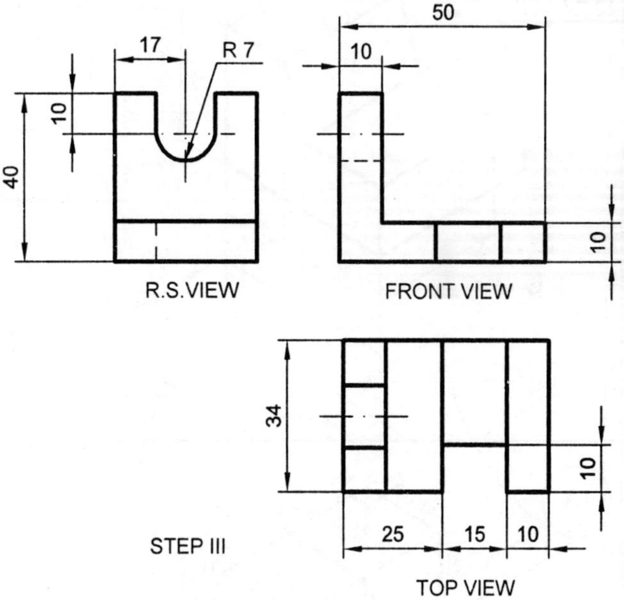

STEP III

Isometric View: Mark the directions of Front, Top and Right Side Views.

STEP I : Draw rectangles using the overall dimensions L, W and H.
Mark centre lines (axes) at the appropriate places.

STEP II : Draw the details simultaneously in all 3 views in the following order : Circles / Arcs, Visible lines of major shape of the object, Minor details.

STEP III : Darken all the visible lines and curves.
Mark the invisible lines. Add the dimensions. Errase the projectors.

2. TYPICAL PRACTICAL PROBLEMS - ENGINEERING COMPONENTS

Draw the following orthographic views for the given isometric views : (a) Front View , (b) Top View and (c) Side View as viewed from the side available for view. Mark the Dimensions.

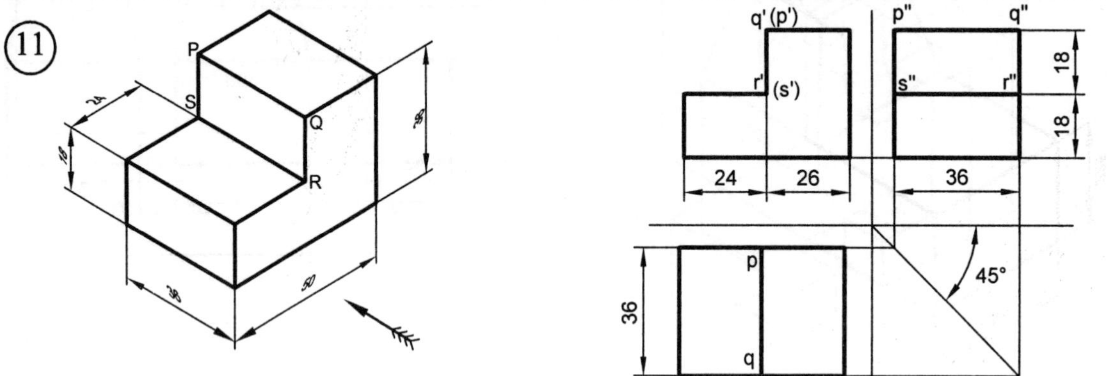

FIG. 11

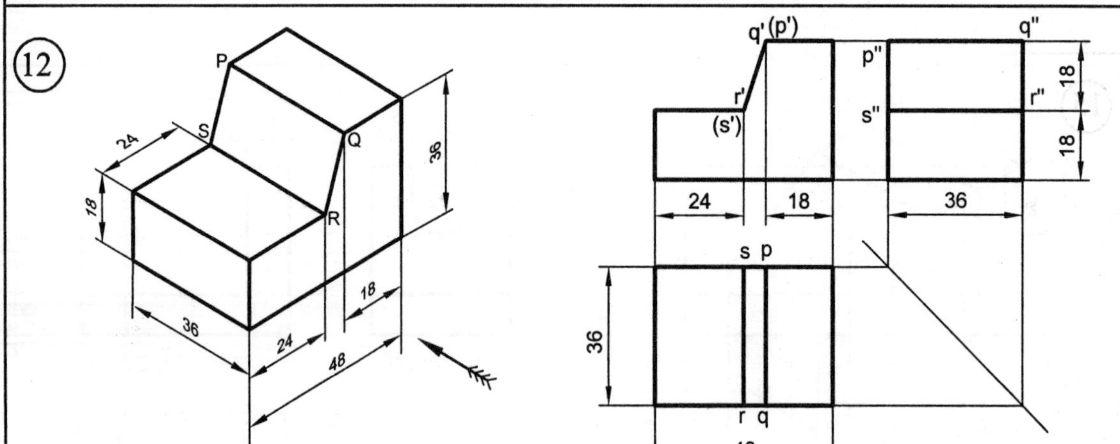

FIG. 12

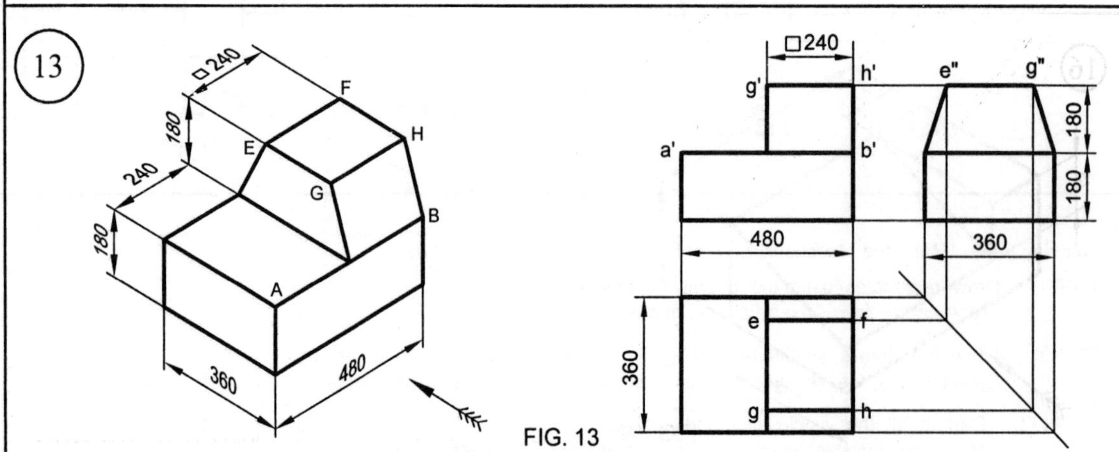

FIG. 13

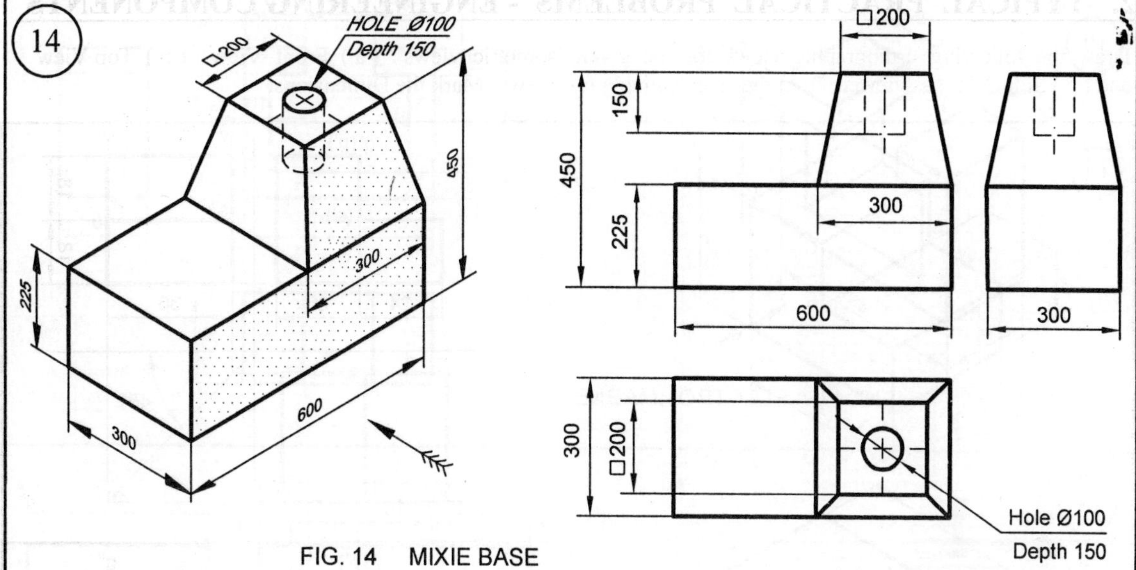

FIG. 14 MIXIE BASE

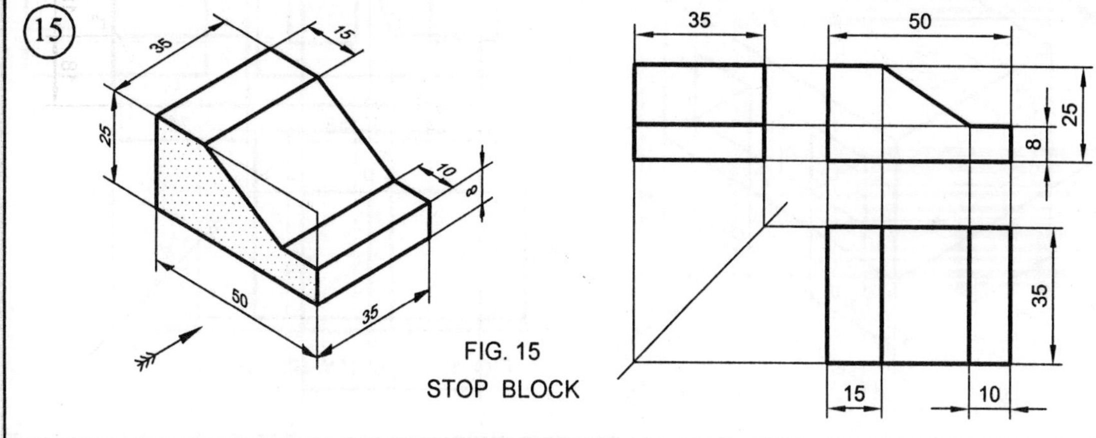

FIG. 15
STOP BLOCK

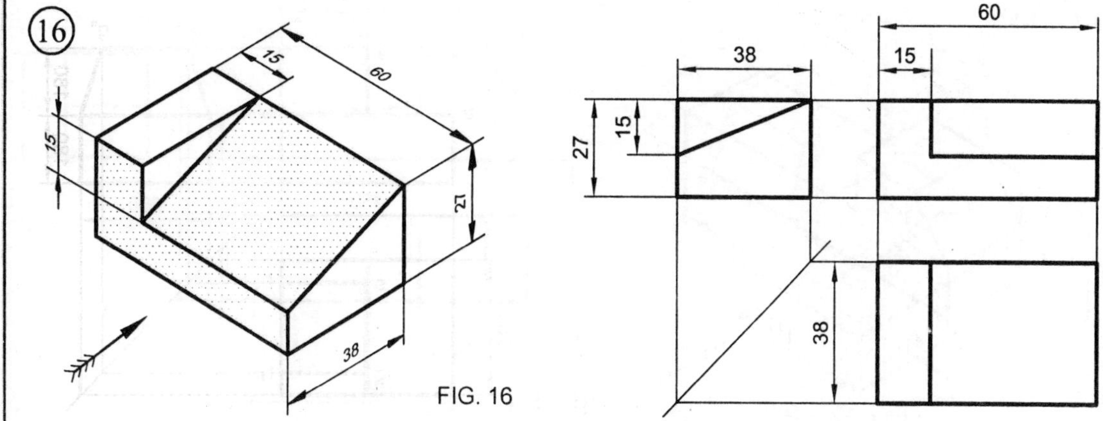

FIG. 16

(17)

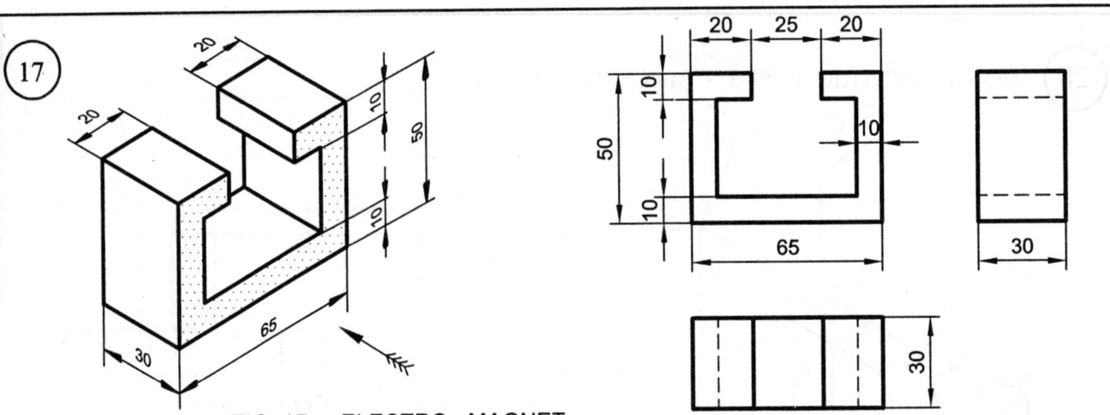

FIG. 17 ELECTRO - MAGNET

(18)

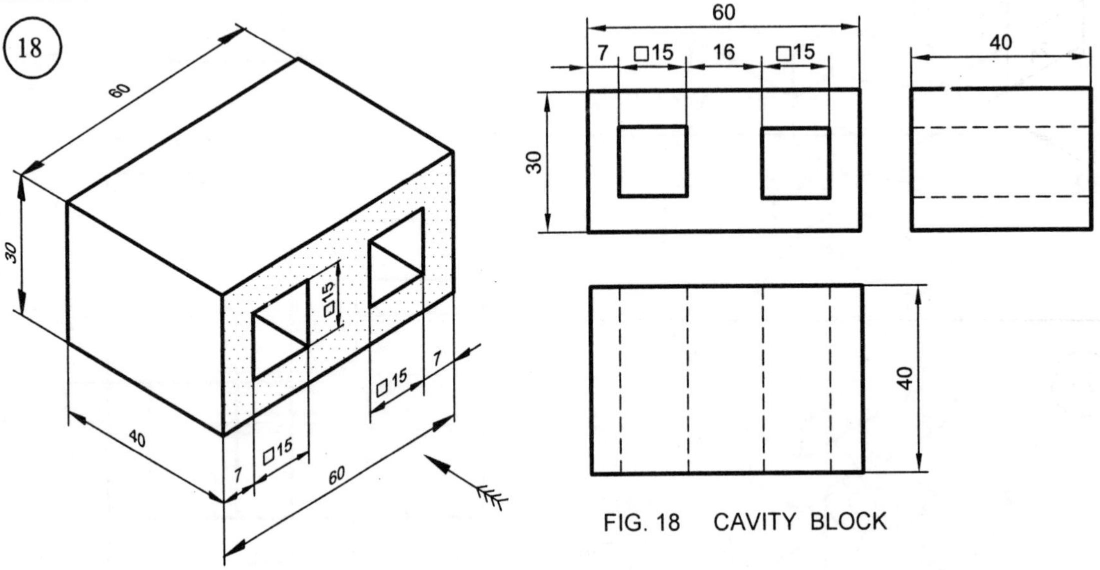

FIG. 18 CAVITY BLOCK

(19) (UQ)

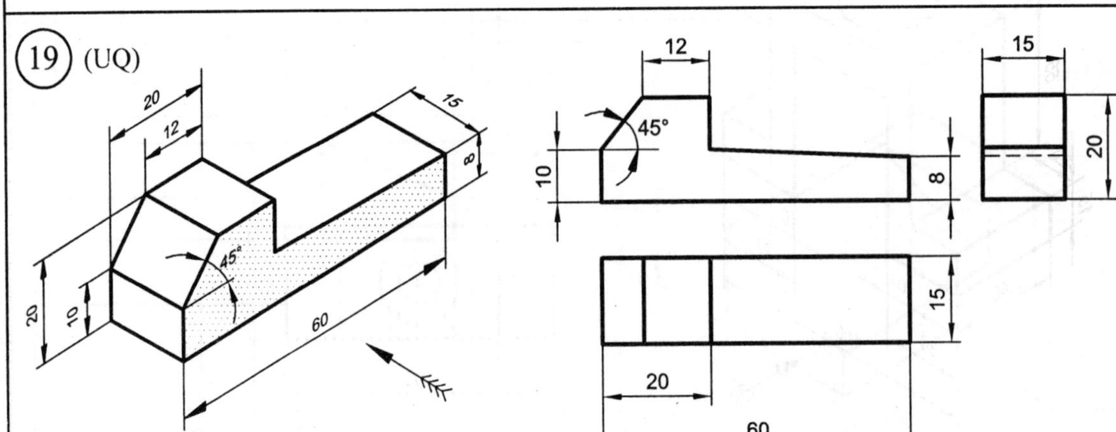

FIG. 19 GIB - HEADED KEY

(20) NON-ISOMETRIC LINES - MACHINE BLOCK (UQ)

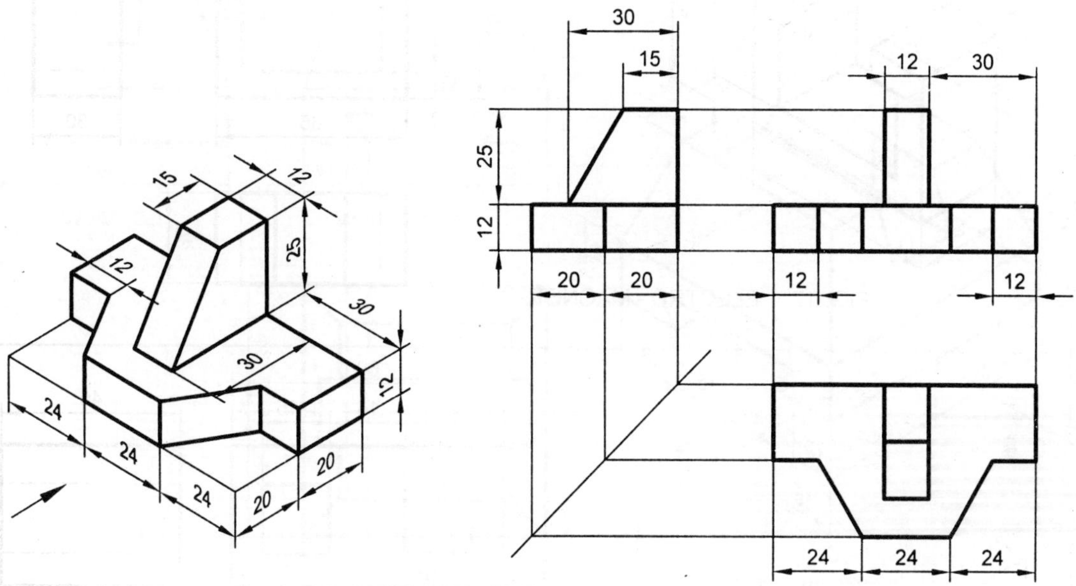

FIG. 20

(21)

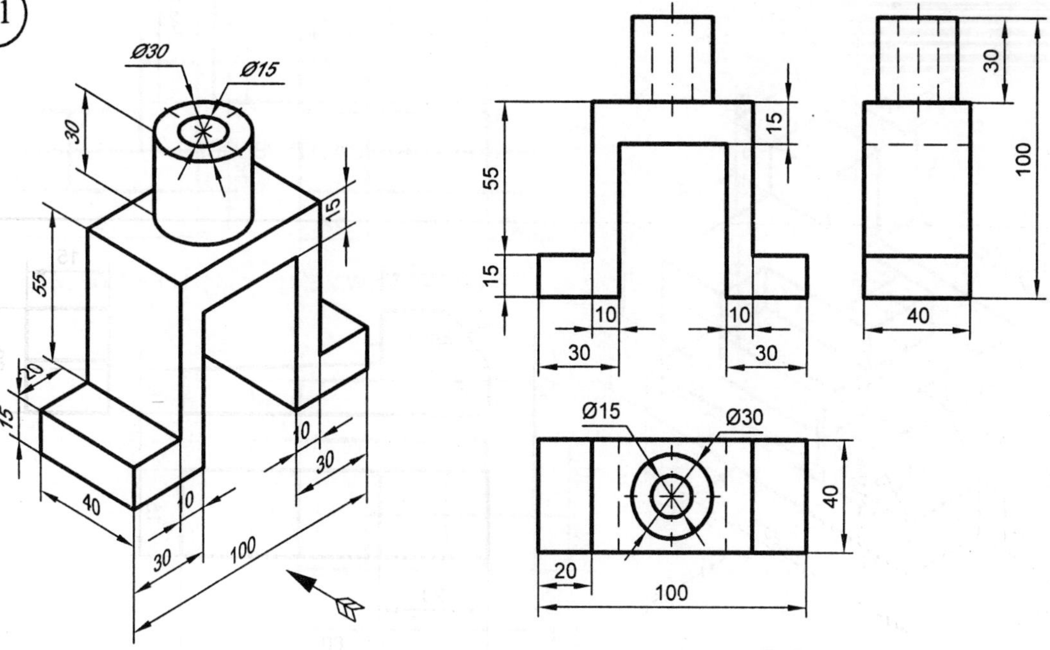

FIG. 21 SUPPORT (UQ)

(22) TAPER BLOCK (UQ)

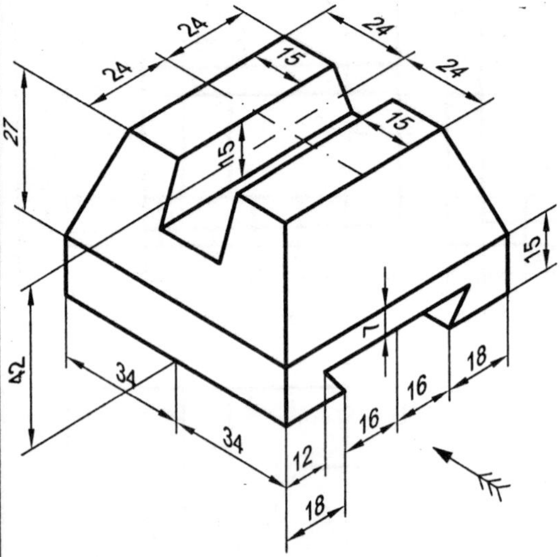

FIG. 22

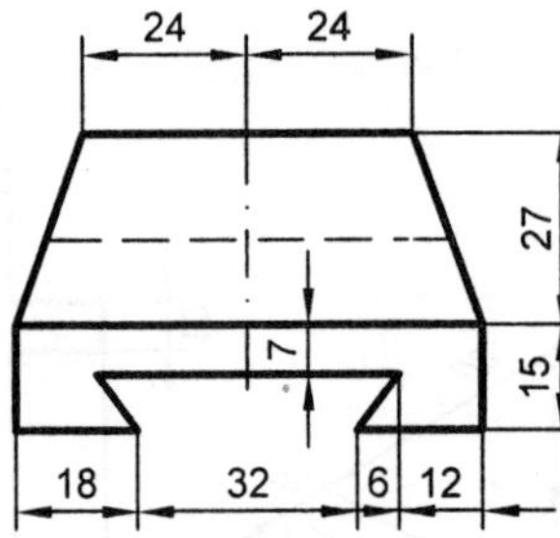

FRONT VIEW

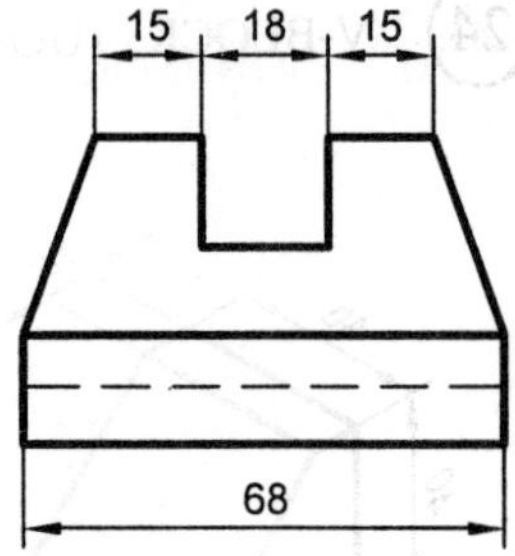

L.S. VIEW

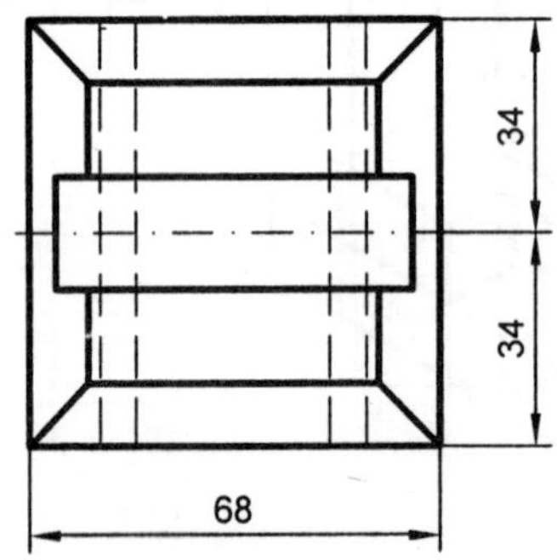

TOP VIEW

(23) BUILDING BLOCK (UQ)

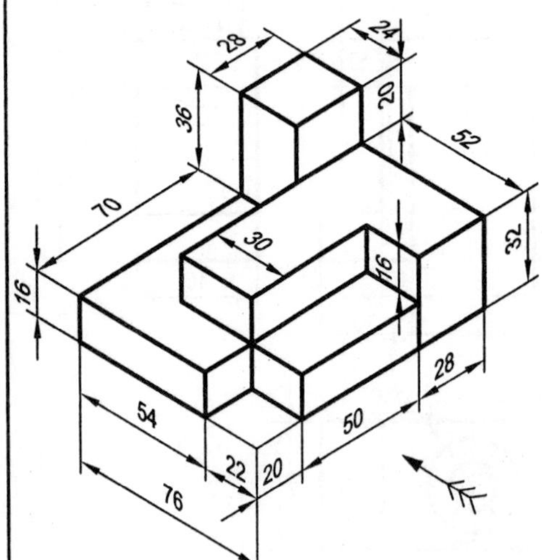

Scale 1 : 2

FIG. 23

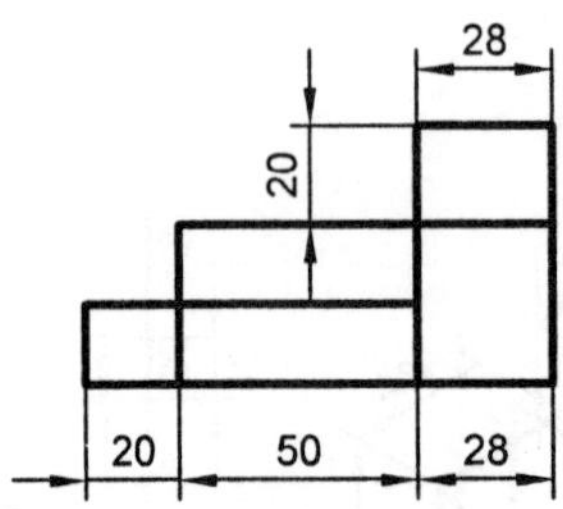

FRONT VIEW

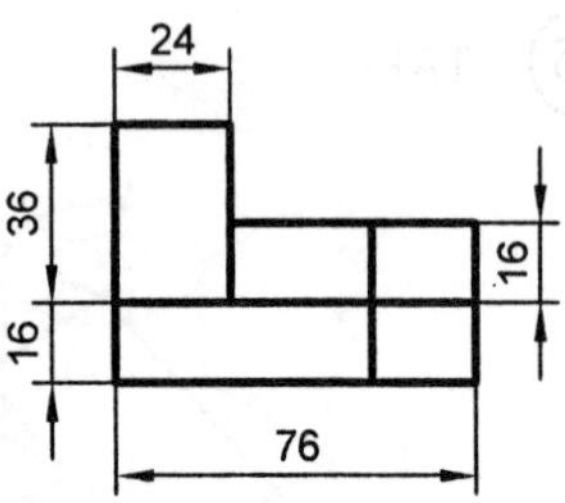

LEFT SIDE VIEW

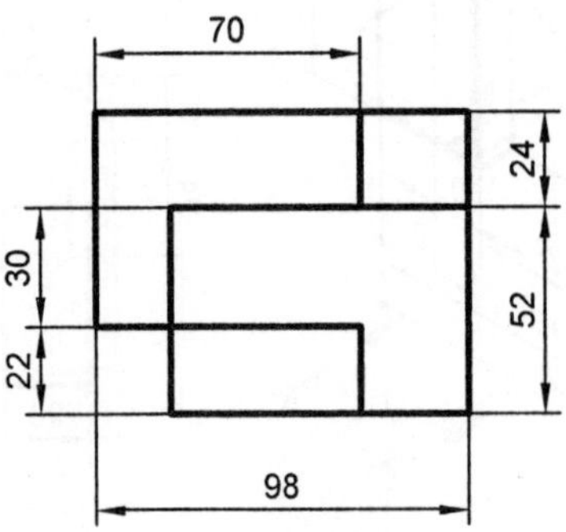

TOP VIEW

(24) V-BLOCK (UQ)

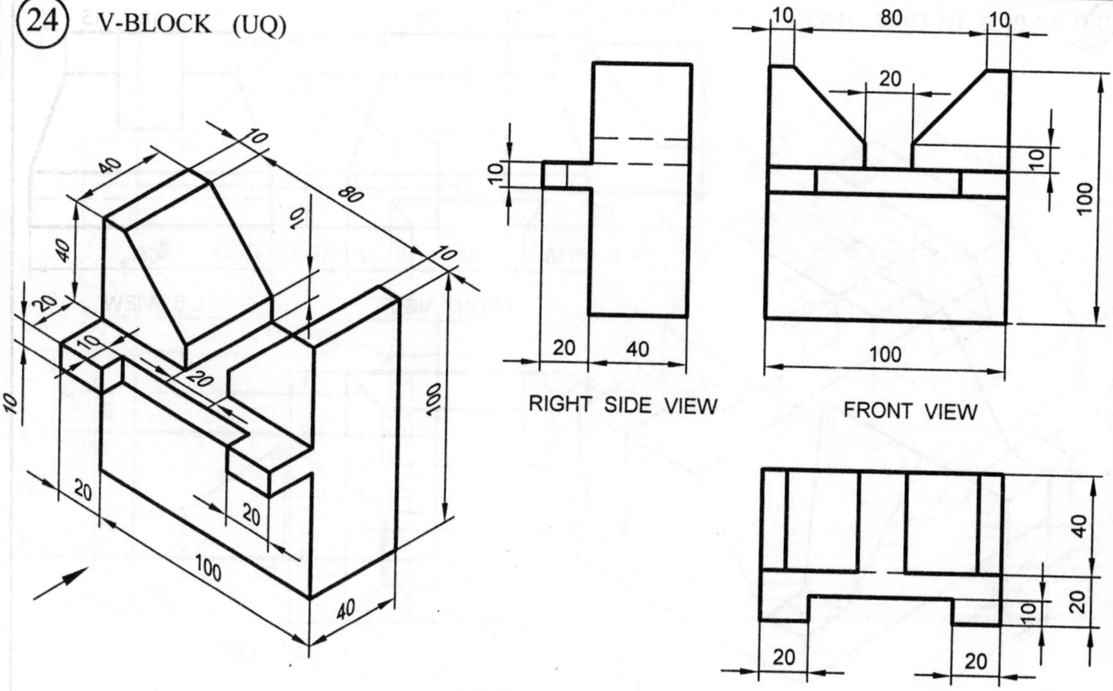

FIG. 24

(25) TAPER BLOCK (UQ)

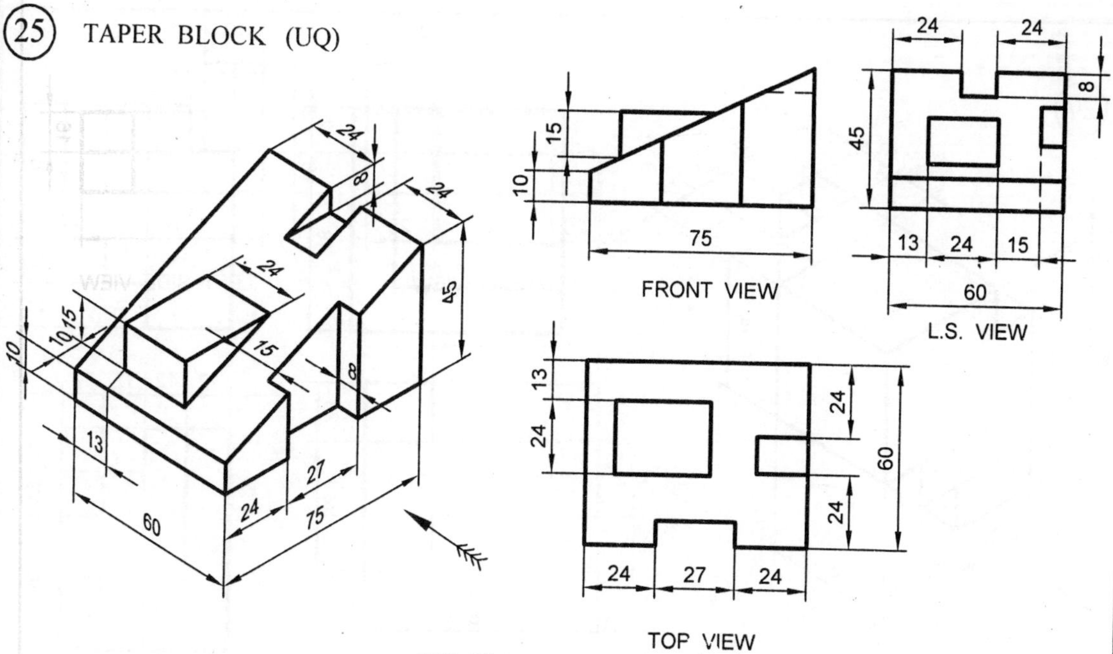

FIG. 25

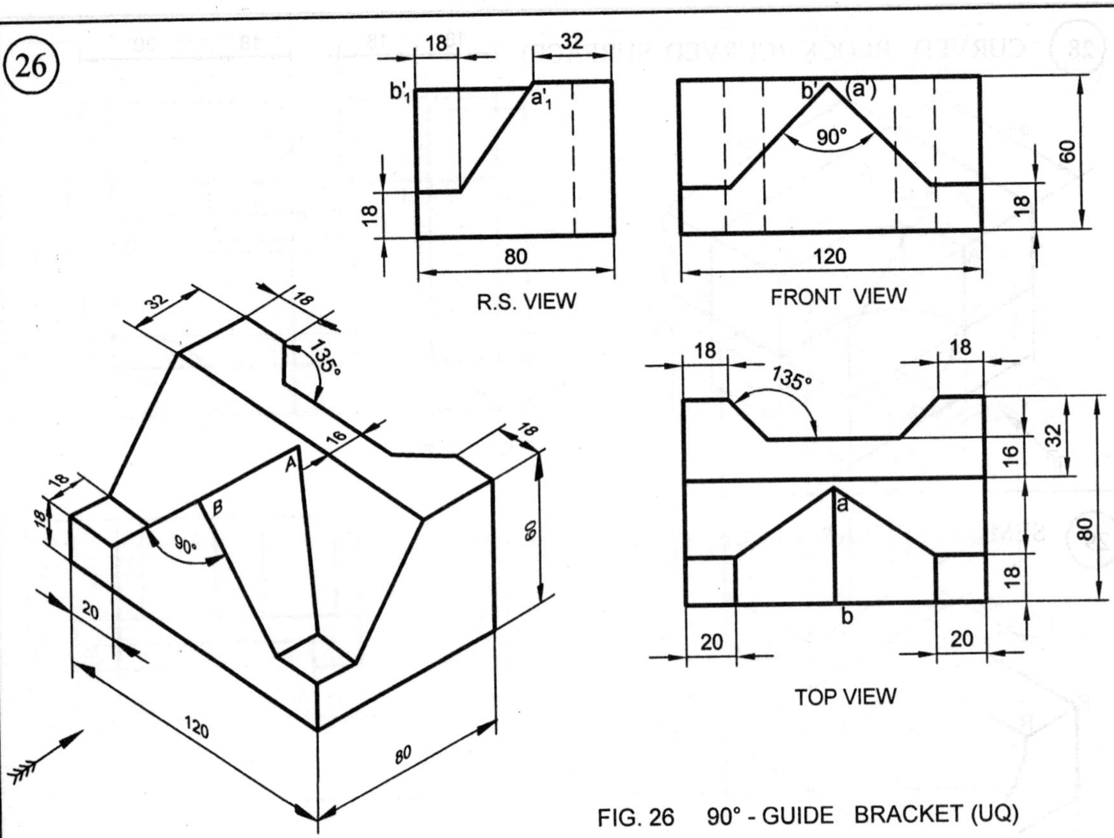

FIG. 26 90° - GUIDE BRACKET (UQ)

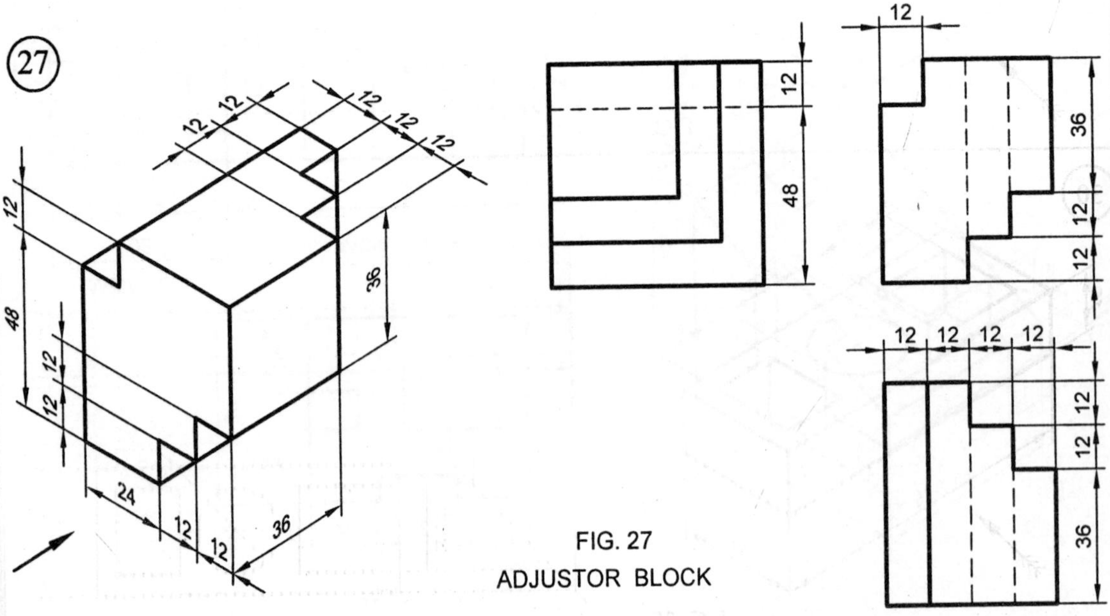

FIG. 27
ADJUSTOR BLOCK

(28) CURVED BLOCK (CURVED SURFACE)

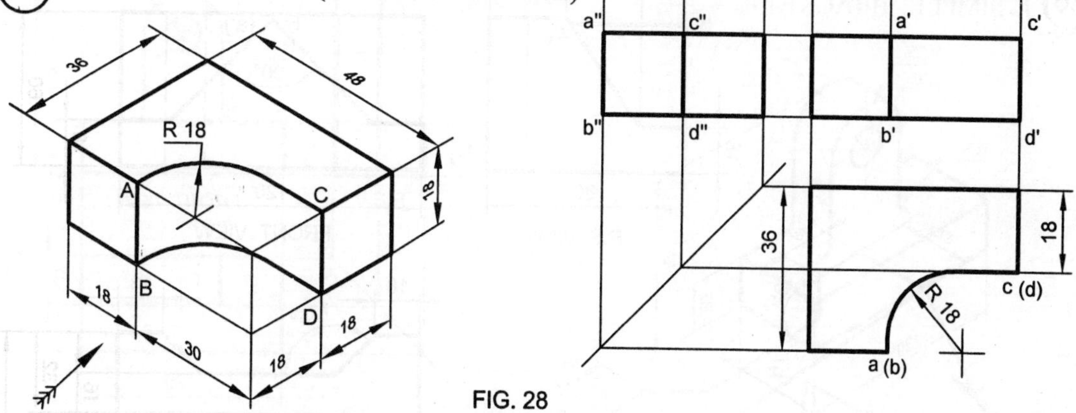

FIG. 28

(29) SEMI - CIRCULAR BLOCK (CURVED SURFACE)

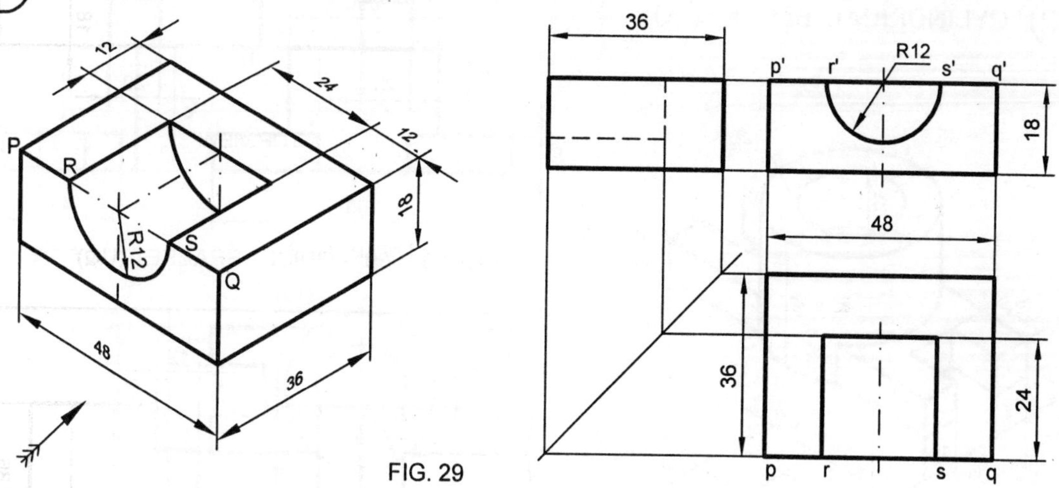

FIG. 29

(30) (UQ)

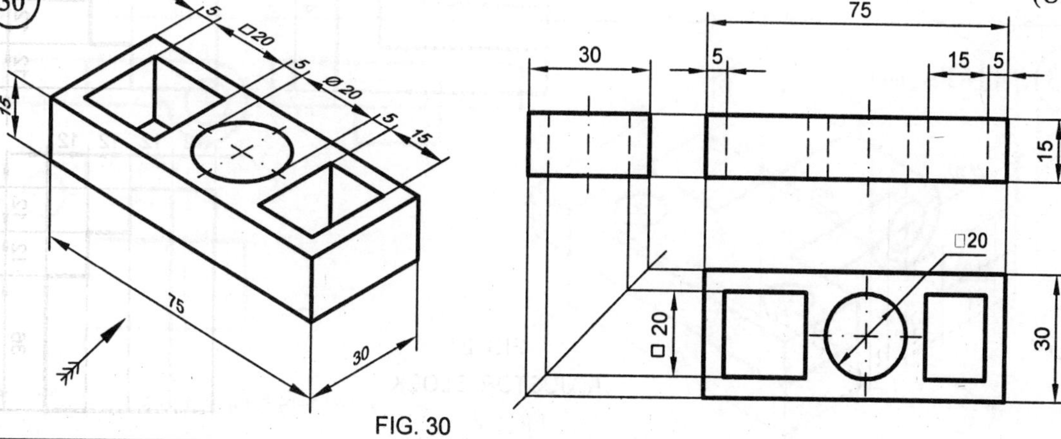

FIG. 30

31 SIMPLE BRACKET

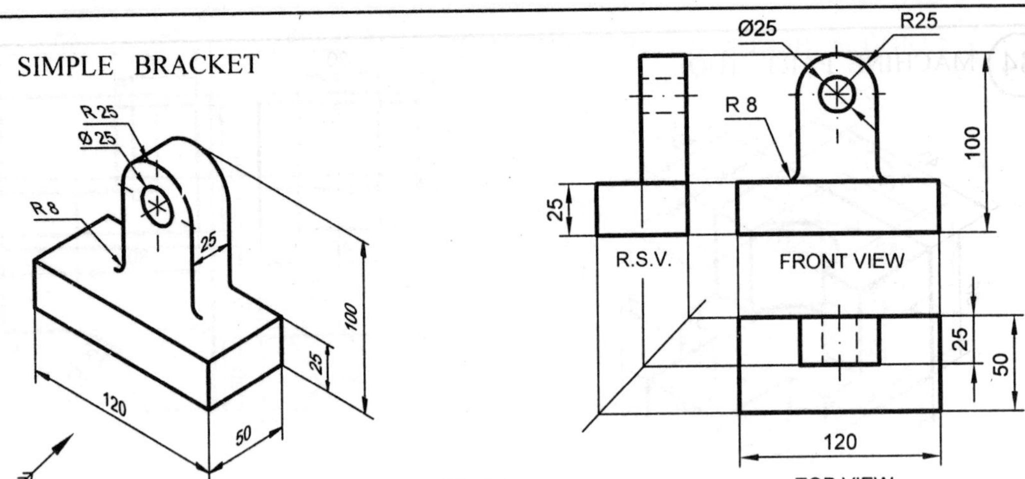

FIG. 31

32 CYLINDRICAL BLOCK (UQ)

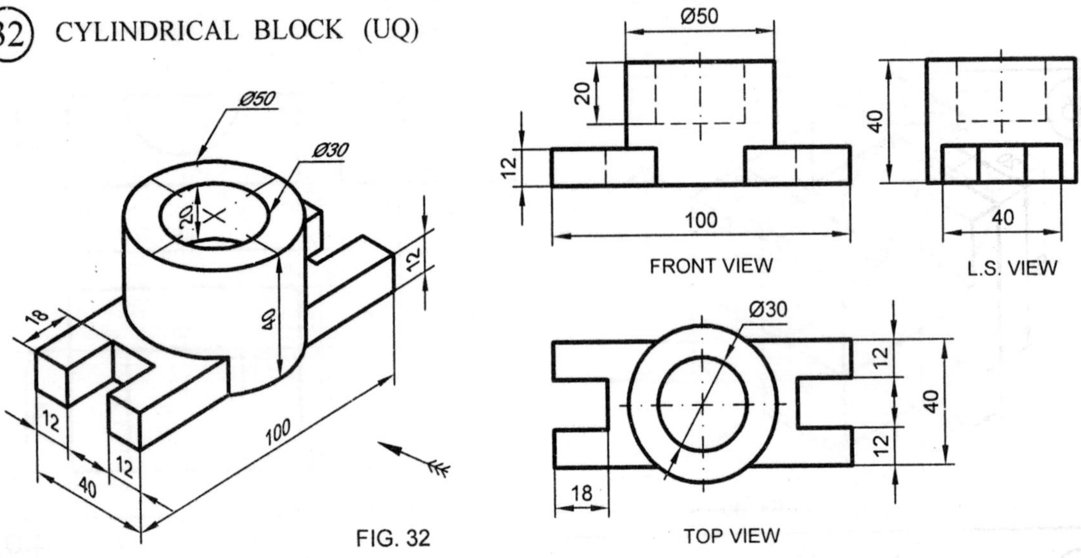

FIG. 32

33 HINGE (UQ)

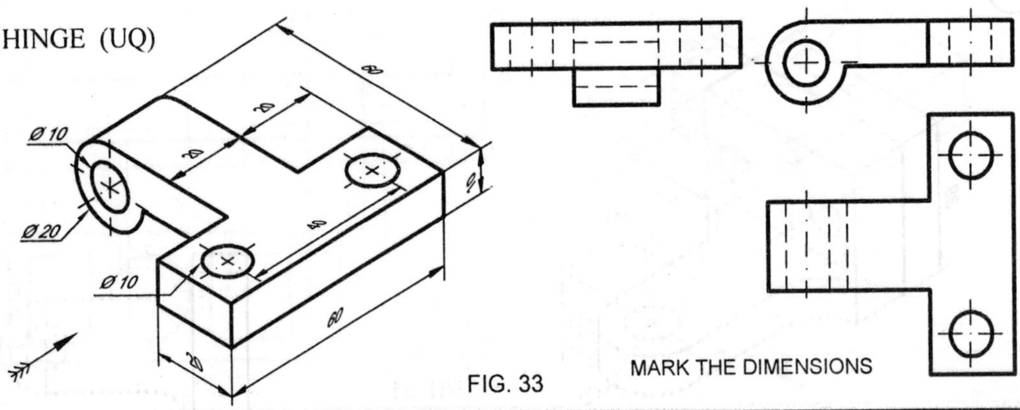

FIG. 33

(34) MACHINE PART (UQ)

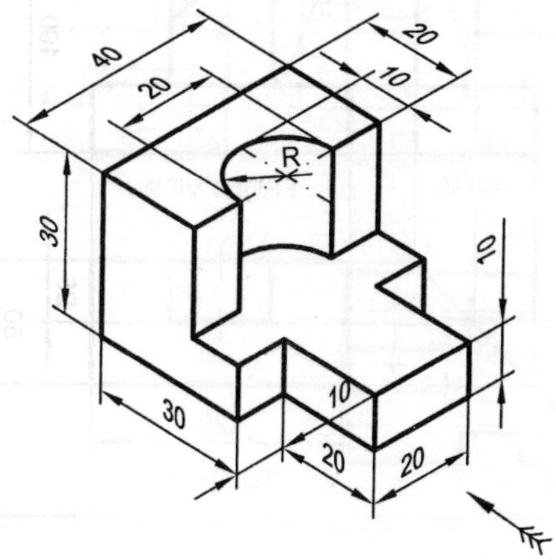

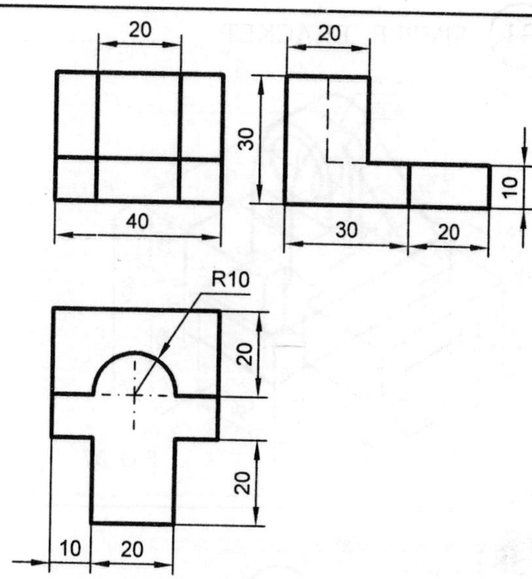

FIG. 34

(35)

(UQ)

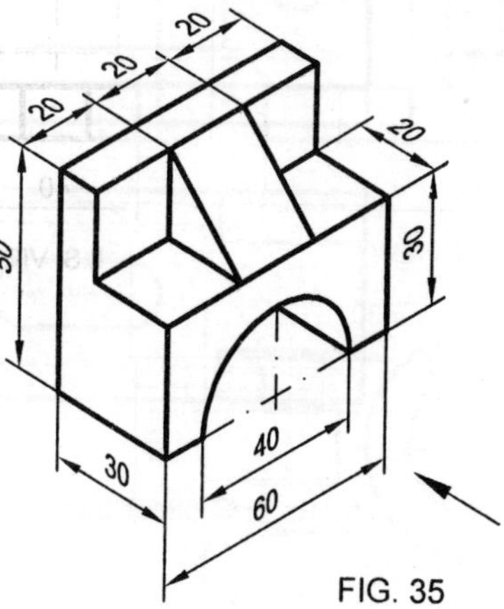

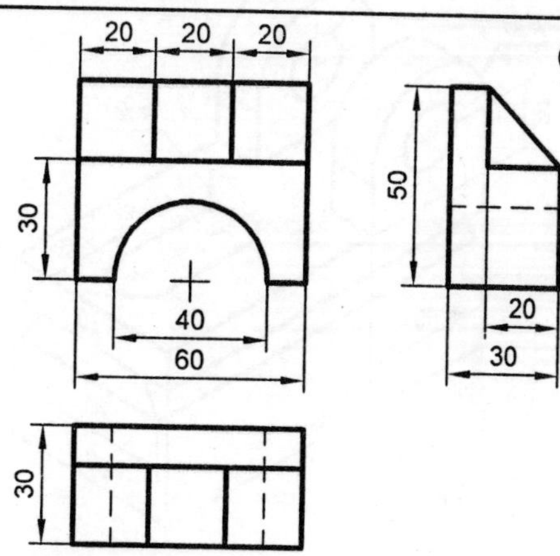

FIG. 35

(36) FIXTURE (UQ)

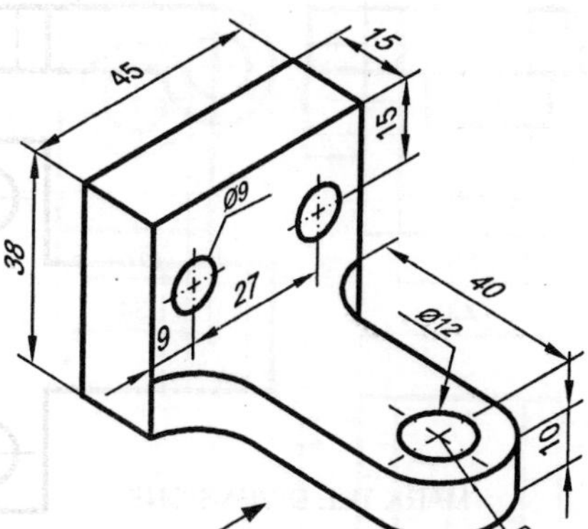

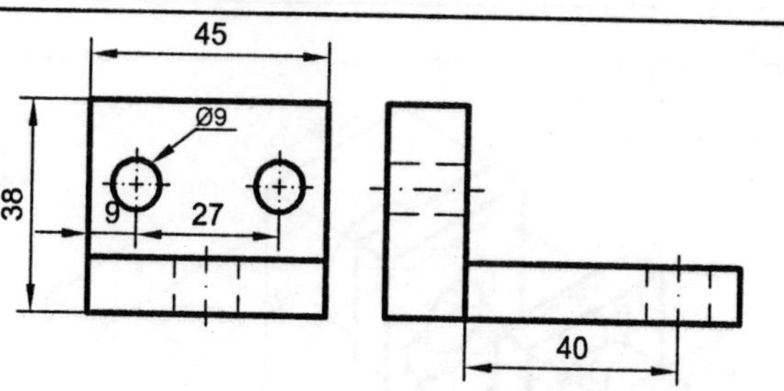

FIG. 36

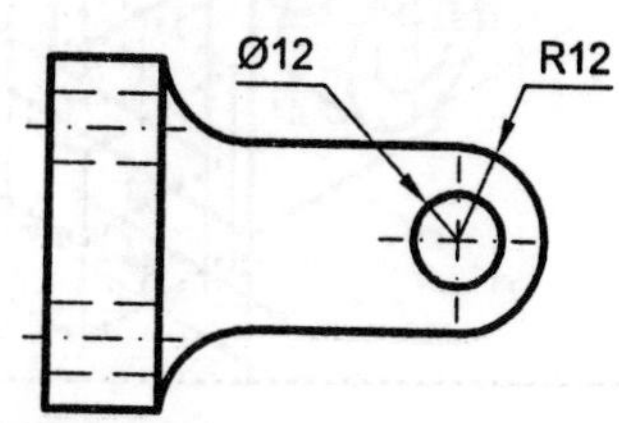

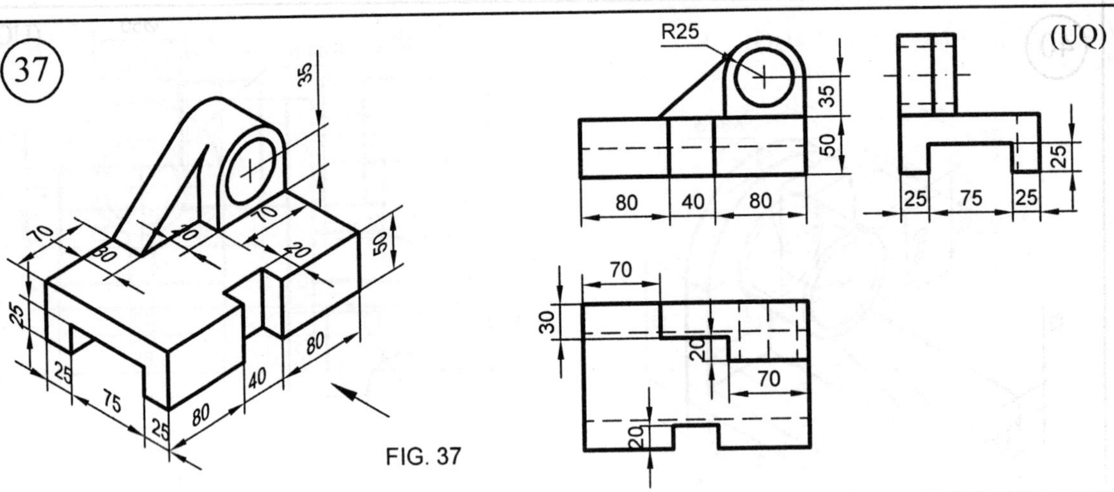

FIG. 37

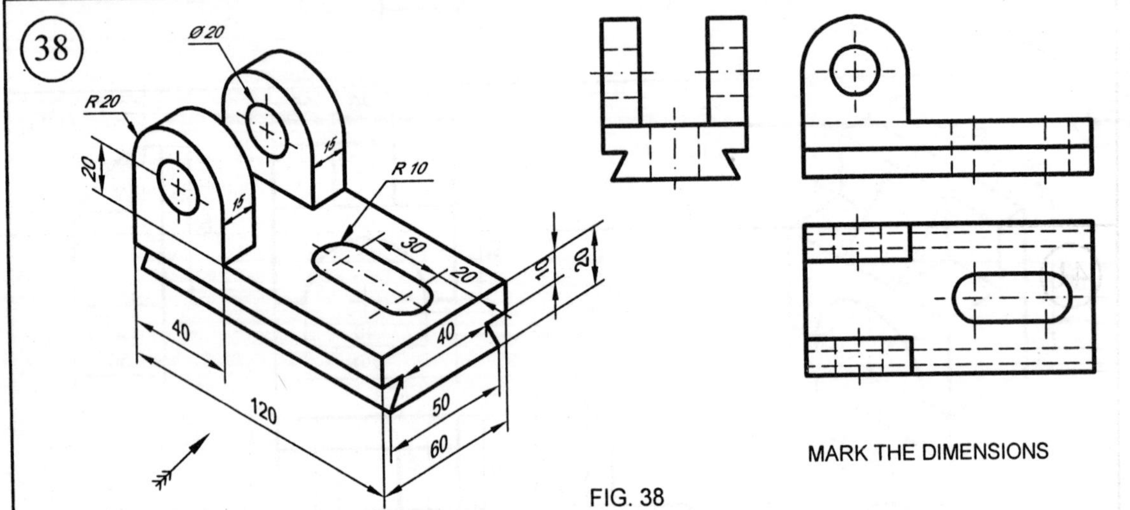

MARK THE DIMENSIONS

FIG. 38

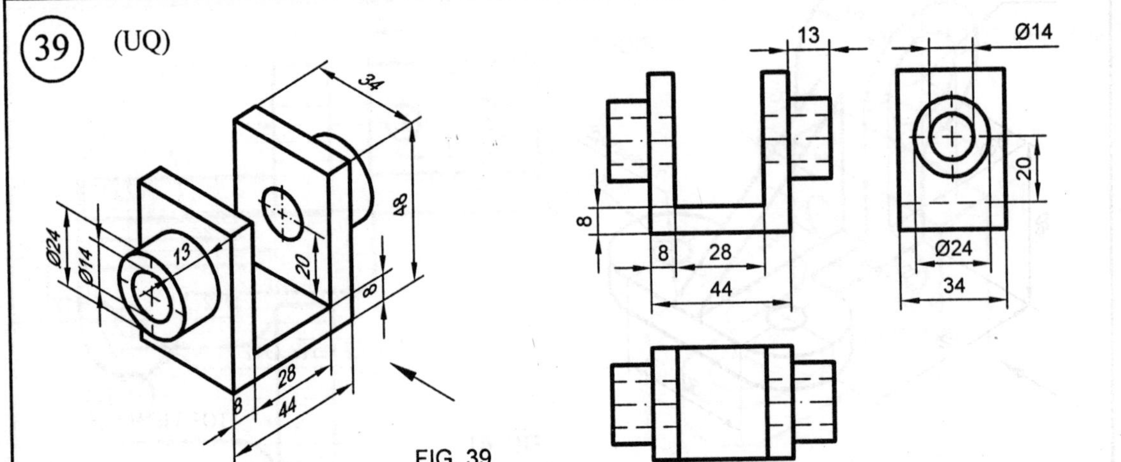

FIG. 39

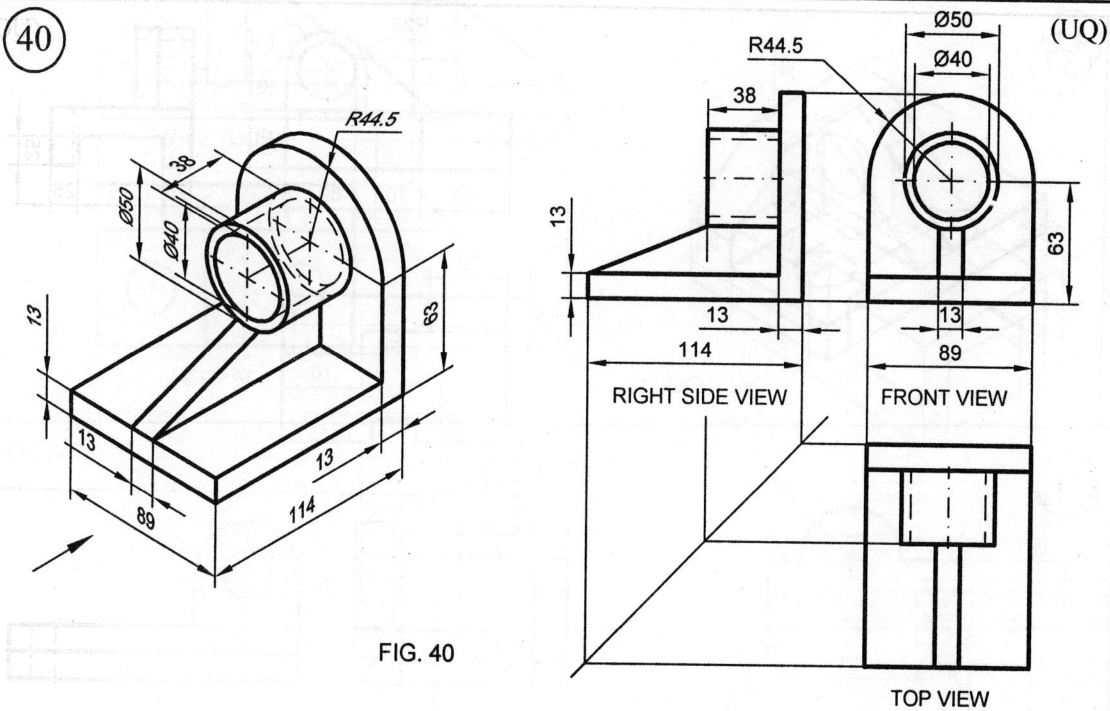
40
(UQ)
R44.5
38
Ø50
Ø40
13
63
13
13
89
114
Ø50
R44.5
Ø40
38
13
63
13
13
114
89
RIGHT SIDE VIEW
FRONT VIEW
TOP VIEW

FIG. 40

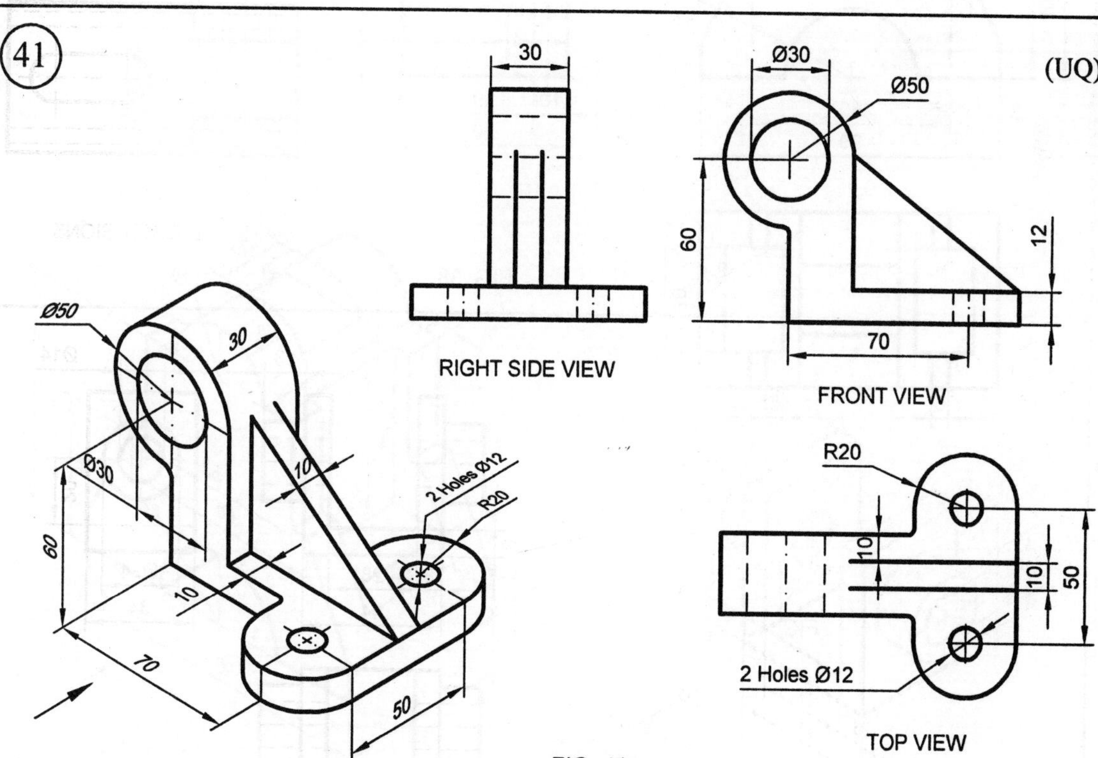
41
(UQ)
30
Ø30
Ø50
60
12
70
RIGHT SIDE VIEW
FRONT VIEW
Ø50
30
Ø30
10
2 Holes Ø12
R20
60
10
70
50
R20
10
10
50
2 Holes Ø12
TOP VIEW

FIG. 41

(UQ)

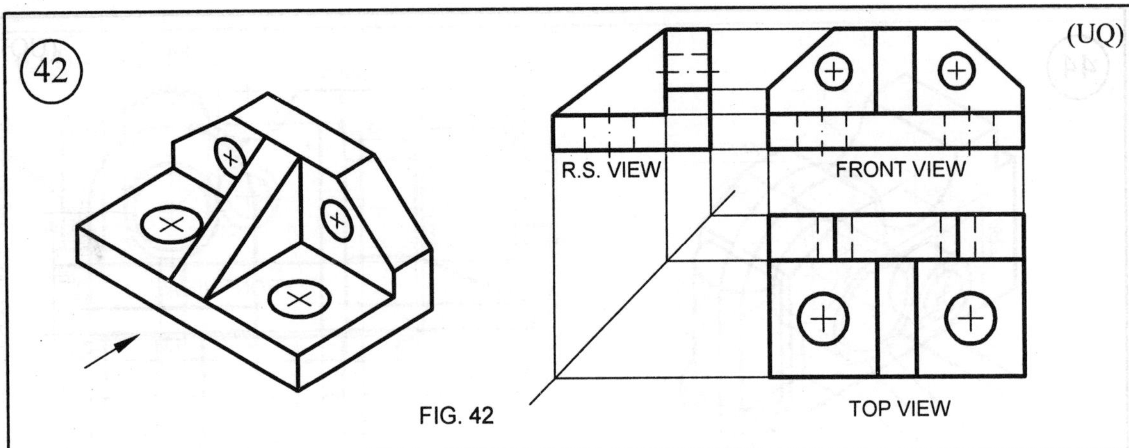

FIG. 42

(UQ)

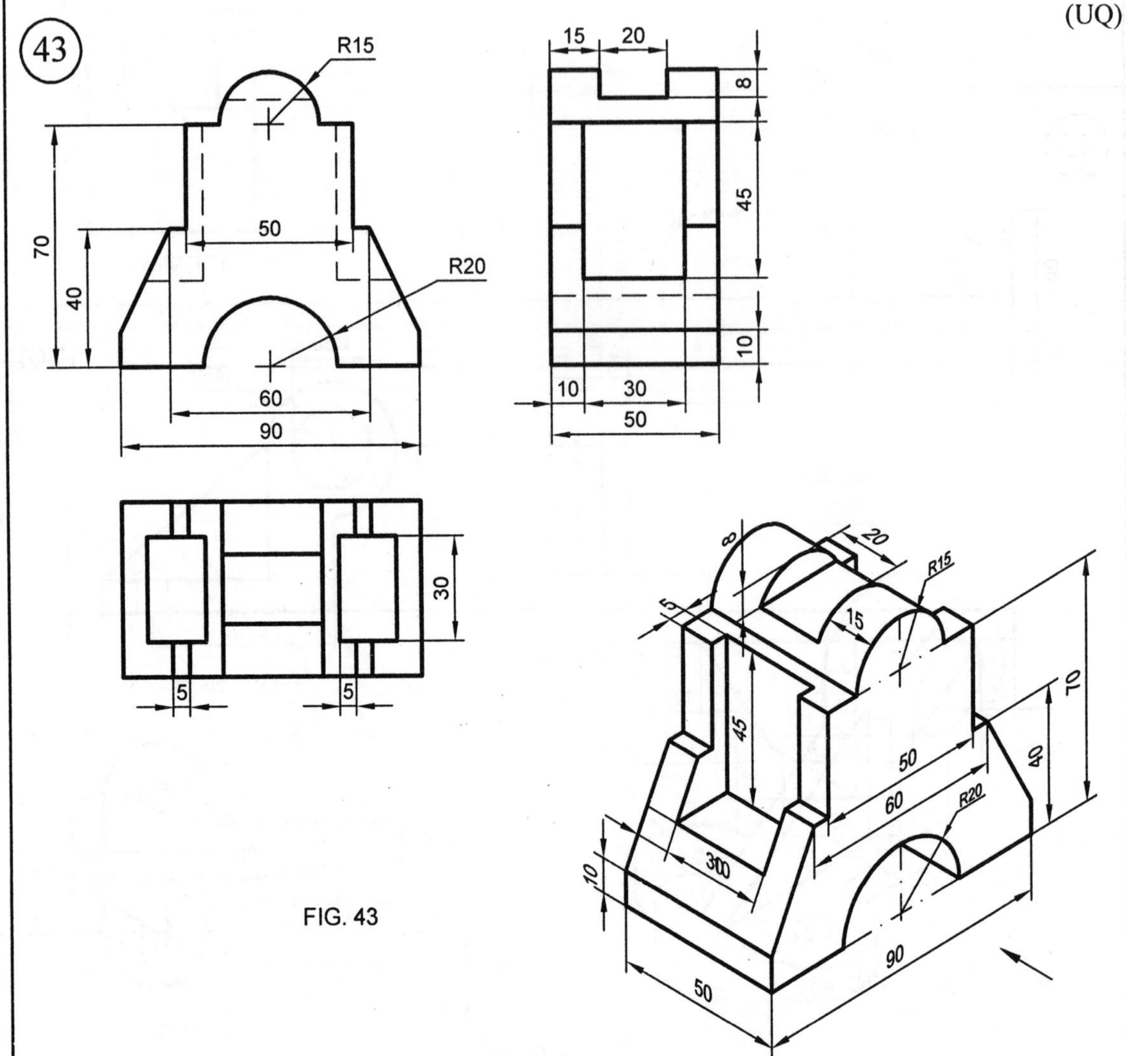

FIG. 43

44

(UQ)

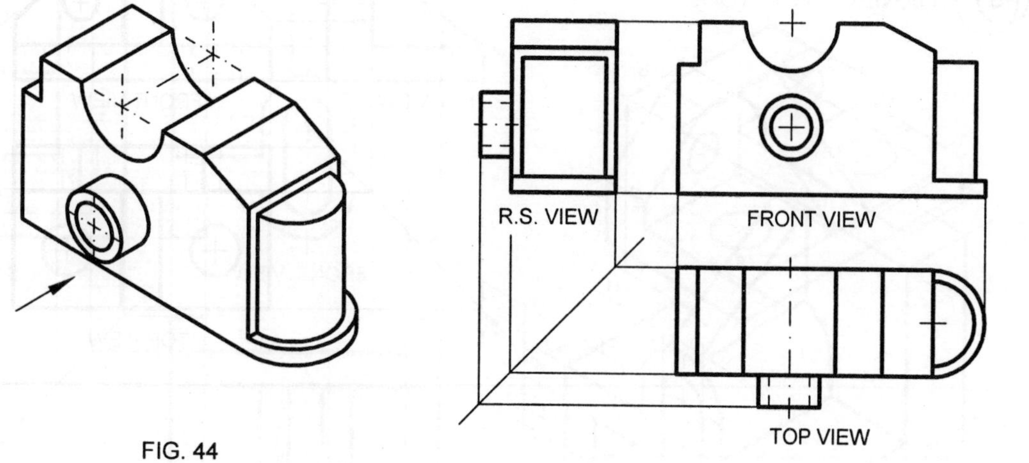

FIG. 44

45

(UQ)

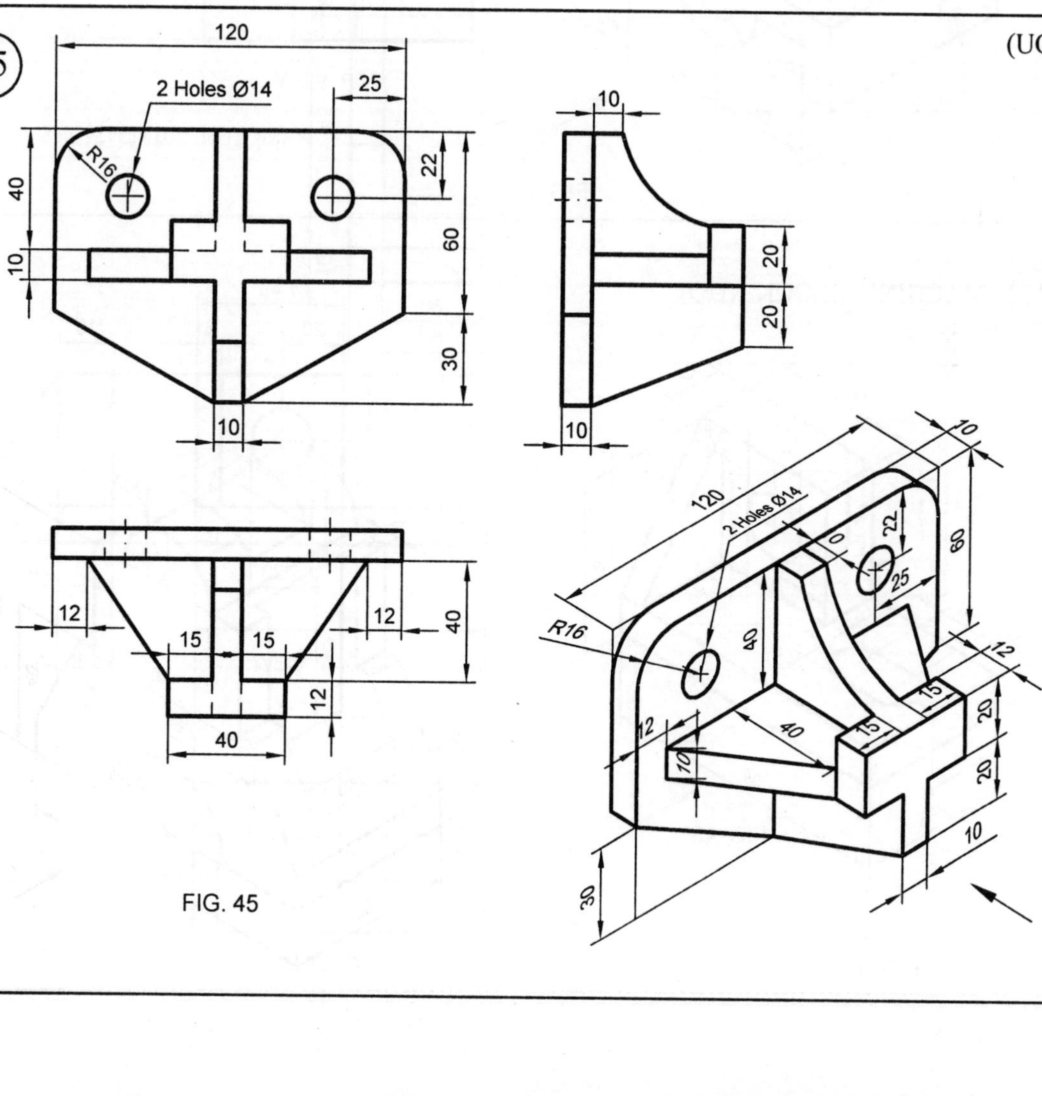

FIG. 45

(46) CORNER STOP (UQ)

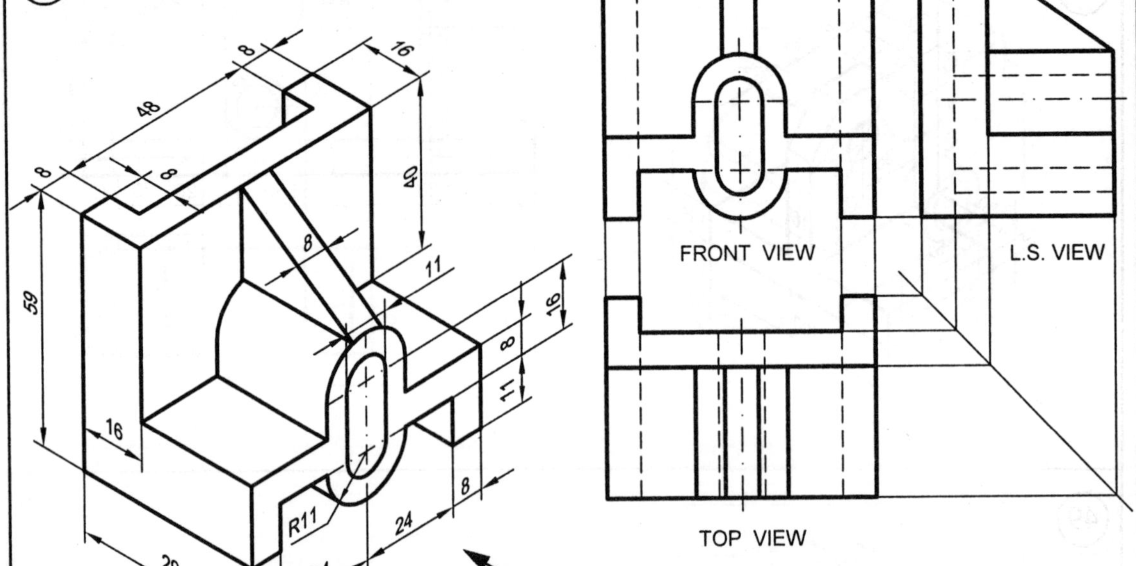

MARK THE DIMENSIONS

FIG. 46

(47) MACHINE BLOCK (UQ)

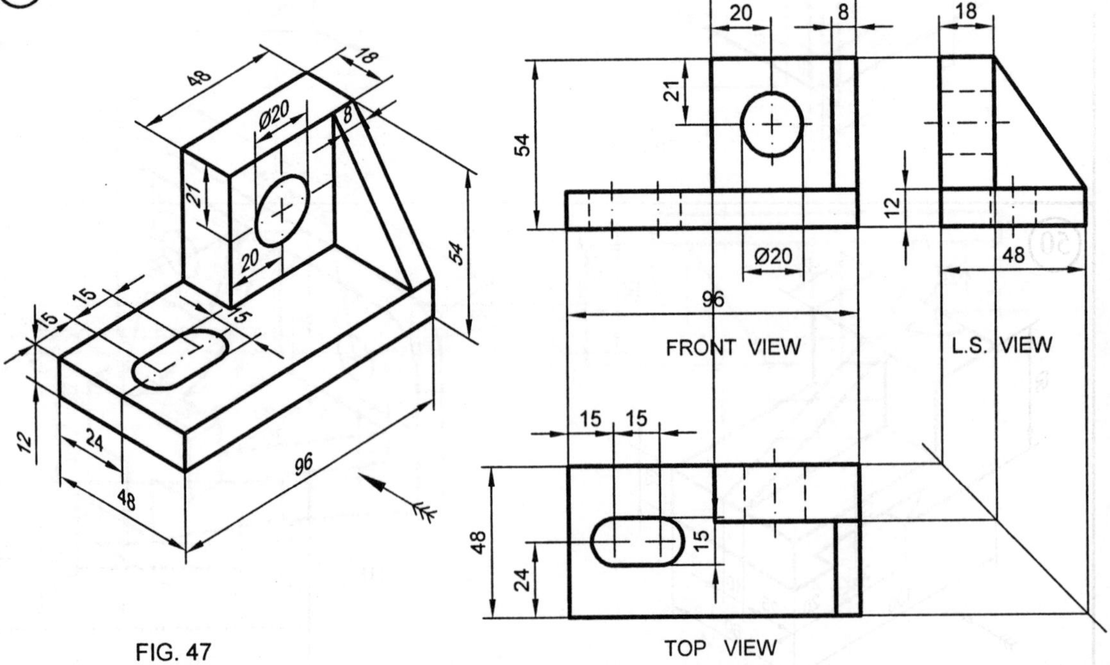

FIG. 47

(48)

(UQ)

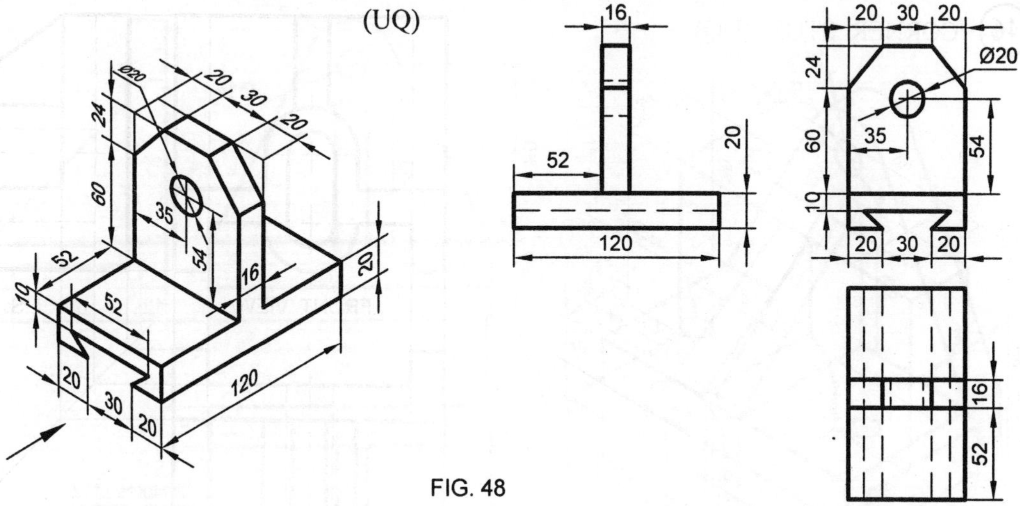

FIG. 48

(49)

(UQ)

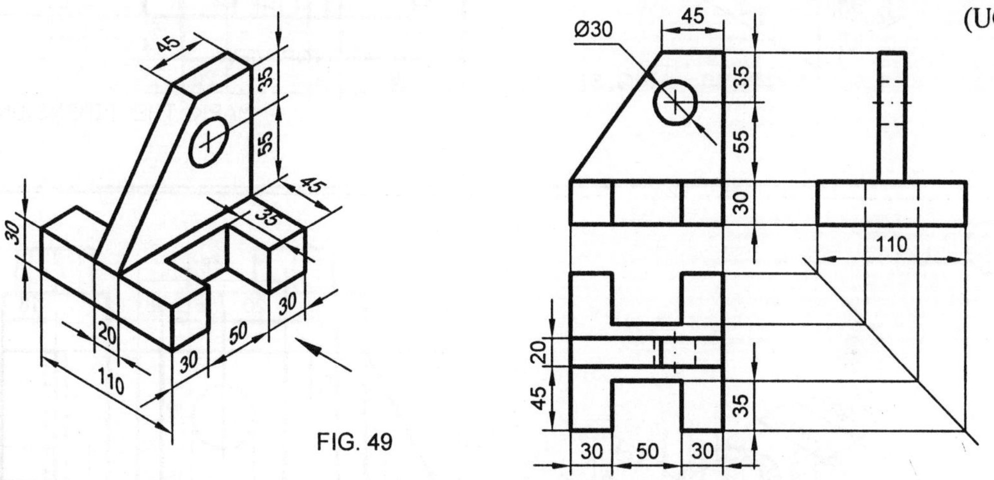

FIG. 49

(50)

(UQ)

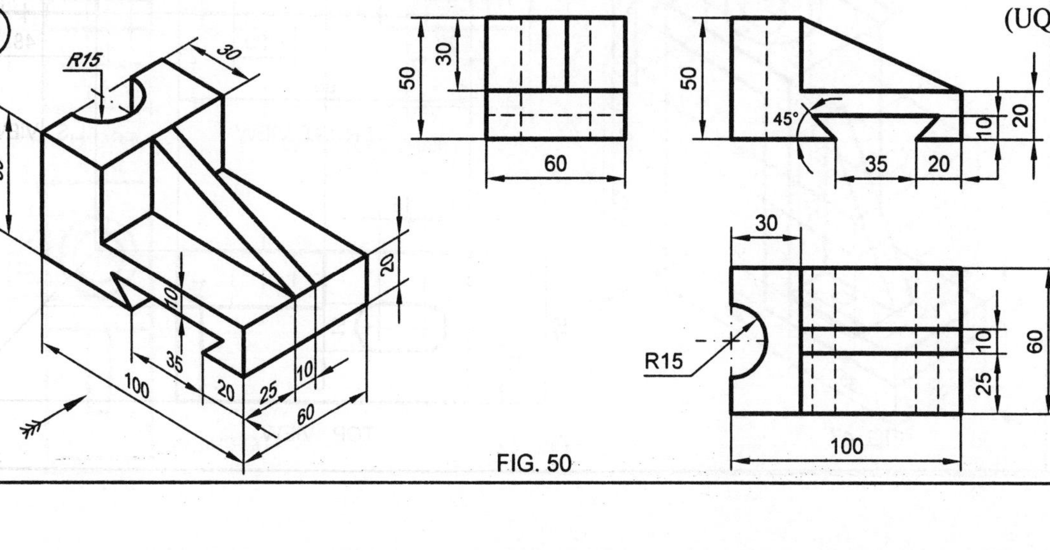

FIG. 50

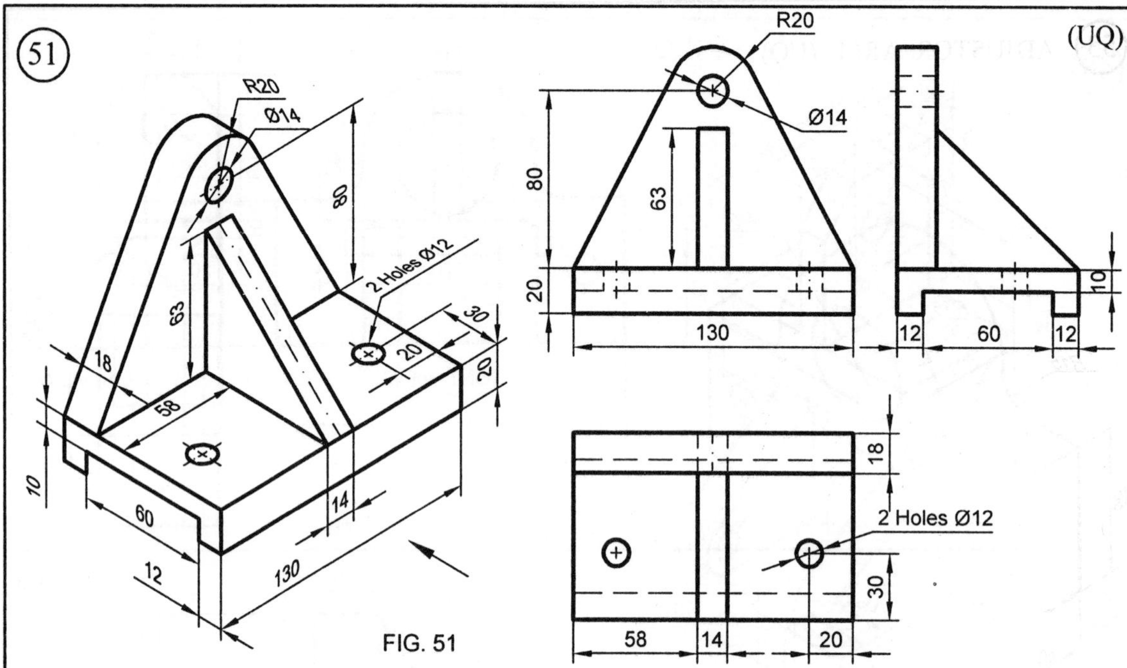
51
(UQ)
R20
Ø14
80
63
2 Holes Ø12
30
20
20
18
58
10
60
14
12
130
20
130
12
60
12
10
18
2 Holes Ø12
30
58
14
20

FIG. 51

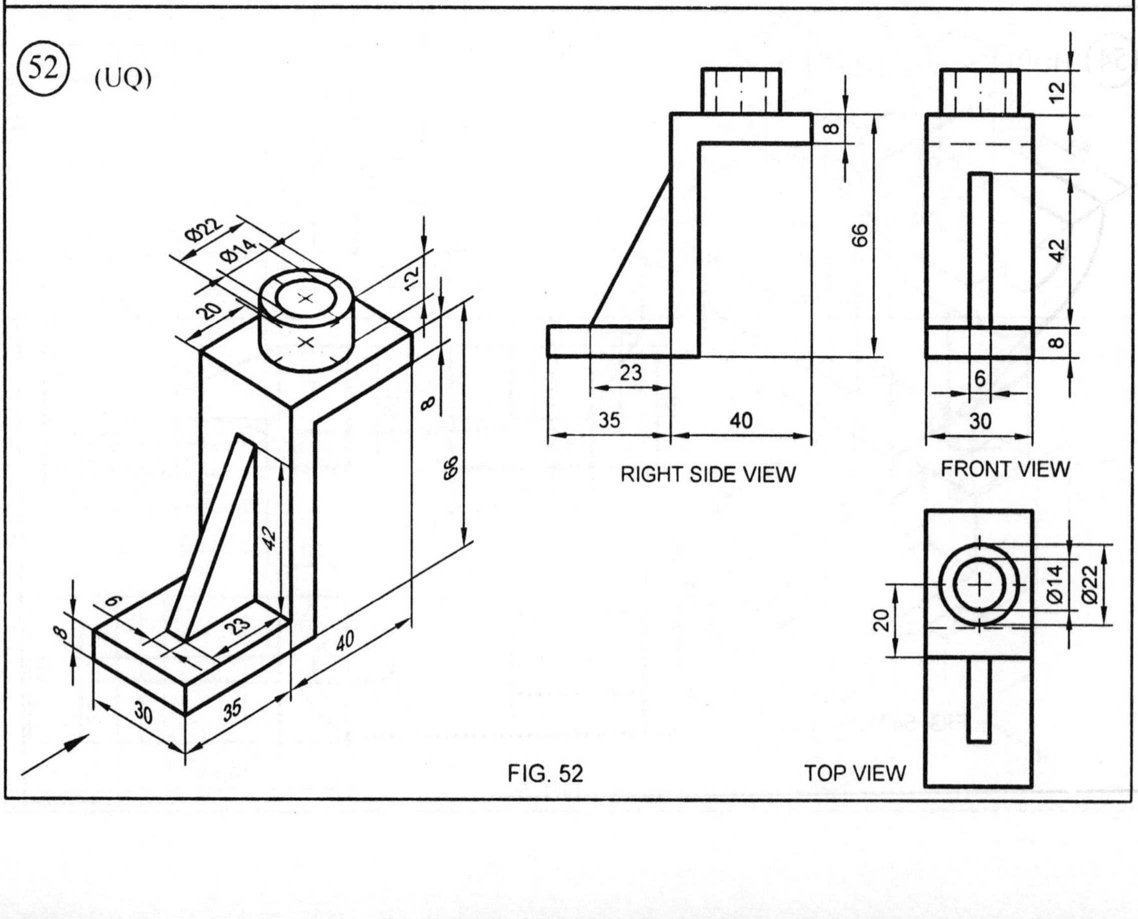
52
(UQ)
Ø22
Ø14
12
20
8
66
42
6
8
23
40
30
35
8
66
23
35
40
RIGHT SIDE VIEW
12
42
8
6
30
FRONT VIEW
Ø14
Ø22
20
TOP VIEW

FIG. 52

(53) ADJUSTOR ARM (UQ)

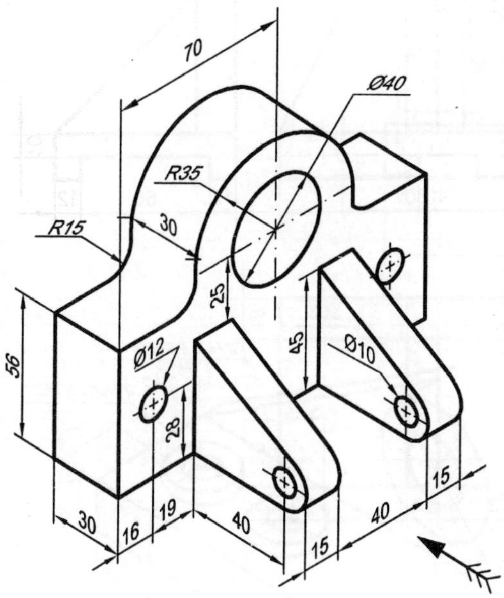

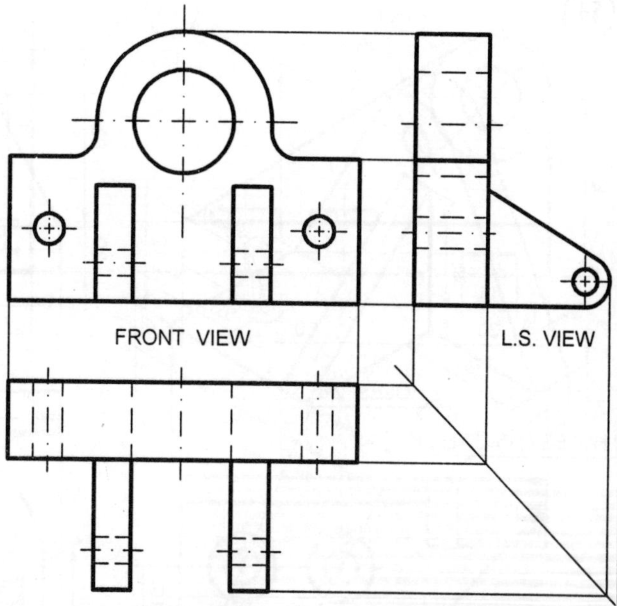

FIG. 53

MARK THE DIMENSIONS

(54) HORIZONTAL SHAFT SUPPORT (UQ)

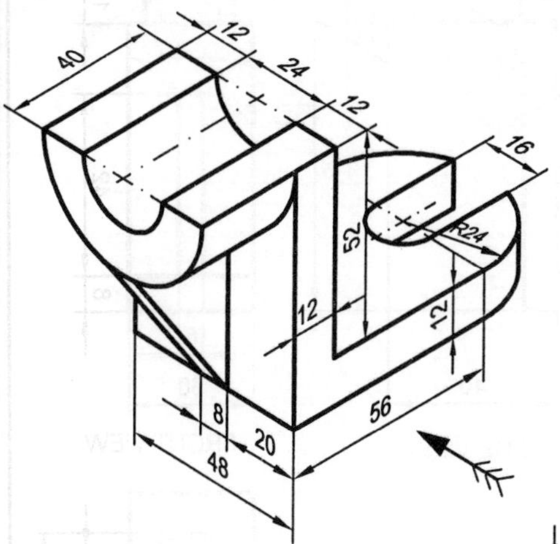

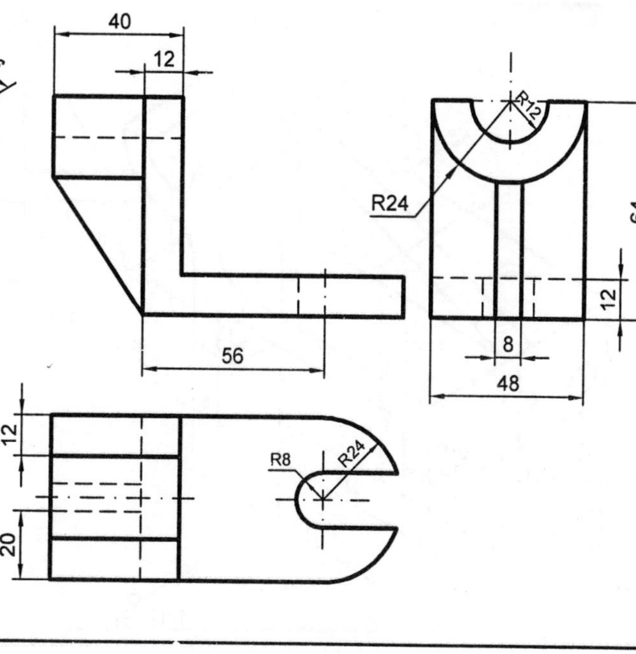

FIG. 54

(55) (UQ)

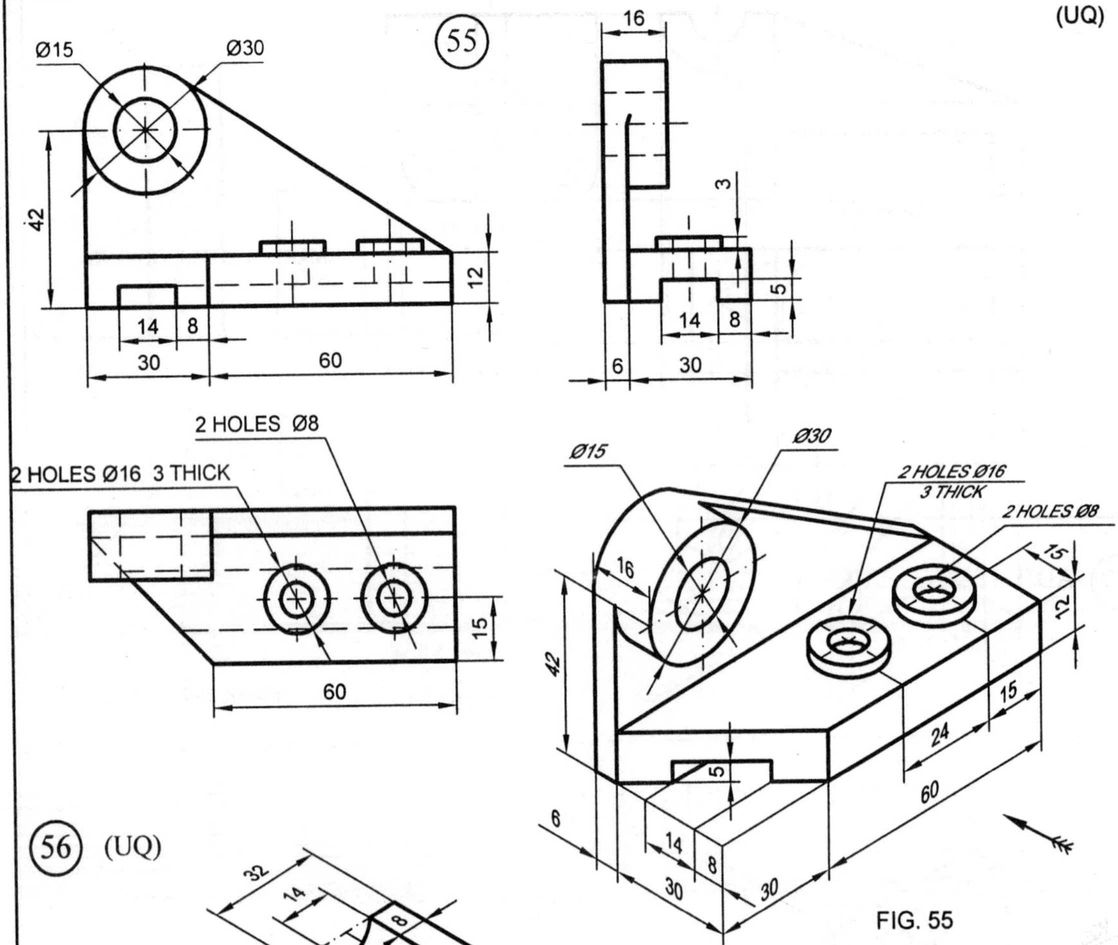

FIG. 55

(56) (UQ)

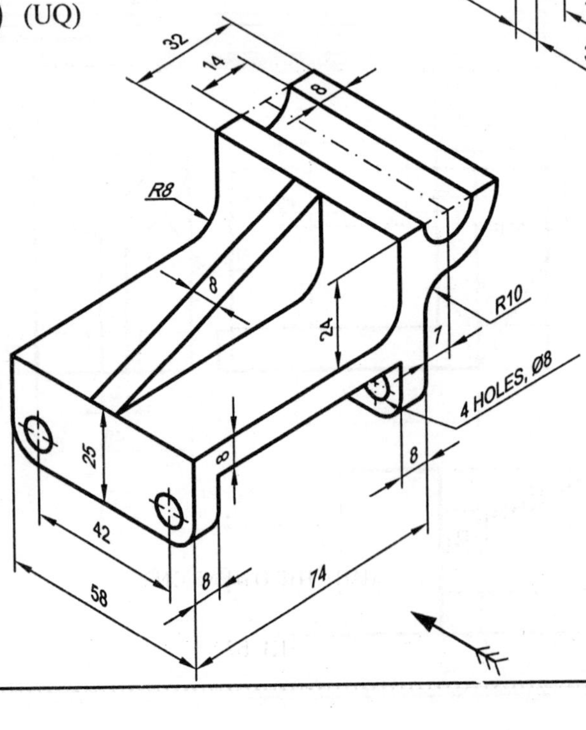

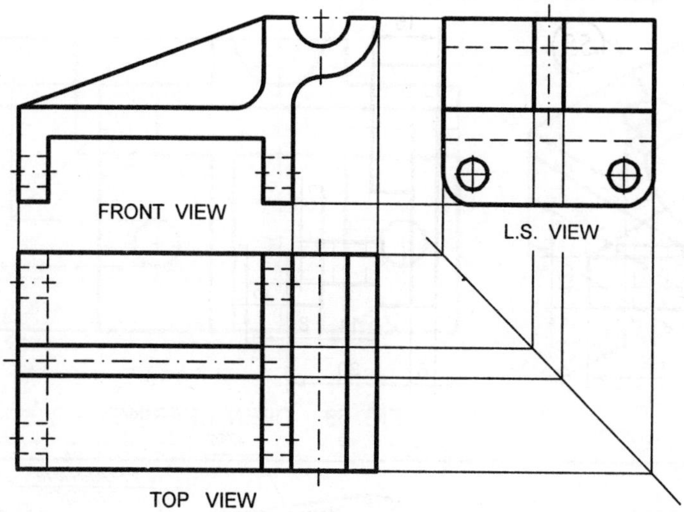

MARK THE DIMENSIONS

FIG. 56

(57) ADJUSTABLE SUPPORT

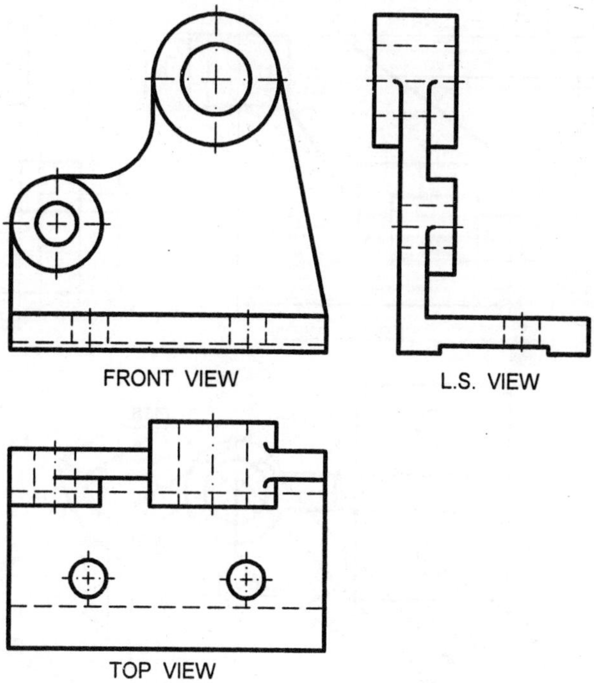

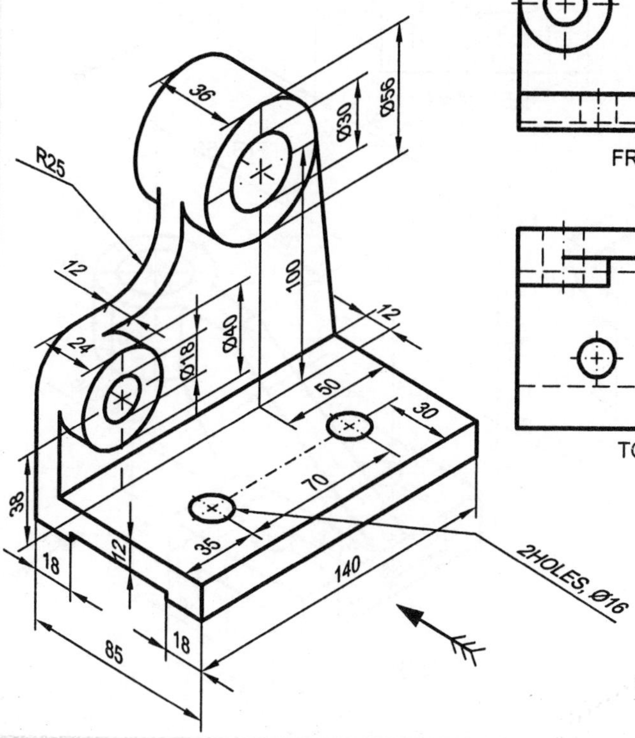

MARK THE DIMENSIONS

FIG. 57

(58)

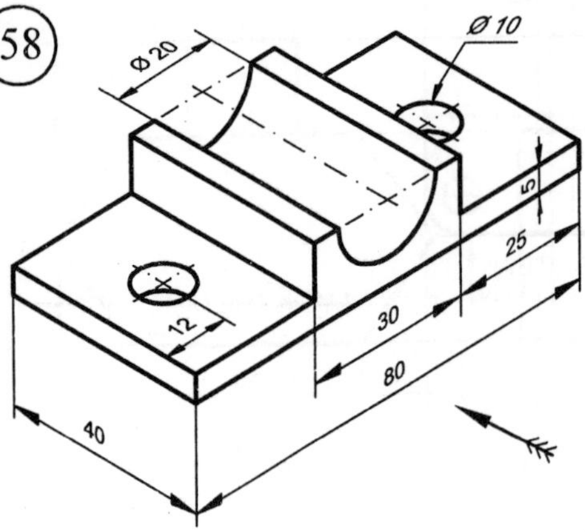

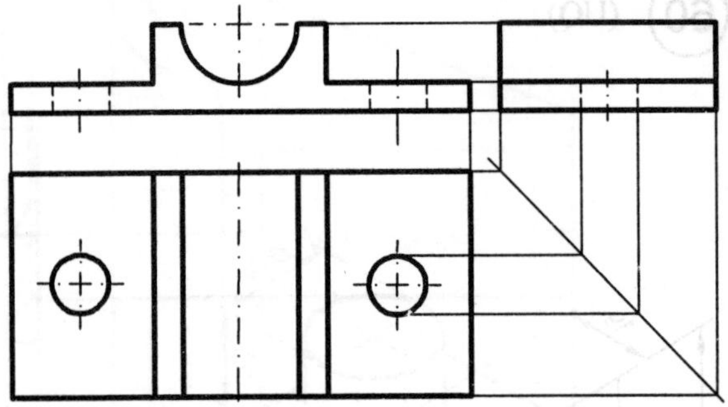

FIG. 58 OPEN BEARING

(59)

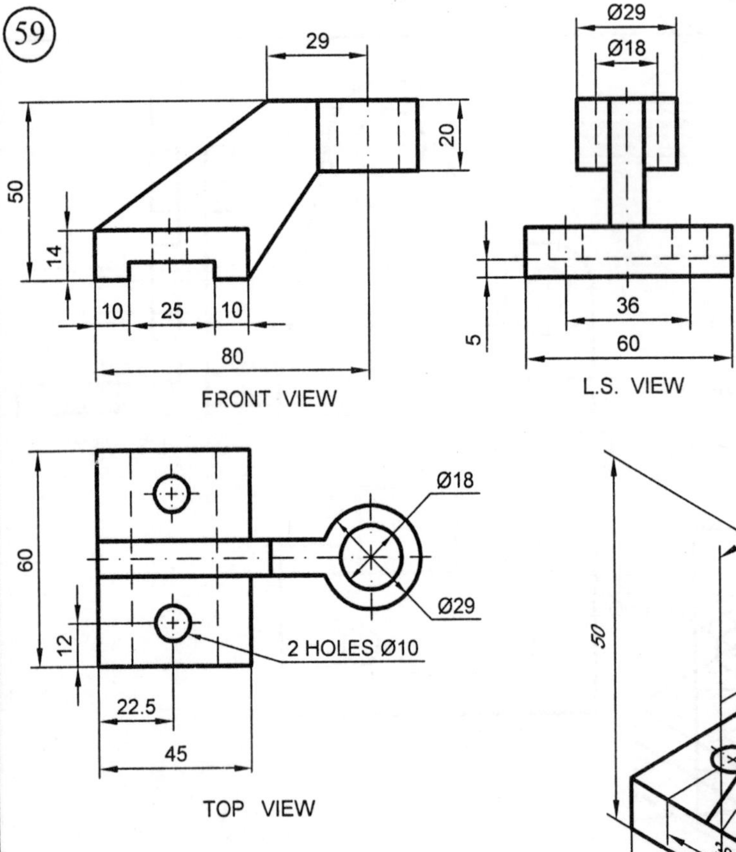

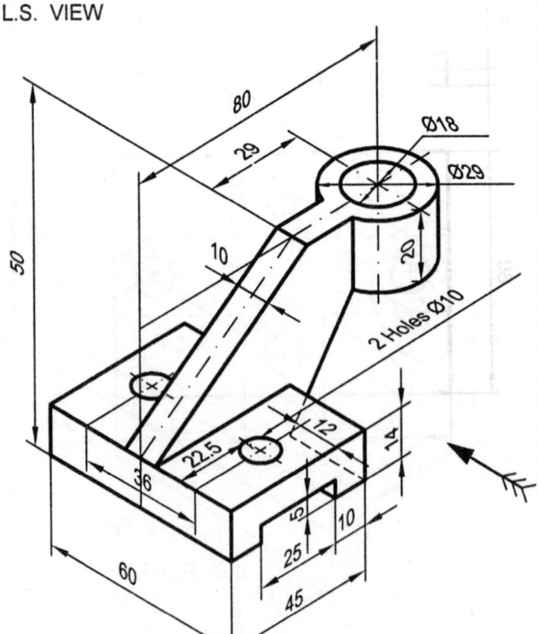

FIG. 59

VERT!CAL SHAFT SUPPORT (UQ)

60 (UQ)

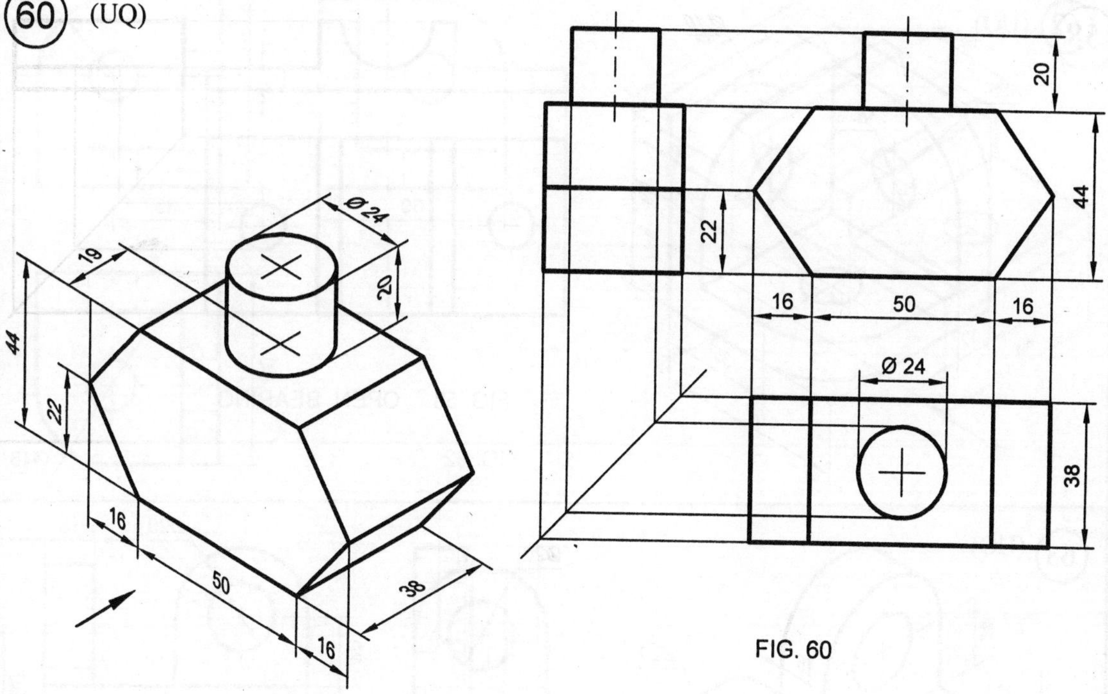

FIG. 60

61 (UQ)

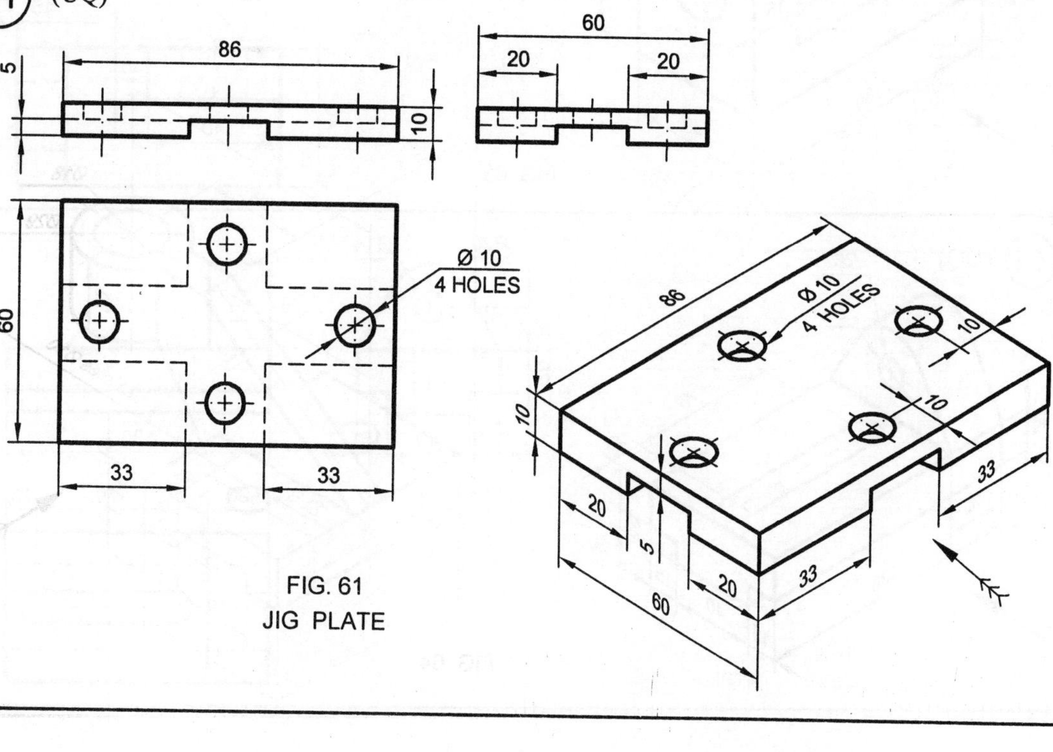

FIG. 61
JIG PLATE

(62) (UQ)

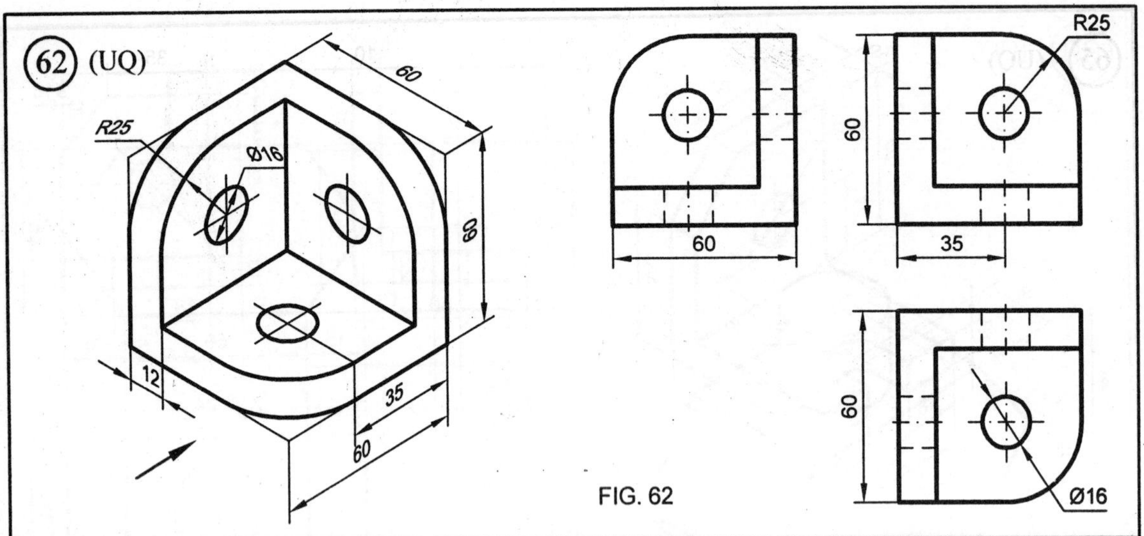

FIG. 62

(63) (UQ)

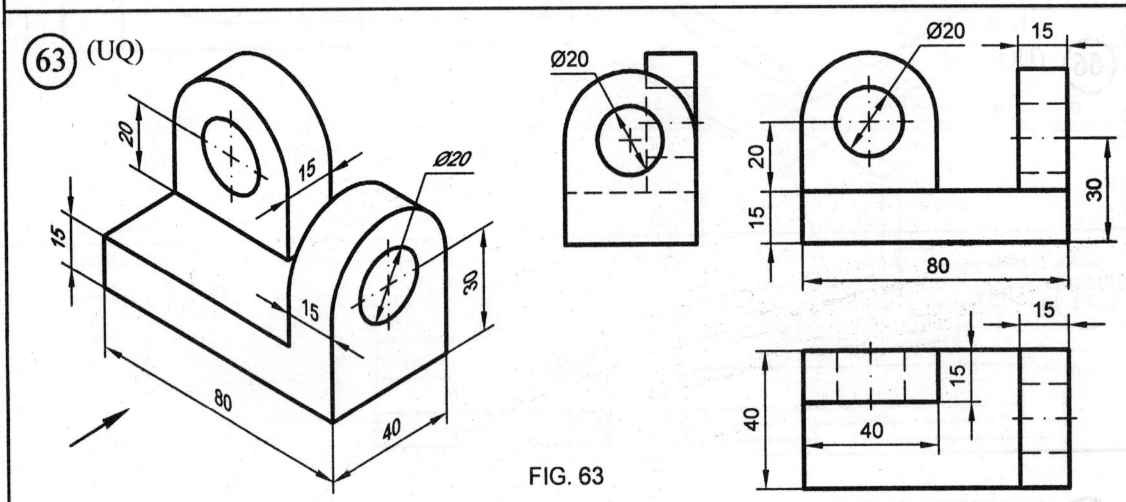

FIG. 63

(64) (UQ)

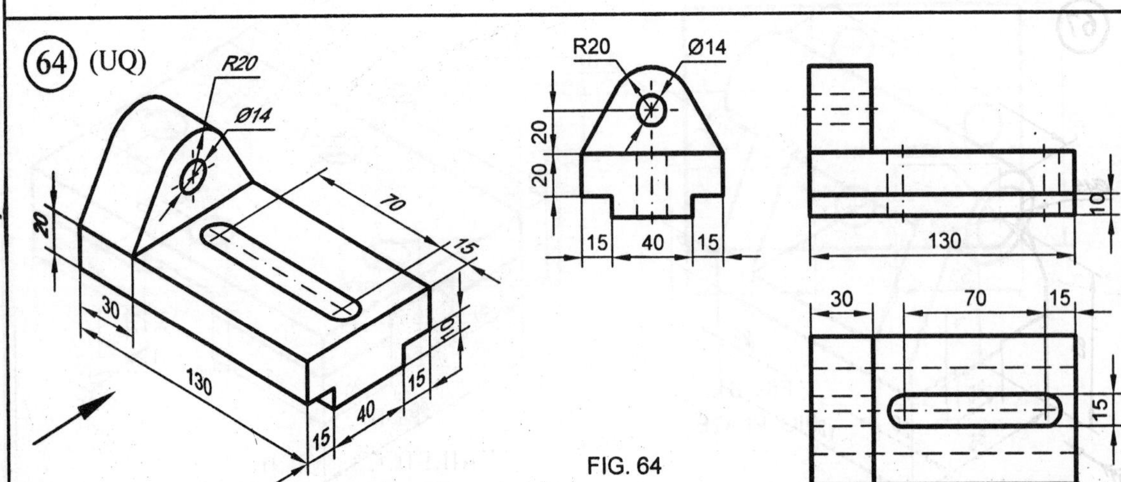

FIG. 64

(65) (UQ)

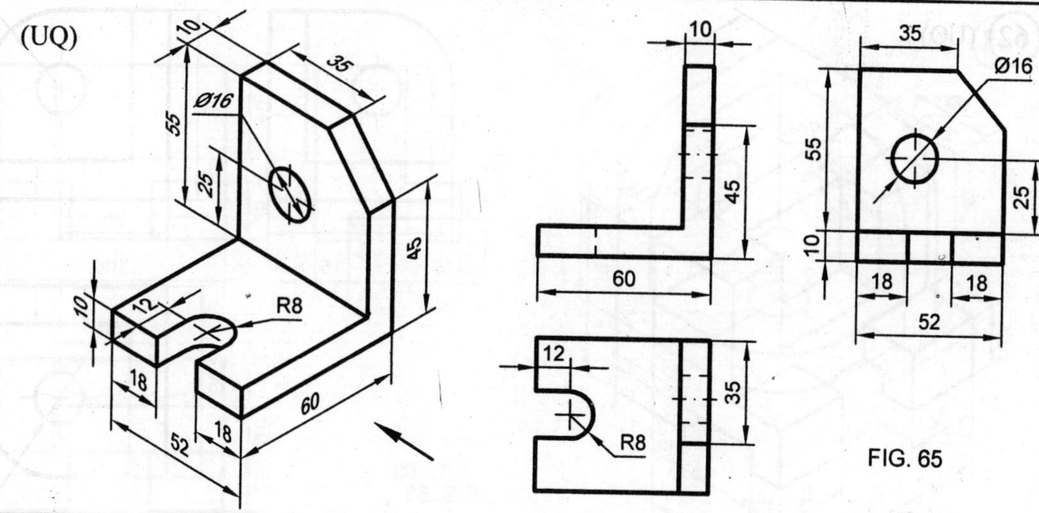

FIG. 65

(66) (UQ)

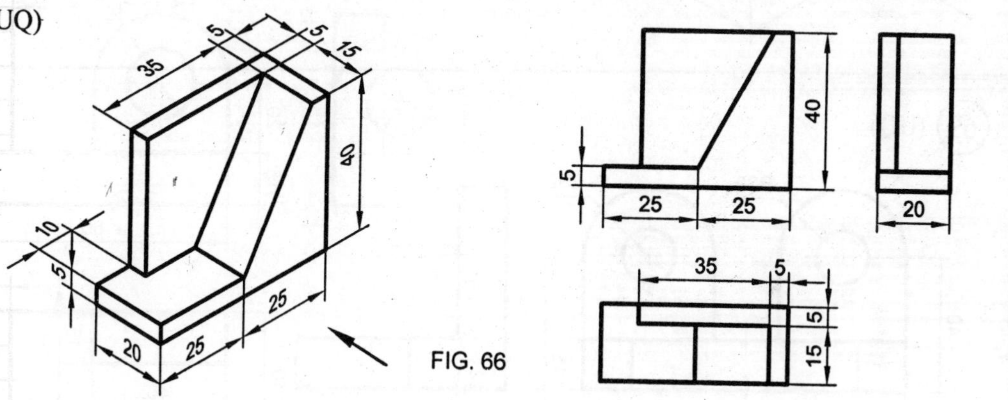

FIG. 66

(67)

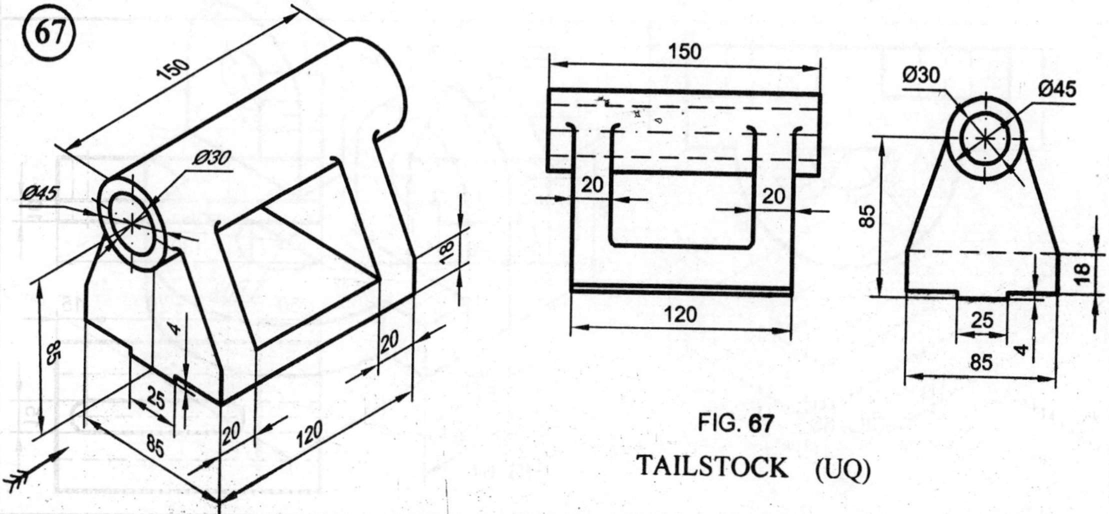

FIG. 67

TAILSTOCK (UQ)

(68) (UQ)

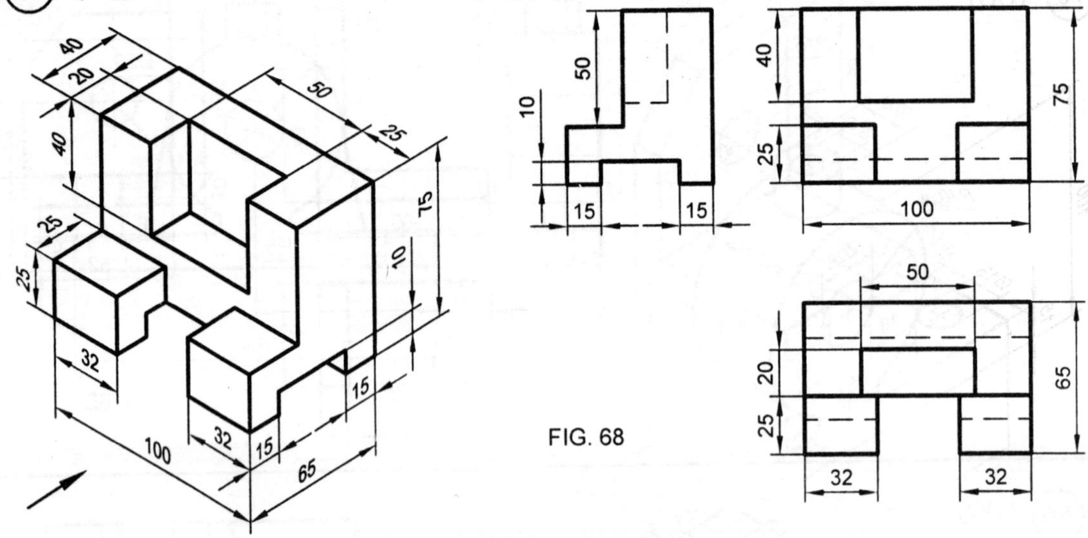

FIG. 68

(69) (UQ)

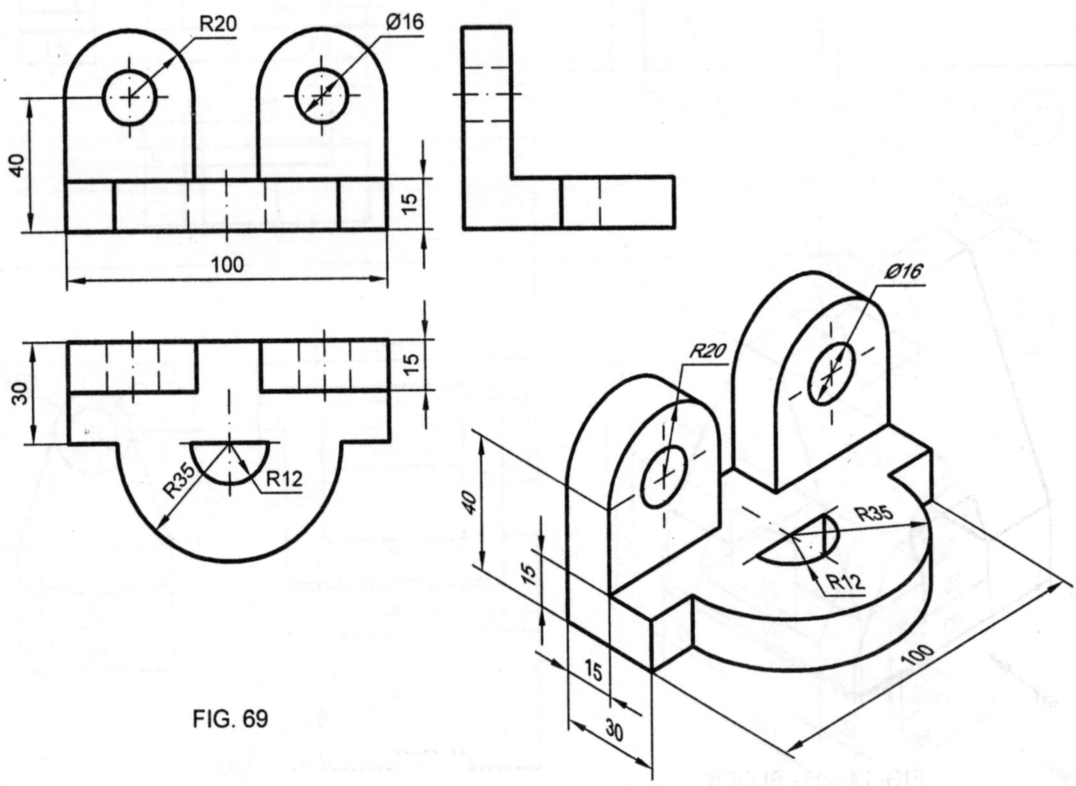

FIG. 69

(70) (UQ)

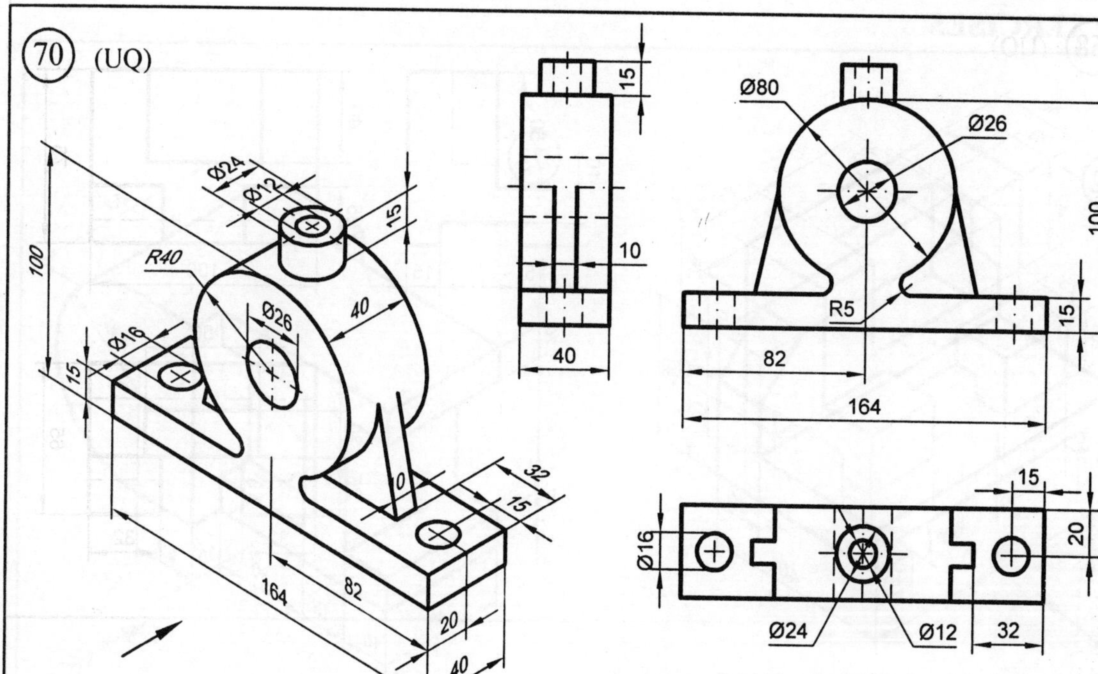

FIG. 70

(71) (UQ)

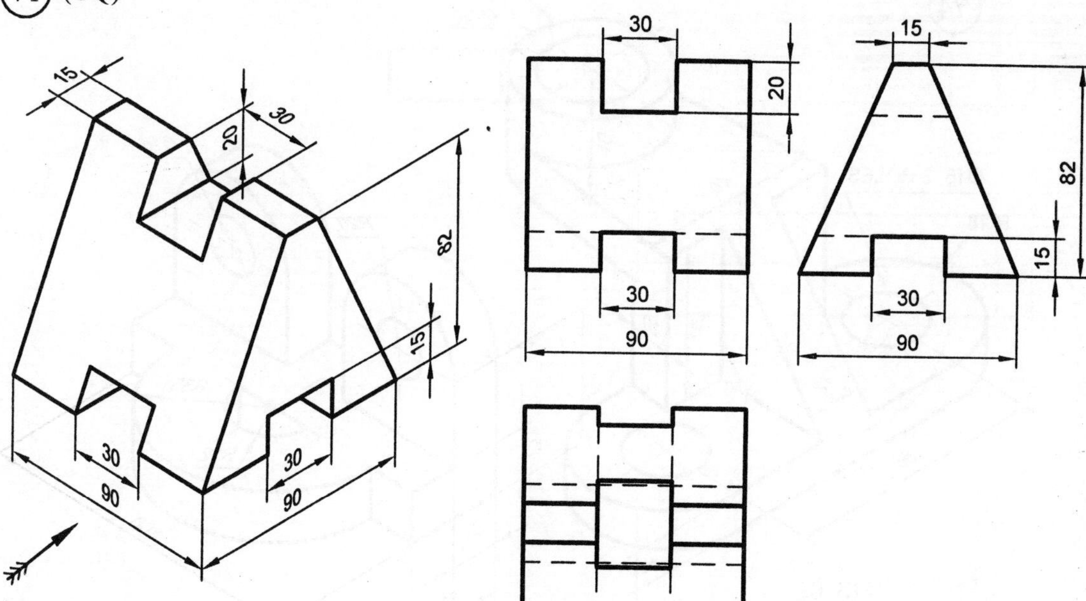

FIG. 71 H - BLOCK

3. EXERCISES

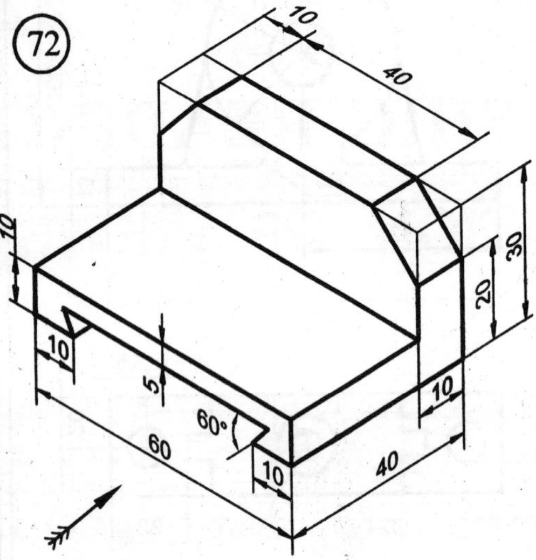

FIG. 72 DOVETAIL GUIDE

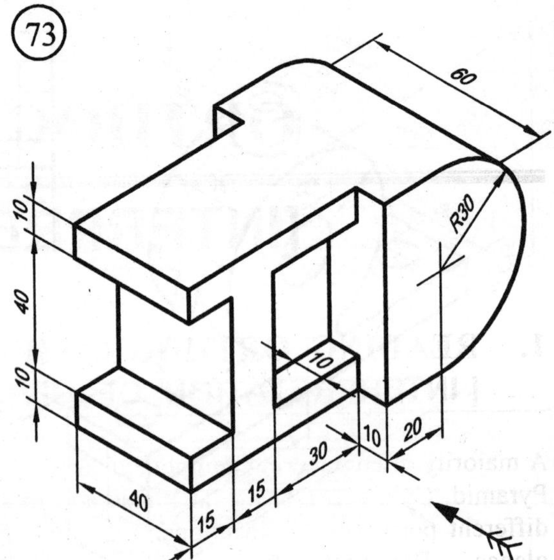

FIG. 73

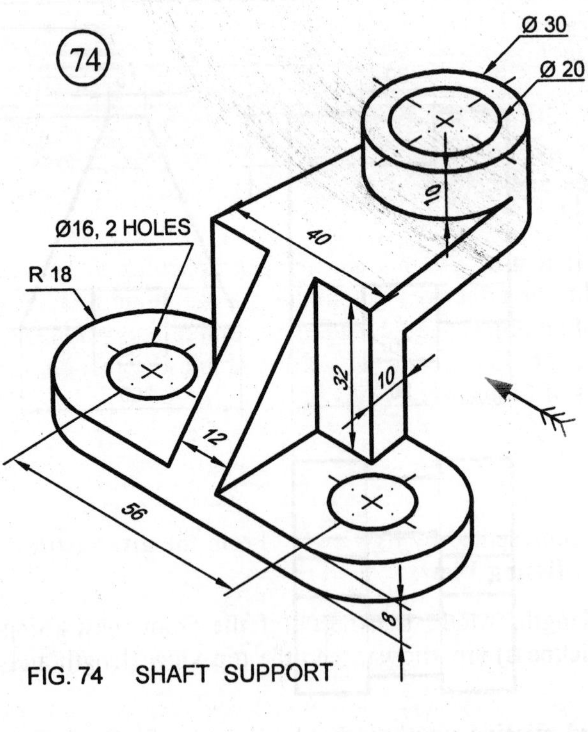

FIG. 74 SHAFT SUPPORT

*

22

ORTHOGRAPHIC READING

[INTERPRETATION OF VIEWS]

1. READING ORTHOGRAPHIC VIEWS – MISSING VIEWS [INTERPRETATION OF GIVEN VIEWS]

A majority of engineering components are combination of common geometrical forms like Prism, Pyramid, Cylinder, Cone, Flat Plates and their Frustum or Truncated Types, joined together in different positions and locations. Do not try to visualize the whole object at a time or at a glance. But, try to visualize piece-by-piece and should keep in memory the shape, size and location of each piece separately. After visualizing all the pieces, combine them together in their respective positions and locations to obtain a picture of the whole object.

Orthographic Reading *or* Interpretation of Given Views

Orthographic Reading means to understand and to interpret or visualize the shape and size of an object from the given orthographic views. Each view gives only two dimensions of a 3-D object. Therefore, at least two views are essential to visualize the shape of the object. Do not try to visualize the shape of the object simply by looking at one view, but *simultaneously* examine all the views. Projections of all the lines and corners in all the views should be noted and the meaning of the same should be visualized.

A language will be effective, if it uses a minimum number of words to convey one's thoughts. Similarly, a drawing is said to be effective, if it contains a minimum number of lines and views. The lines in all the views of an object should lead to one and only one *interpretation.* For this, one needs to have a thorough knowledge of Orthographic Projection, Drawing Conventions, Meaning of Various Types of Lines, Areas, etc.

1.1 MISSING VIEWS

In orthographic projection, an object is represented by its views. From the given pair of views, one must be able to visualize the **Third Missing View**.

Example: In a rectangular plate (length, width, thickness), if the front view (length and thickness) and side view (width and thickness) are known, then the top view (length and width) can be developed.

The ability to visualize the third missing view can be developed only **through practice. Yes,** ***Practice makes us perfect.***

Practice in reading a drawing is done, using Mitre Lines by either projecting one or more views from the given views or by converting one or both the given views into sectional views.

How to identify Lines and Planes in the given Views?

1. When a line is perpendicular to a plane, its projection on that plane is a point, while its projection on the other plane is a line equal to its True Length.
2. When a line is parallel to a plane, its projection on that plane will show its True Length and True Inclination with the other plane.
3. When a line is inclined to both HP and VP, its projections are shorter than the True Length and inclinations to XY are greater than the True Inclinations.
4. When a plane is perpendicular to a reference plane, its projection on that plane is a line.
5. When a plane is parallel to a reference plane, its projection on that plane shows its true shape and size.
6. When a plane is perpendicular to one of the reference planes and inclined to the other, its inclination is shown by the angle, which its projection on the plane to which it is perpendicular, makes with XY. Its projection on the plane to which it is inclined, is smaller than the plane itself.
7. Each line in one view represents a plane in another view, depending upon the position of the plane with reference to the plane of projection. When a line is observed in any two views, it will be a plane in the third view.

Guide Lines to obtain the Missing View

There are *no* fixed Rules to obtain the Missing View. Remember the following Guide Lines:

1. A *point* in one view may represent a corner *or* an edge of the object in the other view.
2. A *line* in one view may represent a line *or* surface of the object in the other view. A *center line* indicates the existence of curved surface *or* hole. It also indicates object's symmetry.
3. *Area* enclosed by one boundary in one view represents a flat *or* curved surface in the other view.
4. From the given orthographic views, try to visualize an object and prepare rough sketch of the pictorial view.
5. From the rough pictorial view, prepare orthographic views and compare with the given views. Read carefully each line in each view. Find out the corresponding projection of each line from another view.
6. For the given two views, it is possible to imagine and interpret different missing views, i.e., different possible objects. *See Fig. 1 and Fig. 2.*

How to obtain the Missing View?

1. Draw the two given views. Draw projectors from one given view to another given view to locate the corresponding points / lines *or* edges / faces in one view to points / lines *or* edges / faces in the other view.
2. ***Mitre Lines:*** Draw projectors from the two given views to the third Missing View using Mitre Lines.
3. Draw the Missing View by locating required points at the intersections of the above projectors.
4. ***Isometric View:*** Visualize the object. Draw or sketch an isometric view for complicated objects. Isometric view will also help to draw the hidden features.

1.2 ORTHOGRAPHIC READING

The method of orthographic reading by recognizing the geometry of the solids forming the object is explained as follows:

Example 1: *See Fig. 1.* It gives one set of front and side views and four different shapes of top views. By reading the front and side views, the shape of the object may be visualized as a rectangular plate having a cut from the front side.

The shape of the cut is fully under-stood by reading the top view. The possible four different forms of cuts are as shown, which are matching the front and side views.

Now, the shape of the object is finalized mentally by the three-view combination. An experienced engineer reads the drawing rapidly by going through the given views and analyzing the geometry of each part subconsciously. The reader shifts his eyes rapidly back and forth from the front view to the top and side views and finalizes the shape of the object in his mind.

Example 2: *See Fig. 2.* The meaning of the lines PQ and RS in the front view can be determined only after referring to the top view. These lines might represent a projection or a cavity as shown in the different top views.

Thus, one front view represents five different types of objects whose top views are different as shown.

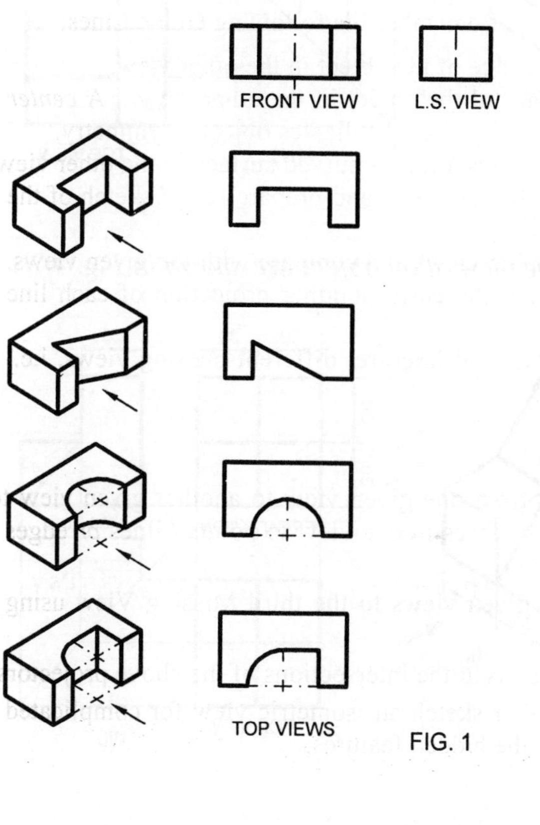

FIG. 1

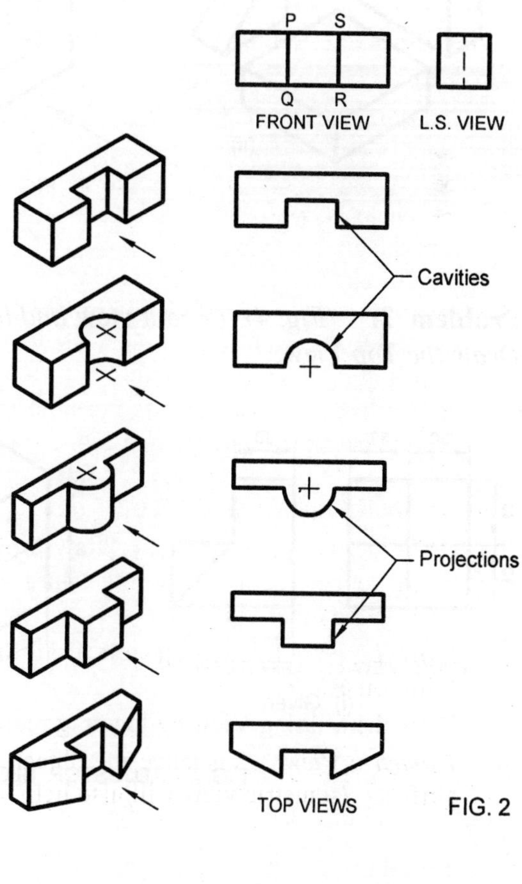

FIG. 2

2. MISSING VIEWS – TYPICAL PROBLEMS

Problem 1: *(Fig. 3) Front view and right side view of an object are shown in Fig. 3(i). Draw the Top View.*

1. Imagine the object from the given two views. For this, let us imagine the object piece-by-piece. Object can be conveniently divided into *three pieces* as shown in Fig. (ii).
2. Correlate the front and side views of each piece into a single object in the form of an *isometric view* as shown in Fig. (iii).
3. Having visualized the full object, draw the missing top view using the *Mitre Line* as shown in Fig. (iv).

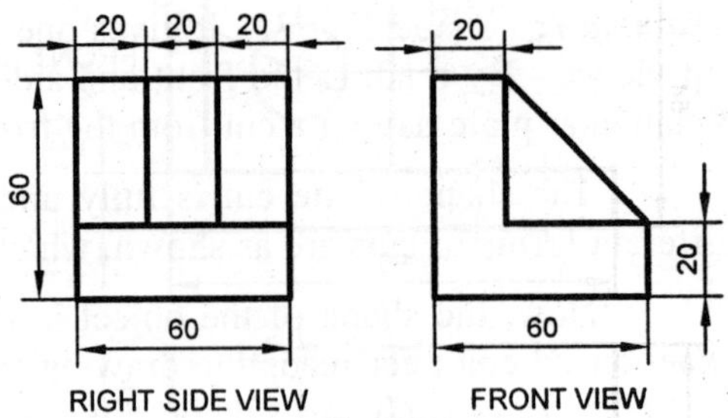

(i) Given

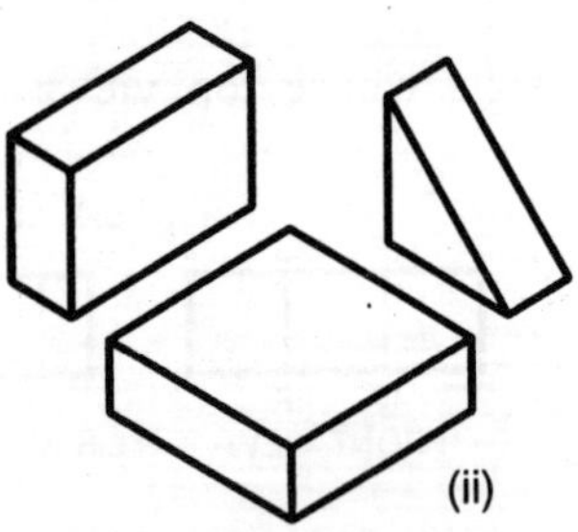

(ii)

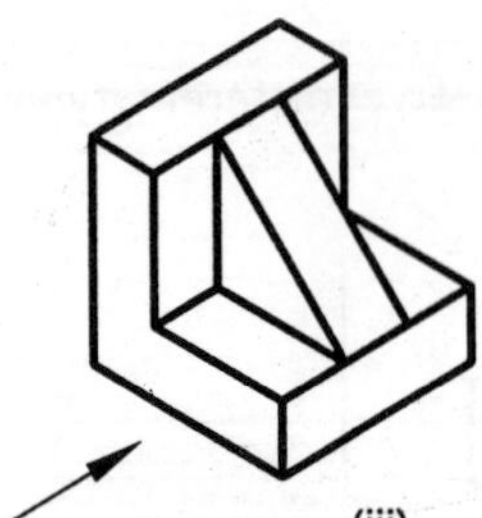

(iii)

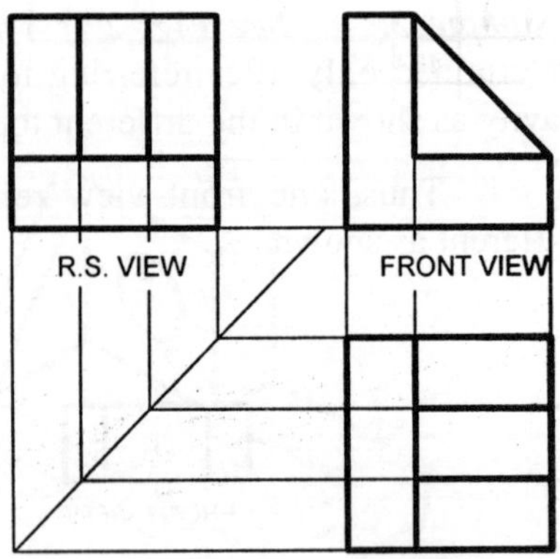

(iv) TOP VIEW

FIG. 3

Problem 2: *(Fig. 4) Front view and left side view of an object are shown in Fig. 4(i). Draw the Top View.*

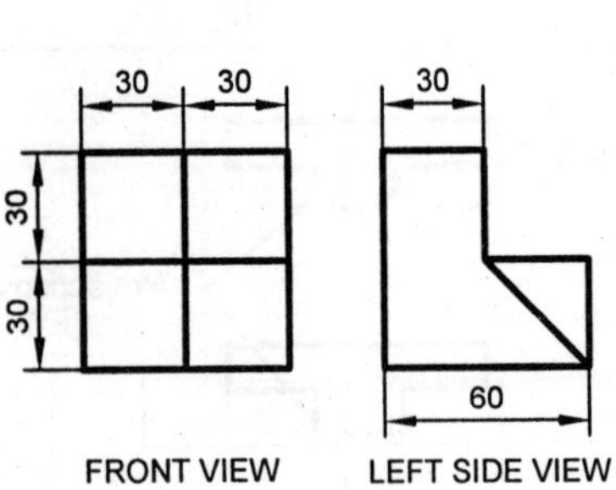

(i) Given

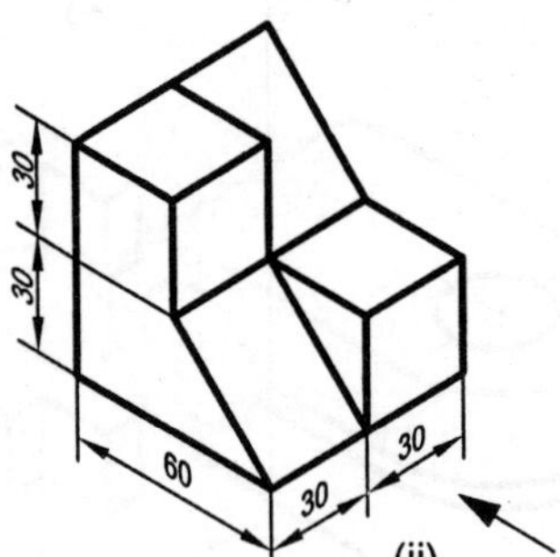

(ii)

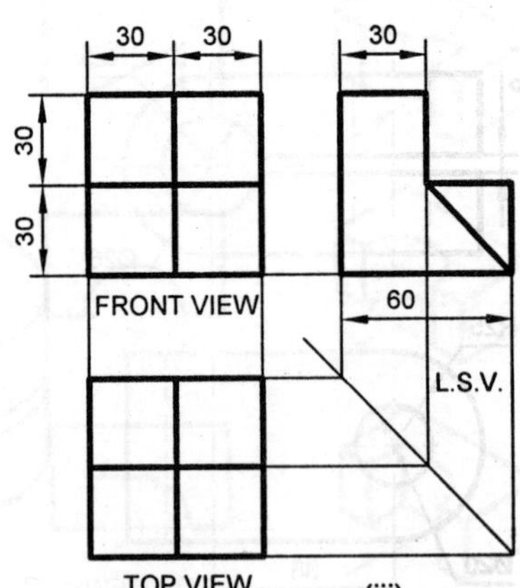

(iii)

FIG. 4 ADJUSTOR BLOCK

Problem 3: *(Fig. 5) Draw the Left Side View of the object shown in Fig. 5(i).*

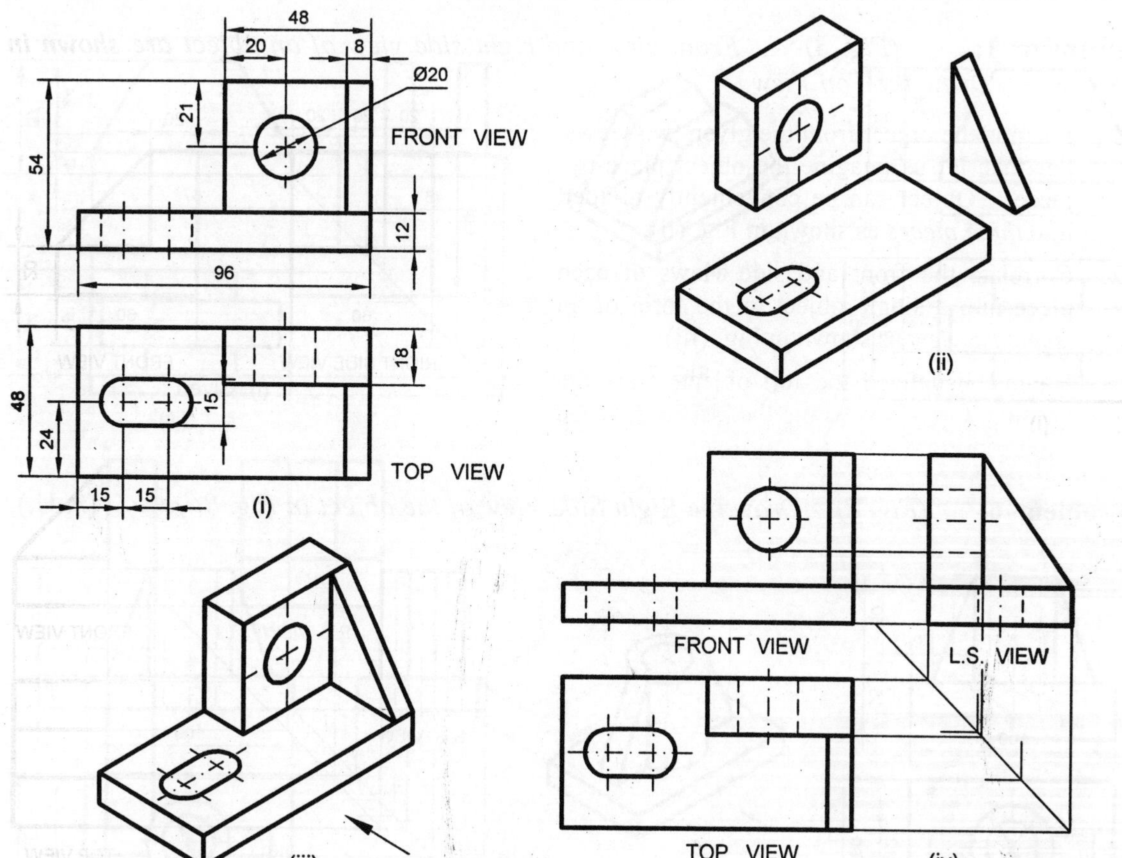

FIG. 5

Problem 4: *(Fig. 6) Draw the Left Side View of the object shown in Fig. 6(i).*

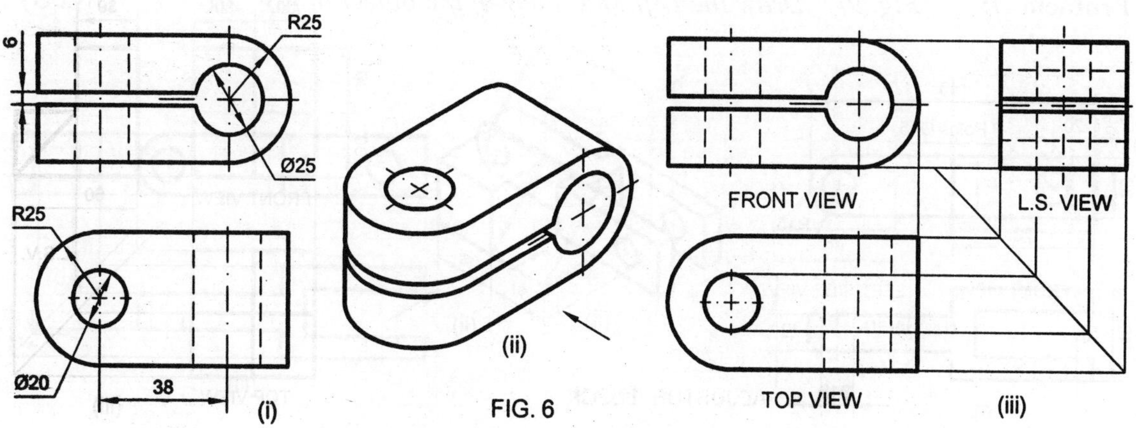

FIG. 6

Problem 5: *(Fig. 7) Draw the Right Side View of the object in Fig. 7(i).* (K.U.)

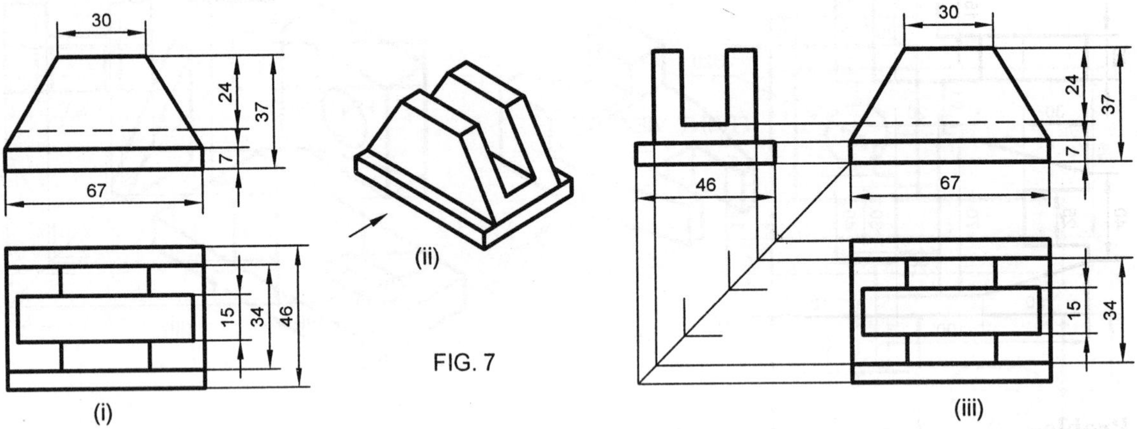

FIG. 7

Problem 6: *(Fig. 8) Draw the Right Side View of the object in Fig. 8(i).* (A.U.)

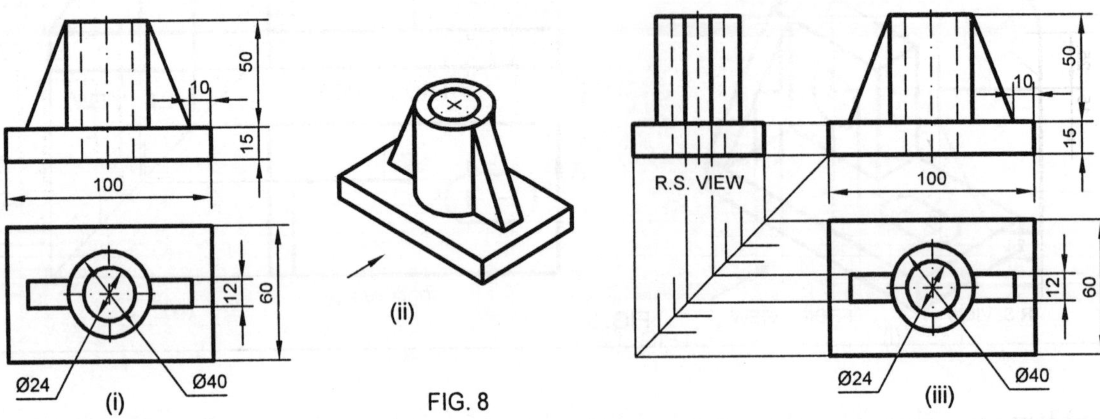

FIG. 8

Problem 7: *(Fig. 9) Draw the Left Side View of the object in Fig. 9(i).* (JNTU)

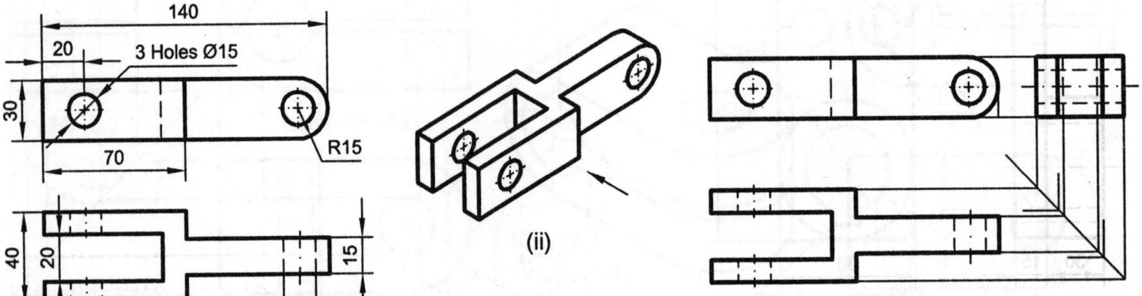

FIG. 9

Problem 8: *(Fig. 10) Draw the Left Side View of the object shown in Fig. 10(i).* (M.U.)

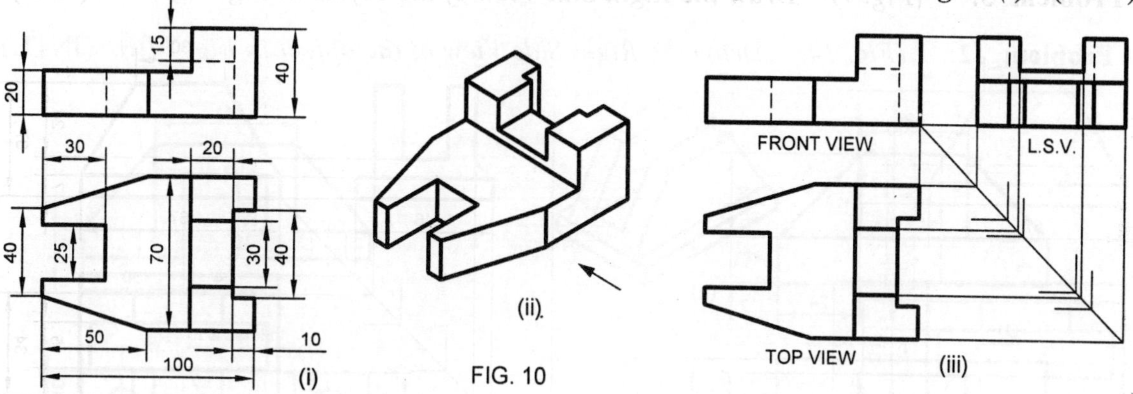

FIG. 10

Problem 9: *(Fig. 11) Draw the Top View of the object shown in Fig. 11(i).* (K.U.)

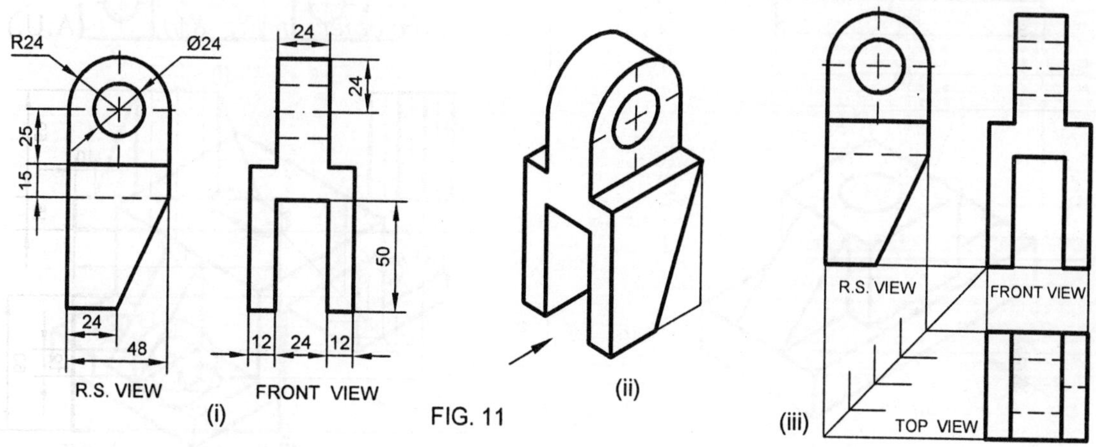

FIG. 11

Problem 10: *(Fig. 12) Draw the Top View of the object shown in Fig. 12(i).* (P.U.)

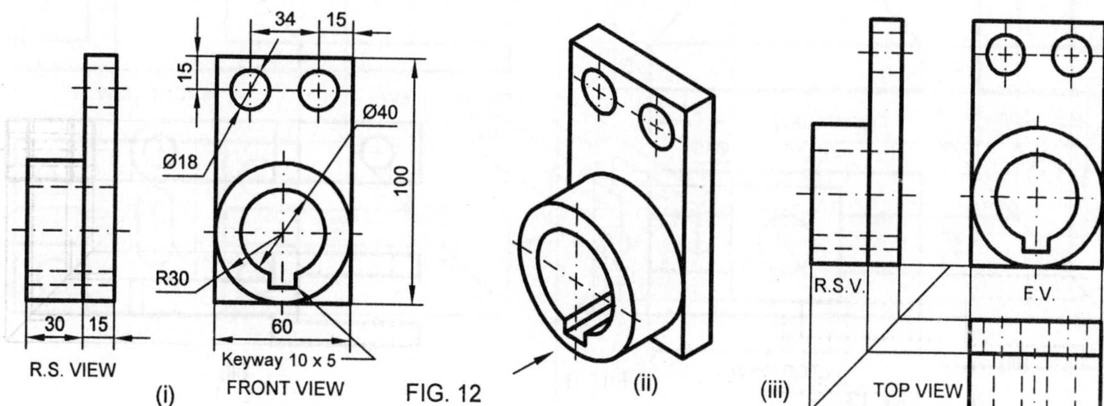

FIG. 12

Problem 11: *(Fig. 13) Draw the Right Side View of the object in Fig. 13(i).* (AU)

Problem 12: *(Fig. 14) Draw the Right Side View of the object in Fig. 14(i).* (JNTU)

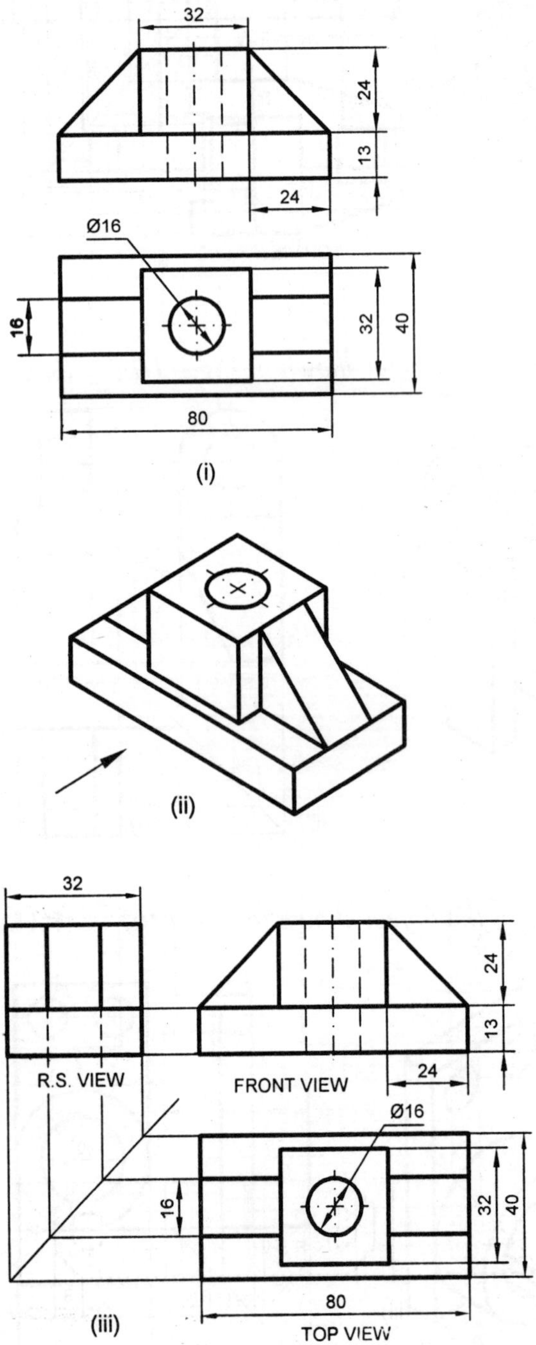

FIG. 13

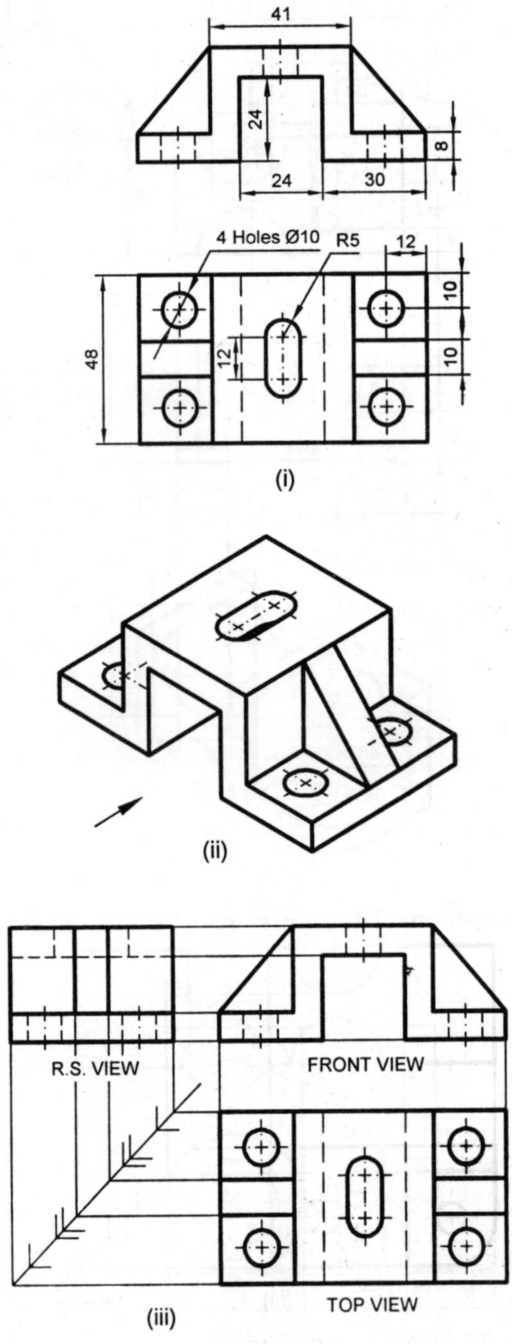

FIG. 14

Problem 13: *(Fig. 15) Draw the Left Side View of the object in Fig. 15(i).* (MU)

Problem 14: *(Fig. 16) Draw the Right Side View of the object shown in Fig. 16(i).* (GU)

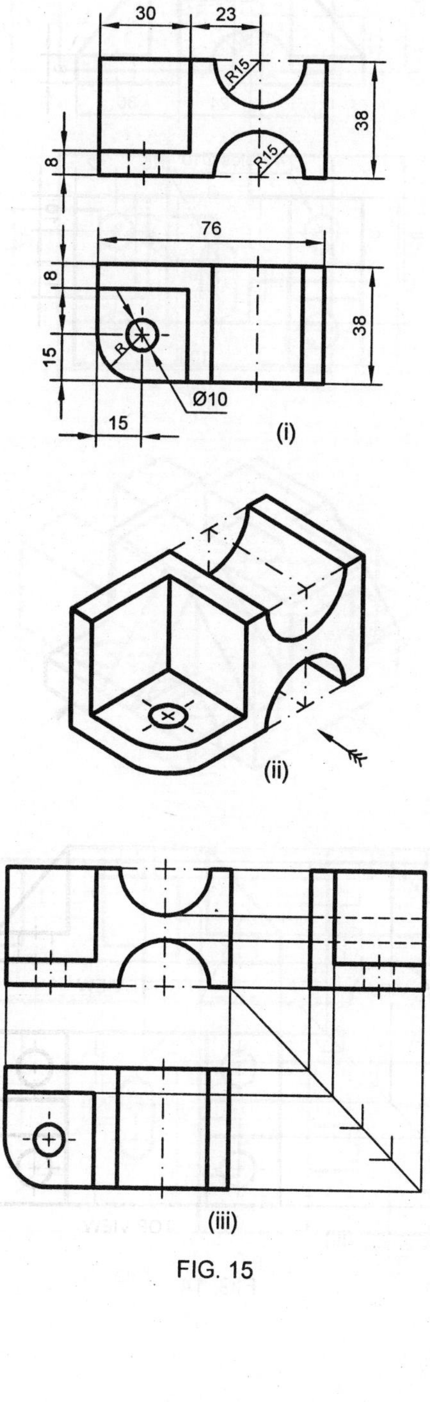

FIG. 15

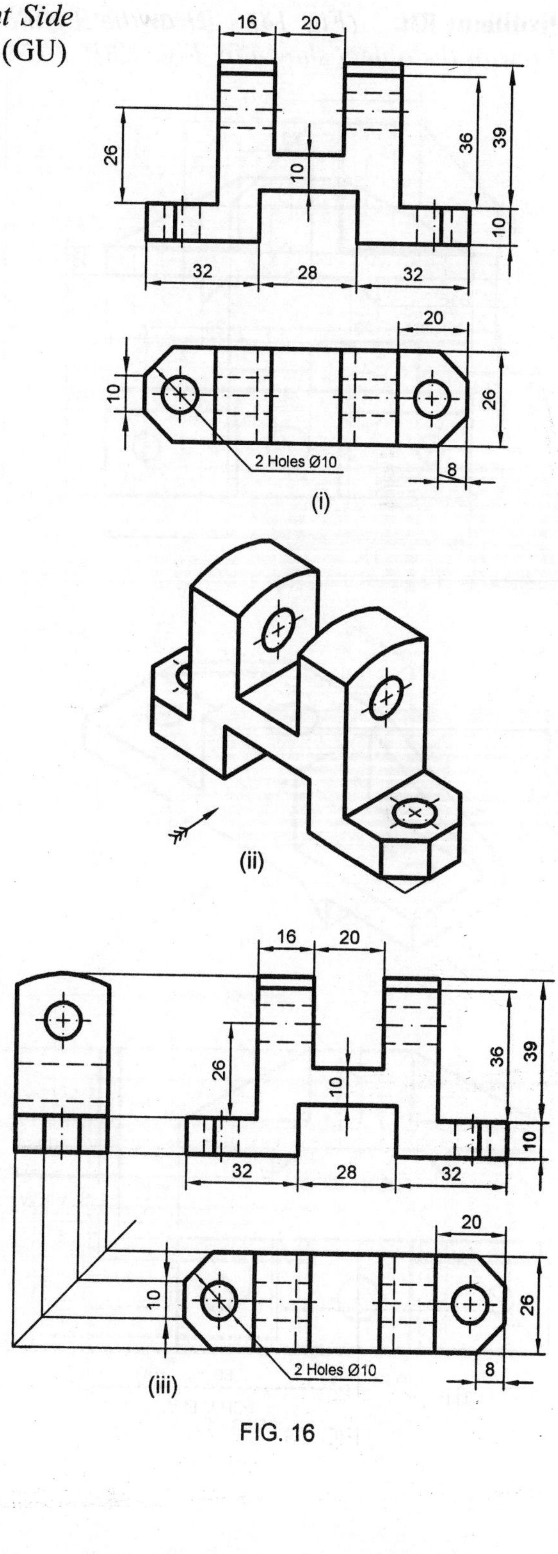

FIG. 16

Problem 15: *(Fig. 17) Draw the Left Side View of the object in Fig. 17(i).* (PU)

Problem 16: *(Fig. 18) Draw the Left Side View of the object shown in Fig. 18(i).* (MU)

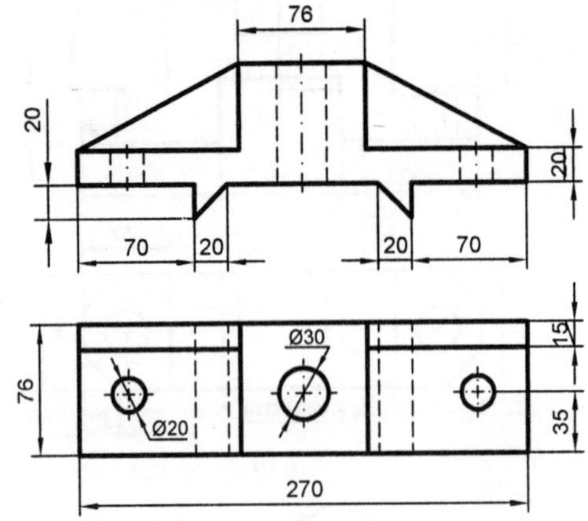

SCALE 1 : 2 (i)

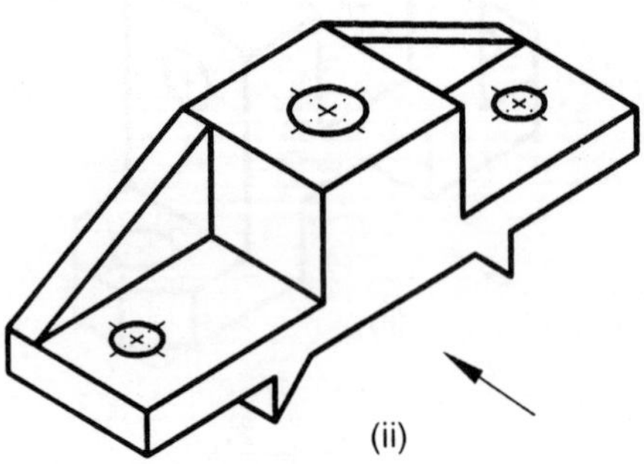

(ii)

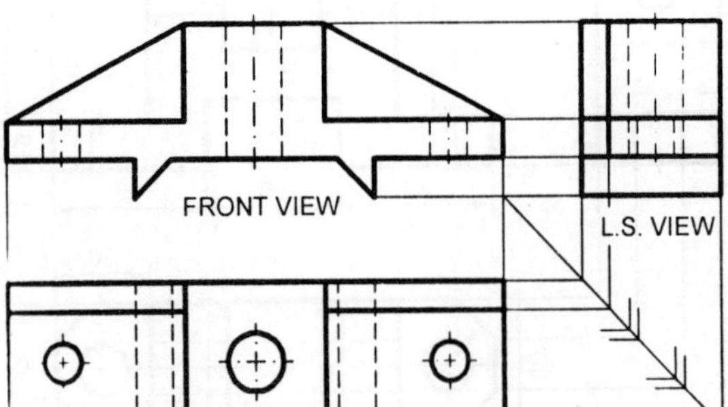

(iii)

FIG. 17

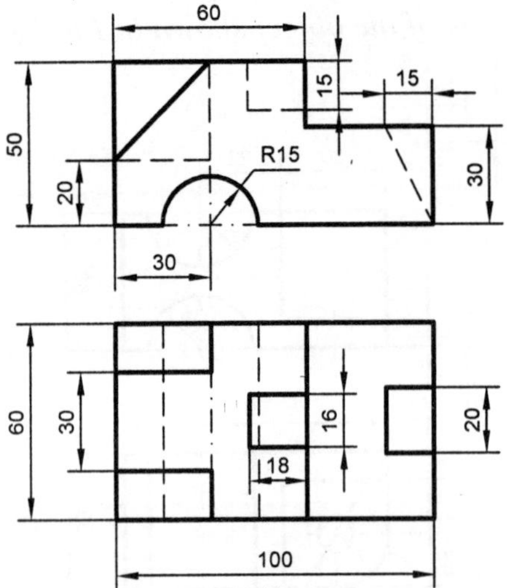

(i)

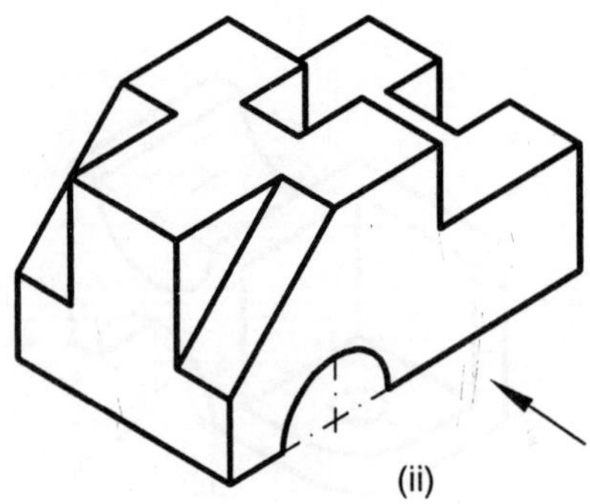

(ii)

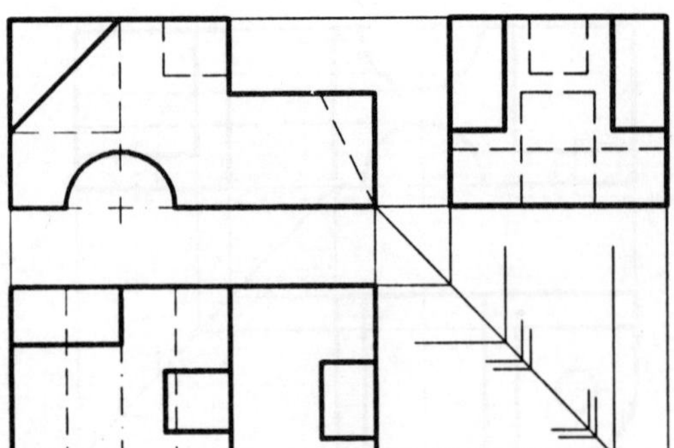

(iii)

FIG. 18

Problem 17: *(Fig. 19) Draw the Left Side View of the object in Fig. 19(i).* (MU)

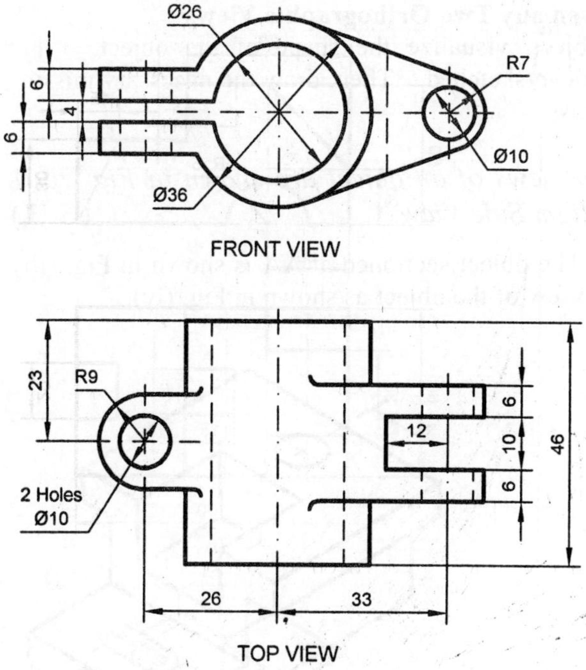

(i)

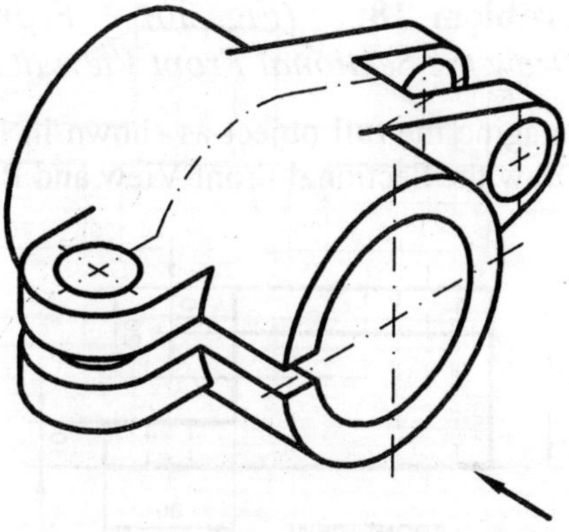

(ii)

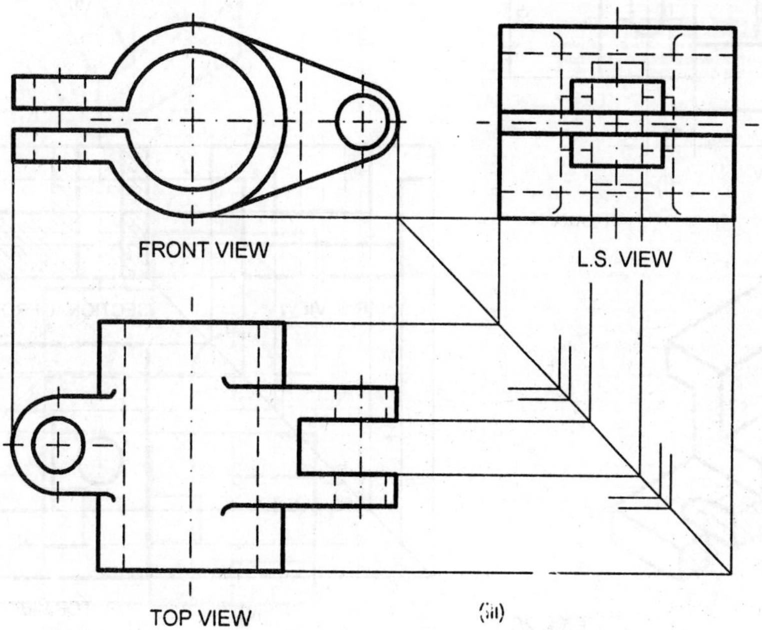

(iii)

FIG. 19

3. MISSING VIEWS WITH SECTIONS – TYPICAL PROBLEMS

To draw the Missing Views with Sections, given any Two Orthographic Views
Given any two orthographic views of an object, visualize the shape of the object. For complicated objects, isometric views may be drawn/sketched. Then, draw the necessary missing view of the object and the required sectional views.

Problem 18: *(Fig. 20) Front and top views of an object are shown in Fig. 20(i). Draw the Sectional Front View at AA and Right Side View.* (G.U.)

Imagine the full object as shown in Fig. (ii). The object sectioned at AA is shown in Fig. (iii). Draw the Sectional Front View and Right Side View of the object as shown in Fig. (iv).

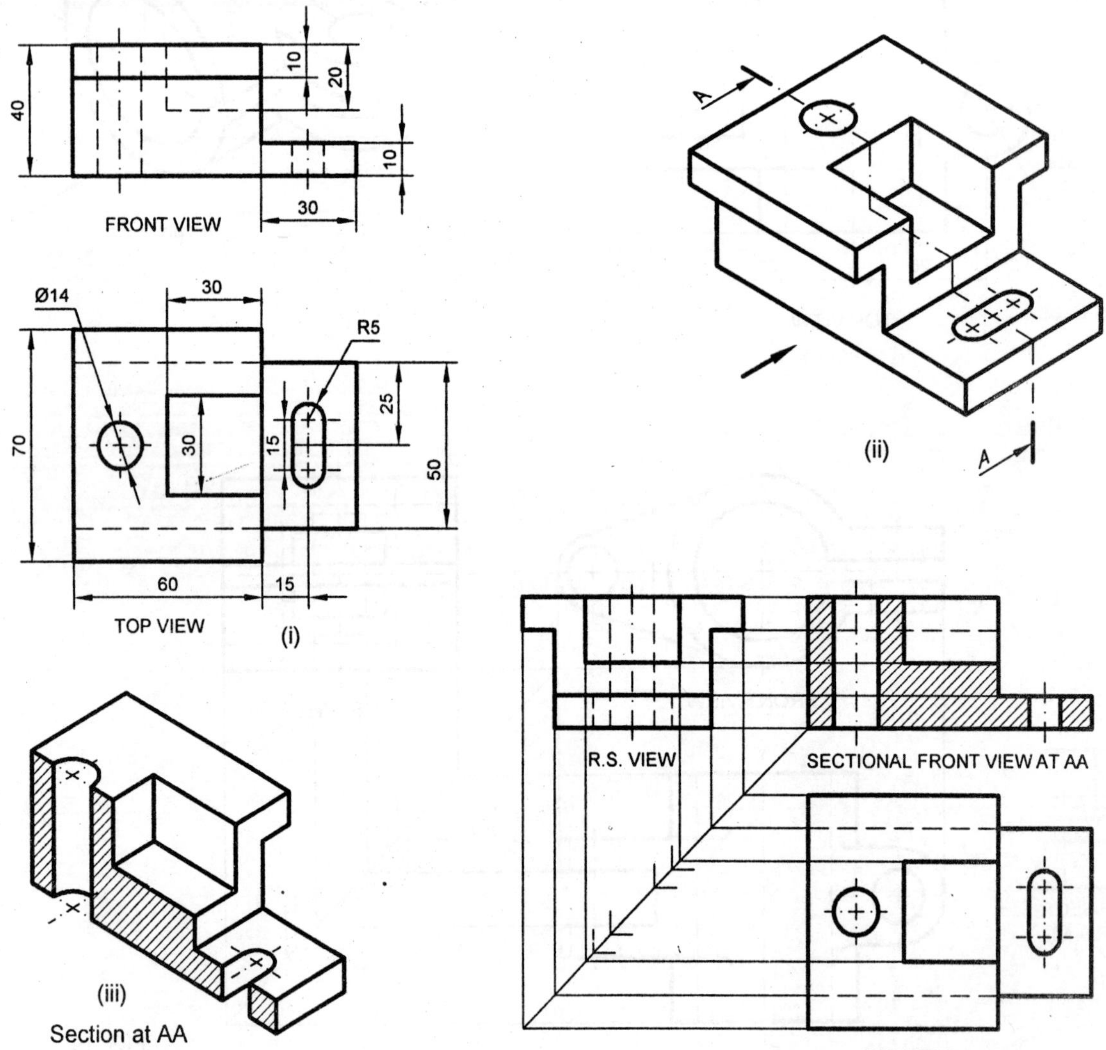

FIG. 20

Problem 19: *(Fig. 21) Front and top views of an object are shown in Fig. 21(i). Draw the Sectional F.V. at AA, Sectional R.S.V. at BB and Sectional T.V. at CC.*

Also, draw the Half-Sectional F.V., Half-Sectional T.V. and Half Sectional R.S.V.

Fig. (i) shows the given views. Fig. (ii) shows visualized Isometric View.

Fig. (iii) shows Sectional F.V., Sectional T.V. and Sectional R.S.V. Note that the curved surface of the hole behind the S.P. line is shown as a rectangle in the S.F.V. and S.R.S.V.

Half-Sectional Views

For a symmetrical object, one half of a full section is to be drawn. Such a view is called *Half-Sectional View.*

To obtain a Half-Sectional View, the object is cut by two imaginary cutting planes perpendicular to each other. $\frac{1}{4}^{th}$ of the object is removed [Fig. (ii)]. Orthographic view of the remaining $\frac{3}{4}^{th}$ of the object is called Half-Sectional View.

Figs. (iv) & (v) show two possibilities of the Half-Sectional Front Views, Half-Sectional Top Views and Half-Sectional Right Side Views.

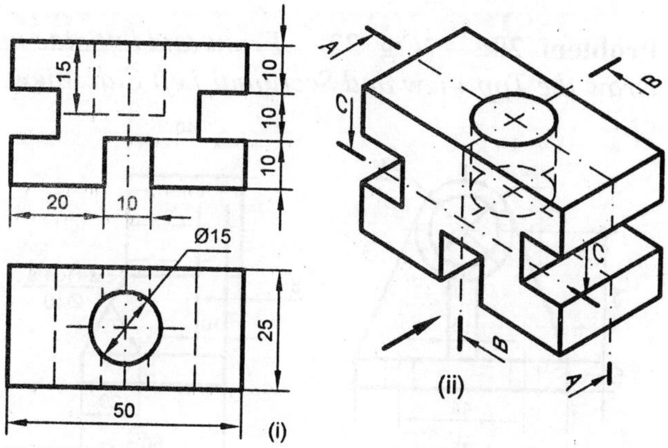

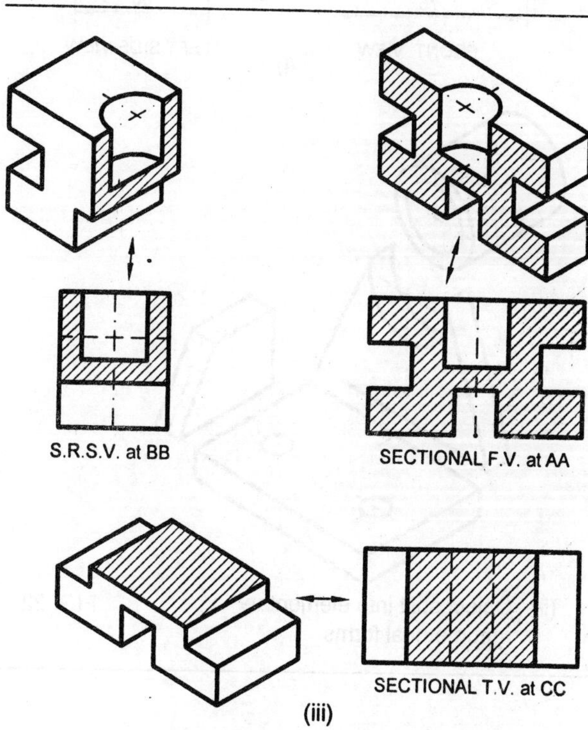

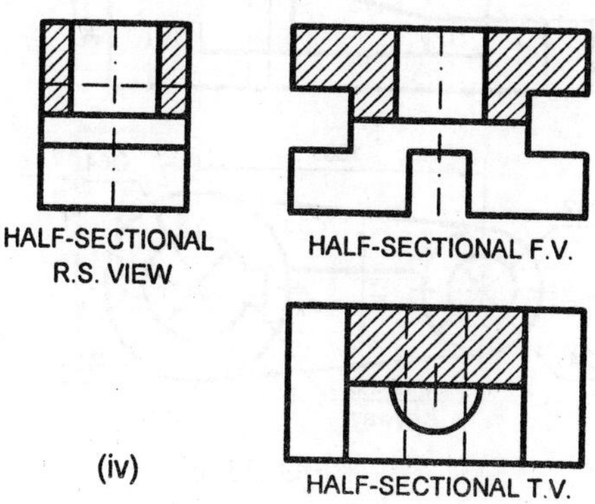

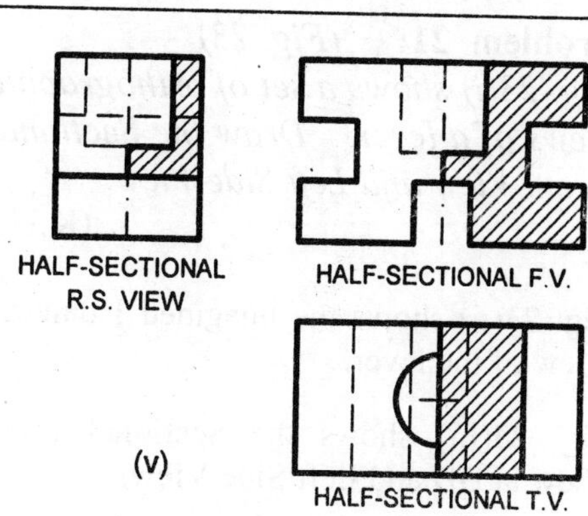

FIG. 21

Problem 20: *(Fig. 22) Front and left side views of an object are shown in Fig. 22(i). Draw the Top View and Sectional Left Side View.* (M.U.)

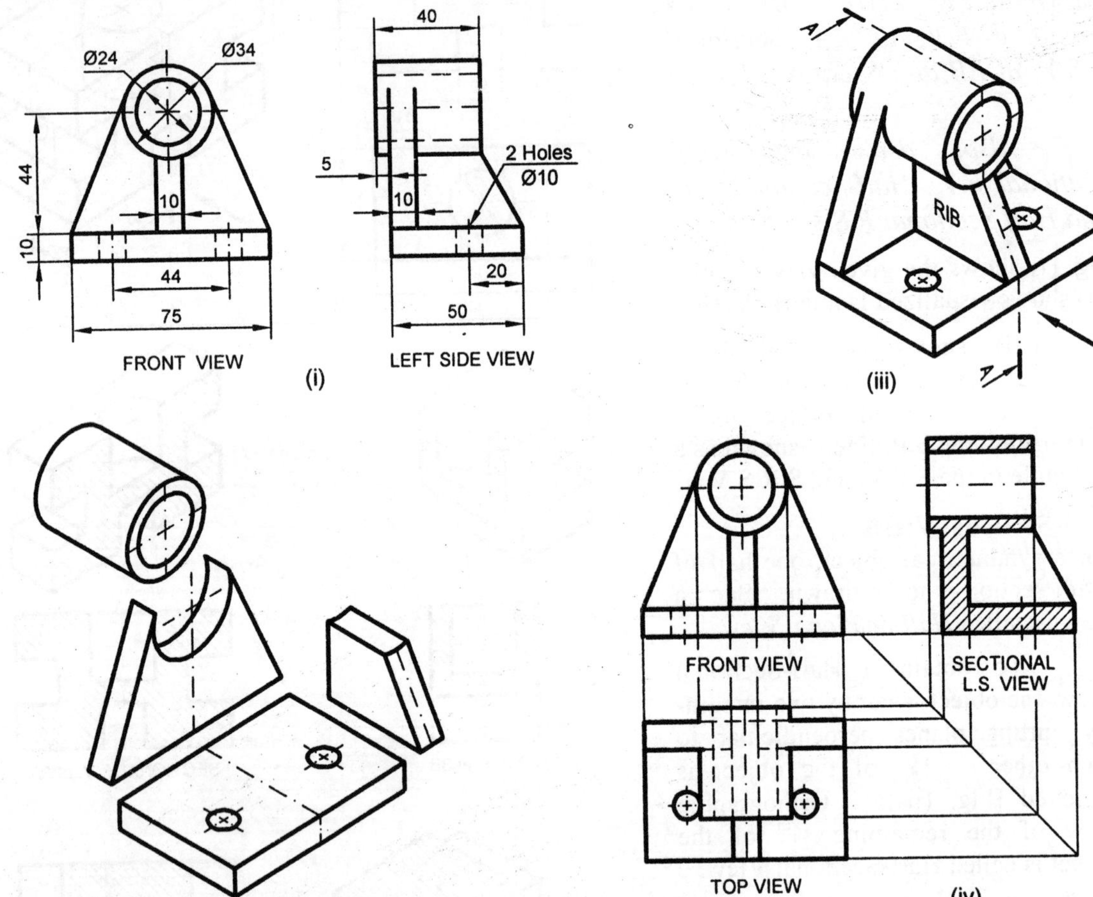

FRONT VIEW (i) LEFT SIDE VIEW

(iii)

(ii) Object split into elementary geometrical forms

FRONT VIEW

SECTIONAL L.S. VIEW

TOP VIEW

(iv)

FIG. 22

Problem 21: *(Fig. 23)*
Fig. 23(i) shows a set of orthographic views of a lever. Draw the Sectional Front View and Left Side View.

(K.U.)

Fig. 23(ii) shows the imagined Isometric View of the lever.

Fig. 23(iii) shows the Sectional Front View at DD and Left Side View.

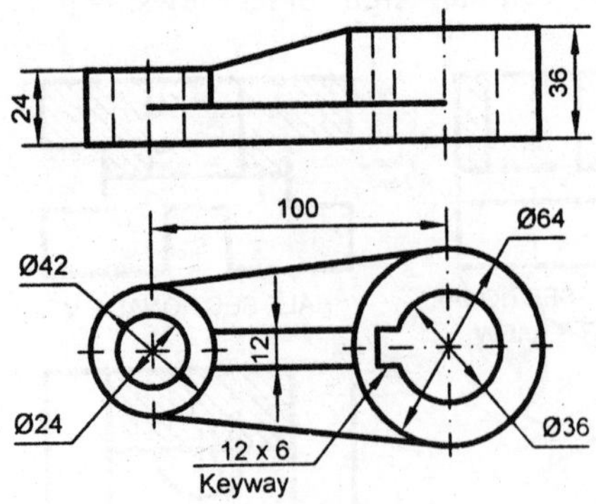

FIG. 23 (i)

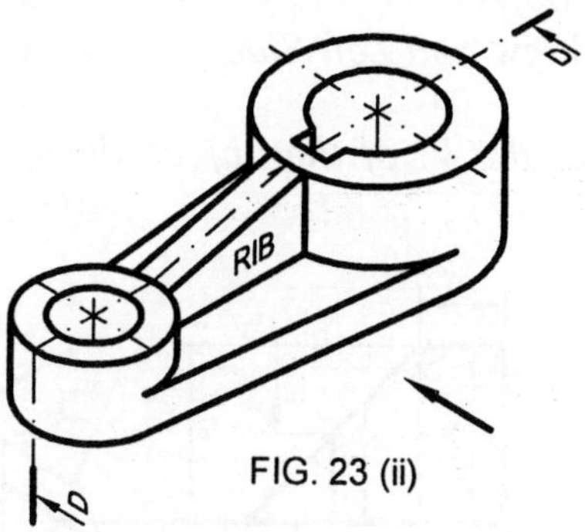

FIG. 23 (ii)

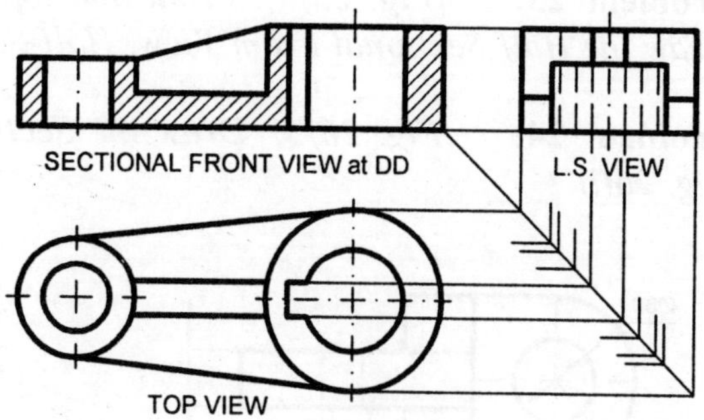

FIG. 23 (iii)

Problem 22: *(Fig. 24)*
Front and top views of an object are shown in Fig. 24(i). Draw the Sectional Front View at EE and Left Side View. (M.U.)

Fig. 24(ii) shows the imagined Isometric View of the object.

Fig. 24(iii) shows the Sectional Front View at EE and Left Side View.

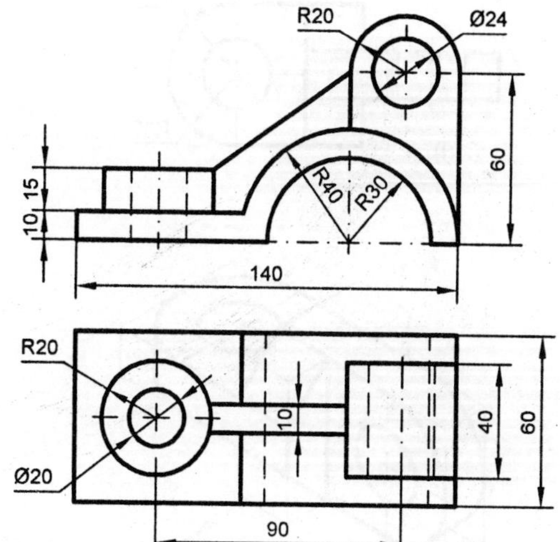

(i)

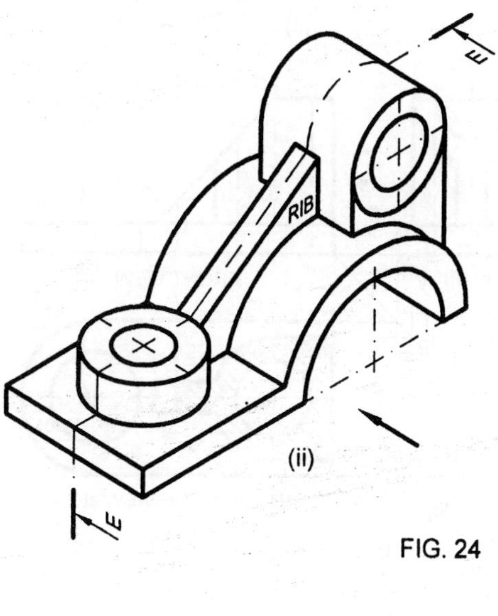

(ii)

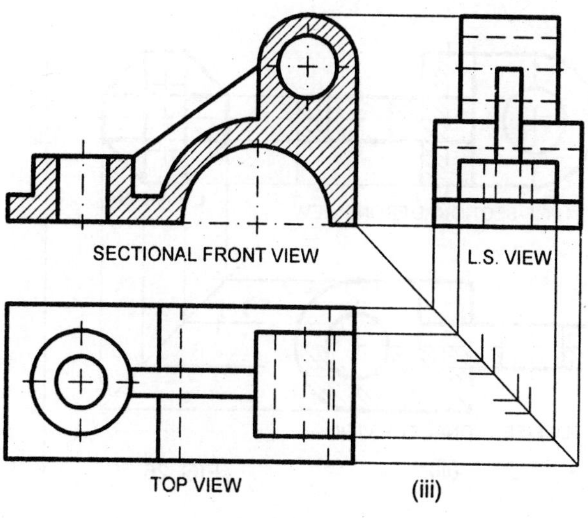

(iii)

FIG. 24

Problem 23: *(Fig. 25) Front and top views of an object are shown in Fig. 25(i). Draw the Half-Sectional Front View, Half-Sectional Top View and Left Side View.* (P.U.)

Problem 24: *(Fig. 26) Draw the Sectional T.V. and R.S.V. of the object shown in Fig. 26(i).* (G.U.)

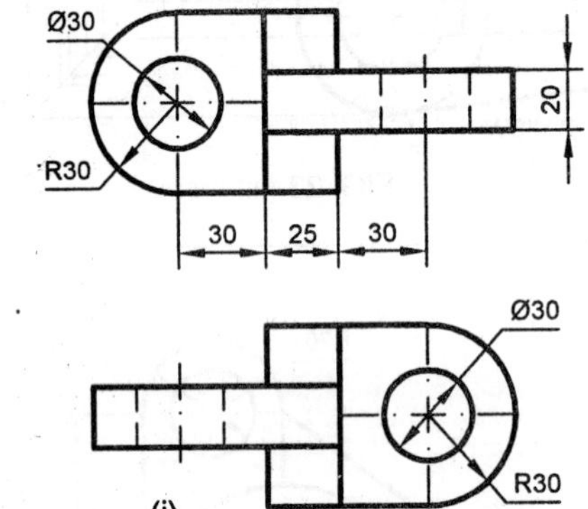

(i)

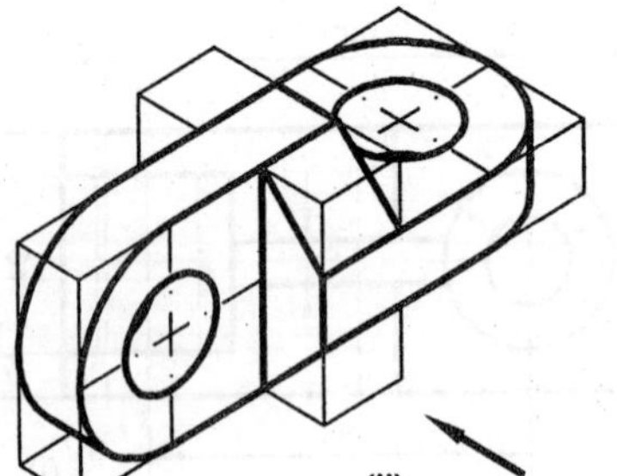
(ii)

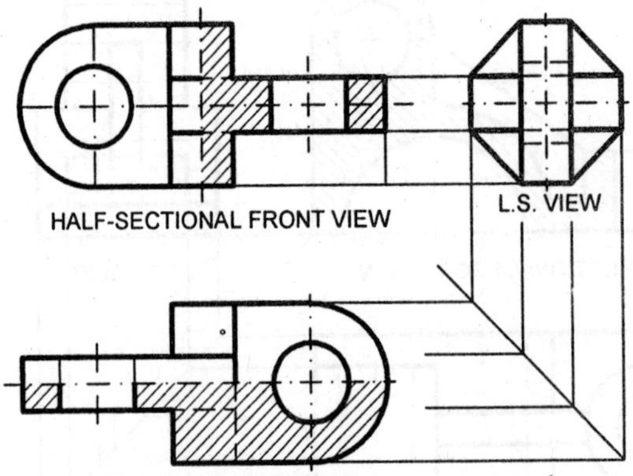

(iii)

FIG. 25

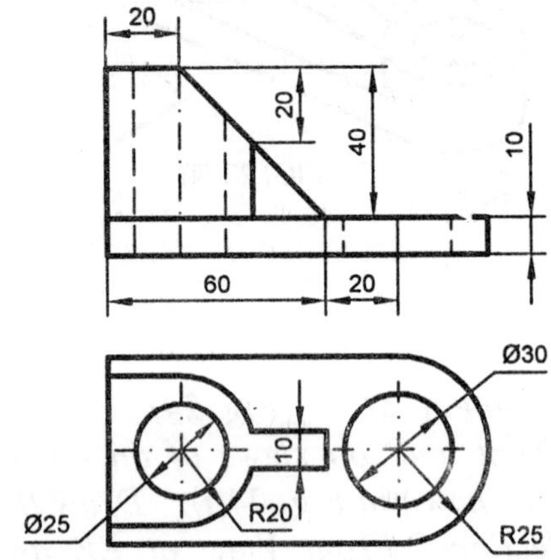

(i)

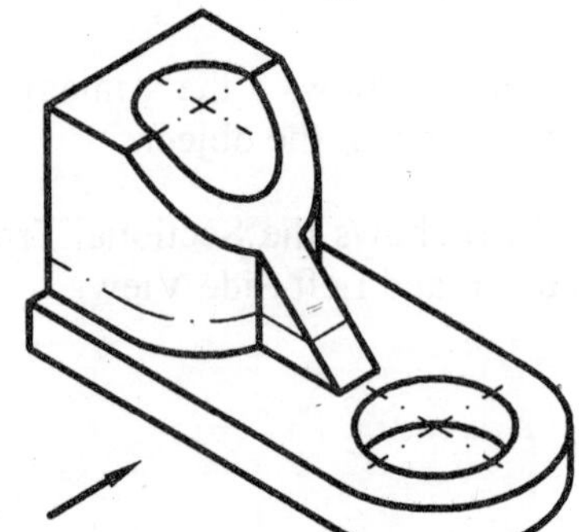
(ii)

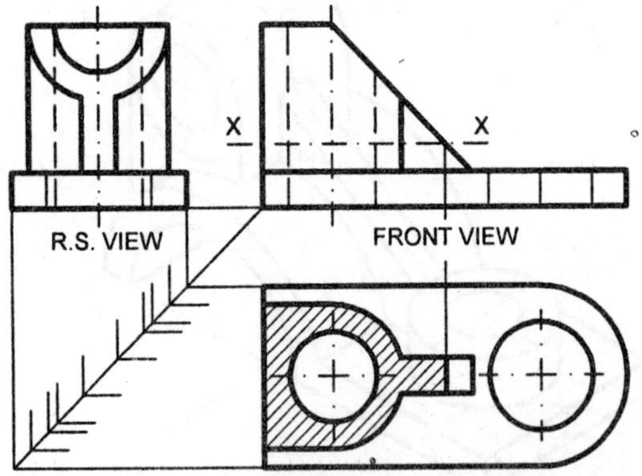

(iii)

FIG. 26

Problem 25: *(Fig. 27) Draw the S.F.V. and S.L.S.V. of the object in Fig. 27(i).* (G.U.)

Problem 26: *(Fig. 28) Draw the Half-Sectional F.V. and Sectional L.S.V. of the object in Fig. 28(i).* (PU)

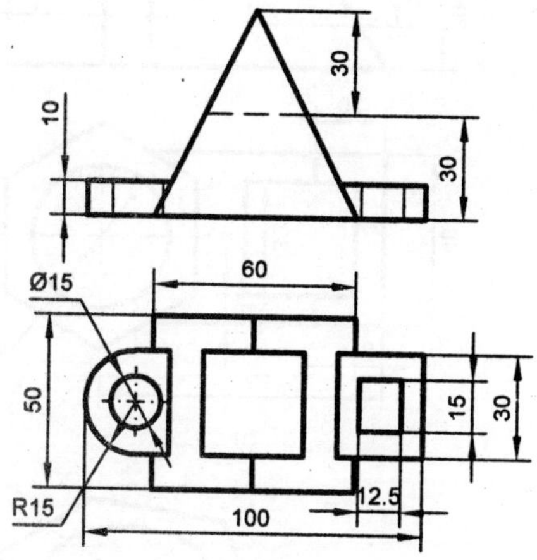

(i)

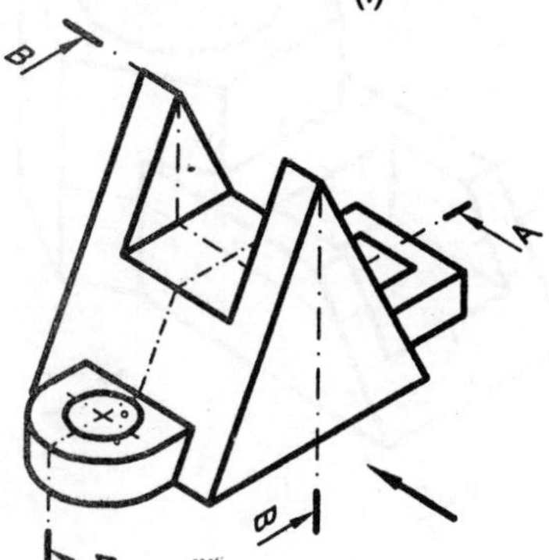

(ii)

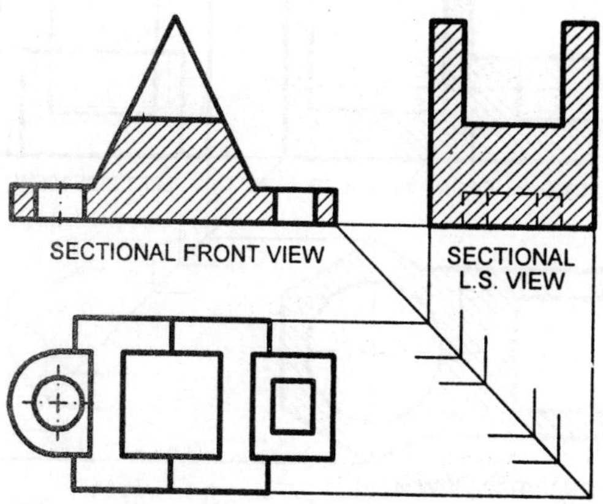

(iii) TOP VIEW

FIG. 27

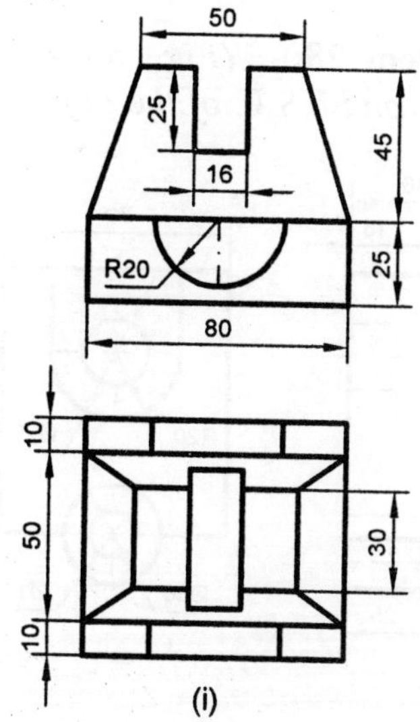

(i)

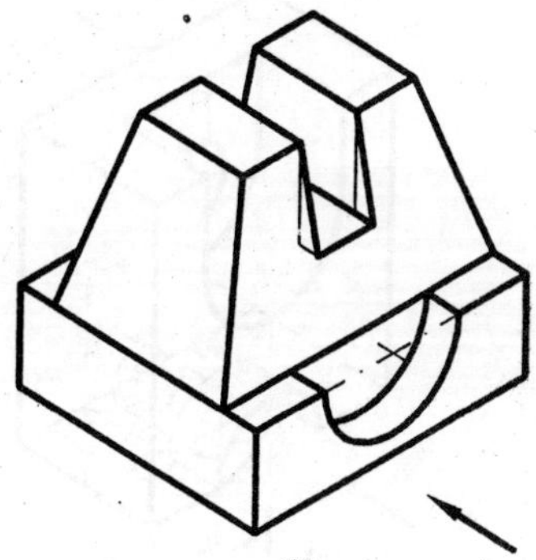

(ii)

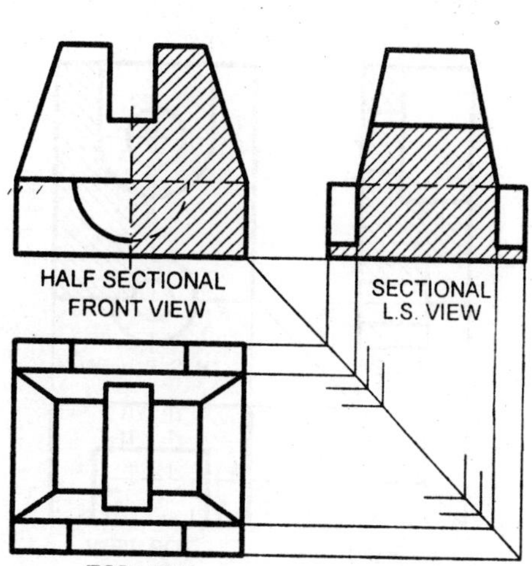

(iii) TOP VIEW

FIG. 28

Problem 27: *(Fig. 29) Draw the Half-Sectional F.V., T.V. and Sectional R.S.V. of the object in Fig. 29(i).* (G.U.)

Problem 28: *(Fig. 30) Draw the S.F.V., S.T.V. and L.S.V. of the object in Fig. 30(i).* (P.U.)

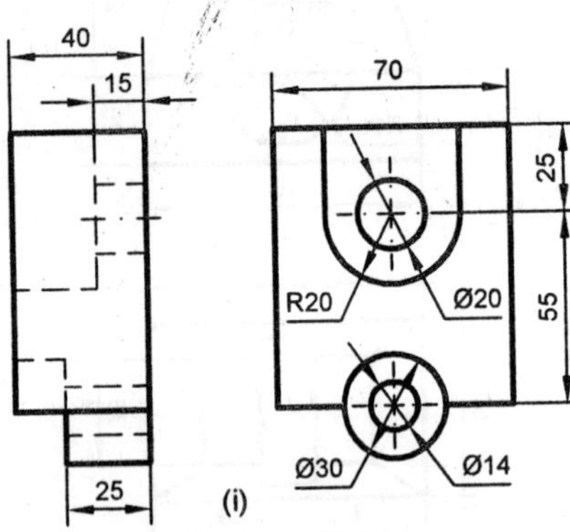

(i)

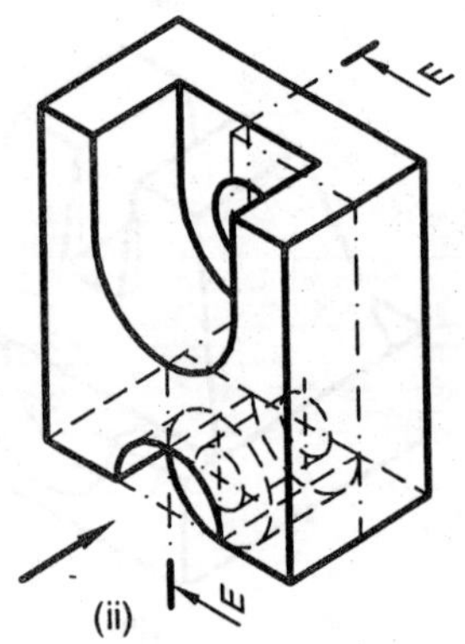

(ii)

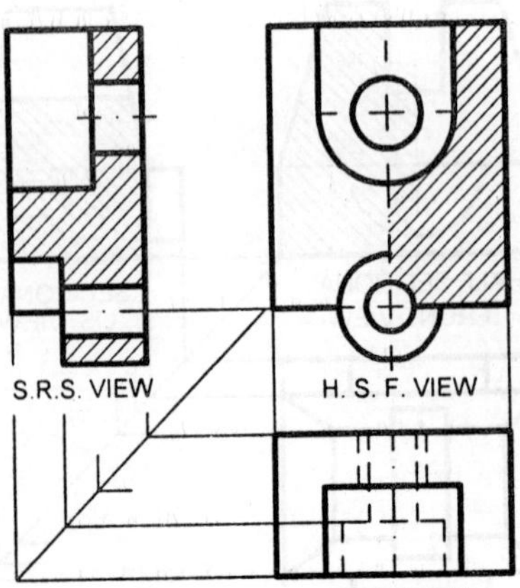

(iii)

FIG. 29

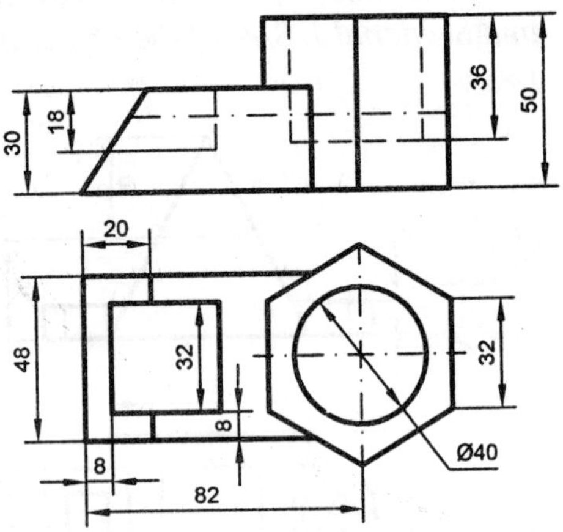

(i)

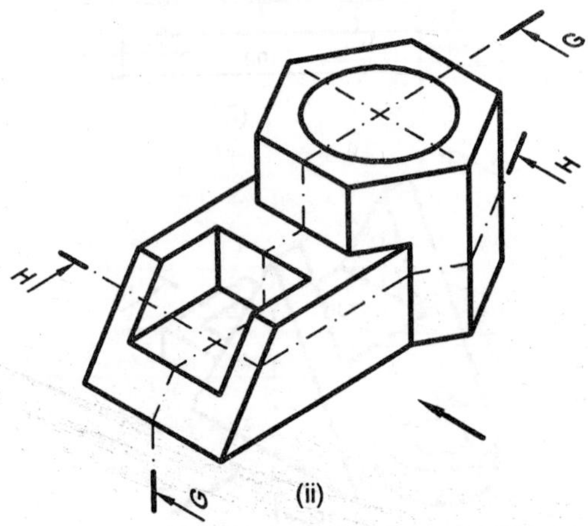

(ii)

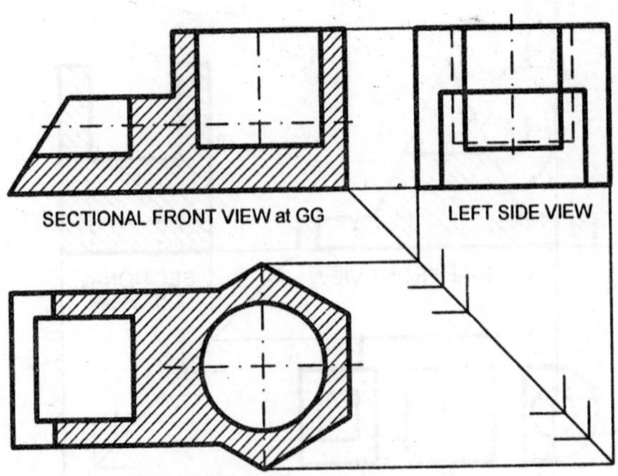

(iii)

FIG. 30

Problem 29: *(Fig. 31) Front and right side views of an object are shown in Fig. 31(i). Draw the Half-Sectional Top View and Half-Sectional Right Side View.*

Problem 30: *(Fig. 32) Front and top views of an object are shown in Fig. 32(i). Draw the Sectional Front View at KK, Half-Sectional Top View and Left Side View.*

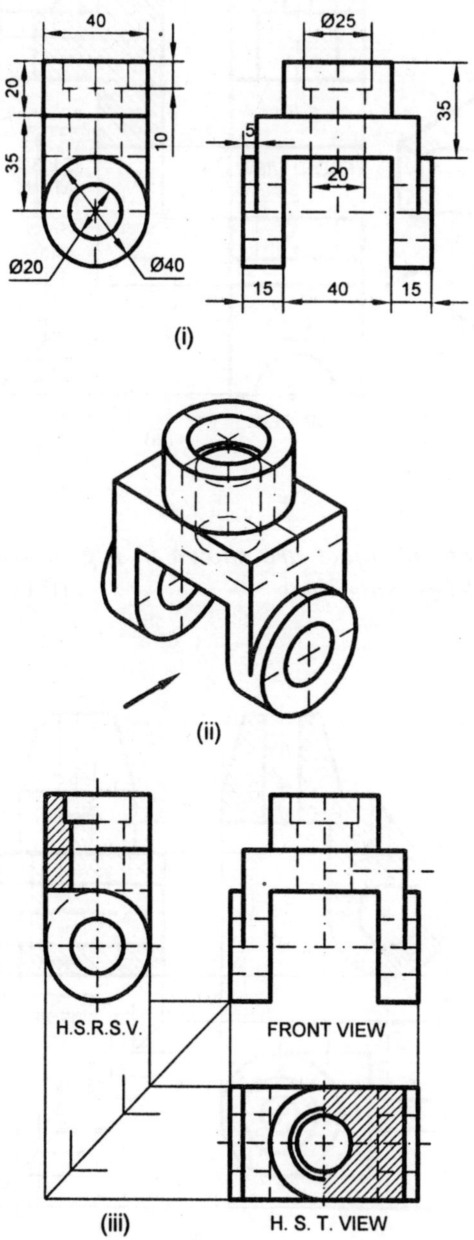

FIG. 31

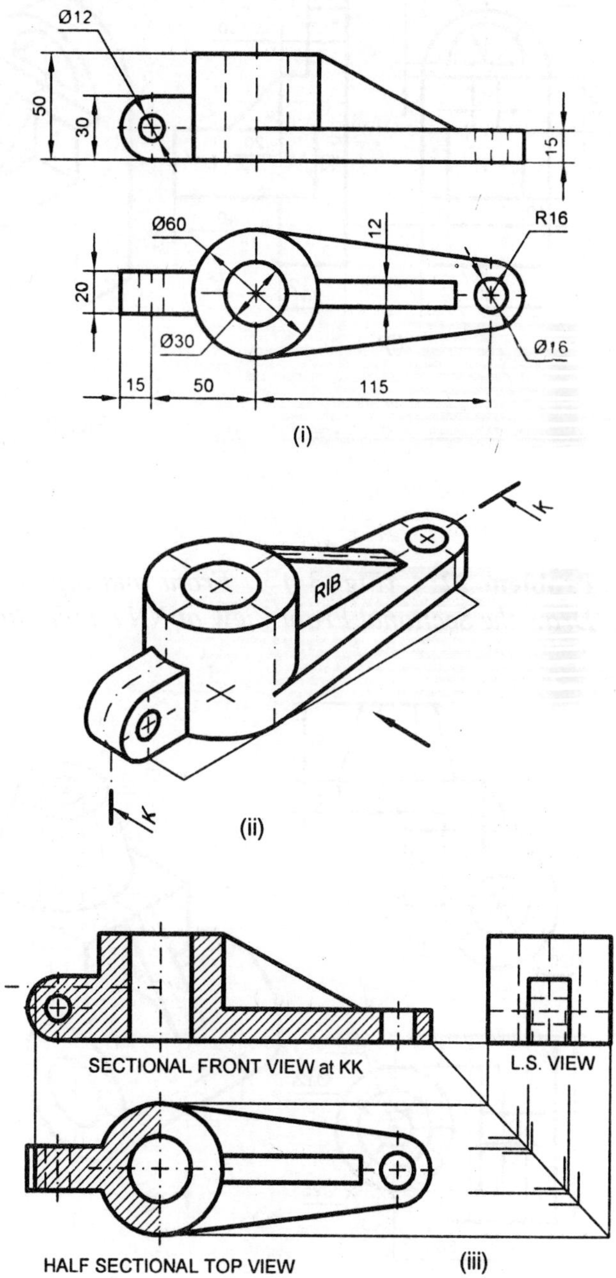

FIG. 32

Problem 31: *(Fig. 33) Front and left side views of an object are shown in Fig. 33(i). Draw the Top View and Sectional Left Side View.* (K.U.)

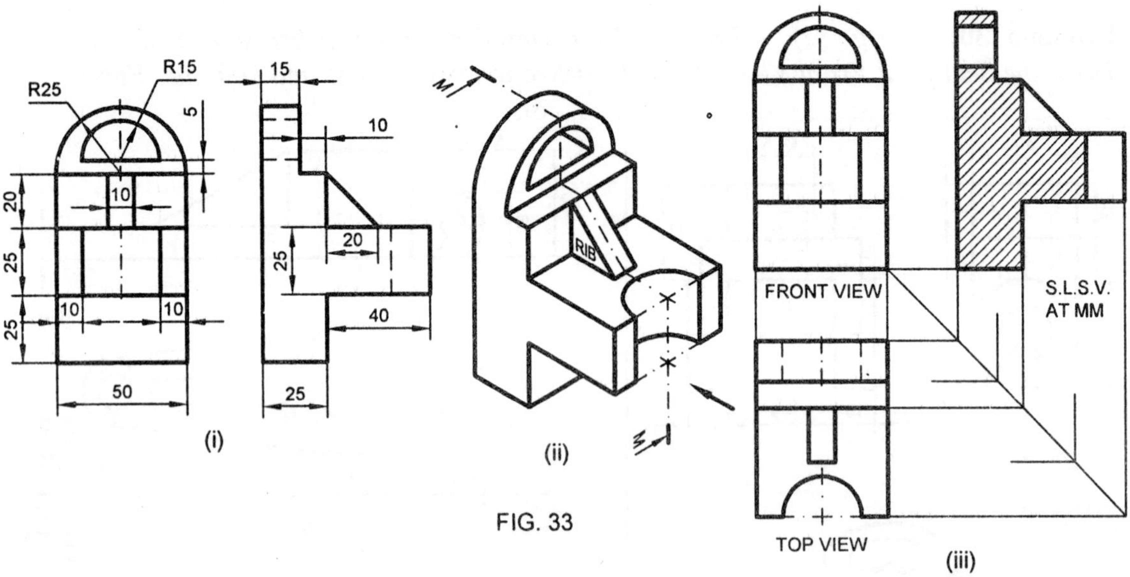

FIG. 33

Problem 32: *(Fig. 34) Front and top views of an object are shown in Fig. 34(i). Draw the Sectional Front View at NN and Sectional Left Side View.* (P.U.)

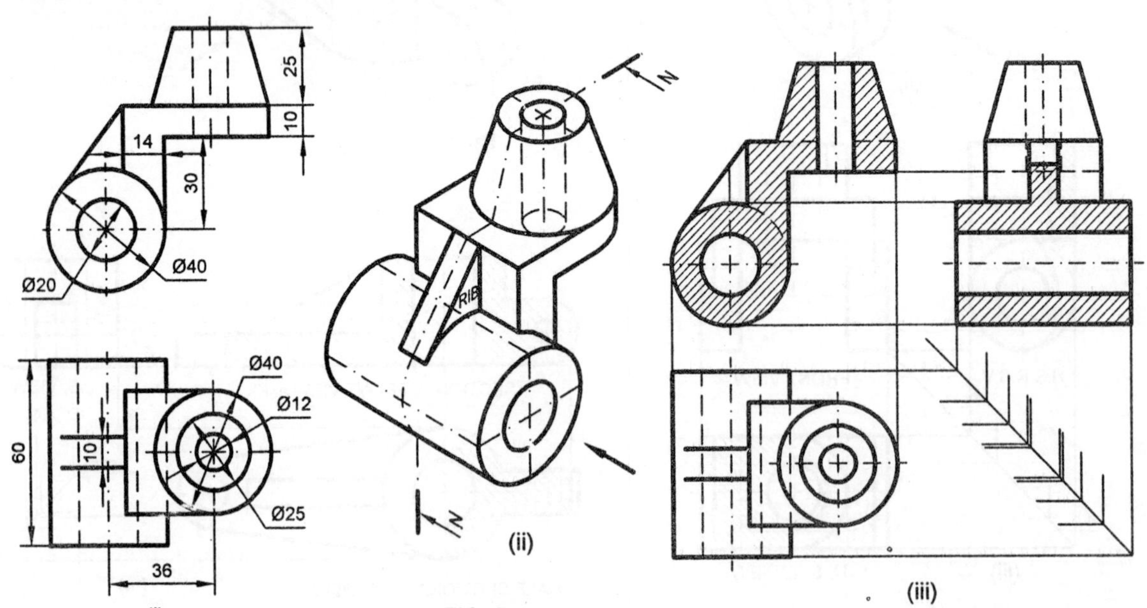

FIG. 34

Problem 33: *(Fig. 35)* *Front and right side views of an object are shown in Fig. 35(i). Draw the Sectional Front View at SS, Top View and Sectional Left Side View at RR.* (M.U.)

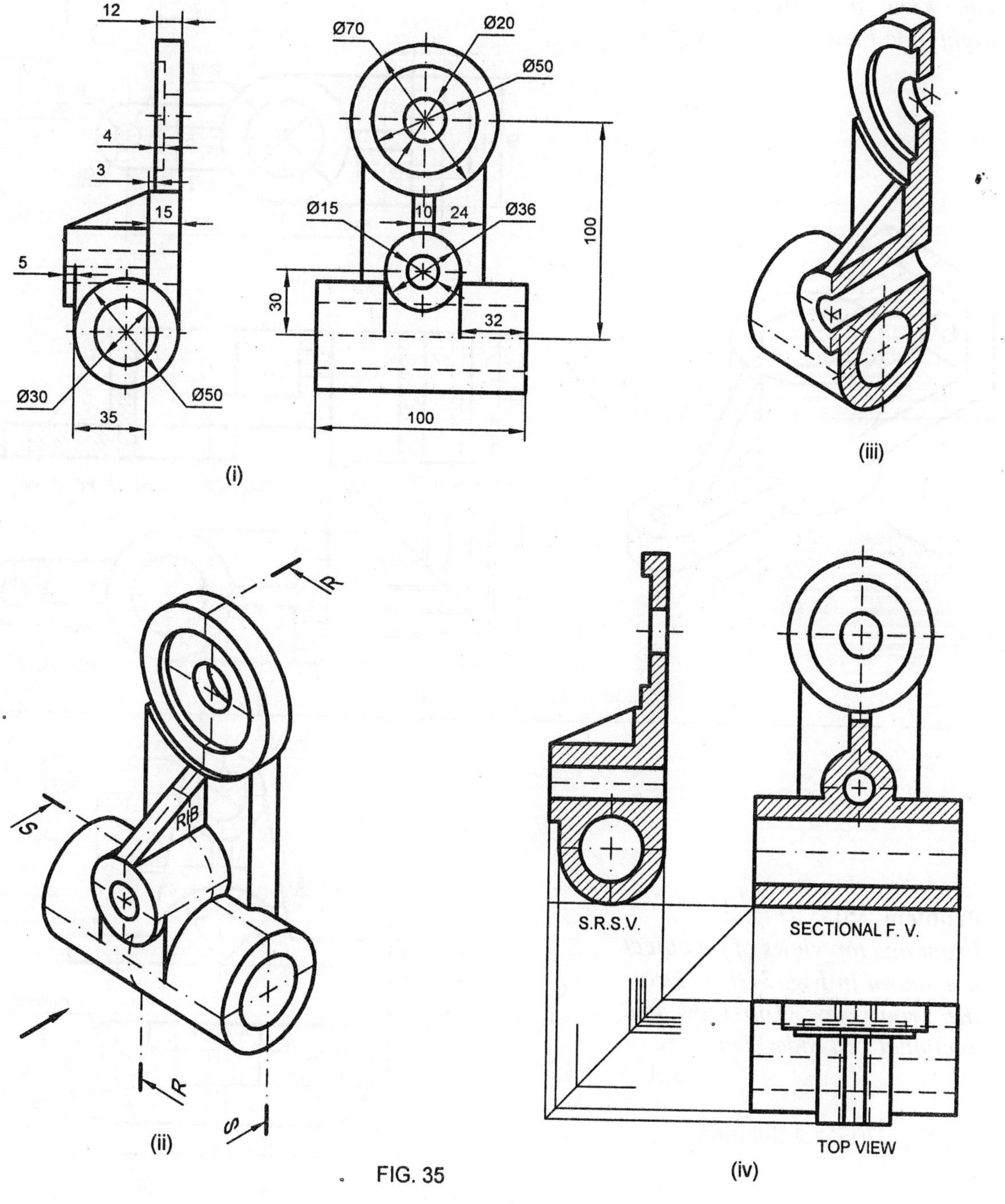

FIG. 35

Problem 34: *(Fig. 36)*
Front and top views of an object are shown in Fig. 36(i). Draw the Half-Sectional Front View, Top View and Half-Sectional Right Side View. (G.U.)

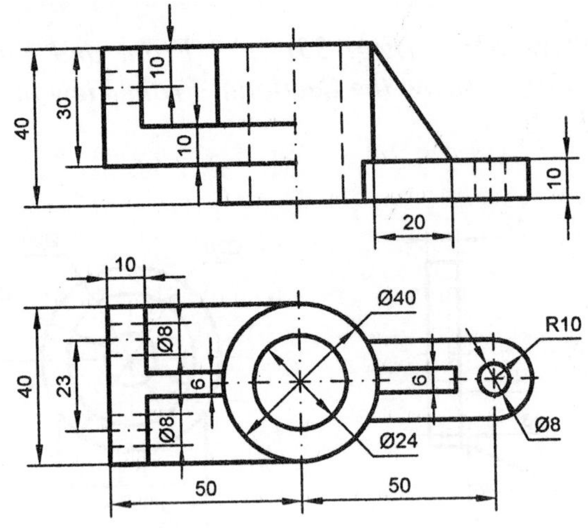

(i)

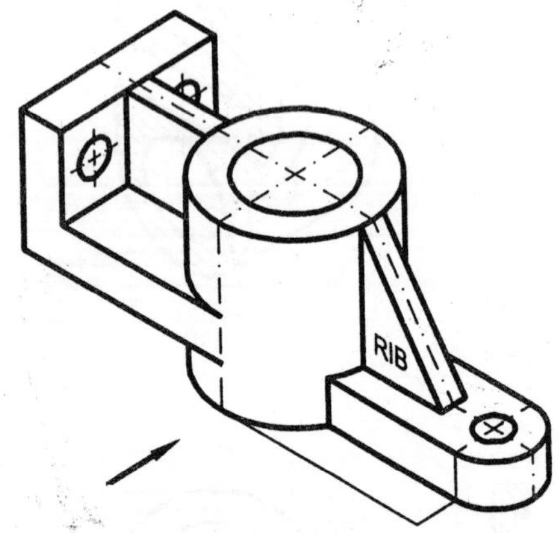

(ii)

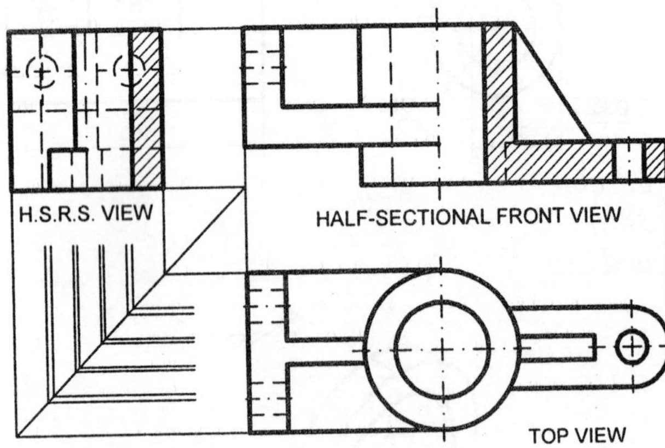

FIG. 36

(iii)

Problem 35: *(Fig. 37)*
Front and top views of an object are shown in Fig. 37(i). Draw the Front View, Top View and Sectional Left Side View.
(K.U.)

See Fig. 37(iii) for Solution.

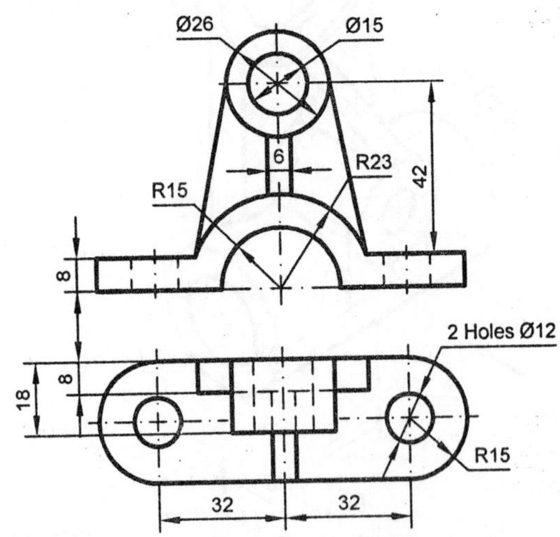

FIG. 37 (i)

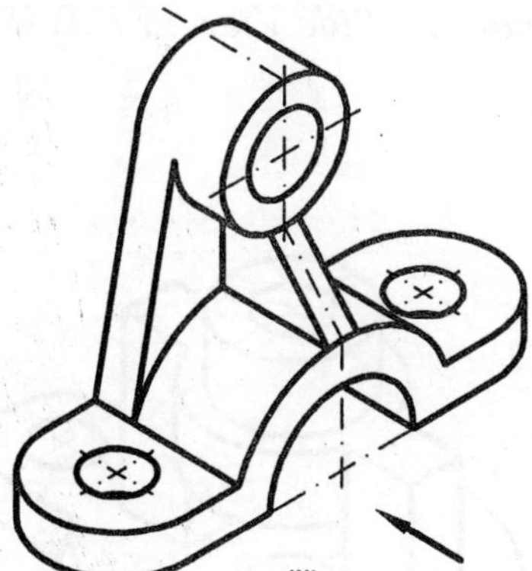

(ii)

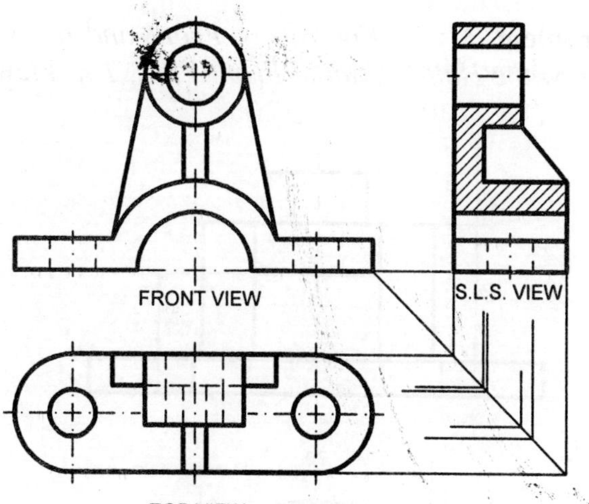

FIG. 37 (iii)

Problem 36: *(Fig. 38)* *Front and left side views of an object are shown in Fig. 38(i). Draw the Top View and Sectional Left Side View.* (M.U.)

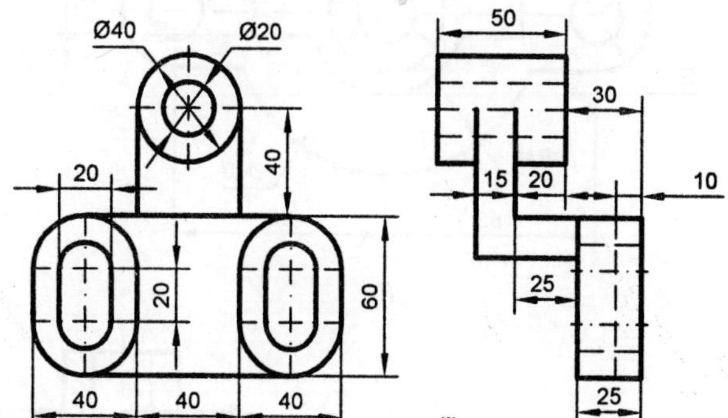

(i)

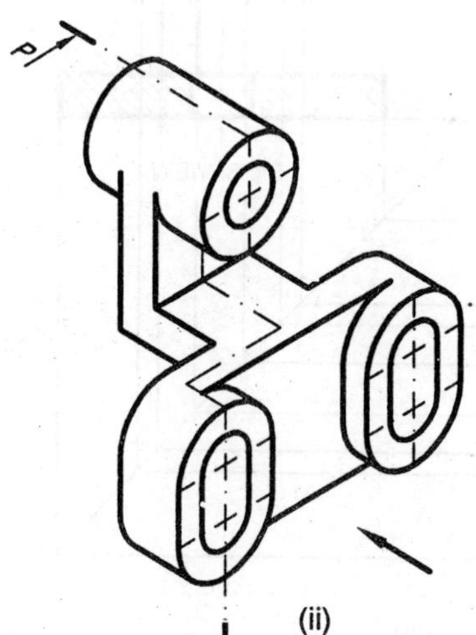

(ii)

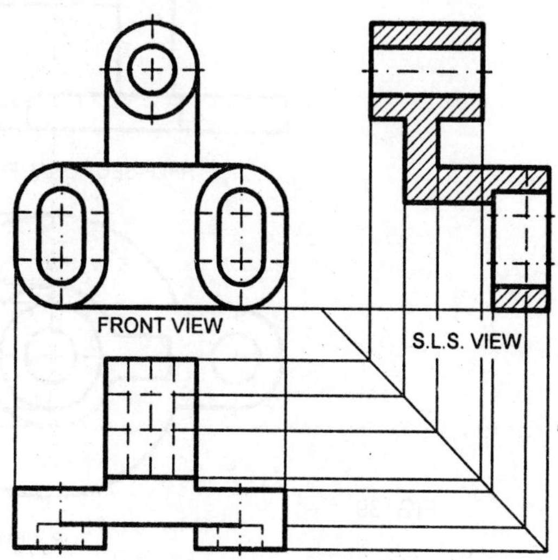

FIG. 38 (iii)

Problem 37: *(Fig. 39) Front and top views of an object are shown in Fig. 39(i). Draw the Half-Sectional Front View, Top View and Sectional Left Side View.* (W.B.U.T.)

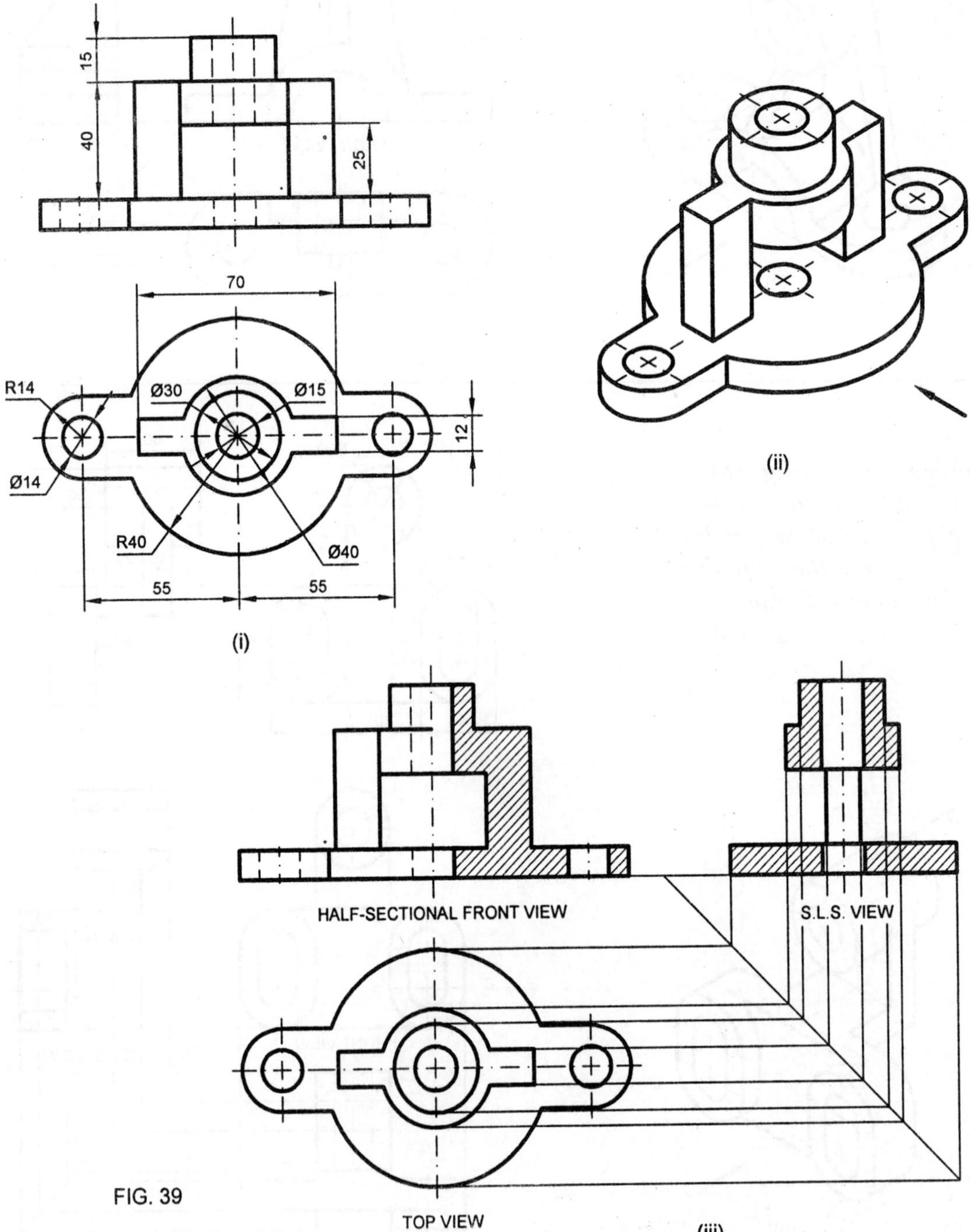

(i) (ii) (iii)

FIG. 39

Problem 38: *(Fig. 40) Front and top views of an object are shown in Fig. 40(i). Draw the Sectional Front View, Top View and Left Side View.* (P.U.)

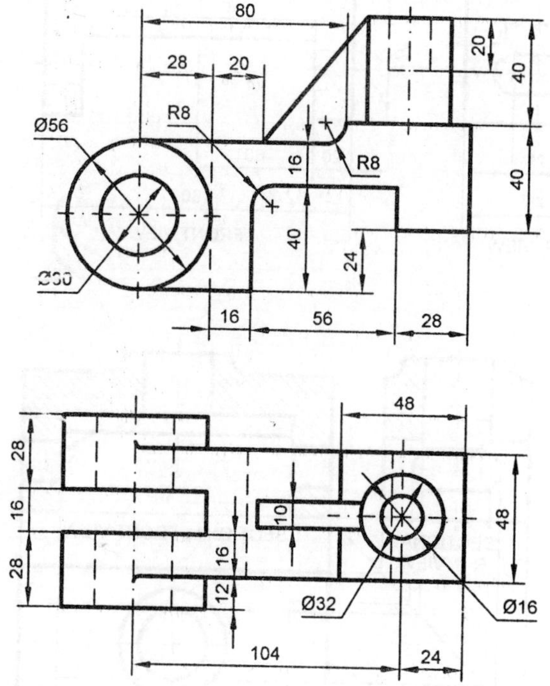

(i)

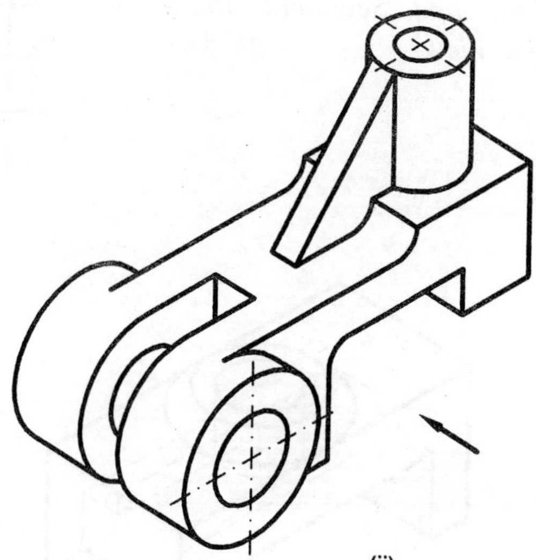

(ii)

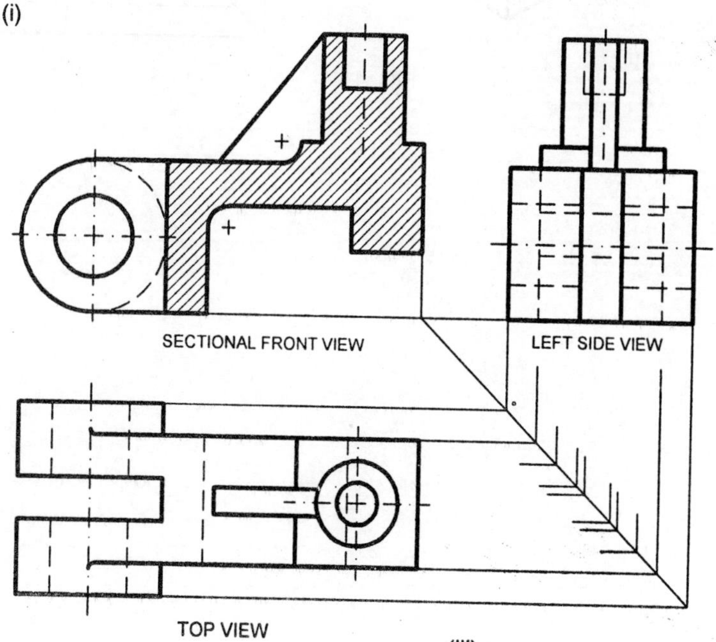

FIG. 40

(iii)

Problem 39: *(Fig. 41)*
Front and right side views of an object are shown in Fig. 41(i). Draw the Sectional Front View, Top View and Sectional Right Side View. (G.U.)

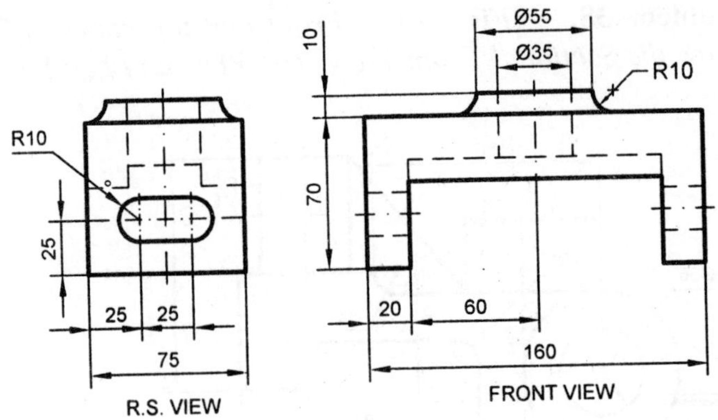

(i)

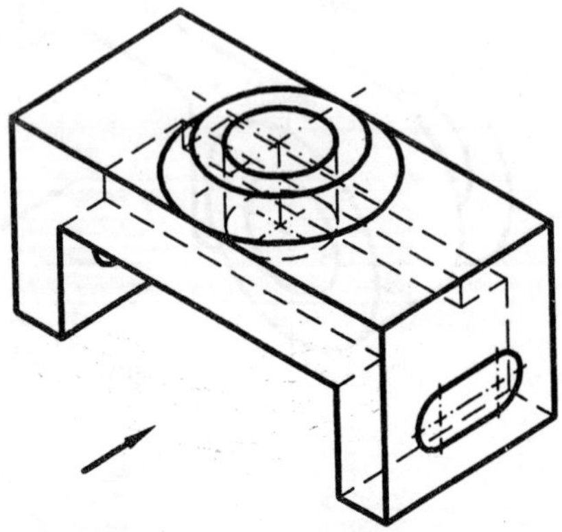

(ii)

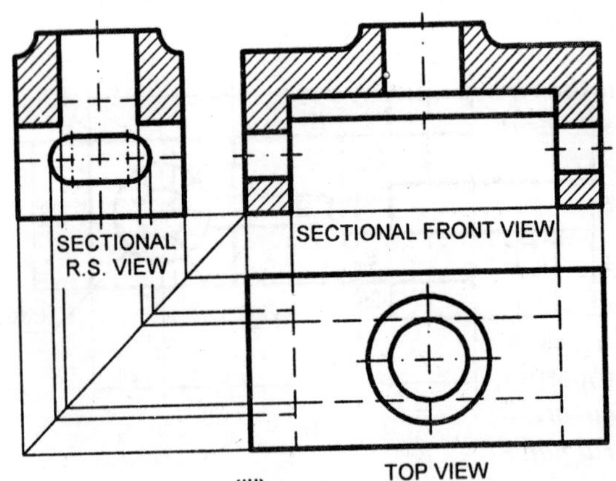

(iii)

FIG. 41

*

23

FREEHAND SKETCHING TECHNIQUES

Freehand sketching means making sketches freehand without the use of drawing instruments. It is not drawn to scale, but should be in *good proportion* by eye judgement. ***In a freehand sketch, all the rules of the graphic language (i.e., B.I.S. Conventions) are as rigidly observed as in a scale-drawing.***

1. PROPORTIONING IN SKETCHING

To sketch rapidly, learn to judge lengths, recognize proportions and compare dimensions by *eye judgment.* ***Until you are able to do this, you may practice using rectangular grid paper or follow the following graphical methods:***

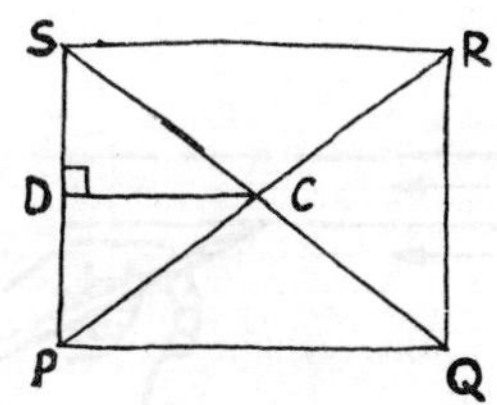

FIG. 1(i)

1.1 To Fix the Midpoint of a Rectangle and its Sides

See Fig. 1(i). Sketch the diagonals of rectangle PQRS. They meet to give center C of the rectangle.

Through C, sketch a line perpendicular to the side SP to obtain its center D.

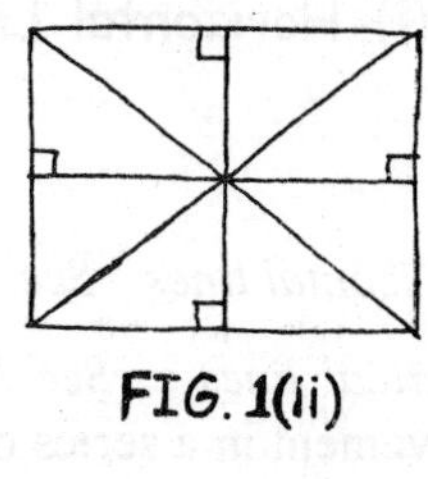
FIG. 1(ii)

1.2 To Subdivide the Rectangle into Quarters

See Fig. 1(ii). Repeat the above midpoint construction to obtain four quadrants or quarters.

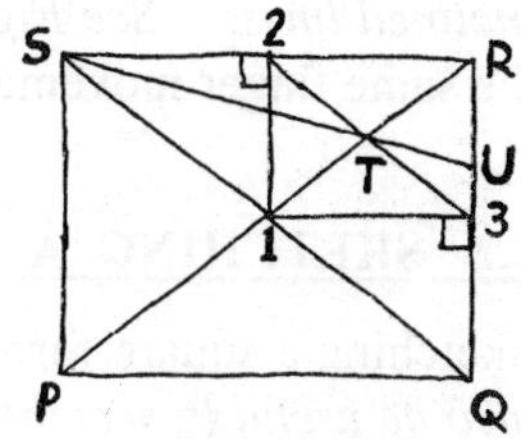

FIG. 1(iii)

1.3 To Subdivide the Rectangle into Thirds

See Fig 1(iii). PQRS is the given rectangle. Sketch the quadrant 12R3. T is the quadrant center. Sketch a line from S through T to meet RQ at U. Now RU = (1/3) RQ.

2. TECHNIQUES OF FREEHAND SKETCHING

2.1 SKETCHING STRAIGHT LINES

See Fig. 2. Mark end points of the line. Make a few trial movements between end points. Move the pencil in 2 or 3 sweeps and sketch a thin line between end points.

Keep your eye in the direction of sketching the line, but not on the pencil point. Waviness, if any, may be corrected.

Darken the finished line in one stroke such that it is uniform and black.

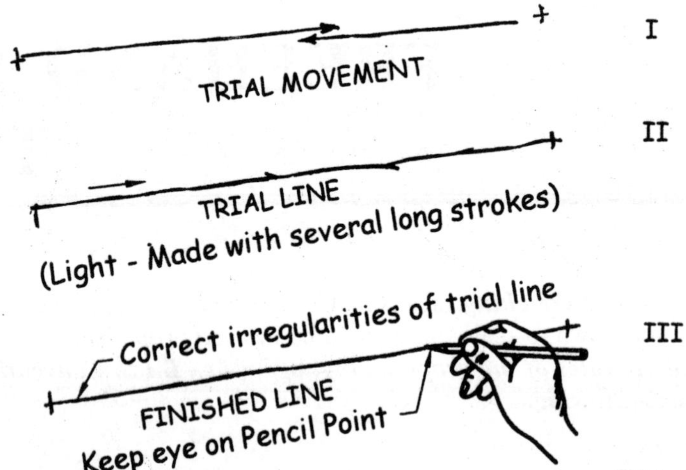

FIG. 2

2.1.1 Sketching Horizontal, Vertical and Inclined Lines

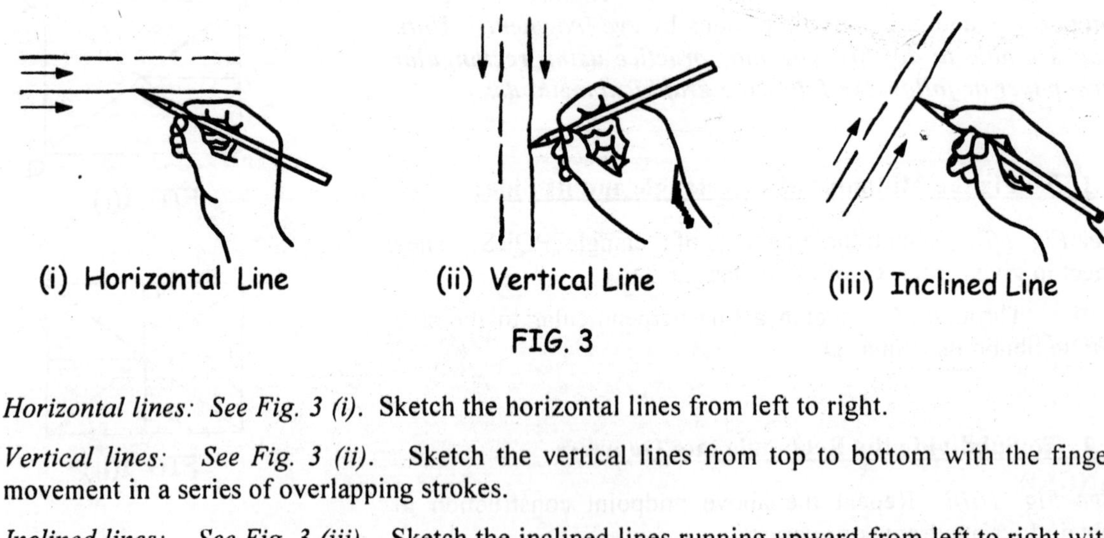

FIG. 3

Horizontal lines: See Fig. 3 (i). Sketch the horizontal lines from left to right.

Vertical lines: See Fig. 3 (ii). Sketch the vertical lines from top to bottom with the finger movement in a series of overlapping strokes.

Inclined lines: See Fig. 3 (iii). Sketch the inclined lines running upward from left to right with the same finger movement used for the horizontal lines.

2.2 SKETCHING A SQUARE

Sketching a square forms an exercise on sketching of straight lines. *Sketching the square will also be useful in sketching circles.*

2.3 SKETCHING CIRCLES

There are *two* methods to sketch a circle freehand.

2.3.1 Sketching CIRCLE in a Square

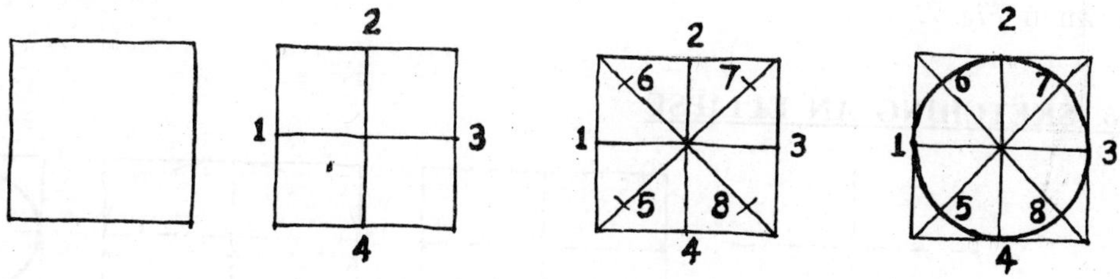

FIG. 4

See Fig. 4. Sketch the enclosing square in *thin* lines. Mark the midpoints of the sides of the square and draw the center lines. Now the center O and radius of the circle are obtained. Also, sketch the diagonals on which mark the points for the diameter by eye judgement. Sketch the arcs tangential to the sides of the square. Sketch a smooth circle passing through the eight points so obtained.

2.3.2 Sketching LARGE CIRCLE using Radial Lines

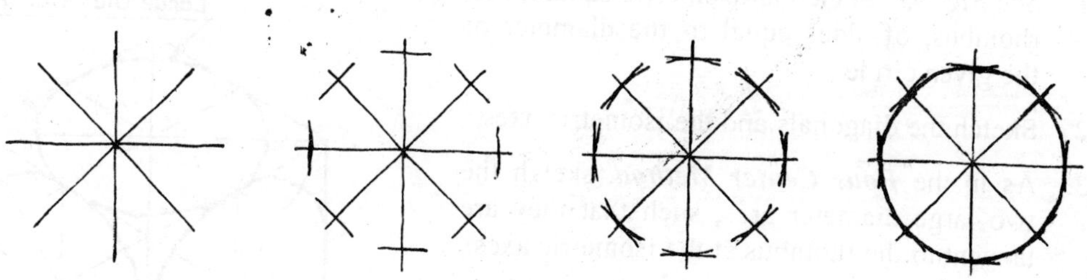

FIG. 5

See Fig. 5. In this method, sketch thin radial lines, say, 8 in number. Sketch short perpendiculars at the end of each radial line at the radial distance from the center. Sketch short arcs tangential to these perpendiculars to approximate the circle. Complete the large circle as shown in *Fig. 5.*

2.4 SKETCHING ANGLES

Figs. 6 (i) & (ii) show sketching of isometric axes.

Fig. 6 (iii) shows sketching of 45° lines.

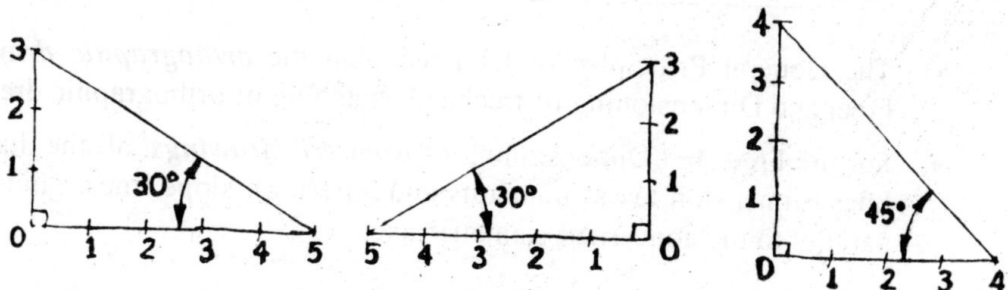

FIG. 6 (i) (ii) (iii)

2.5 SKETCHING ARCS / CURVES

Arcs and curves are sketched by drawing radial lines (similar to circles) as shown in *Fig. 7*.

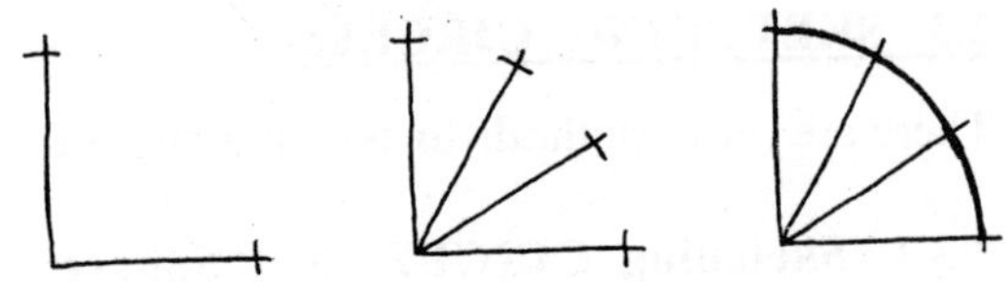

FIG. 7

2.6 SKETCHING AN ELLIPSE

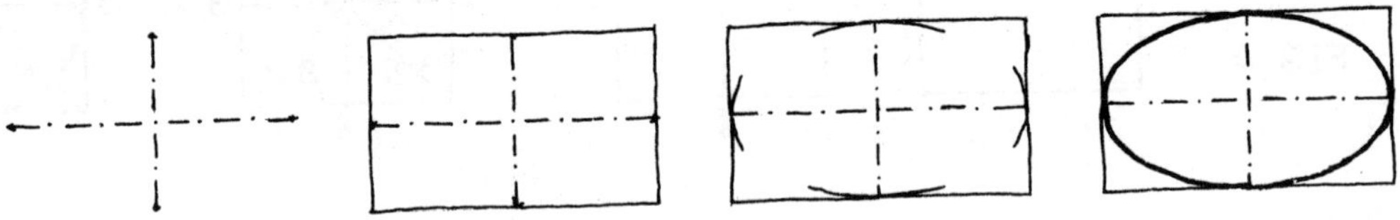

FIG. 8

See Fig. 8. Sketch the major and minor axes of the ellipse as shown. Sketch a rectangle with sides equal to the major and minor axes of the ellipse. Sketch short arcs tangential to the sides of the rectangle, *i.e.*, at the ends of the major and minor axes. Complete the ellipse.

2.7 SKETCHING ELLIPSE IN PICTORIAL VIEW

1. *See Fig. 9.* Sketch an isometric square, *i.e.*, rhombus, of sides equal to the diameter of the given circle.
2. Sketch the diagonals and the isometric axes.
3. As in the ***Four Center Method***, sketch the two large diameter arcs, such that they are tangent to the rhombus at the isometric axes.
4. Sketch the two small diameter arcs, tangent to the rhombus at the isometric axes.
5. *Fig. 9* shows three positions of an isometric circle.

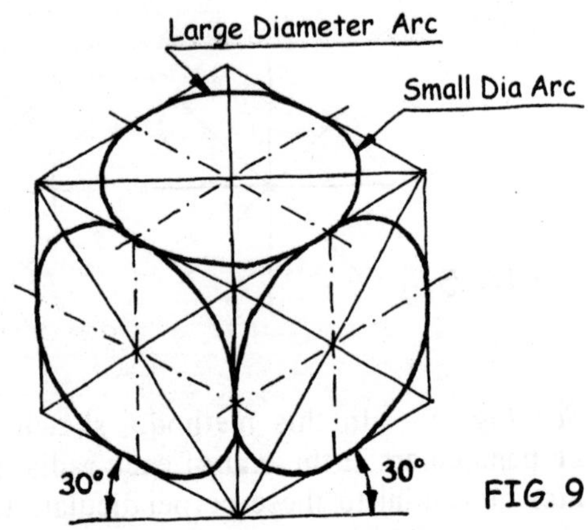

FIG. 9

3. FREEHAND DIMENSIONING

- The General Principles of Dimensioning the ***orthographic drawings*** are applicable to Freehand Dimensioning of freehand sketching of orthographic drawings also.
- For the Freehand Dimensioning of ***isometric drawings***, all the dimension lines, extension lines, dimension texts (numerals and lettering), slope lines, guide lines, etc., are marked parallel to the concerned isometric axes.

*

24

FREEHAND SKETCHING OF ORTHOGRAPHIC VIEWS FROM ISOMETRIC VIEWS

1. TYPICAL PRACTICAL PROBLEMS – Engineering Components

Sketch the following views of the objects shown: (a) Front View, (b) Top View and (c) Side View as viewed from the side available for view. Mark the dimensions.

(1)

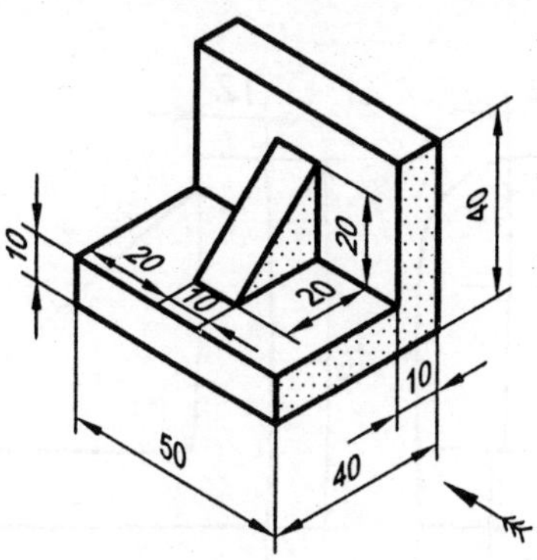

FIG. 1 RIBBED ANGLE PLATE

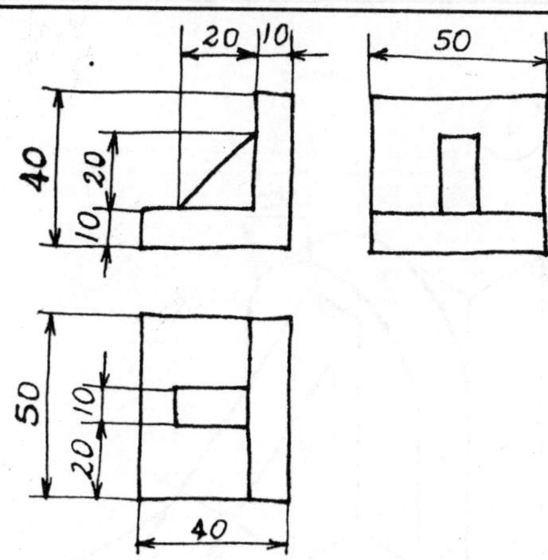

(2)

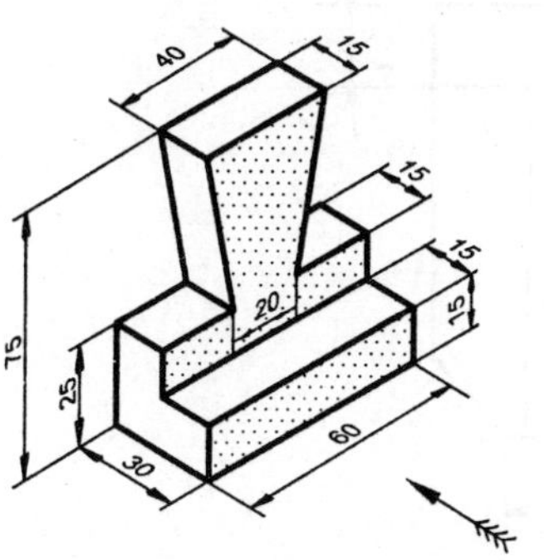

FIG. 2 DOVETAIL BLOCK

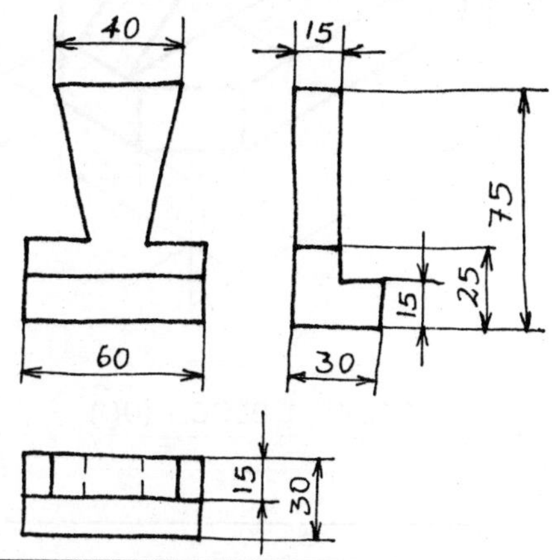

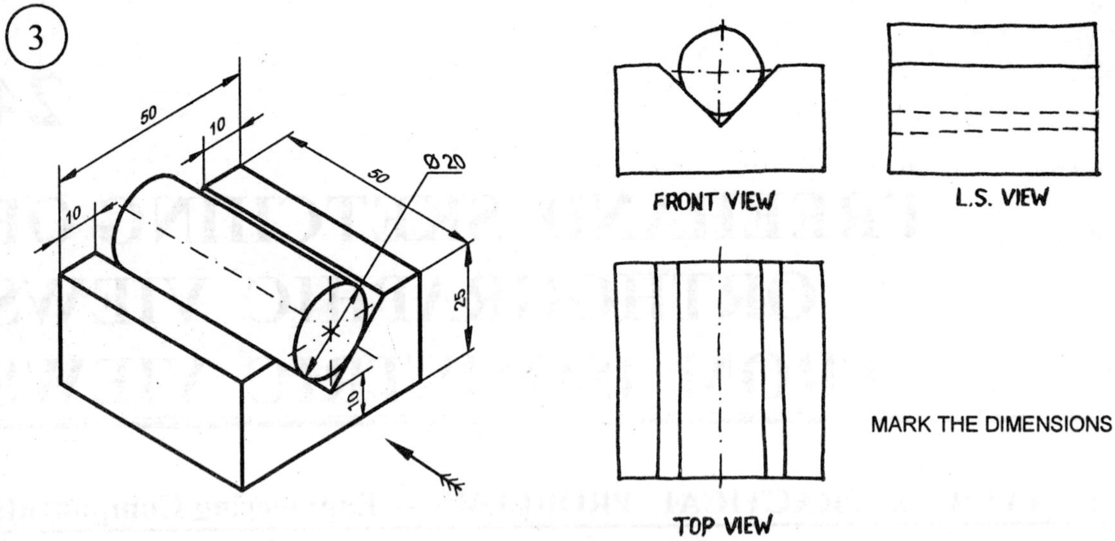

FIG. 3 V - BLOCK WITH CIRCULAR ROD (UQ)

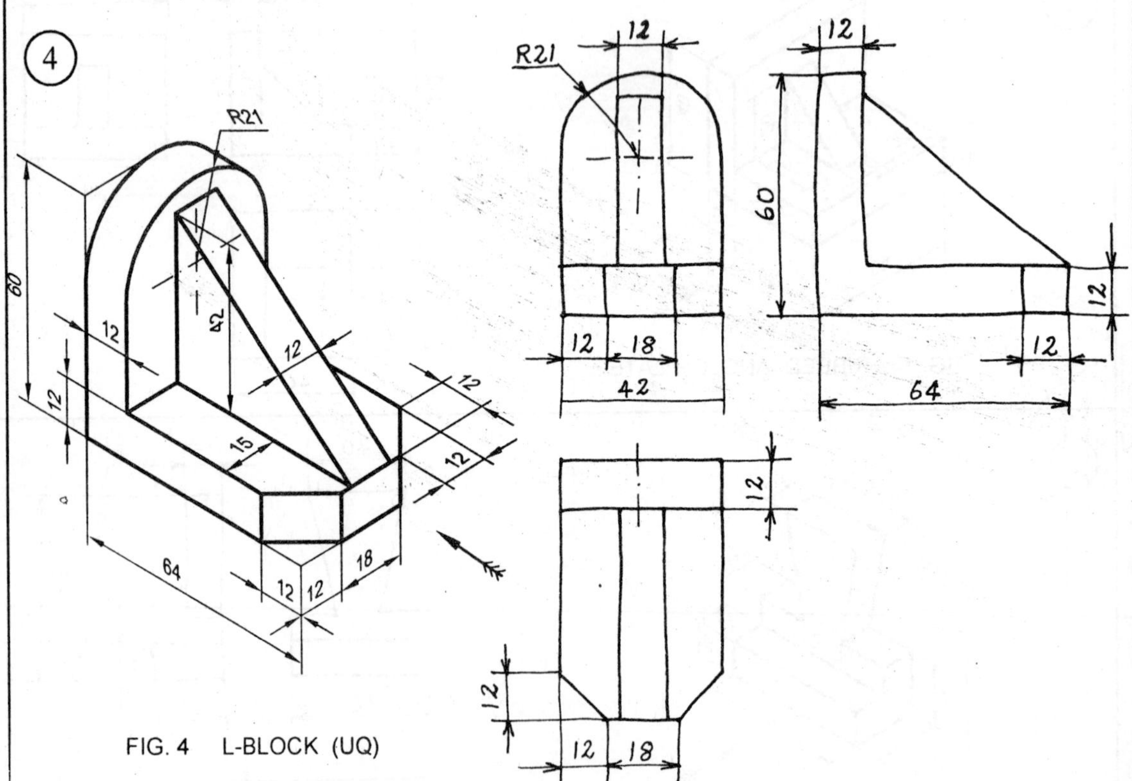

FIG. 4 L-BLOCK (UQ)

(5) BASE BLOCK (UQ)

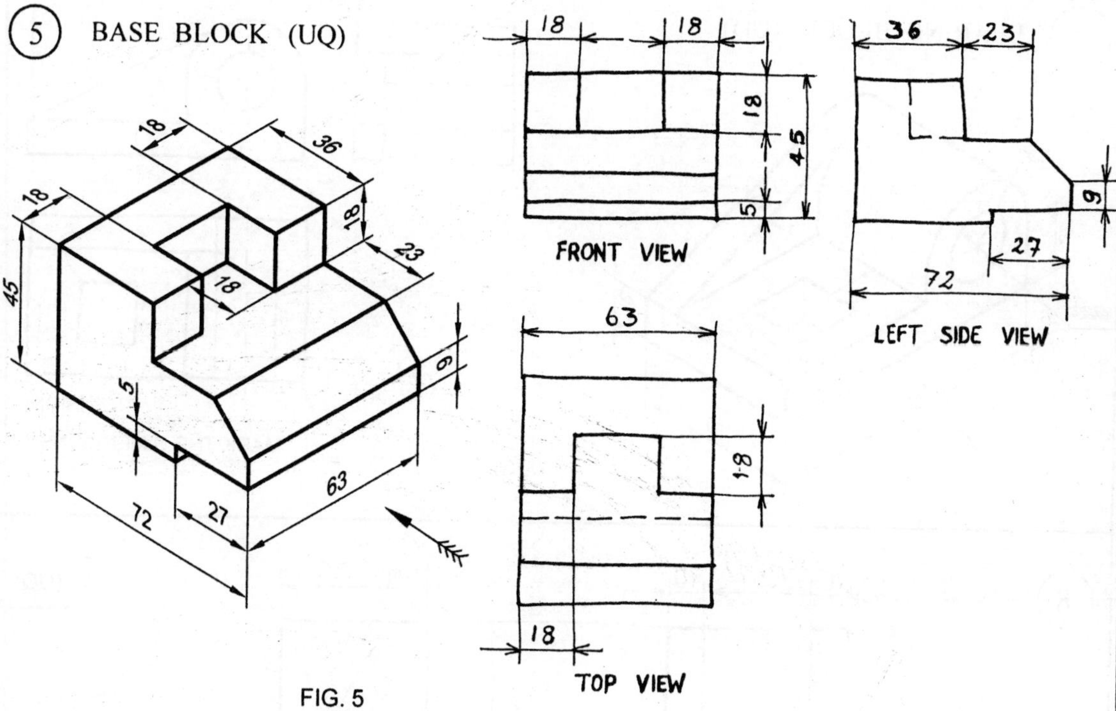

FIG. 5

(6) C.I. BLOCK (UQ)

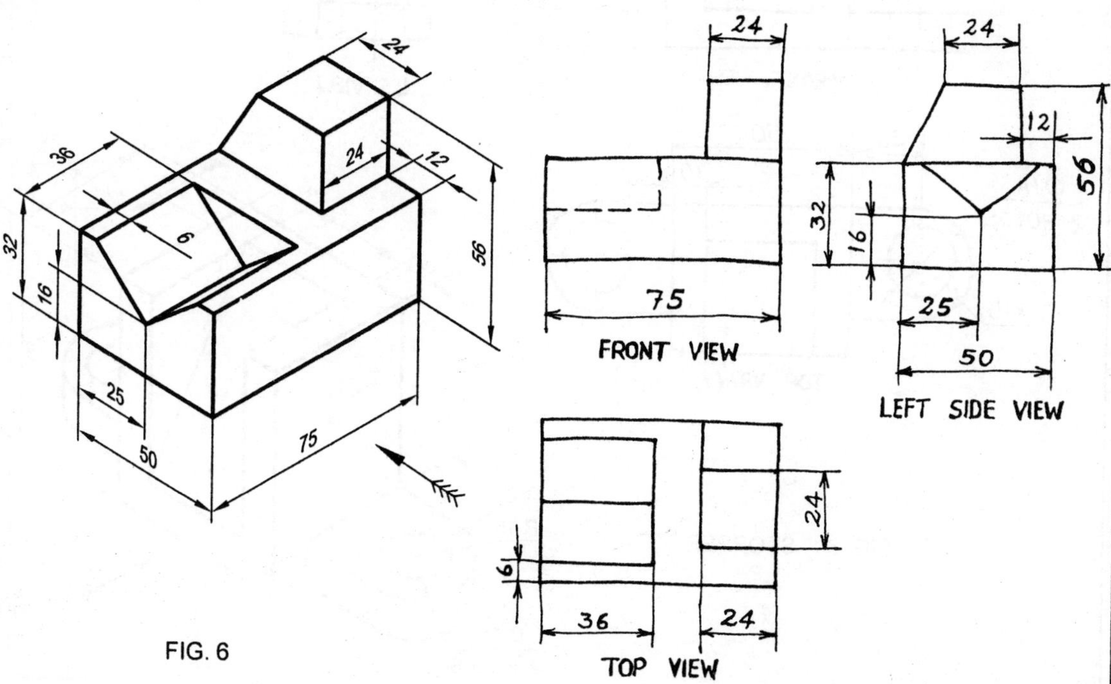

FIG. 6

(7) BEARING BLOCK (UQ)

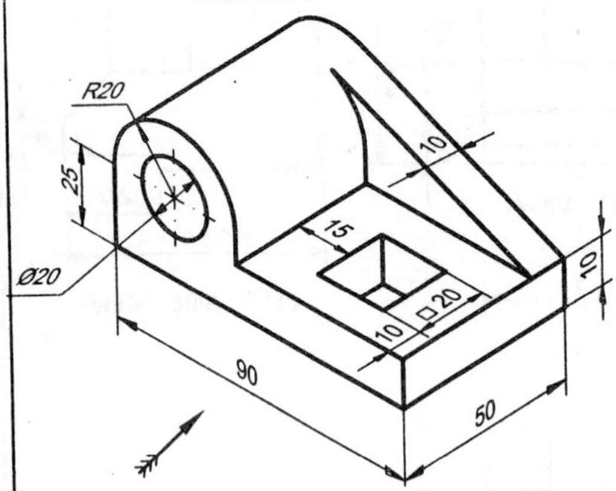

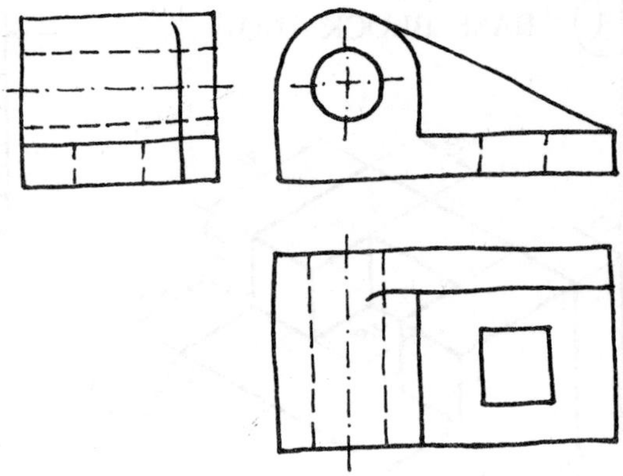

MARK THE DIMENSIONS

FIG. 7

(8) (UQ)

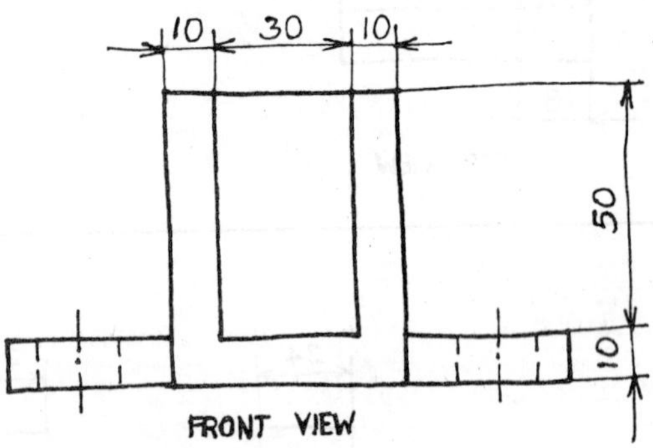

FRONT VIEW

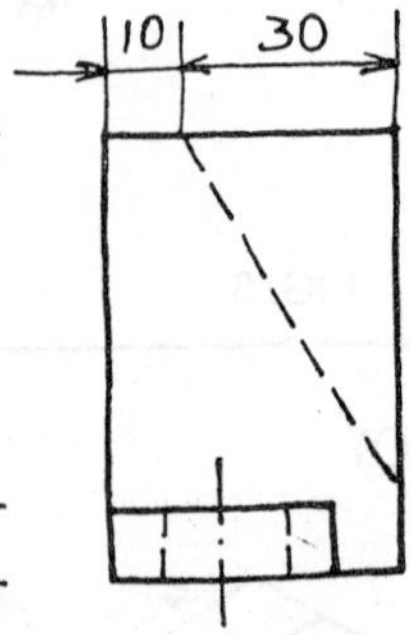

L.S. VIEW

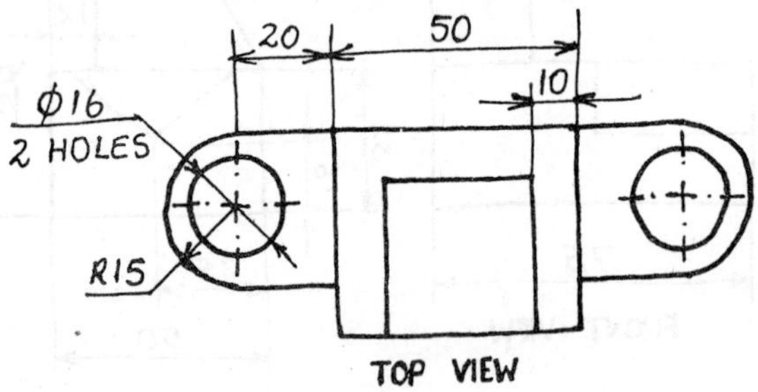

TOP VIEW

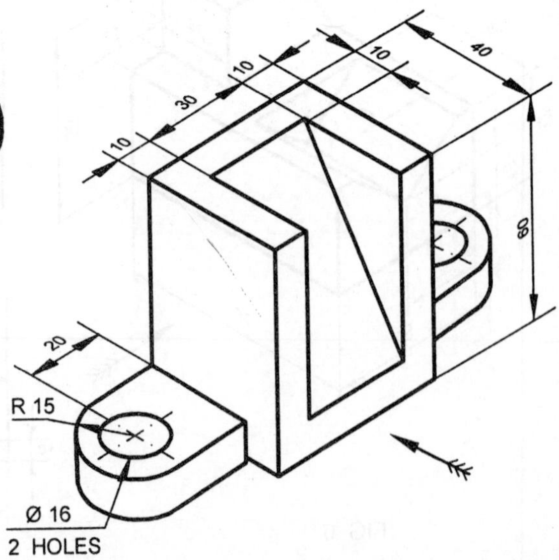

FIG. 8 STOPPER

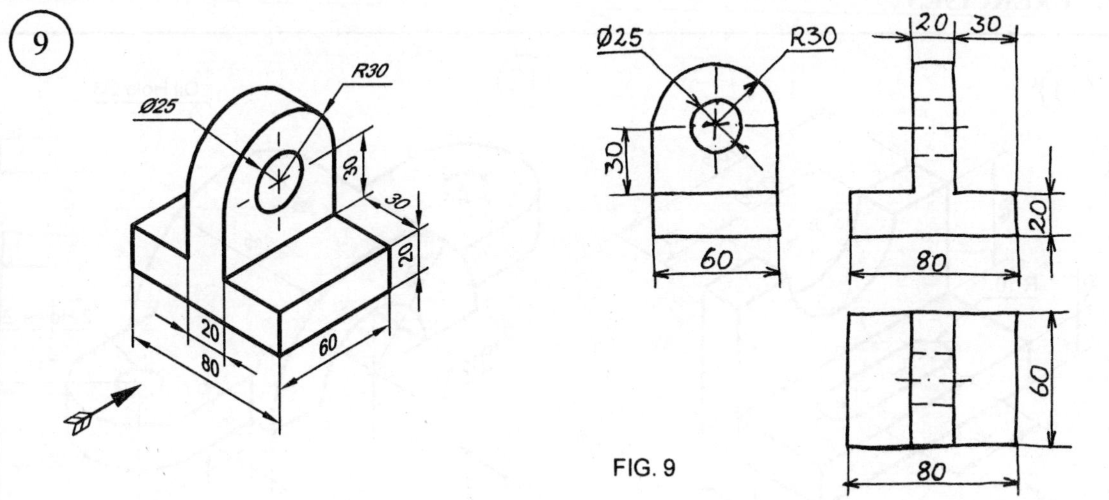

FIG. 9

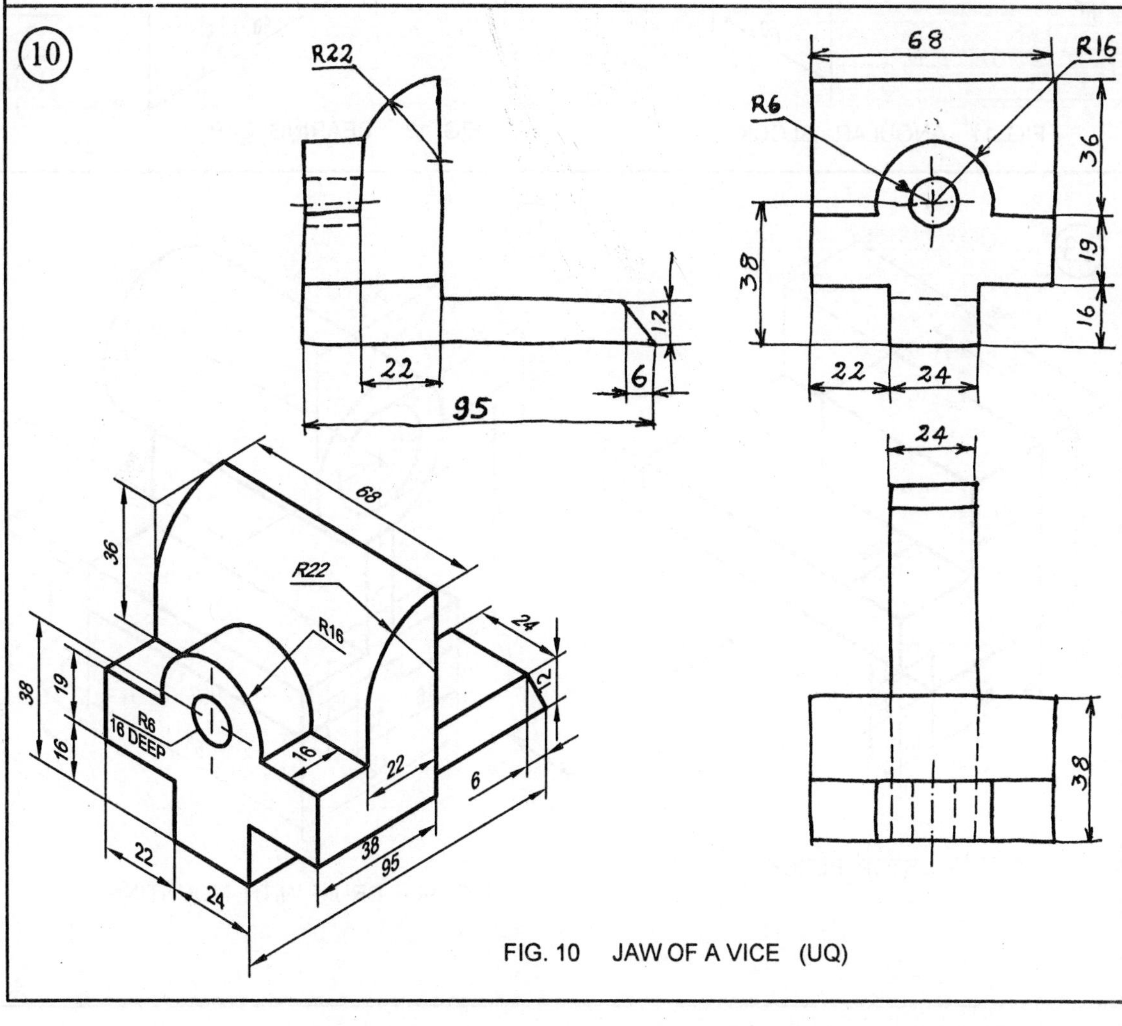

FIG. 10 JAW OF A VICE (UQ)

2. EXERCISES

(11)

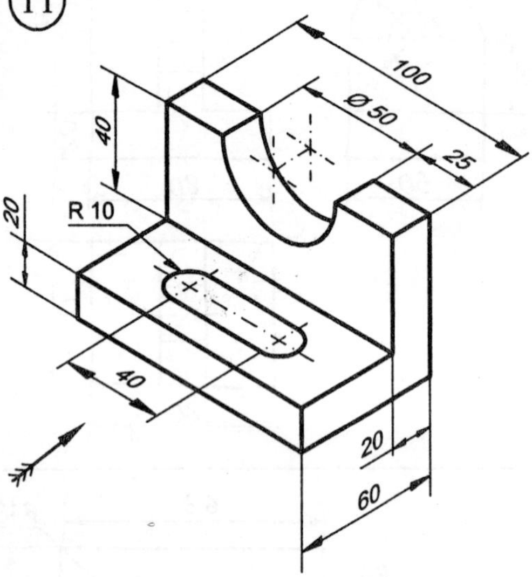

FIG. 11 ANGULAR BLOCK

(12)

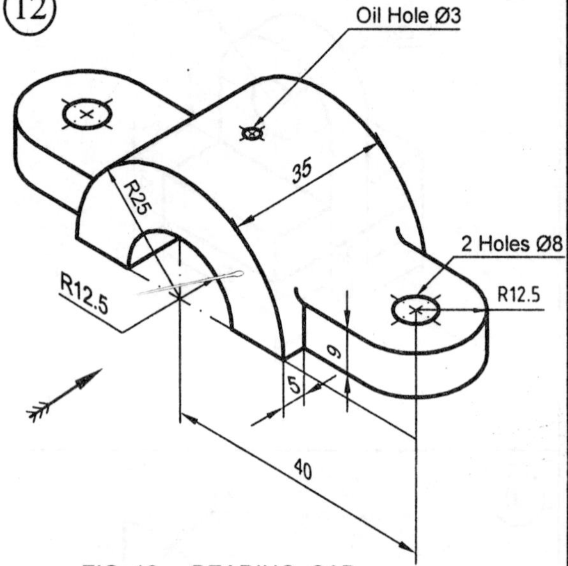

FIG. 12 BEARING CAP

(13)

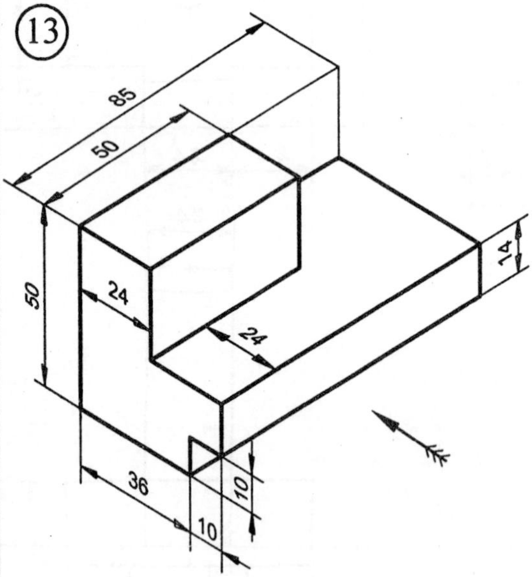

FIG. 13 STOP BLOCK

(14)

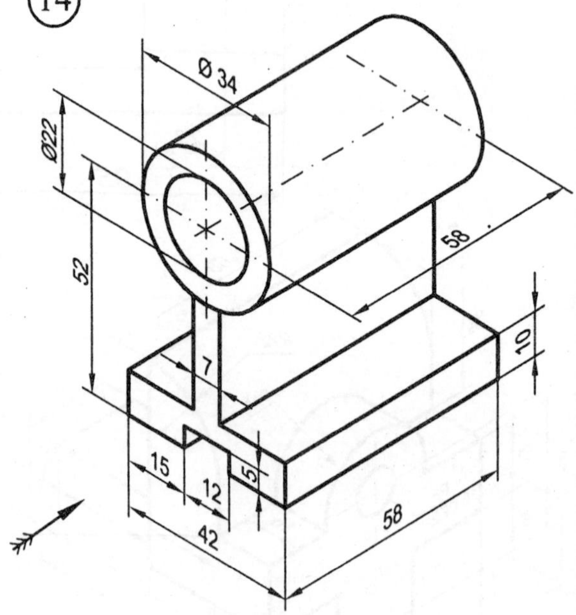

FIG. 14 DRUM WITH MOUNTING

*

25

FREEHAND SKETCHING OF ISOMETRIC VIEWS FROM ORTHOGRAPHIC VIEWS

1. TYPICAL PRACTICAL PROBLEMS – Engineering Components

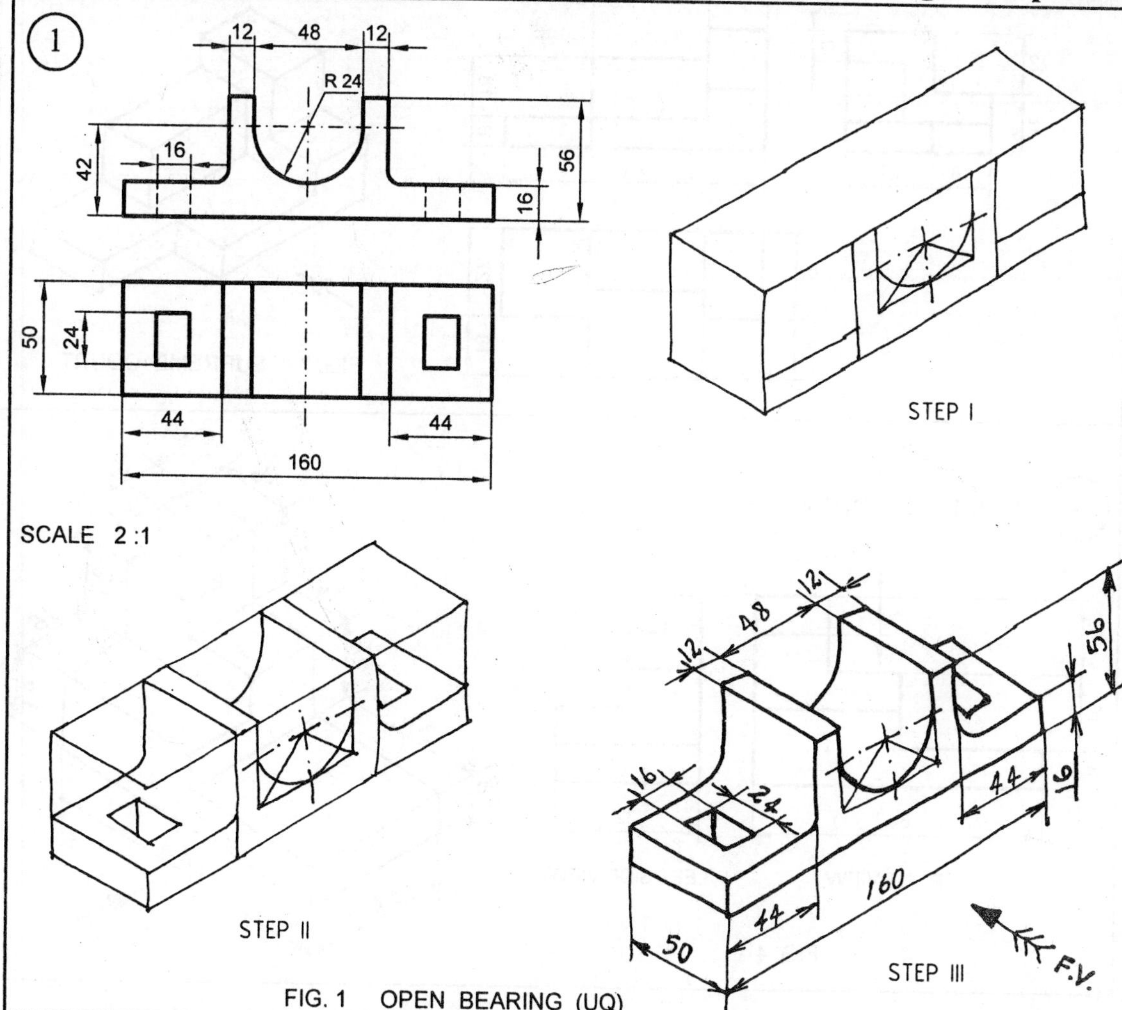

FIG. 1 OPEN BEARING (UQ)

(2) (UQ)

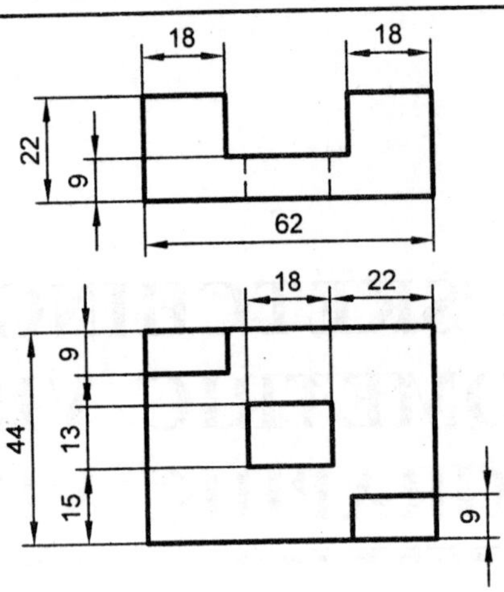

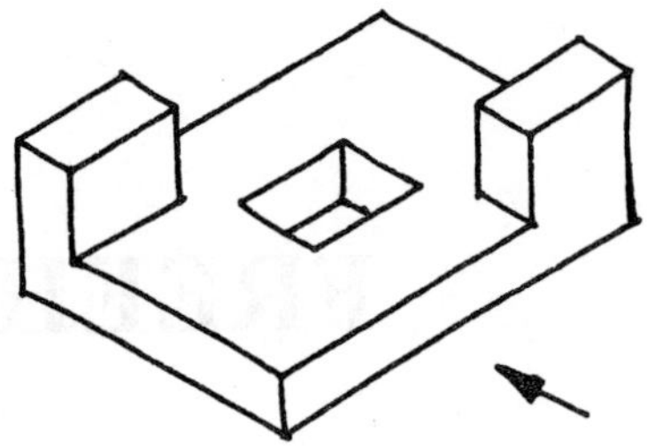

FIG. 2 CENTER GUIDE

(3) (UQ)

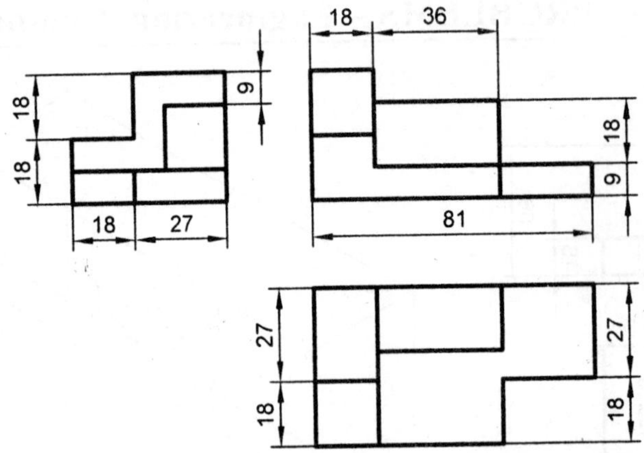

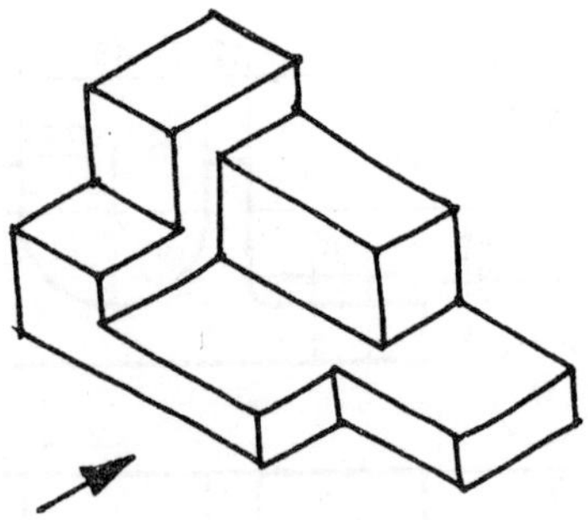

FIG. 3 SUPREME COURT

(4) OFFICE BUILDING (UQ)

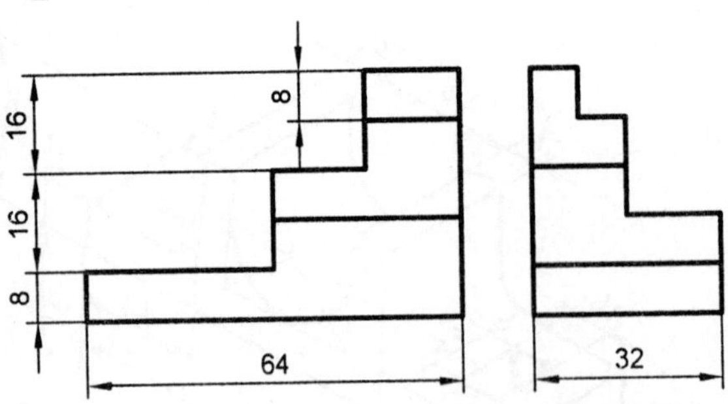

FRONT VIEW LEFT SIDE VIEW

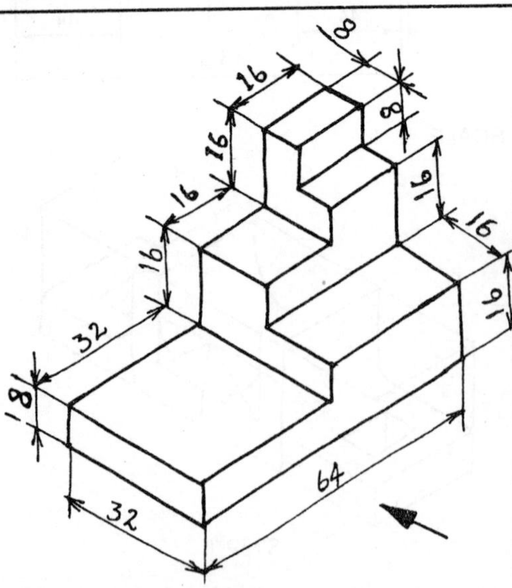

FIG. 4

5 (UQ)

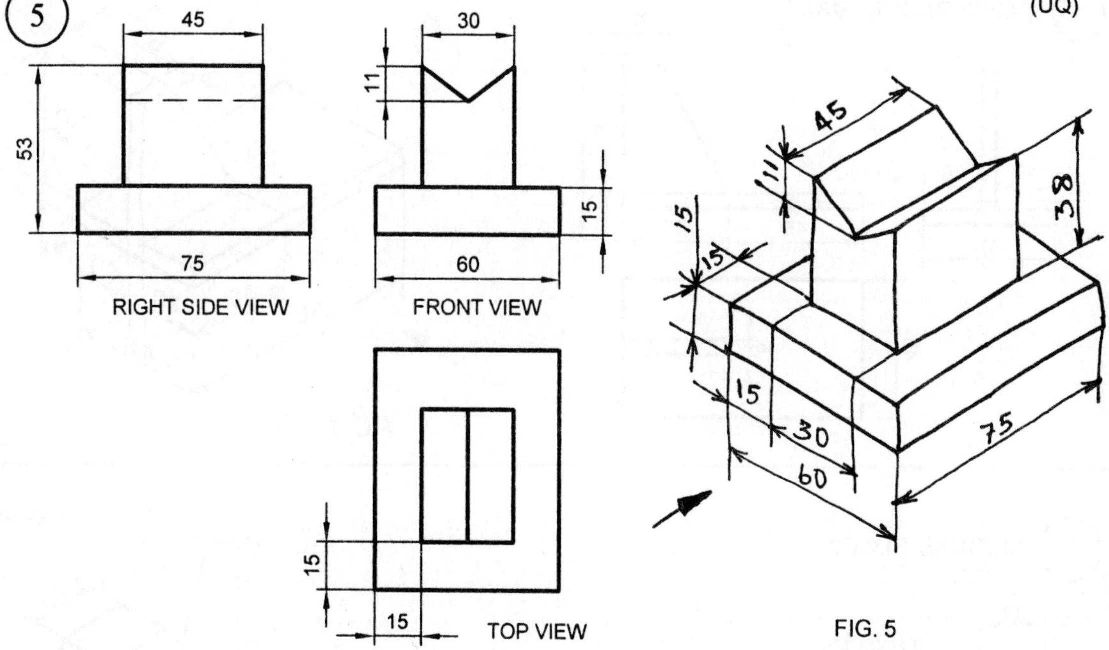

FIG. 5

6 (UQ)

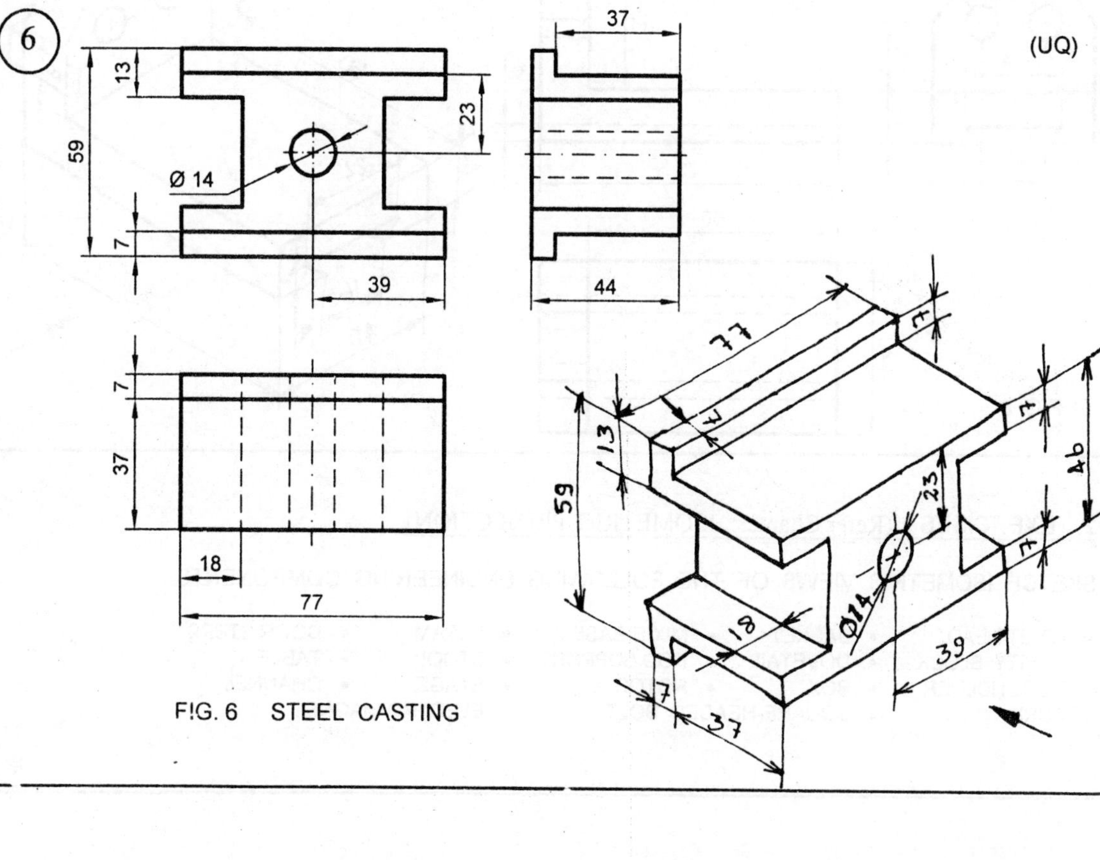

FIG. 6 STEEL CASTING

(7) TAPER BLOCK (UQ)

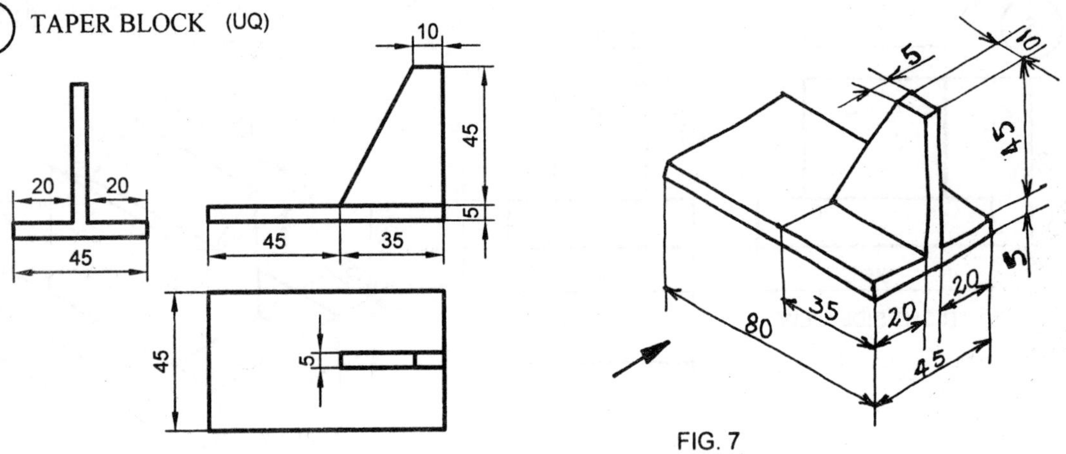

FIG. 7

(8) SLIDING BLOCK (UQ)

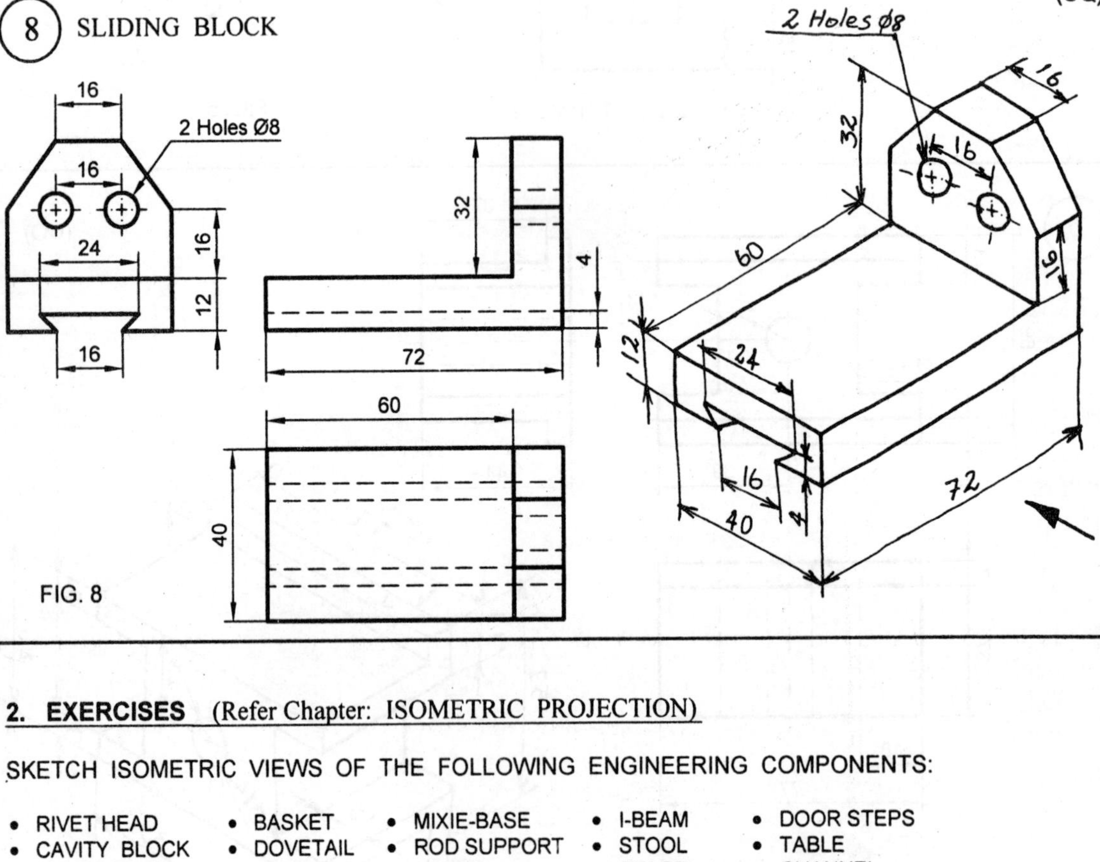

FIG. 8

2. EXERCISES (Refer Chapter: ISOMETRIC PROJECTION)

SKETCH ISOMETRIC VIEWS OF THE FOLLOWING ENGINEERING COMPONENTS:

- RIVET HEAD
- CAVITY BLOCK
- TOOL HOLDER
- FORK
- BASKET
- DOVETAIL
- BOX
- SQUARE-HEADED BOLT
- MIXIE-BASE
- ROD SUPPORT
- ROTOR
- I-BEAM
- STOOL
- STAGE
- ELECTRO-MAGNET
- DOOR STEPS
- TABLE
- CHANNEL

*

26

OBLIQUE PROJECTION

1. PICTORIAL PROJECTIONS – ISOMETRIC AND OBLIQUE

1. **Isometric Projection:** Isometric Projection is a pictorial view in distorted shape with fore-shortened dimensions. See Fig. 1.

2. **Oblique Projection:** ***Oblique*** means ***Slanting/Inclined.*** There are three axes, *viz.*, Vertical, Horizontal and Oblique. Oblique Axis is called *Receding Axis.* It is drawn either at 30° or 45° (Fig. 2). In oblique projection, projectors from various points on the object are drawn parallel to each other and inclined to the Plane of Projection.

   See Fig. 2. The front face of a cube is placed parallel to the plane of projection. Hence, it is shown in its true size and shape.

   The faces of the object *perpendicular* to the plane of projection will be distorted in shape and size. Edges of these perpendicular faces are inclined with the horizontal at 30° or 45°. Oblique drawing is prepared by Box Method or Co-ordinate Method. It is used to sketch pictorial views of machine components for manufacturing purpose.

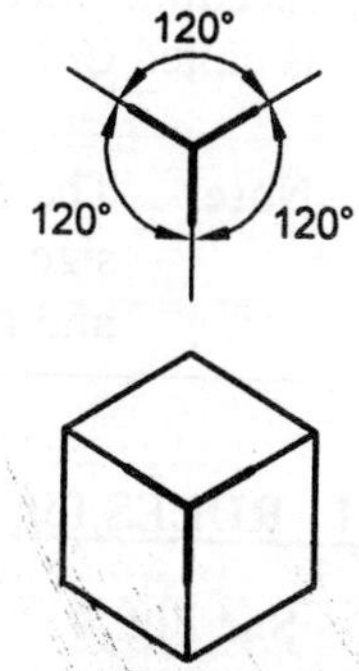

ISOMETRIC OF CUBE
FIG. 1

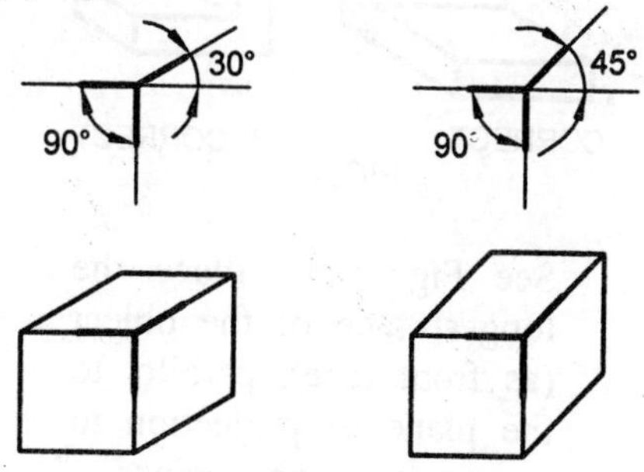

(i) (ii)
FIG. 2
OBLIQUE PROJECTION OF CUBE

Isometric Projection *Vs* Oblique Projection

1. In I.P., three axes are inclined at 120° to each other. In O.P., third axis is inclined at 30° or 45° to the horizontal.
2. In I.P., projectors are parallel to each other and perpendicular to the plane of projection. In O.P., projectors are parallel to each other and inclined to the Plane of Projection.
3. In I.P., object is placed with 3 mutually perpendicular axes making equal angles with plane of projection. Object stands on one of its corners.
   In O.P., object rests on one of its faces. It is placed with a prominent face parallel to plane of projection. Circular or irregular lines on the front face are shown in true shape and size.
4. In I.P., all the faces are distorted in shape and size. In O.P., the faces perpendicular to the Plane of Projection are distorted in shape and size.

2. TYPES OF OBLIQUE PROJECTION

Slope of the lines may be drawn upwards or downwards and either to the left or right. Oblique projection is classified into *two* types, depending on the shortening of the receding axis.

1. **Cavalier Projection:** See Fig. 3(i). When an oblique drawing is prepared to the exact dimensions of an object, it is called a *Cavalier Projection.* In this method, actual depth of the object is shown. Object appears distorted.

2. **Cabinet Projection:** See Fig. 3(ii). To reduce distortion, the dimensions of receding lines are drawn *half-scale.* It is called *Cabinet Projection.*

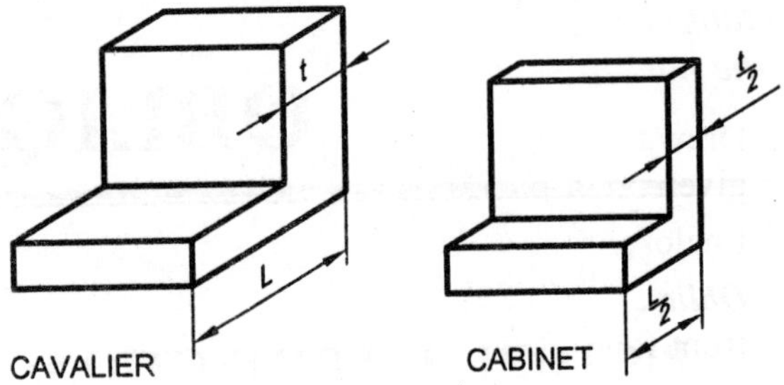

CAVALIER (i) CABINET (ii)

FIG. 3

Note: The receding lines which represent the depth of the object, are to be drawn to half-size. However, mark the actual full-size dimension on the drawing. But, the width and height of the object are to be drawn to full size in the front face.

2.1 RULES for Selecting the Position of an Object for Minimum Distortion

1. See Fig. 4. Place the face with essential contours, (i.e., circles or irregular shapes), parallel to the plane of projection.

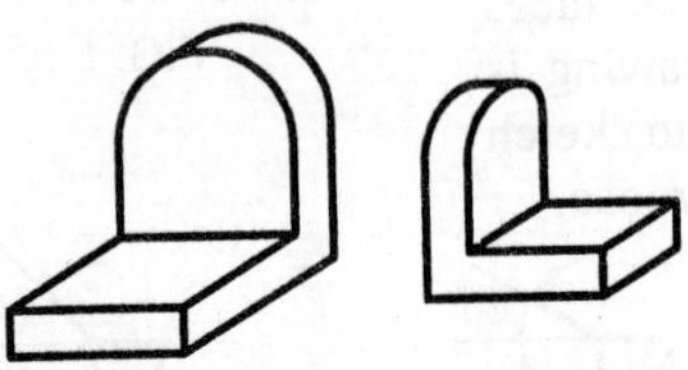
CORRECT INCORRECT

FIG. 4

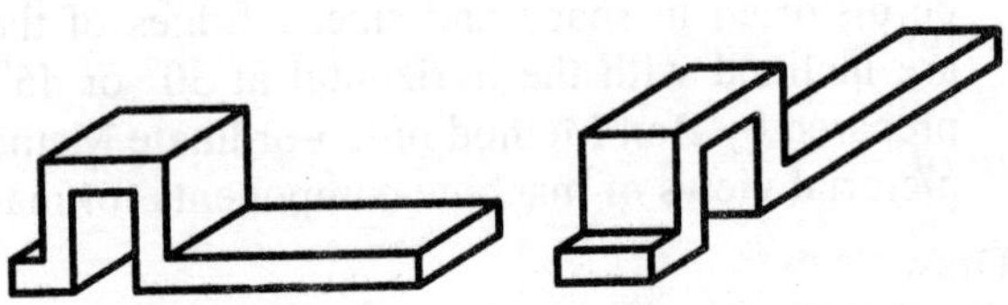
CORRECT INCORRECT

FIG. 5 PLACEMENT OF A LONG OBJECT

2. See Fig. 5. Place the longest face of the object (as front face), parallel to the plane of projection to have realistic appearance.

3. See Fig. 6. For lengthy objects having circular outlines, place the circular face parallel to the plane of projection.

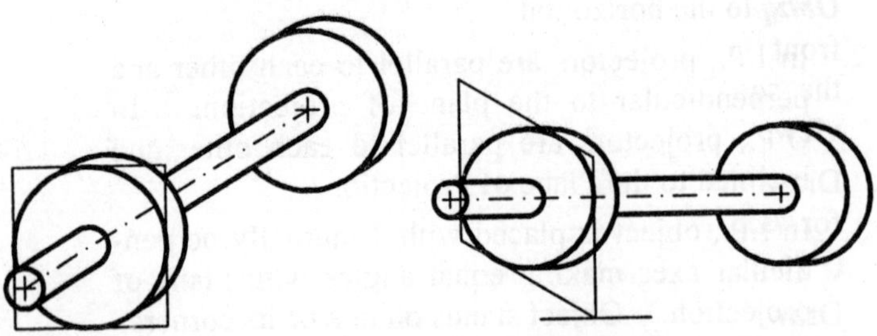
CORRECT INCORRECT

FIG. 6 PLACEMENT OF AN OBJECT HAVING CURVES or CIRCLES

3. OBLIQUE PROJECTION OF SOLIDS

Problem 1: *(Fig. 7) Draw the oblique projection of a hexagonal prism of side of base 20 mm and axis 50 mm long by Cavalier Method. The prism rests on HP on a rectangular face. Its axis is perpendicular to VP. The receding axis is inclined at 45° to the horizontal.*

1. Draw the front and top views of the given prism as shown in Fig. (i).
2. Enclose front view in a rectangle.
3. ***Oblique Projection:*** Draw the front face of the hexagon including the rectangle similar to the front view. See Fig. (ii).
4. Draw the receding lines at 45° for the *same length* of the axis of the prism for Cavalier Method.
5. Draw the rear face of the prism parallel to the front face.
6. Complete the prism as shown.
7. Dimension the oblique projection *similar to **Isometric Projection**.*

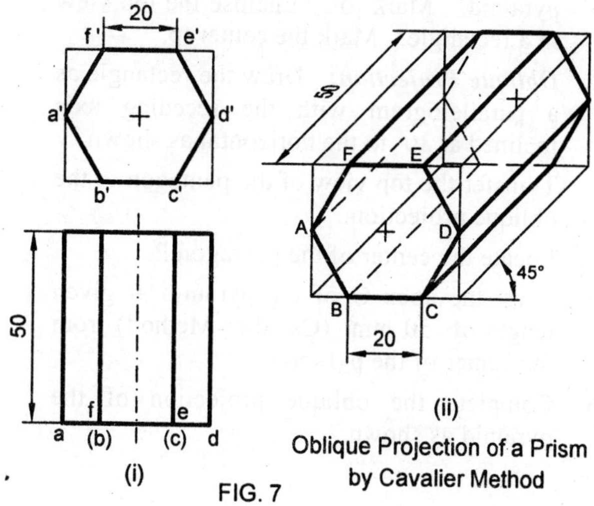

Oblique Projection of a Prism by Cavalier Method

FIG. 7

Problem 2: *(Fig. 8) A hollow cylinder, 46 mm outside diameter, 22 mm inside diameter and axis 76 mm long, is lying on one of its generators on HP. The axis is perpendicular to VP. Draw the oblique projection by Cabinet Method with the receding angle of 30° to the horizontal.*

1. Draw the top and front views of the given cylinder.
2. Enclose the front view in a square.
3. ***Oblique Projection:*** Draw the front face of the cylinder with the square, similar to the front view.
4. Draw the receding lines at 30° for *half the length* of the axis (=76/2) for Cabinet Method.
5. Draw the rear face of the cylinder parallel to the front face. Complete the oblique projection as shown.

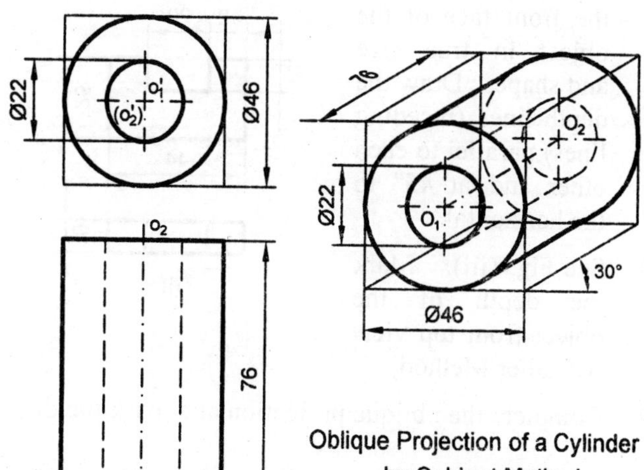

Oblique Projection of a Cylinder by Cabinet Method

FIG. 8

Problem 3: *(Fig. 9) Draw the oblique projection of a pentagonal pyramid of 24 mm side of base and axis 50 mm long by Cavalier Method with the receding axis inclined at 30° to the horizontal. The pyramid rests on its base on HP such that two of its base edges are equally inclined to VP.*

1. Draw the top and front views of the given pyramid. Mark o′. Enclose the top view in a rectangle. Mark the center p.
2. ***Oblique Projection:*** Draw the rectangle as a parallelogram with the receding axis inclined at 30° to the horizontal as shown.
3. Transfer the top view of the pentagon to the oblique projection.
4. Locate the center of the pentagon P.
5. Mark the apex O of the pyramid at given length of 50 mm (Cavalier Method) from the center of the pyramid.
6. Complete the oblique projection of the pyramid as shown.

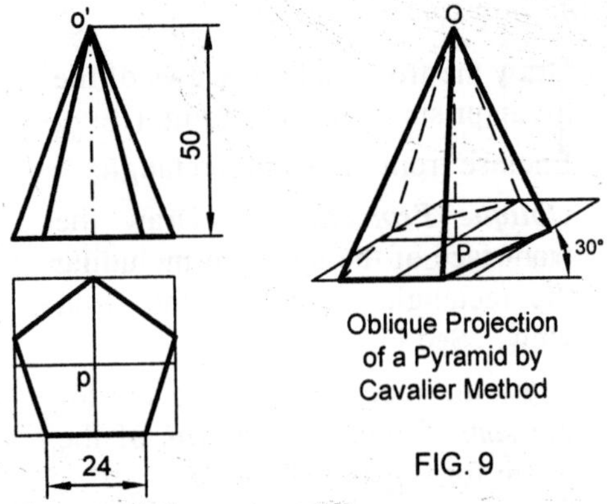

Oblique Projection of a Pyramid by Cavalier Method

FIG. 9

4. OBLIQUE PROJECTION OF OBJECTS

Problem 4: *(Fig. 10) The orthographic views of an object are shown in Fig. (i). Draw its oblique projection with the receding axis inclined at 45° by Cavalier Method.*

1. See Fig. (ii). Draw the front face of the object in true size and shape. Draw the depth lines (receding lines) parallel to each other and at 45° to the horizontal.
2. See Fig. (iii). Mark the depth of the object from top view (Cavalier Method).
3. Complete the oblique projection and mark the dimensions.

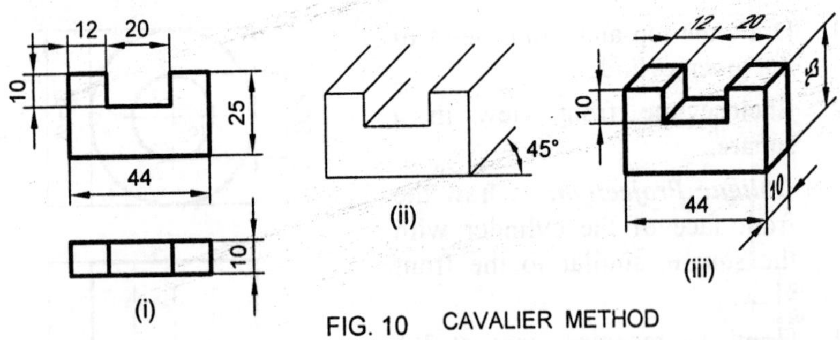

FIG. 10 CAVALIER METHOD

Problem 5: *(Fig. 11) Draw the oblique projection of the object shown in Fig. (i), by Cavalier Method with the receding axis inclined at an angle of 45° to the horizontal.*

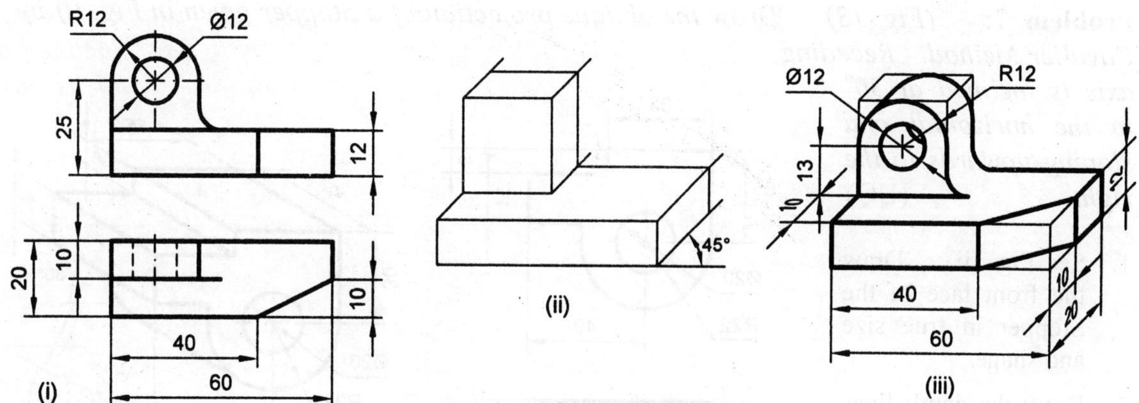

FIG. 11 CAVALIER METHOD

1. See Fig. (ii). Draw the front face of the object in true size and shape by Box Method.
2. Draw the depth lines parallel to each other and inclined at 45° to the horizontal.
3. Complete the construction box. Draw the circle, arcs and taper inside the box as in Fig. (iii).
4. Complete the oblique projection and mark the dimensions.

Problem 6: *(Fig. 12) The front and side views of a V-Block are shown in Fig. (i). Draw the Cavalier Oblique Drawing. Receding axis is inclined at 45° to the horizontal, sloping upwards and to the right.* (UQ)

1. See Fig. (ii). Draw the front face of the object in true size and shape.
2. Draw the depth lines parallel to each other and inclined at 45° to the horizontal.
3. Complete the oblique drawing and mark the dimensions.

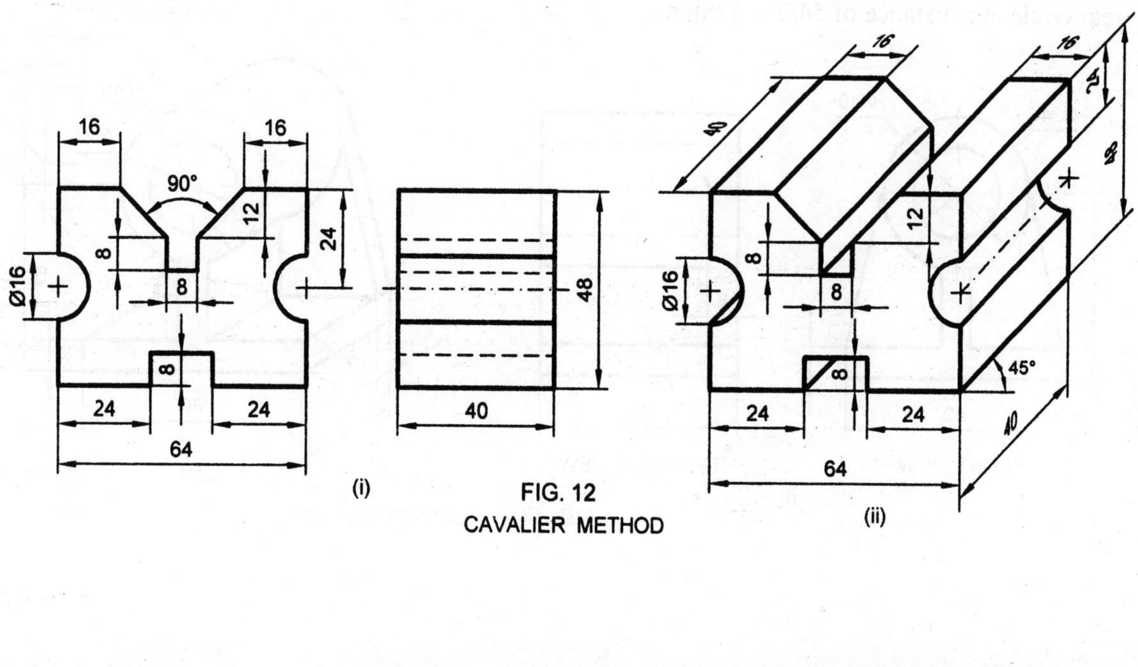

FIG. 12
CAVALIER METHOD

Problem 7: *(Fig. 13) Draw the oblique projection of a Stopper given in Fig. (i) by Cavalier Method. Receding axis is inclined at 30° to the horizontal and sloping upwards to the right.* (UQ)

1. See Fig. (ii). Draw the front face of the Stopper in true size and shape.
2. Draw the depth lines parallel to each other and inclined at 30° to the horizontal.
3. Complete the oblique projection.

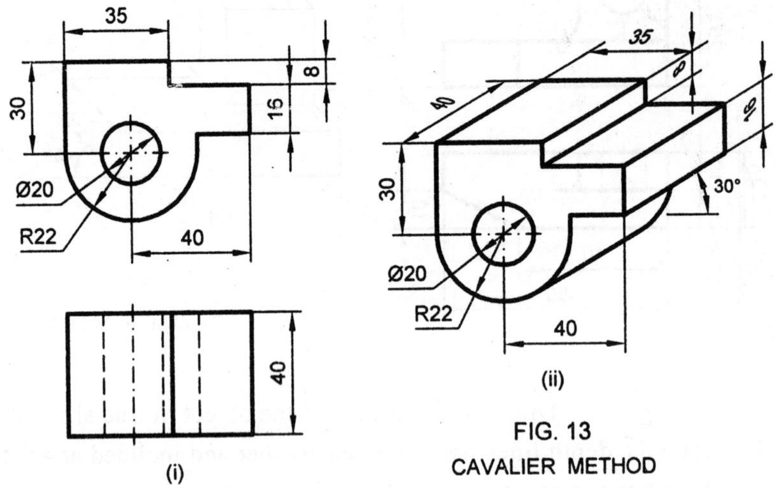

FIG. 13
CAVALIER METHOD

Problem 8: *(Fig. 14) The front and left side views of a Bracket are shown in Fig. (i). Draw the Cabinet Oblique Projection. The receding axis is sloping upwards and to the left, at an angle of 30° to the horizontal.* (UQ)

Note: In Cabinet Oblique Projection, the receding lines which represent the depth of the object, are to be drawn to half-size. However, mark the actual full-size dimension on the drawing.

See Fig. (ii). Mark the receding line at 30° sloping upwards to the left. Mark the center of the rear circle at a distance of 54/2 = 27 mm.

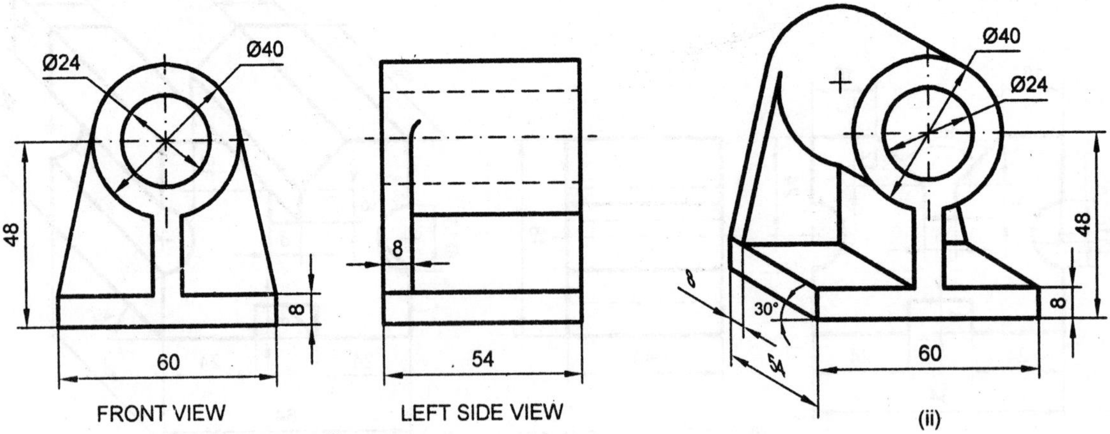

FIG. 14 CABINET METHOD

Problem 9: *(Fig. 15) Orthographic views of an Open Bearing are given in Fig. (i). Draw the Cavalier Oblique View, taking the receding angle as 45° to horizontal.* (UQ)

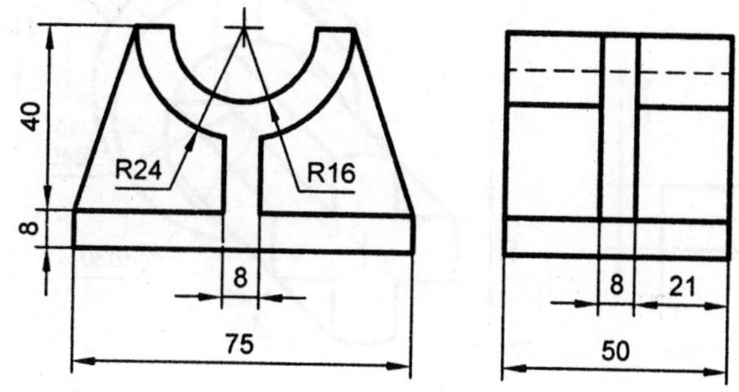

(i)

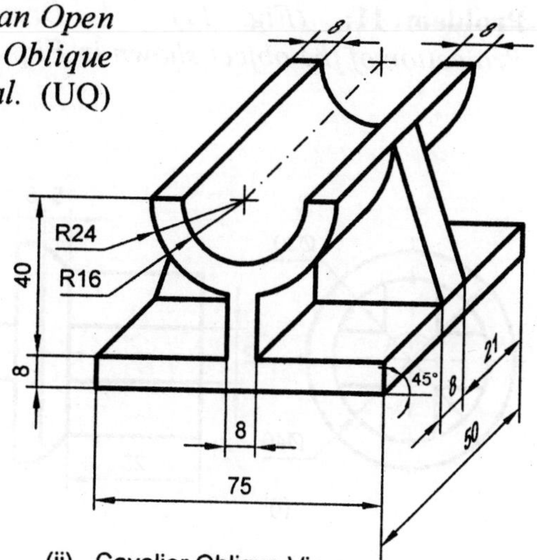

(ii) Cavalier Oblique View

FIG. 15 OPEN BEARING

Problem 10: *(Fig. 16) Draw the Cavalier Oblique View of a Hinge in Fig. (i).* (UQ)

Note: To locate the points on the *ellipse* in Fig. (iii), use the Method of Points followed in Isometric Projection. Draw the diagonals of the square in the top view and draw lines passing through the points of intersection of the circle. Transfer these lines and points to the top face of the oblique view. Draw a smooth curve to get the ellipse.

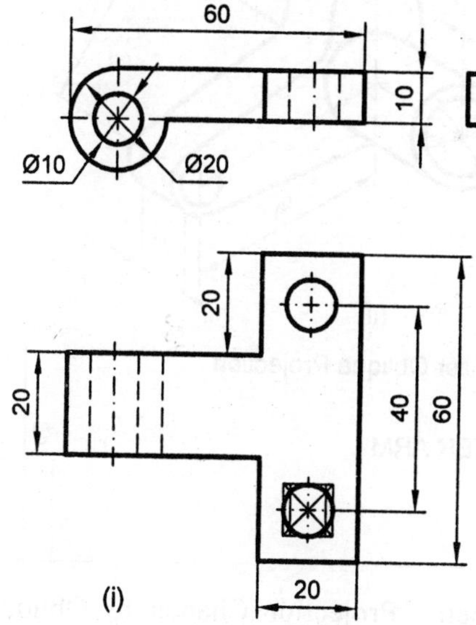

(i)

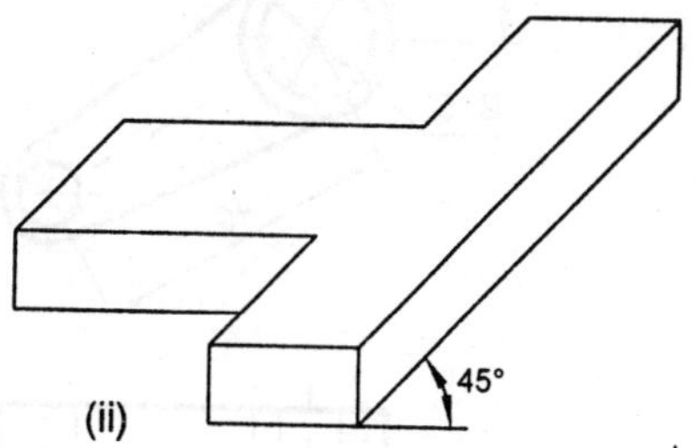

(ii)

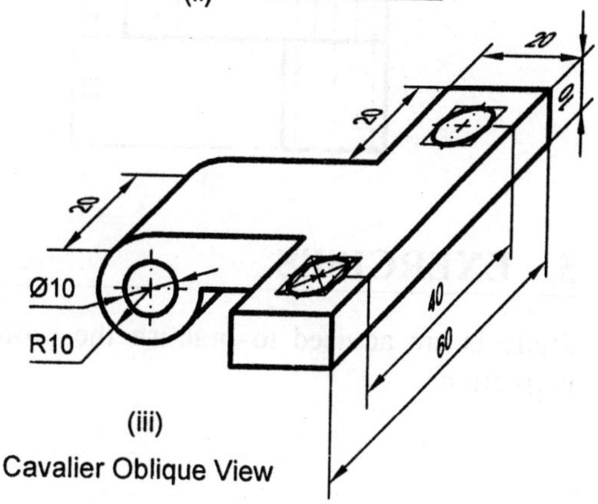

(iii)
Cavalier Oblique View

FIG. 16 HINGE

Problem 11: *(Fig. 17) Draw the Cavalier Oblique Projection of the object shown in Fig. (i).* (UQ)

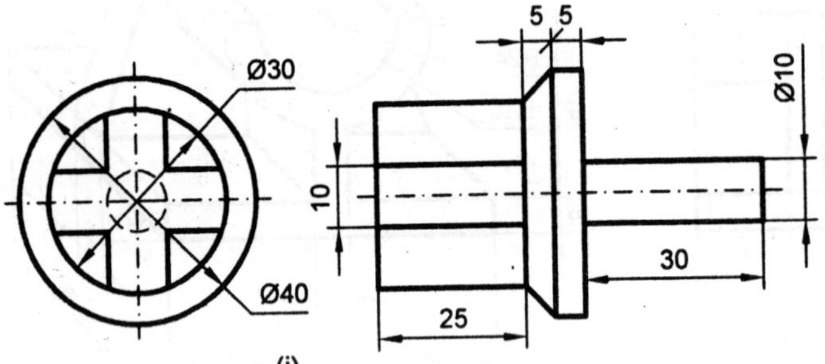

(i)

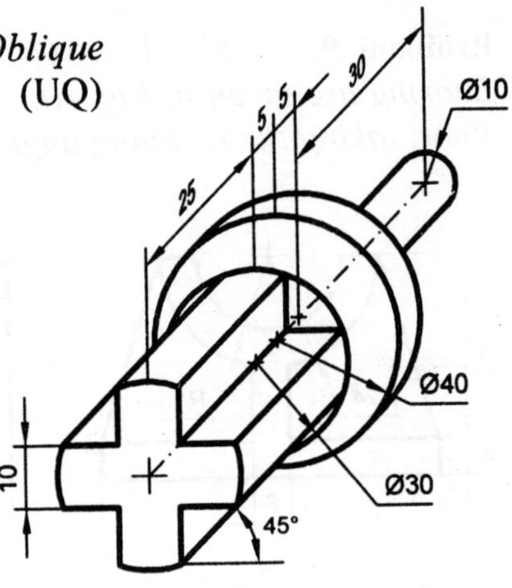

FIG. 17 (ii) Cavalier Oblique Projection

Problem 12: *(Fig. 18) Draw the Cavalier Projection of Rocker Arm in Fig. (i).* (UQ)

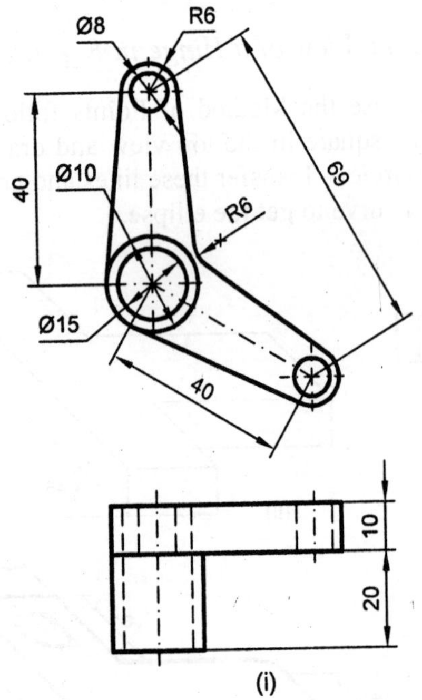

(i)

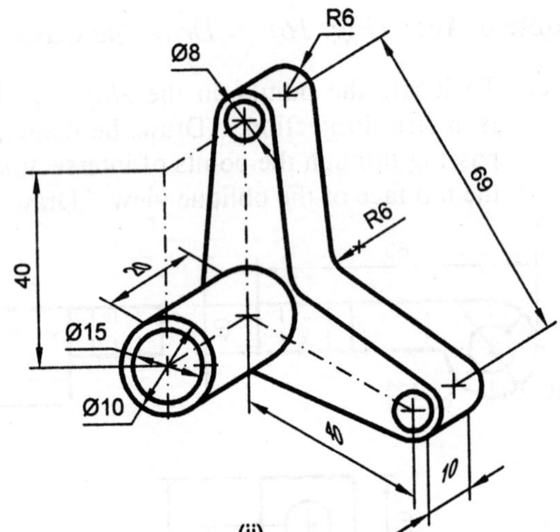

(ii)

Cavalier Oblique Projection

FIG. 18 ROCKER ARM

5. EXERCISES

Students are advised to practise the problems in the Isometric Projection Chapter by Oblique Projection.

*

27

PERSPECTIVE PROJECTION

1. PRINCIPLE OF PERSPECTIVE PROJECTION / VIEW

Perspective Projection is the graphic representation of an object on a single plane called *Picture Plane* (PP), as it appears to an observer. The object is placed behind the PP. The observer is stationed in front of PP. Therefore, visual rays from the eyes of the observer to the object are cut by the PP, *i.e.*, the rays pierce the picture plane and form an image on it. This image is the perspective view of the object.

The arrangement of the eye, picture plane and the object is the same as in the case of ***Third Angle Projection***. Therefore, the top view of the object is drawn above the front view.

Fig. 1 shows the perspective projection of a rail track with electric posts as seen by the observer. Picture taken by a camera is a perspective.

FIG. 1

Applications: As the photographic representation, it is extensively used by architects to show the appearance of proposed buildings, by drafters for illustrations of huge machinery or equipments and also by artists for producing advertisement drawings.

1.1 PERSPECTIVE ELEMENTS – DEFINITIONS

Ground Plane (GP): *See Fig. 2.* Ground Plane is a horizontal plane representing ground on which the object is assumed to be placed.

Picture Plane (PP): It is a vertical transparent plane on which the perspective view of the object is projected. It is placed in between the object and observer. The front view of the perspective elements and of the object (if necessary) may also be projected on the PP.

- Position of PP relative to the object determines the size of the perspective projection.
- When the object is in the PP, its perspective will be in true size.
- When the object is placed in front of PP, its perspective is of enlarged size.
- When the object is placed behind PP, the perspective is in reduced size.

Horizon Plane / Horizontal Plane (HP): It is an imaginary horizontal plane at the level of the observer's eye or S. It is above and parallel to the GP and perpendicular to the PP.

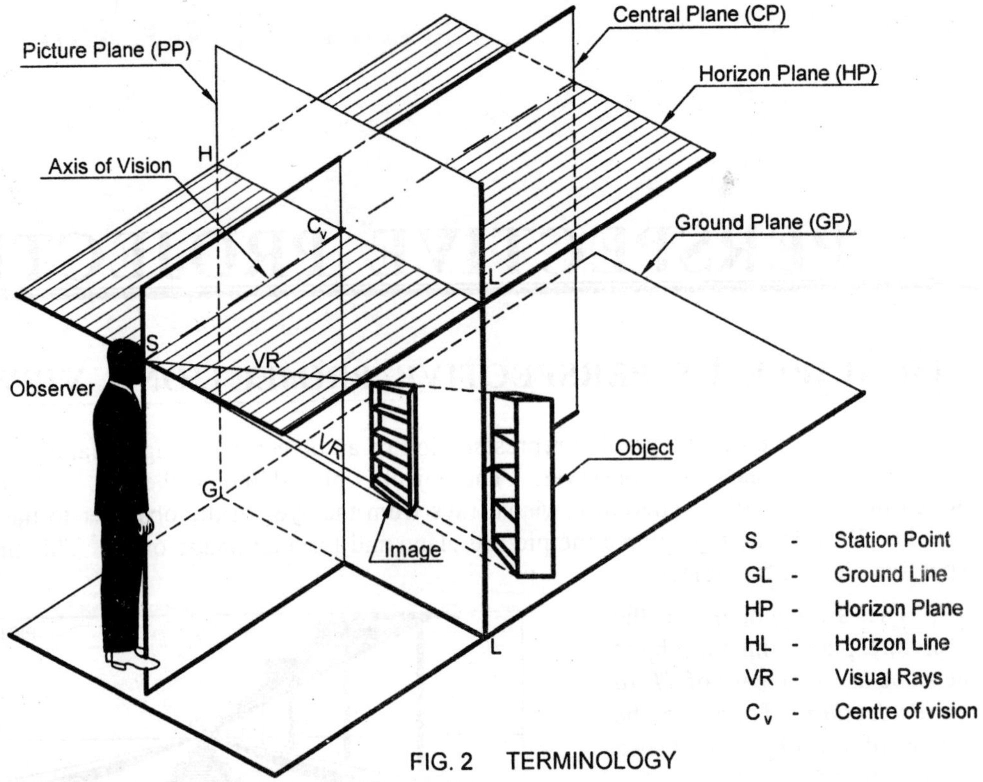

FIG. 2 TERMINOLOGY

Station point (S) *(i.e., eye level)*: Station point is the position of the observer's eye. Its position is fixed at any convenient distance either from the picture plane or from the object. Shape and visibility of edges of perspective projection depends upon the position of S.

- For large objects like buildings, S is taken at eye level of height 1.8 m.
- For small objects, fix S slightly above the top of the object to have a complete view of its top and side surfaces.
- Take the distance of S from PP = twice the width of the object to avoid distorted view.
- Locate S in front of the objects so that CP passes through the center of the object.

Ground Line (GL): It is the line of intersection of the GP with the PP.

Horizon Line (HL): It is the line of intersection of HP with PP. It is parallel to GL.

Axis of Vision (AV) / Perpendicular Axis (PA) / Line of Sight: It is a line drawn perpendicular to the PP from S.

Center of vision (C_v): It is the point at which the axis of vision pierces the PP. It is also the point of intersection of HL with AV.

Central Plane (CP): It is an imaginary vertical plane which passes through the S and C. It is perpendicular to both the PP and the GP. It contains the Axis of Vision (AV).

Auxiliary Ground Plane (AGP) *(Not shown)*: It is a horizontal plane parallel to GP. Top view of the object and of the perspective elements is projected on this plane.

Vanishing Point (VP): All the projections converge to a point known as Vanishing Point.

2. PERSPECTIVE VIEWS OF POINTS, LINES AND PLANES

Problem 1: *(Fig. 3) Draw the perspective projection of a* ***point*** *A situated 20 mm behind the picture plane and 15 mm above the ground plane. The station point is 30 mm in front of the picture plane, 40 mm above the ground plane and lies in a central plane which is 35 mm to the left of the given point.*

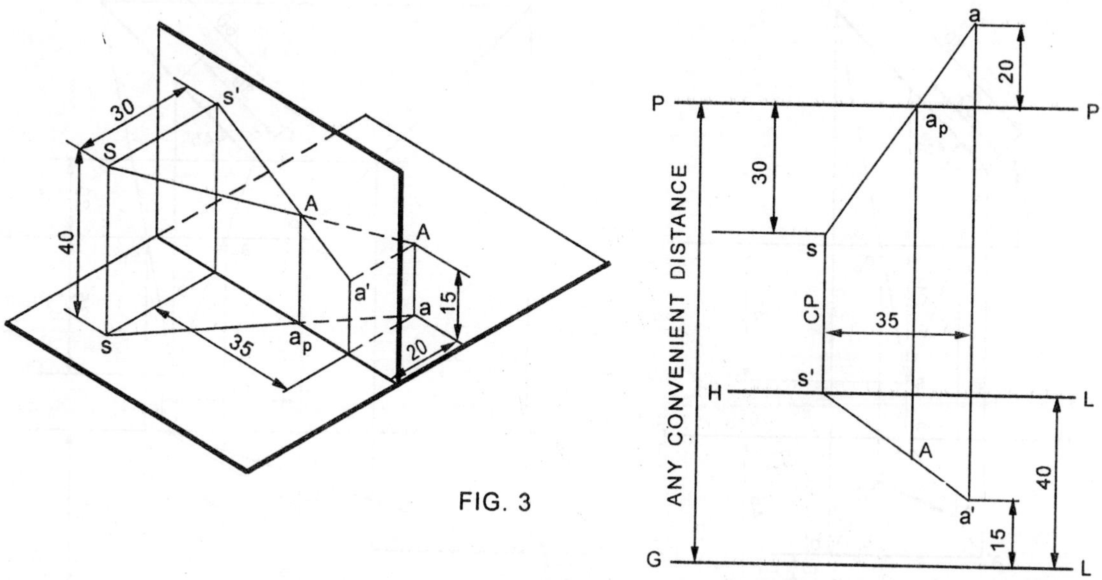

FIG. 3

VISUAL RAY METHOD

In this method, points on the perspective projection are obtained by drawing visual rays from the station point S to both top view and *either* front view *or* side view of the object.

Top view

1. Draw a horizontal line PP to represent the top view of the picture plane.
2. Point A is 20 mm behind PP. Hence mark the point a 20 mm above PP.
3. Station point S lies in a central plane CP which is 35 mm to the left of point A. So, draw a vertical line to represent the top view of CP at 35 mm to the left of a.
4. S is 30 mm in front of PP. Therefore, on CP, mark s 30 mm below PP.
5. Join a and s to represent the top view of the visual ray. It pierces the PP at a_p.

Front view

6. Draw a ground line GL at any convenient distance below PP. Mark a′, the front view of A, 15 mm above GL and in projection with a. To avoid overlap of visual rays and get a clear perspective, select GL such that HL lies below s. S is 40 mm above GP. So, draw HL 40 mm above GL.
7. Further CP also represents front view of CP. Hence mark s′ at the intersection of CP with HL. Join a′s′, the front view of the visual ray. From the piercing point a_p, erect a vertical to cut a′s′ at point A, which is the required perspective projection.

Problem 2: *(Fig. 4) Draw the perspective projection of a* ***line*** *AB, 60 mm long. It is parallel to and 10 mm above the ground plane and inclined at 45° to PP. The end A is 20 mm behind the picture plane. Station point is 35 mm in front of the picture plane and 45 mm above the ground plane and lies in a central plane passing through the mid-point of AB.*

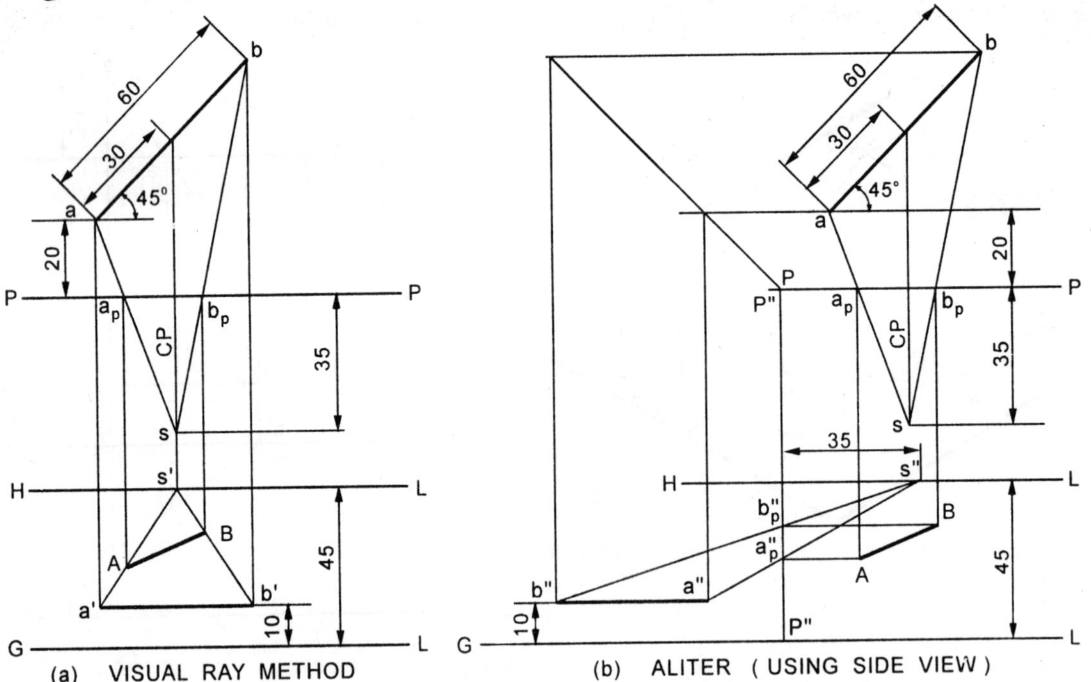

FIG. 4

1. **Top View:** *See Fig. (a).* Draw PP and mark a 20 mm above it. Draw ab = 60 mm (True length of AB) inclined at 45° to PP. From the mid-point of ab erect a vertical line to represent CP. Along the central plane mark s 35 mm below PP. Join a and b with s to represent the top view of the visual rays. Mark the piercing points a_p and b_p on as and bs respectively.
2. **Front View:** Draw GL at any convenient distance below PP. Draw a′ b′ parallel to and 10 mm above GL. Draw HL 45 mm above GL. Mark s′ at the intersection of CP and HL. Join s′ with a′ and b′. From a_p and b_p (piercing points) erect verticals to intersect s′ a′ and s′ b′ (the front view of the visual rays) at A and B respectively. AB is the required perspective projection.

ALITER (using SIDE VIEW): [*Fig. (b)*]

Draw top view as in previous method. Draw GL at any convenient distance below PP. Draw P″ P″, the side view of PP. Draw a″ b″ (Side View of AB) 10 mm above GL. Draw HL 45 mm above GL. As S is 35 mm in front of PP, mark s″ 35 mm from P″ P″ and on HL. Join s″ with a″ & b″ to cut p″ p″ at a''_p & b''_p. From a_p erect vertical to cut a horizontal line drawn from a''_p at A. Similarly obtain B. Join A & B.

Problem 3: *(Fig. 5)* *A* ***square lamina*** *of 40 mm side lies on the ground plane. One of its corners is touching the PP and an edge is inclined at 60° to PP. The station point is 40 mm in front of PP, 60 mm above GP and lies in a central plane which is at a distance of 35 mm to the right of the corner touching the PP. Draw the perspective projection of the lamina.*

VISUAL RAY METHOD : [*Fig. 5(a)*]

Top View

1. Draw the top view of the lamina as a square such that the corner b is touching PP and the edge bc is inclined at 60° to PP.
2. Draw CP, 35 mm from b on right side. Along CP mark s 40 mm below PP.
3. Join s with all the four corners of the square lamina in the top view. Obtain the corresponding piercing points on PP.

Front View

4. Draw GL and obtain the front view of the lamina on it. Draw HL 60 mm above GL and obtain s′ on it. Join s′ with all the corners of the lamina in the front view.

Perspective Projection

5. Since the corner b touches the PP, its perspective will be in its true position.
6. Since the lamina lies on the ground plane, b′ is on GL and is also the perspective projection of B.
7. From a_p draw vertical to intersect a′ s′ at A.
8. Similarly obtain B, C and D. Join ABCD and complete the perspective projection.

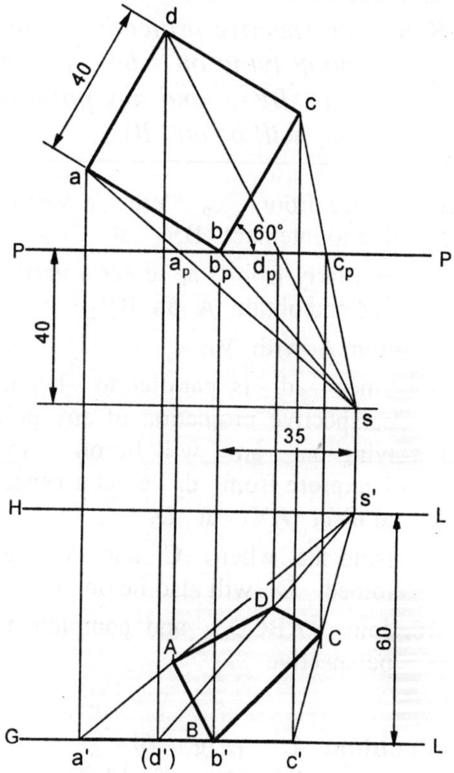

FIG. 5(a) VISUAL RAY METHOD

VANISHING POINT METHOD : [*Fig. 5(b)*]

Vanishing Point is an imaginary point infinite distance away from station point. The point at which the visual ray from the eye to that infinitely distant vanishing point pierces the picture plane is termed as *Vanishing Point*. When the observer views an object, all its parallel edges converge to one / two / three points depending on the locations of the object and the observer.

1. Draw the top view as explained in *Steps 1 to 3* in the above method. Draw GL and HL.

Vanishing Points

2. From s draw a line parallel to bc to cut PP at v_1. Erect vertical from v_1 to cut HL at vanishing point V_1.
3. Similarly from s draw a line parallel to ba to cut PP at v_2. Erect vertical from v_2 and obtain the other vanishing point V_2 on HL.

Perspective Projection

4. Since b touches PP, draw a vertical line from b and obtain B on GL. Join B with V_1 and V_2.

Note: *Perspective projection of any point lying on bc will be on BV_1 and any point on ba will be on BV_2.*

---

5. Hence from c_p erect a vertical line to intersect BV_1 at C.
6. Similarly from a_p erect a vertical line and obtain A on BV_2.
7. Join A with V_1.
8. Since ad is parallel to bc, the perspective projection of any point lying on ad will lie on AV_1. Therefore from d_p erect a vertical to meet AV_1 at D.
9. Note that when C and V_2 are joined, D will also lie on CV_2.
10. Join ABCD and complete the perspective.

Problem 4: *(Fig. 6)* A ***pentagonal lamina*** *of 40 mm side lies on the ground. The corner which is nearest to PP is 15 mm behind it and an edge containing that corner is making 45° with PP.*

The station point is 40 mm in front of PP, 50 mm above GP and lies in a central plane which is at a distance of 70 mm to the left of the corner nearest to the PP.

Draw the perspective projection of the lamina.

Hint: *See Fig. 6.* Perspective projection is drawn by Visual Ray Method using the top and front views.

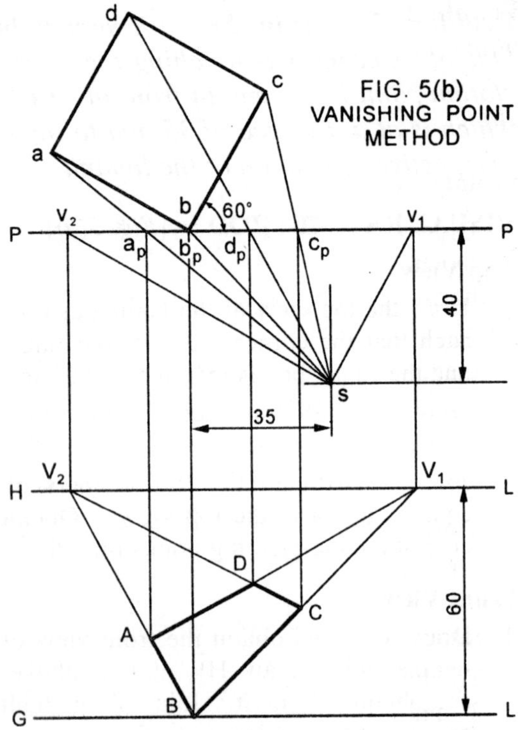

FIG. 5(b) VANISHING POINT METHOD

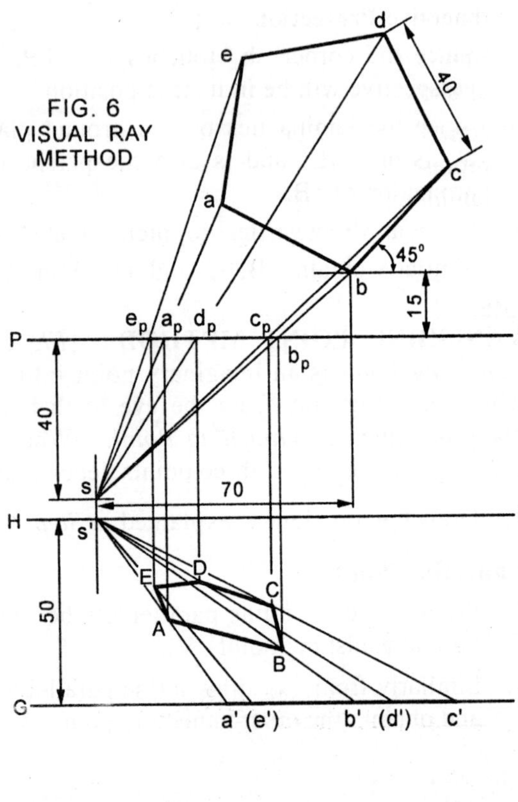

FIG. 6 VISUAL RAY METHOD

Problem 5: *(Fig. 7)* ***A circular lamina** of 45 mm diameter lies on the ground plane and touches the picture plane. The center plane passes through the center of the circle. Station point is 70 mm in front of picture plane and 60 mm above the ground plane. Draw the perspective projection.*

Visual Ray Method

1. Draw the top view of the lamina as a circle touching PP. Draw the vertical and horizontal diameters of the circle.
2. Enclose the circle in a square and join the diagonals.
3. Draw the front view of the lamina. Then draw the perspective projection of the square.
4. Join the diagonals to intersect at O.
5. Through O, draw a vertical and a horizontal line.
6. Obtain the points 1, 5 and 3, 7 on the respective sides of the square in the perspective.
7. To obtain 2 and 4 in the perspective, mark 2′ and 4′.
8. Join 2′(4′) with s′ which intersects the diagonals BD and AC at 2 and 4 respectively.
9. Similarly obtain 6 and 8 in the perspective as shown.
10. Join 1, 2, ..., 8 by a smooth ellipse to represent the perspective projection of the lamina.

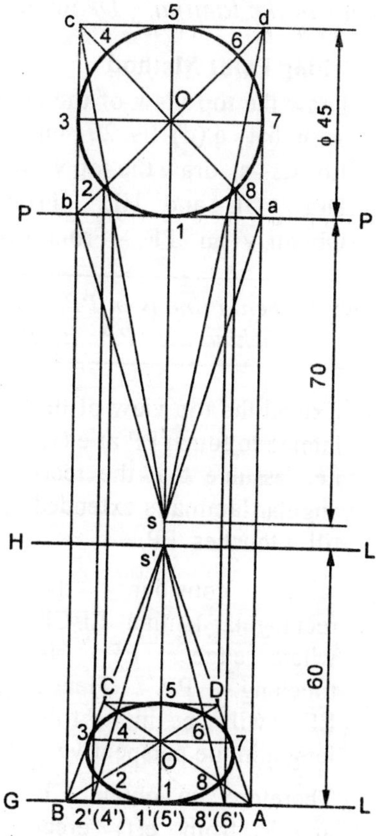

FIG. 7

Note: *1. In the perspective, the center of the ellipse does not coincide with the center of the perspective of the square.*

2. A circle will appear as a circle in its perspective projection, when it is parallel to the PP. In other positions, its perspective will be an ellipse (except when it lies in the CP).

3. Circular features of an object inclined to PP will appear as ellipse in perspective. To obtain the points for the ellipse, enclose the circle in a square. The midpoints of the sides of the square are the four points for the ellipse. Points of intersection of diagonals with the circle are the other four points. Draw lines through these points parallel to the sides of the square. Then draw the perspective of the square along with the parallel lines, locating eight points for drawing the ellipse.

Problem 6: *(Fig. 8)* ***A rectangular lamina*** *of sides 50 mm x 25 mm stands vertically with one of its longer edges on the ground and inclined at 45° to PP. The vertical edge nearest to PP is 20 mm behind it. The station point is 30 mm in front of the PP, 45 mm above the ground and lies in a central plane which passes through the center of the lamina. Draw the perspective projection of the lamina.*

Vanishing Point Method

1. Draw the top view of the lamina as a straight line of length 50 mm inclined at 45° to PP such that a (d) is 20 mm above PP.
2. Locate s, draw the top view of the visual rays and obtain the piercing points.
3. Draw GL and HL. From s draw a line parallel to ab.
4. Obtain v on PP. From v erect a vertical and obtain the vanishing point V on HL.

Rule: *When a line is in PP, true length will be seen in its perspective projection. When a line is behind the PP it will be foreshortened in its perspective.*

---

5. Extend the top view of the lamina to touch PP at e (f). i.e., assume that the rectangular lamina is extended till it touches PP.
6. Now consider the rectangular lamina EBCF where edge EF is touching PP. Hence EF will be in its true length in the perspective.
   Therefore, to obtain E and F from e(f) erect a vertical line to intersect GL. This line is called **measuring line** or **line of heights**.

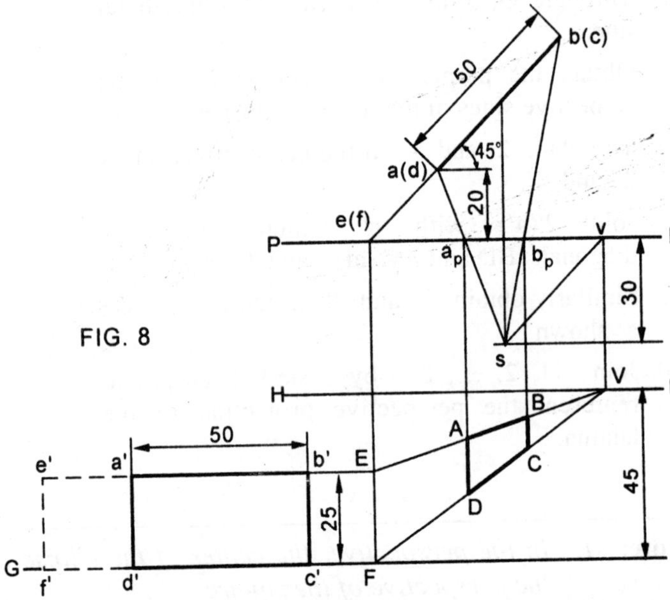

FIG. 8

   On this line the heights of all the points in the lamina can be marked. Hence project e′ and f′ horizontally to intersect the measuring line at E and F. Now EF is the perspective projection and is in its true length = 25 mm.
7. Join F and V. FC is on the ground. Hence, the perspective projection of any point on the edge FC will lie on FV. Hence obtain D and C on FV by drawing vertical lines from the corresponding piercing points.
8. Edge EB is at 25 mm above the ground. Perspective of E has been already marked. Join E and V. Now perspective of any point on edge EB can be obtained on EV. Therefore, obtain A and B on EV by erecting vertical lines from the corresponding piercing points.

9. Join ABCD by *thick* lines and complete the perspective projection of the given lamina. Note that in the perspective, BC is smaller than AD since the edge BC is farther away from PP than AD.

Problem 7: *(Fig. 9)* *A* ***hexagonal lamina*** *of 25 mm side stands vertically on the ground plane and inclined at 50° to PP. The corner nearest to PP is 20 mm behind it. The station point is 45 mm in front of PP, 50 mm above the ground plane and lies in a central plane which passes through the center of the lamina. Draw the perspective projection.*

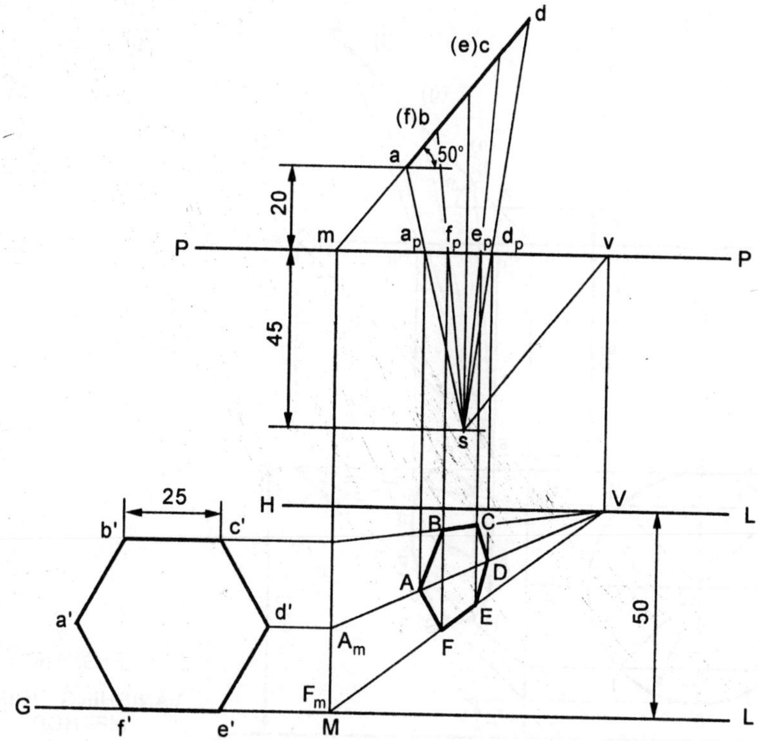

FIG. 9
VANISHING POINT METHOD

1. Draw PP and draw a line inclined at 50° to it.
2. Draw a regular hexagon of side 25 mm such that one of its sides is on GL and name all the corners. Now complete the top view of the lamina with the help of this hexagon.
3. Locate s, draw the visual rays and obtain the piercing points.
4. Draw GL and obtain the **vanishing point V** on HL.
5. Extend the top view of the lamina to meet PP at m. From m draw a vertical line to meet GL at M (mM is the measuring line).
6. The edge FE is on the ground. Project f′ and e′ horizontally to meet the measuring line at F_m. Join F_m with V.
7. From the piercing points f_p and e_p erect verticals to intersect F_mV at F and E respectively.
8. Through a′ and d′ draw a horizontal line and obtain A_m on mM. Join A_m with V.

9. Since the corners A and D of the lamina are at the same height from the ground, obtain the perspective of A and D on A_m V.
10. Similarly obtain B and C and complete the perspective projection.

Problem 8: *(Fig. 10)* *Draw the perspective projection of a* ***circular lamina*** *of 60 mm diameter having its surface vertical and inclined at 45° to the picture plane. The center of the lamina is 40 mm above the ground plane. Station point is 70 mm above ground plane and 55 mm in front of the picture plane. Central plane passes through a point on the circumference of the lamina which is nearest to picture plane and 10 mm behind it.*

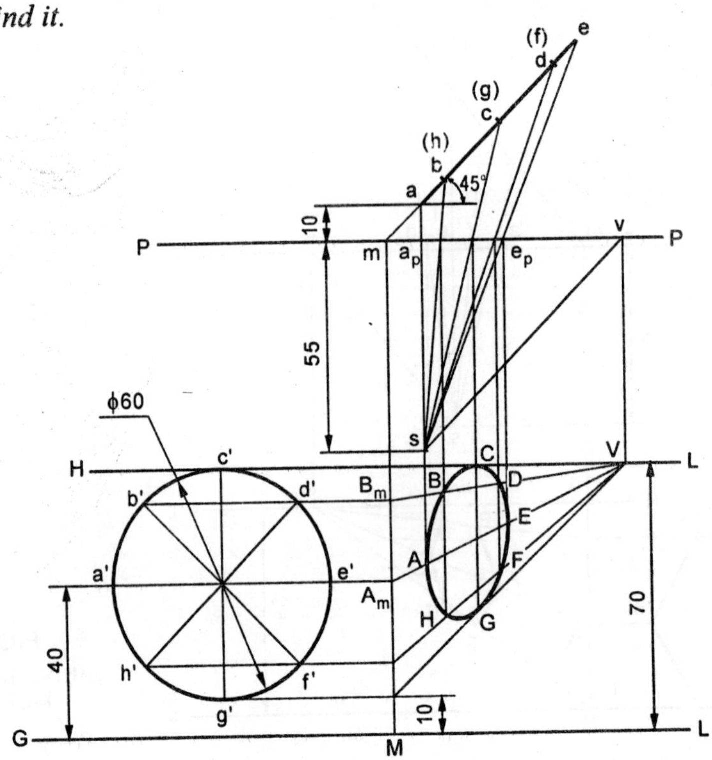

FIG. 10
VANISHING POINT METHOD

1. Draw the top view of the lamina.
2. Draw a circle of 60 mm diameter on the left hand side as shown. Divide it into 8 equal parts and name the division points as a′, b′, etc.
3. Mark these points on the top view also.
4. Mark s and obtain the piercing points on PP.
5. Obtain the **vanishing point V** and draw the measuring line mM.
6. Project the division points horizontally and obtain A_m, B_m, etc., on the measuring line.
7. Join them with V. From the piercing points, erect verticals and obtain A, B, etc.
8. Join these points by an ellipse to represent the perspective projection of circular lamina.

3. PERSPECTIVE PROJECTION OF SOLIDS

Problem 9: *(Fig. 11)* ***A square prism****, side of base 40 mm and height 60 mm rests with its base on the ground such that one of its rectangular faces is parallel to and 10 mm behind the picture plane. The station point is 30 mm in front of PP, 80 mm above the ground plane and lies in a central plane 45 mm to the right of the center of the prism. Draw the perspective projection of the square prism.* (UQ)

VISUAL RAY METHOD

Top View

1. *See Fig. 11.* Draw the top view of the prism as a square of side 40 mm such that ab is parallel to and 10 mm above PP.
2. Locate s. Draw top view of the visual rays. Mark piercing points.

Front View

3. Draw front view of the prism for given position. Locate s′ and draw front view of the visual rays.
4. From piercing points erect vertical lines to cut the corresponding visual rays in the front view. Thus obtain all corners in perspective projection.

To mark the visible and invisible edges in the perspective

5. Draw boundary lines (*thick lines*).
6. The faces $ab(b_1)(a_1)$ and $bc(c_1)(b_1)$ are nearer to s and visible. Hence draw BB_1, BA and BC as *thick lines*.
7. Edge $d(d_1)$ is farther away from s. So draw DD_1, D_1A_1 and D_1C_1 as *dashed* lines.

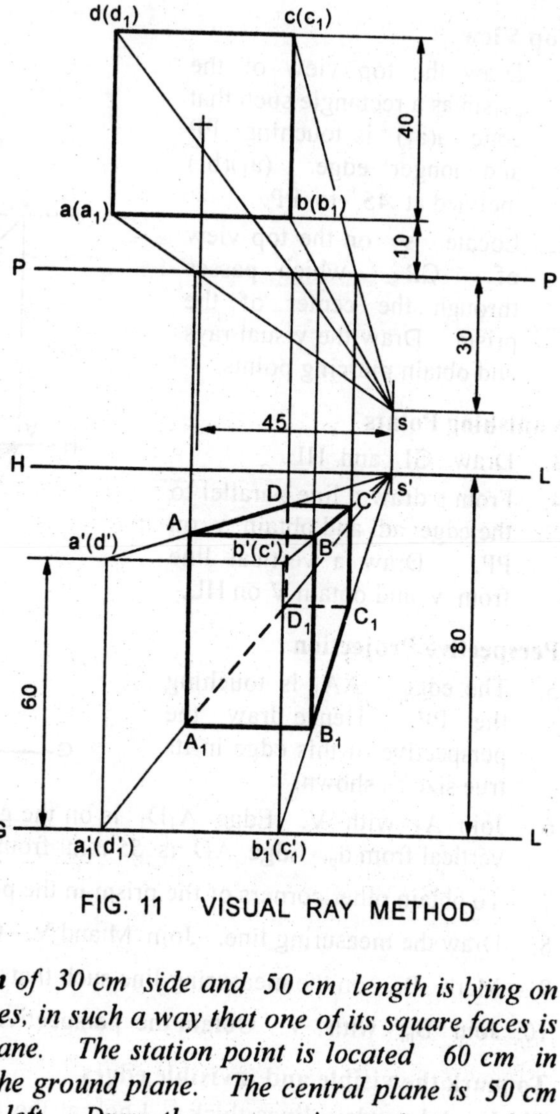

FIG. 11 VISUAL RAY METHOD

Problem 10: *(Exercise)* ***A square prism*** *of 30 cm side and 50 cm length is lying on the ground plane on one of its rectangular faces, in such a way that one of its square faces is parallel to and 10 cm behind the picture plane. The station point is located 60 cm in front of the picture plane and 40 cm above the ground plane. The central plane is 50 cm away from the axis of the prism towards the left. Draw the perspective projection of the prism.*

[In case of large objects, perspective view may partly overlap front view causing confusion.]

Problem 11: *(Fig. 12)* ***A rectangular prism**, sides of base 50 mm x 30 mm and height 55 mm, rests with its base on the ground plane. A vertical edge is in the picture plane and one of the longer edges of its base is inclined at 45° to PP and behind it. The station point is 50 mm in front of PP, 75 mm above the ground plane and lies in a central plane which passes through the center of the prism. Draw the perspective projection of the solid.* (UQ)

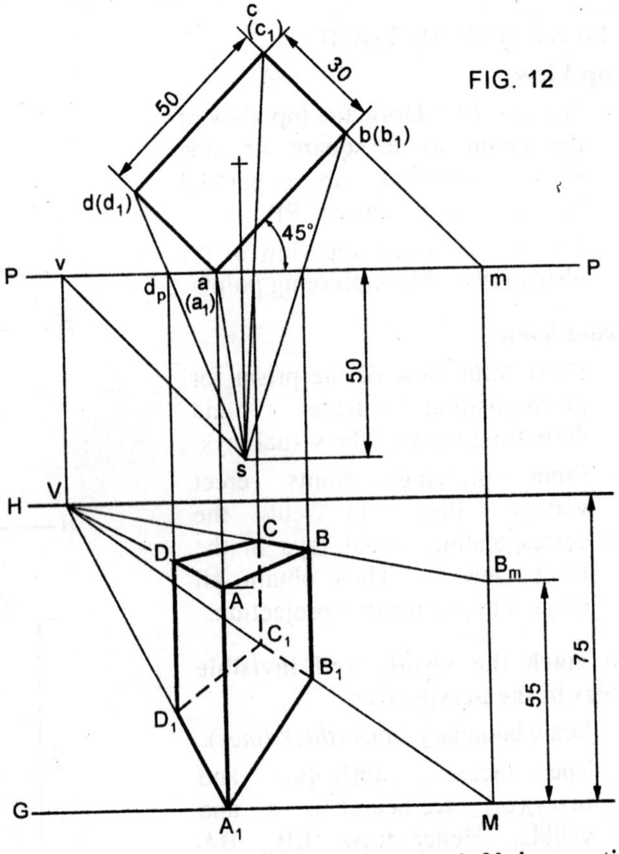

FIG. 12

VANISHING POINT METHOD

Top View

1. Draw the top view of the prism as a rectangle such that edge $a(a_1)$ is touching PP and longer edge $(a_1)(b_1)$ inclined at 45° to PP.
2. Locate s on the top view of CP which passes through the center of the prism. Draw the visual rays and obtain piercing points.

Vanishing Points

3. Draw GL and HL.
4. From s draw a line parallel to the edge ad and obtain v on PP. Draw a vertical line from v and obtain V on HL.

Perspective Projection

5. The edge AA_1 is touching the PP. Hence draw the perspective of this edge in its true size as shown.
6. Join A_1 with V. Edge A_1D_1 is on the ground. Hence obtain D_1 on A_1V by erecting vertical from d_p. Edge AD is 55 mm from ground. Hence obtain D on AV.
7. To obtain other corners of the prism in the perspective, extend cb to meet PP at m.
8. Draw the measuring line. Join M and V. Obtain the perspective of B_1 and C_1 on MV.
9. Mark B_m on the measuring line such that MB_m = 55 mm = height of the prism.
10. Join B_m with V. Obtain the perspective of B and C on B_mV.

To mark the visible and invisible edges

11. Draw boundary lines thick. Look at the top view and visual rays. The edge $a(a_1)$ is nearer to the station point. Hence draw AA_1, AD and AB as thick lines.
12. The edge $c(c_1)$ is farther away from s. Hence draw C_1D_1, C_1B_1 and C_1C as dashed lines.

Problem 12: *(Exercise)* *A **cube** of edge 30 mm rests with one of its faces on the ground plane such that a vertical edge touches the picture plane. The vertical faces of the cube are equally inclined to the PP and behind it. A station point is 40 mm in front of the PP, 50 mm above the ground plane and lies in a central plane 15 mm to the right of the axis of the cube. Draw the perspective projection of the cube.*

Note: ***Distance Points*** *– Vanishing points for all horizontal lines inclined at 45° to PP are known as Distance Points because of their definite positions. They are equidistant from center of vision. The distance of each from the center of vision is equal to the distance of S from PP.*

Problem 13: *(Fig. 13)* *A **square pyramid** of base edge 40 mm and altitude 50 mm, rests with its base on the ground plane such that all the edges of the base are equally inclined to the PP. One of the corners of the base is touching the PP. The station point is 60 mm in front of the PP, 80 mm above the ground plane and lies in a central plane which passes through the axis of the pyramid. Draw the perspective projection.*

(UQ)

VANISHING POINT METHOD

1. Draw the top view of the square pyramid and the visual rays. From s draw a line parallel to ab. Obtain the vanishing point V on HL.
2. Corner A is touching the PP and on the ground. Hence erect a vertical line from a and mark the perspective of A on GL. Join A with V.
3. As edge AB is on the ground, obtain the perspective of B on AV.
4. To obtain the perspective of D and C extend cd to meet PP at m.
5. Draw the measuring line mM.
6. Since the edge DC is on the ground, join M with V.
7. Obtain the perspective of D and C on this line MV.
8. To mark the perspective of the apex O, draw a line parallel to ab and passing through o to meet PP at h.
9. Draw another measuring line hH.
10. On this line mark the height of the apex as O_hH = 50 mm.
11. Join O_h with V. Obtain the perspective of O on this line.
12. Then complete perspective as shown.

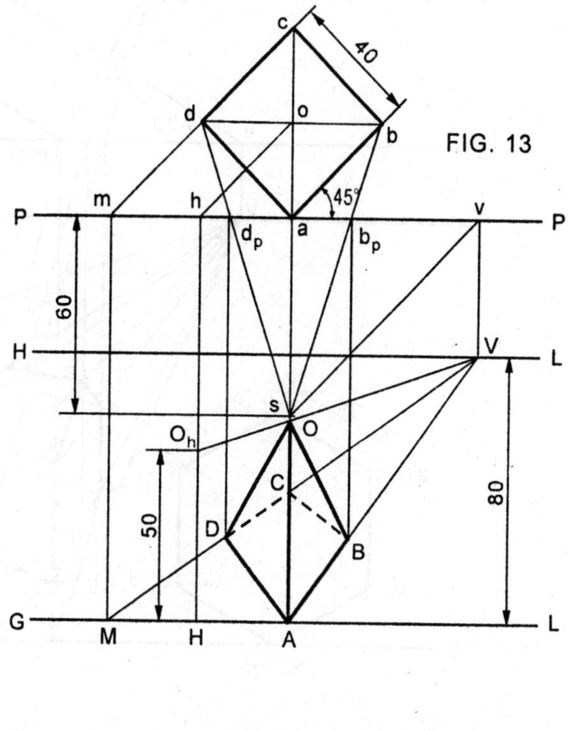

FIG. 13

Problem 14: *(Exercise) A rectangular pyramid, sides of base 55 mm x 20 mm and height 60 mm rests with its base on the ground plane such that one of the longer edges of the base is parallel to and 20 mm behind the picture plane. The station point is 40 mm in front of the picture plane, 60 mm above the ground plane and lies in a central plane which passes through the axis of the pyramid. Draw the perspective projection.* (UQ)

Hint: This problem can be easily solved by Visual Ray Method using the top and side views of the visual rays.

Problem 15: *(Fig. 14) Draw the perspective projection of a* ***hexagonal prism*** *lying on the ground plane on one of its longer edges such that one of its rectangular faces is perpendicular to the ground plane.*

The axis is inclined at 30° to the picture plane and an edge of the base is touching the picture plane. The station point is 110 mm in front of the PP, 95 mm above the ground plane and lies in a central plane which bisects the axis. For the prism, side of base is 25 mm and height 75 mm. (UQ)

VANISHING POINT METHOD

FIG. 14

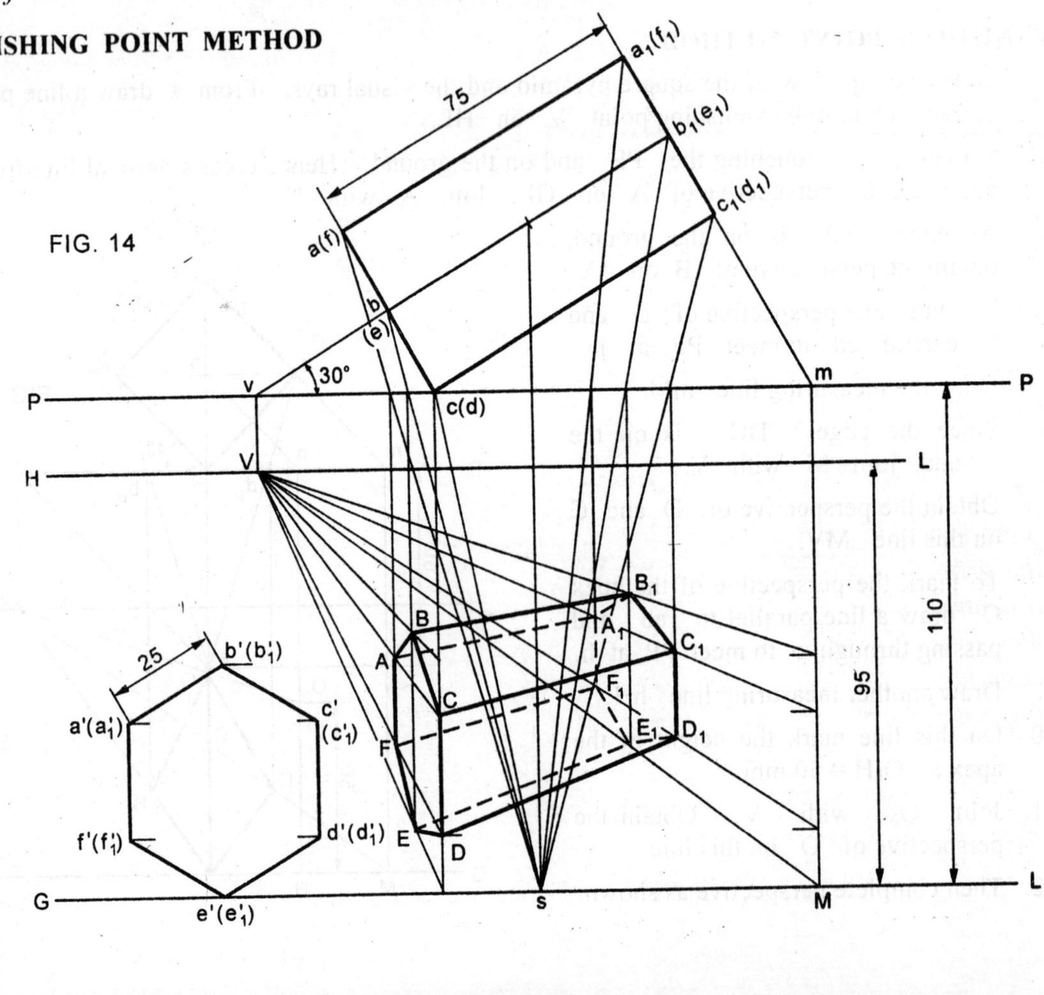

1. Draw the top view of the hexagonal prism, such that edge c(d) is touching the PP and axis inclined at 30° to PP, with the help of the hexagon drawn on left-hand side.
2. Locate s on CP which bisects the axis. Join s with all corners of prism in top view.
3. Draw GL and HL.
4. Consider two parallel lines ca and c_1a_1. Each line contains the top view of six corners. Hence from s draw a line parallel to ca. Obtain V on HL.
5. From c(d) draw a vertical line (measuring line). Mark the heights of all the six corners of the base on it. Since edge CD is touching PP, it will be in true size in its perspective.
6. From all the points on this measuring line, draw vanishing lines to converge at V. Draw vertical lines from the piercing points.
7. Obtain the perspective of E, F, A and B on the corresponding vanishing lines.
8. Obtain the perspective of 6 corners of other end face. Complete the perspective projection.

Problem 16: *(Exercise)* *A **pentagonal prism**, side of base 25 mm and axis 60 mm long, lies with one of its rectangular faces on the ground plane such that a pentagonal face is touching the PP. The station point is 20 mm in front of the picture plane, 55 mm above the ground plane and lies in a central plane which is at 80 mm to the right of the center of the prism. Draw the perspective projection.* (UQ)

Problem 17: *(Fig. 15)*
*A **pentagonal prism**, side of base 30 mm and height 50 mm rests with its base on the ground plane such that one of its rectangular faces is inclined at 45° to PP and the vertical edge nearer to PP is 10 mm behind it.*

The station point is 45 mm in front of the picture plane, 90 mm above the ground plane and lies in a central plane which is 15 mm to the left of the vertical edge nearer to PP. Draw the perspective projection. (UQ)

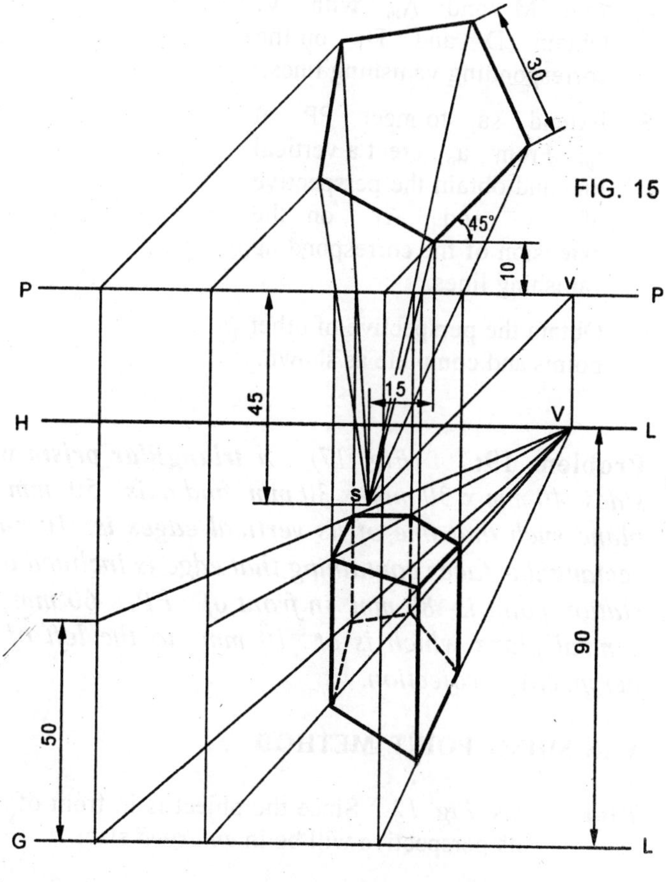

FIG. 15

Problem 18: *(Fig. 16) A* ***cube*** *of 35 mm long edges rests with a square face on the ground plane such that one of the vertical edges of the cube is 8 mm in front of the picture plane and a vertical face containing that edge is inclined at 30° to PP. The station point is 40 mm in front of the PP, 70 mm above the ground plane and lies in a central plane which is 45 mm to the left of the center of the cube. Draw the perspective projection.* (UQ)

1. Draw the top view of the cube for the given position. Locate s and join it with all the corners of the cube in top view.
2. Draw GL and HL. Then obtain V on HL.
3. Note that the picture plane intersects ad at m. Draw the measuring line mM and mark the height of the cube.
4. Join M and A_m with V. Obtain D and D_1 on the corresponding vanishing lines.
5. Extend sa to meet PP at a_p. From a_p erect a vertical line and obtain the perspective of A and A_1 on the extension of the corresponding vanishing lines.
6. Obtain the perspective of other points and complete as shown.

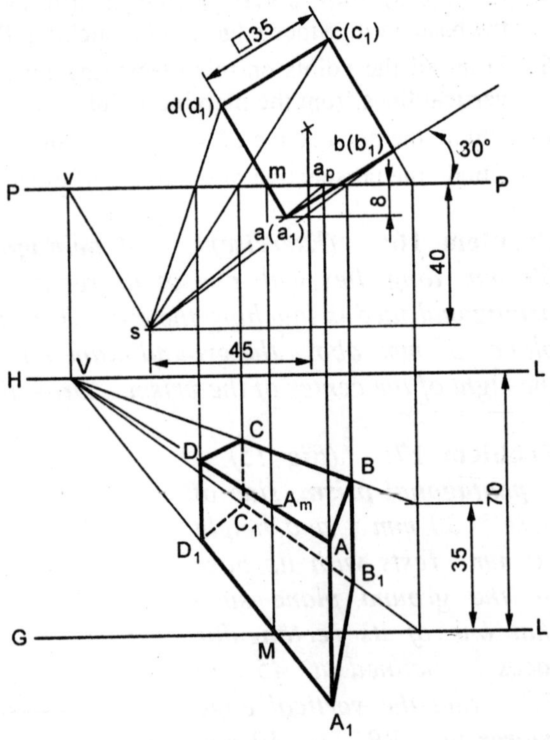

FIG. 16

Problem 19: *(Fig. 17) A* ***triangular prism*** *whose base is an isosceles triangle of sides 40 mm x 30 mm x 30 mm and axis 50 mm long rests with its base on the ground plane such that one of its vertical edges is 10 mm in front of the PP. One of the rectangular faces containing that edge is inclined at 45° to PP and in front of it. The station point is 80 mm in front of PP, 60 mm above the ground plane and lies in a central plane which is at 10 mm to the left of the center of the prism. Draw the perspective projection.* (UQ)

VANISHING POINT METHOD

Hint: *See Fig. 17.* Since the object is in front of PP, (between the PP and the observer) its perspective will be in *enlarged* size.

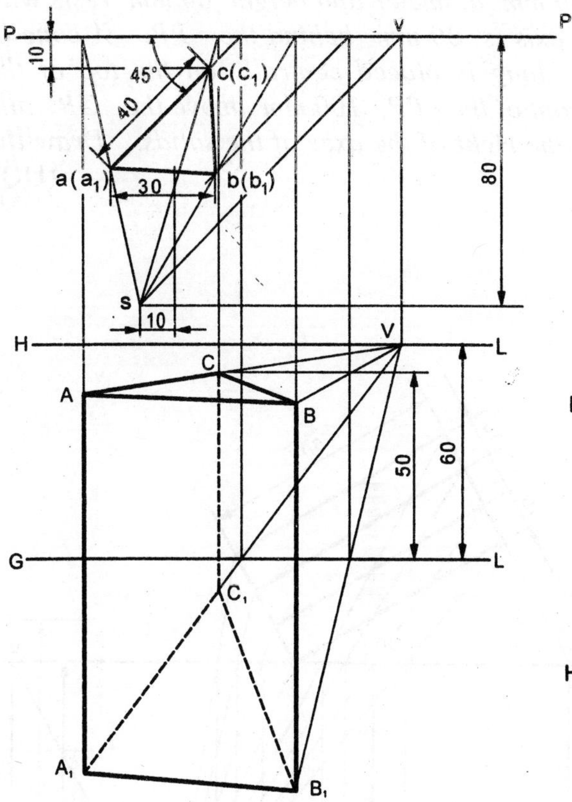

FIG. 17

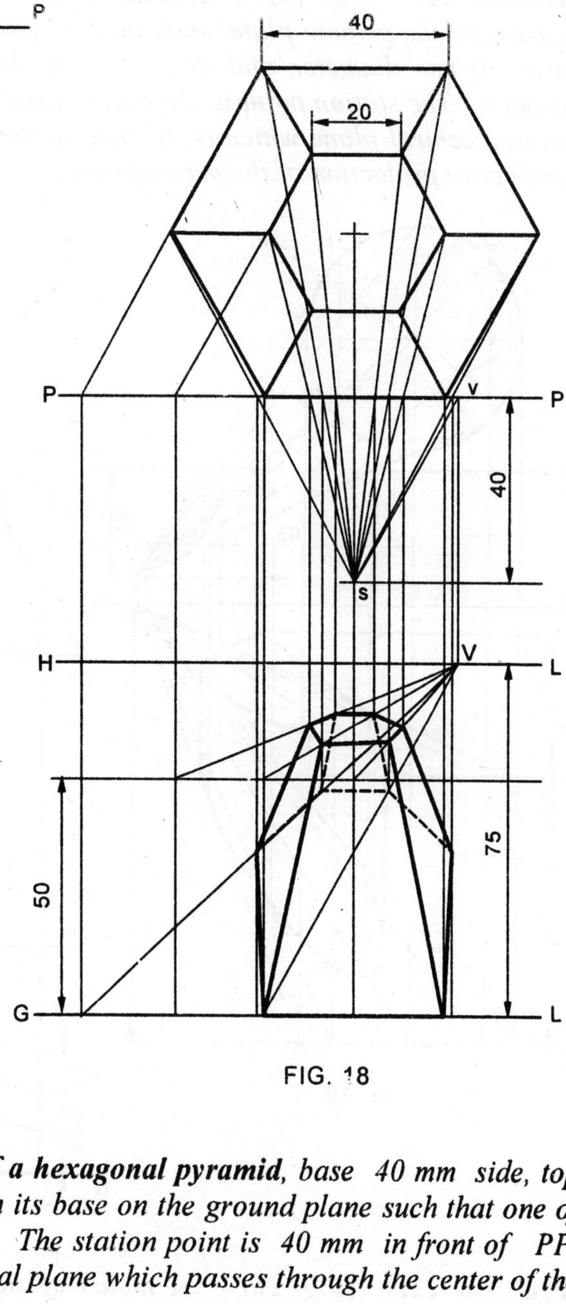

FIG. 18

Problem 20: *(Fig. 18) A **frustum of a hexagonal pyramid**, base 40 mm side, top 20 mm side and height 50 mm rests with its base on the ground plane such that one of the edges of its base is touching the PP. The station point is 40 mm in front of PP, 75 mm above the GP and lies in a central plane which passes through the center of the frustum. Draw the perspective projection.* (UQ)

Hint: *See Fig. 18.* Follow the principle of symmetry to minimize the congestion of projector lines and to have clarity of the figure; that is, locate a corner by drawing the projector from the piercing point on the Picture Plane to cut the corresponding converging line. Mark the other corresponding point by symmetry.

Problem 21: *(Fig. 19) A* ***cylinder*** *of 50 mm diameter and height 60 mm rests with its base on the ground plane such that the axis is 30 mm behind the PP. A cone of base 50 mm diameter and axis 25 mm long is placed centrally on the top of the cylinder. The station point is 25 mm in front of the PP, 100 mm above the GP and lies in a central plane which is 65 mm to the right of the axes of the solids. Draw the perspective projection of the arrangement.* (UQ)

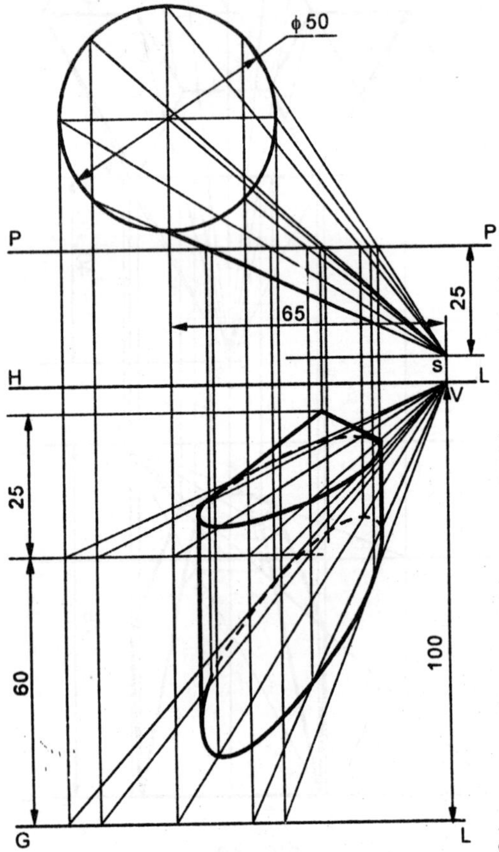

FIG. 19

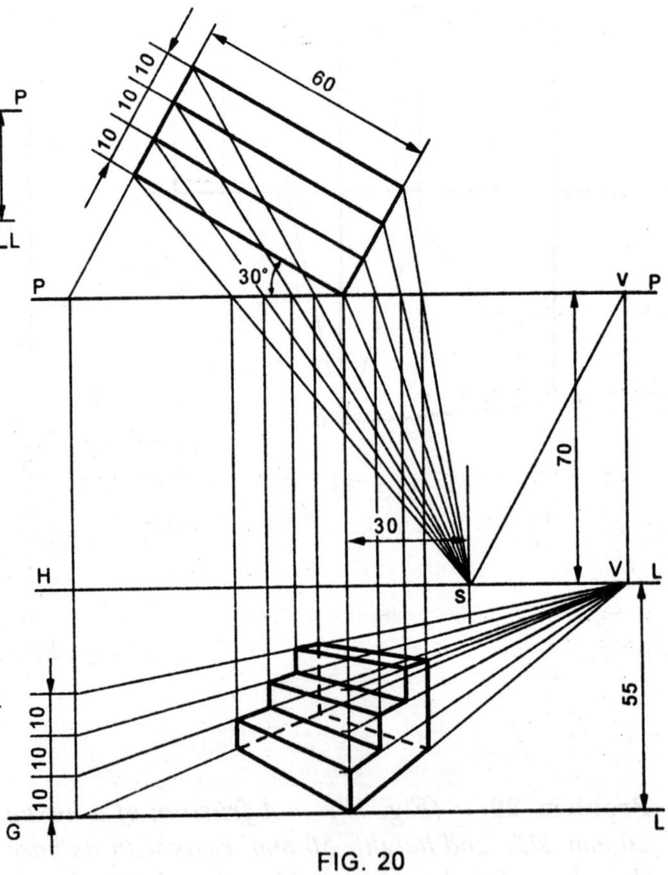

FIG. 20

Problem 22: *(Fig. 20)* ***A model of steps*** *has three steps of 10 mm tread and 10 mm rise. The length of the steps is 60 mm. The model is placed with the vertical edge of the first step touching the PP and its longer edge inclined at 30° to PP. The station point is 70 mm in front of PP, 55 mm above the ground plane and lies in a central plane which is at 30 mm to the right of the vertical edge touching the PP. Draw the perspective projection.* (UQ)

Problem 23: *(Fig. 21) Draw the perspective projection of a* ***shed*** *whose top and side views are given.*

The roof of the shed is inclined at 30° to the ground plane. The horizontal long edges of the shed are inclined at 45° to PP. One of the vertical edges of the wall touches the picture plane and 2.6 m to the left of the central plane. The station point is 9 m in front of PP and 6.5 m above the ground plane. The height and width of the door are 2.2 m and 1.4 m respectively. (UQ)

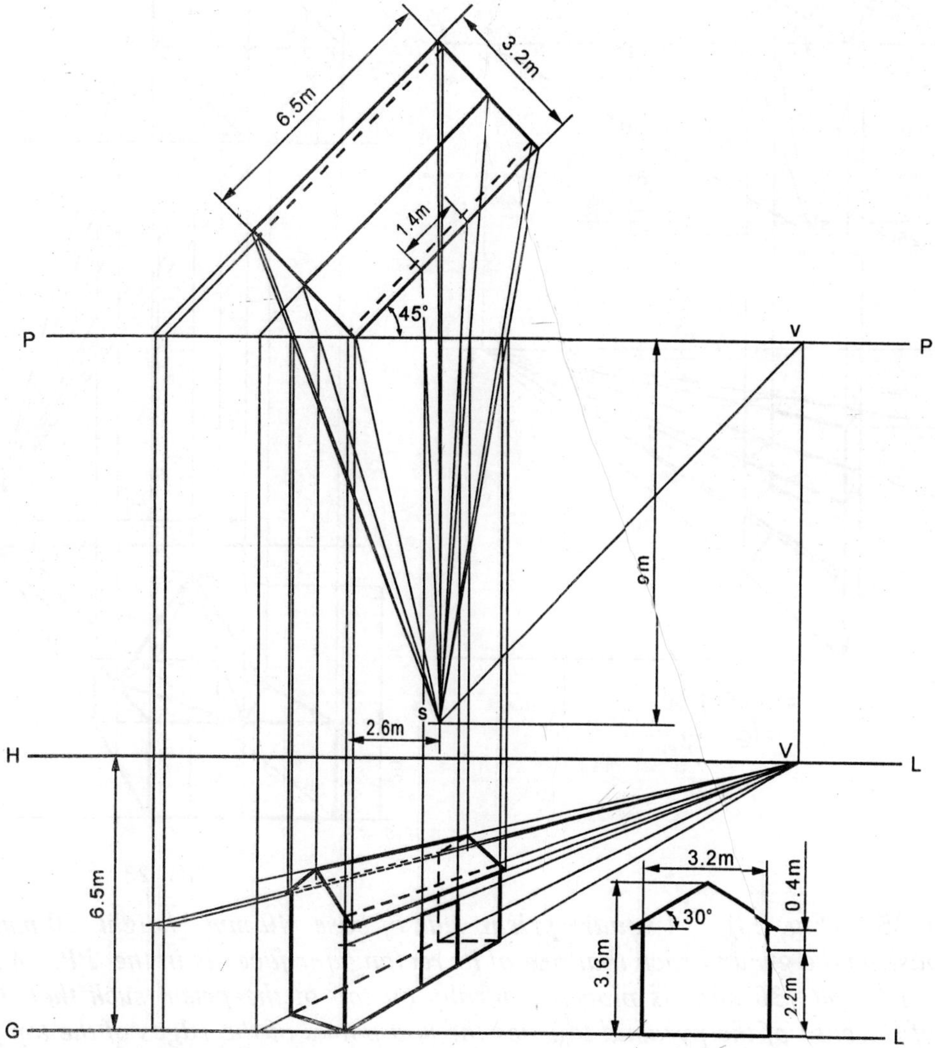

SCALE 1:20

FIG. 21

Problem 24: *(Fig. 22) A **steel rack** of size 1 m x 0.4 m x 1.5 m has 3 shelves equally spaced. It stands on the ground plane such that one of its vertical edges touches the PP and a longer horizontal edge is inclined at 30° to PP. The station point is 1.25 m in front of PP, 2m above the ground plane and lies in a central plane which is at 0.25 m to the right of the edge, touching the picture plane. Draw the perspective projection of the rack when its front portion is open.* (UQ)

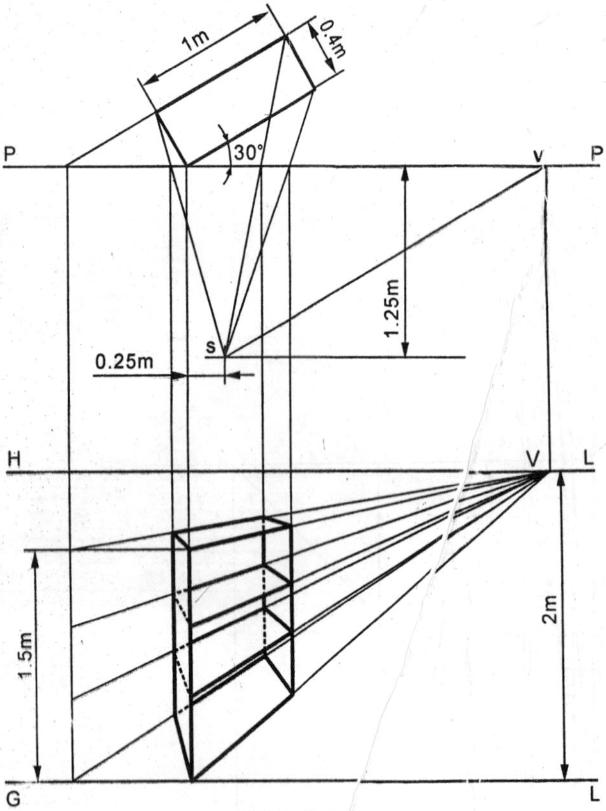

SCALE 1:50

FIG. 22

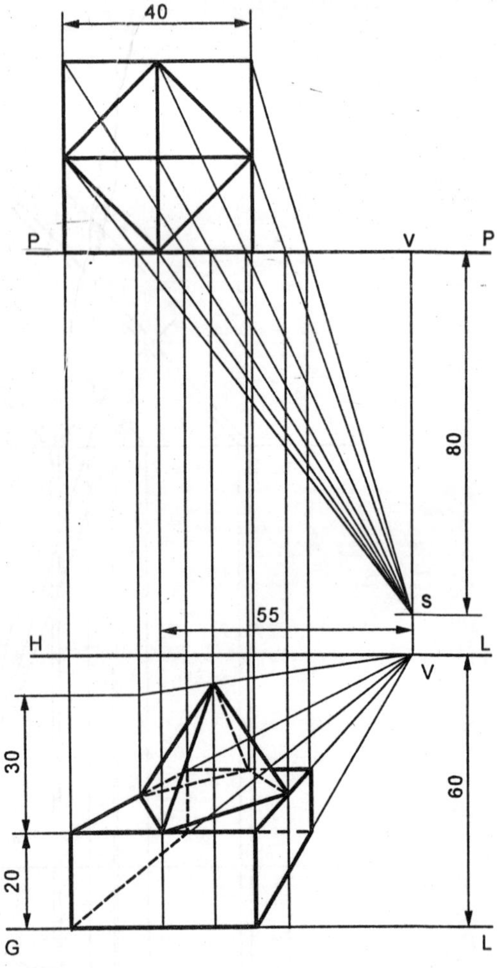

FIG. 23

Problem 25: *(Fig. 23) A **square prism**, side of base 40 mm, height 20 mm rests with its base on the ground such that one of its rectangular faces is in the PP. A **square pyramid** of height 30 mm is placed centrally on top of the prism such that the four corners of the base of the pyramid touches the mid-points of the edges of the top face of the prism. Station point is 80 mm in front of the picture plane, 60 mm above the GP and lies in a central plane which is 55 mm to the right of the axes of the solids. Draw the perspective projection of the **combination of solids.*** (UQ)

Problem 26: *(Exercise) A 25 mm thick* ***octagonal slab****, rests with its base on the ground, supporting a* ***square pyramid*** *of base 40 mm side and 50 mm altitude. The corners of the pyramid coincide with the alternate corners of the slab at the top. A rectangular face of the slab is parallel to and 30 mm behind PP. The station point is 120 mm in front of PP and 40 mm above the ground and 50 mm to the right of apex of the pyramid. Draw the perspective projection.*

4. ADDITIONAL PROBLEMS

27. Draw the perspective view of a cube of 25 mm edge, resting on ground on one of its faces. It has one of its vertical edges in the picture plane and all its vertical faces are equally inclined to the picture plane. The station point is 55 mm in front of the picture plane, 40 mm above the ground and lies in the central plane which is 10 mm to the left of the center of the cube.

(UQ)

28. A cylinder 30 mm diameter and axis 40 mm long is lying on the ground plane with its axis perpendicular to the picture plane. The nearest point of contact with the ground is 60 mm on the left of the station point and 10 mm from the picture plane. The station point is 40 mm above the ground and 60 mm in front of the PP. Draw the perspective projection of the cylinder.

29. A frustum of a square pyramid, base 28 mm side, top 22 mm side and 36 mm height is resting on its base on the GP such that the sides of base are equally inclined to the Picture Plane. The axis of the frustum is 30 mm to the right of the station point. The station point is 45 mm in front of the PP and 50 mm above the GP. The nearest base corner is 10 mm behind the PP. Draw the perspective projection.

30. Draw the perspective projection of a rectangular block of 3 m x 2 m x 1.5 m resting on a horizontal plane with one side of the rectangular plane making an angle of 45° with VP. The observer is at a distance of 6 m from the picture plane. Assume eye level as 1.5 m.

(UQ)

31. A solid is in the form of a square prism of side of base 40 mm upto a height of 50 mm and thereafter tapers into frustum of a square pyramid whose top surface is of 25 mm side. The total height of the solid is 70 mm.

Draw the solid in perspective, given that one side of the base of the solid resting on the ground is inclined at 25° to the PP and the corner containing that side is 40 mm to the right of the eye and is touching the PP. The eye is 100 mm from PP and 90 mm above the ground.

(UQ)

32. A man of 1.8 m height stands at a distance of 5 m from a flight of four stone steps having a width of 2 m, tread 0.3 m and rise 0.2 m. The flight makes an angle of 45° with the PP and touches the same at a distance of 2 m to the right of the center of vision. Draw the perspective projection of the flight.

33. A memento is in the shape of a square prism of side of base 30 mm and altitude 40 mm. Above this height, it tapers uniformly to a square of side 20 mm. The overall height of the memento is 60 mm. The solid is kept on the ground with one edge of the base parallel to the picture plane and 15 mm behind the picture plane. The axis of the solid is 35 mm to the right of the eye of the observer. The eye is 60 mm from the picture plane and 80 mm above the ground.
Draw the perspective projection of the memento. (UQ)

Hint: Draw the perspective projection of three squares, one corresponding to the bottom end, the second corresponding to the top end of the square prism and the third square corresponding to the topmost end.

34. A hexagonal nut of side of base 3.2 cm and axis 3.2 cm with a cylindrical hole of 3.2 cm diameter through it has the axis of the hole coinciding with the axis of the prism. The base of the prism is on the ground and the nearest vertical edge is in the picture plane. The face containing the nearest edge is inclined at 10° to the picture plane. The height of the station point is 5 cm. The station point is opposite to the center of the nut and its distance from the picture plane is 15 cm.
Draw the perspective projection of the nut.

35. A semi-circular arched opening in a wall has a length of 4 m, thickness 0.5 m and height 3.5 m. The opening is 1.5 m wide and the springing points are at a height of 2 m. The wall makes an angle of 45° with the picture plane.
Select a suitable position of the observer and draw the perspective projection. (UQ)

36. A pillar is in the form of a frustum of a pyramid 2 m high, 0.5 m square bottom and 0.3 m square top. It is mounted centrally on a base which is a square prism of height 0.7 m and side of base 0.8 m.
Draw the pillar in perspective, given that one side of the base is perpendicular to the picture plane and the lower corner of the base nearest to the eye is 0.3 m behind the picture plane and 0.6 m to the right of observer. The eye is 3 m from the picture plane and 2.5 m above the ground. (Scale 1 : 20) (UQ)

*